Ciprian Foias
Arthur E. Frazho

The Commutant Lifting Approach to Interpolation Problems

1990

Springer Basel AG

Authors' addresses:

Prof. Ciprian Foias
Department of Mathematics
Swain Hall East
Indiana University
Bloomington, IN 47405
USA

Prof. Arthur Frazho
School of Aeronautics
and Astronautics
Purdue University
West Lafayette, IN 47907
USA

Library of Congress Cataloging-in-Publication Data

Foiaş, Ciprian:
  The commutant lifting approach to interpolation problems / Ciprian
Foias, Arthur E. Frazho.
    p.   cm. – – (Operator theory, advances and applications ; v. 44)
  Includes bibliographical references.
  ISBN 978-3-0348-7714-5   ISBN 978-3-0348-7712-1 (eBook)
  DOI 10.1007/978-3-0348-7712-1

  1. Interpolation.   2. Lifting theory.   I. Frazho, Arthur E.,
1950–   .   II. Title.   III. Series.
QA281.F65   1990
511'. 42 – – dc20

Deutsche Bibliothek Cataloguing-in-Publication Data

Foias, Ciprian:
The commutant lifting approach to interpolation problems /
Ciprian Foias ; Arthur E. Frazho.

  (Operator theory ; Vol. 44)
  ISBN 978-3-0348-7714-5
NE: Frazho, Arthur E.:; GT

ISBN 978-3-0348-7714-5

TO OUR PARENTS WITH LOVE AND ADMIRATION.

PREFACE

Classical H^∞ interpolation theory was conceived at the beginning of the century by C. Carathéodory, L. Fejér and I. Schur. The basic method, due to Schur, in solving these problems consists in applying the Möbius transform to peel off the data. In 1967, D. Sarason encompassed these classical interpolation problems in a representation theorem of operators commuting with special contractions. Shortly after that, in 1968, B. Sz.-Nagy and C. Foias obtained a purely geometrical extension of Sarason's results. Actually, their result states that operators intertwining restrictions of co-isometries can be extended, by preserving their norm, to operators intertwining these co-isometries; starting with R. G. Douglas, P. S. Muhly and C. Pearcy, this is referred to as the commutant lifting theorem.

In 1957, Z. Nehari considered an L^∞ interpolation problem which in turn encompassed the same classical interpolation problems, as well as the computation of the distance of a function f in L^∞ to H^∞. At about the same time as Sarason's work, V. M. Adamjan, D. Z. Arov and M. G. Krein proved that the solution to the Nehari interpolation problem could be deduced from a representation theorem for Hankel operators. Moreover, Adamjan, Arov and Krein gave a full description of all suboptimal solutions of their representation theorem for Hankel operators. Their description is an elegant generalization of Schur's fractional representation of the solutions to the Carathéodory interpolation problem. In 1970, Adamjan, Arov and Krein, and L. B. Page also noted that the commutant lifting theorem provides the general representation theorem for Hankel operators, including block Hankel operators. At this stage one could assert that the commutant lifting theorem provided the natural geometric framework for classical and modern H^∞ interpolation problems. The extension of the Adamjan-Arov-Krein theory to the general framework of the commutant lifting theorem was elaborated later in the second half of the seventies by the Romanian chapter of the Sz.-Nagy school of dilation theory. Surprisingly in the 1960's, geophysicists in seismic oil prospecting independently arrived at the Carathéodory interpolation problem and rediscovered the Schur algorithm (e.g., see [Ro 2]). In their terminology, the Schur numbers which parameterize the solutions to the Carathéodory interpolation problem are known as reflection coefficients for the interfaces of the layered medium. The extension of the Adamjan, Arov and Krein theory to the commutant lifting theorem was influenced by the geophysical interpretation of the Carathéodory interpolation problem. In particular, this suggested a layered medium picture of the Schäffer-Sz.-Nagy matrix form of isometric dilations, and consequently led to the first parameterization of the set of all solutions in the commutant lifting theorem, by adequate sequences of arbitrary contractions (choice sequences). In the classical Carathéodory interpolation problem, the contractions in the choice sequence are the Schur numbers (or the reflection coefficients). A bird's eye view of this connection was given in 1978 (see [Fo 10]).

In the 1980's, there was a renewed interest by mathematicians and engineers in the Nehari problem along with its connections to the theory of Hankel and Toeplitz operators, reproducing kernels, signal processing and control theory. This is well illustrated by the achievements of the Gohberg and Kailath schools as well as the success of the Mathematical Theory of Network and Systems (MTNS) symposia. In particular, successful approaches to interpolation problems and control theory include the band extension method by H. Dym and I. Gohberg, the Beurling - Lax - Halmos theorem for Krein spaces by J. A. Ball and J. W. Helton, the state space approach by K. Glover, and the de Branges reproducing kernel method by Dym. For these developments see [DyG 1,2,3], [BH], [Gl 1,2], [GoKW 1,2] and the monographs [BGR], [Dy 1], [Helt 5]. Meanwhile, the work on the general framework on the commutant lifting theorem continued to grow mainly in Romania, the U.S.A. and Venezuela. It is the purpose of this book to present a unified approach, based on the geometric framework of the commutant lifting theorem, to solve many classical and modern interpolation problems arising in mathematics, engineering, and geophysics. Therefore, we tried to make this book as self-contained as possible. The prerequisites for Chapter I to III are some elementary real and complex analysis as well as linear algebra. In Chapters IV and V we assume some knowledge of elementary functional analysis or at least linear algebra. Starting with Chapter VI we require some familiarity with Hilbert spaces, particularly with infinite orthogonal sums of Hilbert spaces. To keep these chapters as self contained as possible, we tried to make the reading of each chapter depend on only a few preceding sections (see the chart at the end of the introduction). For this reason previously obtained propositions are restated on occasion, especially when the results are reproved by new methods.

In Chapter I we present a rather complete theory of the Carathéodory interpolation problem, with emphasis on the computational aspects of the Schur method. In Chapter II we develop the connection between the Carathéodory interpolation problem and the Levinson algorithm from filtering theory. Chapter III presents a geophysical interpretation of many of the results in the preceding two chapters. The geophysical layered medium models also provide a useful intuitive background for direct and inverse scattering algorithms in interpolation theory. Many of the developments in this monograph were motivated by the material in the first three chapters. In particular, throughout the book we will continue to return to the Carathéodory interpolation problem as the standard test case.

In Chapter IV we present some geometric results concerning the extension and completion of matrix contractions, which are useful in the analysis of the commutant lifting theorem. (It is interesting to note that these theorems were obtained after the commutant lifting theorem.) Chapter V is devoted to contractive one step intertwining liftings, which are the basic building blocks in the analysis of the commutant lifting theorem. Chapter VI is an introduction to the Sz.-Nagy dilation theory. We present only the standard facts of this theory concerning the isometric and the unitary dilations of contractions. In addition, we give a complete classification of non-minimal isometric liftings inspired by the developments in the commutant lifting theorem. The commutant lifting theorem is introduced and studied in Chapter VII. Because of its central role in our work, several different proofs are presented, each of which illuminates another facet of the theorem. The first applications of the commutant lifting theorem are given in Chapter VIII. The Carathéodory and Nevanlinna-Pick interpolation problems, and the representation of Hankel operators are given a geometric treatment. Other applications, such as the extensions of operators commuting with the restrictions of isometries, the characterization of left and right inverses of Toeplitz operators, and Toeplitz Corona theorems, are also presented. Chapter IX is the first place where we introduce and develop the connection between operator valued function theory and the commutant lifting theorem. In particular, we present the Beurling-Lax-Halmos Theorem, some aspects of the Sz.-Nagy-Foias model theory, the Douglas-Shapiro-Shields factorization of certain operator valued functions, and the Adamjan-Arov-Krein representation of Hankel operators. We also introduce several block Nehari type problems occurring in control theory (see also Chapter XII). In Chapter X we give the explicit way in which one uses the commutant lifting theorem to obtain a solution, and characterize uniqueness for several classical interpolation problems. These include the block versions of the Carathéodory, Nevanlinna-Pick, and Hermite-Fejér interpolation problems in both their classical and tangential forms. This chapter also presents the connection between these interpolation problems and the representation theory of Hankel operators. It is shown that these problems can be solved by using a very nice application of the commutant lifting theorem to a general momentum problem discovered by M. Rosenblum and J. Rovnyak in 1980. Since their approach involves a new geometric framework we devoted Chapter XI to their theory.

Some of the most important applications of the previous interpolation problems arise in control theory. Therefore, in Chapter XII we present a bird's eye view of the connection between those problems and robust control theory pioneered by J. W. Helton, A. Tannenbaum and G. Zames. In Chapter XIII we present the two basic approaches of describing and parameterizing all solutions in the commutant lifting theorem. We begin by introducing the parameterization provided by the choice sequences of contractions. Then we obtain a general Schur type fractional representation of the solutions in the commutant lifting theorem. After that, we develop several scattering algorithms connecting the solutions in the commutant lifting theorem with their choice sequences. These algorithms, when specialized to the classical Carathéodory interpolation problem, reduce to the Robinson algorithm (in Chapter III), and the Schur and Adamjan-Arov-Krein algorithms (in Chapter I). Finally, by using the functional model of contractions, we obtain a new Schur type fractional representation in the commutant lifting theorem, which is more akin to the classical Schur and the modern Adamjan-Arov-Krein representations. In Chapter XIV we introduce the Redheffer cascading theory as the natural systems theory framework for the commutant lifting theorem. This yields a direct proof of the commutant lifting theorem and of the Schur type fractional representation of its solutions. By incorporating all the data in one system, we obtain a new Schur type fractional representation of the solutions in the commutant lifting theorem. This third type of Schur representation is precisely the Redheffer scattering matrix obtained by cascading a fixed two port system with an arbitrary passive one port system. The last Schur representations in Chapters XIII and XIV are explicitly computed for Hankel operators. (The formulas in Chapter XIII applied to strictly contractive Hankel operators yield the fractional representations in [AAK 4] and [GoKW 1], while the formulas in Chapter XIV correspond to cascading of a fixed two port to an arbitrary passive one port in network theory. The connection between these two types of formulas is also discussed in Chapter XIV.) In Chapter XV we use the geometry of the Naimark dilation to study positive definite sequences of operators and their corresponding choice sequences, and to obtain the Levinson algorithm for block Toeplitz matrices. Chapter XVI first establishes a complete characterization of all 2 by 2 and 3 by 3 positive block matrices. These results are used to solve the positive block version of the classical Carathéodory interpolation problem, and at the same time provide some deeper insight in the block Levinson algorithm. Chapter XVII uses the wave equation to show how some of the layered medium models in Chapter III naturally arise in geophysics.

We remark that even though most of the mathematical arguments in this book are elementary, because of the many facets of the commutant lifting theorem the underlying geometric structures and their computational counterparts are often complex. Our hope is that this book will encourage researchers to find other ways of understanding the commutant lifting theorem which will make all its facets transparent.

We warmly thank our good friend I. Gohberg for his constant encouragement and support in writing this book. We also thank our colleagues D. Augenstein, H. Bercovici, R. L. Ellis, D. C. Lay, D. K. Le, M. Ostoja-Starzewski, P. J. Sherman and A. Tannenbaum who read parts of the manuscript and provided us with valuable remarks. Many of the results in this book to which the authors contributed were obtained in collaboration with other mathematicians to whom we also express our thanks. This manuscript was typed on the Unix system using troff with pic and mm macros command package. Finally, this work was supported in part by grants from the research fund of Indiana University, the National Science Foundation Grant ECS-8419354 and the Office of Naval Research Grant N00014-89-J-1747.

Indiana University	and	Purdue University, September 1989	C. F. and A. E. F.
Bloomington		West Lafayette	

CHART

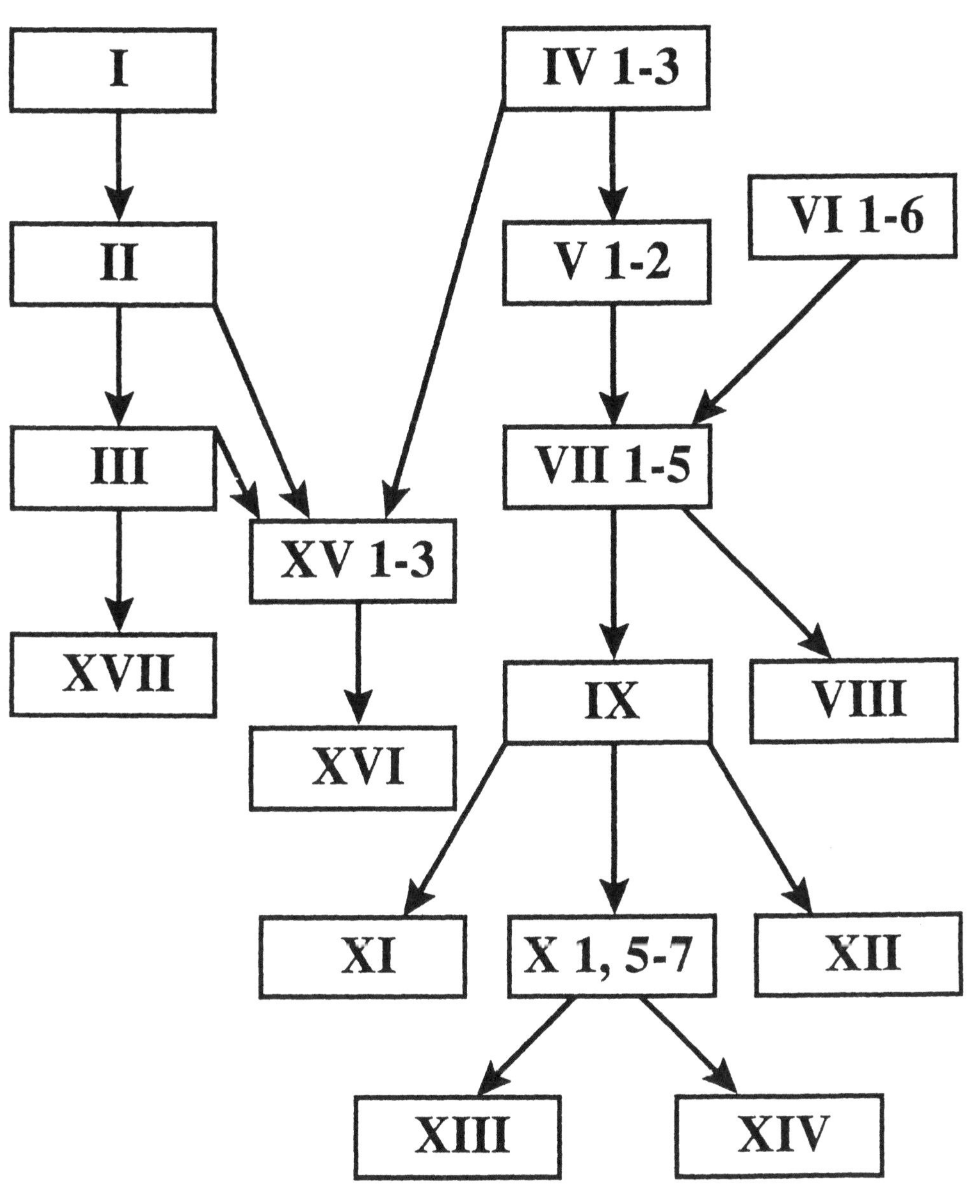

The numbers represent the basic sections required for further reading.

TABLE OF CONTENTS

TABLE OF CONTENTS

Page

Page

CHAPTER I

ANALYSIS OF THE CARATHEODORY INTERPOLATION PROBLEM

In this chapter we present the Schur method for solving the Carathéodory interpolation problem. This method relies on a sequence of numbers, known as the Schur numbers associated with the data for the interpolation problem. It is shown that unique solutions to the Carathéodory interpolation problem are connected with Blaschke products and stable polynomials. As an application we obtain the Jury test for the stability of a polynomial. Next, we give the Schur representation for the set of all solutions to the Carathéodory interpolation problem. The part specified by the data in this representation is a fractional transformation with polynomial coefficients. We introduce elementary techniques from operator theory to give some interesting geometric interpretations of these polynomials coefficients. Using these techniques we also prove that the Carathéodory interpolation problem is solvable if and only if a certain matrix is a contraction. From this we obtain the Schur-Cohn test for the stability of a polynomial. However, due to the importance of Schur numbers in applications, we also present whenever possible several algorithms to recursively compute the Schur numbers from the data (inverse algorithms). It will be easy for the reader to convert these algorithms to recursively compute the data from the Schur numbers (direct algorithms).

1. THE CARATHEODORY INTERPOLATION PROBLEM
AND SCHUR NUMBERS

In this section we present Schur's solution to the classical Carathéodory interpolation problem. From this we derive a "layer peeling algorithm" to recursively compute the Schur numbers associated with the Carathéodory interpolation problem.

Throughout, the open unit disc is denoted by D. The Hardy space H^∞ is the set of all analytic functions f in D whose norm

$$\|f\|_\infty \overset{\Delta}{=} \sup_{z \in D} |f(z)|$$

is finite. The closed unit ball in H^∞ is denoted by H^∞_1, that is, H^∞_1 is the set of all functions f in H^∞ satisfying $\|f\|_\infty \leq 1$. The following Carathéodory interpolation problem plays an important role in this book:

1.1. PROBLEM. Given a polynomial

$$a_o + a_1 z + \cdots + a_{n-1} z^{n-1} \tag{1.1}$$

find necessary and sufficient conditions for the existence of an $f \in H_1^\infty$ such that $f(z)$ has a series expansion of the form:

$$f(z) = a_o + a_1 z + \cdots + a_{n-1} z^{n-1} + 0(z^n) \quad (z \in D) . \tag{1.2}$$

This naturally leads to the following question, raised and solved by Schur.

1.2 PROBLEM. Give an explicit description of all solutions (if there are any) to Problem 1.1.

To present Schur's solution it is convenient to call the string

$$a = (a_o, a_1, ..., a_{n-1}) \tag{1.3}$$

of n complex numbers *the data of Problem 1.1* (or the *Carathéodory data of length n) for the function* $f \in H_1^\infty$ *in* (1.2). Let us recall the following elementary result.

1.3 LEMMA. *Let* λ, μ, *and* a *be complex numbers with* $|a| < 1$ *and*

$$\mu = \omega_a(\lambda) \stackrel{\Delta}{=} (\lambda+a)(1+\overline{a}\lambda)^{-1}. \tag{1.4}$$

Then $\mu \in D$ *if and only if* $\lambda \in D$. *Furthermore,* λ *and* μ *uniquely determine each other by (1.4) and*

$$\lambda = \omega_{-a}(\mu) . \tag{1.5}$$

We proceed now with the presentation of Schur's solution. To this end, let us solve Problem 1.2 for a single point $a = (a_o)$. The general case will follow from this by induction. First, the maximum principle implies that Problem 1.1 has no solution if $|a_o| > 1$, and that it has the unique solution $f(z) = a_o$ if $|a_o| = 1$. Now assume that $|a_o| < 1$. Let $f \in H_1^\infty$ satisfy $f(0) = a_o$ and consider the analytic function f_1 in D defined for $z \in D$ and $z \neq 0$ by

$$f_1 = \frac{1}{z}\, \omega_{-a_o}(f(z)) = \frac{1}{z}\, \frac{f(z)-a_o}{1-\overline{a}_o f(z)}\,.\tag{1.6}$$

By the classical Schwarz Lemma the function f_1 is in H^∞. The map σ taking any such f to the corresponding f_1 is clearly invertible:

$$f(z) = \omega_{a_o}(zf_1(z)) = \frac{a_o + zf_1(z)}{1+\overline{a}_o zf_1(z)}\quad (z \in D)\,.\tag{1.7}$$

Since $|a_o| < 1$, the map σ is one-to-one by Lemma 1.3. This lemma, the maximum principle and the fact that $|a_o| < 1$, imply that $f \in H_1^\infty$ if and only if $f_1 \in H_1^\infty$. Concluding we have the following result.

    1.4 LEMMA. *Let* $a = (a_o)$. *If* $|a_o| > 1$, *then Problem 1.1 has no solution. If* $|a_o| = 1$, *then Problem 1.1 has a unique solution* $f = a_o$. *If* $|a_o| < 1$, *then the set of all solutions to Problem 1.1 is given by (1.7), where* $f_1 \in H_1^\infty$ *is arbitrary. In this case, f and f_1 uniquely determine each other.*

    We shall use the convention that $0/0 = 0$ and $\infty \pm \infty = \infty$. So if $f \equiv a_o$ and $|a_o| = 1$, then $f_1 = \sigma f$ in (1.6) is identically 0. Also if $|a_o| = 1$, then (1.7) gives $f \equiv a_o$ for any $f_1 \in H_1^\infty$. Therefore, (1.7) describes all solutions to Problem 1.1 with data $a = (a_o)$ when $|a_o| \leq 1$.

    The general solution to Problems 1.1 and 1.2 can be obtained by recursively applying Lemma 1.4. To this end, let $a = (a_o, a_1, \ldots, a_{n-1})$ be the Carathéodory data for f in H_1^∞. The case $|a_o| \geq 1$ is covered by Lemma 1.4. So we assume that $|a_o| < 1$. Let $f_1 = \sigma f$ be the function defined in (1.6) and let $a^1 = (a_0^1, a_1^1, \ldots, a_{n-2}^1)$ be the data (of length $n - 1$) for f_1. By using $zf_1(1-\overline{a}_o f) = f - a_o$ and matching like coefficients of z^j we have

$$a_o^1 = a_1(1-|a_o|^2)^{-1}$$

$$a_j^1 = (a_{j+1} + \overline{a}_o \sum_{i=1}^{j} a_{j-i}^1 a_i)(1-|a_o|^2)^{-1}\quad (1 \leq j \leq n-2)\,,\tag{1.8}$$

or equivalently,

$$a_1 = a_o^1(1-|a_o|^2)$$

$$a_{j+1} = a_j^1(1-|a_o|^2) - \overline{a}_o \sum_{i=1}^{j} a_{j-i}^1 a_i\quad (1 \leq j \leq n-2)\,.\tag{1.9}$$

By a slight abuse of notation we also define σ by

$$\sigma(a_0, a_1, ..., a_{n-1}) \overset{\Delta}{=} (a_0^1, ..., a_{n-2}^1) .\tag{1.10}$$

Equations (1.8), (1.9) show that a and $(a_0, \sigma a)$ uniquely determine each other. Recall that $f \in H_1^\infty$ if and only if $f_1 = \sigma f \in H_1^\infty$. Therefore, obtaining a solution f to Problem 1.1 with the data a is equivalent to obtaining a solution $f_1 (=\sigma f)$ to Problem 1.1 with data σa. So the strategy in solving Problem 1.1 is to recursively apply σ until we have reduced the data to length one.

To demonstrate how this is done we define σa by (1.10) even if $|a_0| \geq 1$, with the convention that $a_j^1 = \infty$ if $|a_0| = 1$ and the numerator in the right side of (1.8) is nonzero (otherwise if $|a_0| = 1$, then $a_j^1 = 0/0 = 0$). Then

$$\sigma^i a = (a_0^i, a_1^i, ..., a_{n-1-i}^i)$$

is well defined for $0 \leq i \leq n-1$. Let $r_i = a_0^i$ for $0 \leq i \leq n-1$. In particular, $\sigma^0 a = a$ and $r_0 = a_0$. Therefore, if $|r_0| > 1$, Problem 1.1 has no solution, by Lemma 1.4. If $|r_0| = 1$, Problem 1.1 has a unique solution if and only if $\sigma a = (0, 0, ..., 0)$. This in turn obviously is equivalent to $r_1 = r_2 = \cdots = r_{n-1} = 0$. If $|r_0| < 1$, then by the above discussion, Problem 1.1 with data a is solvable if and only if it is also solvable with data σa. Moreover in this case, any solution f_1 of the latter yields a solution f of the former by

$$f = \omega_{r_0}(zf_1) .$$

By induction if $|r_0| < 1$, $|r_1| < 1$, ..., $|r_{i-1}| < 1$ (for $i \geq 1$), then all solutions of Problem 1.1 with data a are given by

$$f = \omega_{r_0}(z\omega_{r_1}(...(z\omega_{r_{i-1}}(zf_i))...)) ,$$

where f_i is a solution of Problem 1.1 with data $\sigma^i a$. Repeating for f_i and $\sigma^i a$ our earlier discussion concerning f and a, we infer that Problem 1.1 for data a has no solution if $|r_i| > 1$, or if $|r_i| = 1$ and $r_j \neq 0$ for any $i < j \leq n-1$. Secondly, we infer that if $|r_i| = 1$ and $r_j = 0$ for all $i < j \leq n-1$, then Problem 1.1 has a unique solution, namely

$$f = \omega_{r_0}(z\omega_{r_1}(...(z\omega_{r_{i-1}}(zr_i))...)) .\tag{1.11}$$

Finally, if $|r_i| < 1$ for all $0 \leq i \leq n-1$, then any solution of Problem 1.1 with data a is of the form

$$f = \omega_{r_0}(z\omega_{r_1}(...(z\omega_{r_{i-1}}(z\omega_{r_i}(zf_{i+1})))...)) ,$$

where f_{i+1} is a solution of Problem 1.1 with data $\sigma^{i+1} a$.

Obviously the "numbers" r_i ($0 \leq i \leq n-1$) play a basic role in this approach. They are called the *Schur numbers* of data a, or of the solution f. Using this terminology, we can conclude the inductive argument with the following theorem which solves both Problems 1.1 and 1.2.

1.5 THEOREM. *Let $(r_0, r_1, ..., r_{n-1})$ be the Schur numbers of the data a.*

(i) If $|r_i| = 1$ for some $0 \le i \le n-1$ and if $r_j = 0$ for all $i < j \le n-1$, then Problem 1.1 has precisely one solution, namely (1.11).

(ii) If $|r_i| < 1$ for all $0 \le i \le n-1$, then the set of all solutions to Problem 1.1 are given by

$$f = \omega_{r_0}(z\omega_{r_1}(...(z\omega_{r_{n-1}}(zf_n))...)), \tag{1.12}$$

where $f_n \in H_1^\infty$ is arbitrary. In this case, f and f_n uniquely determine each other.

(iii) If the hypotheses of either (i) or (ii) do not hold, then there is no solution to Problem 1.1.

1.6 REMARK. The above theorem shows that if the set of all solutions for Problem 1.1 is not a singleton, then it is parameterized by the closed unit ball in H^∞, that is, H_1^∞.

We note that the Schur numbers $(r_0, r_1, ..., r_{n-1})$ and the data $(a_0, a_1, ..., a_{n-1})$ determine each other when all the r_i's are finite. This follows from (1.8) and (1.9). Furthermore, let $f \in H^\infty$ and let $r_0, r_1, ...$ be the Schur numbers of f and $r_0', r_1', ...$ those of $\sigma^i f$. Then $r_0' = r_i$, $r_1' = r_{i+1}, ...$ This follows at once by examining the previous induction argument. Also we note that in case (i) of Theorem 1.5, the formulas (1.11) and (1.12) coincide. Finally, it is noted that $f \in H_1^\infty$ if and only if all its Schur numbers r_i $(0 \le i < \infty)$ are in the closed unit disc.

One can compute the Schur numbers r_i from the data $(a_0, a_1, ..., a_{n-1})$ by using (1.8). This requires approximately $(1/3)\, n^3$ computations. To complete this section, we will present the "layer peeling algorithm" to compute the Schur numbers. This will require approximately n^2 computations. To this end, let $f = f_0$ be a function in H_1^∞ and f_j the function defined inductively by (1.6) or (1.12), that is,

$$f_j = \frac{r_j + zf_{j+1}}{1 + \bar{r}_j zf_{j+1}} \qquad (r_j = f_j(0)). \tag{1.13}$$

Now assume that u_j and y_j are functions in H^∞ satisfying $u_j(0) \ne 0$ and $f_j u_j = y_j$. Obviously u_j and y_j uniquely determine f_j by $f_j = y_j / u_j$ and $r_j = f_j(0) = y_j(0)/u_j(0)$. Using $f_j u_j = y_j$ in (1.13) and solving for f_{j+1} we have

$$f_{j+1}(\bar{r}_j y_j - u_j) = z^{-1}(r_j u_j - y_j) \ .$$

Let u_j and y_j be the functions in H^∞ defined by

$$u_{j+1} = (\bar{r}_j y_j - u_j) \quad \text{and} \quad y_{j+1} = z^{-1}(r_j u_j - y_j) \quad \text{for } z \neq 0 \ . \tag{1.14}$$

This and $r_j u_j(0) = y_j(0)$ imply that $u_{j+1}(0) = (|r_j|^2 - 1)u_j(0)$. So if $|r_j| \neq 1$ and $u_j(0) \neq 0$ then $u_{j+1}(0) \neq 0$. Moreover, (1.14) shows that one can compute u_{j+1} and y_{j+1} recursively from u_j and y_j. In fact, (1.14) is a difference equation with the initial condition u_0 and y_0. Furthermore, if $u_0(0) \neq 0$ and $|r_i| = |f_i(0)| < 1$ for all $0 \leq i \leq j$, then $u_{j+1}(0) \neq 0$ and the next Schur number r_{j+1} is given by $r_{j+1} = y_{j+1}(0)/u_{j+1}(0)$. This leads to the following algorithm for computing the Schur number from f_0: First let u_0 be any function in H^∞ satisfying $u_0(0) \neq 0$ and set $y_0 = f_0 u_0$. (In fact one can choose $u_0 = 1$ and $y_0 = f_0$.) Then use (1.14) to compute u_j and y_j. If $|r_i| \neq 1$ for all $0 \leq i < j$, then the next Schur number is $r_j = y_j(0)/u_j(0)$.

To obtain the Schur numbers from the data $(a_0, ..., a_{n-1})$ let

$$u_j(z) = \sum_{i \geq 0} u_{j,i} z^i \quad \text{and} \quad y_j(z) = \sum_{i \geq 0} y_{j,i} z^i \qquad (z \in D) \tag{1.15}$$

be the power series expansion of u_j and y_j. Substituting (1.15) into (1.14) and equating coefficients of z^i we have

$$\begin{bmatrix} u_{j+1,0} & 0 \\ u_{j+1,1} & y_{j+1,0} \\ u_{j+1,2} & y_{j+1,1} \\ u_{j+1,3} & y_{j+1,2} \\ \vdots & \vdots \end{bmatrix} = \begin{bmatrix} u_{j,0} & y_{j,0} \\ u_{j,1} & y_{j,1} \\ u_{j,2} & y_{j,2} \\ u_{j,3} & y_{j,3} \\ \vdots & \vdots \end{bmatrix} \begin{bmatrix} -1 & r_j \\ \bar{r}_j & -1 \end{bmatrix} \ . \tag{1.16}$$

By (1.2) and the fact that $y_0 = f_0 u_0$ we have

$$y_{0,i} = \sum_{m=0}^{i} a_{i-m} u_{0,m} \ . \tag{1.17}$$

Thus $(y_{0,0}, ..., y_{0,n-1})$ is uniquely determined by the data $(a_0, ..., a_{n-1})$ and "the input" $(u_{0,0}, ..., u_{0,n-1})$. This and (1.16) shows that $(u_{0,0}, ..., u_{0,n-1})$ and $(y_{0,0}, ..., y_{0,n-1})$ uniquely determine $(u_{j,0}, ..., u_{j,n-1-j})$ and $(y_{j,0}, ..., y_{j,n-1-j})$ for all $0 \leq j \leq n-1$. So we can use (1.16) to recursively determine the Schur numbers

$$r_j = y_{j,0}/u_{j,0} \quad (\text{for } 0 \leq j < n) \tag{1.18}$$

from the data $(a_0, ..., a_{n-1})$. This recursive algorithm is summarized in the following procedure known as the *layer peeling algorithm*. The name "layer peeling" comes from its interpretation in Geophysics discussed in Chapter III, Section 4.

1.7 PROCEDURE. Let $(a_o, ..., a_{n-1}) = (y_{0,0}, ..., y_{0,n-1})$ be the data for Problem 1.1. Set $(u_{o,o}, ..., u_{o,n-1}) = (1, 0, 0, ..., 0)$ and $r_o = a_o$. For $j = 1, 2, ..., n-1$ recursively compute $u_{j,i}$ and $y_{j,i}$ (for $0 \leq i < j$) and r_j by interlacing (1.16) and (1.18).

The previous procedure provides a fast recursive algorithm for computing all the Schur numbers r_j when $|r_j| \neq 1$ for all j. If at any point in this procedure $|r_j| = 1$, then this algorithm stops. In this case there is a unique solution or no solution to Problem 1.1. A finer analysis is needed to distinguish between these two possibilities. This analysis is presented in Proposition 4.1 in Section 4.

2. BLASCHKE PRODUCTS, UNIQUE EXTENSIONS AND THE JURY TEST

In this section we will show that if there is a unique solution to the Carathéodory interpolation problem, then this solution is given by a Blaschke product. Then we will use Blaschke products to obtain the Jury test, which gives necessary and sufficient conditions for the zeros of a polynomial p to lie in the open unit disc, and does not involve factoring p.

To begin, we say that f is a *finite Blaschke product if*

$$f = \gamma \prod_{i=1}^{n} \left(\frac{z - \alpha_i}{1 - \overline{\alpha}_i z} \right) \tag{2.1}$$

where $|\alpha_i| < 1$ for all i and γ is a constant of modulus one. The *order* of the Blaschke product is n. Obviously f is analytic in D. It is easy to verify that $|z - \alpha_i|^2 = |1 - \overline{\alpha}_i z|^2$ when $z = e^{it}$. Thus $|f(e^{it})| = 1$ for all t in $[0, 2\pi]$. By the maximum principle f is in H_1^∞. Hence, a finite Blaschke product is a rational function f in H_1^∞ satisfying $|f(e^{it})| = 1$ for all t in $[0, 2\pi]$. Moreover, one can show that f is a finite Blaschke product if and only if f is a rational analytic function in D satisfying $|f(e^{it})| = 1$ for all t in $[0, 2\pi]$, or equivalently, f is a rational function in H_1^∞ satisfying $|f(e^{it})| = 1$ for all t in $[0, 2\pi]$.

If p(z) is a polynomial of degree less than or equal to n, then the *reverse polynomial* $J_{n+1}p$ is defined by

$$J_{n+1}p \overset{\Delta}{=} z^n \overline{p\left(\frac{1}{\overline{z}}\right)}.$$

The operator J_{n+1} reverses the order and conjugates the coefficients of z^i in p(z). The reverse polynomial $J_{n+1}p$ is also denoted by $p^{\#} = J_{n+1}p$. Finally, notice that if $m \geq n$, then $J_{m+1}p = z^{m-n}J_{n+1}p$.

A polynomial p is called *stable* if all the zeros of p are in the open unit disc D. The following result provides another characterization of finite Blaschke products.

2.1 LEMMA. *A rational function f(z) is a Blaschke product of order n if and only if $f = p/p^{\#}$, where p is a stable polynomial of degree n and $p^{\#}(=J_{n+1}p)$ is its reverse polynomial.*

PROOF. If p is a stable polynomial of degree n, then

$$p(z) = c \prod_{i=1}^{n} (z - \alpha_i) \tag{2.2}$$

where $|\alpha_i| < 1$ for all i and the coefficient c of z^n is not zero. A simple calculation shows that

$$p^{\#}(z) = \overline{c} \prod_{i=1}^{n} (1 - \overline{\alpha}_i z) . \tag{2.3}$$

Thus $f = p/p^{\#}$ is the Blaschke product of order n. On the other hand, if f is the Blaschke product of order n given by (2.1), we choose a complex number c such that $c^2 = \gamma$ and we define the stable polynomial p by (2.2), where α_i are the zeros of f. Then $p^{\#}$ is given by (2.3) and $f = p/p^{\#}$ because $|c| = 1$ and $c/\overline{c} = c^2 = \gamma$. This completes the proof.

2.2 REMARK. Let p be a polynomial of degree n and assume that $f = p/p^{\#}$ is in H^{∞}. Since $|f(e^{it})| = 1$ for all t we conclude by an earlier remark that f is a Blaschke product of order less than or equal to n. If f is a Blaschke product of order $j (\leq n)$, this means that $n - j$ poles and zeros in $p/p^{\#}$ cancel. Consulting (2.2) and (2.3), we see that α is a zero of p if and only if $1/\overline{\alpha}$ is a zero of $p^{\#}$. Therefore, we have a pole-zero cancellation in $p/p^{\#}$ if and only if p has a zero on the unit circle, or p has zeros at α and $1/\overline{\alpha}$.

Now as in the proof of Theorem 1.5, we consider the functions f_j recursively defined by

$$f_j = \frac{r_j + zf_{j+1}}{1 + \overline{r}_j zf_{j+1}} \quad \text{and} \quad f_{j+1} = \frac{1}{z}[\frac{f_j - r_j}{1 - \overline{r}_j f_j}] \tag{2.4}$$

where $r_j = f_j(0)$ is the corresponding Schur number. The following result plays an important role in our approach to the uniqueness question in the Carathéodory problem.

2.3 LEMMA. *The function f_j is a Blaschke product of order $i \geq 1$ if and only if* $|f_j(0)| = |r_j| < 1$ *and f_{j+1} is a Blaschke product of order $i-1$.*

PROOF. Assume that $f_j = p_i/p_i^{\#}$ where p_i is a stable polynomial of degree i and $p_i^{\#} = J_{i+1}p_i$. If

$$p_i(z) = c_o + c_1 z + c_2 z^2 + \cdots + c_i z^i \ , \tag{2.5}$$

then $c_i \neq 0$ and $r_j = f_j(0) = c_o/\overline{c}_i$ satisfies $|r_j| < 1$, because f_j is not a constant. Substituting this into the second equation in (2.4) we see that $f_{j+1} = p_{i-1}/p_{i-1}^{\#}$, where p_{i-1} is a polynomial of degree less than or equal to $i-1$ given by

$$p_{i-1}(z) = z^{-1}(p_i(z) - r_j p_i^{\#}(z)) = \tag{2.6}$$
$$(c_1 - r_j\overline{c}_{i-1}) + (c_2 - r_j\overline{c}_{i-2})z + \cdots + (c_i - r_j\overline{c}_o)z^{i-1}$$

and $p_{i-1}^{\#} = J_i p_{i-1}$ is the reverse polynomial for p_{i-1}. Because f_{j+1} is in H^{∞} Remark 2.2 implies that f_{j+1} is a Blaschke product of order less than or equal to $i-1$. Conversely, assume that $f_{j+1} = p_{i-1}/p_{i-1}^{\#}$ where p_{i-1} is a stable polynomial of degree $i-1$, namely

$$p_{i-1}(z) = d_o + d_1 z + \cdots + d_{i-1} z^{i-1}$$

and $p_{i-1}^{\#} = J_i p_{i-1}$; also assume that the coefficient r_j in (2.4) is in D. Then a simple computation shows that $f_j = p_i/p_i^{\#}$ with

$$p_i(z) = r_j p_{i-1}^{\#}(z) + z p_{i-1}(z) =$$
$$r_j\overline{d}_{i-1} + (r_j\overline{d}_{i-2} + d_o)z + \cdots + (r_j\overline{d}_o + d_{i-2})z^{i-1} + d_{i-1}z^i \ .$$

Since f_j is in H^{∞} an application of Lemma 2.1 verifies that f_j is a Blaschke product of order less than or equal to i. The two established facts also readily show that when both f_j and f_{j+1} are Blaschke products their order differs by one. This completes the proof.

We are now ready for the main result of this section.

2.4 THEOREM. *Let $r_o, r_1, r_2, \cdots$ be the Schur numbers for a function f in H_1^{∞}. Then f is a Blaschke product of order n if and only if $|r_n| = 1$. Moreover, if the Schur numbers r_i for the Carathéodory data $(a_o, a_1, ..., a_{n-1})$ satisfy $|r_i| < |r_j| = 1$ for all $0 \leq i < j$ and $r_i = 0$ for $i > j$, then there is a unique solution to Problem 1.1 and this solution is a Blaschke product of order j.*

PROOF. Let $f = f_0$ be a Blaschke product of order n. By applying the previous lemma recursively we see that f_j defined in (2.4) is a Blaschke product of order $n - j$ and $|r_j| = |f_j(0)| < 1$ for all $j < n$. Furthermore, f_n is a Blaschke product of order zero, or equivalently, $f_n = r_n$ is a constant of modulus one. Conversely if $|r_n| = 1$, then $f_n = r_n$ is a Blaschke product of order zero. As above by recursively using the previous lemma we see that $f = f_0$ is a Blaschke product of order n. The second statement of the theorem now follows from the first statement already proven and from Theorem 1.5. This completes the proof.

To complete this section, we will obtain the Jury test to check whether or not the zeros of a polynomial lie in D. This is the analogue of the Routh test for the half plane. To this end, recall that Lemmas 2.1 and 2.3 show that p_n is a stable polynomial of degree n if and only if $f_0 = p_n/p_n^{\#}$ is a Blaschke product of order n, or equivalently, $|r_0| = |p_n(0)/p_n^{\#}(0)| < 1$ and $f_1 = p_{n-1}/p_{n-1}^{\#}$ is a Blaschke product of order $n - 1$, where p_n and p_{n-1} are given by (2.5) and (2.6) with $n = i$ and $j = 0$. By Lemma 2.3 this implies that p_n is a stable polynomial of degree n if and only if $|r_0| = |p_n(0)/p_n^{\#}(0)| < 1$ and p_{n-1} is a stable polynomial of degree $n - 1$. By repeating the above analysis we see that p_{n-1} is a stable polynomial of degree $n - 1$ if and only if $|r_1| = |p_{n-1}(0)/p_{n-1}^{\#}(0)| < 1$ and p_{n-2} is a stable polynomial of degree $n - 2$ where p_{n-1} and p_{n-2} are given by (2.5) and (2.6) with $n - 1 = i$ and $j = 1$. Continuing in this fashion, we see that p_n is a stable polynomial of degree n if and only if $|r_i| < 1$ for all $0 \leq i \leq n - 2$ and p_1 is a stable polynomial of degree one. However, $p_1 = az + b$ is a stable polynomial of degree one if and only if $|r_{n-1}| = |b/\bar{a}| < 1$. Therefore, p_n is a stable polynomial of degree n if and only if $|r_i| < 1$ for all $0 \leq i \leq n - 1$. Summarizing the above analysis produces the following result due to E.I. Jury.

2.5 THEOREM. *Let* p_n *be a polynomial of degree* n *and* p_{n-j} *the polynomials of degree less than or equal to* $n - j$ *recursively defined by*

$$p_{n-j-1}(z) = z^{-1}(p_{n-j}(z) - r_j p_{n-j}^{\#}(z)) \tag{2.7}$$

where $p_{n-j}^{\#} = J_{n-j+1} p_{n-j}^{\#}$ *is the reverse polynomial of* p_{n-j} *and the Schur number* r_j *is given by* $p_{n-j}(0)/p_{n-j}^{\#}(0)$. *Then* p_n *is a stable polynomial of degree* n *if and only if* $|r_j| < 1$ *for all* $0 \leq j \leq n - 1$. *Moreover,* p_{n-j} *is a stable polynomial of degree* $n - j$ *if and only if* $|r_j| < 1$ *and* p_{n-j-1} *is a stable polynomial of degree* $n - j - 1$.

Using the fact that r_j is the j–th Schur number for $f_0 = p_n/p_n^{\#}$ we immediately obtain the following result.

2.6 COROLLARY. *A polynomial p_n of degree n is stable if and only if the Schur numbers r_j for $f_o = p_n/p_n^\#$ satisfy $|r_j| < 1$ for all $0 \le j \le n-1$.*

3. THE SCHUR REPRESENTATION

In this section we shall give a closed and more explicit form to the parameterization by H_1^∞ of all solutions to Problem 1.1, in the case when this Problem has a solution.

We begin with the following result.

3.1 LEMMA. *Let $(r_o, r_1, ..., r_{n-1})$ be a string of complex numbers in the closed unit disc. Let $n > 1$,*

$$\begin{bmatrix} u_n & x_n \\ v_n & y_n \end{bmatrix} \triangleq \begin{bmatrix} 1 & zr_{n-1} \\ \bar{r}_{n-1} & z \end{bmatrix} \cdots \begin{bmatrix} 1 & zr_1 \\ \bar{r}_1 & z \end{bmatrix} \begin{bmatrix} 1 & r_o \\ \bar{r}_o & 1 \end{bmatrix} \tag{3.1}$$

and

$$\begin{bmatrix} u_1 & x_1 \\ v_1 & y_1 \end{bmatrix} \triangleq \begin{bmatrix} 1 & r_o \\ \bar{r}_o & 1 \end{bmatrix}. \tag{3.2}$$

Then u_n, v_n, x_n, and y_n are polynomials of degree less than or equal to $n-1$ such that $u_n(0) = 1$,

$$\det \begin{bmatrix} u_n & x_n \\ v_n & y_n \end{bmatrix} = z^{n-1}(1 - |r_o|^2) \cdots (1 - |r_{n-1}|^2) \quad (z \in \mathbb{C}) \tag{3.3}$$

and

$$u_n = J_n y_n, \qquad v_n = J_n x_n. \tag{3.4}$$

PROOF. Equation (3.3) and the fact that u_n, v_n, x_n, y_n are polynomials of degree less than or equal to $n - 1$, follows from (3.1) and (3.2). So assume that (3.4) is true for u_{n-1}, v_{n-1}, x_{n-1} and y_{n-1}. Using

$$\begin{bmatrix} u_n & x_n \\ v_n & y_n \end{bmatrix} = \begin{bmatrix} 1 & zr_{n-1} \\ \bar{r}_{n-1} & z \end{bmatrix} \begin{bmatrix} u_{n-1} & x_{n-1} \\ v_{n-1} & y_{n-1} \end{bmatrix} \tag{3.5}$$

we have by induction

$$u_n(z) = u_{n-1}(z) + z r_{n-1} v_{n-1}(z) = J_{n-1} y_{n-1}(z) + z r_{n-1} J_{n-1} x_{n-1}(z) =$$

$$z^{n-2} \overline{y_{n-1}(\frac{1}{\overline{z}})} + z^{n-1} r_{n-1} \overline{x_{n-1}(\frac{1}{\overline{z}})} = z^{n-1} \overline{[\frac{1}{\overline{z}} y_{n-1}(\frac{1}{\overline{z}}) + \overline{r}_{n-1} x_{n-1}(\frac{1}{\overline{z}})]} =$$

$$z^{n-1} \overline{y_n(\frac{1}{\overline{z}})} = J_n y_n(z) \qquad (z \in C \text{ and } z \neq 0).$$

A similar calculation yields the second equation in (3.4). This completes the proof.

The main result of this section is

3.2   THEOREM. *Assume that Problem 1.1 has a solution with data* $a = (a_0, a_1, ..., a_{n-1})$, *and let* $(r_0, r_1, ..., r_{n-1})$ *be the Schur numbers of* a. *Define recursively the polynomials* x_j *and* y_j *for* $1 \leq j \leq n$ *by*

$$\begin{bmatrix} x_j \\ y_j \end{bmatrix} = \begin{bmatrix} 1 & z r_{j-1} \\ \overline{r}_{j-1} & z \end{bmatrix} \begin{bmatrix} x_{j-1} \\ y_{j-1} \end{bmatrix} \quad (z \in C \text{ and } 2 \leq j \leq n) \tag{3.6}$$

where

$$x_1(z) = r_0 \quad \text{and} \quad y_1(z) = 1 . \tag{3.7}$$

Then x_j *and* y_j *are polynomials of degree less than or equal to* $j-1$. *Moreover, any solution* f *to Problem 1.1 is of the form*

$$f(z) = \frac{x_n(z) + z y_n(z) f_n(z)}{y_n^{\#}(z) + z x_n^{\#}(z) f_n(z)} , \tag{3.8}$$

where $f_n \in H_1^{\infty}$ *is arbitrary, and* $x_n^{\#} \overset{\Delta}{=} J_n x_n$ *and* $y_n^{\#} \overset{\Delta}{=} J_n y_n$. *The solution to Problem 1.1 is unique if and only if*

$$x_n x_n^{\#} - y_n y_n^{\#} = 0 \quad (\text{for all } z \in C) . \tag{3.9}$$

Finally, if (3.9) does not hold, then formula (3.8) provides a one-to-one correspondence between a solution $f \in H_1^{\infty}$ *to Problem 1.1 and* $f_n \in H_1^{\infty}$.

PROOF. Lemma 3.1 and (3.5) imply that x_j and y_j are polynomials of degree less than or equal to $j-1$. By Theorem 1.5 the set of all solutions f to Problem 1.1 is given by (1.12). We claim that

$$f = \frac{x_n + zy_n f_n}{u_n + zv_n f_n} \,, \tag{3.10}$$

where $f_n \in H_1^\infty$ and x_n, y_n, u_n, v_n are the polynomials in (3.1) and (3.2). Equations (1.7) and (3.2) show that (3.10) holds when $n = 1$. Now use induction and assume that (3.10) holds for $n-1$, that is,

$$f = \frac{x_{n-1} + zy_{n-1} f_{n-1}}{u_{n-1} + zv_{n-1} f_{n-1}} \,. \tag{3.11}$$

Substituting

$$f_{n-1} = \omega_{r_{n-1}}(zf_n) = \frac{r_{n-1} + zf_n}{1 + \bar{r}_{n-1} zf_n} \tag{3.12}$$

into (3.11) with (3.5) yields (3.10). Equation (3.10) and Lemma 3.1 give (3.8). Furthermore, part (ii) of Theorem 1.5, and (3.11), (3.12) show that f and f_n in (3.10) uniquely determine each other, when the solution to Problem 1.1 is not unique.

To complete the proof, we show that (3.9) is equivalent to a unique solution of Problem 1.1. This follows from (3.3), (3.4) and the fact that the solution is unique if and only if one of the Schur numbers has modulus one. This completes the proof.

The formula (3.8) is called the *Schur representation* of the solution of Problem 1.1. We emphasize once again that in this representation the polynomials x_n and y_n depend only on the data, while f_n is arbitrary in H_1^∞. To complete this section we will present several useful corollaries of the Schur representation.

3.3 COROLLARY. *Let the* x_j *and* y_j *in Lemma 3.1 be of the form*

$$\begin{aligned}
x_j(z) &= x_{j,0} + x_{j,1} z + \cdots + x_{j,j-1} z^{j-1} \\
y_j(z) &= y_{j,0} + y_{j,1} z + \cdots + y_{j,j-1} z^{j-1} \,.
\end{aligned} \tag{3.13}$$

Then

$$\begin{aligned}
x_{j,0} &= r_0, \qquad x_{j,j-1} = r_{j-1}, \qquad y_{j,j-1} = 1 \quad (1 \le j \le n) \\
y_{j,0} &= \bar{r}_{j-1} r_0, \qquad (1 < j \le n)
\end{aligned} \tag{3.14}$$

and

$$\begin{aligned}
x_{j,i} &= x_{j-1,i} + r_{j-1} y_{j-1,i-1}, \\
y_{j,i} &= \bar{r}_{j-1} x_{j-1,i} + y_{j-1,i-1} \quad (2 \le j,\ 1 \le i \le j-2) \,.
\end{aligned} \tag{3.15}$$

PROOF. Equation (3.1) gives for $1 < j \le n$

$$\begin{bmatrix} u_j(0) & x_j(0) \\ v_j(0) & y_j(0) \end{bmatrix} = \begin{bmatrix} 1 & r_o \\ \bar{r}_{j-1} & \bar{r}_{j-1}r_o \end{bmatrix}.$$

This and (3.4) yield (3.14), and (3.6) yields (3.15). This completes the proof.

3.4 COROLLARY. *If Problem 1.1 has a solution, then*

$$a_j = a_j^o + r_j t_j \tag{3.16}$$

where

$$t_j = (1 - |r_o|^2) \cdots (1 - |r_{j-1}|^2) \tag{3.17}$$

and $a_1^o = 0$,

$$a_j^o = -(a_{j-1}\bar{y}_{j,j-2} + \cdots + a_1\bar{y}_{j,o}) \quad (1 < j \le n). \tag{3.18}$$

PROOF. Equation (1.9) shows that the corollary holds for $j = 1$. So let $j > 1$. Then by the Schur representation (3.11) with n replaced by $j+1$ and (3.3) with n replaced by j we have:

$$u_j f = \frac{x_j(u_j + zv_j f_j) + z(u_j y_j - x_j v_j)f_j}{u_j + zv_j f_j} = x_j + \frac{z^j f_j t_j}{u_j + zv_j f_j},$$

where t_j is given by (3.17). Matching like coefficients of z^j in this equation, and using $f_j(0) = r_j$ and the fact that x_j is a polynomial of degree $j-1$ produces

$$a_j u_{j,o} + a_{j-1} u_{j,1} + \cdots + a_1 u_{j,j-1} = r_j t_j.$$

This with $u_j = y_j^\#$ and the definition of a_j^o in (3.18) yields (3.16). This completes the proof.

Equation (3.16) says that a_j is contained in a closed ball $a_j^o + \bar{D}t_j$ with center a_j^o and radius t_j. The center a_j^o and radius t_j only depend upon the data $(a_o, a_1, ..., a_{j-1})$. In other words, if there exists a solution to Problem 1.1 with data $(a_o, ..., a_{n-2})$, then there is a solution to Problem 1.1 with data $(a_o, ..., a_{n-1})$ if and only if a_{n-1} is contained in the ball $a_{n-1}^o + \bar{D}t_{n-1}$. This observation readily produces the following result.

3.5 COROLLARY. *The solution to Problem 1.1 with data* $a' = (a_o, ..., a_{n-2})$ *is unique if and only if all* $f \in H_1^\infty$ *solving this problem have data of length n of the form* $(a_o, a_1, ..., a_{n-1})$ *with the same* a_{n-1}, *namely* $a_{n-1} = a_{n-1}^o$.

The following is a recursive algorithm which is used to determine when there is a unique solution to Problem 1.1.

3.6 COROLLARY. *Let* $(r_0, r_1, ..., r_{n-1})$ *be the Schur numbers for the data* $(a_0, a_1, ..., a_{n-1})$ *and assume that* $|r_i| < |r_{j-1}| = 1$ *for all* $0 \leq i < j - 1$ *and some* $j < n$. *Then there exists a (unique) f in* H_1^∞ *solving Problem 1.1 if and only if*

$$a_k = -(a_{k-1}\overline{y}_{j,j-2} + \cdots + a_{k+1-j}\overline{y}_{j,o}) \quad (j \leq k < n) \tag{3.19}$$

PROOF. Theorem 3.2 implies that $f_j = 0$ and $f = x_j/y_j^{\#}$ is the unique solution to Problem 1.1 with data $(a_0, a_1, ..., a_{j-1})$. Therefore Problem 1.1 with the Carathéodory data $(a_0, a_1, ..., a_{n-1})$ is solvable if and only if $(a_0, a_1, ..., a_{n-1})$ is the Carathéodory data of f.

Assume that Problem 1.1 is solvable. Then $(a_0, a_1, ..., a_{n-1})$ is the Carathéodory data of f. Recall that the degree of x_j is less than or equal to $j-1$. Using $y_j^{\#}f = x_j$ with $y_{j,j-1} = 1$ and matching like coefficients of z^k yields (3.19). Moreover, the a_k's in (3.19) are uniquely determined by $(a_0, a_1, ..., a_{j-1})$. On the other hand, if the data $(a_0, a_1, ..., a_{n-1})$ happens to satisfy (3.19), then the coefficients $a_j, ..., a_{n-1}$ are recursively determined by $(a_0, ..., a_{j-1})$ in the same way as the data for f. Therefore they must coincide with the Carathéodory data of f. Thus f is a solution to Problem 1.1. This completes the proof.

As demonstrated by the previous corollaries the numbers $t_1 = 1 - |r_0|^2$, t_2, t_3, ... play an important role in interpolation theory. Here is another example of their importance.

3.7 COROLLARY. *If Problem 1.1 has a solution, then*

$$1 - |f(e^{it})|^2 = (1 - |f_n(e^{it})|^2) \frac{t_n}{|y_n^{\#}(e^{it}) + e^{it}f_n(e^{it})x_n^{\#}(e^{it})|^2} \tag{3.20}$$

for almost all $t \in [0, 2\pi)$.

PROOF. From the Schur representation (3.8) we obtain

$$1 - |f(z)|^2 = [(|y_n^{\#}(z)|^2 - |x_n(z)|^2) - |zf_n(z)|^2(|y_n(z)|^2 - |x_n^{\#}(z)|^2) +$$

$$2\mathrm{Re}(zf_n(z)(x_n^{\#}(z)\overline{y_n^{\#}(z)} - y_n(z)\overline{x_n(z)}))](|y_n^{\#}(z) + zx_n^{\#}(z)f_n(z)|)^{-2} \ .$$

For $z = e^{it}$

$$|y_n^\#(z)|^2 - |x_n(z)|^2 = (y_n y_n^\# - x_n x_n^\#)(z)\bar{z}^{n-1} = t_n = |y_n(z)|^2 - |x_n^\#(z)|^2$$

and

$$x_n^\#(z)\overline{y_n^\#(z)} - y_n(z)\overline{x_n(z)} = 0$$

(see (3.3) and (3.4)). Formula (3.20) follows now from the fact that since f and f_n are both in H^∞, the values

$$f(e^{it}) = \lim_{r \to 1-} f(re^{it}) \text{ and } f_n(e^{it}) = \lim_{r \to 1-} f_n(re^{it})$$

are defined almost everywhere in $[0, 2\pi)$; see [Ho].

4. COMPUTING SCHUR NUMBERS

The reader who is anxiously awaiting an application can proceed to Chapter III, to see how Schur numbers naturally arise in layered medium problems in Geophysics. However, one should not skip the rest of this section. Recall that in Section 1 we briefly introduced an algorithm to compute the Schur numbers from the Carathéodory data. In this section and the next section we will present in a more detailed way several other algorithms to compute the Schur numbers from the data $(a_0, a_1, ..., a_{n-1})$. These algorithms may also play an important role in certain aspects of Mathematics, Electrical Engineering and Geophysics.

To this end, let us recall some elementary facts concerning the functional calculus for linear operators on C^j. Let T be any operator on C^j and let $p(z) = c_0 + c_1 z + c_2 z^2 + \cdots$ be any polynomial with complex coefficients. Then

$$P(T) \overset{\Delta}{=} c_0 I + c_1 T + c_2 T^2 + \cdots$$

where I is the identity operator on C^j, that is, $I\xi = \xi$ for all $\xi \in C^j$. For example, if T_j is the operator given by the multiplication with the j by j matrix with 1's immediately below the diagonal and zero's elsewhere, that is,

$$T_j \overset{\Delta}{=} \begin{bmatrix} 0 & 0 & 0 & \ldots & 0 & 0 & 0 \\ 1 & 0 & 0 & \ldots & 0 & 0 & 0 \\ 0 & 1 & 0 & \ldots & 0 & 0 & 0 \\ \cdot & \cdot & \cdot & & \cdot & \cdot & \cdot \\ \cdot & \cdot & \cdot & & \cdot & \cdot & \cdot \\ 0 & 0 & 0 & \ldots & 1 & 0 & 0 \\ 0 & 0 & 0 & \ldots & 0 & 1 & 0 \end{bmatrix} \tag{4.1}$$

then $p(T_j)$ is the multiplication with the matrix

$$p(T_j) = \begin{bmatrix} c_o & 0 & 0 & \cdots & 0 & 0 \\ c_1 & c_o & 0 & \cdots & 0 & 0 \\ c_2 & c_1 & c_o & \cdots & & \cdot \\ \cdot & \cdot & \cdot & \cdots & \cdot & \cdot \\ \cdot & \cdot & \cdot & \cdots & \cdot & \cdot \\ \cdot & \cdot & \cdot & \cdots & c_o & 0 \\ c_{j-1} & c_{j-2} & c_{j-3} & \cdots & c_1 & c_o \end{bmatrix} . \tag{4.2}$$

In the sequel we shall not distinguish between matrices and the corresponding linear operators. So the matrices in (4.1) and (4.2) will also be denoted by T_j and $p(T_j)$. Let us also notice that the formula (4.2) for $p(T_j)$ also makes sense when $p = c_o + c_1 z + \ldots$ is the power expansion of a function analytic in a neighborhood of the origin. It is easy to check that for f and g in H^∞:

$$(f+g)(T_j) = f(T_j) + g(T_j) \quad \text{and} \quad (fg)(T_j) = f(T_j)g(T_j) . \tag{4.3}$$

Since $f(T_j)$ is a lower triangular matrix with $f(0)$ as its diagonal entry, $f(T_j)$ is invertible if and only if $f(0)$ is nonzero.

Finally, for fixed Carathéodory data $a = (a_o, a_1, \ldots, a_{n-1})$ we shall need the matrix A_j defined by

$$A_j \triangleq \begin{bmatrix} a_o & 0 & 0 & \cdots & 0 & 0 \\ a_1 & a_o & 0 & \cdots & 0 & 0 \\ a_2 & a_1 & a_o & \cdots & \cdot & \cdot \\ \cdot & \cdot & \cdot & \cdots & \cdot & \cdot \\ \cdot & \cdot & \cdot & \cdots & \cdot & \cdot \\ \cdot & \cdot & \cdot & \cdots & a_o & 0 \\ a_{j-1} & a_{j-2} & a_{j-3} & \cdots & a_1 & a_o \end{bmatrix} \tag{4.4}$$

for $1 \le j \le n$. Notice that if f is any solution of Problem 1.1 with data a, then

$$A_j = f(T_j) . \tag{4.5}$$

As an illustration of this approach, we present the following as a practical test for the existence of a solution to Problem 1.1 when one of the Schur numbers is of modulus one.

4.1 **PROPOSITION.** *Let* $(r_o, r_1, \ldots, r_{n-1})$ *be the Schur numbers for the data* $(a_o, a_1, \ldots, a_{n-1})$ *and assume that* $|r_i| < |r_{j-1}| = 1$ *for all* $0 \le i < j-1$ *and some* $j < n$. *Then there exists a (unique)* f *in* H_1^∞ *solving Problem 1.1 if and only if* $y_j^\#(T_n)A_n = x_j(T_n)$, *or equivalently, (writing out these n by n matrices)*

$$\begin{bmatrix} \overline{y}_{j,j-1} & 0 & \cdots & \cdots & 0 \\ \cdot & \overline{y}_{j,j-1} & \cdots & \cdots & \cdot \\ \cdot & \cdot & \cdots & \cdots & \cdot \\ \cdot & \cdot & \cdots & \cdots & \cdot \\ \overline{y}_{j,o} & \cdot & \cdots & \cdots & \cdot \\ 0 & \overline{y}_{j,o} & \cdots & \cdots & \cdot \\ \cdot & 0 & \cdots & \cdots & \cdot \\ \cdot & \cdot & \cdots & \cdots & \cdot \\ \cdot & \cdot & \cdots & \cdots & 0 \\ 0 & 0 & \cdots & \cdots & \overline{y}_{j,j-1} \end{bmatrix} \quad A_n = \begin{bmatrix} x_{j,o} & 0 & \cdots & \cdots & 0 \\ \cdot & x_{j,o} & \cdots & \cdots & \cdot \\ \cdot & \cdot & \cdots & \cdots & \cdot \\ \cdot & \cdot & \cdots & \cdots & \cdot \\ x_{j,j-1} & \cdot & \cdots & \cdots & \cdot \\ 0 & x_{j,j-1} & \cdots & \cdots & \cdot \\ \cdot & 0 & \cdots & \cdots & \cdot \\ \cdot & \cdot & \cdots & \cdots & \cdot \\ \cdot & \cdot & \cdots & \cdots & 0 \\ 0 & 0 & \cdots & \cdots & x_{j,o} \end{bmatrix} \tag{4.6}$$

PROOF. Theorem 3.2 implies that $f = x_j/y_j^\#$ is the unique solution to Problem 1.1 with data $(a_o, ..., a_{j-1})$. Hence, Problem 1.1 with data $(a_o, ..., a_{n-1})$ has a unique solution (namely $f = x_j/y_j^\#$) if and only if $f(T_n) = A_n$. For this f we have by (4.5),

$$y_j^\#(T_n)f(T_n) = x_j(T_n) \tag{4.7}$$

where $y_j^\#(T_n)$ and $x_j(T_n)$ are the first and third matrices in (4.6). Equation (3.14) implies that $y_{j,j-1} = 1$. Hence $y_j^\#(T_n)$ is invertible. So the matrix $f(T_n)$ in (4.7) is uniquely determined by $y_j^\#$ and x_j. Therefore (4.6) holds if and only if $f(T_n) = A_n$, or equivalently, $f = x_j/y_j^\#$ is the unique solution to Problem 1.1 with data $(a_o, ..., a_{n-1})$. This completes the proof.

The preceding considerations will allow us to follow a matrix approach to develop some algorithms to compute the Schur numbers. We start with the following result.

4.2 PROPOSITION. *Assume that all the Schur numbers* $(r_0, r_1, ..., r_{n-1})$ *of the data* $(a_0, a_1, ..., a_{n-1})$ *are contained in the unit disc* D *and* $n \geq 2$. *Then*

$$r_o = a_o, \quad r_1 = a_1(1 - |a_o|^2)^{-1} \tag{4.8}$$

$$r_j = \frac{a_j + a_1\overline{y}_{j,o} + \cdots + a_{j-1}\overline{y}_{j,j-2}}{1 - a_o\overline{x}_{j,o} - \cdots - a_{j-1}\overline{x}_{j,j-1}} = \frac{a_j - a_j^o}{t_j} \quad (2 \leq j \leq n-1) \tag{4.9}$$

where x_j *and* y_j *are given in (3.6), (3.7) and (3.13) while* a_j^o *is in (3.18).*

PROOF. Equation (4.8) follows from (1.8). Taking $f_n = 0$ and $j+1$ instead of n in the Schur representation (3.8), we have $y_{j+1}^\# f = x_{j+1}$, where $f \in H_1^\infty$ is a Carathéodory function for the data $(a_o, ..., a_j)$, and x_{j+1} and y_{j+1} are defined in (3.1). Thus by (4.3) and

(4.5) we have $y_{j+1}^{\#}(T_{j+1})f(T_{j+1}) = x_{j+1}(T_{j+1})$, or equivalently, in matrix form:

$$
\begin{bmatrix}
\overline{y}_{j+1,j} & 0 & \cdots & 0 \\
\cdot & \overline{y}_{j+1,j} & \cdots & \cdot \\
\cdot & \cdot & \cdots & \cdot \\
\cdot & \cdot & \cdots & \cdot \\
\cdot & \cdot & \cdots & \cdot \\
\overline{y}_{j+1,o} & \overline{y}_{j+1,1} & \cdots & \overline{y}_{j+1,j}
\end{bmatrix}
A_{j+1} =
\begin{bmatrix}
x_{j+1,o} & 0 & \cdots & 0 \\
\cdot & x_{j+1,o} & \cdots & \cdot \\
\cdot & \cdot & \cdots & \cdot \\
\cdot & \cdot & \cdots & \cdot \\
\cdot & \cdot & \cdots & 0 \\
x_{j+1,j} & x_{j+1,j-1} & \cdots & x_{j+1,o}
\end{bmatrix}
\tag{4.10}
$$

where T_j and A_j are defined in (4.1) and (4.4). Matching the two south west entries yields

$$\overline{y}_{j+1,o}a_o + \cdots + \overline{y}_{j+1,j}a_j = x_{j+1,j} .$$

This and (3.14), (3.15) give

$$a_o r_j \overline{x}_{j,o} + a_1(r_j \overline{x}_{j,1} + \overline{y}_{j,o}) + \cdots + a_{j-1}(r_j \overline{x}_{j,j-1} + \overline{y}_{j,j-2}) + a_j = r_j . \tag{4.11}$$

Substituting (3.18) into (4.11) gives

$$a_j = a_j^o + r_j(1 - a_o \overline{x}_{j,o} - \cdots - a_{j-1}\overline{x}_{j,j-1}) . \tag{4.12}$$

Notice that $x_{j,i}$ only depends on the data $(a_o, ..., a_{j-1})$. So the next a_j is given by (4.12), where r_j is any complex number in D. Comparing this to (3.16) we see that

$$t_j = 1 - a_o \overline{x}_{j,o} - \cdots - a_{j-1}\overline{x}_{j,j-1} \tag{4.13}$$

where t_j is defined in (3.17). By the hypothesis all the Schur numbers are in D. So $t_j \neq 0$. Now (4.9) follows from (3.18), (4.12) and (4.13). This completes the proof.

As noted earlier, one can use (1.8) to compute the Schur numbers $(r_0, r_1, ..., r_{n-1})$ from the data $(a_o, a_1, ..., a_{n-1})$. This requires about $(1/3) n^3$ computations. The following presents a far more efficient method of finding r_j and requires approximately $4n^2$ computations. In other words, the above proposition leads to the following recursive algorithm to compute the Schur numbers from the data.

4.3 PROCEDURE. Let $(a_o, a_1, ..., a_{n-1})$ be the Carathéodory data and set $r_o = a_o$. For $j = 1, ..., n-1$ recursively compute x_j, y_j and r_j by interlacing formulas (3.13), (3.14), (3.15) with formula (4.9). Notice that the $(j+1)$ Schur number is given by $r_j = (a_j - a_j^o)/t_j$ where for a_j^o we use formula (3.18) while for t_j we can use either (3.17) or (4.13). If at any step $|r_j| \geq 1$, then the algorithm stops. In this case, there is no solution if $|r_j| > 1$. If $|r_j| = 1$, then one checks (3.19) or (4.6) to determine if there is a solution, necessarily unique.

5. COMPUTING THE SCHUR NUMBERS FROM $I-A_nA_n^*$

In this Section we present a slightly more robust variation of Procedure 4.3 by using the entries of the matrices $1-A_jA_j^*$ for $1 \le j \le n$. In order to fix the framework throughout the present Section, we shall assume that Problem 1.1 with data $(a_0, a_1, ..., a_{n-1})$ is solvable. We begin with the following result.

5.1 PROPOSITION. *If Problem 1.1 has a solution, then*

$$(I-A_jA_j^*) \begin{bmatrix} x_{j+1,1} \\ x_{j+1,2} \\ \cdot \\ \cdot \\ \cdot \\ x_{j+1,j} \end{bmatrix} = \begin{bmatrix} a_1 \\ a_2 \\ \cdot \\ \cdot \\ \cdot \\ a_j \end{bmatrix} \tag{5.1}$$

where the $x_{j+1,i}$'s are defined in (3.14) and (3.15) and $1 \le j \le n-1$.

PROOF. If $j = 1$, then (5.1) reduces to (4.8), so we assume $j > 1$. Theorem 3.2 gives

$$y_{j+1}^{\#}(T_{j+1})f(T_{j+1}) + T_{j+1}x_{j+1}^{\#}(T_{j+1})f(T_{j+1})f_{j+1}(T_{j+1}) =$$

$$\tag{5.2}$$

$$x_{j+1}(T_{j+1}) + T_{j+1}y_{j+1}(T_{j+1})f_{j+1}(T_{j+1})$$

where $f_{j+1} \in H_1^{\infty}$, while $f \in H_1^{\infty}$ is a solution to Problem 1.1 with data $(a_0, a, ..., a_j)$. As before, using $f(T_{j+1}) = A_{j+1}$ and $f_{j+1} = 0$ this yields

$$y_{j+1}^{\#}(T_{j+1})A_{j+1} = x_{j+1}(T_{j+1}) . \tag{5.3}$$

Now choosing $f_{j+1} = 1$ in (5.2) and subtracting (5.3), we find that

$$T_{j+1}x_{j+1}^{\#}(T_{j+1})A_{j+1} = T_{j+1}y_{j+1}(T_{j+1}) . \tag{5.4}$$

Thus,

$$
\begin{bmatrix}
\overline{x}_{j+1,j} & 0 & \cdots & 0 \\
\overline{x}_{j+1,j-1} & \overline{x}_{j+1,j} & \cdots & \cdot \\
\cdot & \cdot & \cdots & \cdot \\
\cdot & \cdot & \cdots & \cdot \\
\cdot & \cdot & \cdots & 0 \\
\overline{x}_{j+1,o} & \overline{x}_{j+1,1} & \cdots & \overline{x}_{j+1,j}
\end{bmatrix} A_{j+1} -
\begin{bmatrix}
y_{j+1,o} & 0 & \cdots & 0 \\
\cdot & y_{j+1,o} & \cdots & \cdot \\
\cdot & \cdot & \cdots & \cdot \\
\cdot & \cdot & \cdots & \cdot \\
\cdot & \cdot & \cdots & 0 \\
y_{j+1,j} & y_{j+1,j-1} & \cdots & y_{j+1,o}
\end{bmatrix} =
$$

$$
\begin{bmatrix}
0 & 0 & . & . & . & . & . & 0 & 0 \\
0 & 0 & . & . & . & . & . & 0 & 0 \\
\cdot & \cdot & \cdot & \cdot & \cdot & \cdot & \cdot & \cdot & \cdot \\
\cdot & \cdot & \cdot & \cdot & \cdot & \cdot & \cdot & \cdot & \cdot \\
\cdot & \cdot & \cdot & \cdot & \cdot & \cdot & \cdot & \cdot & \cdot \\
0 & 0 & . & . & . & . & . & 0 & 0 \\
* & * & . & . & . & . & . & * & *
\end{bmatrix}
\tag{5.5}
$$

where * is possibly a nonzero entry. Equating the row above the last one in (5.5), we have

$$
[\overline{x}_{j+1,1}, \ldots, \overline{x}_{j+1,j}]\, A_j = [y_{j+1,j-1}, \ldots, y_{j+1,o}]\,.
\tag{5.6}
$$

Now using (5.3) and the fact that A_{j+1} and $y_{j+1}^{\#}(T_{j+1})$ commute, we have

$$
A_{j+1}
\begin{bmatrix}
\overline{y}_{j+1,j} & 0 & \cdots & 0 \\
\cdot & \overline{y}_{j+1,j} & \cdots & \cdot \\
\cdot & \cdot & \cdots & \cdot \\
\cdot & \cdot & \cdots & \cdot \\
\cdot & \cdot & \cdots & 0 \\
\overline{y}_{j+1,o} & \overline{y}_{j+1,1} & \cdots & \overline{y}_{j+1,j}
\end{bmatrix} =
\begin{bmatrix}
x_{j+1,o} & 0 & \cdots & 0 \\
\cdot & x_{j+1,o} & \cdots & \cdot \\
\cdot & \cdot & \cdots & \cdot \\
\cdot & \cdot & \cdots & \cdot \\
\cdot & \cdot & \cdots & 0 \\
x_{j+1,j} & x_{j+1,j-1} & \cdots & x_{j+1,o}
\end{bmatrix}.
\tag{5.7}
$$

The first column in (5.7) with $y_{j+1,j} = 1$ yields

$$
A_j
\begin{bmatrix}
\overline{y}_{j+1,j-1} \\
\overline{y}_{j+1,j-2} \\
\cdot \\
\cdot \\
\overline{y}_{j+1,o}
\end{bmatrix}
+
\begin{bmatrix}
a_1 \\
a_2 \\
\cdot \\
\cdot \\
a_j
\end{bmatrix}
=
\begin{bmatrix}
x_{j+1,1} \\
x_{j+1,2} \\
\cdot \\
\cdot \\
x_{j+1,j}
\end{bmatrix}.
\tag{5.8}
$$

Finally, taking the *adjoint,* that is, complex conjugate transpose of (5.6) and substituting it into (5.8) gives (5.1) and completes the proof.

A similar identity to (5.1), which will play an important role in the general interpolation theory developed in later chapters, is given by the following result.

5.2 PROPOSITION. *If Problem 1.1 has a solution, then*

$$(I - A_j A_j^*) \begin{bmatrix} y_{j+1,o} \\ \cdot \\ \cdot \\ \cdot \\ y_{j+1,j-1} \end{bmatrix} = A_j \begin{bmatrix} \bar{a}_j \\ \cdot \\ \cdot \\ \cdot \\ \bar{a}_1 \end{bmatrix} \tag{5.9}$$

where the $y_{j+1,i}$'s are defined in (3.14) and (3.15) and $1 \le j \le n-1$.

PROOF. Comparing the last row in (4.10) yields

$$[\bar{y}_{j+1,o}, \, ..., \, \bar{y}_{j+1,j-1}]A_j + [a_j, \, ..., \, a_1] = [x_{j+1,j}, \, ..., \, x_{j+1,1}] \, ,$$

since $y_{j+1,j} = 1$. Taking the adjoint and multiplying on the left by A_j gives

$$A_j \begin{bmatrix} \bar{x}_{j+1,j} \\ \cdot \\ \cdot \\ \cdot \\ \bar{x}_{j+1,1} \end{bmatrix} = A_j A_j^* \begin{bmatrix} y_{j+1,o} \\ \cdot \\ \cdot \\ \cdot \\ y_{j+1,j-1} \end{bmatrix} + A_j \begin{bmatrix} \bar{a}_j \\ \cdot \\ \cdot \\ \cdot \\ \bar{a}_1 \end{bmatrix} . \tag{5.10}$$

Moving the A_{j+1} matrix on the left of the $[x_{j,i}]$ matrix in (5.5) and comparing the first column yields

$$A_j \begin{bmatrix} \bar{x}_{j+1,j} \\ \cdot \\ \cdot \\ \cdot \\ \bar{x}_{j+1,1} \end{bmatrix} = \begin{bmatrix} y_{j+1,o} \\ \cdot \\ \cdot \\ \cdot \\ y_{j+1,j-1} \end{bmatrix} .$$

Substituting this into (5.10) produces (5.9) and completes the proof.

5.3 PROPOSITION. *If Problem 1.1 has a solution, then*

$$a_j = r_j t_j - [a_{j-1}, ..., a_1] A_{j-1}^* \begin{bmatrix} x_{j,1} \\ \cdot \\ \cdot \\ \cdot \\ x_{j,j-1} \end{bmatrix} = r_j t_j + a_j^o \qquad (5.11)$$

PROOF. This follows by substituting the adjoint of (5.6) with $j+1$ replaced by j into (3.18) and using (3.16).

Finally, we notice that a_j^o is given by (3.18) and the last term in (5.11).

Denote the last row of the matrix $I - A_j A_j^*$ by

$$[d_{j,1}, d_{j,2}, ..., d_{j,j}]$$

and note that

$$d_{j,i} = -(a_{j-1}\bar{a}_{i-1} + a_{j-2}\bar{a}_{i-2} + \cdots + a_{j-i}\bar{a}_o) \quad (1 \le i \le j-1)$$
$$d_{j,j} = 1 - |a_o|^2 - \cdots - |a_{j-1}|^2 . \qquad (5.12)$$

The Schur coefficient r_j can be computed directly from these coefficients as shown by the following result.

5.4 PROPOSITION. *Assume that all the Schur numbers* $(r_o, r_1, ..., r_{n-1})$ *of data* $(a_o, a_1, ..., a_{n-1})$ *are in the unit disk* D *and* $n \ge 3$. *Then*

$$r_j = \frac{a_j - (d_{j,1}x_{j,1} + \cdots + d_{j,j-1}x_{j,j-1})}{d_{j,j} + d_{j,1}y_{j,o} + \cdots + d_{j,j-1}y_{j,j-2}} \quad (2 \le j \le n-1) \qquad (5.13)$$

where $x_{j,i}$ *and* $y_{j,i}$ *are given by (3.14) and (3.15).*

PROOF. Notice that (3.14), (3.15) and (5.1) give

$$a_j = (d_{j,1}x_{j,1} + \cdots + d_{j,j-1}x_{j,j-1}) + r_j(d_{j,j} + d_{j,1}y_{j,o} + \cdots + d_{j,j-1}y_{j,j-2}) . \qquad (5.14)$$

Notice that $x_{j,i}$, $d_{j,i}$ and $y_{j,i}$ only depend upon the data $(a_o, ..., a_{j-1})$. Equation (5.14) says that when r_j runs through $\overline{D}$ the next a_j runs through a specified closed disk. On the other hand, equation (3.16) says that a_j runs through a closed disc with center a_j^o and radius t_j. Both discs must have the same center and radius. Thus

$$a_j^o = d_{j,1}x_{j,1} + \cdots + d_{j,j-1}x_{j,j-1}$$

and

$$t_j = d_{j,j} + d_{j,1} y_{j,o} + \cdots + d_{j,j-1} y_{j,j-2} \, . \tag{5.15}$$

Now (5.13) follows from (5.14) and the fact that $t_j \neq 0$. This completes the proof.

The preceding proposition provides the following recursive algorithm for calculating the Schur numbers from the data.

5.6 PROCEDURE. Let $(a_0, a_1, ..., a_{n-1})$ be the Carathéodory data. First, notice that the $d_{j,i}$'s in (5.12) can be computed recursively for $i \geq 2$ by the following difference equation

$$d_{j,1} = -a_{j-1} \bar{a}_0, \ \ d_{j,2} = -a_{j-1} \bar{a}_1 + d_{j-1,1} \, ,$$

$$d_{j,3} = -a_{j-1} \bar{a}_2 + d_{j-1,2} \, , \ ..., \ d_{j,j-1} = -a_{j-1} \bar{a}_{j-2} + d_{j-1,j-2} \, , \tag{5.16}$$

$$d_{j,j} = d_{j-1,j-1} - |a_{j-1}|^2$$

with the initial conditions

$$[d_{2,1}, d_{2,2}] = [-a_1 \bar{a}_0, \ 1 - |a_o|^2 - |a_1|^2] \, . \tag{5.17}$$

Now set $r_o = a_0$. For $j = 1, 2, ..., n-1$ recursively compute x_j, y_j and r_j by interlacing (3.14), (3.15) with (5.13) (notice that one can compute the $d_{j,i}$'s in parallel with x_j, y_j and r_j.) As before if $|r_j| \geq 1$ for any j, then the solution to Problem 1.1 does not exist, or is unique.

This procedure takes approximately $5n^2$ calculations to compute r_j. Obviously, it takes more calculations than our previous procedures. However, in this algorithm the terms $d_{j,i}$ are more "robust" to "noise" than in our previous algorithms. Finally, we can replace the denominator in the formulas (4.9) and (5.13) by t_j. These coefficients can be recursively computed by $t_{j+1} = (1 - |r_j|^2) t_j$.

We complete this section with the following basic observation.

5.7 REMARK. All the formulas in this chapter hold when $r_0, r_1, ..., r_{n-2}$ are in the unit disc, regardless of the value of r_{n-1}.

6. CONTRACTIONS AND THE CARATHEODORY-FEJER THEOREMS

In this section we will show that there exists a solution to the Carathéodory interpolation problem if and only if $\|A_n\| \leq 1$, where A_n is the matrix in (4.4) with $j = n$.

To begin, let C^j be endowed with the usual scalar product

$$(\xi, \eta) = \xi_1 \bar{\eta}_1 + \cdots + \xi_j \bar{\eta}_j \qquad (\xi = [\xi_i]_1^j \text{ and } \eta = [\eta_i]_1^j \in \mathbf{C}^j)$$

and with the corresponding norm $\|\xi\| = (\xi, \xi)^{1/2}$. Let T be a linear operator on $\mathbf{C}^j$. The norm of T is defined by

$$\|T\| = \sup\{\|T\xi\| : \xi \in \mathbf{C}^j \text{ and } \|\xi\| \le 1\} .$$

An operator T is a *contraction* if $\|T\| \le 1$. If $\|T\| < 1$, then T is a *strict contraction*. For such an operator the functional calculus considered in Section 4 for polynomials can be extended to power series

$$p(z) = c_o + c_1 z + c_2 z^2 + \cdots$$

provided that

$$|c_o| + |c_1| + |c_2| + \cdots \; < \infty \; . \tag{6.1}$$

Indeed, in this case,

$$\|c_o I\| + \|c_1 T\| + \|c_2 T^2\| + \cdots \le |c_o| + |c_1| + |c_2| + \cdots < \infty$$

Therefore the operator

$$p(T) \overset{\Delta}{=} c_o I + c_1 T + c_2 T^2 + \cdots$$

makes sense. (Notice that absolute convergence of a series in norm implies the absolute convergence of the entries of the corresponding series of matrices.) A simple computation also shows that

$$(p + q)(T) = p(T) + q(T) \quad \text{and} \quad (pq)(T) = p(T)q(T) \tag{6.2}$$

where p and q are power series satisfying (6.1). In particular, if $|a| < 1$, then

$$p(z) = (1 + \bar{a}z)^{-1} = \sum_{0}^{\infty} (-\bar{a})^n z^n \qquad (z \in D)$$

satisfies (6.1). Thus $p(T)$ makes sense and (6.2) shows that indeed

$$p(T) = (I + \bar{a}T)^{-1} = \sum_{0}^{\infty} (-\bar{a})^n T^n \; .$$

Also for $|a| < 1$ we have

$$\omega_a(T) = (T + a)(I + \bar{a}T)^{-1} \tag{6.3}$$

where ω_a is defined in (1.4).

6.1 LEMMA. *Let* $|a| < 1$. *If* T *is a contraction, respectively a strict contraction, then* $\omega_a(T)$ *is a contraction, respectively a strict contraction.*

PROOF. If ξ is in C^j, then

$$\|(I + \bar{a}T)\xi\|^2 - \|\omega_a(T)(I + \bar{a}T)\xi\|^2 = \|(I + \bar{a}T)\xi\|^2 - \|(T + a)\xi\|^2 =$$

$$(6.4)$$

$$(1 - |a|^2)(\|\xi\|^2 - \|T\xi\|^2) \geq (1 - |a|^2)(1 - \|T\|^2)\|\xi\|^2 \geq 0 .$$

Setting $\eta = (I + \bar{a}T)\xi$ we have

$$\|\eta\| = \|(I + \bar{a}T)\xi\| \leq \|\xi\| + |a|\,\|\xi\| \leq 2\|\xi\| .$$

Substituting this into (6.4) yields

$$\|\eta\|^2 - \|\omega_a(T)\eta\|^2 \geq (1 - |a|^2)(I - \|T\|^2)\frac{\|\eta\|^2}{4} ,$$

and the lemma is now obvious.

Let T_n be the n by n matrix in (4.1) with $j = n$. Obviously T_n is a contraction. Let $p(z)$ be analytic in a neighborhood of the origin and assume that $p(0) \neq -1/\bar{a}$. Then $\omega_a(p(z))$ is analytic in a neighborhood of the origin and the operator $(\omega_a(p))(T_n)$ makes sense, according to our functional calculus in Section 4. Since $\omega_a(p)(1+\bar{a}p) = p+a$ we deduce from our functional calculus for T_n that $(\omega_a(p))(T_n)(I + \bar{a}p(T_n)) = p(T_n) + aI$. Because $(1 + \bar{a}p)(0)$ is not zero, the operator $I + \bar{a}p(T_n)$ is invertible. Moreover, if $p(T_n)$ is a contraction, then

$$\omega_a(p(T_n)) = (p(T_n) + aI)(I + \bar{a}p(T_n))^{-1} = \omega_a(p)(T_n) \qquad (6.5)$$

where $\omega_a(p(T_n))$ is defined by our functional calculus. By virtue of (6.5) and the previous lemma, we can now state the following result.

6.2 LEMMA. *Let* $p(z)$ *be analytic in a neighborhood of the origin and assume that* $p(0) = -1/\bar{a}$ *where* $|a| < 1$. *If* $p(T_n)$ *is a contraction, respectively strict contraction, then* $\omega_a(p(T_n))$ *is a contraction, respectively strict contraction.*

The solution to the Carathéodory problem will be based on the following matrix.

$$A_n = \begin{bmatrix} a_o & 0 & 0 & \cdots & 0 & 0 \\ a_1 & a_o & 0 & \cdots & 0 & 0 \\ a_2 & a_1 & a_o & \cdots & \cdot & \cdot \\ \cdot & \cdot & \cdot & \cdots & \cdot & \cdot \\ \cdot & \cdot & \cdot & \cdots & \cdot & \cdot \\ \cdot & \cdot & \cdot & \cdots & a_o & 0 \\ a_{n-1} & a_{n-2} & a_{n-3} & \cdots & a_1 & a_o \end{bmatrix} \tag{6.6}$$

where $(a_o, a_1, ..., a_{n-1})$ is the data for Problem 1.1. Recall that this matrix is $f(T_n)$, where $f(z) = a_o + a_1 z + a_2 z^2 + \cdots$ is a solution to the Carathéodory interpolation problem.

6.3 LEMMA. *If Problem 1.1 is solvable, then A_n is a contraction. Moreover, if there is more than one solution, then A_n is a strict contraction.*

PROOF. If there is more than one solution, then according to Theorem 1.5 part (ii), the Schur numbers $(r_0, ..., r_{n-1})$ are all in the open unit disc and a solution is given by

$$f = x_n / y_n^\# = \omega_{r_o}(z\omega_{r_1}(\cdots (z\omega_{r_{n-1}}(g)) \cdots)) \tag{6.7}$$

where $g = 0$. By applying Lemma 6.2 iteratively we see that $A_n = f(T_n)$ is a strict contraction because $g(T_n) = 0$ is a strict contraction. In case there is a unique solution, we refer to Theorem 1.5 part (i) and use formula (1.11) (recall that $|r_i| = 1$) in place of (6.7) to deduce that $A_n = f(T_n)$ is a contraction. This completes the proof.

To obtain the converse of Theorem 6.3 we need a supplemental result to Section 5. Recall that $(r_0, r_1, ..., r_{n-1})$ are the Schur numbers for the data $(a_0, a_1, ..., a_{n-1})$. It is assumed that the Schur numbers $r_0, r_1, ..., r_{n-2}$ are in the unit disc D. By Remark 5.7 the vectors $x_{j,i}$ and $y_{j,i}$ defined in (3.14) and (3.15) for $1 \le j \le n$, $0 \le i \le n-1$ make sense, and the formulas (4.13), (5.5) and (5.7) hold for $j = n-1$. This sets the stage for the following result.

6.4 PROPOSITION. *If the Schur numbers $(r_0, r_1, ..., r_{n-2})$ are in the unit disc D, then*

$$(I - A_n A_n^*) \begin{bmatrix} y_{n,o} \\ . \\ . \\ . \\ y_{n,n-1} \end{bmatrix} = \begin{bmatrix} 0 \\ . \\ . \\ 0 \\ t_n \end{bmatrix} \tag{6.8}$$

where t_n is defined in (3.17).

PROOF. From (5.5) for $j = n-1$ we easily deduce

$$A_n \begin{bmatrix} \overline{x}_{n,n-1} \\ . \\ . \\ . \\ \overline{x}_{n,o} \end{bmatrix} = \begin{bmatrix} y_{n,o} \\ . \\ . \\ . \\ y_{n,n-1} \end{bmatrix} + \begin{bmatrix} 0 \\ . \\ . \\ . \\ 0 \\ t \end{bmatrix} \tag{6.9}$$

where

$$t = (\overline{x}_{n,o} a_o + \cdots + \overline{x}_{n,n-1} a_{n-1}) - y_{n,n-1} . \tag{6.10}$$

In a similar way from (5.7) we deduce

$$A_n^* \begin{bmatrix} y_{n,o} \\ . \\ . \\ . \\ y_{n,n-1} \end{bmatrix} = \begin{bmatrix} \overline{x}_{n,n-1} \\ . \\ . \\ . \\ \overline{x}_{n,o} \end{bmatrix} . \tag{6.11}$$

Applying A_n to (6.11) and using (6.9) we obtain

$$(I - A_n A_n^*) \begin{bmatrix} y_{n,o} \\ . \\ . \\ . \\ y_{n,n-1} \end{bmatrix} = - \begin{bmatrix} 0 \\ . \\ . \\ . \\ 0 \\ t \end{bmatrix}$$

Comparing (4.13) with j replaced by n to (6.10) and using $y_{n,n-1} = 1$ we see that $t_n = -t$. This completes the proof.

We can now give a partial converse to Lemma 6.3.

6.5 THEOREM. *The matrix* A_n *is a strict contraction if and only if all the Schur numbers* r_0, r_1, ..., r_{n-1} *are in* D.

PROOF. By virtue of Lemma 6.3 we only have to prove that if A_n is a strict contraction then all the Schur numbers are in D. To this end we proceed by induction on n. The theorem is obviously true for n = 1. Assume it is true for n–1. Observing that A_{n-1} may be regarded as the restriction of A_n to the space $\{0\} \oplus C^{n-1}$, we can infer that r_0, ..., r_{n-2} belong to D. Therefore formulas (6.8) holds and taking the scalar product with the vector $\xi = [y_{n,j}]_{j=0}^{n-1}$, we obtain

$$t_n = ((1 - A_n A_n^*)\xi, \xi) > 0$$

Hence $1 - |r_{n-1}|^2 > 0$; see (3.17). This completes the proof.

Theorem 6.5 and the following provides a converse to Lemma 6.3.

6.6 LEMMA. *There exists a unique solution to Problem 1.1 if and only if the norm of* A_n *is one.*

PROOF. Assume that the solution to Problem 1.1 is unique. Lemma 6.3 shows that A_n is a contraction. If A_n is a strict contraction, then the solution is not unique by Theorems 1.5 and 6.5. Hence, the norm of A_n is one.

Now assume that $\|A_n\| = 1$. Let j be the first integer satisfying $\|A_{j-1}\| < \|A_j\| = 1$. Theorem 6.5 shows that $(r_0, r_1, ..., r_{j-2})$ are in the unit disc D. Using $\xi = [y_{j,i}]_{i=0}^{j-1}$ and $y_{j,j-1} = 1$ in Proposition 6.4 yields

$$((I - A_j A_j^*)\xi, \xi) = t_j \geq 0 . \qquad (6.12)$$

Since $t_{j-1} > 0$, by (3.17) and Theorem 6.5, this gives $|r_{j-1}| \leq 1$. Notice that $|r_{j-1}| = 1$. Since otherwise $\|A_j\| < 1$ by Theorem 6.5. Consulting (3.17) t_j in (6.12) is zero and ξ is a nonzero vector satisfying $\|A_j^* \xi\| = \|\xi\|$. Let ξ' be the vector in C^n with the first n–j components zero and ξ as the last j components. By using the upper triangular form of A_n^* we have

$$\|\xi\|^2 \geq \|A_n^* \xi'\|^2 = \sum_{k=j}^{n-1} |\bar{a}_k y_{j,j-1} + \cdots + \bar{a}_{k-j+1} y_{j,o}|^2 + \|A_j^* \xi\|^2 =$$

$$\sum_{k=j}^{n-1} |a_k \overline{y}_{j,j-1} + \cdots + a_{k-j+1}\overline{y}_{j,o}|^2 + \|\xi\|^2 \geq \|\xi\|^2 .$$

Therefore all the terms in the sum are zero i.e.,

$$a_k = -(a_{k-1}\overline{y}_{j,j-2} + \cdots + a_{k+1-j}\overline{y}_{j,o}) \qquad (\text{if } j \leq k < n) .$$

This and Corollary 3.6 shows that the solution to Problem 1.1 is unique. This completes the proof.

Combining the previous results yields the main result of this section.

6.7 THEOREM. *Problem 1.1 has a solution if and only if* A_n *is a contraction. Moreover, the solution is unique if and only if the norm of* A_n *is one.*

From this theorem we immediately deduce that if $f(z) = a_o + a_1 z + a_2 z^2 + \cdots$ is in H_1^∞ then the matrix A_n in (6.6) is a contraction for all $n \geq 1$, and therefore $|a_j| \leq 1$ for all j. This also establishes one part of the following result.

6.8 COROLLARY. *Let* $f(z) = a_o + a_1 z + a_2 z^2 + \cdots$ *be analytic for* $|z| < 1$. *Then* $f \in H_1^\infty$ *if and only if the matrix* A_n *in (6.6) is a contraction for all* $n \geq 1$.

PROOF. It remains to prove that if A_n is a contraction for all $n \geq 1$, then $|f(z)| \leq 1$ for all $|z| < 1$. To this end, we observe first that for $\rho < 1$ we have $|a_j| \leq 1 \leq 1/\rho^j$ for all $j \geq 0$. By virtue of Theorem 6.7 for every n there exists a function f_n in H_1^∞ such that

$$f_n(z) = a_o + a_1 z + \cdots + a_{n-1}z^{n-1} + a_n^{(n)}z^n + \cdots$$

By a remark made above $|a_j^{(n)}| \leq 1$ for all $j \geq n$. It follows that if $|z| < \rho$, then

$$|f(z) - f_n(z)| \leq \sum_{j=n}^{\infty} |a_j - a_j^{(n)}| \, |z|^j \leq (\frac{|z|}{\rho})^n \frac{2}{1 - |z|/\rho} \to 0$$

as $n \to \infty$. Since $|f_n(z)| \leq 1$ it follows that $|f(z)| \leq 1$. The desired condition follows now by letting $\rho \to 1$. This completes the proof.

7. THE SCHUR-COHN TEST

In this section we will present the Schur-Cohn test, which states that a polynomial of degree n is stable if and only if a certain n by n matrix is strictly positive.

We say that a n by n matrix A with entries $[a_{i,j}]$ for $1 \leq i \leq n$ and $1 \leq j \leq n$ is *positive* (notation ≥ 0) if

$$(Ax,x) = \sum_{1,1}^{n,n} a_{i,j} x_j \bar{x}_i \geq 0 \qquad (\text{for all } x = [x_i]_1^n \text{ in } C^n).$$

The matrix A is *strictly positive* (notation > 0) if $(Ax,x) > 0$ for all nonzero x in C^n. Let

$$p_n = c_0 + c_1 z + \cdots + c_n z^n \tag{7.1}$$

be a polynomial of degree n and $p_n^\# \overset{\Delta}{=} J_{n+1} p$ be its reverse polynomial. The n by n matrices P_n and Q_n are defined by

$$P_n \overset{\Delta}{=} p_n(T_n) = \begin{bmatrix} c_0 & 0 & \cdots & 0 \\ c_1 & c_0 & \cdots & 0 \\ c_2 & c_1 & \cdots & 0 \\ \vdots & \vdots & \vdots & \vdots \\ c_{n-1} & c_{n-2} & \cdots & c_0 \end{bmatrix} \text{ and } Q_n \overset{\Delta}{=} p_n^\#(T_n) = \begin{bmatrix} \bar{c}_n & 0 & \cdots & 0 \\ \bar{c}_{n-1} & \bar{c}_n & \cdots & 0 \\ \bar{c}_{n-2} & \bar{c}_{n-1} & \cdots & 0 \\ \vdots & \vdots & \vdots & \vdots \\ \bar{c}_1 & \bar{c}_2 & \cdots & \bar{c}_n \end{bmatrix} \tag{7.2}$$

Notice that because c_n is nonzero, Q_n is invertible and the operator $P_n Q_n^{-1}$ is well defined. We supplement the Jury test in Section 2 with the following result, known as the Schur-Cohn test for the stability of a polynomial.

7.1 THEOREM. *Let p_n be a polynomial of degree n. Then the following are equivalent.*

(i) *The polynomial p_n is stable.*

(ii) *The operator $P_n Q_n^{-1}$ is a strict contraction.*

(iii) *The operator $Q_n^* Q_n - P_n^* P_n$ is strictly positive.*

PROOF. Corollary 2.6 shows that p_n is a stable polynomial of degree n if and only if the Schur numbers r_j for $p_n/p_n^\#$ are in the open unit disc D for all $0 \leq j \leq n-1$. This and Theorem 6.5 implies that p_n is a stable polynomial of degree n if and only if $(p_n/p_n^\#)(T_n)$ is a strict contraction. By using our functional calculus in Section 4 we see that $(p_n/p_n^\#)(T_n) = P_n Q_n^{-1}$. Therefore p_n is stable if and only if $P_n Q_n^{-1}$ is a strict contraction. This proves that (i) and (ii) are equivalent. To prove that (ii) and (iii) are equivalent set $x = Q_n y$. Then

$$\|x\|^2 - \|P_n Q_n^{-1} x\|^2 = \|Q_n y\|^2 - \|P_n y\|^2 = ((Q_n^* Q_n - P_n^* P_n) y, y).$$

Because Q_n is invertible, this shows that $P_n Q_n^{-1}$ is a strict contraction if and only if $Q_n^* Q_n - P_n^* P_n$ is strictly positive. This completes the proof.

I.8. NOTES AND COMMENTS

The work of Schur [Schur 1,2], which is the basic content of this chapter, has had a significant impact in operator theory (see [Go 1]). It foreshadowed many of the mathematical methods used in signal processing [Ka 2] and geophysics [RoT 1]. In fact, a major portion of the book is motivated by his work. The first part of Section 1 is due to Schur [Schur 1]. The layer peeling algorithm is taken from [Ka 2]; for related results see [DewVK], [LeK 1,2]. The Jury test in Section 2 is due to Jury [Jury], [JuryB]. However, this test is also implicitly contained in [Schur 2]. Left half plane versions of the Jury test, which is essentially variations of the famous Routh test, are contained in [FrzL]. The results in Sections 3 and 4 are adapted from Schur's papers. The algorithms in Section 5 are special cases of some of the results in [AAK 3]. This fact will be demonstrated in Section XIII.9. Our proof of the Carathédory - Fejér theorem is based on one of the ideas of J. von Neumann in [Neu 4]. Theorem 6.7 and Corollary 6.8 is due to Carathéodory and Fejér [CaFe]. The Schur-Cohn test is due to Schur [Schur 2] and Cohn [Co]. For further results on the Schur-Cohn test see [Par], [PtY 1], [VK], [Y 1]. Some of these results will also be given in Section V.5 and X.8. In connection with this chapter see [AAK 1], [AlpD], [Br 3], [Ca], [Er], [Fejer], [Ge], [Ma] and [Szegö].

CHAPTER II

ANALYSIS OF THE CARATHEODORY INTERPOLATION PROBLEM FOR POSITIVE-REAL FUNCTIONS

In this chapter we will show that the Carathéodory interpolation problem is equivalent to a certain interpolation problem involving positive-real functions and Toeplitz matrices. Then we develop and use the Levinson algorithm from filtering theory to give another recursive algorithm to compute the Schur numbers from the Carathéodory data. Finally, we will present the maximal entropy solution to the Carathéodory interpolation problem.

1. POSITIVE-REAL FUNCTIONS

This section is devoted to positive-real functions and Toeplitz matrices.

A function $\hat{f}$ is *positive-real* if $\hat{f}$ is analytic in D and $\mathrm{Re}(\hat{f}(z)) \geq 0$ for all $z \in$ D. A function F is of *positive type* if $F(z) = \mathrm{Re}(\hat{f}(z))$ where $\hat{f}(z)$ is positive-real. A function F of positive type admits an expansion of the form

$$F(z) = \sum_{o}^{\infty} b_n z^n + \sum_{-1}^{-\infty} b_n \overline{z}^{|n|} \qquad (z \in D) \tag{1.1}$$

which converges in D; indeed this readily follows if we denote the power series expansion of $\hat{f}$ by

$$\hat{f} = b_o + 2 \sum_{1}^{\infty} b_n z^n \quad .$$

We call F a *positive extension* of $(b_o, b_1, ..., b_{n-1})$ if F is of positive type and $(b_o, b_1, ..., b_{n-1})$ are the first n coefficients in (1.1). Finally, notice that if F is of positive type, then $b_n = \overline{b}_{-n}$ and $F(0) = b_o \geq 0$.

Let f be a function in H_1^∞ such that f(0) is real and $f \not\equiv 1$. Consider the function F defined by

$$F(z) = \text{Re } \frac{1+f(z)}{1-f(z)} = \frac{1-|f(z)|^2}{|1-f(z)|^2} \geq 0 \quad (z \in D) \tag{1.2}$$

The *Cayley transform* $\hat{f}$ of f is defined by

$$\hat{f}(z) = \frac{1+f(z)}{1-f(z)} \text{ and } f(z) = \frac{\hat{f}(z)-1}{\hat{f}(z)+1} \quad (z \in D) \tag{1.3}$$

Clearly $\hat{f}$ is analytic in D, and f and $\hat{f}$ uniquely determine each other. Furthermore, $F(z) = $ Re $\hat{f}(z) \geq 0$ for all z in D, that is, $\hat{f}$ is positive-real. Hence F is of positive type.

Now let F be the function of positive type given in (1.1). Consider the function $\hat{f}$ analytic in D given by

$$\hat{f}(z) = \sum_{o}^{\infty} \hat{a}_n z^n = b_o + 2\sum_{1}^{\infty} b_n z^n \quad (z \in D) . \tag{1.4}$$

Obviously $F(z) = $ Re $\hat{f}(z)$ and $\hat{f}$ is positive-real. Let f be the unique analytic function in D defined by (1.3). Using the fact that Re $\hat{f}(z) \geq 0$ in D along with (1.2) and (1.3), it is easy to verify that f is in H_1^∞ and f(0) is real. Note f(0) is real because $b_o \geq 0$. Thus by (1.1) to (1.4) each function F of positive type uniquely determines and is uniquely determined by a function $f \in H_1^\infty$ such that f(0) is real. Throughout this section f, $\hat{f}$ and F are always the functions related by (1.1) to (1.4).

Let f admit a power series expansion of the form

$$f(z) = \sum_{o}^{\infty} a_n z^n \quad (z \in D \text{ and } a_o \text{ real}) . \tag{1.5}$$

Equation (1.3) gives $\hat{f}(1-f) = 1+f$ and $f(1+\hat{f}) = \hat{f}-1$. Expanding these relations into power series and using $b_o = \hat{a}_o$ and $2b_n = \hat{a}_n$ if $n \geq 1$ we have

$$2b_n(1-a_o) = [a_n + b_o a_n + 2\sum_{i=1}^{n-1} b_{n-i} a_i] \qquad (\text{if } n \geq 1)$$

$$a_n(1+b_o) = [2b_n(1-a_o) - 2\sum_{i=1}^{n-1} b_{n-i} a_i] \qquad (\text{if } n \geq 1) \tag{1.6}$$

$$b_o = \frac{1+a_o}{1-a_o} \text{ and } a_o = \frac{b_o-1}{b_o+1} .$$

(Both sums in (1.6) are taken to be zero if $n = 1$.) To avoid trouble, we assume that $a_o \neq 1$ and $b_o \neq -1$. (If $a_o = 1$, then Problem I.1.1 is trivial. If $b_o = -1$, there exists no positive extension of $(b_o, ..., b_{n-1})$. Recall that b_o must be positive. So $a_o = 1$ or $b_o = -1$ is not of interest to us). Equation (1.6) shows that $(a_o, a_1, ..., a_{n-1})$ in (1.5) and

$(b_o, b_1, ..., b_{n-1})$ in (1.1) uniquely determine each other. Combining this with our above analysis shows that f is a solution to Problem I.1.1 with data $(a_o, a_1, ..., a_{n-1})$ if and only if its corresponding F is a positive extension of $(b_o, b_1, ..., b_{n-1})$. In other words, Problem 1.1 in Chapter I is equivalent to the following Carathéodory positive extension problem: Given the data $(b_o, ..., b_{n-1})$, find necessary and sufficient conditions for the existence of a positive extension F of $(b_o, ..., b_{n-1})$.

Throughout the rest of this chapter $(a_o, a_1, ..., a_{n-1})$ and $(b_o, b_1, ..., b_{n-1})$ are related by (1.6), with a_o real and f, $\hat{f}$, F by (1.1) to (1.4). Summing up the previous analysis proves the following result.

1.1 LEMMA. *A function F is of positive type if and only if F is given by (1.2) where f $\in$ H_1^∞ and f(0) is real. Furthermore, this f and F uniquely determine each other by (1.1) to (1.5). A function f is a solution to Problem I.1.1 with data $(a_o, a_1, ..., a_{n-1})$ if and only if its corresponding F defined by (1.1) and (1.6) is a positive extension of $(b_o, b_1, ..., b_{n-1})$. Finally, Problem I.1.1 with data $(a_o, a_1, ..., a_{n-1})$ has a unique solution if and only if there exists a unique positive extension of $(b_o, b_1, ..., b_{n-1})$.*

To complete this section we will show that the data $(b_o, b_1, ..., b_{n-1})$ admits a positive extension if and only if its n by n Toeplitz matrix B_n defined by

$$B_n = \begin{bmatrix} b_o & \overline{b}_1 & \overline{b}_2 & \cdots & \overline{b}_{n-1} \\ b_1 & b_o & \overline{b}_1 & \cdots & \overline{b}_{n-2} \\ b_2 & b_1 & b_o & \cdots & \overline{b}_{n-3} \\ . & . & . & \cdots & . \\ . & . & . & \cdots & . \\ . & . & . & \cdots & . \\ b_{n-1} & b_{n-2} & b_{n-3} & \cdots & b_o \end{bmatrix} \tag{1.7}$$

is positive. Here is the appropriate place to mention that a n by n matrix C with entries $c_{i,j}$ for $1 \le i \le n$ and $1 \le j \le n$ is called *Toeplitz* if $c_{i,j} = c_{i-j}$ for all i and j. Notice that besides the matrix B_n all the matrices in Section I.4 are Toeplitz matrices. If F is a positive extension of $(b_o, b_1, ..., b_{n-1})$, then we set $B_n = T_n(F)$. The following plays a fundamental role in positive extension theory.

1.2 THEOREM *A function F is of positive type if and only if $T_n(F) \ge 0$ for all* $n \ge 1$.

PROOF. Assume F is of positive type. Lemma 1.1, formula (1.2) and the classical Fatou lemma (see [Ho]) imply that $F(e^{it}) = \lim_{r \to 1} F(re^{it}) \geq 0$ a.e. For any $x = [x_k]_0^{n-1}$ in C^n define $x(e^{it})$ by

$$x(e^{it}) = \sum_{k=0}^{n-1} x_k e^{ikt} . \tag{1.8}$$

Then we have

$$(T_n(F)[x_k]_0^{n-1}, [x_k]_0^{n-1}) =$$

$$\frac{1}{2\pi} \int_0^{2\pi} F(e^{it}) x(e^{it}) \bar{x}(e^{it}) dt = \frac{1}{2\pi} \int_0^{2\pi} F(e^{it}) |x(e^{it})|^2 dt \geq 0 . \tag{1.9}$$

To prove the other half assume that $T_n(F) \geq 0$ for all n. This implies that

$$\begin{bmatrix} b_o & \bar{b}_i \\ b_i & b_o \end{bmatrix} \geq 0 \qquad \text{(for all i).}$$

Hence $|b_i| \leq b_o$ and the series in (1.1) converges absolutely in D. For covenience we denote $\bar{b}_j$ by b_{-j} for all $j \geq 1$. Then for $|z| < 1$ we have

$$\frac{F(z)}{1-|z|^2} = \sum_{i=0}^{\infty} z^i \bar{z}^i [\sum_{j=0}^{\infty} b_j z^j + \sum_{j=1}^{\infty} b_{-j} \bar{z}^j] =$$

$$\sum_{i \geq 0 \, j \geq 0} b_j z^{i+j} \bar{z}^i + \sum_{i \geq 0 \, j \geq 1} b_{-j} z^i \bar{z}^{i+j} =$$

$$\sum_{j=0}^{\infty} \sum_{i=j}^{\infty} b_{i-j} z^i \bar{z}^j + \sum_{i=0}^{\infty} \sum_{j=i+1}^{\infty} b_{i-j} z^i \bar{z}^j =$$

$$\sum_{i \geq 0 \, j \geq 0} b_{i-j} z^i \bar{z}^j = \lim_{n \to \infty} \sum_{i=0, \, j=0}^{n,n} b_{i-j} z^i \bar{z}^j \geq 0 .$$

Thus $F(z) \geq 0$ for all $z \in D$ and the proof is complete.

The following shows that the role of A_n in Problem I.1.1 is played by the Toeplitz matrix B_n in the positive extension Carathéodory interpolation problem.

1.3 THEOREM. *Let* B_n *be the Toeplitz matrix defined in (1.7). Then* $B_n \geq 0$ *if and only if* $(b_o, ..., b_{n-1})$ *admits a positive extension.*

PROOF. If $(b_0, ..., b_{n-1})$ admits a positive extension, then by virtue of Theorem 1.2, the matrix $B_n = T_n(F)$ is positive. Now assume that $B_n \geq 0$. Let $\hat{f}_n$ be the polynomial defined by $\hat{f}_n = b_0 + 2b_1 z + 2b_2 z + \cdots + 2b_{n-1} z^{n-1}$. For any x in C^n

$$\mathrm{Re}(\hat{f}_n(T_n)x, x) = (B_n x, x) \geq 0 \tag{1.10}$$

where T_n is the matrix on C^n defined in (I.4.1) with j = n. We claim that -1 is not an eigenvalue of $\hat{f}_n(T_n)$. If $\hat{f}_n(T_n)x = -x$ for some x in C^n, then (1.10) gives $-\|x\|^2 \geq 0$, or equivalently, x = 0. Thus $(\hat{f}_n(T_n) + I)^{-1}$ exists. Using this and the functional calculus in Section I.4 we have

$$(\hat{f}_n(T_n) - I)(\hat{f}_n(T_n) + I)^{-1} = f_n(T_n) \tag{1.11}$$

where f_n is the function defined in (1.3) for $\hat{f} = \hat{f}_n$. By (1.6) we see that $(a_0, a_1, ..., a_{n-1})$ is the Carathéodory data for f_n. Thus $f_n(T_n) = A_n$. To complete the proof, we will show that A_n is a contraction since then $(b_0, b_1, ..., b_{n-1})$ has a positive extension by Lemma 1.1 and Theorem I.6.7. To this end we use (1.11) and (1.10) to obtain

$$\|(\hat{f}_n(T_n) + I)x\|^2 - \|f_n(T_n)(\hat{f}_n(T_n) + I)x\|^2 =$$

$$\|(\hat{f}(T_n) + I)x\|^2 - \|(\hat{f}(T_n) - I)x\|^2 = 4\,\mathrm{Re}(\hat{f}_n(T_n)x, x) \geq 0 \,.$$

Since -1 is not an eigenvalue of $\hat{f}_n(T_n)$, this shows that $f_n(T_n) = A_n$ is a contraction and completes the proof.

Obviously we have the following result.

1.4 COROLLARY. *Problem I.1.1 with data* $(a_0, a_1, ..., a_{n-1})$ *has a solution if and only if* $B_n \geq 0$, *where* B_n *is the Toeplitz matrix in (1.7). (Recall that* $(a_0, a_1, ..., a_{n-1})$ *and* $(b_0, b_1, ..., b_{n-1})$ *are related by (1.6).)*

Solving Problem I.1.1 with data $(a_0, a_1, ..., a_{n-1})$ is equivalent to finding all positive extensions F of $(b_0, b_1, ..., b_{n-1})$. The set of all positive extensions F of $(b_0, b_1, ..., b_{n-1})$ is given by (1.2), where f is given by the Schur representation (I.3.8) in Theorem I.3.2 with data $(a_0, a_1, ..., a_{n-1})$. The role of A_n in Problem I.1.1 is played by B_n in the positive extension problem.

2. THE LEVINSON ALGORITHM

This section is devoted to the Levinson algorithm. This algorithm can be used to recursively determine the Schur numbers $(r_0, ..., r_{n-1})$ from the data $(b_0, ..., b_{n-1})$. It can also be used to test when B_n is strictly positive.

This algorithm is based on the following fact.

2.1 THEOREM. *Let* B_n *be the Toeplitz matrix given by (1.7). Then* $B_n > 0$ *if and only if* $b_o > 0$ *and for every* $j = 2, 3, ..., n$ *there exists a solution* $\alpha_j = [\alpha_{j,k}]_{k=1}^{j}$ *of the equation*

$$[\alpha_{j,1}, ..., \alpha_{j,j-1}, 1]B_j^{tr} = [0,0, ..., 0, \alpha_{j,j}] . \tag{2.1}$$

such that $\alpha_{j,j} > 0$.

Here and in the sequel M^{tr} denotes the transpose of the square or rectangular matrix M.

PROOF. Assume that $B_n > 0$. Let e_j for $1 \leq j \leq n$ be the standard basis for C^n, that is, e_j is one in the j-th coordinate and zero elsewhere. We endow C^n with the new inner product $<x,y> = (B_n x,y)$, where x and y are in C^n and $(\cdot, \cdot)$ denotes the standard inner product on C^n; see Section I.6. Then $b_o = <e_1, e_1> > 0$. Let H_j denote the subspace of C^j formed by the vectors in C^n which have the last n−j coordinates zero. Furthermore, for $j > 1$, let e_j' be the orthogonal projection of e_j onto H_{j-1}. This means that

$$e_j' = -\sum_{i=1}^{j-1} \alpha_{j,i} e_i \tag{2.2}$$

and $e_j - e_j'$ is orthogonal to H_{j-1}, that is,

$$0 = <(\sum_{i=1}^{j-1} \alpha_{j,i} e_i) + e_j, e_k> \qquad (\text{for } 1 \leq k \leq j-1) . \tag{2.3}$$

Hence

$$<(\sum_{i=1}^{j-1} \alpha_{j,i} e_i) + e_j, e_j> = <e_j - e_j', e_j - e_j'> . \tag{2.4}$$

Using the relations $< e_j, e_i > = b_{i-j}$ or $\bar{b}_{j-i}$ according to whether $i \geq j$ or $i \leq j$, we easily see that (2.1) coincides with (2.3), (2.4) because

$$\alpha_{j,j} = < e_j - e_j', e_j - e_j' > . \tag{2.4a}$$

Finally, $\alpha_{j,j} > 0$ because B_j is strictly positive and e_j' does not equal e_j. From this (2.1) and $\alpha_{j,j} > 0$ readily follows.

Now assume that (2.1) holds ($b_0 > 0$) and $\alpha_{j,j} > 0$. For $j = 2$ we have

$$[\alpha_{2,1},\ 1]\begin{bmatrix} b_o & b_1 \\ \overline{b}_1 & b_o \end{bmatrix} = [0,\ \alpha_{2,2}]\ ,$$

or equivalently,

$$\alpha_{2,1} = \frac{-\overline{b}_1}{b_o} \text{ and } 0 < \alpha_{2,2} = b_o - \frac{|b_1|^2}{b_o}\ . \tag{2.5}$$

Taking the determinant of B_2 and using the fact that $b_o > 0$, we see that $0 < \alpha_{22}$ is equivalent to $B_2 > 0$. Now we use induction. Assume that $B_{n-1} > 0$ and (2.1) holds for j $= n$ where $\alpha_{n,n} > 0$. We also define e_n' by

$$e_n' = -\sum_{i=1}^{n-1} \alpha_{n,i} e_i\ .$$

Then using (2.1) several times we obtain for x in H_{n-1} the following relations

$$(B_n(x+e_n),(x+e_n)) = (B_n x,x)+2\mathrm{Re}(B_n e_n,x) + (B_n e_n,e_n) =$$

$$(B_{n-1}x,x) + 2\mathrm{Re}(B_n e_n',x) + (B_n e_n',\ e_n) + \alpha_{n,n} =$$

$$(B_{n-1}x,\ x) + 2\mathrm{Re}(B_{n-1} e_n',\ x) + (B_n e_n,\ e_n') + \alpha_{n,n} =$$

$$(B_{n-1}x,\ x) + 2\mathrm{Re}(B_{n-1} e_n',\ x) + (B_{n-1} e_n',\ e_n') + \alpha_{n,n} =$$

$$(B_{n-1}(x+e_n'),\ (x+e_n')) + \alpha_{n,n} > 0\ .$$

This clearly implies that B_n is strictly positive and the proof is complete.

2.2   REMARK. Since B_j is Toeplitz, equation (2.1) is equivalent to

$$[1,\ \overline{\alpha}_{j,j-1},\ ...,\ \overline{\alpha}_{j,1}]B_j^{\mathrm{tr}} = [\overline{\alpha}_{j,j},\ 0,\ 0...,\ 0] \tag{2.6}$$

This follows by inverting the order of the components in (2.1) and conjugating them.

The following is the Levinson algorithm.

2.3   COROLLARY. *Let* $B_n > 0$. *Then the* $\alpha_{j,i}$'*s in Theorem 2.1 for* $2 \le j \le n$ *can be recursively computed starting with (2.5) by*

$$[\alpha_{j+1,1},\ ...,\ \alpha_{j+1,j}] = [0,\alpha_{j,1},\ ...,\ \alpha_{j,j-1}] - \frac{\Delta j}{\alpha_{j,j}}\ [1,\ \overline{\alpha}_{j,j-1},\ ...,\ \overline{\alpha}_{j,1}] \quad (for\ j < n) \tag{2.7}$$

and

$$\alpha_{j+1,j+1} = \alpha_{j,j} - \frac{|\Delta_j|^2}{\alpha_{j,j}} \qquad\qquad (for\ j < n) \qquad\qquad (2.8)$$

where

$$\Delta_j = \overline{b}_j + \sum_{i=1}^{j-1} \alpha_{j,i}\overline{b}_i . \qquad\qquad (for\ j < n) \qquad\qquad (2.9)$$

PROOF. Let $2 \leq j < n$. Formulas (2.1), (2.6), the Toeplitz structure of B_{j+1} and the fact that $\alpha_{j,j}$ is positive give:

$$[0,\ \alpha_{j,1},\ ...,\ \alpha_{j,j-1},\ 1]B_{j+1}^{tr} = [\Delta_j,\ 0,\ 0,\ ...,\ \alpha_{j,j}]$$

and

$$[1,\ \overline{\alpha}_{j,j-1},\ ...,\ \overline{\alpha}_{j,1},\ 0]B_{j+1}^{tr} = [\alpha_{j,j},\ 0,\ 0,\ ...,\ 0,\ \overline{\Delta}_j]$$

A simple subtraction yields:

$$\{[0,\ \alpha_{j,1},\ ...,\ \alpha_{j,j-1},\ 1] - \frac{\Delta_j}{\alpha_{j,j}}\ [1,\ \overline{\alpha}_{j,j-1},\ ...,\ \overline{\alpha}_{j,1},\ 0]\}B_{j+1}^{tr} =$$

$$[0,\ 0,\ ...,\ 0,\ \alpha_{j,j} - \frac{|\Delta_j|^2}{\alpha_{j,j}}] .$$

Comparing this to (2.1) and using Theorem 2.1 with the fact that B_{j+1} is nonsingular, we obtain (2.7) and (2.8). This completes the proof.

Theorem 2.1, Corollary 2.3 and its proof immediately yields the following result

2.4 COROLLARY. *Given a Toeplitz matrix* B_n. *Set* $\alpha_{1,1} = b_0$ *and compute* $\alpha_{j,j}$ *according to Corollary 2.3. Then* $\alpha_{j,j} > 0$ *for all* $1 \leq j \leq n-1$ *if and only if* $B_n > 0$.

We conclude this section with the following basic remark, which is the analogue of Remark I.5.7.

2.5 REMARK. Let B_n be a Toeplitz matrix and assume that B_{n-1} is strictly positive. Then one can use the Levinson algorithm to compute the $\alpha_{n,i}$'s satisfying

$$[\alpha_{n,1},\ ...,\ \alpha_{n,n-1},\ 1]B_n^{tr} = [0,\ 0,\ ...,\ 0,\ \alpha_{n,n}] , \qquad\qquad (2.10)$$

that is, the Levinson algorithm can be used one step beyond B_{n-1} when $B_{n-1} > 0$. On the other hand, formulas (2.8) and (2.9) yield

$$\alpha_{n,n} = \alpha_{n-1,n-1} - |\bar{b}_{n-1} + \sum_{i=1}^{n-2} \alpha_{n-1,i}\bar{b}_i|^2 \cdot \alpha_{n-1,n-1}^{-1} \ .$$

The above sum

$$\sigma = -\sum_{1}^{n-2} \alpha_{n-1,i}\bar{b}_i$$

as well as $\alpha_{n-1,n-1}$ depend only on $(b_0, ..., b_{n-2})$, so for every b_{n-1} in the open (closed) disc with center σ and radius $\alpha_{n-1,n-1}$ the corresponding $\alpha_{n,n}$ occurring in (2.8) is strictly positive (positive), respectively. Therefore the set of all b_{n-1} such that B_n is strictly positive (positive) is in the open (closed) disc, with center σ and radius $\alpha_{n-1,n-1}$, respectively.

3. UNIQUENESS AND POSITIVE EXTENSIONS

In this section we will show that the data $(b_0, ..., b_{n-1})$ admits a unique positive extension if and only if its Toeplitz matrix B_n is positive and singular. Since B_n plays the role of A_n, this is equivalent to the fact that Problem I.1.1 admits a unique solution if and only if the norm of A_n is one.

We begin with the following result.

3.1 LEMMA. *Let* B_j *be a Toeplitz matrix satisfying* $B_{j-1} > 0$, *where* $j > 1$. *Then* B_j *is positive and singular if and only if* $\alpha_{j,j} = 0$, *or equivalently,*

$$[\alpha_{j,1}, ..., \alpha_{j,j-1}, 1]B_j^{tr} = [0, 0, ..., 0] \tag{3.1}$$

where the coefficients $\alpha_{j,i}$ *are computed from the Levinson algorithm.*

PROOF. Assume that (3.1) holds. Let $\xi_j = [\alpha_{j,1}, ..., \alpha_{j,j-1}, 1]^{tr}$ and let x be any vector in C^j with its last component zero. For γ in C^1 we have

$$(B_j(x+\gamma\xi_j), x+\gamma\xi_j) = (B_{j-1}x, x) \geq 0.$$

Since vectors of the form $x + \gamma\xi_j$ span C^j, this and (3.1) imply that B_j is positive and B_j is singular.

Now assume that B_j is positive and singular. This and $B_{j-1} > 0$ implies that

$$B_j[\alpha'_{j,1}, ..., \alpha'_{j,j-1}, 1]^{tr} = [0, 0, ..., 0]^{tr}$$

for some suitable $\alpha'_{j,i}$. Subtracting this from the transpose of (2.10), with n replaced by j, yields

$$B_j[\alpha_{j,1} - \alpha'_{j,1}, \ ..., \ \alpha_{j,j-1} - \alpha'_{j,j-1}, \ 0]^{tr} = [0, \ 0, \ ..., \ 0, \ \alpha_{j,j}]^{tr} \ .$$

Since $B_{j-1} > 0$ this gives $\alpha_{j,i} = \alpha'_{j,i}$ for $1 \leq i \leq j-1$. Hence $\alpha_{j,j} = 0$. This completes the proof.

The main result of this section is the following theorem.

3.2 THEOREM. *There exists a unique positive extension of* $(b_0, b_1, ..., b_{n-1})$ *if and only if* B_n *is positive and singular.*

PROOF. Assume there exists a unique positive extension of $(b_0, b_1, ..., b_{n-1})$. Theorem 1.3 shows that $B_n \geq 0$. For the moment assume that B_n is strictly positive, Remark 2.5 with n replaced by n+1 shows that B_n has more than one positive extension. Therefore B_n has to be singular.

Now assume that B_n is positive and singular. Let j be the first integer such that B_j is singular. For $m \geq j$ we set

$$\xi_m = [0, \ 0, \ ..., \ 0, \ \alpha_{j,1}, \ ..., \ \alpha_{j,j-1}, \ 1] \in \mathbb{C}^m$$

where the $\alpha_{j,i}$'s are given in Lemma 3.1. Let F be any positive extension of $(b_0, b_1, ..., b_{j-1})$; see Theorem 1.3. By Theorem 1.2 this F determines a positive Toeplitz matrix B_m containing B_j in its lower right hand corner. Equation (3.1) gives $(B_m \xi_m, \xi_m) = (B_j \xi_j, \xi_j) = 0$. Since B_m is positive, this implies that $B_m \xi_m = 0$. (Notice that if A is a positive matrix satisfying $(Ax, x) = 0$, then $A^{1/2}x = 0$ where $A^{1/2}$ is the positive square root of A. Hence $Ax = 0$.) Thus

$$b_k = -(\overline{\alpha}_{j,j-1} b_{k-1} + \cdots + \overline{\alpha}_{j,1} b_{k-j+1}) \qquad (k \geq j-1) . \tag{3.2}$$

Therefore all the b_k's for $k \geq j$ are uniquely determined by B_j, or equivalently, by $\{b_i\}_0^{j-1}$. So the extension is unique. This completes the proof.

3.3 COROLLARY. *The Toeplitz matrix* B_n *is positive and singular, or equivalently,* $(b_0, b_1, ..., b_{n-1})$ *admits a unique positive extension if and only if for some* $j \leq n$ *we have* $\alpha_{i,i} > 0$ *for* $1 \leq i < j$ *and* $\alpha_{j,j} = 0$ *and (3.2) holds for* $j-1 \leq k \leq n-1$, *where* $\alpha_{k,k}$ *for* $1 \leq k \leq j$ *are computed by the Levinson algorithm.*

PROOF. If B_n is positive and singular, then Corollary 2.4, Lemma 3.1 and the last paragraph in the proof of Theorem 3.2 shows that (3.2) holds.

Now assume that (3.2) holds, $\alpha_{i,i} > 0$ for $1 \leq i < j$ and $\alpha_{j,j} = 0$. Then Corollary 2.4 and Lemma 3.1 imply that $B_{j-1} > 0$, and B_j is positive and singular. Theorem 3.2 implies that $(b_0, b_1, ..., b_{j-1})$ admits a unique positive extension F. Moreover, the proof of Theorem 3.2 shows that the data $(b_0, b_1, ..., b_{n-1})$ of F satisfies (3.2). (The b_k's for $k \geq j$ in (3.2) are uniquely determined by $(b_0, b_1, ..., b_{j-1})$). Thus $(b_0, b_1, ..., b_{n-1})$ admits a unique positive extension. By Theorem 3.2 again, B_n is positive and singular. This completes the proof.

Combining the above with Lemma 1.1 and the results in Section 6 in Chapter I, we obtain the following result.

3.4 COROLLARY. *Let $(a_0, a_1, ..., a_{n-1})$ and $(b_0, b_1, ..., b_{n-1})$ be related by (1.6). Problem I.1.1 with data $(a_0, a_1, ..., a_{n-1})$ has a solution if and only if B_n is positive. Moreover, Problem I.1.1 has a unique solution if and only if B_n is positive and singular.*

4. COMPUTING SCHUR NUMBERS FROM THE LEVINSON ALGORITHM

In this section we show how the Levinson algorithm can be used to compute the Schur numbers of the data $(a_0, a_1, ..., a_{n-1})$. Throughout the remaining part of this chapter, $(a_0, a_1, ..., a_{n-1})$ is the data to Problem I.1.1 with Schur numbers $(r_0, r_1, ..., r_{n-1})$ and t_n is defined in (I.3.17). The data $(b_0, b_1, ..., b_{n-1})$ for the positive extension problem is given by (1.6).

Throughout B_j is the j by j matrix in the northwest corner of the Toeplitz matrix B_n in (1.7). Obviously B_j is a Toeplitz matrix.

4.1 LEMMA. *Let B_j be strictly positive. Then $B_j = T_j(F_0)$ where*

$$F_0(e^{it}) = t_j \, | \, y_j^\#(e^{it}) - x_j(e^{it}) \, |^{-2} \qquad (4.1)$$

and x_j and y_j are the polynomials (I.3.6), (I.3.7) occurring in the Schur representation (I.3.8). (Recall that $y_j^\# = J_j y_j$.) Furthermore, the polynomials $y_j^\# - x_j$ and $y_j^\#$ have no zeros in $\overline{D}$ and $\|x_j / y_j^\#\|_\infty < 1$. Finally, $F_0 \in L^\infty$ and there exists a $\gamma > 0$ such that

$$F_0(e^{it}) \geq \gamma > 0 \quad \text{a.e.} \qquad (4.2)$$

PROOF. By Corollary 3.4 all the Schur numbers $(r_0, r_1, ..., r_{j-1})$ of the data $(a_0, a_1, ..., a_{j-1})$ are contained in the unit disc D. Theorem I.3.2 shows that $f = x_j / y_j^\#$ is a solution to Problem I.1.1 with data $(a_0, a_1, ..., a_{j-1})$. By Lemma 1.1 and (1.2), (I.3.3) and (I.3.4) for $j = n$, the function F_0 defined by

$$F_o(e^{it}) = \frac{1-|x_j(e^{it})/y_j^{\#}(e^{it})|^2}{|1-x_j(e^{it})/y_j^{\#}(e^{it})|^2} = \frac{|y_j^{\#}(e^{it})|^2-|x_j(e^{it})|^2}{|y_j^{\#}(e^{it})-x_j(e^{it})|^2} = \tag{4.3}$$

$$t_j|y_j^{\#}(e^{it})-x_j(e^{it})|^{-2}$$

is a positive extension of $(b_o, b_1, ..., b_{j-1})$ and $B_j = T_j(F_o)$.

For the second part recall that $f = x_j/y_j^{\#}$ is in H_1^{∞} and $f(0) = x_{j,o}/\overline{y}_{j,j-1} = r_o$ by virtue of equations (I.3.13) and (I.3.14). These facts with the maximum principle yield $|f(z)| < 1$ for all z in D, that is, $|x_j(z)| < |y_j^{\#}(z)|$ for all z in D. So $y_j^{\#} - x_j$ has no zeros in D. Also it is clear that $|x_j(z)| \leq |y_j^{\#}(z)|$ for all z in $\overline{D}$. If $y_j^{\#}(z_o) = 0$ for some z_o of modulus one, then $x_j(z_o) = 0$ and thus $y_j^{\#}(z_o) - x_j(z_o) = 0$. However, by (I.3.4) the determinant in (I.3.3) evaluated at z_o is zero, that is,

$$0 = z_0^{j-1}(1-|r_o|^2)(1-|r_1|^2)\cdots(1-|r_{j-1}|^2). \tag{4.4}$$

Since $|r_i| < 1$ for all $i < j$ we have that $t_j = (1-|r_o|^2)\cdots(1-|r_{j-1}|^2)$ is strictly positive. This and the fact that $|z_o| = 1$ contradicts (4.4). We conclude that the two polynomials $y_j^{\#}$ and $y_j^{\#} - x_j$ have no zeros in $\overline{D}$. By (4.1) and $t_j > 0$ we obtain (4.2). The proof is now complete.

The main result of this section is given by the following theorem.

4.2 THEOREM. *The following are equivalent.*

(i) B_j is strictly positive

(ii)   There exists a $F \in L^{\infty}$ such that $B_j = T_j(F)$ and $F(e^{it}) \geq \gamma$ a.e. for some $\gamma > 0$.

(iii)   There exists a $f \in H^{\infty}$ such that $\|f\|_{\infty} < 1$, $f(0)$ is real and $B_j = T_j(F)$, where

$$F(e^{it}) = \mathrm{Re}\, \frac{1+f(e^{it})}{1-f(e^{it})} . \tag{4.5}$$

Finally, if any one of the above conditions hold, then

$$r_o = \frac{b_o-1}{b_o+1} \quad \text{and} \quad r_{j-1} = -\overline{\alpha}_{j,1} \tag{4.6}$$

where $\alpha_{j,1}$ is computed by the Levinson algorithm.

PROOF. Lemma 4.1 with $F = F_o$ show that (i) implies (ii). Assume that (ii) holds. Following the proof of Theorem 1.2 we associate to $[x_k]_0^{j-1}$ in C^j the function $x(e^{it})$ in L^2

defined by (1.8) where $j = n$. Equation (1.9) and $B_j = T_j(F)$ give

$$(B_j[x_k]_0^{j-1}, [x_k]_0^{j-1}) = \frac{1}{2\pi} \int_0^{2\pi} F(e^{it})\,|x(e^{it})|^2 dt \geq$$

$$\frac{\gamma}{2\pi} \int_0^{2\pi} |x(e^{it})|^2 dt = \gamma\|[x_k]_0^{j-1}\|^2 \ .$$

Therefore B_j is strictly positive, and (i) and (ii) are equivalent.

If (i) holds, then all the Schur numbers are in the unit disc. Thus $t_j > 0$. Setting $f = x_j/y_j^\#$ in (I.3.20) where $j = n$ and $f_n = 0$ yields $1 - |f(e^{it})|^2 = t_j\,|y_j^\#(e^{it})|^{-2}$. Since $y_j^\#$ is a polynomial with no zeros in $\overline{D}$ and $t_j > 0$, this implies that $1 - |f(e^{it})|^2 \geq \gamma_1 > 0$. Hence $\|f\|_\infty < 1$. This and (4.3) gives (iii). On the other hand if (iii) holds, then Theorem I.2.4 and $\|f\|_\infty < 1$ show that the solution to Problem I.1.1 with data $(a_0, a_1, ..., a_{j-1})$ is not unique. (If the solution f were unique, then f would be a Blaschke product. In particular, $\|f\|_\infty = 1$.) Corollary 3.4 (with $n=j$) implies that B_j is strictly positive. So (i), (ii) and (iii) are equivalent.

The first equation in (4.6) follows form $r_0 = a_0$ and (1.6). To complete the proof assume that the above conditions hold. Consider the polynomial $\beta_j^\#(z)$ defined by

$$\beta_j^\#(z) = \frac{y_j^\# - x_j}{1 - r_0} = 1 + \overline{\beta}_{j,j-1}z + \cdots + \overline{\beta}_{j,2}z^{j-2} + \overline{\beta}_{j,1}z^{j-1} \ . \tag{4.7}$$

The 1 appears in (4.7) because (I.3.14) implies that $y_j^\#(0) - x_j(0) = 1 - r_0$. By virtue of Lemma 4.1, the function $(y_j^\# - x_j)^{-1}$ is analytic in a neighborhood of $\overline{D}$. This, the equation $y_j^\#(0) - x_j(0) = 1 - r_0$ and the fact that $r_0 = a_0$ is real (see Section 1) give:

$$F_0(e^{it})\beta_j^\#(e^{it}) = \frac{t_j}{(1-r_0)(y_j^\#(e^{it}) - x_j(e^{it}))} =$$

$$\frac{t_j}{(1-r_0)(y_j^\#(0) - x_j(0))} + \sum_{k=1}^{\infty} \gamma_k e^{-ikt} = \beta_{j,j} + \sum_{k=1}^{\infty} \gamma_k e^{-ikt} \tag{4.8}$$

where

$$\beta_{j,j} = t_j(1-r_0)^{-2} = b_0(1-|r_1|^2)(1-|r_2|^2) \cdots (1-|r_{j-1}|^2) \ . \tag{4.9}$$

The second equality follows from the first equation in (4.6) or (1.6). Using (4.8) and $B_j = T_j(F_0)$ with the standard basis $\{e_k\}_0^{j-1}$ in C^j yields

$$(B_j[1, \bar{\beta}_{j,j-1}, ..., \bar{\beta}_{j,1}]^{tr}, e_k) = \frac{1}{2\pi} \int_0^{2\pi} F_0(e^{it})\beta_j^{\#}(e^{it})e^{-ikt}dt = \beta_{j,j} \text{ if } k = 0$$

$$= 0 \text{ if } 1 \leq k \leq j-1. \qquad (4.10)$$

In matrix form (4.10) becomes

$$B_j[1, \bar{\beta}_{j,j-1}, ..., \bar{\beta}_{j,1}]^{tr} = [\beta_{j,j}, 0, 0, ..., 0]^{tr}. \qquad (4.11)$$

By Remark 2.2 we have $\beta_{j,i} = \alpha_{j,i}$ for $1 \leq i \leq j$ where the coefficients $\alpha_{j,i}$ are computed by the Levinson algorithm. Now the second equation in (4.6) follows from (I.3.14) and (4.7). This completes the proof.

The proof of the previous theorem readily establishes the following remarkable fact.

4.3 COROLLARY. *Let* B_j *be strictly positive and assume that*

$$\alpha_j(z) = \alpha_{j,1} + \alpha_{j,2}z + \cdots + \alpha_{j,j-1}z^{j-2} + z^{j-1}$$

where $\alpha_{j,i}$ *for* $1 \leq i < j$ *are computed recursively by the Levinson algorithm. Then*

$$y_j^{\#} - x_j = (1 - r_0)\alpha_j^{\#} \qquad (4.12)$$

where $\alpha_j^{\#} = J_j\alpha_j$.

4.4 REMARK. The previous analysis shows that we can use the Levinson algorithm to construct a positive extension of the data $(b_0, ..., b_{n-1})$. This is done by first computing the $\alpha_{j,i}$ in the Levinson algorithm. If $\alpha_{j,j} > 0$ for all $1 \leq j \leq n-1$ then B_n is strictly positive. Moreover, $\alpha_j^{\#}$ has no zeros in $\bar{D}$. Lemma 4.1 formulas (4.1), (4.9), (4.7) and Corollary 4.3 show that

$$F_0(e^{it}) = \frac{\alpha_{n,n}}{|\alpha_n(e^{it})|^2} \qquad (4.13)$$

is a positive extension of $(b_0, ..., b_{n-1})$. Furthermore, F_0 is in L^{∞} and there exists a $\gamma > 0$ such that $F_0(e^{it}) \geq \gamma > 0$ for all t. On the other hand, if $\alpha_{i,i} > 0$ for $1 \leq i < j$ and $\alpha_{j,j} = 0$ for some j, then one can use Corollary 3.3 to determine whether or not there exists a positive extension of $(b_0, ..., b_{n-1})$. If there exists a unique positive extension F to $(b_0, ..., b_{n-1})$, then this F is given by (1.1), where b_k for $k \geq n$ is computed according to equation (3.2).

The above analysis readily produces the following procedure for computing the Schur numbers $(r_0, ..., r_{n-1})$ from the data $(b_0, ..., b_{n-1})$.

4.5 PROCEDURE. Assume that there are multiple solutions to Problem I.1.1, or equivalently, that B_n is strictly positive. Then the Schur numbers $(r_0, ..., r_{n-1})$ are given by $r_{j-1} = -\overline{\alpha}_{j,1}$ where $\alpha_{j,i}$ (for $1 \le i \le j$) is computed recursively by the Levinson algorithm for $j = 1, 2, ..., n$.

The next corollary relates the Schur numbers to the determinant of B_j.

4.6 COROLLARY. *If* B_n *is strictly positive, then*

$$r_j = \frac{-(-1)^j}{\det B_j} \det \begin{bmatrix} b_1 & b_2 & \cdots & b_{j-1} & b_j \\ b_0 & b_1 & \cdots & b_{j-2} & b_{j-1} \\ \overline{b}_1 & b_0 & \cdots & b_{j-3} & b_{j-2} \\ . & . & \cdots & . & . \\ . & . & \cdots & . & . \\ . & . & \cdots & . & . \\ \overline{b}_{j-2} & \overline{b}_{j-3} & \cdots & b_0 & b_1 \end{bmatrix} \qquad (1 \le j < n) \qquad (4.14)$$

where $(r_0, r_1, ..., r_{n-1})$ *are the Schur numbers for* $(a_0, a_1, ..., a_{n-1})$.

PROOF. Dividing out the $\alpha_{j+1,j+1}$ in (2.6) where we replace j by j + 1 yields

$$B_{j+1}[\gamma_1, \gamma_2, ..., \gamma_{j+1}]^{tr} = [1, 0, 0, ..., 0]^{tr} \text{ and } -r_j = \overline{\alpha}_{j+1,1} = \frac{\gamma_{j+1}}{\gamma_1} \ . \qquad (4.15)$$

Using the determinant with the classical adjoint to invert B_j in (4.15) gives (4.14) and completes the proof.

5. MAXIMAL ENTROPY

In this section we will show that the positive extension F_0 in (4.1), or equivalently, F_0 given by the Levinson algorithm in (4.13) is the maximal entropy extension of $(b_0, ..., b_{n-1})$.

We begin with the following result due to Szegö-Kolmogorov-Krein.

5.1 THEOREM. *Let* $F(e^{it})$ *be a positive* (≥ 0) *Lebesgue intergable function on* $(0, 2\pi)$. *Then*

$$\inf_{p} \frac{1}{2\pi} \int_{0}^{2\pi} |1-p(e^{it})|^2 F(e^{it})dt = \exp \left[\frac{1}{2\pi} \int_{0}^{2\pi} \log F(e^{it})dt\right] \qquad (5.1)$$

where the infimum is taken over the set of all polynomials p satisfying $p(0) = 1$.

For an elegant proof of Theorem 5.1 see Chapter 4 of [Ho].

It is emphasized that the right hand side of (5.1) is zero if $\log F(e^{it})$ is not integrable, or equivalently, $\int \log F \, dt = -\infty$. Let us recall that a positive extension F of $(b_0, b_1, \ldots, b_{n-1})$ corresponds to a solution f in H_1^∞ of the Caratheodory interpolation Problem I.1.1, with data $(a_0, a_1, \ldots, a_{n-1})$; see Section 1. For convenience, we say that the Schur numbers $(r_0, r_1, \ldots, r_{n-1})$ associated with f corresponding to the positive extension F are the *Schur numbers corresponding to* F. We also notice that the Schur numbers $(r_0, r_1, \ldots, r_{n-1})$ are uniquely determined by $(a_0, a_1, \ldots, a_{n-1})$, or equivalently, by $(b_0, b_1, \ldots, b_{n-1})$. This sets the stage for the following result.

5.2 THEOREM. *Assume that* B_n *is strictly positive and F is a positive extension of* $(b_0, \ldots, b_{n-1})$. *Then*

$$\exp \left[\frac{1}{2\pi} \int_{0}^{2\pi} \log F(e^{it})dt\right] = b_0 \prod_{i=1}^{\infty} (1 - |r_i|^2) \qquad (5.2)$$

where r_j *for* $j \geq 1$ *are the Schur numbers corresponding to* F. *In particular, if* F_0 *is the function defined in (4.13), then*

$$\exp \frac{1}{2\pi} \int_{0}^{2\pi} \log F_0(e^{it})dt = b_0 \prod_{i=1}^{n-1} (1 - |r_i|^2) \, . \qquad (5.3)$$

PROOF. Let p be a polynomial and assume that F is a positive extension of $(b_0, \ldots, b_{n-1})$. Using (1.8), (1.9), (2.4a), (4.9) with $T_n(F) = B_n$, and the fact that the coefficients $\alpha_{n,i}$ in the Levinson algorithm come from the orthogonal projection e_n' of e_n onto H_{n-1} (see the proof of Theorem 2.1) we have

$$\inf \{\frac{1}{2\pi} \int_{0}^{2\pi} |1-p(e^{it})|^2 F(e^{it})dt: \quad \deg p < n \quad \text{and} \quad p(0) = 0\} =$$

$$\inf \{\frac{1}{2\pi} \int_{0}^{2\pi} |e^{i(n-1)t} - p^{\#}(e^{it})|^2 F(e^{it})dt: \quad \deg p^{\#} \leq n-2\} = \qquad (5.4)$$

$$\inf (B_n(e_n - \sum_{1}^{n-1} x_k e_k), \ (e_n - \sum_{1}^{n-1} x_k e_k)) = \|e_n - e_n'\|^2 = \alpha_{n,n} = \beta_{n,n} = b_0 \prod_{i=1}^{n-1} (1 - |r_i|^2)$$

where the $\{e_i\}$ are the standard basis for C^n. Equation (5.2) follows from Theorem 5.1 and (5.4). Equation (5.3) follows from the fact that the Schur numbers r_j corresponding to F_0 are all zero if $j \geq n$. (Recall that the f corresponding to F_0 is $f = x_n/y_n^{\#}$ and $f_n = 0$. Thus $r_j = 0$ if $j \geq n$.) This completes the proof.

The previous theorem with the fact that a function f in H_1^{∞} uniquely determines and is uniquely determined by its Schur numbers gives the following result.

5.3   COROLLARY. *Assume that B_n is strictly positive and F is a positive extension of* $(b_0, ..., b_{n-1})$. *Then*

$$\frac{1}{2\pi} \int_0^{2\pi} \log F(e^{it})dt \leq \frac{1}{2\pi} \int_0^{2\pi} \log F_0(e^{it})dt \tag{5.5}$$

where F_0 is the positive extension in (4.13) *of* $(b_0, ..., b_{n-1})$ *obtained by the Levinson algorithm. Moreover, there is equality in* (5.5) *if and only if* $F = F_0$.

If F is a positive Lebesgue integrable function, then

$$\frac{1}{2\pi} \int_0^{2\pi} \log F(e^{it})dt$$

is the *entropy integral* of F. Equation (5.5) says that F_0 is a *maximal entropy* extension of $(b_0, ..., b_{n-1})$. In other words, if F is any other extension of $(b_0, ..., b_{n-1})$, then the entropy integral of F is less than or equal to the entropy integral of F_0.

II.6. NOTES AND COMMENTS

All the results in this section are classical; see [DudM], [Mar], [Ge], [GrS] and [Szegö] for further results and historical comments. Here we have tried to emphasize the connection between Schur numbers, the Levinson algorithm, and positive extensions. Let us also mention that Theorem 1.2 is due to Toeplitz [To 4] and later Schur [Schur 1] presented an alternate proof. Our proof was taken from [KovP]. The Levinson algorithm [Lev] in Corollary 2.3 plays an important role in linear prediction and signal processing (see the monographs [DudM] and [Mar] for further details). A matrix version of the Levinson algorithm was first obtained by Whittle [Wh], and Wiggins and Robinson [WigR]. This algorithm is also discussed in [DudM], [MoVK] and Chapters XV, XVI of this book. The fact that $y_j^{\#} - x_j$ and $y_j^{\#}$ have no zeros in $\overline{D}$ is a classical result in orthogonal polynomials (see [Ge] and [Szegö] for related results). Finally, it is noted that Schur numbers and maximal entropy realizations naturally occur in prediction theory; see [DewD], [DewVK], [DudM], [MoVK] and [Mar]. The product $b_o \Pi (1 - |r_i|^2)$ is known as the *prediction error*.

Let us present two sample problems in prediction theory intimately related to the content in Chapters I and II. To begin, let $\{\xi_n : -\infty < n < \infty\}$ be a sequence of functions in $L^2(\Omega, \mu)$, where $(\Omega, A \mu)$ is a probability space. The sequence $\{\xi_n : -\infty < n < \infty\}$ is a *weakly stationary* random process if

$$(\xi_n, \xi_m) \overset{\Delta}{=} \int_\Omega \xi_n \overline{\xi}_m d\mu = (\xi_{n+1}, \xi_{m+1}) \qquad \text{(for all n and m)} \qquad (6.1)$$

and

$$(\xi_n, 1) = \int_\Omega \xi_n d\mu = (\xi_{n+1}, 1) \qquad \text{(for all n) .} \qquad (6.2)$$

By replacing ξ_n with $\xi_n - (\xi_n, 1)$ we can assume that the means in (6.2) are all zero. In this case, the sequence $\{b_n : -\infty < n < \infty\}$ defined by $b_n = (\xi_n, \xi_o)$ is called the *correlation sequence* of the process ξ_n. The stationarity conditions in (6.1) imply that

$$b_{n-m} = (\xi_n, \xi_m) \qquad \text{(for all n and m) .} \qquad (6.3)$$

Moreover, $\overline{b}_n = b_{-n}$ for all n.

A simple version of the Kolmogorov - Wiener prediction problem is: Given a weakly stationary random process $\{\xi_n : -\infty < n < \infty\}$, what is the best approximation $\hat{\xi}_n$ in the L^2 norm of ξ_n by linear combination of $\{\xi_m : m < n\}$ and what is the variance of the prediction error $\xi_n - \hat{\xi}_n$, that is, $\|\xi - \hat{\xi}_n\|^2$. Obviously, $\hat{\xi}_n$ must be the orthogonal projection of ξ_n onto the closed linear subspace H_n of $L^2(\Omega, \mu)$ spanned by $\{\xi_m : m < n\}$. If the Toeplitz matrices B_n in (1.7) satisfies $B_n - \gamma I_{C^n} \geq 0$ for all n with the same positive constant $\gamma > 0$, then (by using Theorem 4.2) there exists a function F in L^∞ such that $B_n = T_n(F)$ for all $n \geq 0$ and $F(e^{it}) \geq \gamma > 0$ a.e. Consulting (1.1) and (6.3) we have

$$(\xi_n, \xi_o) = b_n = \frac{1}{2\pi}\int_0^{2\pi} e^{-int}F(e^{it})dt . \qquad (6.4)$$

From Theorem 5.1 it follows that

$$\|\xi_n - \hat{\xi}_n\|^2 = \inf\{\|\xi_n - \sum_{-\infty}^{n-1} c_i\xi_i\|^2 : c_{n-1}, c_{n-2}, \dots \text{ in } C^1 \text{ has finite support}\} =$$

$$\inf \left\{ \frac{1}{2\pi} \int_0^{2\pi} |1 - p(e^{it})|^2 F(e^{it}) dt : p(z) \text{ is a polynomial such that } p(0) = 0 \right\} =$$

$$\exp \left[\frac{1}{2\pi} \int_0^{2\pi} \log F(e^{it}) dt \right] .$$

Not let $\psi(z)$ be the outer spectral factor of F defined by

$$\psi(z) = \exp \left[\frac{1}{2\pi} \int_0^{2\pi} \frac{e^{it} + z}{e^{it} - z} \log F(e^{it}) dt \right] \qquad (z \in D)$$

and let

$$\phi(z) \stackrel{\Delta}{=} 1 - \frac{\psi(0)}{\psi(z)} = \phi_1 z + \phi_2 z^2 + \cdots$$

be the Taylor series expansion of ϕ. Then

$$\hat{\xi}_n = \phi_1 \xi_{n-1} + \phi_2 \xi_{n-2} + \phi_3 \xi_{n-3} + \cdots$$

Moreover, $\|\xi_n - \hat{\xi}_n\| = |\psi(0)|$.

A very interesting and useful complement to this problem was considered by Burg [Bur 1,2,3]. Namely, assume that only $(b_o, b_1, ..., b_{n-1})$ and $\{\xi_i : 1 \le i < n\}$ are actually known. Besides the orthogonal projection $\hat{\xi}_n$ also consider the orthogonal projection η_n of ξ_n onto the space generated by $\{\xi_i : 1 \le i < n\}$. Then obviously,

$$\|\xi_n - \eta_n\|^2 = \|\xi_n - \hat{\xi}_n\|^2 + \|\hat{\xi}_n - \eta_n\|^2 \ge \|\xi_n - \hat{\xi}_n\|^2$$

and $\hat{\xi}_n = \eta_n$ if and only if $\|\xi_n - \hat{\xi}_n\|^2$ attains its maximal possible value $\|\xi_n - \eta_n\|^2$. By virtue of (5.3) and (5.4) we have

$$\|\xi_n - \eta_n\|^2 = \exp \left[\frac{1}{2\pi} \int_0^{2\pi} \log F_o(e^{it}) dt \right] .$$

In this case (4.3) shows that $\phi(z) = 1 - \alpha_n^{\#}(z)$ where $\alpha_n^{\#}(z)$ is the polynomial provided by the Levinson algorithm and thus

$$\hat{\xi}_n = \eta_n = -(\overline{\alpha}_{n,n-1} \xi_{n-1} + \overline{\alpha}_{n,n-2} \xi_{n-2} + \cdots + \overline{\alpha}_{n,1} \xi_1)$$

(see (4.12), (4.13) and Lemma 4.1). This is essentially the geometrical basis of Burg's solution to this prediction problem. Many of the mathematical ideas involved in these approaches to prediction theory are more or less explicitly present in the classical treatise [Szegö].

In connection with this chapter see [ArsCo], [Ba], [BrK], [Con 1], [DeGK], [Des], [DewVK], [Frz 7], [FrzK], [Ge], [GoS], [GrS], [Ka 2], [KaBM], [KaVM], [KrK], [KrN], [LeK], [Mar], [Mas], [Mo], [RoT 2], [Szegö] and [Tr].

CHAPTER III

SCHUR NUMBERS, GEOPHYSICS AND INVERSE SCATTERING PROBLEMS

Schur numbers occur in one of the simplest mathematical models in geophysics, which is used by the petroleum industry to process seismic data. In this chapter we will give an introductory explanation of how Schur numbers naturally arise in certain layered medium models in geophysics. Then it is shown that computing the Schur numbers from the data $(a_0, ..., a_{n-1})$ is equivalent to solving an inverse scattering problem.

1. THE LAYERED MEDIUM MODEL

We begin by developing the basic layered medium model in which Schur numbers play the role of reflection coefficients.

Consider a plane wave traveling through a layered medium, orthogonally to the interfaces between the layers. We assume that the upper layer labeled by 0 extends to ∞. The interface between the j-th and j+1 layer is called the j-th interface; see Figure 1.

<table>
<tr><td>0-layer</td><td></td></tr>
<tr><td colspan="2">————————————————————————————————</td></tr>
<tr><td>1st-layer</td><td>0-interface</td></tr>
<tr><td colspan="2">————————————————————————————————</td></tr>
<tr><td>2nd-layer</td><td>1st interface</td></tr>
<tr><td colspan="2">————————————————————————————————</td></tr>
<tr><td>3rd-layer</td><td>2nd interface</td></tr>
<tr><td colspan="2">————————————————————————————————</td></tr>
<tr><td>⋮ ⋮ ⋮</td><td>3rd interface</td></tr>
</table>

Figure 1

Each interface is characterized by the following properties:

(i) If a downgoing wave of amplitude a strikes the j-th interface, then a wave of amplitude ar_j will be reflected upward into the j-th layer, and a wave of amplitude at_j will cross the j-th interface into the lower j+1 layer; see Figure 2. The numbers (without physical dimension) r_j and t_j are called the *reflection and transmission* coefficients of the j-th interface, respectively. They enjoy the following properties

$$t_j = 1 + r_j \quad \text{and} \quad -1 \le r_j \le 1 . \tag{1.1}$$

(ii) If an upgoing wave of amplitude a strikes the j-th interface, then a wave of amplitude $-ar_j$ is reflected downward into the j+1 layer and a wave of amplitude $a(1-r_j)$ will be transmitted upward into the j-th layer, see Figure 2.

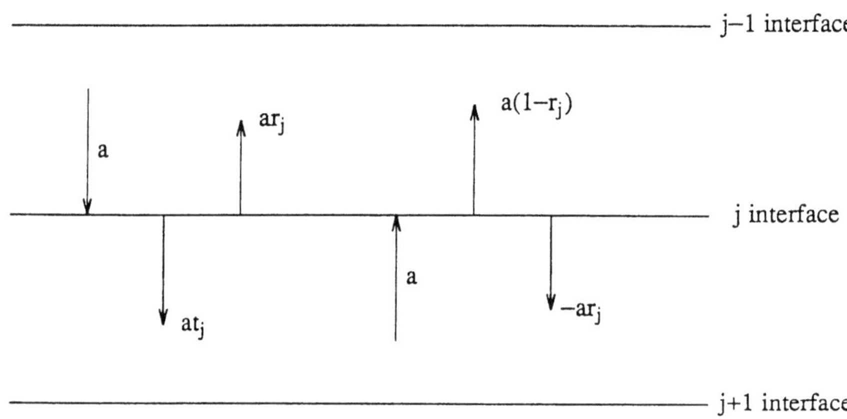

Figure 2

A *unit impulse* is a downgoing wave of amplitude one which strikes an interface only once, at time zero. (The amplitude is zero at all other times). We always assume that the time it takes for a downgoing (or upgoing) wave to travel the distance between the j and j+1 (or the j+1 and j) interface is T/2, where T is the same positive number for all $j \ge 0$. So if a (downgoing) unit impulse strikes the zero interface, then we obtain upgoing waves at the zero interface of amplitudes

$$(a_0, a_1, a_2, a_3, \ldots) \tag{1.2}$$

at times

$$(0, T, 2T, 3T, 4T, \ldots) . \tag{1.3}$$

It is assumed that our system is linear, stationary and causal. This means that if we strike the zero interface by downgoing waves of amplitudes

$$(\alpha_0, \alpha_1, \alpha_2, \alpha_3, \dots) \tag{1.4}$$

at times (1.3), then at the zero interface we obtain an output of upgoing waves of amplitudes

$$(a_0, a_1, a_2, a_3, \dots) * (\alpha_0, \alpha_1, \alpha_2, \alpha_3, \dots) \triangleq$$

$$(a_0\alpha_0, \; a_1\alpha_0 + a_0\alpha_1, \; a_2\alpha_0 + a_1\alpha_1 + a_0\alpha_2, + \dots) = \tag{1.5}$$

$$(\beta_0, \beta_1, \beta_2, \beta_3, \dots)$$

at times (1.3). Here $*$ denotes the convolution. By taking the z-transform, that is, by identifying the time nT with z^n, we see that (1.5) is formally equivalent to

$$\beta(z) = R_0(z)\alpha(z) \quad \text{where}$$

$$R_0(z) = \sum_0^\infty a_n z^n, \quad \alpha(z) = \sum_0^\infty \alpha_n z^n \text{ and } \beta(z) = \sum_0^\infty \beta_n z^n. \tag{1.6}$$

$R_0(z)$ is the *scattering function* of the layered medium below the 0-interface. For convenience we shall call it the scattering function at the 0-interface. Analogously we define $R_j(z)$ as the *scattering function at the* j-th interface by

$$R_j(z) = \sum_{i=0}^\infty a_i^j z^i \tag{1.7}$$

where $(a_0^j, a_1^j, a_2^j, \dots)$ are the amplitudes of the upgoing waves at times (1.3) at the j-th interface, when excited by a (downgoing) unit impulse at the j-th interface and the $j-1$ layer extends to infinity.

2. SCHUR NUMBERS AND REFLECTION COEFFICIENTS

In this section we show that the reflection coefficients are the Schur numbers when one views $a_0, a_1, a_2, \dots$ as the data in the Carathéodory interpolation problem.

We begin with the following result, which displays the relation between two scattering functions at successive interfaces.

2.1 PROPOSITION. *As power series the scattering functions* R_j *and* R_{j+1} *are related by*

$$R_j(z) = \omega_{r_j}(zR_{j+1}(z)) \triangleq \frac{r_j + zR_{j+1}(z)}{1 + r_j zR_{j+1}(z)}. \tag{2.1}$$

PROOF. It suffices to consider the case $j = 0$. Let a (downgoing) unit impulse hit the zero interface. Clearly

$$r_o \text{ is reflected upward at time zero.} \tag{2.2}$$

Using properties (i) and (ii) in Section 1 we have upgoing waves (starting at time T) just above the zero interface of amplitudes

$$(1+r_o)(1-r_o)a_o^1(\text{at time T}), \qquad (1+r_o)(1-r_o)a_1^1(\text{at time 2T}), \; \cdots \;,$$

or equivalently,

$$(1+r_o)(1-r_o)(0, \; a_o^1, \; a_1^1, \; a_2^1, \; ...) \;.$$

By (1.7) this is the z-transform of

$$(1 - r_o^2)zR_1(z) \;. \tag{2.3}$$

Analogously, the waves with z-transform

$$z(1+r_o)(-r_o)R_1(z) \tag{2.4}$$

are reflected downward just below the zero interface (see Figure 3) starting at time T. This produces upgoing waves just above the zero interface (starting at time 2T) whose z-transform is

$$z^2(1+r_o)(1-r_o)(-r_o)R_1(z)^2 \;. \tag{2.5}$$

Also

$$z^2(1+r_o)(-r_o)^2R_1(z)^2 \tag{2.6}$$

is the z-transform of the waves which starts at time 2T and are reflected downward just below the zero interface. This in turn produces upward waves just above the zero interface (starting at time 3T) whose z-transform is

$$z^3(1+r_o)(1-r_o)(-r_o)^2R_1(z)^3 \;. \tag{2.7}$$

In general the above analysis yields upgoing waves just above the zero interface (starting at time nT) whose the z-transform is

$$z^n(1+r_o)(1-r_o)(-r_o)^{n-1}R_1(z)^n \;. \tag{2.8}$$

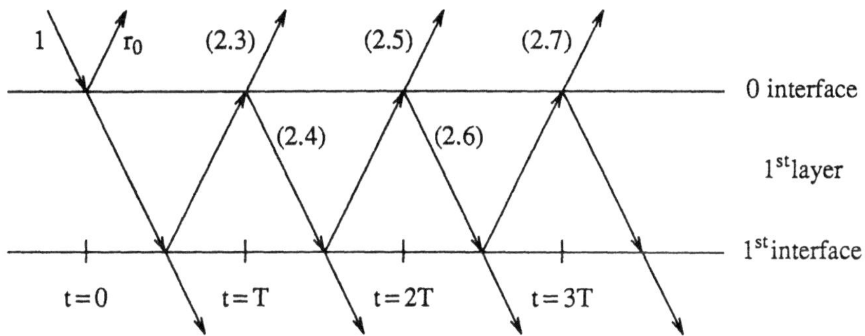

Figure 3

In Figure 3 the vertical axis represents depth while the horizontal one represents time. Figure 3 also displays the starting points of the waves in (2.3) to (2.7). Notice that in space as indicated by Figures 1 and 2 the waves are moving vertically.

Finally, summing up all the upgoing waves just above the zero interface (see (2.2) and (2.8)) we obtain

$$R_o(z) = r_o + \sum_0^\infty z^{n+1}(1+r_o)(1-r_o)(-r_o)^n R_1(z)^{n+1} =$$

$$r_o + z(1-r_o^2)R_1(z)(1+zr_oR_1(z))^{-1} = \frac{r_o+zR_1(z)}{1+zr_oR_1(z)} \; . \tag{2.9}$$

The second equality follows from using the fact that as power series

$$\sum_0^\infty z^n(-r_o)^n R_1^n(z) = (1+zr_oR_1(z))^{-1} \; .$$

This completes the proof.

Comparing (I.1.13) and (2.1) we see that r_o is the 0-th Schur number of R_o and

$$(a_o^1, a_1^1, ..., a_{n-2}^1) = \sigma(a_o, a_1, ..., a_{n-1})$$

where σ is defined in (I.1.10) and a_i, a_i^1 are given in (I.1.8) and (I.1.9). We conclude with the following facts.

2.2 COROLLARY. *In the preceding model, let the amplitudes*

$$(a_0, \ a_1, \ a_2, \ ..., \ a_{n-1}) \tag{2.10}$$

of the (upward) output at the zero interface be the data of Problem I.1.1. Then the reflection coefficients $(r_0, \ r_1, \ ..., \ r_{n-1})$ *of the layered medium are precisely the Schur numbers of the data.*

PROOF. This readily follows from (2.1) (for j = 0, 1, 2, ...) and the definition of the Schur members given in Section I.1.

2.3 COROLLARY. *Problem I.1.1 with the data in (2.10) is solvable.*

PROOF. To see this, notice that $(a_0, \ a_1, \ ..., \ a_{n-1})$ is not affected by the values of $r_n, \ r_{n+1}, \ r_{n+2}, \ ...$ (This follows because it takes nT units of time for the effect of the n-th interface to appear on the surface.) Without loss of generality, assume that $r_m = 0$ for all $m \geq n$, or equivalently, $R_n = 0$. Notice that this actually means that the n-th layer is extended downward to infinity. (In the case when one of the reflection coefficients is of modulus one we can set $r_m = 0$ for all $m \geq j + 1$, where j is the smallest integer such that $|r_j| = 1$. This follows because if $|r_j| = 1$, then the j-th interface acts as a barrier. If $r_j = 1$, then no wave below the j-th interface can enter the j-th layer, or if $r_j = -1$, then no wave in the j-th layer can cross the j-th interface; see properties (i) and (ii) in Section 1.) Hence $R_{n-1}(z) = r_{n-1}$ and $R_{n-2}, \ R_{n-3}, \ ..., \ R_0$ are by virtue of (2.1) and Theorem I.1.5 the Taylor expansions of $z = 0$ of functions in H_1^∞. Thus

$$R_0(z) = a_0 + a_1 z + a_2 z^2 + \cdots + a_{n-1} z^{n-1} + 0(z) \quad (z \in D)$$

and $R_0 \in H_1^\infty$. This completes the proof.

2.4 COROLLARY. *If the j-th reflection coefficient is of modulus one, then the scattering function* R_0 *is a Blaschke product of order j.*

PROOF. This follows from Theorem I.2.4, Proposition 2.1 and Corollary 2.3

3. COMPUTING WITH THE LAYERED MEDIUM MODEL

In this section we will use the layered medium model to obtain another recursive algorithm for computing the Schur numbers, or reflection coefficients from the data $(a_0, \ ..., \ a_{n-1})$.

We begin by establishing some notation. Throughout, the layered medium is hit by a (downgoing) unit impulse at the zero interface. We define $X_j^{(n)}$, $Y_j^{(n)}$ as the amplitude of the downgoing, upgoing wave which strikes the j-th interface at time $(n - \frac{j}{2})T$, respectively (see Figure 4). (The waves $X_j^{(n)}$ and $Y_j^{(n)}$ are different from the $x_{i,j}$ and $y_{i,j}$ in Section I.3). The following boundary conditions are a simple consequences of (i), (ii) in Section 1 and the fact that the time it takes a wave to travel between two adjoining layers is $T/2$:

$$X_n^{(n)} = (1 + r_{n-1})X_{n-1}^{(n-1)} = (1 + r_{n-1}) \cdots (1 + r_0), \qquad X_0^{(0)} = 1$$

$$\tag{3.1}$$

$$X_0^{(n)} = 0, \; Y_j^{(n)} = 0 \text{ and } X_{j+1}^{(n)} = 0 \qquad (\text{if } j \geq n \geq 1).$$

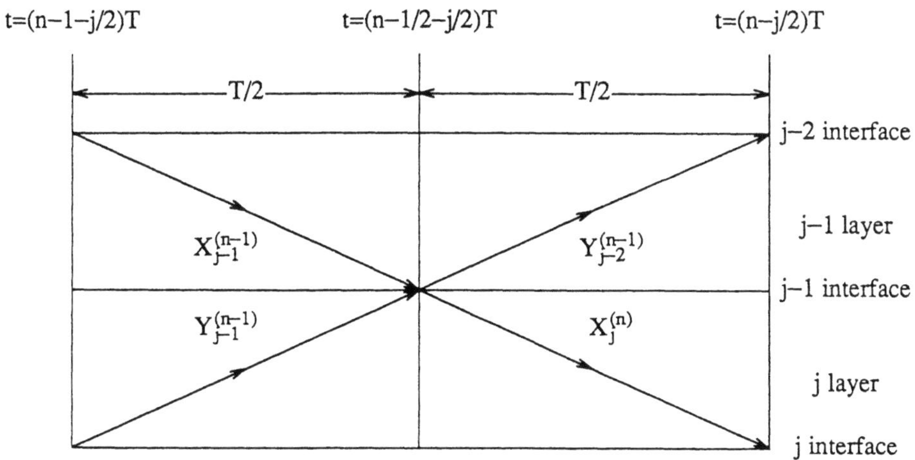

Figure 4

By consulting Figure 4 we readily obtain the following result.

3.1 LEMMA. *For* $n \geq 2$ *the string* $(X_1^{(n)}, X_2^{(n)}, ..., X_n^{(n)})$ *can be obtained recursively from the string* $(X_1^{(n-1)}, X_2^{(n-1)}, ..., X_{n-1}^{(n-1)})$ *by (3.1) and the equation*

$$\begin{bmatrix} Y^{(n-1)}_{j-2} \\ X^{(n)}_j \end{bmatrix} = \begin{bmatrix} r_{j-1} & 1-r_{j-1} \\ 1+r_{j-1} & -r_{j-1} \end{bmatrix} \begin{bmatrix} X^{(n-1)}_{j-1} \\ Y^{(n-1)}_{j-1} \end{bmatrix} \quad (j = n,\ n-1,\ ...,\ 1) . \tag{3.2}$$

3.2 LEMMA. *With the previous notations we have*

$$a_n = \sum_{j=1}^{n} (1-r_o)(1-r_1)\cdots(1-r_{j-1})r_j X^{(n)}_j \qquad (n \geq 1). \tag{3.3}$$

PROOF. The amplitude of the upgoing wave reflected in the j-th layer (at time $(n - j/2)T$) is $r_j X^{(n)}_j$. Since the upgoing wave must also travel through the interfaces j–1, j–2, ..., 0, the amplitude of its upgoing component immediately above the 0-th interface at time nT is given by

$$X^{(n)}_j r_j (1-r_{j-1})(1-r_{j-2})\cdots(1-r_o) . \tag{3.4}$$

Since at time $\dfrac{nT}{2}$ the downgoing wave has only reached the n-th interface, we have $X^{(n)}_j = 0$ for $j > n$. Summing (3.4) over j yields (3.3) and completes the proof.

Using Lemma 3.1 and 3.2 we obtain the following recursive algorithm for computing the reflection coefficients (or Schur numbers) from the output (or data) $(a_0,\ a_1,\ a_2,\ ...)$.

3.3 PROCEDURE. For $i \geq 1$ let

$$\begin{aligned} Z_i &= (1-r_{i-1})Z_{i-1} = (1-r_o)(1-r_1)\cdots(1-r_{i-1}) \\ t_i &= (1-|r_{i-1}|^2)t_{i-1} = (1-|r_o|^2)(1-|r_1|^2)\cdots(1-|r_{i-1}|^2) . \end{aligned} \tag{3.5}$$

Equations (3.1),(3.3) and $a_o = r_o$ yield

$$r_1 = a_1(1-|a_o|^2)^{-1} \text{ and } r_n = \frac{a_n - \sum_{j=1}^{n-1} Z_j r_j X^{(n)}_j}{t_n} \quad (\text{if } n \geq 2) \tag{3.6}$$

where $X^{(n)}_j$ and Z_j, t_j are recursively defined by interlacing (3.1), (3.2) and (3.5) with (3.6). Notice that in using (3.2), for given n, one starts with $j = n$ and finishes with $j = 1$. Equation (3.6) recursively gives r_n. If at any step $|r_n| = 1$, then the solution to Problem I.1.1 with data of length greater than or equal to n is unique. Since the data comes from a layered medium model, the solution to Problem I.1.1 always exists by Corollary 2.3.

This algorithm is just as efficient as Procedure I.4.3. It takes approximately $4n^2$ calculations to obtain the reflection coefficients. However, it is not as robust as

Procedure I.5.6 which takes approximately $5n^2$ calculations. Finally, we notice that the determination of the output $(a_0, a_1, a_2, ...)$ resulting from the input $(1, 0, 0, ...)$ when the reflection coefficients are known is called the *direct problem*. The reverse problem of computing the reflection coefficients from the input and output data is, of course, called the *inverse problem or inverse scattering problem*. Obviously Procedures I.1.7, I.4.3, I.5.6 and 3.3 are inverse scattering algorithms which solve the inverse problem. They can easily be adapted to solve the direct problem. The details are left to the reader.

4. THE SCHUR REPRESENTATION AND THE LAYERED MEDIUM MODEL

In this section we will use the layer medium model to obtain a compelling interpretation of the Schur representation and of the layer peeling algorithm.

We begin by considering the layered medium scheme in Section 1, where the waves hit the zero- interface at times nT (for $n=0, \pm 1, \pm 2, \cdots$). Obviously these waves hit the j-th interface at times $(n + j/2)T$ (for $n = 0 \pm 1, \pm 2, \cdots$). In defining the z–transform of the waves at the j-th interface it is convenient (as will be made clear below) to shift the origin of time to $(j/2)T$. This means that the z–transform of waves reaching the j-th interface at times

$$(..., (\frac{j}{2} -1)T, \frac{jT}{2}, (\frac{j}{2} + 1)T, ...)$$

with corresponding amplitudes $(..., a_{-1}, a_0, a_1, ...)$ is given by

$$\sum_{j=-\infty}^{\infty} a_j z^j \ .$$

Now let

$$g_0 = \sum_{i=-\infty}^{\infty} g_{0,i} z^i \text{ and } f_0 = \sum_{i=-\infty}^{\infty} f_{0,i} z^i$$

be the z–transform of the downgoing, respectively upgoing waves in the zero-layer (immediately above the zero interface). In the j-th layer these waves correspond to the waves given by

$$g_j' = \sum_{i=-\infty}^{\infty} g_{j,i}' z^i \text{ and } f_j' = \sum_{i=-\infty}^{\infty} f_{j,i}' z^i$$

at the upper interface and

$$g_j = \sum_{i=-\infty}^{\infty} g_{j,i} z^i \quad \text{and} \quad f_j = \sum_{i=-\infty}^{\infty} f_{j,i} z^i \tag{4.1}$$

at the lower interface. We emphasize that f_j and f_j' are the z–transforms of upgoing waves while g_j and g_j' are the z–transforms of downgoing waves. To help the reader we represented in Figure 5 the amplitude of these waves striking the j-th interface at time $(n + j/2)T$.

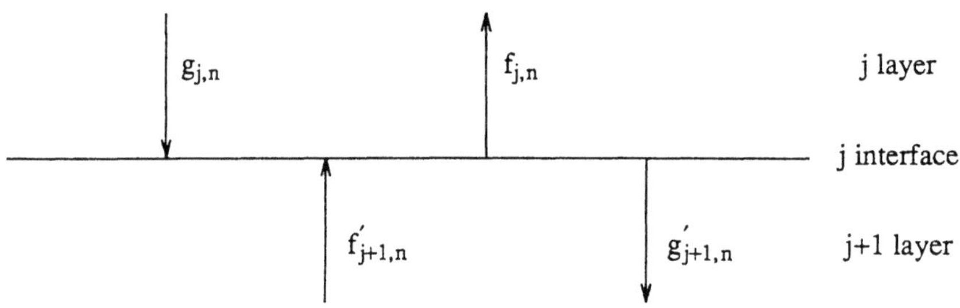

Figure 5

The *scattering matrix* for the j-th interface is the matrix which expresses the output at the j-th interface in terms of the input at the j-th interface. By Figure 5 and properties (i) and (ii) in Section 1, this matrix is given by the 2 by 2 matrix in the following equation

$$\begin{bmatrix} f_j \\ g_{j+1}' \end{bmatrix} = \begin{bmatrix} r_j & 1-r_j \\ 1+r_j & -r_j \end{bmatrix} \begin{bmatrix} g_j \\ f_{j+1}' \end{bmatrix}. \tag{4.2}$$

It is interesting to note that if S_j is the scattering matrix in (4.2), then $S_j^2 = I$, or the inverse of S_j is S_j. By rearranging (4.2) we have

$$[g_j, f_j] = [g_{j+1}', f_{j+1}'] \begin{bmatrix} 1 & r_j \\ r_j & 1 \end{bmatrix} \frac{1}{1+r_j}. \tag{4.3}$$

Using $f_{j+1}' = z f_{j+1}$ and $g_{j+1}' = g_{j+1}$ in (4.3) yields

$$[g_j, f_j] = [g_{j+1}, zf_{j+1}] \begin{bmatrix} 1 & r_j \\ r_j & 1 \end{bmatrix} \frac{1}{1+r_j} \tag{4.4}$$

if $r_j \neq -1$. This with (I.3.1) implies that

$$[g_0, f_0] = [g_j, zf_j] \begin{bmatrix} u_j & x_j \\ v_j & y_j \end{bmatrix} \frac{1}{(1+r_0)(1+r_1)...(1+r_{j-1})} . \tag{4.5}$$

Now let $g_0 = 1$ be the unit impulse. Then $f_0 = R_0$ is the scattering function at the zero interface. As before, R_n is the scattering function at the n-th interface. Thus $f_n = R_n g_n$. Substituting $g_0 = 1$, $f_0 = R_0$ and $f_n = R_n g_n$ into (4.5) for $j = n$ produces

$$[1, R_0] = [g_n, zR_n g_n] \begin{bmatrix} u_n & x_n \\ v_n & y_n \end{bmatrix} \frac{1}{(1+r_0)(1+r_1)...(1+r_{n-1})} . \tag{4.6}$$

Eliminating g_n in (4.6) gives

$$R_0 = \frac{x_n + zy_n R_n}{u_n + zv_n R_n} . \tag{4.7}$$

By virtue of (I.3.4) this is precisely the Schur representation (I.3.8) of the solution $f = R_0$ to Problem I.1.1 with the data $(a_0, a_1, ..., a_{n-1})$. Summing up the previous analysis proves the following result.

4.1 PROPOSITION. *Let $r_0, r_1, ..., r_{n-1}$ be the first n reflection coefficients different from -1 of the layered medium. Then the Schur representation (I.3.8) is representing the scattering function $R_0 = f_0$ at the zero interface in terms of the scattering function $R_n = f_n$ at the n-th interface.*

At this point it is appropriate to mention that the matrix

$$\frac{1}{1+r_j} \begin{bmatrix} 1 & r_j \\ r_j & 1 \end{bmatrix} \tag{4.8}$$

in (4.3) is called the *transmission matrix of the j-th interface,* because it transforms the waves below the interface into the waves above the interface. The transmission matrix plays an important role in inverse scattering theory, interpolation theory, signal processing, etc. In this paragraph we will use this matrix to obtain the layer peeling algorithm to compute the reflection coefficients from the data.

To complete this section, we use the layered medium model to present another method for obtaining the layer peeling algorithm in Procedure I.1.7. To begin, we assume that the input of the zero interface g_o starts at time zero, that is, $g_{o,i} = 0$ for all $i < 0$. Since these waves reach the j-th layer at times larger than or equal to $(j/2)T$, we deduce from (4.1) that both $g_{j,i}$ and $f_{j,i}$ are also zero for $i < 0$. This means that g_j and f_j admit a power series expansion of the form:

$$g_j = \sum_{i=0}^{\infty} g_{j,i} z^i \quad \text{and} \quad f_j = \sum_{i=0}^{\infty} f_{j,i} z^i . \tag{4.9}$$

By inverting the transmission matrix in (4.4) and matching like coefficients of z^i we have

$$\begin{bmatrix} g_{j,o} & 0 \\ g_{j,1} & f_{j,o} \\ g_{j,2} & f_{j,1} \\ g_{j,3} & f_{j,2} \\ \cdot & \cdot \\ \cdot & \cdot \\ \cdot & \cdot \end{bmatrix} = \begin{bmatrix} g_{j-1,o} & f_{j-1,o} \\ g_{j-1,1} & f_{j-1,1} \\ g_{j-1,2} & f_{j-1,2} \\ g_{j-1,3} & f_{j-1,3} \\ \cdot & \cdot \\ \cdot & \cdot \\ \cdot & \cdot \end{bmatrix} \begin{bmatrix} 1 & -r_{j-1} \\ -r_{j-1} & 1 \end{bmatrix} \frac{1}{1-r_{j-1}} . \tag{4.10}$$

From the definition of $f_{j,o}$ and $g_{j,o}$ with $f_j = R_j g_j$ it follows that

$$r_j = f_{j,o}/g_{j,o} \qquad (j \geq 0) \tag{4.11}$$

Equations (4.10) and (4.11) lead to the following inverse scattering algorithm.

4.2 PROCEDURE. Let g_o be the input function at the zero interface satisfying $g_{j,i} = 0$ if $i < 0$ and f_o the corresponding output function. Set $r_o = f_{o,o}/g_{o,o}$. For $j=1, 2, 3, \cdots$ recursively compute $g_{j,i}$ and $f_{j,i}$ (for $i \geq 0$) and r_j by interlacing (4.10) with (4.11).

If one only wants to compute the first n reflection coefficients $(r_o, r_1, ..., r_{n-1})$, then the previous procedure only needs the amplitudes $f_{o,i}$ and $g_{o,i}$ for $0 \leq i \leq n-1$. This follows from (4.10), or from the physical fact that it takes $nT/2$ units of time for a wave to travel from the zero interface to the n-th interface. This algorithm obviously necessitates approximately $(3/2) n^2$ arithmetical calculations.

Finally, notice that formulas (4.10) and (I.1.16) differ by a multiplicative constant on the right. By virtue of (4.11) and (I.1.18) such factors do not effect the computation of the reflection coefficients. Therefore Procedure I.1.7 actually coincides with Procedure 4.2. This procedure is called the *layer peeling algorithm* because it peels off the waves in the first j−1 layers to arrive at the waves g_j and f_j immediately above the j-th interface.

5. THE LEVINSON ALGORITHM AND MARINE SEISMOLOGY

In this section we use the Levinson algorithm to compute the reflection coefficients in a marine seismology problem.

In marine seismology one has the same layered medium structure described by Figures 1 and 2 in Section 1, with one additional constraint. The zero interface air-water acts like a perfect reflector from below, that is, $r_o = -1$; see Figure 6. (There are no constraints on the other reflection coefficients r_j for $j \geq 1$, except of course $|r_j| \leq 1$.) Here as in Section 4, the functions g_1', f_1', g_1, f_1 are the downgoing, upgoing, downgoing, upgoing waves just below, below, above, above the zero, zero, first, first interface, respectively (see Figure 6).

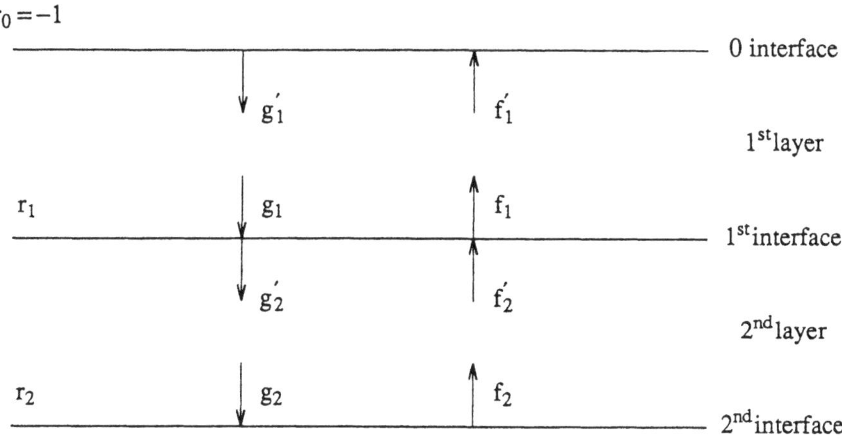

Figure 6

To compute the reflection coefficients r_i for $i \geq 1$ in marine seismology one sets off a single explosion $g_{o,o}' = 1$ immediately below the zero interface. Since $r_o = -1$ the properties (i) and (ii) in Section 1 give

$$1 + f_1' = g_1' . \tag{5.1}$$

Using (4.1), we have

$$f_1' = zf_1, \quad g_1' = g_1 \text{ and } f_1 = R_1 g_1 \tag{5.2}$$

where R_1 is the scattering function at the first interface. Substituting (5.2) into (5.1) yields $1 + zR_1 g_1 = g_1$, or equivalently, $g_1 = (1 - zR_1)^{-1}$. From this and the fact that $f_1' = zR_1 g_1$ it follows that

$$f_1' = \frac{zR_1}{1 - zR_1} = \sum_{i=1}^{\infty} b_i z^i \, . \tag{5.3}$$

where b_i is the amplitude of f_1' at time iT for $i = 1, 2, \ldots$ We define the function F by

$$F = \overline{f_1'} + 1 + f_1' = \frac{1 - |zR_1|^2}{|1 - zR_1|^2} \geq 0 \qquad (z \in D) \, . \tag{5.4}$$

The last equality follows from (5.3). The inequality holds because R_1 is in H_1^{∞}.

The above analysis shows that F is the function of positive type uniquely determined by $f = zR_1 \in H_1^{\infty}$; see Section II.1. This shows that $0, r_1, r_2, r_3, \ldots$ are the Schur numbers for f where $r_1, r_2, r_3, \ldots$ are the reflection coefficients of the layered medium below the first layer. Equations (5.3) and (5.4) imply that the Toeplitz matrix $B_n = T_n(F)$ generated by F is given by

$$B_n = T_n(F) = \begin{bmatrix} 1 & b_1 & b_2 & \cdots & b_{n-1} \\ b_1 & 1 & b_1 & \cdots & b_{n-2} \\ b_2 & b_1 & 1 & \cdots & b_{n-3} \\ . & . & . & \cdots & . \\ . & . & . & \cdots & . \\ . & . & . & \cdots & . \\ b_{n-1} & b_{n-2} & b_{n-3} & \cdots & 1 \end{bmatrix} . \tag{5.5}$$

By Theorem II.4.2 or Procedure II.4.5, the reflection coefficients or Schur numbers r_j are recursively given by

$$r_j = -\alpha_{j+1, 1} \qquad (\text{if } j \geq 1) \tag{5.6}$$

where the $\alpha_{j,i}$'s are the constants obtained from the Levinson algorithm applied to B_n in (5.5); see Sections II.2 and II.4. (Since the entries in the Toeplitz matrix B_n are real, the $\alpha_{j,i}$'s computed by the Levinson algorithm are real and one does not have to take the complex conjugate in (5.6).) Summing up this analysis produces the following algorithm for computing the reflection coefficients in our marine seismology problem ($r_o = -1$).

5.1 PROCEDURE. Set off a single explosion $g'_{0,0} = 1$ immediately below the zero interface and collect the resulting amplitudes b_i for f'_1 at times iT for $i \geq 1$. Then $r_1, r_2, \cdots$ is given by (5.6), where the $\alpha_{j,i}$'s are recursively computed from the Levinson algorithm applied to B_n.

The layer peeling algorithm in Section 4 can be used to recursively obtain the reflection coefficients r_j in marine seismology. The set up is the same as before. A single explosion $g'_{0,0} = 1$ is set off immediately below the zero interface and the resulting amplitudes b_i of f'_1 at times iT for $i \geq 1$ are collected. Then the second equation in (5.2), with (5.1) implies that $g_1 = 1 + f'_1$. This and (5.3) show that the input g_1 immediately above the first interface is

$$g_1 = 1 + \sum_{i=1}^{\infty} b_i z^i . \tag{5.7}$$

The first equation in (5.2), with (5.3) show that the output f_1 immediately above the first interface is

$$f_1 = \sum_{i=0}^{\infty} b_{i+1} z^i . \tag{5.8}$$

Equations (5.7) and (5.8) provide us with the following initial conditions immediately above the first interface

$$\begin{aligned} g_{1,o} = 1 \text{ and } g_{1,i} = b_i \quad &(\text{if } i \geq 1) \\ f_{1,i} = b_{i+1} \quad &(\text{if } i \geq 0) . \end{aligned} \tag{5.9}$$

These "initial conditions" follow from the fact that the zero interface acts like a perfect reflector from below. So the output $f_{1,i-1} = b_i$ get reflected into the input $g_{1,i} = b_i$ one time unit T later.

The conditions in (5.9) show that the marine seismology setting is a special case of the usual layered medium setting in Section 4. By ignoring the zero interface the marine seismology structure is equivalent to the usual layered medium structure in Section 4, under the additional constraint that (5.9) holds at the top interface. In particular, one can compute the reflection coefficients r_j in marine seismology by computing the reflection coefficients r_j for $j \geq 1$ in the usual layered medium setting assuming that (5.9) holds. Consulting Section 4 these reflection coefficients are given by (see (4.11))

$$r_1 = b_1 \text{ and } r_j = f_{j,o}/g_{j,o} \quad (\text{if } j \geq 2) \tag{5.10}$$

where $f_{j,i}$ and $g_{j,i}$ are computed recursively by (4.10) starting with $j = 2$ and using the initial conditions (5.9), as in Procedure 4.2. Summing up gives us another algorithm for computing the reflection coefficients r_j in marine seismology.

5.2 PROCEDURE. Set of a single explosion $g'_{o,o} = 1$ immediately below the zero interface, and collect the resulting amplitudes b_i for f'_1 at times it for $i \geq 1$. Define $g_{1,i}$ and $f_{1,i}$ for $i \geq 0$ according to (5.9) and set $r_1 = b_1$. For $j = 2, 3, \cdots$ recursively compute $g_{j,i}$ and $f_{j,i}$ (for $i \geq 0$), and r_j by interlacing (4.10) with (5.10).

6. SCATTERING AND ROTATION MATRICES

In this section we will show that the 2 by 2 scattering matrix in (4.2) is unitarily equivalent to a rotation matrix on C^2. (Recall that two operators A and B are *unitarily equivalent* if there exists a unitary operator U acting between the appropriate spaces such that $A = U^*BU$.) In the last chapter of the monograph we will show that the reflection coefficient r_j is given by

$$r_j = \frac{Z_j - Z_{j+1}}{Z_j + Z_{j+1}} \tag{6.1}$$

where $Z_j \,(>0)$ is a physical parameter called the *impedance* of the j-th layer. The *energy* of a wave of amplitude a in the j-th layer (with speed 1) is $a^2 Z_j$.

Let u, y, u', y', be the waves represented in Figure 7.

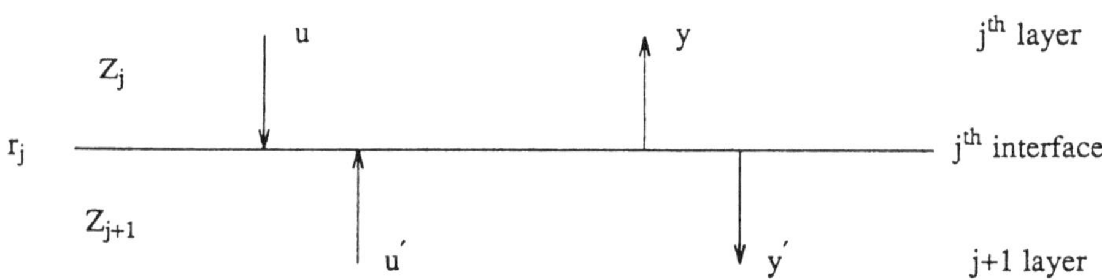

Figure 7

By the properties (i) and (ii) in Section 1,

$$\begin{bmatrix} y \\ y' \end{bmatrix} = \begin{bmatrix} r_j & 1-r_j \\ 1+r_j & -r_j \end{bmatrix} \begin{bmatrix} u \\ u' \end{bmatrix}. \tag{6.2}$$

Recall that the matrix S_j in (6.2) defined by

$$S_j \triangleq \begin{bmatrix} r_j & 1-r_j \\ 1+r_j & -r_j \end{bmatrix} \tag{6.3}$$

is the *scattering matrix* of the j-th interface. It transforms the input into the output at the j-th interface. This matrix plays a fundamental role in scattering and interpolation theories. The matrix has already played an important role in calculating the reflection coefficients in Procedure 3.3; see (3.2). It was also used in Section 4 to obtain the Schur representation (see (4.2)).

Using (6.1) and (6.2) a straight forward calculation shows that

$$Z_j y^2 + Z_{j+1} y'^2 = Z_j u^2 + Z_{j+1} u'^2 . \tag{6.4}$$

This says that the energy of the output of the j-th interface equals the energy of the input of the j-th interface. In other words, the energy is conserved by the scattering produced at the j-th interface. We denote by H the vector space C^2 endowed with the inner product

$$(x,y)_H = Z_j x_1 \bar{y}_1 + Z_{j+1} x_2 \bar{y}_2 = (Qx, Qy) \qquad (x_1, x_2, y_1, y_2 \in C^1) \tag{6.5}$$

where Q is the diagonal matrix whose diagonal entries are $\sqrt{Z_j}$ and $\sqrt{Z_{j+1}}$ and $(\cdot, \cdot)$ is the standard inner product on C^2. Equations (6.4) and (6.5) imply that the operator S_j on H is unitary, that is, it preserves the inner product on H. Consulting (6.5) we see that Q is a unitary operator, from H onto C^2, that is, it transforms the inner product on H into that of C^2. So using $Q^* = Q^{-1}$ the operator $R_{r_j} \triangleq Q S_j Q^{-1}$ on C^2 is unitarily. This and (6.1), (6.3) give

$$R_{r_j} = Q S_j Q^{-1} = \begin{bmatrix} r_j & (1-r_j^2)^{1/2} \\ (1-r_j^2)^{1/2} & -r_j \end{bmatrix} . \tag{6.6}$$

6.1 PROPOSITION. *The scattering matrix S_j on H in (6.3) is unitarily equivalent to the rotation matrix R_{r_j} on C^2 in (6.6).*

The matrix R_{r_j} on C^2 in (6.6) is called the *rotation matrix* for r_j. Rotation matrices play an important role in solving many inverse scattering and interpolation problems. In particular, they will play a major role in our work starting with the next chapter.

III.7. NOTES AND COMMENTS

All the results in this section concerning layered medium and inverse scattering are well known in geophysics; see for instance [Cl], [Kan], [Ro] and [RoT]. The connection between geophysics, Schur numbers and Carathéodory interpolation is implicitly contained in these books and explicitly presented in [Fo 10]. Here we have tried to emphasize this connection. Proposition 2.1 and its proof is taken from [Ro 2]. Finally, it is noted that Procedure 3.3 is due to Robinson [Ro 1]. The layer peeling algorithm is taken from [Ka 2]; for related results see [DewD], [DewVK] and [LeK]. The connection between the Levinson algorithm and marine seismology can be found in [Cl]. In connection with this chapter see [BrK], [Cl], [Kan], [Ro] and [RoT].

CHAPTER IV

CONTRACTIVE EXPANSIONS ON EUCLIDIAN AND HILBERT SPACE

In this chapter we consider matrices with operator or matrix entries, often called block matrices. First we obtain a complete characterization of all 2 by 2 contractive block matrices. Then we develop a Levinson type algorithm to determine whether or not a block matrix is contractive. As an application we will present a new method for solving the Carathéodory interpolation problem. Although this chapter is presented in a Hilbert space framework, we recommend to the reader who is unfamiliar with elementary functional analysis to consider all spaces as Euclidian spaces, that is, different C^n spaces with their standard inner product.

1. CONTRACTIVE 2 by 1 COLUMN MATRICES

To begin we recall some elementary facts concerning operators on Hilbert spaces. Let H and H' be two Hilbert spaces and T a linear operator mapping H into H'. The norm of T is defined by

$$\|T\| = \sup\{\|Th\|{:}h \in H \text{ and } \|h\| \le 1\}$$

where the norms on both H and H' are denoted by $\|\cdot\|$. Recall that the norm of an element f in a Hilbert space is defined from the inner product in that Hilbert space in the following way $\|f\| = (f,f)^{\frac{1}{2}}$. An operator T is *bounded* if $\|T\|$ is finite, and T is a *contraction* if $\|T\| \le 1$. The operator T is a *strict contraction* $\|T\| < 1$. Throughout this book, when we say that T is an operator we mean that T is a bounded linear operator. The *adjoint* T^* of T is the operator mapping H' into H defined by

$$(Th, h') = (h, T^*h') \qquad (\text{where } h \in H \text{ and } h' \in H').$$

Recall that $\|T\| = \|T^*\|$. In particular, T is a contraction if and only if T^* is a contraction. The operator T is an *isometry* if $\|Th\| = \|h\|$ for all h in H. The operator T is *unitary* if T is an isometry and the range of T is H'. It is easy to show that T is unitary if and only if T and T^* are isometric. The indentity, respectively zero operator on any Hilbert space is always denoted by I, respectively 0. The orthogonal projection onto a subspace F of H is denoted by P_F.

An operator Q on H is *positive* (notation $Q \geq 0$) if $(Qh, h) \geq 0$ for all $h \in H$. Moreover, if Q is positive, there exists a unique (bounded linear) positive operator $Q^{1/2}$ such that $(Q^{1/2})^2 = Q$. This positive square root of Q is the norm limit of polynomials in Q. Notice that T is a contraction if and only if $I - T^*T$ is a positive operator on H. We say that Q is *strictly positive* if $Q - \delta I$ is positive for some real $\delta > 0$. We mention that if Q is positive, then Q is strictly positive if and only if its square root is strictly positive. Here and throughout this text, we say that an operator is invertible if it has a bounded inverse, that is, its inverse is also an operator. We also mention that if Q is positive, then Q is invertible if and only if Q is strictly positive.

Throughout this monograph for any contraction T we define its *defect operator* D_T by the positive operator $(I - T^*T)^{1/2}$ and by D_T the closure of its range $(D_T = \overline{D_T H})$. (Notice that in the space D_T the D is italicized, while in the operator D_T the D is not italicized.) Obviously

$$T(I - T^*T) = (I - TT^*)T ,$$

and consequently

$$Tp(I - T^*T) = p(I - TT^*)T$$

for any polynomial $p(z) = a_o + a_1 z + \cdots$. Therefore by continuity we have

$$TD_T = D_{T^*} T . \tag{1.1}$$

It is emphasized that (1.1) is a useful property that will be used time and time again. We shall often use the following identity

$$\|h\|^2 = \|Th\|^2 + \|D_T h\|^2 \quad (h \in H) . \tag{1.2}$$

This follows from

$$\|h\|^2 - \|Th\|^2 = ((I - T^*T)h, h) = (D_T^2 h, h) = \|D_T h\|^2 .$$

Using these relations it is easy to check that T is a strict contraction if and only if $I - T^*T$, or equivalently, $I - TT^*$ is strictly positive. It follows that T is a strict contraction if and only if D_T, or equivalently, D_{T^*} is invertible.

Let H, H', H_1 and H_2 be (complex) Hilbert spaces and T a column matrix of the form

$$T = \begin{bmatrix} A \\ C \end{bmatrix} : \ H \to H' \oplus H_2 \tag{1.3}$$

where A and C are operators acting between the appropriate spaces. Recall that in formula (1.3) we identify the column vector $[h', h_2]^{tr}$ with the vector $h' \oplus h_2$ in $H' \oplus H_2$. The following simple lemma plays a basic role in this chapter.

1.1 LEMMA. *The operator* T *in (1.3) is a contraction if and only if* A *is a contraction and there exists a contraction* Y *mapping* D_A *into* H_2 *such that*

$$C = YD_A . \tag{1.4}$$

In this case, the formula

$$WD_T = D_Y D_A \tag{1.5}$$

defines a unitary operator W *from* D_T *onto* D_Y. *In particular,* T *is an isometry if and only if* Y *is an isometry.*

PROOF. Assume that A is a contraction and (1.4) holds where Y is a contraction. Equation (1.2) gives

$$\|D_T h\|^2 = \|h\|^2 - \|Th\|^2 = \|h\|^2 - \|Ah\|^2 - \|YD_A h\|^2 =$$

$$\|D_A h\|^2 - \|YD_A h\|^2 = \|D_Y D_A h\|^2 \geq 0 \qquad (h \in H) .$$

Hence T is a contraction. This relation also yields (1.5). On the other hand, if T is a contraction, then

$$\|Ch\|^2 \leq \|h\|^2 - \|Ah\|^2 = \|D_A h\|^2 \qquad (h \in H) .$$

So there exists a contraction Y mapping $D_A H$ into H_2 such that (1.4) holds. This Y extends by continuity to a contraction from D_A into H_2. This proof is now complete.

1.2 COROLLARY. *Let* A *be a strict contraction. Then* T *in (1.3) is a strict contraction if and only if* Y *is a strict contraction.*

PROOF. Since D_A is invertible, (1.5) shows that D_T is invertible if and only if D_Y is invertible. This completes the proof.

Consider the operator

$$T = [A,\ B] : H \oplus H_1 \to H' \tag{1.6}$$

where A and B are operators acting between the appropriate spaces. Applying Lemma 1.1 to the adjoint of T in (1.6) produces the following result.

1.3 COROLLARY. *The operator* T *in (1.6) is a contraction if and only if* A *is a contraction and there exists a contraction* X *mapping* H_1 *into* D_{A^*} *such that*

$$B = D_{A^*} X. \tag{1.7}$$

1.4 COROLLARY. *Let* T *be a contraction from* H *into* H'. *Then the operator*

$$R_T = \begin{bmatrix} T & D_{T^*} \\ D_T & -T^* \end{bmatrix} :\ H \oplus D_{T^*} \to H' \oplus D_T \tag{1.8}$$

is unitary.

PROOF. Taking the adjoints in (1.1), we infer that

$$T^* D_{T^*} \subseteq D_T. \tag{1.9}$$

Therefore (1.8) makes sense. By Lemma 1.1 the columns of R_T are isometries. From (1.1) it follows that these columns are orthogonal. Therefore R_T is an isometry. But its adjoint $R_T^* = R_{T^*}$ is a matrix of the same form and R_T^* is also an isometry. Thus R_T is unitary and the proof is complete.

The operator R_T will be called the *rotation matrix* of T.

1.5 COROLLARY. *Let* T *be a contraction of the form (1.6) and* X *be the contraction defined by (1.7). Then the operator* W *from* D_T *into* $D_A \oplus D_X$ *defined by*

$$WD_T = \begin{bmatrix} D_A & -A^* X \\ 0 & D_X \end{bmatrix} \tag{1.10}$$

is unitary.

PROOF. By Corollary 1.4 and Lemma 1.1

$$\begin{bmatrix} A & D_{A^*} X \\ D_A & -A^* X \\ 0 & D_X \end{bmatrix} = \begin{bmatrix} A & D_{A^*} & 0 \\ D_A & -A^* & 0 \\ 0 & 0 & I \end{bmatrix} \begin{bmatrix} I & 0 \\ 0 & X \\ 0 & D_X \end{bmatrix} \tag{1.11}$$

is an isometry from $H \oplus H_1$ into $H' \oplus D_A \oplus D_X$. Applying Lemma 1.1 with A replaced by $[A, D_{A}\cdot X]$ to the isometry in (1.11), we obtain that W corresponding to Y from D_T into $D_A \oplus D_X$ is an isometry. Notice that the right hand side of (1.10) is upper triangular and the diagonal entries have dense range. So W is onto and the proof is complete.

Throughout we do not differentiate between H, H_1 and $H \oplus \{0\}$, $\{0\} \oplus H_1$, respectively.

1.6 PROPOSITION. *Let* T *be a contraction of the form (1.6) and* X *be the contraction defined by (1.7). Then the following properties are equivalent:*

(i) X *is an isometry;*

(ii) $D_T = \overline{D_T(H \oplus \{0\})}$;

(iii) $H_1 \oplus D_{T^\cdot} = \{P_{H_1} T^* h \oplus D_{T^\cdot} h : h \in H'\}^-$.

PROOF. Corollary 1.5 gives

$$W(D_T \ominus \overline{D_T(H \oplus \{0\})}) = WD_T \ominus (D_A \oplus \{0\}) = \{0\} \oplus D_X .$$

Thus (i) and (ii) are equivalent. By applying Lemma 1.1 to T^* we see that the operator W_* from $D_{T^\cdot}$ to $D_{X^\cdot}$ defined by $W_* D_{T^\cdot} = D_{X^\cdot} D_{A^\cdot}$ is unitary. Notice that

$$\begin{bmatrix} I & 0 \\ 0 & W_* \end{bmatrix}[(H_1 \oplus D_{T^\cdot}) \ominus \{P_{H_1} T^* h \oplus D_{T^\cdot} h : h \in H'\}^-] =$$

$$\tag{1.12}$$

$$(H_1 \oplus D_{X^\cdot}) \ominus \left\{ \begin{bmatrix} X^* D_{A^\cdot} \\ D_{X^\cdot} D_{A^\cdot} \end{bmatrix} H' \right\}^- = (H_1 \oplus D_{X^\cdot}) \ominus \begin{bmatrix} X^* \\ D_{X^\cdot} \end{bmatrix} D_{A^\cdot} .$$

Now Corollary 1.4 applied to $R_{X^\cdot}$ yields

$$(H_1 \oplus D_{X^\cdot}) \ominus \begin{bmatrix} X^* \\ D_{X^\cdot} \end{bmatrix} D_{A^\cdot} = \begin{bmatrix} D_X \\ -X \end{bmatrix} D_X .$$

From this and (1.12) it follows that (iii) holds if and only if $D_X = \{0\}$, or equivalently, X is an isometry, that is, (i) holds. This completes the proof.

1.7 REMARK. There is a geometric interpretation of Corollary 1.3, which is quite useful. Let $F \subseteq K$ and set $G = K \ominus F$. Let Γ be a contraction from F to H'. Then the set of all contractions C_1 from K to H' satisfying $C_1 | F = \Gamma$ is given by the formula

$$C_1 = \Gamma P_F + D_{\Gamma^*} \Gamma_1 P_G \tag{1.13}$$

where Γ_1 is an arbitrary contraction from G to D_{Γ^*}.

2. CONTRACTIVE 2 by 2 LOWER TRIANGULAR MATRICES

In this section we obtain a complete characterization of all contractive 2 by 2 lower triangular matrices. To begin let T be a matrix of the form

$$T = \begin{bmatrix} A & 0 \\ C & D \end{bmatrix} : \; H \oplus H_1 \to H' \oplus H_2 \tag{2.1}$$

where A, C and D are operators acting between the appropriate spaces.

2.1 LEMMA. *The operator* T *defined in (2.1) is a contraction if and only if* A *and* D *are contractions, and* C *is of the form*

$$C = \Delta_* \Gamma D_A \tag{2.2}$$

where Γ *is contraction from* D_A *into* D_{D^*} *and* Δ_* *is the positive square root of* $I - DD^*$.

PROOF. The condition that A and D are contractions is obviously a necessary one for T to be a contraction. Henceforth we assume that this condition is satisfied. Lemma 1.1 implies that T in (2.1) is a contraction if and only if the row matrix $Y = [Y_1, Y_2]$ from $D_{[A,0]} = D_A \oplus H_1$ into H_2 defined by

$$[C,D] = Y D_{[A,0]} = [Y_1, Y_2] \begin{bmatrix} D_A & 0 \\ 0 & I \end{bmatrix} \tag{2.3}$$

is a contraction. But (2.3) shows that $D = Y_2$. By Corollary 1.3 the row matrix $[Y_1, D]$ is a contraction if and only if $Y_1 = \Delta_* \Gamma$ defines a contraction Γ from D_A into D_{D^*}. The Lemma follows by observing that (2.3) also yields

$$C = Y_1 D_A = \Delta_* \Gamma D_A \; .$$

2.2 COROLLARY. *Let* T *be a contraction of the form (2.1) and* Γ *the contraction defined by (2.2). Then the operators* W *from* D_T *into* $D_\Gamma \oplus D_D$ *and* W_* *from* D_{T^*} *into* $D_{A^*} \oplus D_{\Gamma^*}$ *defined by*

$$W D_T = \begin{bmatrix} D_\Gamma D_A & 0 \\ -D^* \Gamma D_A & \Delta \end{bmatrix} \tag{2.4}$$

and

$$W_* D_{T^*} = \begin{bmatrix} D_{A^*} & -A\Gamma^*\Delta_* \\ 0 & D_{\Gamma^*}\Delta_* \end{bmatrix} . \tag{2.5}$$

are both unitary where Δ is the positive square root of $I - D^* D$.

PROOF. According to Lemma 1.1 and Corollary 1.5 the formulas

$$W_1 D_T = D_{[Y_1, Y_2]} \begin{bmatrix} D_A & 0 \\ 0 & I \end{bmatrix} \tag{2.6}$$

and

$$W_2 D_{[Y_1, Y_2]} = \begin{bmatrix} D_\Gamma & 0 \\ -D^*\Gamma & \Delta \end{bmatrix}$$

define unitary operators between the appropriate spaces. (Recall that $Y_2 = D$ and $Y_1 = \Delta_* \Gamma$.) This and the fact that $W = W_2 W_1$ is unitary yields (2.4). Applying the previous result to T^*, one easily obtains that W_* is also unitary. This completes the proof.

2.3 COROLLARY. *The operator* T *defined in (2.1) is a contraction if and only if* A *is a contraction and the formula* $C = YD_A$ *defines an operator* Y *from* D_A *into* H_2 *such that* [Y, D] *is a contraction from* $D_A \oplus H_1$ *into* H_2.

PROOF. This corollary follows at once from Lemma 2.1 and Corollary 1.3.

3. CONTRACTIVE 2 by 2 MATRICES

In this section we present the general form for a 2 by 2 matrix contraction. Our previous results are special cases of Theorem 3.1 below.

Let T be the block matrix defined by

$$T = \begin{bmatrix} A & B \\ C & D \end{bmatrix} : H \oplus H_1 \to H^{'} \oplus H_2 \tag{3.1}$$

where A, B, C and D are operators acting between the appropriate spaces.

3.1 THEOREM. *The operator* T *defined in (3.1) is a contraction if and only if* A *is a contraction and* B, C, D *are operators of the form*

$$B = D_{A^*} X , \quad C = YD_A \quad and \quad D = D_{Y^*}\Gamma D_X - YA^* X \tag{3.2}$$

where X from H_1 into D_{A^*} and Y from D_A into H_2 and Γ from D_X into D_{Y^*} are all contractions.

PROOF. The condition that the first row and the first column of T are contractions is obviously a necessary one for T to be a contraction. Henceforth we assume that this condition is satisfied. Thus by Corollary 1.3 and Lemma 1.1 the operators X and Y defined by the first two equations in (3.2) are contractions. From Corollary 1.5 we infer that the operator W_1 from $D_{[A,B]}$ into $D_A \oplus D_X$ defined by

$$W_1 D_{[A,B]} = \begin{bmatrix} D_A & -A^*X \\ 0 & D_X \end{bmatrix} \tag{3.3}$$

is unitary. Therefore, Lemma 1.1 shows that T is a contraction if and only if the operator Y' defined by

$$Y' D_{[A,B]} = [YD_A, D]$$

is a contraction from $D_{[A,B]}$ into H_2. Setting $[Y_1, Y_2] = Y' W_1^*$ we obtain

$$[Y_1, Y_2] \begin{bmatrix} D_A & -A^*X \\ 0 & D_X \end{bmatrix} = Y' D_{[A,B]} = [YD_A, D] \ . \tag{3.4}$$

This shows that $Y_1 = Y$. Moreover, by Corollary 1.3 the row matrix $[Y, Y_2]$ is a contraction if and only if $Y_2 = D_{Y^*}\Gamma$ defines a contraction Γ from D_X into D_{Y^*}. Therefore the theorem follows by observing that (3.4) and $Y_1 = Y$ also yield the last equation in (3.2).

3.2 COROLLARY. *Let T be a contraction of the form (3.1) and X, Y and Γ be the contractions defined in (3.2). Then the operators W from D_T into $D_Y \oplus D_\Gamma$ and W_* from D_{T^*} into $D_{X^*} \oplus D_{\Gamma^*}$ defined by*

$$WD_T = \begin{bmatrix} D_Y D_A & -(D_Y A^*X + Y^*\Gamma D_X) \\ 0 & D_\Gamma D_X \end{bmatrix} \tag{3.5}$$

and

$$W_* D_{T^*} = \begin{bmatrix} D_{X^*} D_{A^*} & -(D_{X^*} AY^* + X\Gamma^* D_{Y^*}) \\ 0 & D_{\Gamma^*} D_{Y^*} \end{bmatrix} \tag{3.6}$$

are both unitary.

PROOF. Let Y' be the operator in the previous proof and let W' be the unitary operator from D_T onto $D_{Y'}$ defined according to Lemma 1.1, with $[A, B]$ instead of A. Let W_2 be the unitary operator from $D_{[Y, Y_2]}$ onto $D_Y \oplus D_\Gamma$ defined according to Corollary 1.5 with $[Y, Y_2] = [Y, D_Y \cdot \Gamma]$ instead of T. Using $[Y, Y_2] = Y'W_1^*$ we see that

$$D_{[Y,Y_2]}^2 = W_1 D_{Y'}^2 W_1^* .$$

So by forming polynomials and passing limits we obtain

$$D_{Y'} = W_1^* D_{[Y,Y_2]} W_1 .$$

This and (3.3) gives

$$W' D_T = D_{Y'} D_{[A,B]} = W_1^* D_{[Y,Y_2]} \begin{bmatrix} D_A & -A^*X \\ 0 & D_X \end{bmatrix} =$$

$$W_1^* W_2^* \begin{bmatrix} D_Y & -Y^*\Gamma \\ 0 & D_\Gamma \end{bmatrix} \begin{bmatrix} D_A & -A^*X \\ 0 & D_X \end{bmatrix} = W_1^* W_2^* \begin{bmatrix} D_Y D_A & -(D_Y A^*X + Y^*\Gamma D_X) \\ 0 & D_\Gamma D_X \end{bmatrix} .$$

and hence $W = W_2 W_1 W'$ is unitary. Applying this conclusion to the adjoint T^* of T, we obtain that W_* is also unitary. This completes the proof.

3.3 COROLLARY. *Let* A, X *and* Y *all be strict contractions. Then* T *in (3.1) is a strict contraction if and only if* Γ *is a strict contraction.*

PROOF. From formula (3.5) we deduce that D_T is invertible if and only if both $D_Y D_A$ and $D_\Gamma D_X$ are invertible. Since A, X and Y are all strict contractions this implies that D_T is invertible if and only if D_Γ is invertible. This completes the proof.

3.4 REMARK. Corollary 1.2 shows that A, X and Y are all strict contractions if and only if both $[A, B]$ and $[A^*, C^*]$ are strict contractions. Thus if $[A, B]$ and $[A^*, C^*]$ are both strict contractions, then T in (3.1) is a strict contraction if and only if Γ is a strict contraction.

3.5 COROLLARY. *If* T *in (3.1) is a contraction, then* T *admits a factorization of the form:*

$$T = \begin{bmatrix} I & 0 & 0 \\ 0 & Y & D_{Y^*} \end{bmatrix} \begin{bmatrix} A & D_A^* & 0 \\ D_A & -A^* & 0 \\ 0 & 0 & \Gamma \end{bmatrix} \begin{bmatrix} I & 0 \\ 0 & X \\ 0 & D_X \end{bmatrix} \tag{3.7}$$

where X, Y and Γ are contractions defined by (3.2) acting between the appropriate spaces.

PROOF. This follows from Theorem 3.1 and (3.2), (3.1).

The following immediate consequence of Theorem 3.1 has an important role in many applications.

3.6 COROLLARY. *Let T be the matrix (3.1) where A, B and C are fixed. Then there exists an operator D such that T is a contraction if and only if*

$$[A,B] \quad and \quad \begin{bmatrix} A \\ C \end{bmatrix} \tag{3.8}$$

are contractions. In this case, formula (3.2) establishes a one to one correspondence between the set of all contractions T and the set of all contractions Γ from D_X into D_{Y^*}.

3.7 REMARK. In Corollary 3.6 the contraction T is uniquely determined by the fixed contractions in (3.8) if and only if X or Y^* is an isometry. Otherwise one can choose $\Gamma = 0$ and a nonzero Γ in (3.2).

4. COMPUTING ALL CONTRACTIVE 2 by 2 MATRICES

In this section we present a method to obtain the set of all contractions T in (3.1) when the operators A, B, and C are fixed.

To avoid technical problems, throughout this section, it is assumed that A is a strict contraction, that is,

$$\|A\| < 1 . \tag{4.1}$$

This implies that both D_A and D_{A^*} are invertible. We make no assumptions on B and C except that they are known operators. The following identity which is a consequence of (1.1) will play an important role in this section

$$D_A^{-1} A^* = A^* D_{A^*}^{-1} . \tag{4.2}$$

Finally, it is emphasized that our procedure will also determine when [A, B] and $[A^*, C^*]$ are strict contractions.

To obtain the set of all contractions T directly from Theorem 3.1 one must first calculate the operators X and Y defined by

$$X = D_A^{-1} \cdot B \quad \text{and} \quad Y = CD_A^{-1} \, . \tag{4.3}$$

If X and Y are both contractions, then the set of all contractions T in (3.1) are given by (3.2), where Γ is an arbitrary contraction from D_X into D_{Y^*}. Otherwise [A, B] or [A*, C*] will not be a contraction, and there is no contraction T of the form (3.1). In this method one must calculate both D_A^{-1} and $D_{A^*}^{-1}$. In this section we will present an approach for obtaining all contractive T in (3.1), by only calculating $D_A^{-2} = (I - A^*A)^{-1}$ or $D_{A^*}^{-2} = (I - AA^*)^{-1}$, and thus avoiding the calculation of the square roots D_A or D_{A^*}. In general it is cumbersome to compute the square root of an operator.

To this end let $\hat{X}$ and $\hat{Y}$ be the operators defined by

$$\hat{X} = D_A^{-2} A^* B = D_A^{-1} A^* X \quad \text{and} \quad \hat{Y} = D_A^{-2} C^* = D_A^{-1} Y^* \, . \tag{4.4}$$

The second and fourth equalities follow from the definition of the operators (not necessarily contractions) X and Y defined in (4.3). We claim that

$$YA^*X = C\hat{X} = \hat{Y}^* A^* B \, . \tag{4.5}$$

The first equality follows from (4.2), (4.3) and (4.4). The second equality follows from (1.1), (4.3) and (4.4):

$$YA^*X = \hat{Y}^* D_A A^* X = \hat{Y}^* A^* D_{A^*} \cdot X = \hat{Y}^* A^* B \, .$$

The following identities are also important in our procedure for determining all contractive T

$$\begin{aligned} I - YY^* &= I - C\hat{Y} = I - \hat{Y}^* C^* \\ I - X^*X &= I - B^*B - B^*A\hat{X} = I - B^*B - \hat{X}^* A^* B \, . \end{aligned} \tag{4.6}$$

The first equality is a simple consequence of the definition of Y and $\hat{Y}$. The third equality in (4.6) follows from (4.2), (4.4) and the following calculation

$$I - X^*X = I - X^* D_{A^*}^2 \cdot X - X^* AA^* X =$$

$$I - B^*B - X^* D_{A^*} \cdot D_A^{-1} AA^* X = I - B^*B - B^* A\hat{X} \, .$$

The second and fourth equalities in (4.6) are consequences of the self adjointness of $I - YY^*$ and $I - X^*X$.

The following allows us to test when [A*, C*] is a contraction by only calculating D_A^{-2}.

4.1 LEMMA. *Let* A *be a strict contraction. Then* $[A^*, C^*]$ *is a contraction (strict contraction) if and only if*

$$I - C\hat{Y} \quad \text{is positive} \tag{4.7}$$

(strictly positively).

PROOF. Equations (4.3) and (4.6) with Lemma 1.1 show that $[A^*, C^*]$ is a contraction if and only if (4.7) holds. Corollary 1.2 shows that $[A^*, C^*]$ is a strict contraction if and only if Y is a strict contraction, or equivalently, that $I - C\hat{Y}$ is strictly positive. This completes the proof.

A similar argument also proves the next result.

4.2 LEMMA. *Let* A *be a strict contraction. Then* $[A, B]$ *is a contraction (strict contraction) if and only if*

$$I - B^*B - B^*A\hat{X} \quad \text{is positive} \tag{4.8}$$

(strictly positive).

Now we are ready to present a procedure involving only D_A^{-2} for calculating all contractions T in (3.1), when A, B, and C are fixed (and of course A is a strict contraction).

4.3 PROCEDURE. First calculate D_A^{-2}. Then calculate $\hat{X}$, $\hat{Y}$ and the operators in (4.6). If (4.7) and (4.8) hold, then the set of all contractions T in (3.1) is given by

$$D = D_Y \cdot \Gamma D_X - C\hat{X} = D_Y \cdot \Gamma D_X - \hat{Y}^* A^* B \tag{4.9}$$

where Γ is an arbitrary contraction from D_X into $D_Y\cdot$ and

$$D_Y\cdot = (I - C\hat{Y})^{1/2} \quad \text{and} \quad D_X = (I - B^*B - B^*A\hat{X})^{1/2} . \tag{4.10}$$

If (4.7) or (4.8) does not hold, then there is no contraction T of the form (3.1).

Finally, it is noted that (4.9) gives two different methods for computing the contraction D. Although this procedure involves the square roots of $D_Y\cdot$ and D_X, in many cases these operators act on spaces of low dimension. This facilitates the computation of $D_Y\cdot$ and D_X. In the case of the Carathéodory interpolation problem these spaces are of dimension less than or equal to one, as will be seen in Section 7.

To complete this section, we present a dual of the previous algorithms, that is, a procedure that only involves $D_{A^*}^{-2}$. Both methods are useful because in some cases A is not a square matrix and therefore the rank of $D_{A^*}^2$ may be much smaller than the rank of D_A^2 or vice versa. So inverting $D_{A^*}^2$ may be easier than inverting D_A^2 or vice versa. To begin let $\hat{X}_*$ and $\hat{Y}_*$ be the operators defined by

$$\hat{X}_* = D_{A^*}^{-2} B = D_{A^*}^{-1} X \quad \text{and} \quad \hat{Y}_* = D_{A^*}^{-2} AC^* = D_{A^*}^{-1} AY^* . \tag{4.11}$$

By performing calculations similar to our previous ones we have

$$YA^*X = \hat{Y}_*^* B = CA^* \hat{X}_*$$

$$I - YY^* = I - CC^* - CA^* \hat{Y}_* = I - CC^* - \hat{Y}_*^* AC^* \tag{4.12}$$

$$I - X^*X = I - B^* \hat{X}_* = I - \hat{X}_*^* B .$$

By altering the proofs of Lemmas 4.1 and 4.2 using (4.12), we obtain the following results.

4.4 COROLLARY. *Let* A *be a strict contraction. Then* $[A^*, C^*]$ *is a contraction (strict contraction) if and only if*

$$I - CC^* - CA^* \hat{Y}_* \quad \text{is positive} \tag{4.13}$$

(strictly positive).

4.5 COROLLARY. *Let* A *be a strict contraction. Then* $[A, B]$ *is a contraction (strict contraction) if and only if*

$$I - B^* \hat{X}_* \quad \text{is positive} \tag{4.14}$$

(strictly positive).

4.6 PROCEDURE. First calculate $D_{A^*}^{-2}$. Then calculate $\hat{X}_*$, $\hat{Y}_*$ and the operators in (4.12). If (4.13) and (4.14) hold, then the set of all contractions T in (3.1) is given by

$$D = D_{Y^*} \Gamma D_X - \hat{Y}_*^* B = D_{Y^*} \Gamma D_X - CA^* \hat{X}_* \tag{4.15}$$

where Γ is an arbitrary contraction from D_X into D_{Y^*} and

$$D_{Y^*} = (I - CC^* - CA^* \hat{Y}_*)^{1/2} \quad \text{and} \quad D_X = (I - B^* \hat{X}_*)^{1/2} . \tag{4.16}$$

If (4.13) or (4.14) does not hold, then there is no contraction T of the form (3.1). As expected (4.15) gives two different methods for computing D.

Finally, it is noted that $\hat{X}$, $\hat{X}_*$ and $\hat{Y}$, $\hat{Y}_*$ are related by

$$\hat{X} = A^*\hat{X}_* \text{ and } \hat{Y}_* = A\hat{Y} . \tag{4.17}$$

This follows from (4.2), (4.4) and (4.11).

5. CONTRACTIVE LEVINSON SYSTEMS

In this section we show that the operators $\hat{X}$, $\hat{Y}$, and $\hat{X}_*$, $\hat{Y}_*$ solve a Levinson type equation. As before throughout this section it is always assumed that A is a strict contraction, and B, C are fixed operators.

To begin consider the following linear system of equations

$$[I-[A,B]^*[A,B]]\begin{bmatrix} F \\ G \end{bmatrix} = \begin{bmatrix} D_A^2 & -A^*B \\ -B^*A & I-B^*B \end{bmatrix}\begin{bmatrix} F \\ G \end{bmatrix} = \begin{bmatrix} R \\ Q \end{bmatrix} \tag{5.1}$$

where F, G, R and Q are operators acting between the appropriate spaces. We begin with the following result.

5.1 LEMMA. *Let A be a strict contraction, and G, R be known operators. Then there exists a unique solution F, Q to (5.1), namely*

$$F = D_A^{-2}A^*BG + D_A^{-2}R$$
$$Q = (I - B^*B - B^*AD_A^{-2}A^*B)G - B^*AD_A^{-2}R . \tag{5.2}$$

PROOF. By substituting (5.2) into (5.1) it is easy to verify that (5.2) is a solution to (5.1). (This solution was obtained by solving for F in the first row of (5.1) and then by substituting this F into the last row of (5.1) to find Q.) To prove uniqueness, because the solution set is linear, we assume without loss of generality, that the known operators G and R are both zero. Since A is a strict contraction this implies that $F = 0$. Hence $Q = 0$ and the proof is complete.

5.2 COROLLARY. *Let A be a strict contraction. Then $F = \hat{X}$ and $Q = I - X^*X$ is the unique solution to*

$$(I - [A, B]^*[A, B])\begin{bmatrix} F \\ I \end{bmatrix} = \begin{bmatrix} 0 \\ Q \end{bmatrix} . \tag{5.3}$$

In this case [A, B] is a contraction (strict contraction) if and only if Q is positive (strictly positive).

PROOF. The first statement follows from (4.6) and (5.2) along with the definitions in (4.4). The second statement follows from (4.6) and Lemma 4.2. This completes the proof.

5.3 COROLLARY. *Let* A *be a strict contraction. Then* $F' = \hat{Y}$ *and* $Q' = -X^* A Y^*$ *is the unique solution to*

$$(I - [A, B]^* [A, B]) \begin{bmatrix} F' \\ 0 \end{bmatrix} = \begin{bmatrix} C^* \\ Q' \end{bmatrix}. \tag{5.4}$$

PROOF. This follows from (4.4), (4.5) and (5.2).

The next two Corollaries are the analogue of Corollaries 5.2 and 5.3 for $[A^*, C^*]$.

5.4 COROLLARY. *Let* A *be a strict contraction. Then* $F_* = \hat{Y}_*$ *and* $Q_* = I - Y Y^*$ *is the unique solution to*

$$\left(I - \begin{bmatrix} A \\ C \end{bmatrix} [A^*, C^*]\right) \begin{bmatrix} F_* \\ I \end{bmatrix} = \begin{bmatrix} 0 \\ Q_* \end{bmatrix}. \tag{5.5}$$

In this case $[A^*, C^*]$ *is a contraction (strict contraction) if and only if* Q_* *is positive (strictly positive).*

PROOF. This proof is similar to Corollary 5.2 and omitted.

5.5 COROLLARY. *Let* A *be a strict contraction. Then* $F'_* = \hat{X}_*$ *and* $Q'_* = -Y A^* X$ *is the unique solution to*

$$\left(I - \begin{bmatrix} A \\ C \end{bmatrix} [A^*, C^*]\right) \begin{bmatrix} F'_* \\ 0 \end{bmatrix} = \begin{bmatrix} B \\ Q'_* \end{bmatrix}. \tag{5.6}$$

PROOF. The proof is similar to Corollary 5.3 and is omitted.

The solution to (5.3), (5.4) (or (5.5), (5.6)) is a contractive version of the matrix expression for the Levinson algorithm in Section II.2. Moreover, the solution to (5.3) and (5.4) provides a test for determining whether or not both [A,B] and $[A^*, C^*]^*$ are contractions. Since $Q = I - X^* X$ the row matrix [A, B] is a contraction if and only if Q is positive. Because (4.6) gives $I - C F' = I - Y Y^*$ the column matrix $[A^*, C^*]^*$ is a contraction if and only if $I - C F'$ is positive. These tests are similar to the test using the

Levinson algorithm for determining the positivity of a Toeplitz matrix (see Corollary II.2.4). If both Q and $I - CF'$ are positive, then by Theorem 3.1 the set of all D such that the block matrix T in (3.1) is a contraction is given by

$$D = Q'^* + (I - CF')^{1/2}\Gamma Q^{1/2} \tag{5.7}$$

where Γ is an arbitrary contraction acting between the appropriate spaces. It is emphasized that one can check for contractions and obtain all these D's by solving the Levinson system in (5.3), (5.4).

In a similar manner the Levinson system in (5.5), (5.6) provides a test for determining whether or not both [A, B] and $[A^*, C^*]^*$ are contractions. Since $Q_* = I - YY^*$ the column matrix $[A^*, C^*]^*$ is a contraction if and only if Q_* is positive. Because (4.12) gives $I - B^*F'_* = I - X^*X$ the row matrix [A,B] is a contraction if and only if $I - B^*F'_*$ is positive. If both Q_* and $I - B^*F'_*$ are positive, then by Theorem 3.1 the set of all D such that the block matrix T in (3.1) is a contraction is given by

$$D = Q'_* + Q_*^{1/2}\Gamma(I - B^*F'_*)^{1/2} \tag{5.8}$$

where Γ is a contraction acting between the appropriate spaces. As before it is emphasized that one checks for contractions and obtains all these D's by solving the Levinson system in (5.5), (5.6).

Each of these methods requires a solution to one matrix equation and shows exactly how D_{Y^*}, D_X and $-YA^*X$ naturally occur in (5.3) to (5.6).

6. EXPANDING MATRIX CONTRACTIONS

In this section we apply some of the proceeding results to an interesting particular case, which includes the expansion of a n by n contractive matrix to a n+1 by n+1 contractive matrix.

We begin with a simple application of Theorem 3.1 and Corollary 3.6 to a matrix T from $H \oplus C^1$ to $H' \oplus C^1$ of the form

$$T = \begin{bmatrix} A & B \\ C & D \end{bmatrix}$$

$$\tag{6.1}$$

$$\text{where} \quad B\gamma = \gamma b \text{ and } C^*\gamma = \gamma c_* \qquad \text{(for all } \gamma \text{ in } C^1) .$$

In this case A maps H into H' and b is a "column" vector in H' and c_* is a "column" vector in H. Here A, B and C are specified. Finally, D = [d] where d is in C^1.

6.1 PROPOSITION. *The matrix* T *in (6.1) is a contraction for an adequate choice of* d *if and only if the matrices*

$$[A, B] \quad and \quad \begin{bmatrix} A \\ C \end{bmatrix} \tag{6.2}$$

are contractions. In this case, there exists an α *in* C^1 *and a* $\tau \geq 0$ *such that the set of all contractive* T *(with the same* A, B, C*) is given by*

$$d = \alpha + \tau\rho \qquad (with \ \rho \in \overline{D}) \ . \tag{6.3}$$

PROOF. The first assertion is a restatement of part of Corollary 3.6. If the operators in (6.2) are contractions, and X and Y are the contractions defined in (3.2), then $-YA^{*}X = [\alpha]$. Obviously the spaces D_X and $D_{Y^{\cdot}}$ are contained in C^1. If $D_X = \{0\}$ or $D_Y = \{0\}$, then we set $\tau = 0$. Otherwise $D_X = C^1 = D_{Y^{\cdot}}$. In this case, $D_X = [d_X]$ and $D_{Y^{\cdot}} = [d_{Y^{\cdot}}]$ where d_X and $d_{Y^{\cdot}}$ are both strictly positive real numbers. Moreover, the contraction Γ has the form $[\rho]$ for some ρ in $\overline{D}$. Setting $\tau = d_X d_{Y^{\cdot}}$ with (3.2) yields the representation in (6.3). This completes the proof.

6.2 REMARK. Assume that the operators in (6.2) are fixed contractions. The representation (6.3) expresses the fact that T is a contraction if and only if d belongs to the closed unit disc $\alpha + \tau\overline{D}$ with center at α and radius τ. In particular, this implies that α and τ are uniquely determined.

Obviously the contraction T is uniquely determined by A, B and C if and only if $\tau = 0$, that is, $d_X = 0$ or $d_{Y^{\cdot}} = 0$. The next proposition gives an equivalent characterization of $d_X = 0$. Throughout, the range of an operator R is denoted by ran R.

6.3 PROPOSITION. *Let* R = [A,B] *be a contraction from* $H \oplus C^1$ *into* H' *and* X *be the contraction from* C^1 *into* $D_{A^{\cdot}}$ *defined by* B = $D_{A^{\cdot}}$X. *Then the following properties are equivalent:*
   (i) X *is an isometry, that is,* $d_X = 0$*;*
   (ii) b $\notin$ ran $D_{R^{\cdot}}$*;*
   (iii) R^{*}b $\notin$ ran D_R.
Moreover, if A *is a strict contraction, then the properties (i) to (iii)*
are equivalent to
   (iv) $\|[A,B]\| = 1$.

PROOF. If $b = D_R \cdot h$ for some h in H', then $R^* b = D_R R^* h$ and $R^* b$ is in the range of D_R. On the other hand, if $R^* b = D_R x$ for some x in $H \oplus C^1$, then

$$D_R \cdot (Rx + D_R \cdot b) = R D_R x + D_R^2 \cdot b = b$$

and b is in the range of $D_R \cdot$. Therefore (ii) and (iii) are equivalent.

Proposition 1.6 implies that $d_X \neq 0$ if and only if there exists a nonzero $\lambda \oplus h$ in $C^1 \oplus D_R \cdot$ which is orthogonal to $\{B^* h' \oplus D_R \cdot h' : h' \in H'\}$, or equivalently, there exists a nonzero $\lambda \oplus h$ satisfying $b\lambda + D_R \cdot h = 0$. We claim that $\lambda \neq 0$. If $\lambda = 0$, then $D_R \cdot h = 0$ and h in $D_R \cdot$ yields h = 0. Therefore $d_X \neq 0$ if and only if b is in the range of $D_R \cdot$. This proves the equivalence of (i) and (ii).

Assume that A is a strict contraction. If $d_X = 0$, then choosing $h = D_A^{-1} A^* X 1$ and $\lambda = 1$ gives $D_R (h \oplus \lambda) = 0$, by Corollary 1.5 and (1.10) for R = T. Thus (iv) holds. If (iv) holds, then there exists a sequence of unit vectors $h_n \oplus \lambda_n$ in $H \oplus C^1$ such that

$$D_R (h_n \oplus \lambda_n) \to 0 \ \text{ as } \ n \to \infty . \tag{6.4}$$

Since λ_n is a bounded sequence of complex numbers, without loss of generality, we assume that $\lambda_n \to \lambda$. We claim that $\lambda \neq 0$. If $\lambda = 0$, then (6.4) and (1.10) imply that $D_A h_n \to 0$ as $n \to \infty$. This along with the fact that $\|h_n\| \to 1$ contradicts the invertibility of D_A. Hence $\lambda \neq 0$. Applying (6.4) and (1.10) again yields

$$\|D_A h_n - A^* X \lambda_n\|^2 + d_X |\lambda_n|^2 \to 0 \ \text{ as } \ n \to \infty .$$

Since λ is nonzero this implies that $d_X = 0$. Therefore (i) and (iv) are equivalent. The proof is now complete.

Proposition 6.3 readily gives the following

6.4 COROLLARY. *Let* $R_* = [A^*, C^*]^*$ *be a contraction from H into* $H' \oplus C^1$ *and* Y *the contraction from* D_A *into* C^1 *defined by* $C = Y D_A$. *Then the following properties are equivalent:*

(i) Y^* *is an isometry, that is,* $d_{Y \cdot} = 0$;

(ii) $c_* \notin \operatorname{ran} D_{R.}$;

(iii) $R_* c_* \notin \operatorname{ran} D_{R:}$.

Moreover, if A *is a strict contraction, then the properties (i) to (iii) are equivalent to*

(iv) $\|[A^*, C^*]\| = 1$.

Proposition 6.3 and Corollary 6.4 have the following immediate consequences.

6.5 COROLLARY. *The contraction* T *in (6.1) is uniquely determined by the contractions* [A,B] *and* [A*, C*] *if and only if at least one of the following conditions hold*

$$b \notin \operatorname{ran}D_{[A,B]^*} \ \ or \ \ c_* \notin \operatorname{ran}D_{[A^*,C^*]^*} \ .$$

6.6 COROLLARY. *Let* A *be a strict contraction. The contraction* T *in (6.1) is uniquely determined by the contractions* [A,B] *and* [A*, C*] *if and only if the norm of* [A,B] *or* [A*, C*] *is one.*

6.7 REMARK. If $\|A\| = 1$, then the conclusion of Corollary 6.6 might not be true. Indeed if

$$A = \begin{bmatrix} 1 & 0 \\ 0 & 0 \end{bmatrix}, \quad B = \begin{bmatrix} 0 \\ 1/2 \end{bmatrix}, \quad C = [0, \ 1/2] \ ,$$

then any d satisfying $|d| \leq .75$ will make

$$\begin{bmatrix} A & B \\ C & D \end{bmatrix} = \begin{bmatrix} 1 & 0 & 0 \\ 0 & 0 & .5 \\ 0 & .5 & d \end{bmatrix}$$

a contraction (see Lemma 2.1).

6.8 COROLLARY. *If the contractions* A,B *and* C *uniquely determine a contraction* T *of the form (6.1), then* $\|T\| = 1$.

PROOF. If $\|A\| = 1$, then $\|T\| = 1$. If A is a strict contraction, then Corollary 6.6 implies that the norm of [A,B] or [A*, C*] is one. Thus $\|T\| = 1$ and the proof is complete.

Remark 6.7 shows that the converse to Corollary 6.8 is not true.

We conclude this section by showing how one can use the Levinson system in Section 5 to construct the set of all contractive T in (6.1) when A is a matrix.

6.9 PROPOSITION. *Let* A *be a strictly contractive matrix from* $H = \mathbb{C}^m$ *to* $H' = \mathbb{C}^n$ *and set* $R = [\, A \, , \, B \,]$ *where* B *is a column vector in* $\mathbb{C}^n$. *Then there exists an* $f = [\, f_i \,]_1^m$ *and* $f' = [\, f'_i \,]_1^m$ *in* $\mathbb{C}^m$, *and scalars* q, q' *satisfying*

$$D_R^2 \begin{bmatrix} -f \\ 1 \end{bmatrix} = \begin{bmatrix} 0 \\ q \end{bmatrix} \quad and \quad D_R^2 \begin{bmatrix} f' \\ 0 \end{bmatrix} = \begin{bmatrix} C^* \\ q' \end{bmatrix} \tag{6.5}$$

where $C = [\, c_1 \,, c_2 \,, \ldots , c_m \,]$ *is a row vector. Moreover, the operators* R *and* $[\, A^* \,, C^* \,]$ *are a contractions if and only if* $q \geq 0$ *and* $1 \geq Cf'$, *respectively. In this case, the set of all* d *such that the operator* T *in (6.1) is a contraction is given by*

$$d = \sum_1^m f_i c_i + \sqrt{q}\, \left[1 - \sum_1^m f_i' c_i \right]^{1/2} r \tag{6.6}$$

where r *is a scalar in the closed unit disc.*

PROOF.   The first part follows from Corollaries 5.2 and 5.3 along with $F = -f$, $F' = f'$, $Q = q$ and $Q' = q'$ (see also end of Section 5). The third term in (6.6) corresponds to $(I - CF')^{1/2}\Gamma Q^{1/2}$ in (5.7). Since (5.7) describes all D such that T is a contraction it remains to verify that $Cf' = \bar{q}'$. To this end notice that (5.3) and Corollary 5.3 give

$$CF = YD_A F = YD_A D_A^{-2} A^* B = YA^* D_A^{-1} D_A X = YA^* X = -Q'^* \,. \tag{6.7}$$

Thus $\bar{q}' = Cf$. Now (6.6) follows from (5.7). This completes the proof.

By a similar argument employing Corollaries 5.4 and 5.5 along with (5.8) we obtain the following result.

6.10 PROPOSITION. *Let* A *be a strictly contractive matrix from* $H = C^m$ *to* $H' = C^n$ *and set* $R_* = [\, A^* \,, C^* \,]^*$ *where* C *is a row vector in* C^m. *Then there exists an* $f_* = [\, f_{*i} \,]_1^n$ *and* $f_*' = [\, f_{*i}' \,]_1^n$ *in* C^n, *and scalars* q_*, q_*' *satisfying*

$$D_{R_*}^2 \begin{bmatrix} -f_* \\ 1 \end{bmatrix} = \begin{bmatrix} 0 \\ q_* \end{bmatrix} and \; D_{R_*}^2 \begin{bmatrix} f_*' \\ 0 \end{bmatrix} = \begin{bmatrix} B \\ q_*' \end{bmatrix} \tag{6.8}$$

where $B = [\, b_1 \,, b_2 \,, \ldots , b_n \,]^{tr}$ *is a column vector. Moreover, the operators* R_* *and* $[\, A \,, B \,]$ *are contractions if and only if* $q_* \geq 0$ *and* $1 \geq B^* f_*$ *respectively. In this case, the set of all* d *such that the operator* T *in (6.1) is a contraction is given by*

$$d = \sum_1^n \bar{f}_{*i} b_i + \sqrt{q_*}\, (1 - \sum_1^n \bar{f}_{*i}' b_i)^{1/2} r \tag{6.9}$$

where r *is a scalar in the closed unit disc.*

As expected both (6.6) and (6.9) express the set of all d such that T is a contraction, by a closed disc with center

$$\sum_1^m f_i c_i = \sum_1^n \overline{f}_{*i} b_i \tag{6.10}$$

and radius

$$\sqrt{q}\left[1 - \sum_1^m f_i' c_i\right]^{1/2} = \sqrt{q_*}\left[1 - \sum_1^n \overline{f}_{*i}' b_i\right]^{1/2} \ ; \tag{6.11}$$

since both sets describe the same closed disc, they must have the same center and radius.

7. THE CARATHEODORY INTERPOLATION PROBLEM REVISITED

In this section we will present an instructive application of the previous results to the Carathéodory interpolation problem. This will provide a new role for the Schur numbers, namely as contractions Γ appearing in formula like (3.1) and (3.2).

For convenience we shall study matrices of the form

$$A_j' = \begin{bmatrix} 0 & 0 & 0 & \ldots & 0 & a_o \\ \cdot & \cdot & \cdot & \cdot\,\cdot\,\cdot & \cdot & \cdot \\ \cdot & \cdot & \cdot & \cdot\,\cdot\,\cdot & \cdot & \cdot \\ \cdot & \cdot & \cdot & \cdot\,\cdot\,\cdot & \cdot & \cdot \\ 0 & 0 & a_o & \ldots & a_{j-4} & a_{j-3} \\ 0 & a_o & a_1 & \ldots & a_{j-3} & a_{j-2} \\ a_o & a_1 & a_2 & \ldots & a_{j-2} & a_{j-1} \end{bmatrix} \tag{7.1}$$

where $(a_o, a_1, ..., a_{j-1})$ are j complex numbers. Unlike the Toeplitz matrix A_j in (I.4.4) the matrix A_j' in (7.1) is a j by j *Hankel matrix*. Indeed a matrix C with entries $C_{i,j}$ for $1 \le i \le n$ and $1 \le j \le n$ satisfying $C_{i,j} = C_{i+j}$ for some sequence $\{C_m\}$ is referred to in the literature as a *Hankel matrix*. It is easy to show that

$$A_j = A_j' \begin{bmatrix} 0 & 0 & \ldots\ldots & 0 & 1 \\ 0 & 0 & \ldots\ldots & 1 & 0 \\ \cdot & \cdot & \ldots\ldots & \cdot & \cdot \\ \cdot & \cdot & \ldots\ldots & \cdot & \cdot \\ \cdot & \cdot & \ldots\ldots & \cdot & \cdot \\ 0 & 1 & \ldots\ldots & 0 & 0 \\ 1 & 0 & \ldots\ldots & 0 & 0 \end{bmatrix} \tag{7.2}$$

where A_j is the j by j Toeplitz matrix defined in equation (I.4.4). The matrix on the right in (7.2) is unitary. It has one's on the south west to north east diagonal and zero's elsewhere. Obviously $\|A_j\| = \|A'_j\|$. Therefore according to Theorem I.6.7, the Carathéodory interpolation problem with data $(a_0, a_1, ..., a_{j-1})$ is solvable if and only if A'_j is a contraction. In this case the solution is unique if and only if the norm of A'_j is one.

Assume that there exists a solution to the Carathéodory interpolation problem with data $(a_0, a_1, ..., a_{j-1})$, or equivalently, that A'_j is a contraction. From Corollary I.6.8 we see that $f(z) = \sum_0^\infty a_n z^n$ is a solution to the Carathéodory interpolation problem with data $(a_0, a_1, ..., a_{j-1})$ if and only if the n by n Hankel matrix

$$A'_n = \begin{bmatrix} 0 & 0 & 0 & . & . & . & 0 & a_0 \\ . & . & . & . & . & . & . & . \\ . & . & . & . & . & . & . & . \\ . & . & . & . & . & . & . & . \\ 0 & 0 & a_0 & . & . & . & a_{n-4} & a_{n-3} \\ 0 & a_0 & a_1 & . & . & . & a_{n-3} & a_{n-2} \\ a_0 & a_1 & a_2 & . & . & . & a_{n-2} & a_{n-1} \end{bmatrix} \tag{7.3}$$

is a contraction for all $n > j$. In other words, the Carathéodory interpolation problem is equivalent to the following: Given a j by j contractive Hankel matrix A'_j, find all sequences of complex numbers $(a_j, a_{j+1}, a_{j+2}, \cdots)$ such that the n by n Hankel matrix A'_n is a contractive expansion of A'_j for all $n > j$. An important special case of this problem is the following *one step Carathéodory interpolation problem:* Given a j by j contractive Hankel matrix find all complex numbers a_j such that the j+1 by j+1 Hankel matrix A'_{j+1} in (7.3) with $n = j+1$ is a contraction. A recursive solution to this one step Carathéodory interpolation problem solves the Carathéodory interpolation problem. The one step problem is solved in the next proposition by only using the results in this chapter.

7.1 PROPOSITION. *Let A'_j be a fixed contraction. Then*

(i) There exists a constant a_j in C^1 such that A'_{j+1} is a contractive expansion of A'_j.

(ii) The set of all a_j such that A'_{j+1} is a contraction is given by

$$a_j = \alpha_j + \tau_j \rho_j \quad (with \ |\rho_j| \leq 1) \tag{7.4}$$

where α_j is in C^1 and $\tau_j \geq 0$ are constants which depend only on A'_j.

(iii) If A'_j is a strict contraction, then $\|A'_{j+1}\| = 1$ if and only if $|\rho_j| = 1$.

(iv) There exists a unique contractive expansion A'_{j+1} of A'_j if and only if the norm of A'_j is one.

PROOF. We set

$$A'_{j+1} = \begin{bmatrix} A & B \\ C & D \end{bmatrix} \tag{7.5}$$

where

$$A = \begin{bmatrix} 0 & 0 \\ 0 & A'_{j-1} \end{bmatrix}, \quad B = \begin{bmatrix} a_0 \\ a_1 \\ \cdot \\ \cdot \\ \cdot \\ a_{j-1} \end{bmatrix}, \tag{7.6}$$

$$C = [a_0, a_1, ..., a_{j-1}], \text{ and } D = [a_j] \ .$$

Notice that

$$\|[A, B]\| = \|A'_j\| = \|[A^*, C^*]\| \ . \tag{7.7}$$

Proposition 6.1 along with $d = a_j$ produces parts (i) and (ii). Remark 3.4, equation (7.7) and $\Gamma = [\rho_j]$ (see the proof of Proposition 6.1) give part (iii).

To prove half of (iv), assume that A'_j has a unique contractive expansion A'_{j+1}. If A'_j is a strict contraction, then Remark 3.4 and (7.7) shows that both X and Y are strict contractions. Thus $\tau_j = d_X d_{Y^*}$ is nonzero (see the proof of Proposition 6.1). Proposition 6.1 (or part (ii) above) shows that the entries a_j of all contractive expansions A'_{j+1} of A'_j can be arbitrarily chosen in the closed disc with center α_j and radius τ_j. This contradicts the uniqueness of A'_{j+1}. Therefore the norm of A'_j is one.

Now assume that the norm of A'_j is one. If A is a strict contraction, then Corollary 6.6 and (7.7) show that A'_j admits a unique contractive expansion A'_{j+1}. If the norm of A is one, then there exists a vector $x = [x_i]_0^{j-1}$ such that $\|A^* x\| = \|x\| = 1$. Since the first column of A^* is zero, $x_0 = 0$. From the structure of our Hankel matrices it follows that $x' = [x_1, x_2, ..., x_{j-1}, 0]$ satisfies

$$\|D_{[A,B]^*} x'\|^2 = \|x'\|^2 - \|[A,B]^* x'\|^2 = \|x\|^2 - \|A^* x\|^2 = 0 \ .$$

So if $b = B1$ is in the range of $D_{[A,B]^*}$ then b is orthogonal to x', that is,

$$\bar{a}_o x_1 + \bar{a}_1 x_2 + \cdots + \bar{a}_{j-2} x_{j-1} = 0. \tag{7.8}$$

Obviously

$$A^* x = \begin{bmatrix} 0 \\ \bar{a}_o x_{j-1} \\ \cdot \\ \cdot \\ \cdot \\ \bar{a}_o x_2 + \bar{a}_1 x_3 + \cdots + \bar{a}_{j-3} x_{j-1} \\ \bar{a}_o x_1 + \bar{a}_1 x_2 + \cdots + \bar{a}_{j-2} x_{j-1} \end{bmatrix}$$

Equation (7.8) shows that the last component of $A^* x$ is zero. Thus

$$1 = \|A^* x\|^2 \le \|A_{j-2}^{'*}[x_i]_2^{j-1}\|^2 \le \|[x_i]_2^{j-1}\|^2 = 1 - |x_1|^2 .$$

Hence $x_1 = 0$. By repeating this argument it is easy to show that $x_2 = 0$, ..., $x_{j-1} = 0$. This contradicts the fact that $\|x\| = 1$. Therefore b is not in the range of $D_{[A,B]}^*$. Corollary 6.5 implies that A_j' admits a unique contractive expansion. Finally, the proof is complete.

7.2 REMARK. One can also use ideas similar to those in the proof of Lemma I.6.6 to verify the following statement in Proposition 7.1 part (iv): If the norm of A_n' is one, then A_n' admits a unique contractive expansion A_{n+1}'. (The converse follows from Corollary 6.8.)

To prove this let $j \le n$ be the first integer such that $\|A_j'\| = 1$ and A_{n+1}' be any contractive expansion of A_j'. Then A is a strict contraction. Proposition 6.3 and (7.7) show that $D_X = 0$. By Corollary 5.2 and (5.3) there exists a vector $x' = [x_i]_0^j$ with $x_j = 1$ satisfying $\|x'\| = \|[A,B]x'\|$. Since the first column of A is zero, $x_o = 0$ and $\|x\| = \|A_j'x\|$ where $x = [x_i]_1^j$. Let $x_n = 0 \oplus x$ be the vector in C^n, where the first $n-j$ components are zero and x forms the last j components. Using the triangular form of A_n' we have

$$\|x\|^2 \ge \|A_n' x_n\|^2 = \sum_{i=j}^{n-1} |a_i x_j + a_{i-1} x_{j-1} + \cdots + a_{i-j+1} x_1|^2 + \|A_j' x\|^2 =$$

$$\sum_{i=j}^{n-1} |a_i x_j + a_{i-1} x_{j-1} + \cdots + a_{i-j+1} x_1|^2 + \|x\|^2 .$$

So all the terms in the sums are zero, that is,

$$a_i = -(a_{i-1}x_{j-1} + \cdots + a_{i-j+1}x_1) \quad (j \le i < n) \tag{7.9}$$

because $x_j = 1$. Corollary 5.2 and (5.3) show that $[x_i]_1^{j-1}$ are uniquely determined by A_j'. Therefore the set $\{a_i\}_j^n$ corresponding to any contractive expansion A_{n+1}' of A_j' are uniquely determined by A_j'. In fact these a_i's are given in (7.9). This implies that A_n' admits a unique contractive expansion. The proof is now complete.

Assume that the Hankel matrix A_n' in (7.3) formed by $\{a_j\}_0^{n-1}$ is a contraction. Proposition 7.1 part (ii) implies that each a_j is in the closed disc, with center α_j and radius τ_j. The first "data" point a_0 is in the closed unit disc, that is, $\alpha_0 = 0$ and $\tau_0 = 1$. By convention we set $\rho_j = 0$ if $\tau_j = 0$ in equation (7.4). Recall that α_j and τ_j are uniquely determined by A_j'; see Proposition 7.1. This and $a_0 = \rho_0$ implies that the complex numbers $\{\rho_j\}_0^{n-1}$ in (7.4) are uniquely determined by A_n'. In other words, the set $\{a_j\}_0^{n-1}$ uniquely determines a set of n complex numbers $\{\rho_j\}_0^{n-1}$ in the closed unit disc by (7.4). We will call $\{\rho_j\}_0^{n-1}$ the *choice sequence* for A_n'. We are ready for the next result.

7.3 PROPOSITION. *Let A_n' be a contractive Hankel matrix of the form (7.1) with choice sequence $\{\rho_i\}_0^{n-1}$. Then*

(i) The matrix A_n' is a strict contraction if and only if $|\rho_j| < 1$ for all $0 \le j < n$.

(ii) The matrix A_n' has norm one if and only if there exists an integer j such that

$$|\rho_i| < |\rho_j| = 1 \text{ for } 0 \le i < j \text{ and } \rho_i = 0 \text{ for } j < i < n. \tag{7.10}$$

PROOF. Obviously A_j' is a strict contraction for all $j \le n$ if and only if A_n' is a strict contraction. Thus (i) follows from Proposition 7.1 (iii). For part (ii) assume that $\|A_n'\| = 1$. Let $j+1$ be the first integer such that $\|A_{j+1}'\| = 1$. Then (i) and part (iii) of Proposition 7.1 imply that $|\rho_i| < |\rho_j| = 1$ for $0 \le i < j$. By Proposition 7.1 (iv), the Hankel matrix A_{j+1}' has a unique contractive expansion of the same special Hankel form; this expansion must be A_n'. Hence $\tau_i = 0$ for $i > j$. By our convention $\rho_i = 0$ for $i > j$ and (7.10) holds. On the other hand, if (7.10) holds, then (i) and Proposition 7.1 (iii) imply that $\|A_{j+1}'\| = 1$. So $\|A_n'\| = 1$ and the proof is complete.

Using $\alpha_0 = 0$ and $\tau_0 = 1$, one can show that there is a one to one correspondence between the set of all contractive Hankel matrices A_n', and the set of all choice sequences $\{\rho_j\}_0^{n-1}$ satisfying either $|\rho_j| < 1$ for all $0 \le j < n$, or (7.10).

7.4 COROLLARY. *Let A_n' be a contractive Hankel matrix of the form (7.3). Let $\{\rho_i\}_0^{n-1}$ be the choice sequence for A_n' and $\{r_i\}_0^{n-1}$ be the Schur numbers for the data $\{a_i\}_0^{n-1}$. Then $r_j = \rho_j$ for $0 \le j < n$.*

PROOF. We proceed by induction. Since $\rho_o = a_o = r_o$ we assume that $\rho_i = r_i$ for $0 \leq i \leq j-1$ and $j \geq 1$. Let $\{a_i\}_0^{j-1}$ be the data for the Carathéodory interpolation problem. According to Corollary I.3.4 and Theorem I.6.7, the set of all a_j such that A'_{j+1} is a contraction is given by the closed disc with center a_j^0 and radius t_j. By Proposition 7.1 (iii) the set of all a_j such that A'_{j+1} is a contraction is closed disc with center α_j and radius τ_j. In both cases these a_j's describe the same set, that is, $\alpha_j = a_j^0$ and $\tau_j = t_j$ for $0 \leq j < n$. Comparing (7.4) with equation (I.3.16) gives $\rho_j = r_j$ and completes the proof.

The following result shows how the Levinson system in Section 5 can be used to construct the next a_j from A'_j.

7.5 PROPOSITION. *Let A'_{j-1} be strictly contractive and set $R = [\, A\, ,\, B\,]$ where A, B and C are defined in (7.6). Then there exists an $f = [\, f_i\,]_1^j$ in C^j and q in C^1 satisfying*

$$D_R^2 \begin{bmatrix} -f \\ 1 \end{bmatrix} = \begin{bmatrix} 0 \\ q \end{bmatrix}. \tag{7.11}$$

Moreover, A'_j is a contraction if and only if $q \geq 0$. In this case, the set of all a_j such that A'_{j+1} is a contraction is given by

$$a_j = \sum_1^j f_i a_{i-1} + qr \tag{7.12}$$

where r is a scalar in the closed unit disc.

PROOF. The first statement follows from Proposition 6.9. To prove the second part notice that

$$I - X^* X = I - B^* D_A^{-2} B \text{ and } I - YY^* = I - CD_A^{-2}C^* . \tag{7.13}$$

Since B and $D_{A^*}^2$ is the transpose of C and D_A^2 respectively, (7.13) implies that $I - X^* X = I - YY^*$. Corollaries 1.2 and 5.2 with $A'_j = [\, A^*\, ,\, C^*\,]^*$ show that A_j is a contraction if and only if $q \geq 0$. If $q \geq 0$, then $d_{Y^*} = d_X = q$. Now equation (7.12) follows from (7.5), (7.6), (6.6) and Theorem 3.1. This completes the proof.

Finally, it is noted that equation (7.11) is a contractive version of the Levinson system in (II.2.1). At this point one can use (7.11) to compute the set of all contractive A'_{j+1} from A'_j recursively, and develop fast recursive inverse scattering algorithms. In Chapter XIII we will show how one can use these Levinson type systems to recursively compute all contractive A'_{j+1} from A'_j and develop inverse scattering algorithms in a very general setting.

IV.8. NOTES AND COMMENTS

The characterization of the row $[A, B]$ or column $[A, C]^{tr}$ contraction is usually referred to as the Douglas Lemma [Do 1]. Previously, Julia [Ju 1-3] already noticed that a contraction A can be extented to an isometry $[A, C]^{tr}$. It was Halmos who introduced, and made use of the rotation matrix of a contraction [Ha 1,5]. The description of all contractions on a Hilbert space K extending a contraction on a subspace H also discussed in Section 1 is explicitly presented in [DaKW]. The characterization of all lower triangular contractions in Section 2 is taken from [Sz.-NF 6]. The characterization of all 2 by 2 matrix contractions is due to [DaKW] and later by [ArsG]. The basic Corollary 3.6 is usually referred to as the Parrott Lemma [Parr 2] and was circulated by Parrott in the mid seventies. The computational formulas given in Sections 4 and 5 were inspired by the Levinson algorithm for block Toeplitz matrices (see also Chapters XV and XVI). Finally, it is important to notice that in this chapter, as well as the rest of the monograph, the role of the defect operator D_A can be played by any solution Φ of $I = A^*A + \Phi^*\Phi$. This type of equation is usually referred to as a *Lyapunov equation*. In connection with this chapter see [Da], [ElL], [FoT 3], [GroJSW], [JoR] and [Woe].

CHAPTER V

CONTRACTIVE ONE STEP INTERTWINING LIFTINGS

This chapter is devoted to a thorough analysis of contractive one step intertwining liftings. This analysis will play a very important role in the following chapters in the commutant lifting theorem and interpolation problems. An application of this theory to the Carathéodory interpolation problem will also be presented in Section 3.

1. CHARACTERIZING ALL CONTRACTIVE ONE STEP INTERTWINING LIFTINGS

In this section we give a complete characterization of the set of all contractive one step intertwining liftings.

Let A be an operator from H into H'. An operator A_1 from $K(\supseteq H)$ to $K'(\supseteq H')$ is called *a lifting of* A if $P_{H'}A_1 = AP_H$, or equivalently, A_1 has a matrix representation of the form

$$A_1 = \begin{bmatrix} A & 0 \\ B & C \end{bmatrix} : K = H \oplus D \to K' = H' \oplus D' . \tag{1.1}$$

Throughout T on H and T' on H' are contractions. The *contractive one step lifting* T_1 on $H \oplus D_T$ and T'_1 on $H' \oplus D_{T'}$ of T and T' is defined by

$$T_1 = \begin{bmatrix} T & 0 \\ D_T & 0 \end{bmatrix} \text{ on } H \oplus D_T \text{ and } T'_1 = \begin{bmatrix} T' & 0 \\ D_{T'} & 0 \end{bmatrix} \text{ on } H' \oplus D_{T'} , \tag{1.2}$$

respectively. Lemma IV.2.1 guarantees that T_1 and T'_1 is a contractive lifting of T and T', respectively. The one step liftings T_1 and T'_1 will play an important role throughout this monograph.

The set of all operators, respectively contractions, A from H to H' intertwining T and T', that is,

$$AT = T'A \tag{1.3}$$

is denoted by $I(T, T')$, respectively $I_1(T, T')$. Let U on K and U' on K' be contractive liftings of T and T', respectively and assume that A is a contraction. A basic problem in operator theory is the existence and characterization of all (contractive) liftings A_1 of A in $I_1(U, U')$. Even in very simple cases this problem has no solution, as the following example shows:

$$T = 0, \ T' = 0 \text{ on } H = H' = C^1, \ A = 1 \text{ and}$$

$$U = \begin{bmatrix} 0 & 0 \\ 0 & 1 \end{bmatrix}, \quad U' = \begin{bmatrix} 0 & 0 \\ 1 & 0 \end{bmatrix} \text{ on } C^2 .$$

In this example one of the liftings, namely U', is a contractive one step lifting while the other U is not. Moreover, there is no intertwining lifting of A. This emphasizes the significance in the next lemma, which yields a quite surprising result, that is, every A in $I_1(T, T')$ admits a contractive lifting A_1 in $I_1(T_1, T_1')$ of A. For reasons which will become clear later any operator A_1 in $I_1(T_1, T_1')$ lifting A in $I_1(T, T')$ will be called a *contractive one step intertwining liftings* of A.

Let A be a contraction in $I(T, T')$. Let the spaces F and G be defined by

$$F = \{D_A Th \oplus D_T h : h \in H\}^- \text{ and } G = (D_A \oplus D_T) \ominus F . \tag{1.4}$$

Throughout this section Γ_o is the contraction mapping F into $D_{T'}$ defined by

$$\Gamma_o(D_A Th \oplus D_T h) = D_{T'} Ah \qquad (h \in H) . \tag{1.5}$$

To prove that Γ_o is a contraction first notice that

$$\|D_A Th\|^2 + \|D_T h\|^2 = \|Th\|^2 - \|ATh\|^2 + \|h\|^2 - \|Th\|^2 =$$

$$\|h\|^2 - \|ATh\|^2 = \|h\|^2 - \|T'Ah\|^2 = \qquad (h \in H) \tag{1.6}$$

$$\|h\|^2 - \|Ah\|^2 + \|Ah\|^2 - \|T'Ah\|^2 = \|D_A h\|^2 + \|D_{T'} Ah\|^2 .$$

In particular,

$$\|D_{T'} Ah\|^2 \le \|D_A Th \oplus D_T h\| \qquad (h \in H).$$

From this it follows that the operator Γ_o defined in (1.5) is a contraction. Throughout this section Π_H is the operator from $H \oplus D_T$ onto H defined by

$$\Pi_H(h \oplus d) = h \qquad\qquad (h \oplus d \in H \oplus D_T) .$$

The operator Π_H picks out the H component in $h \oplus d$. Now the stage is set for the following basic result.

1.1 LEMMA. *Let A be in* $I_1(T, T')$. *The set of all contractive one step intertwining liftings* A_1 *of A is given by*

$$A_1 = \begin{bmatrix} I & 0 \\ 0 & \Gamma_0 P_F + D_{\Gamma_0^*} \Gamma_1 P_G \end{bmatrix} \begin{bmatrix} A\Pi_H \\ D_A \oplus I \end{bmatrix} \tag{1.7}$$

where Γ_0 *is the contraction mapping F into* $D_{T'}$ *defined in (1.5) and* Γ_1 *is an arbitrary contraction mapping G into* $D_{\Gamma_0^*}$. *In particular, there is a one to one correspondence between the set of all contractive one step intertwining liftings* A_1 *of A and the set of all contractions* Γ_1 *mapping G into* $D_{\Gamma_0^*}$. *Furthermore, the relation*

$$\alpha D_{\Gamma_0}(D_A Th \oplus D_T h) = D_A h \qquad (h \in H) \tag{1.8}$$

defines a unitary operator α *mapping* D_{Γ_0} *onto* D_A.

PROOF. (The notation $D_A \oplus I$ makes sense in this setting, because we always identify $[h, d]^{tr}$ with $h \oplus d$ for all h in H and d in D_T. Thus $(D_A \oplus I)[h, d]^{tr} = D_A h \oplus d$.) By Corollary IV.2.3, an operator A_1 from $H \oplus D_T$ to $H' \oplus D_{T'}$ is a contractive lifting of A if and only if A_1 admits a matrix representation of the form

$$A_1 = \begin{bmatrix} A & 0 \\ XD_A & Y \end{bmatrix} : H \oplus D_T \to H' \oplus D_{T'} \tag{1.9}$$

where $[X, Y]$ is a contraction mapping $D_A \oplus D_T$ into $D_{T'}$. By (1.2) the operator A_1 in (1.9) is in $I(T_1, T_1')$ if and only if

$$[X, Y] \begin{bmatrix} D_A Th \\ D_T h \end{bmatrix} = D_{T'} Ah \qquad (h \in H) , \tag{1.10}$$

or equivalently, by the definition of F in (1.4)

$$[X, Y] | F = \Gamma_0 . \tag{1.11}$$

Therefore A_1 is a contractive one step intertwining lifting of A if and only if A_1 admits a matrix representation of the form (1.9) where $[X, Y]$ is a contraction satisfying (1.11). Remark IV.1.7 shows that the set of all contractions $[X, Y]$ satisfying (1.11) is given by

$$[X, Y] = \Gamma_0 P_F + D_{\Gamma_0^*} \Gamma_1 P_G .$$

Substituting this into (1.9) yields (1.7) and proves the first part of the lemma.

To complete the proof, notice that (1.5) and (1.6) give

$$\|D_{\Gamma_0}(D_A Th \oplus D_T h)\|^2 =$$

$$\|D_A Th \oplus D_T h\|^2 - \|\Gamma_0(D_A Th \oplus D_T h)\|^2 = \|D_A h\|^2$$

where h is in H. From this it follows that the operator α mapping D_{Γ_0} onto D_A defined in (1.8) is unitary. This finishes the proof.

Now we will define some important spaces and operators, which will be used in a more explicit characterization of the set of all contractive one step intertwining liftings. As before $A \in I_1(T, T')$. By (1.6) the formula

$$\omega(D_A Th \oplus D_T h) = D_{T'} Ah \oplus D_A h \qquad (h \in H) \tag{1.12}$$

defines an isometric operator from F into $D_{T'} \oplus D_A$. By introducing the spaces

$$F' = \{D_{T'} Ah \oplus D_A h : h \in H\}^- \text{ and } G' = (D_{T'} \oplus D_A) \ominus F', \tag{1.13}$$

then ω is an unitary operator from F onto F'. We mention that unlike the spaces F and F' the dimension of the *residual spaces* G and G' may be different.

First let us verify the following useful relation

$$\overline{P_{G'}(D_{T'} \oplus \{0\})} = G'. \tag{1.14}$$

To see this let $f \oplus g$ live in $G'(\subseteq D_{T'} \oplus D_A)$ and be orthogonal to the left hand side of (1.14). Thus $f \oplus g$ is orthogonal to $D_{T'} \oplus \{0\}$; hence $f = 0$. Equation (1.13) and g in D_A imply that $g = 0$. Therefore (1.14) holds.

Throughout this section Π' from $D_{T'} \oplus D_A$ onto $D_{T'}$ and Π'_A from $D_{T'} \oplus D_A$ onto D_A are the operators defined by

$$\Pi'(d' \oplus d_A) = d' \text{ and } \Pi'_A(d' \oplus d_A) = d_A \qquad (d' \oplus d_A \in D_{T'} \oplus D_A). \tag{1.15}$$

The operator Π' picks out the first component of $D_{T'} \oplus D_A$, while Π'_A picks out the second component. Finally, we are ready for the main result of this section.

1.2 THEOREM. *Let A be in $I_1(T, T')$. There exists a one to one correspondence between the set of all contractive one step intertwining lifting A_1 of A, and the set of all contractions Γ mapping G into G'. In fact, this correspondence is provided by the relation*

$$A_1 = \begin{bmatrix} I & 0 \\ 0 & \Pi'(\omega P_F + \Gamma P_G) \end{bmatrix} \begin{bmatrix} A\Pi_H \\ D_A \oplus I \end{bmatrix}. \tag{1.16}$$

PROOF. The definition of Γ_o in (1.5) and ω in (1.12) give

$$\Gamma_o = \Pi'\omega. \tag{1.17}$$

For f in $D_{T'}$ we have

$$\|D_{\Gamma_o^*}f\|^2 = \|f\|^2 - \|\Gamma_o^*f\|^2 = \|f\|^2 - \|\omega^*P_F\Pi'^*f\|^2 = \|f\|^2 - \|P_F\Pi'^*f\|^2 = \|P_G\Pi'^*f\|^2.$$

It follows that the relation $\gamma_1 D_{\Gamma_o^*} = P_G\Pi'^*$ defines an isometry γ_1 mapping $D_{\Gamma_o^*}$ into G'. Equation (1.14) shows that γ_1 is onto, or equivalently, its adjoint γ is unitary. Thus by passing to the adjoints the equation

$$D_{\Gamma_o^*}\gamma = \Pi' \mid G' \tag{1.18}$$

defines a unitary operator γ from G' onto $D_{\Gamma_o^*}$. Substituting the Γ_o in (1.17) and $D_{\Gamma_o^*}\gamma$ in (1.18) into the A_1 in (1.7), produces the A_1 in (1.16) where $\gamma\Gamma = \Gamma_1$. This and Lemma 1.1 completes the proof.

1.3 REMARK. The proof of the previous Theorem shows that the contractive intertwining liftings A_1 of A in (1.7) and (1.16) are equal if and only if $\Gamma_1 = \gamma\Gamma$.

Let M mapping N' into M and N mapping N into N' be contractions. We call the factorization M·N *regular* if

$$D_M \oplus D_N = \{D_M Nf \oplus D_N f : f \in N\}^-, \tag{1.19}$$

or equivalently, if $G(M\cdot N) = \{0\}$ where $G(M\cdot N)$ is the *residual space* defined by

$$F(M\cdot N) \triangleq \{D_M Nf \oplus D_N f : f \in N\}^-$$
$$G(M\cdot N) \triangleq (D_M \oplus D_N) \ominus F(M\cdot N). \tag{1.20}$$

With these definitions we clearly have

$$F = F(A\cdot T), \ F' = F(T'\cdot A), \ G = G(A\cdot T) \text{ and } G' = G(T'\cdot A). \tag{1.21}$$

From the previous theorem we readily obtain the following result.

1.4 COROLLARY. *Let A be in $I_1(T, T')$. There is a unique contractive one step intertwining lifting A_1 of A if and only if A·T or T'·A is a regular factorization.*

1.5 LEMMA. *Let* A *be in* $I_1(T,T')$. *Let* Γ_0 *be the contraction,* α *and* γ *be the unitary operators defined in (1.5), (1.8) and (1.18), respectively. Then*

$$\alpha D_{\Gamma_0} = \Pi'_A \omega \ and - \alpha \Gamma_0^* \gamma = \Pi'_A \mid G' . \tag{1.22}$$

PROOF. By (1.8) and the definition of ω we have

$$\alpha D_{\Gamma_0}(D_A Th \oplus D_T h) = D_A h = \Pi'_A \omega (D_A Th \oplus D_T h) \qquad (h \in H).$$

This yields the first equation in (1.22). The second equation follows by using (1.14), (1.17), (1.18) the first part of (1.22) and (1.5) in the following calculation

$$\alpha \Gamma_0^* \gamma P_{G'} \Pi'^* = \alpha \Gamma_0^* D_{\Gamma_0^*} = \alpha D_{\Gamma_0} \Gamma_0^* =$$

$$\Pi'_A \omega \omega^* P_{F'} \Pi'^* = \Pi'_A (I - P_{G'}) \Pi'^* = - \Pi'_A P_{G'} \Pi'^* .$$

This completes the proof.

1.6 COROLLARY. *Let* A *be in* $I_1(T,T')$. *Let* A_1 *be the same contractive intertwining lifting of* A *in (1.7) and (1.16) where the contraction* $\Gamma_1 = \gamma\Gamma$ *and* γ *is the unitary operator defined in (1.18). Then the relation*

$$WD_{A_1} = \begin{bmatrix} \Pi'_A(\omega P_F + \Gamma P_G) \\ D_\Gamma P_G \end{bmatrix}(D_A \oplus I) = \begin{bmatrix} \alpha D_{\Gamma_0} P_F - \alpha \Gamma_0^* \Gamma_1 P_G \\ D_{\Gamma_1} P_G \end{bmatrix}(D_A \oplus I) . \tag{1.23}$$

defines a unitary operator W *mapping* D_{A_1} *onto* $D_A \oplus D_\Gamma$ *where* α *is the unitary operator defined in (1.8).*

PROOF. The second equality in (1.23) follows from (1.22) in Lemma 1.5 and $D_{\Gamma_1} = D_\Gamma$. Since Γ is a contraction from G into G', the operator V defined by the following product of two isometries

$$V = \begin{bmatrix} I & 0 \\ 0 & (\omega P_F + \Gamma P_G) \\ 0 & D_\Gamma P_G \end{bmatrix} \begin{bmatrix} A\Pi_H \\ D_A \oplus I \end{bmatrix}$$

is also an isometry mapping $H \oplus D_T$ into $H' \oplus (D_{T'} \oplus D_A) \oplus D_\Gamma = (H' \oplus D_{T'}) \oplus (D_A \oplus D_\Gamma)$. The previous theorem shows that the contractive intertwining lifting A_1 of A is $A_1 = P_{H' \oplus D_T} V$. For f in $H \oplus D_T$ this gives

$$\|D_{A_1} f\|^2 = \|Vf\|^2 - \|A_1 f\|^2 = \|P_{D_A \oplus D_\Gamma} Vf\|^2 .$$

From this it follows that the first equality in (1.23) defines an isometry W from D_{A_1} to $D_A \oplus D_\Gamma$. For h in H the equation (1.23) gives

$$WD_{A_1}(Th \oplus D_T h) = \Pi'_A \omega(D_A Th \oplus D_T) \oplus 0 = D_A h \oplus 0 .$$

Therefore the range of W contains $D_A \oplus \{0\}$. From (1.23) we infer that $\{0\} \oplus D_\Gamma G$ is also contained in the range of the isometry W. Hence the isometry W is onto, that is, W is unitary. This completes the proof.

Later we will show how the following result can be used in computing certain contractive intertwining liftings, or equivalently, solving certain interpolation problems.

1.7 COROLLARY. *Let A be a contraction in $I(T, T')$. Let A_1 be the contractive one step intertwining lifting of A in (1.16). Then $G(A_1 \cdot T_1)$, respectively $G(T'_1 \cdot A_1)$, can be identified with D_Γ, respectively D_{Γ^*}. In particular, $A_1 \cdot T_1$, respectively, $T'_1 \cdot A_1$, is a regular factorization if and only if Γ is an isometry, respectively co-isometry.*

PROOF. A simple calculation shows that $D_{T_1} = 0 \oplus I$ on $H \oplus D_T$ and $D_{T_1} = \{0\} \oplus D_T$ (which is identified with D_T). Using the unitary operator W in (1.23) with the definition of ω in (1.12) we have

$$(W \oplus I)[G(A_1 \cdot T_1)] =$$

$$(W \oplus I)[(D_{A_1} \oplus D_T) \ominus \{D_{A_1} T_1 (h \oplus d) \oplus D_{T_1}(h \oplus d): (h \oplus d) \in H \oplus D_T\}^-] =$$

$$(D_A \oplus D_\Gamma \oplus D_T) \ominus \{D_A h \oplus 0 \oplus d: (h \oplus d) \in H \oplus D_T\}^- = \{0\} \oplus D_\Gamma \oplus \{0\}$$

where I is the identity on $D_{T_1} = \{0\} \oplus D_T$. This proves that $G(A_1 \cdot T_1)$ can be identified with D_Γ.

To prove the other half of the corollary notice that (1.16), (1.23) and the definition of ω gives

$$(I' \oplus W)[G(T'_1 \cdot A_1)] =$$

$$(I' \oplus W)[(D_{T'_1} \oplus D_{A_1}) \ominus \{D_{T'_1} A_1 (h \oplus d) \oplus D_{A_1}(h \oplus d): h \oplus d \in H \oplus D_T\}^-] =$$

$$[D_{T'} \oplus D_A \oplus D_\Gamma)] \ominus \{((\omega P_F \oplus \Gamma P_G)(D_A h \oplus d)) \oplus D_T P_G(D_A h \oplus d): h \oplus d \in H \oplus D_T\}^- =$$

$$(G' \oplus D_\Gamma) \ominus \{\Gamma g \oplus D_\Gamma g: g \in G\}^- = [D_{\Gamma^*}, -\Gamma]^* D_{\Gamma^*} \qquad (1.24)$$

where I' is the natural identification of $D_{T'_1} = \{0\} \oplus D_{T'}$ to $D_{T'}$. The last equality in

(1.24) follows from the fact that the rotation operator

$$R_\Gamma = \begin{bmatrix} \Gamma & D_{\Gamma^*} \\ D_\Gamma & -\Gamma^* \end{bmatrix} : \ G \oplus D_{\Gamma^*} \to G' \oplus D_\Gamma$$

is unitary (see Corollary IV.1.4). In other words, the orthogonal complement of the range of the first column $[\Gamma^*, D_\Gamma]^*$ of R_Γ is precisely the range of the second column $[D_{\Gamma^*}, -\Gamma]^*$ of R_Γ. Because $[D_{\Gamma^*}, -\Gamma]^*$ is an isometry, (1.24) shows that $G(T_1' \cdot A_1)$ can be identified with D_{Γ^*}. This completes the proof.

2. REGULAR FACTORIZATION

Regular factorizations play an important role in lifting theory. Here we obtain some further results on regular factorizations.

A contraction A from H to H' is *pure* if $\|Ah\| < \|h\|$ for all nonzero h in H. We say that an operator $A = A_o \oplus A_1$ maps $H = H_o \oplus H_1$ into $H' = H_o' \oplus H_1'$ if A maps H_o into H_o' and A maps H_1 into H_1'. Moreover, A_o is the operator from H_o to H_o' and A_1 the operator from H_1 to H_1' defined by $A_o \triangleq A|H_o$ and $A_1 \triangleq A|H_1$. We begin with the following result.

2.1 LEMMA. *Let* Z *be a contraction from* H *to* H'. *Then* Z *admits a unique decomposition of the form* $Z = Z_o \oplus Z_u$ *mapping* $H_o \oplus H_u$ *into* $H_o' \oplus H_u'$ *where* Z_o *is a pure contraction and* Z_u *is unitary. (By unique we mean that if* $Z = \bar{Z}_o \oplus \bar{Z}_u$ *is any other decomposition where* $\bar{Z}_o$ *is a pure contraction and* $\bar{Z}_u$ *is unitary, then* $Z_o = \bar{Z}_o$ *and* $Z_u = \bar{Z}_u$.) *In fact* $Z_o = Z|D_Z$ *from* $H_o = D_Z$ *to* $H_o' = D_{Z^*}$, *and* $Z_u = Z|H_u$ *from* $H_u = H \ominus D_Z$ *onto* $H_u' = H' \ominus D_{Z^*}$.

PROOF. Using $ZD_Z = D_{Z^*}Z$ and $Z^*D_{Z^*} = D_Z Z^*$ we see that

$$ZD_Z \subseteq D_{Z^*} \quad \text{and} \quad Z^*D_{Z^*} \subseteq D_Z . \tag{2.1}$$

The second relation in (2.1) shows that $Z(H \ominus D_Z)$ is contained in $H' \ominus D_{Z^*}$. So setting

$$H_o = D_Z , \ H_u = H \ominus D_Z , \ \ H_o' = D_{Z^*} \text{ and } H_u' = H' \ominus D_{Z^*}$$

Z admits a decomposition of the form $Z = Z_o \oplus Z_u$ mapping $H_o \oplus H_u$ into $H_o' \oplus H_u'$ where $Z_o = Z|H_o$ and $Z_u = Z|H_u$. Notice that h is in H_u if and only if $D_Z h = 0$, or equivalently, $\|h\|^2 = \|Zh\|^2$. This implies that Z_u is an isometry and Z_o is pure. (If $\|Z_o h_o\| = \|h_o\|$ for some h_o in D_Z, then $\|Zh_o\|^2 = \|h_o\|^2$ and h_o is orthogonal to D_Z. Hence $h_o = 0$.) In particular, Z_u is an isometry for any contraction Z. Since

$$Z_u^* = Z^* \,|\, H_u' = Z^* \,|\, (H \ominus D_{Z^*}) = (Z^*)_u$$

we see that Z_u^* is an isometry. Hence Z_u is unitary.

To complete the proof let $Z = \bar{Z}_o \oplus \bar{Z}_u$ mapping $\bar{H}_o \oplus \bar{H}_u$ into $\bar{H}_o' \oplus \bar{H}_u'$ be any decomposition where $\bar{Z}_o$ is pure and $\bar{Z}_u$ is unitary. For h in $\bar{H}_u$ we have $\|h\|^2 = \|\bar{Z}_u h\|^2 = \|Zh\|^2$, or equivalently, $D_Z h = 0$, and h is in H_u. Thus $\bar{H}_u \subseteq H_u$ and $L = H_u \ominus \bar{H}_u \subseteq \bar{H}_o$. Since $Z | \bar{H}_o$ is pure and $Z | H_u$ is unitary this implies that $L = \{0\}$, that is, $H_u = \bar{H}_u$. The proof is now complete.

The operator Z_o [Z_u] in Lemma 2.1 is called the *pure [unitary] part* of Z. As in Section 1 (see (1.19), (1.20)) we consider the contractions M mapping N' into M and N mapping N into N'. Lemma IV.2.1 implies that the operator Z mapping $D_M \oplus D_N$ into $D_{M^*} \oplus D_{N^*}$ defined by

$$Z = \begin{bmatrix} M \,|\, D_M & 0 \\ -D_{N^*} D_M \,|\, D_M & N \,|\, D_N \end{bmatrix} \tag{2.2}$$

is a contraction. The operator Z in (2.2) is well defined because $M D_M \subseteq D_{M^*}$ and $N D_N \subseteq D_{N^*}$. (Note that in this case the Γ given by Lemma IV.2.1 is $\Gamma = -P_{D_{N^*}} \,|\, D_M$). We are ready for the following result.

2.2 LEMMA. *Let Z be the contraction in (2.2). Then $G(M{\cdot}N)$ and $G(N^*{\cdot}M^*)$ are the orthogonal complements of D_Z and D_{Z^*} in $D_M \oplus D_N$ and $D_{M^*} \oplus D_{N^*}$, respectively.*

PROOF. Let $f \oplus g$ be in $D_M \oplus D_N$. Using $D_{N^*} N = N D_N$ we have

$$\|D_Z(f \oplus g)\|^2 = \|f\|^2 + \|g\|^2 - \|Z(f \oplus g)\|^2 =$$

$$\|f\|^2 + \|g\|^2 - \|Mf\|^2 - \|Ng\|^2 - \|D_{N^*} D_M f\|^2 + 2\mathrm{Re}(D_{N^*} D_M f, Ng) =$$

$$\|D_M f\|^2 + \|D_N g\|^2 - \|D_M f\|^2 + \|N^* D_M f\|^2$$

$$+ 2\mathrm{Re}(N^* D_M f, D_N g) = \|N^* D_M f + D_N g\|^2 .$$

Hence $f \oplus g$ is orthogonal to D_Z if and only if $N^* D_M f + D_N g = 0$, or equivalently, $f \oplus g$ is orthogonal to $F(M{\cdot}N)$ by (1.20). Therefore $G(M{\cdot}N)$ is the orthogonal complement of D_Z. A similar calculation gives the statement concerning $G(N^*{\cdot}M^*)$. This completes the proof.

2.3 COROLLARY. *The spaces $G(M \cdot N)$ and $G(N^* \cdot M^*)$ have the same dimension. In particular, the factorization $M{\cdot}N$ is regular if and only if $N^*{\cdot}M^*$ is* *regular.*

PROOF. Lemmas 2.1 and 2.2 show that Z_u maps $G(M{\cdot}N)$ onto $G(N^*{\cdot}M^*)$ where Z_u is the unitary part of the operator Z in (2.2). Thus $G(M{\cdot}N) = \{0\}$ if and only if $G(N^*{\cdot}M^*) = \{0\}$. This completes the proof.

The next result is a useful complement of Theorem 1.2.

2.4 LEMMA. *If A is a contraction in $I(T, T^{'})$, then*

$$\dim G(A{\cdot}T) \le \dim D_{T^*} \,, \tag{2.3}$$

$$\dim G(T^{'}{\cdot}A) \le \dim D_{T^{'}} \,. \tag{2.4}$$

PROOF. Equation (2.4) follows from (1.13) and (1.14). Replacing $T^{'}{\cdot}A$ with $T^*{\cdot}A^*$ in (1.14) gives

$$G(T^*{\cdot}A^*) = \overline{P_{G(T^*{\cdot}A^*)}(D_{T^*} \oplus \{0\})} \tag{2.5}$$

This and Lemma 2.2 yields (2.3) and completes the proof.

2.5 PROPOSITION. *Let A be a contraction in $I(T, T^{'})$.*

(i) If the factorization $T^{'}{\cdot}A$ is regular, D_A and $D_{T^{'}}$ are finite dimensional and $T^{'}$ is not an isometry, then $\ker D_A \ne \{0\}$.

(ii) If the factorization $A{\cdot}T$ is regular, H is finite dimensional and T is not an isometry, then $\ker D_A \ne \{0\}$.

(iii) If $T^{'}{\cdot}A$ is not a regular factorization, $\ker D_A \ne \{0\}$ and the dimension of $D_{T^{'}}$ is one, then $\ker D_A$ is invariant to T and $T\,|\,(\ker D_A)$ is an isometry.

PROOF. The hypothesis in (i) implies

$$D_{T^{'}} \oplus D_A = \{D_{T^{'}}Ah \oplus D_A h{:}h \in H\} \,. \tag{2.6}$$

For any nonzero d in $D_{T^{'}}$ there exists a nonzero h in H such that

$$d \oplus 0 = D_{T^{'}}Ah \oplus D_A h \,.$$

Hence $D_A h = 0$ and the proof of (i) is finished.

To prove (ii) we notice that the regularity of the factorization $A{\cdot}T$ and $D_T \ne 0$ implies that

$$\dim D_A < \dim D_A + \dim D_T \leq \dim H < \infty .$$

Therefore $\ker D_A \neq \{0\}$.

Finally, we assume the hypothesis in (iii). Let $D_A h = 0$ for some in h in H. We claim that $D_{T'} Ah = 0$. If $D_{T'} Ah \neq 0$, then $D_{T'} \oplus \{0\} \subsetneq F'$, because the dimension of $D_{T'}$ is one. Hence $D_{T'} \oplus D_A = F'$ and the factorization $T' \cdot A$ is regular. This is a contradiction. Thus we must have $D_{T'} Ah = 0$. Recall that

$$\omega(D_A Th \oplus D_T h) = D_{T'} Ah \oplus D_A h \quad (h \in H)$$

is a unitary operator from F onto F'; see Section 1. By what we have just shown $D_{T'} Ah \oplus D_A h = 0$ when $D_A h = 0$. Thus $D_A Th$ and $D_T h$ are both zero. Therefore $\ker D_A$ is invariant to T and $T | \ker D_A$ is an isometry. This completes the proof.

2.6 COROLLARY. *Let H be finite dimensional and A be in $I_1(T, T')$. Assume that the dimension of $D_{T'}$ is one and all the eigenvalues of T are contained in the open unit disc. Then A admits a unique contractive one step intertwining lifting A_1 if and only if the norm of A is one.*

PROOF. Assume that A admits a unique contractive intertwining one step lifting. Corollary 1.4 shows that $T' \cdot A$ or $A \cdot T$ is a regular factorization. If $T' \cdot A$ is a regular factorization, then part (i) of Proposition 2.5 implies that the norm of A is one. On the other hand, if $A \cdot T$ is a regular factorization, then part (ii) of Proposition 2.5 shows that the norm of A is one, since $D_T \neq \{0\}$.

Now assume that the norm of A is one. We claim that $T' \cdot A$ is a regular factorization. If not, then part (iii) of Proposition 2.5 implies that $\ker D_A$ is invariant to T and $T | \ker D_A$ is an isometry. This contradicts the fact that all the eigenvalues of T are in the open unit disc. So $T' \cdot A$ is a regular factorization. Corollary 1.4 shows that A admits a unique contractive one step lifting. The proof is now complete.

2.7 COROLLARY. *Let A be a finite dimensional contraction in $I(T, T')$ where the eigenvalues of both T and T' are contained in the open unit disc and $\delta_T = \delta_{T'} = 1$. The following statements are equivalent:*

(i) The factorization $A \cdot T$ is regular.

(ii) The factorization $T' \cdot A$ is regular.

(iii) The norm of A is one.

Moreover, if any one of these conditions hold, then A admits a unique contractive one step intertwining lifting.

PROOF. The proof of the previous corollary shows that (ii) and (iii) are equivalent. Because ω is unitary we have $\dim F = \dim F'$. This implies that $\delta_A + \delta_T = \dim F$ if and only if $\delta_A + \delta_{T'} = \dim F'$. Therefore A·T is a regular factorization if and only if T'·A is a regular factorization. The last statement follows from the previous corollary. This completes the proof.

We conclude this section with two characterizations for regular factorizations, which are of independent interest.

2.8 PROPOSITION. *Let* M *mapping* N' *into* M *and* N *mapping* N *into* N' *be two contractions. Then the following are equivalent:*

(i) *The factorization* M · N *is regular.*

(ii) $\{D_M f \oplus D_{N^*} f : f \in N'\}^- = D_M \oplus D_{N^*}$

(iii) $D_M N' \cap D_{N^*} N' = \{0\}$.

Since we do not use this result we refer the reader to [Sz.-NF 13] for its proof.

3. APPLICATIONS OF CONTRACTIVE ONE STEP LIFTINGS TO CARATHEODORY INTERPOLATION

In this section we will use the previous results on contractive one step intertwining liftings to solve the one step Carathéodory interpolation problem considered in Section IV.7.

To begin let us recall the one step Carathéodory interpolation problem: Given the contractive Hankel matrix A_j' in (IV.7.1), find all a_j in C^1 such that the Hankel matrix A_{j+1}' in (IV.7.3) with $n = j+1$ is a contractive expansion of A_j'.

To solve this problem let S_j and S_j' be the contractions on C^j defined by

$$
S_j = \begin{bmatrix}
0 & 1 & 0 & \ldots & \ldots & 0 \\
0 & 0 & 1 & \ldots & \ldots & 0 \\
0 & 0 & 0 & \ldots & \ldots & \cdot \\
\cdot & \cdot & \cdot & \cdot & \ldots & \cdot \\
\cdot & \cdot & \cdot & \cdot & \ldots & \cdot \\
\cdot & \cdot & \cdot & \cdot & \ldots & 1 \\
0 & 0 & 0 & \ldots & \ldots & 0
\end{bmatrix}
\quad \text{and} \quad
S_j' = \begin{bmatrix}
0 & 0 & \ldots & \ldots & 0 & 0 \\
1 & 0 & \ldots & \ldots & 0 & 0 \\
0 & 1 & \ldots & \ldots & 0 & 0 \\
\cdot & \cdot & \cdot & \ldots & \cdot & \cdot \\
\cdot & \cdot & \cdot & \ldots & \cdot & \cdot \\
\cdot & \cdot & \cdot & \ldots & \cdot & \cdot \\
0 & 0 & \ldots & \ldots & 1 & 0
\end{bmatrix}
\tag{3.1}
$$

The matrix S_j has 1's immediately above the diagonal and zero's elsewhere. The matrix S_j' is the transpose of S_j. It is easy to show that an operator B on C^j intertwines S_j and S_j', that is, $S_j' B = B S_j$ if and only if $B = A_j'$ where A_j' is a Hankel matrix of the form (IV.7.1). Let $T = S_j$ and $T' = S_j'$. A simple calculation show that

$$D_{S_j} = P_1 \text{ and } D_{S_j'} = P_j \qquad (3.2)$$

where P_i is the orthogonal projection onto the i-th component of C^j for $1 \le i \le j$. By using this, the contractive one step lifting T_1 and T_1' of S_j and S_j' can be identified with S_{j+1} and S_{j+1}' respectively. So without loss of generality we set $T_1 = S_{j+1}$ and $T_1' = S_{j+1}'$. In this case, an operator B is in $I_1(T_1, T_1')$ if and only if $B = A_{j+1}'$ where A_{j+1}' is a $(j+1) \times (j+1)$ contractive Hankel matrix of the form (IV.7.3), with $j = n+1$. In particular, A_{j+1}' is a contractive one step intertwining lifting of A_j' if and only if A_{j+1}' is a contractive Hankel expansion of A_j'. Therefore constructing all contractive one step liftings A_{j+1}' of A_j' is equivalent to finding all a_j in C^1 such that A_{j+1}' is a contractive Hankel expansion of A_j'. Finally, Theorem 1.2 guarantees that there exists a contractive Hankel expansion A_{j+1}' of A_j'.

By identifying $D_{S_j'}$ with C^1, Theorem 1.2 shows that A_{j+1}' is a contractive Hankel expansion of $A_j' = A$ if and only if the last row is

$$[a_0, a_1, a_2, ..., a_j] = \Pi'(\omega P_F + \Gamma P_G)(D_A \oplus 1) \qquad (3.3)$$

where Γ mapping G into G' is a contraction. Here the range of Π' is taken to be C^1 by the identification of $[0, 0, \cdots, 0, \lambda]^{tr}$ with λ for all λ in C^1. In particular, if $\Gamma = 0$ the last row becomes

$$[a_0, a_1, ..., a_{j-1}, \alpha_j] = \Pi' \omega P_F (D_A \oplus 1) \qquad (3.4)$$

where clearly α_j is uniquely determined by A_j', or equivalently, by $(a_0, a_1, ..., a_{j-1})$. In case A_j' has a unique one step intertwining lifting A_{j+1}' this A_{j+1}' is given by (IV.7.3) where $a_j = \alpha_j$ and $n = j+1$.

On the other hand, if A_{j+1}' is not unique, then by subtracting (3.4) from (3.3) we obtain

$$[0, 0, ..., 0, b] = \Pi' \Gamma P_G (D_A \oplus 1) \qquad (3.5)$$

where $b \in C^1$ is uniquely determined by Γ. Corollary 1.4 and Lemma 2.4 show that $\dim G = \dim G' = 1$. Therefore $\Gamma \lambda g = \rho_j \lambda g'$ where $g \in G$ and $g' \in G'$ are unit vectors, $\rho_j \in C^1$ satisfies $|\rho_j| \le 1$, while λ runs over C^1. Using this in (3.5) yields

$$\rho_j [0, 0, ..., 0, \tau_j] = \Pi' \Gamma P_G (D_A \oplus 1) \qquad (3.6)$$

where $\tau_j \ge 0$ is in C^1 and is uniquely determined by

$$\tau_j = \Pi' g' (e_{j+1}, g) \qquad (g' \in G', \|g'\| = 1, \tau_j \ge 0), \qquad (3.7)$$

where e_{j+1} is the j+1 unit vector in C^{j+1}. (One can always choose $\tau_j \ge 0$ by replacing g'

by $\zeta g'$ and ρ by $\rho\bar{\zeta}$ for an appropriate choice of ζ in C^1 with $|\zeta| = 1$.) Note we choose $\tau_j = 0$ if the expansion A'_{j+1} is unique. Otherwise $\tau_j > 0$.

Summing up the previous remarks we have: The set of all contractive Hankel expansions A'_{j+1} of A'_j is given by (IV.7.3) where $j+1 = n$,

$$a_j = \alpha_j + \tau_j \rho_j \quad (|\rho_j| \le 1) \tag{3.8}$$

and α_j, τ_j are uniquely determined by A'_j in (3.4), (3.7), that is, the set of all a_j's such that A'_{j+1} is a contractive Hankel expansion of A'_j is given by the closed disc with center α_j and radius τ_j. (As before $\alpha_o = 0$ and $\tau_o = 1$) The contractive Hankel expansion A'_{j+1} is unique if and only if $\tau_j = 0$. Corollary I.3.4 shows that these a_j's are also in the closed disc $a_j^o + t_j \bar{D}$ with center a_j^o and radius t_j. Since the center and radius of a closed disc are unique $a_j^o = \alpha_j$ and $t_j = \tau_j$. Moreover, $\rho_j = r_j$ the j-th Schur number for $(a_o, a_1, ..., a_{n-1})$. Finally, Corollary 2.6 shows that A'_j admits a unique contractive expansion A'_{j+1} if and only if the norm of A'_j is one.

In this way we reconstructed, using only the properties of contractive one step intertwining liftings, all the results obtained in a more laborious way in Theorem I.6.7.

4. A 2 by 2 MATRIX CHARACTERIZATION OF ALL CONTRACTIVE ONE STEP INTERTWINING LIFTINGS

In this section we use Theorem IV.3.1 to to give a another construction for the set of all contractive one step liftings A_1 of A. As before A is in $I_1(T, T')$.

Consulting (1.2) we see that $T'_1 A_1 = A_1 T_1$ if and only if A_1 admits a matrix representation of the form

$$A_1 = \begin{bmatrix} A & 0 \\ G & H \end{bmatrix} : H \oplus D_T \to H' \oplus D_{T'} \tag{4.1}$$

where

$$[G, H] \begin{bmatrix} T \\ D_T \end{bmatrix} = D_{T'} A . \tag{4.2}$$

Recall that the rotation operator R_T defined in (IV.1.8) is unitary. Hence, the operator A_1 in (4.1) satisfying (4.2) is a contraction if and only if the operator

$$A_1 R_T = \begin{bmatrix} AT & AD_{T^*} \\ D_{T'}A & D \end{bmatrix} \text{ with } D = [G, H] \begin{bmatrix} D_{T^*} \\ -T^* \end{bmatrix}$$

is a contraction. Since $R_T^* = R_{T^*}$ we see that A_1 is the product of the above matrix with R_{T^*}. Conversely let the matrix

$$\begin{bmatrix} AT & AD_{T^*} \\ D_{T'}A & D \end{bmatrix}$$

be a contraction. If we define A_1 by the contraction

$$A_1 = \begin{bmatrix} AT & AD_{T^*} \\ D_{T'}A & D \end{bmatrix} R_{T^*} \tag{4.3}$$

then using $TD_T = D_{T^*}T$ it follows that A_1 is a contraction of the form (4.1). Applying $R_T^* = R_{T^*}$ again we see that this A_1 also satisfies (4.2). Therefore the set of all contractive one step intertwining liftings A_1 of A is given by (4.3) where D is chosen such that

$$\begin{bmatrix} AT & AD_{T^*} \\ D_{T'}A & D \end{bmatrix} : H \oplus D_{T^*} \to H' \oplus D_{T'} \tag{4.4}$$

is a contraction. This is precisely the set up in Theorem IV.3.1.

Using $T'A = AT$ we have

$$\|D_{AT}h\|^2 = \|h\|^2 - \|ATh\|^2 = \|h\|^2 - \|Ah\|^2 + \|Ah\|^2 - \|T'Ah\|^2 =$$
$$\|D_A h\|^2 + \|D_{T'}Ah\|^2 \qquad (h \in H). \tag{4.5}$$

This implies that $\|D_{T'}Ah\| \le \|D_{AT}h\|$ for all h in H. So there exists a contraction Y mapping D_{AT} into $D_{T'}$ satisfying

$$D_{T'}A = YD_{AT}. \tag{4.6}$$

The following calculation is the dual to (4.5):

$$\|D_{T^*A^*}h\|^2 = \|h\|^2 - \|T^*A^*h\|^2 = \|D_{A^*}h\|^2 + \|D_{T^*}A^*h\|^2 \qquad (h \in H'). \tag{4.7}$$

This implies that $\|D_{T^*}A^*h\| \le \|D_{T^*A^*}h\|$ for all h in H'. So there exists a contraction X mapping D_{T^*} into $D_{T^*A^*}$ satisfying $D_{T^*}A^* = X^*D_{T^*A^*}$, or equivalently,

$$AD_{T^*} = D_{(AT)^*}X. \tag{4.8}$$

Theorem IV.3.1 along with (4.4), (4.6), (4.8) and the above analysis readily yields the following result.

4.1 PROPOSITION. *Let A be in $I_1(T,T')$. Let Y and X be the contractions defined in (4.6) and (4.8), respectively. Then the set of all contractive one step intertwining liftings A_1 of A is given by (4.3), where*

$$D = D_{Y^*} \Gamma D_X - Y T^* A^* X \tag{4.9}$$

and Γ is an arbitrary contraction from D_X into D_Y.

4.2 LEMMA. *Let A be a strict contraction in $I(T,T')$. Then both X and Y are strict contractions, or equivalently, both D_X and D_Y are invertible.*

PROOF. Equations (4.5) and (4.6) give:

$$\|D_Y D_{AT} h\|^2 = \|D_{AT} h\|^2 - \|Y D_{AT} h\|^2 = \|D_A h\|^2 \qquad (h \in H) \, .$$

We claim that Y is a strict contraction, or equivalently, D_Y is invertible. If D_Y is not invertible, then there exists a sequence h_n in H such that $\|D_{AT} h_n\| = 1$ and

$$\|D_A h_n\| = \|D_Y D_{AT} h_n\| \to 0 \text{ as } n \to \infty \, .$$

Since A is a strict contraction this implies that $h_n \to 0$. Thus $D_{AT} h_n \to 0$ which is a contradiction. Therefore D_Y is invertible, or equivalently, Y is a strict contraction. A similar argument involving (4.7) and (4.8) shows that X is a strict contraction. This completes the proof.

4.3 COROLLARY. *Let A be a strict contraction in $I(T,T')$ and A_1 a contractive one step intertwining lifting of A. Then A_1 is a strict contraction if and only if the contraction Γ defined in (4.9) is a strict contraction.*

PROOF. Obviously AT is a strict contraction. By Lemma 4.2 both X and Y are strict contractions. Taking into account Proposition 4.1, we can apply Corollary IV.3.3 to AT, X and Y. It follows that A_1 is a strict contraction if and only if Γ is a strict contraction. This completes the proof.

To complete this section, we will convert (4.9) to a more usable form involving the unitary operator ω defined in Section 1. To this end notice that (4.7) implies that there exists a unitary operator γ_* mapping $F_* \overset{\Delta}{=} F(T^* A^*)$ onto $D_{T^* A^*}$ satisfying

$$\gamma_*^* D_{T^* A^*} = \begin{bmatrix} D_{T^* A^*} \\ D_{A^*} \end{bmatrix} \, ,$$

or equivalently,

$$D_{T^* A^*} \gamma_* = [A D_{T^*}, D_{A^*}] \,|\, F_* \, . \tag{4.10}$$

(Here we set $F_* = F(T^* \cdot A^*)$ and $G_* = G(T^* \cdot A^*)$; see (1.20).) This and (4.8) give

$$\gamma_*^* X = P_{F_*} | D_{T^*} \ . \tag{4.11}$$

For d in D_{T^*} this yields:

$$\|D_X d\|^2 = \|d\|^2 - \|Xd\|^2 = \|d\|^2 - \|P_{F_*} d\|^2 = \|P_{G_*} d\|^2 \ . \tag{4.12}$$

By (4.12) there exists an isometry α mapping D_X into G_* satisfying

$$\alpha D_X = P_{G_*} | D_{T^*} \ . \tag{4.13}$$

Recall that $G_* = G(T^* \cdot A^*)$ in (2.5), namely $\overline{P_{G_*} D_{T^*}} = G_*$. Therefore α is onto G_*, that is, α is unitary.

Since R_T is unitary, the following product of two isometries

$$\begin{bmatrix} AT & AD_{T^*} \\ D_A T & D_A D_{T^*} \\ D_T & -T^* \end{bmatrix} = \begin{bmatrix} A & 0 \\ D_A & 0 \\ 0 & I \end{bmatrix} \begin{bmatrix} T & D_{T^*} \\ D_T & -T^* \end{bmatrix} \tag{4.14}$$

is also an isometry. This and Corollary IV.1.5 for $[AT, AD_{T^*}]$ imply that there exists a unitary operator γ mapping $D_{AT} \oplus D_X$ onto $D_A \oplus D_T$ satisfying

$$\gamma \begin{bmatrix} D_{AT} & -T^* A^* X \\ 0 & D_X \end{bmatrix} = \begin{bmatrix} D_A T & D_A D_{T^*} \\ D_T & -T^* \end{bmatrix} \ . \tag{4.15}$$

Identifying D_{AT} and D_X with $D_{AT} \oplus \{0\}$ and $\{0\} \oplus D_X$, respectively we have

$$\gamma D_{AT} = \begin{bmatrix} D_A T \\ D_T \end{bmatrix} \ . \tag{4.16}$$

Therefore $\gamma D_{AT} = F$ and $\gamma D_X = G$ where F and G are defined in (1.4). Equations (4.6), (4.16) and the definition of the unitary operator ω in (1.12) give

$$Y \gamma^* | F = \Pi' \omega | F \tag{4.17}$$

where Π' is defined in (1.15). For d in $D_{T'}$ this yields:

$$\|D_{Y^*} d\|^2 = \|d\|^2 - \|Y^* d\|^2 = \|d\|^2 - \|P_{F'} d\|^2 = \|P_{G'} d\|^2$$

where $G' = G(T'A)$ is defined in (1.13) or (1.20). Equation (1.14) implies that there exists a unitary operator α_* mapping G' onto D_{Y^*} such that $\alpha_*^* D_{Y^*} = P_{G'} | D_{T'}$, or equivalently,

$$D_{Y^*} \alpha_* = \Pi' | G' \ . \tag{4.18}$$

Finally $\gamma^* F = D_{AT}$ and (4.17), (4.15) give

$$-YT^* A^* X = [Y,\ 0] \begin{bmatrix} -T^* A^* X \\ D_X \end{bmatrix} =$$

$$- [Y,\ 0] \gamma^* \gamma \begin{bmatrix} -T^* A^* X \\ D_X \end{bmatrix} = \Pi' \omega P_F \begin{bmatrix} D_A D_{T^*} \\ -T^* \end{bmatrix} \tag{4.19}$$

From Proposition 4.1 and (4.9), (4.13), (4.18), (4.19) we infer the following result.

4.4 THEOREM. *Let* A *be in* $I_1(T,\ T')$. *Then the set of all contractive one step intertwining liftings* A_1 *of* A *is given by* (4.3), *where*

$$D = \Pi' \Gamma_* P_G. \,| D_{T^*} + \Pi' \omega P_F \begin{bmatrix} D_A D_{T^*} \\ -T^* \end{bmatrix} \tag{4.20}$$

and Γ_* *is an arbitrary contraction from* G_* *into* G'. *In particular, there is a one to one correspondence between the set of all contractive one step intertwining liftings* A_1 *of* A *and the set of all contractions from* G_* *to* G'.

Corollary 2.3 shows that A·T is a regular factorization if and only if $A^* \cdot T^*$ is a regular factorization. This and the previous theorem also proves Corollary 1.4, that is, A *admits a unique contractive one step intertwining lifting if and only if* A·T *or* $T' \cdot A$ *is a regular factorization.*

We notice that

$$\left\{ P_G \begin{bmatrix} D_A D_{T^*} \\ -T^* \end{bmatrix} D_{T^*} \right\}^- = G. \tag{4.21}$$

To see this recall that the unitary operator γ defined in (4.15) is onto $D_A \oplus D_T$. Since the first column in the second matrix in (4.15) spans F and $F \oplus G = D_A \oplus D_T$ the closure of P_G on the second column must give G, that is, (4.21) holds.

Let β be the unitary operator $\gamma \alpha^*$ mapping G_* onto G. Then

$$\beta P_{G_*} .\,| D_{T^*} = P_G \begin{bmatrix} D_A D_{T^*} \\ -T^* \end{bmatrix} |\, D_{T^*}. \tag{4.22}$$

Indeed using $\gamma D_X = G$ with (4.15), (4.13) and d in D_{T^*} we have

$$P_G(D_A D_{T^\cdot} d \oplus -T^* d) = \gamma P_{D_X}(-T^* A^* X d \oplus D_X d) = \gamma(0 \oplus D_X d) = \gamma \alpha^* P_{G^\cdot} d \, .$$

This proves (4.22).

To complete this section we will show that the two representations of the set of all contractive one step intertwining liftings A_1 of A in Theorem 1.2 and Theorem 4.4 are identical up to a unitary operator on the contractions Γ and Γ_*.

4.5 PROPOSITION. *Let A_1 be a contractive one step intertwining lifting of A. Let Γ and Γ_* be the contractions corresponding to the representation of A_1 in (1.16) and (4.20) respectively. Then $\Gamma\beta = \Gamma_*$ where β is the unitary operator from G_* onto G defined in (4.22).*

PROOF. Since (1.16) and (4.3) represent the same A_1 we have

$$\Pi'(\omega P_F + \Gamma P_G)(D_A \oplus I)R_T = [D_{T'}A, \, D] \tag{4.23}$$

where D is defined in (4.20). The Equation (4.23) with the definition of ω in (1.12) and h in H give

$$D_{T'}Ah = \Pi'(\omega P_F + \Gamma P_G)\,(D_A \oplus I)R_T(h \oplus 0) \, .$$

So using d in $D_{T^\cdot}$ in (4.23) along with (4.22) we obtain

$$Dd = \Pi'(\omega P_F + \Gamma P_G)\,(D_A \oplus I)R_T(0 \oplus d) =$$

$$\Pi'(\omega P_F + \Gamma P_G)(D_A D_{T^\cdot} d \oplus -T^* d) =$$

$$\Pi'\Gamma\beta P_{G^\cdot} d + \Pi'\omega P_F(D_A D_{T^\cdot} d \oplus -T^* d) \, .$$

Comparing this with the D in (4.20) yields

$$\Pi'\Gamma\beta P_{G^\cdot}\,|D_{T^\cdot} = \Pi'\Gamma_* P_{G^\cdot}\,|D_{T^\cdot} \, . \tag{4.24}$$

Equations (1.14) and (2.5) imply that $\Pi'\,|G'$ is one to one and the range of $P_{G^\cdot}\,|D_{T^\cdot}$ is onto a dense set in G_*. Thus (4.24) implies that $\Gamma\beta = \Gamma_*$. This completes the proof.

5. THE SCHUR-COHN TEST REVISITED

In this section we will use some simple properties of contractions to obtain and generalize the Schur-Cohn test for the stability of a polynomial in Theorem I.7.1.

Throughout this section

$$p_n(z) = c_o + c_1 z + \cdots + c_n z^n \tag{5.1}$$

is a polynomial of degree n and $p_n^{\#} \triangleq J_{n+1} p_n$ is its reverse polynomial defined in Section I.2. Recall that p_n is a *stable* polynomial if all its zeros are in the open unit disc. If T is an operator on H, then $p(T)$ is the operator on H obtained by replacing z by T in (5.1). This sets the stage for the following result due to Pták and Young [PtY 1].

5.1 THEOREM. *Let* p_n *be a polynomial of degree* n. *Let* T *be a contraction on* $\mathbb{C}^n$ *whose defect index* $\delta_T = 1$. *If*

$$p_n^{\#}(T)^* p_n^{\#}(T) - p_n(T)^* p_n(T) > 0 , \tag{5.2}$$

then p_n *is a stable polynomial.*

PROOF. Without loss of generality we can assume that $c_n = 1$. Dividing through by c_n does not effect the zeros of $p_n(z)$ and does not change the inequality in (5.2). In this case

$$p_n(z) = \prod_{i=1}^{n}(z - \alpha_i) \quad \text{and} \quad p_n^{\#}(z) = \prod_{i=1}^{n}(1 - \bar{\alpha}_i z) \tag{5.3}$$

where α_i are the zeros of $p_n(z)$. Now let

$$B_i = (I - \bar{\alpha}_i T) \quad \text{and} \quad N_i = (T - \alpha_i I) \quad (\text{for } 1 \le i \le n) \tag{5.4}$$

and notice that for all x in $\mathbb{C}^n$

$$\|B_i x\|^2 - \|N_i x\|^2 = \|(I - \bar{\alpha}_i T)x\|^2 - \|(T - \alpha_i I)x\|^2 =$$

$$\|x\|^2 - 2\operatorname{Re}(\bar{\alpha}_i Tx, x) + |\alpha_i|^2 \|Tx\|^2 - \|Tx\|^2 + 2\operatorname{Re}(Tx, \alpha_i x) - \|\alpha_i x\|^2 =$$

$$(1 - |\alpha_i|^2)(\|x\|^2 - \|Tx\|^2) = (1 - |\alpha_i|^2)\|D_T x\|^2 .$$

Thus

$$\|B_i x\|^2 - \|N_i x\|^2 = (1 - |\alpha_i|^2)\|D_T x\|^2 \quad (x \in \mathbb{C}^n) . \tag{5.5}$$

If n=1, then (5.5) is equivalent to the following formulas which plays a central role in the proof:

$$\|p_n^{\#}(T)x\|^2 - \|p_n(T)x\|^2 = \sum_{1}^{n}(1 - |\alpha_i|^2)\|C_i x\|^2 \quad (x \in \mathbb{C}^n) \tag{5.6}$$

where

$$C_i = D_T N_1 N_2 \cdots N_{i-1} B_{i+1} B_{i+2} \cdots B_n \qquad (1 \le i \le n) . \tag{5.7}$$

One obtains (5.6) by recursively applying (5.5). To see this it is sufficient to consider the case when n=3. The general case follows by a similar argument. So using (5.5) with the fact that N_i and B_i commute we have for all x in $\mathbb{C}^n$

$$\|p_3^\#(T)x\|^2 - \|p_3(T)x\|^2 = \|B_1 B_2 B_3 x\|^2 - \|N_1 N_2 N_3 x\|^2 =$$

$$\|B_1 B_2 B_3 x\|^2 - \|N_1 B_2 B_3 x\|^2 + \|N_1 B_2 B_3 x\|^2 - \|N_1 N_2 N_3 x\|^2 =$$

$$(1 - |\alpha_1|^2)\|D_T B_2 B_3 x\|^2 + \|B_2 N_1 B_3 x\|^2 - \|N_2 N_1 B_3 x\|^2 + \|N_2 N_1 B_3 x\|^2 - \|N_1 N_2 N_3 x\|^2 =$$

$$(1 - |\alpha_1|^2)\|D_T B_2 B_3 x\|^2 + (1 - |\alpha_2|^2)\|D_T N_1 B_3 x\|^2 + \|B_3 N_1 N_2 x\|^2 - \|N_3 N_1 N_2 x\|^2 =$$

$$\sum_1^3 (1 - |\alpha_i|^2)\|D_T C_i x\|^2 .$$

As noted earlier a similar argument establishes (5.6).

If (5.2) holds, then (5.6) is strictly positive for all nonzero x. Since $\delta_T = 1$ the rank of C_i is at most one. So we can choose a nonzero x such that $C_j x = 0$ for all $j \ne i$. Since (5.6) is strictly positive this implies that $(1 - |\alpha_i|^2)\|C_i x\|^2 > 0$. Therefore α_i is in the open unit disc for all i, or equivalently, p_n is stable. This completes the proof.

If $T = T_n$ is the lower shift matrix on $\mathbb{C}^n$ in (I.4.1) with $j = n$, then T is a contraction of rank one. Therefore the previous theorem proves part of the Schur-Cohn test in Theorem I.7.1. Finally, it is noted that when using Theorem 5.1 one must always choose a contraction T such that all the eigenvalues of T are in the open unit disc D. If $Tx = \lambda x$ where $|\lambda| = 1$ and x is nonzero, then

$$\|p_n^\#(T)x\| = \|p_n^\#(\lambda)x\| = \|p_n(\lambda)x\| = \|p_n(T)x\|$$

and (5.2) does not hold. To complete this section we will obtain a converse to the previous theorem. This begins with the following result which a consequence of the definition of a regular factorization ($\delta_{AT} = \dim F(A{\cdot}T)$) and Corollary 2.7.

5.2 LEMMA. *Let A and T be commuting contractions on a finite dimensional space. Assume that all the eigenvalues of T are in D, its defect index $\delta_T = 1$ and the norm of A is one. Then $\delta_{AT} = \delta_T + \delta_A$.*

Now we are ready to prove the following converse to Theorem 5.1.

5.3 THEOREM. *Let p_n be a polynomial of degree n. Let T be a contraction on C^n such that $\delta_T = 1$ and all the eigenvalues of T are in D. Then p_n is a stable polynomial of degree n if and only if (5.2) holds.*

PROOF. Assume that p_n is a polynomial of degree n. Without loss of generality, we assume that the coefficient of z^n is one and p_n is given by (5.3) where all the α_i's are in D. Because α_i is in D we see that B_i^{-1} exists. So we set $W_i = N_i B_i^{-1}$. Notice that $W_i = b_i(T)$ where b_i is the Blaschke product of order one whose zero is at α_i. Moreover, the spectral mapping theorem with the fact that $b_i(z)$ maps D onto D tells us that the eigenvalues of W_i are in D. Using $y = B_i x$ in (5.5) we have

$$\|D_{W_i} y\|^2 = \|y\|^2 - \|W_i y\|^2 = (1 - |\alpha_i|^2)\|D_T B_i^{-1} y\|^2 \quad (y \in C^n) . \tag{5.8}$$

This implies that $\delta_{W_i} = 1$. Therefore W_i is a contraction with eigenvalues in D and $\delta_{w_i} = 1$. Obviously W_i and W_j commute. Applying Lemma 5.2 we see that $\delta_{W_1 W_2} = \delta_{W_1} + \delta_{W_2} = 2$. If $2 < n$, then this shows that there exists a nonzero x such that $D_A x = 0$, or equivalently, $\|x\| = \|Ax\|$ where $A = W_2 W_3$. Moreover, by the spectral mapping theorem the eigenvalues of A are in D. Applying Lemma 5.2 again $\delta_{W_1 W_2 W_3} = \delta_{W_1} + \delta_{W_2 W_3} = 3$. Continuing in this fashion the defect index for $W_1 W_2 \cdots W_n$ is n. Therefore

$$\|(p_n/p_n^\#)(T)x\| = \|W_1 W_2 \cdots W_n x\| < \|x\| \quad \text{(for all } x \neq 0 \text{ in } C^n) . \tag{5.9}$$

Replacing x by $p_n^\#(T)y$ produces (5.2). This completes the proof.

By setting $T = T_n$ the lower shift on C^n the previous theorem reduces to the Schur-Cohn test in Theorem I.7.1. Finally, it is noted that the proof of the previous theorem also yields the following result.

5.4 COROLLARY. *Let T be a contraction on a finite dimensional space H such that $\delta_T = 1$ and all the eigenvalues of T are in D. Let B_m be a Blaschke product of order m where $m \leq \dim H$. Then the defect index of $B_m(T)$ is m.*

V.6. NOTES AND COMMENTS

The existence of a contractive one step intertwining lifting was established implicitly in [Sz.-NF 7] and explicitly in [DoMP]. The representations of all these liftings presented in Lemma 1.1 and Theorem 1.2 were given in [CeF 2] and [ArsCF 1,2]. The uniqueness characterization in Corollary 1.4 was obtained in [AnCF]. Originally regular factorizations were introduced in the study of the connection between proper invariant subspaces for a contraction and the factorization of its characteristic function [Sz.-NF 5]. The representation of all contractive intertwining liftings given in Theorem 4.4 is new and constitutes an elaboration of some of the results in [Frz 8]. Theorem 5.1 as well as its proof is taken from [PtY 1]. For related results concerning the Schur-Cohn test see [Co], [Par], [PtY 2], [VK] and [Y 1,2]. In connection with this chapter see the monographs [Ni] and [Po].

CHAPTER VI

ISOMETRIC AND UNITARY DILATIONS

In this chapter we study isometric and unitary dilations. These concepts are due to Sz.-Nagy and they have no analogue in linear algebra, that is, they are genuinely infinite dimensional concepts. Indeed their natural framework is that of infinite dimensional Hilbert spaces. From now on we assume that the reader is acquainted with elementary facts concerning operators in Hilbert space, as presented in [RieszSz.-N].

1. THE WOLD DECOMPOSITION

In this section we present some elementary facts concerning isometric operators, in particular, their Wold decomposition.

To begin we establish some notation. The closed linear span of a set of subspaces $\{M_n : i \leq n \leq j\}$ where i and j are integers is denoted by

$$\bigvee_i^j M_n \, .$$

The integers i or j can be $-\infty$ or $+\infty$, respectively. The orthogonal direct sum of these spaces is denoted by

$$\bigoplus_i^j M_n$$

This means that if $\bigoplus_i^j m_n$ is in $\bigoplus_i^j M_n$ then the square of its norm is given by

$$\left\| \bigoplus_i^j m_n \right\|^2 = \sum_i^j \|m_n\|^2 \, .$$

Finally, we say that two operators T on H and T' on H' are *unitarily equivalent* if there exists a unitary operator W from H onto H' satisfying $WT = T'W$.

Let U be an isometry on K. A subspace L is *wandering* for U if $U^n L$ is orthogonal to $U^m L$ for all $n \geq 0$ and $m \geq 0$ when $n \neq m$. Since $U^*U = I$, this is equivalent to L being orthogonal to $U^n L$ for all $n \geq 1$. If L is wandering for U, then $M_+(L)$ is the following subspace generated by L

$$M_+(L) = \overset{\infty}{\underset{0}{\oplus}} U^n L = \overset{\infty}{\underset{0}{\bigvee}} U^n L \ .$$

We say that S on K is a *unilateral shift* if S is an isometry and S has a cyclic wandering subspace L, that is, $K = M_+(L)$. (Recall that a subspace L is *cyclic* for an operator C on K if $K = \bigvee \{C^n L : n \geq 0\}$.) Notice that

$$K \ominus SK = (\overset{\infty}{\underset{0}{\oplus}} S^n L) \ominus (\overset{\infty}{\underset{1}{\oplus}} S^n L) = L \ .$$

So if S is a unilateral shift, then it has a unique cyclic wandering subspace L, and this subspace is given by $L = K \ominus SK$. The dimension of L is the *multiplicity* of S. The simplest example of a unilateral shift is the following lower shift matrix S on $l^2(L)$, the space of all square summable unilateral sequences with entries in L

$$S = \begin{bmatrix} 0 & 0 & 0 & 0 & 0 & . & . & . & . \\ I & 0 & 0 & 0 & 0 & . & . & . & . \\ 0 & I & 0 & 0 & 0 & . & . & . & . \\ 0 & 0 & I & 0 & 0 & . & . & . & . \\ 0 & 0 & 0 & I & 0 & . & . & . & . \\ 0 & 0 & 0 & 0 & I & . & . & . & . \\ \vdots & \vdots & \vdots & \vdots & \vdots & \vdots & \vdots & \vdots & \vdots \end{bmatrix} \text{ on } l^2(L) \ . \tag{1.1}$$

Two unilateral shifts S on K and S' on K' are unitarily equivalent if and only if they have the same multiplicity. In particular, S on $K = M_+(L)$ is unitarily equivalent to the lower shift matrix in (1.1).

To prove this result assume that S and S' are unitarily equivalent. Then the cyclic wandering subspace $L = K \ominus SK$ and $L' = K' \ominus S'K'$ have the same dimension. On the other hand, if S and S' have the same multiplicity, then there exists a unitary operator ϕ mapping L onto L', where L and L' are the cyclic wandering subspaces for S and S', respectively. Let us extend ϕ by linearly to an operator W mapping $K = M_+(L)$ onto $K' = M_+(L')$ by setting $WS^n l = S'^n \phi l$, where $n \geq 0$ and l is in L. Using $WSS^n l = S'^{n+1} \phi l = S'WS^n l$ with linearly it is easy to see that $S'W = WS$. Moreover, for l_n in L we have

$$\left\| W \overset{\infty}{\underset{0}{\sum}} S^n l_n \right\|^2 = \left\| \overset{\infty}{\underset{0}{\sum}} S'^n \phi l_n \right\|^2 = \overset{\infty}{\underset{0}{\sum}} \| l_n \|^2 = \left\| \overset{\infty}{\underset{0}{\sum}} S^n l_n \right\|^2 \ .$$

Therefore W is a unitary operator mapping $M_+(L)$ onto $M_+(L')$ and S and S' are unitarily equivalent. This completes the proof.

Let S be a unilateral shift on $K = M_+(L)$. We claim that $S^{*n} \to 0$ strongly as $n \to \infty$. To prove this first notice that $S^{*n} S^j L = 0$ for all $n > j$. This follows from the wandering property of L:

$$(S^{*n} S^j l, \, m) = (S^j l, \, S^n m) = 0 \, . \qquad \text{(for all } l \in L \text{ and } m \in M_+(L))$$

Now let l_j be in L for $j \geq 0$. Then

$$\|S^{*n} \sum_0^\infty S^j l_j\|^2 = \sum_{j \geq n} \|S^{*n} S^j l_j\|^2 = \sum_{j \geq n} \|l_j\|^2 \to 0 \quad \text{(as } n \to \infty) \, .$$

Since $\{l_j\}$ is square summable and $K = M_+(L)$ this implies that $S^{*n} \to 0$ strongly, which proves our claim.

Recall that a subspace H_0 *reduces* an operator A on H if H_0 is invariant to both A and A^*. Obviously H_0 reduces A if and only if its orthogonal complement $H_1 = H \ominus H_0$ reduces A. The notation $A = A_0 \oplus A_1$ on $H_0 \oplus H_1$ means that H_0 or equivalently H_1 reduces A and $H = H_0 \oplus H_1$; moreover, A_0 on H_0 and A_1 on H_1 are the operators defined by $A_0 = A|H_0$ and $A_1 = A|H_1$, respectively. The notation $A_0 \oplus A_1$ is equivalent to A admitting a matrix representation of the form:

$$A = \begin{bmatrix} A_0 & 0 \\ 0 & A_1 \end{bmatrix} \quad \text{on } H_0 \oplus H_1 \, .$$

The following theorem is known as the *Wold decomposition* of an isometry. It plays a fundamental role in operator theory and stochastic processes.

1.1 THEOREM. *(Wold Decomposition). Let* U *be an isometry on* K. *Then* U *admits a unique decomposition of the form* $U = S \oplus W$ *on* $K_s \oplus K_u$ *where* S *is a unilateral shift and* W *is unitary. By unique we mean that if* $U = S' \oplus W'$ *on* $K'_s \oplus K'_u$ *where* S' *is a unilateral shift and* W' *is unitary, then* $K_s = K'_s$ *and* $K_u = K'_u$, *that is,* $S = S'$ *and* $W = W'$. *Moreover, the spaces* K_s *and* K_u *are given by*

$$K_s = M_+(L) \text{ where } L = K \ominus UK \text{ and } K_u = \bigcap_0^\infty U^n K \, . \tag{1.2}$$

Finally, the spaces K_s *or* K_u *may be zero.*

PROOF. Let $L = K \ominus UK$. Obviously L is orthogonal to UK. Since UK contains $U^n L$ for all $n \geq 1$, this implies that L is orthogonal to $U^n L$ for all $n \geq 1$. Hence L is wandering for U. By recursively applying the definition of L we have

$$K = L \oplus UK = L \oplus UL \oplus U^2 K = \cdots = (\overset{n-1}{\underset{0}{\oplus}} U^i L) \oplus (U^n K) \quad (n \geq 1).$$

This shows that h in K is orthogonal to $M_+(L)$ if and only if h is in $U^n K$ for all $n \geq 1$. Because $U^n K$ is a decreasing sequence of subspaces h is orthogonal to $M_+(L)$ if and only if h is in the space K_u defined in (1.2). This yields the decomposition $K = K_s \oplus K_u$ where $K_s = M_+(L)$. Obviously both K_s and K_u are invariant for U. Therefore K_s, or equivalently, K_u reduces U. Clearly $S = U | K_s$ is a unilateral shift. Since

$$UK_u = U(\overset{\infty}{\underset{0}{\cap}} U^n K) = \overset{\infty}{\underset{0}{\cap}} U^{n+1} K = K_u$$

$W = U | K_u$ is unitary. This proves the decomposition of U into $S \oplus W$, where S is a unilateral shift and W is unitary.

Now assume that $U = S' \oplus W'$ on $K_s' \oplus K_u'$ is another decomposition of U, where S' is unilateral shift and W' is unitary. Using $K_s' = M_+(L')$ where L' is the cyclic wandering subspace for S' we have

$$L = K \ominus UK = K_s' \ominus S' K_s' = L'.$$

Thus S and S' have the same cyclic wandering subspace L. Therefore $K_s = K_s'$ and $S = S'$. This completes the proof.

Recall that two operators T on H and T' on H' are *similar* if there exists an invertible operator X from H onto H' satisfying $T'X = XT$. To complete this section we will use the Wold decomposition to show that two isometries which are similar are unitarily equivalent. This begins with the following result.

1.2 LEMMA. *Two unitary operators are similar if and only if they are unitarily equivalent.*

PROOF. Let W on H and V on G be two unitary operators satisfying $VX = XW$ where X is an invertible operator mapping H onto G. Thus $WX^* = X^*V$ and $X^*XW = WX^*X$. For every polynomial p(z) this implies that

$$p(X^*X)W = Wp(X^*X).$$

By passing limits we have $RW = WR$, where R is the positive square root of X^*X. Obviously R is an invertible operator on H. The polar decomposition of X shows that X $= \Omega R$, where Ω is a unitary operator from H onto G. Therefore

$$V\Omega = V\Omega RR^{-1} = \Omega RWR^{-1} = \Omega RR^{-1}W = \Omega W .$$

This completes the proof.

1.3 PROPOSITION. *Two isometries are similar if and only if they are unitarily equivalent.*

PROOF. Let U on K and U' on K' be two (similar) isometries, and X an invertible operator from K onto K' satisfying $U'X = XU$. Let $K = K_s \oplus K_u$ and $K' = K'_s \oplus K'_u$ be the Wold decomposition of U and U', respectively. Notice that

$$X^* \ker(U'^*) = \ker(U^*) .$$

Using $\ker(U'^*) = K' \ominus U'K'$ and $\ker(U^*) = K \ominus UK$ with the Wold decompositions of U and U', we see that the multiplicity of $S = U|K_s$ equals the multiplicity of $S' = U'|K'_s$. Therefore the shifts S and S' are unitarily equivalent. On the other hand, since X is a bijection

$$XK_u = X \bigcap_0^\infty U^n K = \bigcap_0^\infty XU^n K = \bigcap_0^\infty U'^n XK = \bigcap_0^\infty U'^n K' = K'_u .$$

So the unitary parts $W = U|K_u$ and $W' = U'|K'_u$ are also similar. By the previous lemma W and W' are unitarily equivalent. It follows that $U = S \oplus W$ and $U' = S' \oplus W'$ are unitarily equivalent. This completes the proof.

2. BILATERAL SHIFTS

In this section we study bilateral shifts and unitary extensions of isometries.

To this end let U be a unitary operator on K. Since $UU^* = I$ a subspace L is wandering for U if and only if $U^n L$ is orthogonal to $U^m L$ for all integers $n \neq m$. In this case we define the subspace $M(L)$ by

$$M(L) = \bigoplus_{-\infty}^\infty U^n L = \bigvee_{-\infty}^\infty U^n L .$$

Obviously $M(L)$ reduces U. Notice that $M_+(L)$ uniquely determines L by $L = M_+(L) \ominus UM_+(L)$. The space $M(L)$ does not uniquely determine L. For instance $M(L) = M(UL)$. The following proposition shows that $M(L)$ determines the dimension of L.

2.1 PROPOSITION. *Let $M(L) \supseteq M(L')$ where L and L' are wandering subspaces for a unitary operator U on K. Then $\dim(L) \geq \dim(L')$. If L' is finite dimensional and*

$\dim(L) = \dim(L^{'})$, *then* $M(L) = M(L^{'})$.

PROOF. First assume that $\dim L \geq \chi_o$ where χ_o is the dimension of the integers. Then

$$\chi_o \dim(L) = \dim M(L) \geq \dim M(L^{'}) \geq \chi_o \dim(L^{'}) .$$

This implies that $\dim (L) \geq \dim (L^{'})$. Now assume that L is finite dimensional. Let $\{\phi_i\}_1^n$ and $\{\phi_i^{'}\}_1^m$ be a complete orthonormal sequence for L and $L^{'}$, respectively. Notice that $\{U^j\phi_i : 1 \leq i \leq n$ and $-\infty < j < \infty\}$ and $\{U^j\phi_i^{'}: 1 \leq i \leq m$ and $-\infty < j < \infty\}$ is a complete orthonormal sequence for $M(L)$ and $M(L^{'})$ respectively. Using Bessel's inequality and Parseval's equality we have

$$\dim L = \sum_1^n \|\phi_i\|^2 \geq \sum_{i,j,k} |(\phi_i, U^j\phi_k^{'})|^2 =$$

$$\sum_{i,j,k} |(U^{*j}\phi_i, \phi_k^{'})|^2 = \sum_k \|\phi_k^{'}\|^2 = \dim(L^{'}) .$$

This proves the inequality $\dim L \geq \dim L^{'}$. Moreover, equality holds if and only if $\{\phi_i\}_1^n$ is contained in $M(L^{'})$, or equivalently, $M(L) \subseteq M(L^{'})$. This completes the proof.

An operator $\hat{S}$ on M is a *bilateral shift* if $\hat{S}$ is unitary and $M = M(L)$ where L is wandering for $\hat{S}$. The dimension of L is the *multiplicity* of $\hat{S}$. The previous proposition guarantees that the multiplicity is a positive number uniquely determined by $\hat{S}$, that is, independent of the wandering subspace L satisfying $K = M(L)$. The classical example of a bilateral shift is multiplication by e^{it} on $L^2(L)$, the Lebesgue space of all square intergable measurable functions on $[0, 2\pi)$ with values in L. Parseval's equality shows that by replacing all functions in $L^2(L)$ with the bilateral sequence of their Fourier coefficients, we obtain a unitary operator from $L^2(L)$ onto the space $\hat{l}^2(L)$ of all square summable bilateral sequences with entries in L. In this way the previous multiplication operator is unitarily equivalent to $\hat{S}$ on $\hat{l}^2(L)$ given by the infinite by infinite matrix with ones immediately below the main diagonal and zeros elsewhere. Finally, it is noted that two bilateral shifts are unitarily equivalent if and only if they have the same multiplicity. The proof is similar to the corresponding result for unilateral shifts and is omitted.

Let A be an operator mapping H into $H^{'}$. We say that an operator B mapping K into $K^{'}$ is an *extension of* A if $H \subseteq K$ and $H^{'} \subseteq K^{'}$ and $B|H = A$. Obviously B is an extension of A if and only if B admits a matrix representation of the form

$$B = \begin{bmatrix} A & D \\ 0 & C \end{bmatrix} : H \oplus H^\perp \to H' \oplus H'^\perp .$$

We say that two extensions B on K and B' on K' of A are *isomorphic* if there exists a unitary operator ϕ mapping K onto K' satisfying $B'\phi = \phi B$ and $\phi | H = I_H$. Recall that an operator B_* mapping K into K' is a *lifting of* A if $H \subseteq K$ and $H' \subseteq K'$ and $P_{H'} B_* = A P_H$. It is easy to see that B_* is a lifting of A if and only if B_* admits a matrix representation of the form

$$B_* = \begin{bmatrix} A & 0 \\ D & C \end{bmatrix} : H \oplus H^\perp \to H' \oplus H'^\perp .$$

Therefore B is an extension of A if and only if B^* is a lifting of A^*. We say that two liftings B and B' of A are *isomorphic* if B^* and B'^* are *isomorphic* when viewed as extensions of A^*. Our first example of an extension is given in the following proposition.

     2.2 PROPOSITION. *Any unilateral shift can be extended to a bilateral shift of the same multiplicity.*

     PROOF. Let S be a unilateral shift on $M_+(L)$. Consider the space M defined by

$$M = (\overset{-1}{\underset{-\infty}{\oplus}} L) \oplus M_+(L) .$$

Let $\hat{S}$ be the operator on M defined by

$$\hat{S}[(\overset{-1}{\underset{-\infty}{\oplus}} l_n) \oplus (\overset{\infty}{\underset{0}{\oplus}} S^n l_n)] = (\overset{-2}{\underset{-\infty}{\oplus}} l_n) \oplus (l_{-1} \oplus \overset{\infty}{\underset{0}{\oplus}} S^{n+1} l_n) \tag{2.1}$$

where l_n is in L for all n. It is easy to verify that this $\hat{S}$ is a bilateral extension of S. Moreover $M = M(L)$ and S and $\hat{S}$ have the same multiplicity. This completes the proof.

     The bilateral shift extension $\hat{S}$ on M of S in (2.1) satisfies

$$M = \overset{\infty}{\underset{-\infty}{\bigvee}} \hat{S}^n M_+(L) . \tag{2.2}$$

It is easy to show that all bilateral extensions $\hat{S}$ on K of S satisfying the cyclic condition in (2.2) are unitarily equivalent. Moreover, if S has finite multiplicity, then Proposition 2.1 shows that all bilateral extensions of S with the same multiplicity as S are unitarily equivalent.

     Let U be an isometry on H. Then $\hat{U}$ on $\hat{K}$ is a *minimal unitary extension of* U if $\hat{U}$ is an unitary extension of U and

$$\hat{K} = \bigvee_{-\infty}^{\infty} \hat{U}^{n} K . \tag{2.3}$$

The Wold decomposition $U = S \oplus W$ on $M_{+}(L) \oplus K_{u}$ guarantees that any isometry U admits a minimal unitary extension. In fact, this extension is $\hat{U} = \hat{S} \oplus W$ on $M(L) \oplus K_{u}$ where $\hat{S}$ is the bilateral shift extending S on $M_{+}(L)$ in (2.1). This proves the first part of the following result.

2.3 COROLLARY. *Any isometry* U *on* K *admits a minimal unitary extension. Furthermore, all minimal unitary extensions are unique up to an isomorphism.*

PROOF. It only remains to establish the uniqueness result. To this end let $\hat{U}'$ on $\hat{K}'$ be another minimal unitary extension of U satisfying the corresponding minimality condition in (2.3). For any sequence $\{h_{n}\}$ in K with finite support we have

$$\|\sum_{-\infty}^{\infty} \hat{U}^{n} h_{n}\|^{2} = \sum_{n,m} (\hat{U}^{n} h_{n}, \hat{U}^{m} h_{m}) = \sum_{n \geq m} (\hat{U}^{n-m} h_{n}, h_{m}) + \sum_{m > n} (h_{n}, \hat{U}^{m-n} h_{m}) =$$

$$\sum_{n \geq m} (U^{n-m} h_{n}, h_{m}) + \sum_{m > n} (h_{n}, U^{m-n} h_{m}) = \|\sum_{-\infty}^{\infty} \hat{U}'^{n} h_{n}\|^{2} . \tag{2.4}$$

This and the minimality condition in (2.3) implies that there exists a unitary operator ϕ mapping $\hat{K}$ onto $\hat{K}'$ satisfying

$$\phi \sum_{-\infty}^{\infty} \hat{U}^{n} h_{n} = \sum_{-\infty}^{\infty} \hat{U}'^{n} h_{n} \tag{2.5}$$

where h_{n} is a sequence in K with finite support. In particular,

$$\phi \hat{U} \sum_{-\infty}^{\infty} \hat{U}^{n} h_{n} = \sum_{-\infty}^{\infty} \hat{U}'^{n+1} h_{n} = \hat{U}' \phi \sum_{-\infty}^{\infty} \hat{U}^{n} h_{n} .$$

Thus $\phi \hat{U} = \hat{U}' \phi$. Obviously $\phi | H = I$. This completes the proof.

The previous corollary also shows that any co-isometry admits a unitary lifting, which is unique up to an isomorphism. (Recall that an operator C is a co-isometry if its adjoint C^{*} is an isometry.) Let T on H and T' on H' be contractions. Recall that $I(T, T')$ is the set of all (bounded) operators A mapping H into H' satisfying $AT = T' A$. The following shows that $I(U, U')$ can be extended to a subset in $I(\hat{U}, \hat{U}')$.

2.4 COROLLARY. *Let* $\hat{U}$ *on* $\hat{K}$ *be the minimal unitary extension of the isometry* U *on* K *and* $\hat{U}'$ *on* $\hat{K}'$ *be any unitary extension of the isometry* U' *on* K'. *Let* B *be in*

$I(U,U')$. *Then there is only one operator* $\hat{B}$ *in* $I(\hat{U},\hat{U}')$ *extending* B. *This operator is given by*

$$\hat{B} = \text{strong·} \lim_{n \to \infty} \hat{U}'^{*n} B \hat{U}^{n} P_{-n} \tag{2.6}$$

where P_{-n} *is the orthogonal projection on* $\hat{U}^{*n} K$. *Moreover, this extension preserves norm, that is,* $\|\hat{B}\| = \|B\|$. *Finally, if* B *is an isometry (unitary and* $\hat{U}'$ *is the minimal unitary extension of* U'), *then* $\hat{B}$ *is an isometry (unitary).*

PROOF. Let $\hat{B}$ be the operator defined in (2.6). To show that this definition makes sense assume that k is in K and $n > m$. Then

$$\hat{B}\hat{U}^{*m}k = \hat{U}'^{*n} B \hat{U}^{n} P_{-n} \hat{U}^{*m} k = \hat{U}'^{*n} B U^{n-m} k = \hat{U}'^{*m} B k . \tag{2.7}$$

This and the fact that $\{\hat{U}^{*n}K\}_0^\infty$ is dense in $\hat{K}$ implies that the limit in (2.6) exists, and $\hat{B}$ is well defined. Imposing $m = 0$ in (2.7) gives $\hat{B}|K = B$. Using $m+1$ in (2.7) shows that $\hat{B}\hat{U}^{*}\hat{U}^{*m}k = \hat{U}'^{*}\hat{B}\hat{U}^{*m}k$, or equivalently, $\hat{B}$ is in $I(\hat{U},\hat{U}')$. To show uniqueness let C in $I(\hat{U},\hat{U}')$ be another operator extending B. Then

$$\hat{B}\hat{U}^{*m}k = \hat{U}'^{*m} B k = \hat{U}'^{*m} C k = C \hat{U}^{*m} k \qquad (k \in K) .$$

It follows that $C = \hat{B}$ and $\hat{B}$ is unique. Notice that (2.7) also implies that $\|\hat{B}\| = \|B\|$. Finally, if B is an isometry then (2.7) gives

$$\|\hat{B}\hat{U}^{*m}k\| = \|Bk\| = \|\hat{U}^{*m}k\| .$$

Therefore $\hat{B}$ is isometric on the dense set $\{\hat{U}^{*n}K\}_0^\infty$, or equivalently, $\hat{B}$ is an isometry. If $\hat{U}'$ is the minimal unitary extension of U' and if B is unitary, then since B is onto, (2.7) shows that $\hat{B}$ is onto. This completes the proof.

3. MINIMAL ISOMETRIC DILATIONS

In this section we introduce and study the isometric dilation of a contraction.

Let B be an operator on K and H a subspace of K. The *compression of* B *to* H is the operator on H defined by $A = P_H B|H$. We say that B on K is a *dilation* of an operator A on H if

$$A^n = P_H B^n | H \qquad \text{(for all } n \geq 0) . \tag{3.1}$$

It follows that if p(z) is a polynomial, then (3.1) implies that $p(A) = P_H p(B)|H$.

On the other hand, a subspace H is *semi-invariant* for an operator B on K if $H \subseteq K$ and there exists two invariant subspaces M and N for B satisfying $H = N \ominus M$ where $M \subseteq N$. It is easy to show that H is semi-invariant for B if and only if B admits a matrix representation of the form

$$B = \begin{bmatrix} * & * & * \\ 0 & A & * \\ 0 & 0 & * \end{bmatrix} \text{ on } M \oplus H \oplus N^{\perp} \tag{3.2}$$

where A is the compression of B to H. Obviously B is a dilation of A. This proves half of the following lemma.

3.1 LEMMA. *Let* B *be an operator on* K. *Then a subspace* H *is semi-invariant for* B *if and only if* B *is a dilation of* A *where* A *is the compression of* B *to* H.

PROOF. Assume that (3.1) holds. Let N be the following cyclic subspace generated by H

$$N = \bigvee_{0}^{\infty} B^n H .$$

Obviously N is invariant for B. To complete the proof it is sufficient to show that $M = N \ominus H$ is invariant for B. To this end notice that (3.1) yields:

$$P_H BB^n h = AA^n h = AP_H B^n h \qquad \text{(for all n$\geq$0 and h in } H).$$

Hence $P_H B \,|\, N = AP_H \,|\, N$, or equivalently, $A^* = P_N B^* \,|\, H$. Using this along with the fact that $M (\subseteq N)$ is orthogonal to H we have

$$(Bm, h) = (m, B^* h) = (m, P_N B^* h) = (m, A^* h) = 0 \qquad (m \in M \text{ and } h \in H)$$

Therefore BM is orthogonal to H. Since N is invariant for B and $M \subseteq N$, this implies that M is invariant for B. The proof is now complete.

Let T be an operator on H. Let U on K and U' on K' be two dilations of A. Then we say that U and U' are *isomorphic* if there exists a unitary operator W mapping K onto K' satisfying

$$WU = U'W \text{ and } W\,|\,H = I_H . \tag{3.3}$$

If T is a contraction on H, then T admits an isometric dilation. To see this let U be the operator defined by

$$U = \begin{bmatrix} T & 0 & 0 & 0 & 0 & . & . & . & . \\ D_T & 0 & 0 & 0 & 0 & . & . & . & . \\ 0 & I & 0 & 0 & 0 & . & . & . & . \\ 0 & 0 & I & 0 & 0 & . & . & . & . \\ 0 & 0 & 0 & I & 0 & . & . & . & . \\ 0 & 0 & 0 & 0 & I & . & . & . & . \\ \vdots & \vdots & \vdots & \vdots & \vdots & \vdots & \vdots & \vdots & \vdots \end{bmatrix} \quad \text{on } H \oplus l^2(D_T). \tag{3.4}$$

Obviously U is a dilation of T. A simple calculation shows that $U^*U = I$. Therefore U is an isometric dilation of T. An isometric dilation U on K of T is *minimal* if H is cyclic for U, that is,

$$K = \bigvee_{0}^{\infty} U^n H, \tag{3.5}$$

or equivalently, the smallest invariant subspace of U containing H is the whole space K. It is easy to verify that the matrix U in (3.4) is a minimal isometric dilation of T. The following shows that this U in (3.4) is the "only" minimal isometric dilation of T.

    3.2 **THEOREM.** *Any contraction* T *on H admits a minimal isometric dilation. Furthermore, all minimal isometric dilations of* T *are isomorphic.*

    PROOF. As noted earlier U in (3.4) is a minimal isometric dilation of T. The uniqueness proof is similar to the proof of Corollary 2.3. Let U' on K' be another minimal isometric dilation of T. If $\{h_n\}_0^\infty$ is a sequence in H with finite support, then

$$\|\textstyle\sum U^n h_n\|^2 = \sum (U^n h_n, U^m h_m) = \sum_{n \geq m} (U^{n-m} h_n, h_m) + \sum_{n < m} (h_n, U^{m-n} h_m)$$

$$= \sum_{n \geq m} (T^{n-m} h_n, h_m) + \sum_{n < m} (h_n, T^{m-n} h_m) = \|\textstyle\sum U'^n h_n\|^2.$$

The minimality guarantees that the finite sums $\sum U^n h_n$ and $\sum U'^n h_n$ are dense in K and K', respectively. Therefore there exists a unitary operator W mapping K onto K' such that

$$W \sum U^n h_n = \sum U'^n h_n$$

for every finite sum $\sum U^n h_n$. From this it easily follows that (3.3) holds. Thus U is isomorphic to U'. This completes the proof.

    If U on K is the minimal isometric dilation of T on H, then H is invariant for U^* and

$$T^* = U^* \,|\, H, \text{ or equivalently, } P_H U = T P_H \,. \tag{3.6}$$

To see this notice that Lemma 3.1 implies that H is semi-invariant for U. Thus $H = N \ominus M$ where both M and N are invariant for U. By minimality K is the smallest invariant subspace for U containing H. Since $H \subseteq K$ we have $N = K$. Therefore $H = K \ominus M$ is invariant for U^*, or equivalently, the first equation in (3.6) holds. Taking the adjoint yields the second equation. By virtue of the uniqueness of the minimal isometric dilation, equation (3.6) also follows from the matrix form of the minimal isometric dilation in (3.4).

3.3 REMARK. Equation (3.6) shows that the minimal isometric dilation U of T is an isometric lifting of T. If V on K_2 is an isometric lifting of T on H, then V admits a decomposition of the form $V = U \oplus V_1$ on $K \oplus K_1$ where U is the minimal isometric dilation of T and

$$K = \bigvee_0^\infty V^n H \,. \tag{3.7}$$

To see this first notice that K is invariant for V. Using the lifting property $T^* = V^* \,|\, H$ we have

$$V^* K = (V^* H) \bigvee V^* \left(\bigvee_1^\infty V^n H \right) \subseteq H \bigvee \left(\bigvee_0^\infty V^n H \right) = K \,.$$

Therefore K is also invariant for V^* and $U = V \,|\, K$ is the minimal isometric dilation for T. This completes the proof.

3.4 REMARK. If T is a co-isometry on H, then its minimal isometric dilation U is unitary. Indeed in this case it is easy to verify that $UU^* = I$ where U is the matrix minimal isometric dilation of T in (3.4).

The previous remark shows that any co-isometry admits a minimal unitary (dilation) lifting. By taking adjoints this shows that any isometry admits a minimal unitary extension, which is unique up to an isomorphism. This provides another proof of Corollary 2.3.

4. THE GEOMETRY OF MINIMAL ISOMETRIC DILATIONS

In this section we study the Wold decomposition and the geometry of minimal isometric dilations.

To begin recall that for a contraction T the dimension of D_T and $D_T{}^*$ is denoted by δ_T and δ_{T^*} respectively; moreover, δ_T and δ_{T^*} are called the *defect indices of* T.

4.1 PROPOSITION. *Let* U *on* K *be the minimal isometric dilation of* T *on* H. *Then* K *admits a decomposition of the form:*

$$K = H \oplus M_+(L) \text{ where } L = \overline{(U-T)H} \tag{4.1}$$

is a wandering subspace for U. *Moreover, the dimension of* L *is* δ_T.

PROOF. Obviously

$$Uh = Th \oplus P_M Uh \qquad (h \in H) \tag{4.2}$$

where M is the orthogonal complement of H. Because H is invariant for U^* the subspace M is invariant for U. This (4.2) and the minimality of U implies that

$$H \oplus M = K = \bigvee_0^\infty U^n H = H \oplus (\bigvee_0^\infty U^n P_M UH) \,.$$

A simple calculation gives

$$\overline{P_M UH} = \overline{(U-T)H} = L \,. \tag{4.3}$$

So to verify (4.1) it is sufficient to show that $P_M UH$ is wandering for U. Using the invariance of M with h and h$'$ in H we have

$$(U^n P_M Uh, P_M Uh') = (U^n P_M Uh, Uh') = (P_M Uh, U^{*n-1}h') =$$

$$(P_M Uh, T^{*(n-1)}h') = 0$$

for all $n \geq 1$. Therefore L is wandering for U and (4.1) holds. For h in H we have

$$\|(U-T)h\|^2 = \|Uh\|^2 - 2\mathrm{Re}(Uh, Th) + \|Th\|^2 = \|h\|^2 - \|Th\|^2 = \|D_T h\|^2 \,.$$

This implies that there exists a unitary operator ϕ mapping D_T onto L such that

$$\phi D_T h = (U-T)h \qquad (h \in H) \,. \tag{4.4}$$

In particular, the dimension of L equals δ_T. This completes the proof.

The following provides an analysis of the Wold decomposition of the minimal isometric dilation.

4.2 PROPOSITION. *Let* U *on* K *be the minimal isometric dilation of* T *on* H. *The Wold decomposition of* U *is* $S \oplus W$ *on* $M_+(L_*) \oplus K_u$ *where* L_* *is the wandering subspace for* U *defined by*

$$L_* = \overline{(I - UT^*)H} \tag{4.5}$$

and W *is the unitary part of* U. *Moreover, the dimension of* L_* *is* δ_{T^*}.

PROOF. The wandering subspace in the Wold decomposition of U is $L_* = K \ominus UK$. Notice that $U^{*n}L_* = 0$, or equivalently, $P_{L_*}U^n = 0$ for all $n \geq 1$. This and the minimality of U gives

$$L_* = P_{L_*}K = P_{L_*}\left(\bigvee_0^\infty U^n H\right) = \overline{P_{L_*}H} . \tag{4.6}$$

Let h be in H. Then $P_{L_*}h = h - Ug$ where g is in K and $h - Ug$ is orthogonal to UK, or equivalently, $U^*h = g$. Since $U^*|H = T$ we have

$$P_{L_*}h = h - UT^*h \quad (h \in H) . \tag{4.7}$$

This and (4.6) produces (4.5). To complete the proof notice that

$$\|P_{L_*}h\|^2 = \|h - UT^*h\|^2 = \|h\|^2 - 2\mathrm{Re}(h, UT^*h) - \|T^*h\|^2 =$$

$$\|h\|^2 - \|T^*h\|^2 = \|D_{T^*}h\|^2 \quad (h \in H) .$$

Thus there exists a unitary operator ϕ_* mapping D_{T^*} onto L_* satisfying

$$\phi_* D_{T^*}h = P_{L_*}h = (I - UT^*)h \quad (h \in H) . \tag{4.8}$$

This completes the proof.

An operator T on H is *stable (*-stable)* if $T^n \to 0$ ($T^{*n} \to 0$) strongly as $n \to \infty$, respectively. The following shows that T is *-stable if and only if its minimal isometric dilation is a unilateral shift.

4.3 PROPOSITION. *Let* T *be a contraction on* H. *Then* T *is* *-*stable if and only if its minimal isometric dilation* U *on* K *is a unilateral shift.*

PROOF. If U is a unilateral shift, then $T^* = U^*|H$ shows that T is *-stable. On the other hand, let $K = M_+(L_*) \oplus K_u$ be the Wold decomposition of U. If h is in H, then $h = m \oplus h_u$ where m is in $M_+(L_*)$ and h_u is in K_u. Using $T^* = U^*|H$ and $U = S \oplus W$ we have

$$\|T^{*n}h\|^2 = \|S^{*n}m\|^2 + \|W^{*n}h_u\|^2 \to \|h_u\|^2 \quad (\text{as } n \to \infty) .$$

If T is $*$-stable, then $h_u = 0$ and $H \subseteq M_+(L_*)$. Since $M_+(L_*)$ is invariant for U the minimality condition (3.5) gives $K = M_+(L_*)$. Thus U is a unilateral shift. This completes the proof.

To complete this section, we will present a method to construct the Wold decomposition of the minimal isometric dilation directly from a contraction T on H. To this end notice that

$$\|h\| \geq \|T^*h\| \geq \|T^{*2}h\| \geq \|T^{*3}h\| \geq \cdots$$

Therefore $T^n T^{*n}$ forms decreasing sequence of positive self adjoint operators. Hence $T^n T^{*n}$ converges strongly to a positive self adjoint operator Q_1. Now let Q be the positive square root of Q_1. Then

$$\|Qh\|^2 = \lim_{n \to \infty} \|T^{*n}h\|^2 \quad (h \in H) . \tag{4.9}$$

Thus $\|Qh\| = \|QT^*h\|$. So there exists an isometry W_1 on $\overline{QH}$ satisfying $W_1 Q = QT^*$. Let W_2 on K_u, be the minimal unitary extension of W_1. (Corollary 2.3 guarantees that W_2 is unique up to an isomorphism.) Finally, it is noted that $W_2 Q = QT^*$.

For h in H notice that

$$\sum_0^n \|D_{T^*} T^{*i}h\|^2 = \sum_0^n (\|T^{*i}h\|^2 - \|T^{*(i+1)}h\|^2) = \|h\|^2 - \|T^{*n+1}h\|^2 .$$

Using (4.9) and letting $n \to \infty$ this implies that

$$\|h\|^2 = \|Qh\|^2 + \sum_0^\infty \|D_{T^*} T^{*n}h\|^2 . \tag{4.10}$$

Consider the operator Θ mapping H into $l^2(D_{T^*}) \oplus K_u$ defined by

$$\Theta h = (\bigoplus_0^\infty D_{T^*} T^{*n}h) \oplus Qh \qquad (h \in H) . \tag{4.11}$$

Equation (4.10) guarantees that Θ is an isometry. Moreover, $W_2 Q = QT^*$ along with the matrix form for the unilateral shifts in (1.1), where D_{T^*} replaces L gives

$$(S^* \oplus W_2)\Theta = \Theta T^* . \tag{4.12}$$

Therefore T^* can be identified with $(S^* \oplus W_2)|\Theta H$. Thus $S \oplus W$ where $W = W_2^*$ is an isometric dilation for T. One can verify that $S \oplus W$ is minimal. So $S \oplus W$ is the Wold decomposition of the minimal isometric dilation of T.

5. MINIMAL UNITARY DILATIONS

In this section we study the minimal unitary dilation of Sz.-Nagy.

To this end we say that $\hat{U}$ on $\hat{K}$ is a *minimal unitary dilation* of a contraction T on H if $\hat{U}$ is a unitary dilation of T satisfying the following minimality condition

$$\hat{K} = \bigvee_{-\infty}^{\infty} \hat{U}^n H . \tag{5.1}$$

Obviously $\hat{U}$ is a minimal unitary dilation for T if and only if $\hat{U}^*$ is a minimal unitary dilation for T^*. To construct a minimal unitary dilation of T first let U on K be the minimal isometric dilation of T. Then let $\hat{U}$ on $\hat{K}$ be the minimal unitary extension of U. The following shows that $\hat{U}$ is a dilation of T

$$P_H \hat{U}^n | H = P_H P_K U^n | H = P_H U^n | H = T^n \quad \text{(for all } n \geq 0) .$$

The minimality condition in (5.1) follows from the fact that $\hat{U}$ is a minimal unitary extension of U and U is a minimal isometric dilation of T. Therefore $\hat{U}$ is a minimal unitary dilation of T.

One can also construct a minimal unitary dilation $\hat{U}$ on $\hat{K}$ for T in matrix form. To this end let $\hat{U}$ be the matrix defined by

$$\hat{U} = \begin{bmatrix} \ddots & & & \vdots & & \vdots & & \vdots & & \vdots & \\ \cdots & 0 & 0 & 0 & 0 & 0 & \cdots \\ \cdots & I & 0 & 0 & 0 & 0 & \cdots \\ \cdots & 0 & D_{T^*} & T & 0 & 0 & \cdots \\ \cdots & 0 & -T^* & D_T & 0 & 0 & \cdots \\ \cdots & 0 & 0 & 0 & I & 0 & \cdots \\ \cdots & 0 & 0 & 0 & 0 & I & \cdots \\ & \vdots & & \vdots & & \vdots & & \ddots \end{bmatrix}$$

$$\text{on} \quad \cdots D_{T^*} \oplus D_{T^*} \oplus D_{T^*} \oplus H \oplus D_T \oplus D_T \oplus D_T \oplus \cdots \tag{5.2}$$

where T in H appears in the 0-0 position. Using the fact that

$$\begin{bmatrix} D_{T^*} & T \\ -T^* & D_T \end{bmatrix} : D_{T^*} \oplus H \to H \oplus D_T$$

is unitary (see Corollary 1.4 in Chapter IV), it is easy to verify that $\hat{U}$ in (5.2) is a minimal unitary dilation of T.

Following ideas similar to the proof of Theorem 3.2, it is easy to verify that all minimal unitary dilations are unique up to an isomorphism. Summing up the previous analysis yields the main result of this section due to Sz.-Nagy.

5.1 THEOREM. *Any contraction* T *on* H *admits a minimal unitary dilation. Furthermore, all minimal unitary dilations of* T *are isomorphic.*

The previous theorem also yields the following inequality due to von Neumann.

5.2 COROLLARY. *Let* T *be a contraction on* H. *If* $p(z)$ *is a polynomial, then*

$$\|p(T)\| \le \sup_{|z|=1} |p(z)| . \tag{5.3}$$

PROOF. If $\hat{U}$ is the minimal unitary dilation of T, then

$$\|p(T)\| = \|P_H p(\hat{U}) | H\| \le \|p(\hat{U})\| \le \sup_{|z|=1} |p(z)| .$$

The last inequality follows from the spectral representation of a unitary operator. In particular, $\|p(\hat{U})\|$ equals the supremum of $|p(z)|$ where z runs over the spectrum of $\hat{U}$. This completes the proof.

To complete this section, we will study the geometry of minimal unitary dilations. To this end let $\hat{U}$ on $\hat{K}$ be the minimal unitary dilation of T on H. Consider the spaces K and K_* defined by

$$K = \bigvee_{0}^{\infty} U^n H \text{ and } K_* = \bigvee_{-\infty}^{0} \hat{U}^n H . \tag{5.4}$$

Obviously K is invariant for $\hat{U}$ and K_* is invariant for $\hat{U}^*$. It is easy to see that $U = \hat{U} | K$ on K and $U_* = \hat{U}^* | K_*$ on K_* are the minimal isometric dilations of T and T^*, respectively. In particular, $\hat{U}$ and $\hat{U}^*$ is the minimal unitary extension of U and U_* respectively.

If G is wandering for $\hat{U}$, then $M_-(G)$ is the subspace of $\hat{K}$ defined by

$$M_-(G) = \bigoplus_{-\infty}^{-1} \hat{U}^n G = \bigvee_{-\infty}^{-1} \hat{U}^n G .$$

In this case $\hat{U}^* | M_-(G)$ is a unilateral shift and G is its cyclic wandering subspace. Let L and L_* be the wandering subspaces for U defined in (4.3) and (4.5), respectively. Because $\hat{U}$ extends U both L and L_* are wandering for $\hat{U}$. Propositions 4.1 and 4.2 give:

$$K = H \oplus M_+(L) = M_+(L_*) \oplus K_u . \tag{5.5}$$

Since $\hat{U}$ is the minimal unitary extension of U and $M_+(L_*) \oplus K_u$ is the Wold decomposition of K, we have the first equality in:

$$\hat{K} = M(L_*) \oplus K_u = M_-(L_*) \oplus H \oplus M_+(L) \tag{5.6}$$

The second equality follows from (5.5) and the fact that the orthogonal complement of K is $M_-(L_*)$. Finally, Proposition 4.3 shows that T is $*$-stable if and only if $K_u = \{0\}$ or equivalently $\hat{K} = M(L_*)$.

Since $\hat{U}^*$ is a unitary extension of U_* equations (4.3) and (4.5) give

$$\overline{(\hat{U}^* - T^*)H} = \hat{U}^* \overline{(I - \hat{U}T^*)H} = \hat{U}^* L_*$$
$$\overline{(I - \hat{U}^* T)H} = \hat{U}^* \overline{(\hat{U} - T)H} = \hat{U}^* L . \tag{5.7}$$

Using (5.7) and applying Propositions 4.1 and 4.2 to $U_* = \hat{U}^* | K_*$ we have

$$K_* = M_-(L_*) \oplus H = M_-(L) \oplus K_{u*} \tag{5.8}$$

where K_{u*} corresponds to the unitary part in the Wold decomposition of U_*. Because $\hat{U}^*$ is the minimal unitary extension of U_* and $M_-(L) \oplus K_{u*}$ is the Wold decomposition of K_* we have

$$\hat{K} = M(L) \oplus K_{u*} . \tag{5.9}$$

Applying Proposition 4.3 to U_* we see that T is stable if and only if $\hat{K} = M(L)$. Equations (5.6), (5.5) and (5.8) produce the first set of equalities in:

$$\hat{K} \ominus K = M_-(L_*) = K_* \ominus H \text{ and } \hat{K} \ominus K_* = M_+(L) = K \ominus H . \tag{5.10}$$

The second set of equations follows from (5.9), (5.8) and (5.5). In particular, $\hat{U}^* | (\hat{K} \ominus K)$ and $\hat{U} | (\hat{K} \ominus K_*)$ are unilateral shifts with $\hat{U}^* L_*$ and L as their respective cyclic wandering subspaces. Summing up the above analysis yields the first five parts of the following theorem.

5.3 THEOREM. *Let $\hat{U}$ on $\hat{K}$ be the minimal unitary dilation of* T *on H and L and* L_* *be the subspaces defined by*

$$L = \overline{(\hat{U} - T)H} \quad and \quad L_* = \overline{(I - \hat{U}T^*)H} . \tag{5.11}$$

(i) The subspaces L and L_* are wandering for $\hat{U}$ and

$$\hat{K} = M_-(L_*) \oplus H \oplus M_+(L) = M(L) \oplus K_{u*} = M(L_*) \oplus K_u . \tag{5.12}$$

(ii) T is stable (∗-stable) if and only if $\hat{K} = M(L)$ $(\hat{K} = M(L_*))$.

(iii) The Wold decomposition for the minimal isometric dilation U on K of T is

$$(\hat{U}\,|\,M_+(L_*)) \oplus (\hat{U}\,|\,K_u).$$

(iv) The Wold decomposition for the minimal isometric dilation U_* on K_* of T^* is

$$(\hat{U}^*\,|\,M_-(L)) \oplus (\hat{U}^*\,|\,K_{u*}).$$

(v) Equation (5.10) holds. In particular, $\hat{U}^*\,|\,(\hat{K} \ominus K)$ and $\hat{U}\,|\,(\hat{K} \ominus K_*)$ are unilateral shifts with $\hat{U}^* L_*$ and L as their cyclic wandering subspaces, respectively.

(vi)

$$\operatorname*{strong\ lim}_{n\to\infty} \hat{U}^{*n} P_K \hat{U}^n \hat{k} = \hat{k} \quad \textit{(for all $\hat{k}$ in $\hat{K}$)} \tag{5.13}$$

PROOF. Only the proof of part (vi) remains. Part (v) shows that $\hat{U}^*\,|\,(\hat{K} \ominus K) = \hat{U}^*\,|\,M_-(L_*)$ is a unilateral shift. This and (5.5) implies that (5.13) holds for all $\hat{k}$ in

$$(\overset{\infty}{\underset{-n}{\oplus}} \hat{U}^j L_*) \oplus K_u$$

where $n \geq 0$. This and (5.6) yields (5.13). The proof is complete.

6. COMPLETELY NON-UNITARY CONTRACTIONS

In this section we introduce and study completely non-unitary contractions and their H^∞ functional calculus.

A contraction T on H is *completely non-unitary* if there is no nonzero reducing subspace H_u such that $T\,|\,H_u$ is unitary on H_u. For example, the unilateral shift S is completely non-unitary. To see this simply notice that $S^{*n} \to 0$ strongly. Actually the Wold decomposition of an isometry U shows that U is completely non-unitary if and only if it is a unilateral shift. The following shows that any contraction can be uniquely decomposed into a completely non-unitary contraction and a unitary contraction.

6.1 THEOREM. *Let T be a contraction on H. Then T admits a unique (reducing) decomposition of the form $T = T_0 \oplus T_u$ on $H = H_0 \oplus H_u$ where $T_0 \overset{\Delta}{=} T\,|\,H_0$ on H_0 is completely non-unitary and $T_u \overset{\Delta}{=} T\,|\,H_u$ on H_u is unitary. Moreover, the space H_u is given by the set of all h in H satisfying*

$$\|T^n h\| = \|h\| = \|T^{*n} h\| \qquad \textit{(for all $n \geq 1$)} . \tag{6.1}$$

In particular, if T *is an isometry, then this decomposition is precisely the Wold decomposition of* T.

The operator T_o is *the completely non-unitary part* of T and T_u is the *unitary part* of T. In this decomposition $(H = H_o \oplus H_u)$ the space H_o or H_u can be the trivial space $\{0\}$. It is emphasized that the decomposition $T = T_o \oplus T_u$ is unique. By unique we mean that if $T = T_o' \oplus T_u'$ on $H_o' \oplus H_u'$ is another decomposition of T where T_o' is completely non-unitary and T_u' is unitary, then $T_o = T_o'$ and $T_u = T_u'$, or equivalently, $H_o = H_o'$ and $H_u = H_u'$.

PROOF. Let H_u be the subset of H defined by (6.1). Recall that if A is any contraction on H, then $\|Ah\| = \|h\|$ if and only if $D_A h = 0$. So if h satisfies (6.1), then $D_{T^n} h = 0$ and $D_{T^{*n}} h = 0$ for all $n \geq 1$. This implies that

$$H_u = [\bigcap_1^\infty \ker D_{T^n}] \cap [\bigcap_1^\infty \ker D_{T^{*n}}] . \tag{6.2}$$

Therefore H_u is a closed linear subspace of H. We claim that H_u is a reducing subspace for T. If h is in H_u, then (6.1) gives

$$\|T^n Th\| = \|T^{n+1} h\| = \|h\| = \|Th\| \quad (n \geq 1) . \tag{6.3}$$

Recall that $\|Th\| = \|h\|$ produces $T^* Th = h$. Using this in (6.1):

$$\|T^{*n} Th\| = \|T^{*n-1} h\| = \|h\| = \|Th\| \quad (n \geq 1) .$$

Combining this with (6.3) shows that Th is in H_u and H_u is an invariant subspace for T. A similar argument shows that H_u is an invariant subspace for T^*. So H is a reducing subspace. Obviously both $T|H_u$ and $T^*|H_u$ are isometric. Therefore $T_u \triangleq T|H_u$ on H_u is unitary.

We claim that $T_o \triangleq T|H_o$ on $H_o \triangleq H \ominus H_u$ completely non-unitary. If there exists a reducing subspace F of T_o where $T_o|F$ is unitary, then obviously (6.1) holds for all h in F. So $F \subseteq H_u$ and $F = \{0\}$. Thus T_o is completely non-unitary.

Let $T = T_o' \oplus T_u'$ on $H_o' \oplus H_u'$ be another decomposition of T where T_o' is completely non-unitary and T_u' is unitary. Then clearly (6.1) holds for all h in H_u'. This implies that $H_u' \subseteq H_u$. Because both H_u and H_u' are reducing subspaces for T the space $R = H_u \ominus H_u'$ is also a reducing subspace. However $T|R = T_u|R$ is unitary and $T|R = T_o'|R$ is completely non-unitary. Therefore $R = \{0\}$ and $H_u = H_u'$ and $H_o = H_o'$. So our decomposition $T = T_o \oplus T_u$ is unique.

Finally, it is noted that if T is an isometry, then T^n is an isometry $D_{T^n} = 0$ and $\ker D_{T^{*n}} = T^n H$. This implies that H_u in (6.2) is precisely the unitary part of T in the

Wold decomposition of T; see Theorem 1.1 and the equation (1.2). This completes the proof.

Let $\hat{U}$ on $\hat{K}$ be a minimal unitary dilation of T. Moreover, let L and L_* be the wondering subspaces for $\hat{U}$ defined in (5.11). The spaces K_u and K_{u*} are the reducing subspaces for $\hat{U}$ defined in (5.12). This sets the stage for the following result.

6.2 PROPOSITION. *Let $\hat{U}$ on $\hat{K}$ be the minimal unitary dilation of* T *on* H. *If H_u is the reducing subspace corresponding to the unitary part of* T, *then*

$$H_u = \hat{K} \ominus (M(L) \vee M(L_*)) = K_u \cap K_{u*} . \tag{6.4}$$

In particular, T *is a completely non-unitary contraction if and only if*

$$M(L) \vee M(L_*) = \hat{K} . \tag{6.5}$$

PROOF. Because H_u is a reducing subspace for T and $T|H_u$ is unitary, H_u is a reducing subspace for $\hat{U}$. The first equation in (5.12) shows that H_u is orthogonal to both $M_+(L)$ and $M_-(L_*)$. Since H_u is a reducing subspace this implies that H_u is orthogonal to both $M(L)$ and $M(L_*)$, that is

$$H_u \subseteq \hat{K} \ominus (M(L) \vee M(L_*)) = K_u \cap K_{u*} \tag{6.6}$$

The equality follows from (5.12).

We claim that $K_u \cap K_{u*} \subseteq H$. To verify this let h be in $K_u \cap K_{u*}$. Consulting (6.6) we see that H is orthogonal to both $M_+(L)$ and $M_-(L_*)$. By (5.12) this h is in H. Therefore $K_u \cap K_{u*} \subseteq H$. Recall that both K_u and K_{u*} are reducing subspaces for $\hat{U}$. So $K_u \cap K_{u*}$ is a reducing subspace for $\hat{U}$ contained in H. Therefore

$$\|T^n h\| = \|P_H \hat{U}^n h\| = \|\hat{U}^n h\| = \|h\| = \|\hat{U}^{*n} h\| = \|P_H \hat{U}^{*n} h\| = \|T^{*n} h\| \quad (n \geq 1) \tag{6.7}$$

for all h in $K_u \cap K_{u*}$. This and the definition of H_u shows that $K_u \cap K_{u*} \subseteq H_u$. Using this in (6.6) gives (6.4). The last statement (6.5) follows from the fact that T is completely non-unitary if and only if $H_u = \{0\}$. This completes the proof.

If f is any vector in $\hat{K}$, then

$$M(f) = \bigvee_{-\infty}^{\infty} \hat{U}^n f , \quad M_+(f) = \bigvee_{0}^{\infty} \hat{U}^n f \quad \text{and} \quad M_*(f) = \bigvee_{-\infty}^{0} \hat{U}^n f . \tag{6.8}$$

These definitions are used to prove the following result.

6.3 PROPOSITION. *If* T *on* H *is a completely non-unitary contraction and* h *is any nonzero element in* H, *then* $\hat{U}|M(h)$ *contains a bilateral shift.*

PROOF. Observe that the spaces $M_+(h) \ominus \hat{U}M_+(h)$ and $M_*(h) \ominus \hat{U}^* M_*(h)$ are wandering spaces for $\hat{U}|M(h)$. In particular, if one of these spaces is nonzero, $\hat{U}|M(h)$ contains a bilateral shift. So from now on assume that both $M_+(h)$ and $M_-(h)$ are reducing subspaces for $\hat{U}$. Consulting (5.4) we see that $M_+(h) \subseteq K$ and $M_*(h) \subseteq K_*$. Equation (5.10) shows that $M_+(h)$ is orthogonal to $M(L_*)$ and $M_*(h)$ is orthogonal to $M(L)$. By (6.4) this implies that $M_+(h) \cap M_*(h) \subseteq H_u$. Since h is in the reducing subspace $M_+(h) \cap M_*(h)$ it follows that h is in H_u. Because T is completely non-unitary h = 0. This completes the proof.

Let $\hat{U} = \int_o^{2\pi} e^{it}dE_t$ be the spectral representation of $\hat{U}$ where E_t is the resolution of the identity over $[0, 2\pi)$. Let $E(\sigma)$ be the corresponding spectral measure for Borel subsets σ of the unit circle C and $m(\sigma)$ the normalized Lebesgue measure on C. Recall that two measures $\mu(\sigma)$ and $\nu(\sigma)$ are *equivalent* if they are both zero on the same Borel sets of C. This sets the stage for the following result.

6.4 THEOREM. *Let* T *on* H *be a completely non-unitary contraction and* $E(\sigma)$ *the spectral measure for the minimal* $\hat{U}$ *unitary dilation of* T. *Then* $E(\sigma)$ *is equivalent to the normalized Lebesgue measure* $m(\sigma)$ *on* C. *Moreover, if* h *is any nonzero vector in* H, *then* $m(\sigma)$ *is equivalent to* $(E(\sigma)h, h)$.

PROOF. If F is any wandering subspace for $\hat{U}$ and f is in F, then

$$\int_o^{2\pi} e^{int}d(E_tf, f) = (\hat{U}^n f, f) = \|f\|^2 \, \delta(n)$$

where $\delta(0) = 1$ and $\delta(n) = 0$ if $n \neq 0$. By the uniqueness of Fourier series this implies that $(E_t f, f) = \|f\|^2 t/2\pi$ and

$$\|E(\sigma)f\|^2 = (E(\sigma)f, f) = m(\sigma)\|f\|^2 \; . \tag{6.9}$$

In particular, $(E(\sigma)f, f)$ and $m(\sigma)$ are equivalent measures. If $m(\sigma) = 0$, then $E(\sigma)f = 0$. Moreover, using the fact that $\hat{U}$ and $E(\sigma)$ commute $E(\sigma)\hat{U}^n f = 0$ for all n. Therefore if F is any wandering subspace for $\hat{U}$ and $m(\sigma) = 0$, then $E(\sigma)g = 0$ for all g in $M(F)$. Since both L and L_* are wandering subspaces the preceding analysis show that $E(\sigma)g = 0$ for all g in $M(L)$ and all g in $M(L_*)$. Proposition 6.2 implies that $E(\sigma) = 0$. On the other hand, if $E(\sigma) = 0$, then $E(\sigma)h = 0$ for some nonzero h in H. Obviously $E(\sigma)M(h) = 0$. Proposition 6.3 shows that $E(\sigma)f = 0$ for some nonzero wandering vector f in $M(h)$. Applying (6.9) gives $m(\sigma) = 0$. This completes the proof.

To complete this section we will use the proceeding minimal unitary dilation theory to develop the H^∞ functional calculus for completely non-unitary contractions. Let A be the set of all functions whose power series is absolutely convergent, that is, a(z) is in A if and only if

$$a(z) = \sum_0^\infty a_n z^n \quad \text{and} \quad \sum_0^\infty |a_n| < \infty . \tag{6.10}$$

The functions in A are analytic in D and continuous on $\overline{D}$. Let T be a contraction on H. As in Section I.6 if a is in A, then a(T) is the operator on H defined by

$$a(T) \stackrel{\Delta}{=} \sum_0^\infty a_n T^n . \tag{6.11}$$

Because a is absolutely convergent this sum converges in the operator norm. Moreover, since the product of two absolutely convergent series is absolutely convergent we have

$$(\alpha a + \gamma b)(T) = \alpha a(T) + \gamma b(T) \quad \text{and} \quad (ab)(T) = a(T)b(T) \quad (a, b \in A) . \tag{6.12}$$

Therefore the mapping $a \to a(T)$ is an algebra homomorphism from A into $L(H)$, the (bounded linear) operators on H. If u(z) is analytic in D, then $\tilde{u}(z) \stackrel{\Delta}{=} \overline{u(\overline{z})}$. Finally, using (6.10) and (6.11) we have

$$a(T)^* = \tilde{a}(T^*) \qquad (a \in A) . \tag{6.13}$$

Let T on H be a completely non-unitary contraction and $\hat{U}$ on $\hat{K}$ be its minimal unitary dilation. By Theorem 6.4 the spectral measure $E(\sigma)$ for $\hat{U}$ is equivalent to the normalized Lebesgue measure. So if u is in H^∞, then $u(e^{it})$ is integrable with respect to E_t. Moreover, the operator $u^s(\hat{U})$ calculated by the spectral intergal

$$u^s(\hat{U}) = \int_0^{2\pi} u(e^{it}) \, dE_t$$

is well defined and $\|u^s(\hat{U})\| \leq \|u\|_\infty$. Now let $u_r(z) \stackrel{\Delta}{=} u(zr)$ where $0 \leq r < 1$. Obviously $u_r(z)$ is in A and $u_r^s(\hat{U}) = u_r(\hat{U})$ where $u_r(\hat{U})$ is calculated by (6.11) for $a = u_r$. It is well known ([Ho]) that $u_r(e^{it}) \to u(e^{it})$ a.e. as $r \to 1-$. By Lebesgue's bounded convergence theorem and the fact that E_t is absolutely continuous with respect to the Lebesgue measure

$$\|u^s(\hat{U})f - u_r(\hat{U})f\|^2 = \int_0^{2\pi} |u(e^{it}) - u_r(e^{it})|^2 d(E_t f, f) \to 0 \ \text{ as } \ r \to 1-\,.$$

Thus $u_r(\hat{U}) \to u^s(\hat{U})$ strongly as $r \to 1-$.

Let u be in H^∞. The operator $u(T)$ on H is defined by $u(T) \overset{\Delta}{=} P_H u^s(\hat{U})\,|\,H$. If u is in A, then this coincides with our definition of $u(T)$ for functions in A. Since $u_r(z)$ is in A we see that $u_r(T) = P_H u_r(\hat{U})\,|\,H$. By our previous analysis this implies that $u_r(T) \to u(T)$ strongly as $r \to 1-$. Moreover, $\|u(T)\| \le \|u\|_\infty$. In this way for any completely non-unitary contraction T and any function u in H^∞ we can define the operator $u(T)$ as the strong limit of $u_r(T)$ as $r \to 1-$. Through this limiting process the algebraic properties in (6.12) extend from A to H^∞. Since T is completely non-unitary T^* is also completely non-unitary. Using $u_r(T)^* = \bar{u}_r(T^*)$ and passing limits we have $u(T)^* = \bar{u}(T^*)$. Summing up the previous analysis yields the following result, which is the H^∞ functional calculus for completely non-unitary contractions.

6.5 THEOREM. *Let* T *be a completely non-unitary contraction and* $\hat{U}$ *on* $\hat{K}$ *its minimal unitary dilation. The mapping* $u \to u(T)$ *is an algebra homomorphism of* H^∞ *into* $L(H)$ *satisfying the following properties*

(i) $u_r(T) \to u(T)$ *strongly as* $r \to 1-$

(ii) $\|u(T)\| \le \|u\|_\infty$

(iii) $u(T)^* = \bar{u}(T^*)$

(iv) $u(T) = P_H u^s(\hat{U})\,|\,H.$

7. NON-MINIMAL ISOMETRIC DILATIONS

In this section we will introduce the notion of a D minimal isometric lifting of a contraction, which is a special kind of non-minimal isometric dilation. Unlike the minimal isometric dilation, D minimal isometric liftings are not unique. By defining a choice sequence which is an operator version of Schur numbers, we will provide a complete classification of all D minimal isometric liftings. In particular, we will show that there is a one to one correspondence between the set of all D minimal isometric liftings and a certain set of choice sequences.

To begin let V on K be an isometric operator and D be a specified linear subspace orthogonal to H'. We say that V is a D *minimal isometric lifting of* a contraction T' on H', if V is an isometric lifting of T' satisfying the following minimality condition

$$K = \bigvee_0^\infty V^n(H' \oplus D)\,. \tag{7.1}$$

Notice that if $D = \{0\}$, then the D minimal isometric lifting reduces to the usual minimal isometric dilation of T'. In this case, V is unique up to an isomorphism. We say that two D minimal isometric liftings V on K and $\tilde{V}$ on $\tilde{K}$ are *isomorphic* if there exists a unitary operator W from K onto $\tilde{K}$ satisfying $WV = \tilde{V}W$ and $W|H' = I_{H'}$. However, in general D minimal isometric liftings are not unique. In this section we will present a complete classification of all D minimal isometric liftings. We say that $\{\Gamma_i\}_1^\infty$ is a *choice sequence initiated on F to G* if Γ_1 is a contraction mapping F into G and Γ_{i+1} is a contraction mapping D_i into D_{i*} for all $i \geq 1$ where

$$D_i = D_{\Gamma_i}, \quad D_{i*} = D_{\Gamma_i^*}, \quad D_i = D_{\Gamma_i} \text{ and } D_{i*} = D_{\Gamma_i^*} \qquad \text{(for all } i \geq 1). \qquad (7.2)$$

If $F = G$, then $\{\Gamma_i\}_1^\infty$ is a choice sequence *initiated on F*.

7.1 REMARK. Choice sequences initiated on C^1 have played an important role in Chapter I. To see this notice that $\{r_i\}_0^\infty$ is a choice sequence initiated on C^1 if r_i is in D for all i, or $|r_i| < 1$ for $i < j$ and $|r_j| = 1$ and $r_m = 0$ for $m > j$. By Theorem I.1.5 the function $f(z) = \Sigma a_n z^n$ is in H_1^∞ if and only if the Schur numbers $(r_0, r_1, r_2, ...)$ corresponding to the data $(a_0, a_1, a_2, ...)$ is a choice sequence on C^1. In particular, there is a one to one correspondence between the set of all functions f in H_1^∞ and the set of all choice sequences $\{r_i\}_0^\infty$ initiated on C^1.

Corollary 2.3 in Chapter IV shows that $\hat{T}$ on $H' \oplus D$ is a contractive lifting of a contraction T' on H' if and only if $\hat{T}$ admits a matrix representation of the form

$$\hat{T} = \begin{bmatrix} T' & 0 \\ XD_{T'} & Y \end{bmatrix} \text{ on } H' \oplus D \qquad (7.3)$$

where $\Gamma_1 \triangleq [X, Y]$ is a contraction mapping $D_{T'} \oplus D$ into D. Therefore V on K is a D minimal isometric lifting of T' if and only if V is an isometric lifting of T', the compression of V to $H' \oplus D$ is $\hat{T}$ where $\Gamma_1 = [X, Y]$ is a contraction and V satisfies the minimality condition (7.1).

To complete this section we will show that there is a one to one correspondence between the set of all D minimal isometric liftings of T' on H' and the set of all choice sequences initiated from $D_{T'} \oplus D$ to D. In particular, there is a unique D minimal isometric lifting if and only if $D = \{0\}$. To this end let us recall that the rotation matrix

$$R_\Gamma = \begin{bmatrix} \Gamma & D_{\Gamma^*} \\ D_\Gamma & -\Gamma^* \end{bmatrix} : F \oplus D_{\Gamma^*} \rightarrow G \oplus D_\Gamma \qquad (7.4)$$

determined by a contraction Γ from F to G is unitary; see Corollary IV.1.4. In particular,

$$\begin{bmatrix} G \\ D_\Gamma \end{bmatrix} \ominus \begin{bmatrix} \Gamma \\ D_\Gamma \end{bmatrix} F = \begin{bmatrix} D_{\Gamma^*} \\ -\Gamma^* \end{bmatrix} D_{\Gamma^*} \tag{7.5}$$

where $[D_{\Gamma^*}, -\Gamma]^*$ is an isometry. This states that the orthogonal complement of the range of the first column in R_Γ is given by the range of the second column in R_Γ.

Now let V on K be a D minimal isometric lifting of T' on H. The minimality condition (7.1) with the form of $\hat{T}$ in (7.3) implies that V admits a matrix representation of the form:

$$V = \begin{bmatrix} T'P_{H'} & 0 & 0 & 0 & \cdots \\ \Gamma_1(D_{T'}\oplus I) & * & * & * & \cdots \\ B & & * & * & * & \cdots \\ 0 & & * & * & * & \cdots \\ 0 & & 0 & * & * & \cdots \\ 0 & & 0 & 0 & * & \cdots \\ 0 & & 0 & 0 & 0 & \cdots \\ \cdot & & & \cdot & \cdot & \cdot & \cdots \\ \cdot & & & \cdot & \cdot & \cdot & \cdots \\ \cdot & & & \cdot & \cdot & \cdot & \cdots \end{bmatrix} \tag{7.6}$$

$$\text{on} \quad (H'\oplus D) \oplus D_1 \oplus D_2 \oplus D_3 \oplus \cdots \tag{7.6a}$$

where

$$D_n = H_n \ominus H_{n-1} \quad \text{and}$$

$$H_n = \bigvee_{i=0}^{n} V^i(H'\oplus D) = (H\oplus D) \oplus D_1 \oplus D_2 \oplus \cdots \oplus D_n \quad (n\geq 0) . \tag{7.7}$$

Since V is an isometry the columns of V must be isometric and the spaces formed by range of the columns of V must be orthogonal. The first column of V in (7.6) is easy to obtain. The isometric property with $h \oplus f$ in $H' \oplus D$ gives

$$\|B(h\oplus f)\|^2 = \|h\|^2 + \|f\|^2 - \|T'h\|^2 - \|\Gamma_1(D_{T'}h\oplus f)\|^2 =$$

$$\|D_{T'}h \oplus f\|^2 - \|\Gamma_1(D_{T'}h\oplus f)\|^2 = \|D_{\Gamma_1}(D_{T'}h\oplus f)\|^2 .$$

So there exists a unitary ϕ mapping D_{Γ_1} onto $\overline{B(H'\oplus D)}$ satisfying $\phi D_{\Gamma_1}(D_{T'} \oplus I) = B$. Without loss of generality we can choose $\phi = I$. (This allows us to set $D_1 = D_{\Gamma_1}$ and justifies the notation in (7.2).) Therefore the first column in V can be identified with

$$[T'P_{H'}, \ \Gamma_1(D_{T'}\oplus I), \ D_1(D_{T'}\oplus I), \ 0, \ 0, \ 0, \ ...]^{tr} \ .$$

where tr denotes the transpose. Furthermore, $H_1 = (H \oplus D) \oplus D_1$. By applying (7.5) with $\Gamma = \Gamma_1$ we have

$$\begin{bmatrix} D \\ D_1 \end{bmatrix} \ominus \begin{bmatrix} \Gamma_1 \\ D_1 \end{bmatrix} ((D_{T'}\oplus I)(H' \oplus D)) = \begin{bmatrix} D_{1*} \\ -\Gamma_1^* \end{bmatrix} D_{1*} \ . \tag{7.8}$$

Since range of the operator formed by the second column in (7.6) is orthogonal to the range of the operator formed by first column, (7.8) shows that the second column is of the form

$$[0, \ D_1 \cdot \Gamma_2, \ -\Gamma_1^* \Gamma_2, \ C, \ 0, \ 0, \ ...]^{tr}$$

where Γ_2 is an operator from D_1 into D_{1*} and C is an operator from D_1 onto a dense set in D_2. But this column is also an isometry. Therefore Γ_2 must be a contraction and we can identify D_2 with D_{Γ_2} (again justifying the notation in (7.2)). Now the second column of V can be identified with (see Lemma IV.1.1)

$$[0, \ D_1 \cdot \Gamma_2, \ -\Gamma_1^* \Gamma_2, \ D_2, \ 0, \ 0, \ 0, \ ...]^{tr} \ .$$

Clearly $H_2 = H \oplus D_1 \oplus D_2$. Continuing in this fashion, that is, using the minimality condition (7.1) with (7.5) and the fact that all the columns of V must be isometric and with orthogonal ranges, one can show that V can be identified with a matrix of the form

$$V = \begin{bmatrix} T'P_{H'} & 0 & 0 & 0 & \cdots \\ \Gamma_1(D_{T'}\oplus I) & D_{1*}\Gamma_2 & D_{1*}D_{2*}\Gamma_3 & D_{1*}D_{2*}D_{3*}\Gamma_4 & \cdots \\ D_1(D_{T'}\oplus I) & -\Gamma_1^*\Gamma_2 & -\Gamma_1^*D_{2*}\Gamma_3 & -\Gamma_1^*D_{2*}D_{3*}\Gamma_4 & \cdots \\ 0 & D_2 & -\Gamma_2^*\Gamma_3 & -\Gamma_2^*D_{3*}\Gamma_4 & \cdots \\ 0 & 0 & D_3 & -\Gamma_3^*\Gamma_4 & \cdots \\ 0 & 0 & 0 & D_4 & \cdots \\ 0 & 0 & 0 & 0 & \cdots \\ 0 & 0 & 0 & 0 & \cdots \\ \vdots & \vdots & \vdots & \vdots & \cdots \\ & & & & \cdots \\ & & & & \cdots \end{bmatrix} \tag{7.9}$$

$$\text{on } \ K = (H' \oplus D) \oplus D_1 \oplus D_2 \oplus D_3 \oplus \cdots \tag{7.9a}$$

where $\{\Gamma_i\}_1^\infty$ is a choice sequence initiated from $D_{T'} \oplus D$ to D.

The isometry V in (7.9) can be put into a simple form. To see this let V_1 be the isometry on K defined by

$$V_1 = \begin{bmatrix} T'P_{H'} & 0 \\ \Gamma_1(D_{T'} \oplus I) & D_{1*} \\ D_1 & -\Gamma_1^* \end{bmatrix} \oplus I \oplus I \oplus I \oplus \cdots \tag{7.10}$$

For $n \geq 2$, let V_n be the unitary operator defined by

$$I \oplus I \oplus \cdots \oplus I \oplus \begin{bmatrix} \Gamma_n & D_{n*} \\ D_n & -\Gamma_n^* \end{bmatrix} \oplus I \oplus I \oplus \cdots \tag{7.11}$$

where Γ_n is the contraction appearing in the n-th position and I is the identity operator on the appropriate space. A simple calculation shows that the matrix isometry V in (7.9) is given by

$$V = \prod_1^\infty V_n = V_1 V_2 V_3 V_4 \cdots \tag{7.12}$$

where $\{\Gamma_i\}_1^\infty$ is a choice sequence initiated from $D_{T'} \oplus D$ to D. (Actually the product in (7.12) converges strongly to the isometry V in (7.9).) Therefore any D minimal isometric lifting V of T' is unitarily equivalent (isomorphic) to a matrix of the form (7.9), or equivalently, (7.12) where $\{\Gamma_i\}_1^\infty$ is a choice sequence initiated on $D_{T'} \oplus D$ to D uniquely determined by V.

On the other hand, if $\{\Gamma_i\}_1^\infty$ is a choice sequence initiated on $D_{T'} \oplus D$ to D, then (7.10) and (7.11) show that the operator V formed by (7.12) is an isometry. Since this V is also given by (7.9) we see that this V is a D minimal isometric lifting of T'. Thus any choice sequence $\{\Gamma_i\}_1^\infty$ uniquely determines a D minimal isometric lifting of T'. Summing up this analysis readily yields the following result.

7.2 THEOREM. *Let* T' *be a contraction on* H'. *Then* V *is a D minimal isometric lifting of* T' *if and only if* V *is isomorphic to a matrix of the form (7.9) where* $\{\Gamma_i\}_1^\infty$ *is a choice sequence initiated from* $D_{T'} \oplus D$ *to* D. *In particular, there exists a one to one correspondence between the set of all (isomorphic classes of) D minimal isometric liftings of* T' *and the set of all choice sequences initiated from* $D_{T'} \oplus D$ *to* D.

We complete this section, with some supplementary remarks relating the previous theorem to some further developments in this book. In the next chapter we will show that obtaining all contractive liftings in the commutant lifting theorem is equivalent to obtaining all D minimal isometric liftings of T' satisfying $\Gamma_1 | F' = \omega^*$ where ω^* is a unitary operator mapping $F' \subseteq D_{T'} \oplus D$ onto $F \subseteq D$. In this case, the set of all D minimal isometric liftings of T' is given by (7.9), where $\{\Gamma_i\}_1^\infty$ is a choice sequence initiated on $D_{T'} \oplus D$ to D satisfying the constraint $\Gamma_1 | F' = \omega^*$. Moreover, the constraint on $\Gamma_1 | F'$ implies that there is a one to one correspondence between the set of all D minimal isometric liftings of T' satisfying $\Gamma_1 | F' = \omega^*$, and the set of all choices sequences $\{\Gamma_1\}_1^\infty$ initiated on G' to G, where G' is the orthogonal complement of F' and G is the orthogonal complement of F. In Chapter XV we will show that solving the Carathéodory interpolation problem for positive definite functions involves constructing all D minimal isometric lifting of T' in the trivial case when $H' = \{0\}$. In this case there is a one to one correspondence through V in (7.9) between the set of all D minimal isometric liftings of $T' = 0$ and the set of all choice sequences $\{\Gamma_i\}_1^\infty$ initiated on D. Moreover, choosing $T = \Gamma_1$ on $H = D$ (with $H' = \{0\}$) and setting $\Gamma_i = 0$ for all $i \geq 2$ is precisely the matrix form of the minimal isometric dilation of T in (3.4). This is analogous to the maximal entropy solution to the positive definite Carathéodory interpolation discussed in Section II.5.

VI.8. NOTES AND COMMENTS

Probably v. Neumann was the first to be aware of the importance of shift operators in the study of operators on a Hilbert space. The structure Theorem 1.1 for isometries has an important interpretation in the theory of stochastic processes (see [HeL], [WiM] and [Wo]). The concepts of a unitary dilation and an isometric lifting are due to Sz.-Nagy [Sz.-N 1,2,3,4]. The first matrix construction of a unitary dilation was done by Schäffer [Sc]. The geometric structure of these dilations and the occurrence of two shifts with remarkable properties (see Theorem 5.3) made the Sz.-Nagy dilations into an excellent instrument for the study of contractions on Hilbert space (see for instance [Ber] and [Sz.-NF 10]). The concept of semi-invariance and Lemma 3.1 is due to Sarason [Sa 1]. The results in Sections 1 to 6 are taken from Chapters I to III in [Sz.-NF 10]. We selected only those topics which seemed necessary for the further development of this book. We send the interested reader to the Books [Ber] and [Sz.-NF 10] for historical comments as well as many other topics on dilation theory. The elaboration of all nonminimal isometric liftings given is Section 7 is new and was inspired by the results in [ArsCF 2], and will play a basic role in the analysis given in Section VII.8 of a nice proof of the commutant theorem due to Arocena [Ar 1,3,4]. For further details and historical comments on dilation theory see [Ber], [Do 6], [Fi], [LaP 2] and [Sz.-NF 10]. For remarkable, quite different variants on dilation theory see [Ag], [DoP] and [Pau].

CHAPTER VII

THE COMMUTANT LIFTING THEOREM

In this chapter we will introduce the main theorem of this monograph, namely, the Commutant Lifting Theorem and we will give several different proofs for it. Each proof illuminates different features of this theorem. We also use the second proof to discuss the uniqueness question in the commutant lifting theorem. The third proof establishes the connection to Ando's dilation theorem for two commuting contractions, while the fourth proof is based on Arocena's coupling of contractions.

1. THE COMMUTANT LIFTING THEOREM

In this section we present a modified version of the original proof of the commutant lifting theorem and some of its easy corollaries.

Throughout this chapter T on H and T' on H' are contractions. Recall that $I_1(T, T')$ is the set of all contractions A mapping H into H' intertwining T and T', that is, $T'A = AT$. Let U on K and U' on K' be the minimal isometric dilation of T and T', respectively. If A is in $I_1(T, T')$, then B is called a *contractive intertwining lifting of* A if B is in $I_1(U, U')$ and B is a (contractive) lifting of A, that is, $P_{H'}B = AP_H$. A basic problem in this monograph is the existence and characterization of all contractive intertwining liftings. We begin with the following result which will simplify some of our calculations.

1.1 LEMMA. *Let* A *be in* $I_1(T, T')$ *and* U *be the minimal isometric dilation of* T. *Then* AP_H *is in* $I_1(U, T')$. *Moreover,* B *is a contractive intertwining lifting of* A *if and only if* B *is a contractive intertwining lifting of* AP_H. *In particular, the set of all contractive intertwining liftings* B *of* A *and the set of all contractive intertwining liftings* B *of* AP_H *are identical.*

PROOF. Using $TP_H = P_H U$ we have

$$T' A P_H = A T P_H = A P_H U .$$

Therefore $A P_H$ is in $I_1(U, T')$. If B is a contractive intertwining lifting of A, then $U' B = BU$ and $P_{H'} B = A P_H$. Because U is the minimal isometric dilation of U this B is also a contractive intertwining lifting of $A P_H$. On the other hand, if B is a contractive intertwining lifting of $A P_H$, then $U' B = BU$ and $P_{H'} B = A P_H$. So B is also a contractive intertwining lifting of A. This completes the proof.

The following result which proves the existence of a contractive intertwining lifting is referred to in the literature as *the commutant lifting theorem*. Later we will give a complete characterization of all contractive intertwining liftings and several algorithms to compute these liftings.

1.2 THEOREM. *Let A be in $I_1(T, T')$. Then there exists a contractive intertwining lifting B of A.*

PROOF. Due to Lemma 1.1 without loss of generality we assume that $T = U$ is an isometry. In this case B is a contractive intertwining lifting of A if and only if B is a contraction mapping H into K' lifting A and satisfying the intertwining property $U' B = BU$. Let U' on $K' = H' \oplus l^2(D_{T'})$ be the minimal isometric dilation of T' in its matrix form corresponding to (VI.3.4). Since B maps H into K' we see that B admits a matrix representation of the form

$$B = [A^*, Z_1^*, Z_2^*, Z_3^*, \cdots]^* \tag{1.1}$$

where Z_i is an operator mapping H into $D_{T'}$ for all $i \geq 1$. Using $U' B = BU$ along with the matrix form of U', we see that B is a contractive intertwining lifting of A if and only if B is a contraction of the form (1.1) satisfying

$$D_{T'} A = Z_1 U \text{ and } Z_j = Z_{j+1} U \qquad (\text{for all } j \geq 1) . \tag{1.2}$$

To complete the proof we will construct a contraction B satisfying (1.1) and (1.2). Using $T' A = AU$ with the fact that U is an isometry we have

$$\|D_{T'} Ah\|^2 \leq \|D_A h\|^2 + \|D_{T'} Ah\|^2 =$$

$$\|h\|^2 - \|Ah\|^2 + \|Ah\|^2 - \|T' Ah\|^2 = \|Uh\|^2 - \|AUh\|^2 = \|D_A Uh\|^2 \qquad (h \in H) .$$

This implies that there exists a contraction C mapping $\overline{D_A UH}$ into $D_{T'}$ satisfying

$$D_{T'} A = C D_A U = Z_1 U \tag{1.3}$$

where $Z_1 | UH = C D_A | UH$. Now consider the following matrix decomposition of

$[A^*, Z_1^*]^*$

$$B_1 = \begin{bmatrix} A \\ Z_1 \end{bmatrix} = \begin{bmatrix} A|UH & A|G \\ CD_A & D \end{bmatrix} : (UH) \oplus G \to H' \oplus D_{T'} \qquad (1.4)$$

where $G = H \ominus UH$ and D is still undetermined. Using (1.3)

$$\|AUh\|^2 + \|CD_A Uh\|^2 = \|T'Ah\|^2 + \|D_{T'}Ah\|^2 =$$

$$\|T'Ah\|^2 + \|Ah\|^2 - \|T'Ah\|^2 = \|Ah\|^2 \leq \|h\|^2 = \|Uh\|^2 .$$

Therefore the first column of B_1 is a contraction. Obviously the first row of B_1 is a contraction. Corollary 3.6 in Chapter IV implies that there exists a contraction $D = D_o$ such that B_1 in (1.4) is a contraction. Setting $Z_1 = [CD_A, D_o]$ shows that there exists a contraction Z_1 satisfying (1.2), and such that B_1 in (1.4) is a contraction.

Now we use induction and assume that

$$B_n^* = [A^*, Z_1^*, Z_2^*, ..., Z_n^*] \qquad (1.5)$$

is a contraction satisfying (1.2) for $1 \leq j < n$. Let $Z_{n+1} = [Z_n U^*, D]$ be an operator mapping $UH \oplus G$ into $D_{T'}$. Clearly Z_{n+1} satisfies the $(n+1)^{th}$ constraint $Z_{n+1}U = Z_n$ in (1.2). Moreover, B_{n+1} admits a matrix decomposition of the form

$$B_{n+1} = \begin{bmatrix} B_n|UH & B_n|G \\ Z_n U^* & D \end{bmatrix} : (UH) \oplus G \to [H' \oplus (\overset{n}{\underset{1}{\oplus}} D_{T'})] \oplus D_{T'} \qquad (1.6)$$

Using (1.2), $T'A = AU$ and applying B_{n+1} to Uh gives

$$\|B_{n+1}Uh\|^2 = \|AUh\|^2 + \|D_{T'}Ah\|^2 + \sum_2^{n+1} \|Z_i Uh\|^2 =$$

$$\|T'Ah\|^2 + \|Ah\|^2 - \|T'Ah\|^2 + \sum_1^n \|Z_i h\|^2 = \|B_n h\|^2 \leq \|h\|^2 = \|Uh\|^2 .$$

Therefore the first column of B_{n+1} in (1.6) is a contraction. By induction the first row of B_{n+1} in (1.6) is a contraction. Corollary 3.6 in Chapter IV implies that there exists a contraction $D = D_n$ such that B_{n+1} in (1.6) is a contraction. Setting $Z_{n+1} = [Z_n U^*, D_n]$ shows that there exists a Z_{n+1} satisfying (1.2) where B_{n+1} is a contraction. This completes the induction.

We have shown that there exists a sequence of contractions $\{Z_i\}_1^\infty$ satisfying (1.2) where B_n is a contraction for all n. Therefore if B is defined according to (1.1), we have

$$\|Bh\|^2 = \lim_{n \to \infty} (\|Ah\|^2 + \sum_{j=1}^{n} \|Z_j h\|^2) = \lim_{n \to \infty} \|B_n h\|^2 \le \|h\|^2 \qquad (h \in H)$$

and thus B is a contraction from H to K'. Since by construction B satisfies (1.2), it follows that B is an intertwining lifting of A. The proof is now complete.

In Chapter XIII we will present a more comprehensive analysis of the preceding proof to obtain a characterization of all contractive intertwining liftings of A.

1.3 COROLLARY. *Let* A *be in* $I_1(T, T')$ *and* V *on* M *be an isometric lifting of* T *and* W *on* N *a contractive lifting of* T'. *Then there exists a contractive lifting* B *in* $I_1(V, W)$ *of* A.

PROOF. First we prove the corollary under the supplementary condition that W is an isometry. By virtue of Remark VI.3.3

$$V = U \oplus V_1 \quad \text{and} \quad W = U' \oplus W_1$$

where $M = K \oplus M_1$ and $N = K' \oplus N_1$. By the commutant lifting theorem there exists a contractive intertwining lifting B_o in $I_1(U, U')$ of A in $I_1(T, T')$. The operator $B = B_o P_K$ enjoys the desired properties.

Now assume that W is a contractive dilation of T'. Let V' on M' be the minimal isometric dilation of W. By the first part of the proof there exists a contractive lifting B_1 mapping M into M' of A such that $B_1 \in I_1(V, V')$. Finally, $B = P_N B_1$ has the desired properties. This completes the proof.

By taking adjoints in the previous corollary we obtain the following result.

1.4 COROLLARY. *Let* A *be a contraction in* $I(T,T')$ *and* V_* *on* M *a co-isometric extension of* T *and* V'_* *on* M' *a co-isometric extension of* T'. *Then there exists a contraction* B *in* $I(V_*,V'_*)$ *extending* A, *that is,* B *maps* H *into* H' *and* $B\,|\,H = A$.

1.5 COROLLARY. *Let* V *on* V *and* V' *on* V' *be unitary dilations of* T *on* H *and* T' *on* H', *respectively. Let* A *be a contraction in* $I(T,T')$. *Then there exists a contraction* B_1 *in* $I_1(V,V')$ *satisfying* $A = P_{H'} B_1 | H$. *Moreover,* B_1 *can be chosen such that* B_1 *admits a matrix representation of the form:*

$$B_1 = \begin{bmatrix} * & 0 & 0 \\ * & A & 0 \\ * & * & * \end{bmatrix} : K^{\perp} \oplus H \oplus M_+(L) \to K'^{\perp} \oplus H' \oplus M_+(L') \qquad (1.7)$$

where $B \triangleq B_1 | K$ *from* $K = H \oplus M_+(L)$ *to* $K' = H' \oplus M_+(L')$ *is a contractive intertwining lifting of* A.

PROOF. Let K and K' be the spaces defined by

$$K = \bigvee_0^\infty V^n H \text{ and } K' = \bigvee_0^\infty V'^n H' . \tag{1.8}$$

Obviously $U = V | K$ on K and $U' = V' | K'$ on K' are the minimal isometric dilations of T and T', respectively. Consider the spaces $\hat{K}$ and $\hat{K}'$ defined by

$$\hat{K} = \bigvee_{-\infty}^\infty V^n H \text{ and } \hat{K}' = \bigvee_{-\infty}^\infty V'^n H' . \tag{1.9}$$

Clearly $\hat{K}$ and $\hat{K}'$ are reducing subspaces for V and V'. Moreover, $\hat{U} = V | \hat{K}$ on $\hat{K}$ and $\hat{U}' = V' | \hat{K}'$ on $\hat{K}'$ are the minimal unitary dilations of T and T', respectively. The operators $\hat{U}$ and $\hat{U}'$ are also the minimal unitary extensions of U and U', respectively. By the commutant lifting theorem there exists a contractive intertwining lifting B of A. Corollary VI.2.4 shows that B admits a contractive extension $\hat{B}$ in $I_1(\hat{U},\hat{U}')$. Because $\hat{B}$ is an extension of B and B is a lifting of A, the contraction $\hat{B}$ admits a matrix representation of the form (1.7) where $\hat{B}$ replaces B_1. Since $\hat{K}$ and $\hat{K}'$ are reducing subspaces for V and V', respectively the operator $B_1 = \hat{B}P_{\hat{K}}$ is a contraction in $I_1(V,V')$ of the form (1.7). By construction $B = B_1 | K$ is a contractive intertwining lifting of A. This completes the proof.

Let F and F' be subspaces of H and H', respectively. We say that (F,F') is *a hyperinvariant pair of subspaces for* (T,T') if $AF \subseteq F'$ for all operators A in $I(T,T')$. If $F = F'$ and $T = T'$, this reduces to the usual definition of a hyperinvariant subspace, that is, F is *a hyperinvariant subspace for* T if F is an invariant subspace for all operators A in the commutant $I(T, T)$ of T. The concept of a hyperinvariant subspace is used in the following result

1.6 COROLLARY. *Let R be an invariant subspace for T and A a contraction in* $I_1(T|R,T')$. *Let V_* on M and V'_* on M' be co-isometric extensions of T and T' respectively. Assume that (H,H') is a hyperinvariant pair of subspaces for (V_*,V'_*). Then there exists a contraction $\hat{A}$ in $I(T,T')$ extending A, that is, $\hat{A}|R = A$. In this case any contractive intertwining lifting B of $\hat{A}$ satisfies $P_{H'}B|R = A$.*

PROOF. Obviously V_* is a co-isometric extension of $T|R$. By Corollary 1.4 there exists a contraction B_* in $I(V_*,V'_*)$ extending A. Because (H,H') is a hyperinvariant pair of subspaces, $B_*H \subseteq H'$ and $\hat{A} \overset{\Delta}{=} B_*|H$ is a contraction in $I_1(T,T')$. Clearly A =

$\hat{A}|R$. The last statement follows from the lifting property of contractive intertwining liftings and $\hat{A}|R = A$. This completes the proof.

We conclude this section with the following useful remark communicated to us by M. A. Dritschel.

1.7 REMARK. Let U and U' be the minimal isometric dilations of T and T', respectively. Let A be a contraction in $I(T, T')$. Then B is a contractive intertwining lifting of A if and only if B is a contraction in $I(U, U')$ and $A = P_{H'}B|H$.

To prove this result assume that B is in $I_1(U, U')$ and $A = P_{H'}B|H$. Then using $P_{H'}U' = T'P_{H'}$ we have for all h in H

$$P_{H'}BU^n(U - T)h = P_{H'}U'^{n+1}Bh - P_{H'}U'^nBTh =$$

$$T'^{n+1}P_{H'}Bh - T'^nP_{H'}BTh = T'^{n+1}Ah - T'^nATh = 0 .$$

By Proposition VI.4.1 this implies that $P_{H'}B(K \ominus H) = 0$, or equivalently, $P_{H'}B = P_{H'}BP_H = AP_H$. Therefore B is a contractive intertwining lifting of A. Since the other half of this remark is obvious, the proof is now complete.

2. STRONG LIMITS OF CONTRACTIVE LIFTINGS

In this section we will present some elementary lemmas, which will be used in analyzing the commutant lifting theorem.

We begin with a simple lemma concerning the strong limits of contractive liftings. Let A_j be a (contractive) lifting of A_{j-1} for all $j \geq 1$. Clearly A_j is a (contractive) lifting of A_i when $i < j$. The following shows that the limit of the A_j's exist and is also a contractive lifting of all A_i.

2.1 LEMMA. *For all $n \geq 1$, let A_n mapping H_n into H'_n be a contractive lifting of A_{n-1} mapping H_{n-1} into H'_{n-1}. If*

$$K = \bigvee_{n \geq 0} H_n = \{\bigcup_{n=0}^{\infty} H_n\}^- \quad and \quad K' \supseteq \bigvee_{n \geq 0} H'_n \tag{2.1}$$

then the following limit exists

$$B = \text{strong} \lim_{n \to \infty} A_n P_{H_n} k \quad (k \in K) \tag{2.2}$$

and B mapping K into K' is a contractive lifting of all A_n.

PROOF. Clearly

$$A_n = P_{H'_n} A_{n+m} \,|\, H_n \qquad \text{(if n, m} \geq 0) . \tag{2.3}$$

If h_j is in H_j, $n \geq j$ and $m \geq 0$, then

$$\|A_{n+m}h_j - A_n h_j\|^2 = \|(I - P_{H'_n})A_{n+m}h_j\|^2 =$$

$$\|A_{n+m}h_j\|^2 - \|P_{H'_n}A_{n+m}h_j\|^2 = \|A_{n+m}h_j\|^2 - \|A_n h_j\|^2 . \tag{2.4}$$

Thus

$$\|A_j h_j\|^2 \leq \|A_{j+1}h_j\|^2 \leq \|A_{j+2}h_j\|^2 \leq \cdots \leq \|h_j\|^2$$

and $\|A_n h_j\|$ approaches a limit as $n \to \infty$. Equation (2.4) gives (2.2) when $k = h_j \in H_j$. By (2.1) and the fact that $A_n P_{H_n}$ is a contraction for all n it readily follows that the limit in (2.2) exists. Obviously B is a contraction. By letting $m \to \infty$ in (2.3) we have

$$A_n = P_{H'_n} B \,|\, H_n \qquad \text{(for all n} \geq 0) . \tag{2.5}$$

This completes the proof.

2.2 COROLLARY. *For all* $n \geq 1$, *let* A_n *mapping* H_n *into* H'_n *be a contractive lifting of* A_{n-1} *mapping* H_{n-1} *into* H'_{n-1}. *If there is equality for* K' *in (2.1), then*

$$B^* = \text{strong} \lim_{n \to \infty} A_n^* P_{H'_n} \tag{2.6}$$

where B^* *is the adjoint of* B *in (2.2).*

PROOF. Obviously $A_{n+m}^* \,|\, H'_n = A_n^*$ for all $m \geq 0$. So $A_n^* h_j$ converges as $n \to \infty$ for h_j in H'_j. Since A_n is a sequence of contractions and there is equality in (2.1), the right hand side of equation (2.6) exists. Clearly this limit B_* is the adjoint of B in (2.2), since for any k in K and k' in K' one has

$$(B_* k', k) = \lim_{n \to \infty} (A_n^* P_{H'_n} k', k) = \lim_{n \to \infty} (k', A_n P_{H_n} k) = (k', Bk).$$

This completes the proof.

A remarkable particular case of the preceding lemma concerns the isometric dilation of a contraction T on H. The *simple or one step dilation* of T is

$$T_1 = \begin{bmatrix} T & 0 \\ D_T & 0 \end{bmatrix} \text{ on } H_1 = H \oplus D_T . \tag{2.7}$$

Let $T_0 = T$ and T_j be the simple dilation of T_{j-1} for $n \geq j \geq 1$. In this case T_n on H_n is called the *n-step dilation* of T. The following shows that the n-step dilations converge to the minimal isometric dilation.

2.3 COROLLARY. *Let T_n on H_n be the n-step dilation of the contraction T on H. Then the minimal isometric dilation U on K of T is given by*

$$U = \text{strong} \lim_{n \to \infty} T_n P_{H_n} \quad \text{where} \quad K = \bigvee_1^\infty H_n . \tag{2.8}$$

Moreover, U is the minimal isometric dilation of T_n for all $n \geq 0$ and

$$H_n = \bigvee_0^n U^n H . \tag{2.9}$$

PROOF. By Lemma 2.1, equation (2.8) defines a contractive lifting U of $T = T_0$. By (2.5) and (2.7) we have for h_j in H_j and $n > j$

$$\|h_j\|^2 \geq \|Uh_j\|^2 \geq \|P_{H_n} Uh_j\|^2 = \|T_n h_j\|^2 =$$

$$\|T_{n-1} h_j\|^2 + \|D_{T_{n-1}} h_j\|^2 = \|T_{n-1} h_j\|^2 + \|h_j\|^2 - \|T_{n-1} h_j\|^2 = \|h_j\|^2 . \tag{2.10}$$

Therefore $U \,|\, H_j$ is an isometry. This and (2.8) shows that U is an isometry.

Now we prove that U is minimal. Equations (2.5) and (2.10) yield

$$T_n \,|\, H_j = U \,|\, H_j \qquad (\text{if } n > j) . \tag{2.11}$$

This and the definition of the one step dilation in (2.7) give

$$H_n = H_{n-1} \bigvee T_n H_{n-1} = H_{n-1} \bigvee U H_{n-1} . \tag{2.12}$$

From this we readily obtain (2.9) and the minimality condition $K = \bigvee_0^\infty U^n H$. This and Lemma 2.1 completes the proof.

2.4 REMARK. Let T_n on H_n be the n-step dilation of the contraction T on H, and U on K the minimum isometric dilation of T defined by (2.8). Obviously the family $\{T_n\}$ is uniquely determined by T. By the previous corollary

$$T_n = P_{H_n} U \,|\, H_n \quad \text{and} \quad H_n = \bigvee_{i=0}^n U^i H . \tag{2.13}$$

Proposition VI.4.1 along with equation (VI.4.1) show that

$$H_1 = H \bigvee UH = H \oplus L$$

where $L = \overline{(U-T)H}$ is wandering for U. This and (2.13) gives

$$H_n = \bigvee_0^n U^n H = H \oplus L \oplus UL \oplus \cdots \oplus U^{n-1}L . \tag{2.14}$$

In particular,

$$H_n \ominus H_{n-1} = U^{n-1}(H_1 \ominus H_0) . \tag{2.15}$$

Since the minimal isometric dilation is unique up to an isomorphism, the sequence of n-step dilations $\{T_n\}$ is isomorphic to $\{P_{H'_n} U' | H'_n\}$ where U' on K' is any minimal isometric dilation of T and H'_n is the closed linear span of $\{U'^i H\}_0^n$. To be precise there exists a unitary operator W mapping K onto K', and this W also maps H_n onto H'_n and satisfies

$$(P_{H'_n} U' | H'_n)W = WT_n \qquad \text{(for all } n \geq 0) .$$

Due to this isomorphism we say that T_n *is a n-step dilation of* T if T_n is given by (2.13) where U is any minimal isometric dilation of T. By the previous corollary U is also the minimal isometric dilation of T_n for all $n \geq 1$. In particular, if U on K is the matrix minimal isometric dilation of T in (VI.3.4), then the n-step dilation T_n of T is given by

$$T_n = P_{H_n} U | H_n = \begin{bmatrix} T & 0 & 0 & . & . & . & . & . & 0 & 0 \\ D_T & 0 & 0 & . & . & . & . & . & 0 & 0 \\ 0 & I & 0 & . & . & . & . & . & 0 & 0 \\ 0 & 0 & I & . & . & . & . & . & 0 & 0 \\ . & . & . & . & . & . & . & . & . & . \\ . & . & . & . & . & . & . & . & . & . \\ 0 & 0 & 0 & . & . & . & . & . & I & 0 \end{bmatrix} \tag{2.16}$$

$$\text{on } \ H_n = \bigvee_{i=0}^n U^i H = H \oplus [\bigoplus_1^n D_T]$$

2.5 REMARK. The above analysis also yields another proof of the following result in Remark VI.3.4: If T^* is an isometry, then the minimal isometric dilation of T is unitary.

To see this let T_n on H_n be the n-step dilation of T, and U on K its corresponding minimal isometric dilation. Clearly,

$$T_1 = \begin{bmatrix} T & 0 \\ D_T & 0 \end{bmatrix} \text{ on } H \oplus D_T .$$

A direct calculation gives $T_1 T_1^* = I$. Hence T_1^* is an isometry. By the same argument T_n^* is an isometry for all $n \geq 0$. Corollary 2.3 implies that $U^* | H_n = T_n^*$ is isometric for all n. Equation (2.6) adjusted to this setting and the minimality of U proves that U^* is an isometry, or equivalently, U is unitary. This proves our claim.

3. THE SECOND PROOF OF THE COMMUTANT LIFTING THEOREM.

In this section we will use n-step dilations to obtain an alternate proof of the commutant lifting theorem.

As before T on H and T' on H' are contractions and A is in $I_1(T, T')$. The n-step dilations of T on H and T' on H' are denoted by T_n on H_n and T_n' on H_n', respectively. Theorem 1.2 in Chapter V shows that there exists an operator A_1 in $I_1(T_1, T_1')$ which is a contractive lifting of A. Recall that by definition T_n and T_n' are the one step dilations of T_{n-1} and T_{n-1}', respectively for all $n \geq 1$. (Here $T_o = T$ and $T_o' = T'$.) By recursively applying Theorem 1.2 in Chapter V, we conclude that there exists a sequence of operators A_n in $I_1(T_n, T_n')$ such that A_n is a contractive lifting of A_{n-1} for all $n \geq 1$ where $A_o = A$. Lemma 2.1 implies that $A_n P_{H_n} \to A_\infty$ strongly as $n \to \infty$ and A_∞ is a contractive lifting of A_n for all $n \geq 0$. By Corollary 2.3

$$T_n P_{H_n} \to U \text{ and } T_n' P_{H_n'} \to U'$$

strongly as $n \to \infty$ where U and U' is the minimal isometric dilations of T_n and T_n', respectively, for all $n \geq 0$. Since A_n is in $I_1(T_n, T_n')$ we have

$$A_n P_{H_n} U = A_n T_n P_{H_n} = T_n' A_n P_{H_n} .$$

Taking the limit we see that $U' A_\infty = A_\infty U$. Therefore A_∞ is a contractive intertwining lifting of A. This provides an alternate proof of the commutant lifting theorem (Theorem 1.2). We say that A_n is a *contractive n-step intertwining lifting of* A if A_n is in $I_1(T_n, T_n')$ and A_n is a lifting of $A_j = P_{H_j} A_n | H_j$ for all $0 \leq j \leq n$ where $A_o \triangleq A$. For later purposes we sum up the proceeding construction in the following remark.

3.1 REMARK. Let A be in $I_1(T, T')$ and $\{A_n\}_0^\infty$ be a sequence in $I_1(T_n, T_n')$ of contractive n-step intertwining liftings of A satisfying $P_{H_j'} A_n | H_j = A_j$ for all $0 \leq j \leq n$. Then

$$B = A_\infty = \text{strong } \lim_{n \to \infty} A_n P_{H_n} \tag{3.1}$$

is a contractive intertwining lifting of A.

3.2 REMARK. The previous proof of the commutant lifting theorem produces all contractive intertwining liftings B of A in $I_1(T, T')$. To be precise any contractive intertwining lifting B of A is given by (3.1) in the previous remark.

To see this let B be a contractive intertwining lifting of A. Let U on K and U' on K' be the minimal isometric dilations of T and T', respectively. By Remark 2.4 the n-step dilations T_n on H_n and T'_n on H'_n of T and T' are given by

$$T_n^* = U^* \,|\, H_n \text{ and } T_n'^* = U'^* \,|\, H'_n$$
$$H_n = H \oplus (\overset{n-1}{\underset{0}{\oplus}} U^i L) \text{ and } H'_n = H' \oplus (\overset{n-1}{\underset{0}{\oplus}} U'^i L') \tag{3.2}$$

where

$$L = \overline{(U-T)H} \text{ and } L' = \overline{(U'-T')H'}$$

are wandering for U and U', respectively. Since $U \,|\, M_+(L)$ and $U' \,|\, M_+(L')$ are unilateral shifts and $H^\perp = M_+(L)$, we have

$$BH_n^\perp = BU^n H^\perp = U'^n B H^\perp \subseteq U'^n H'^\perp = H_n'^\perp . \tag{3.3}$$

The third relation follows from the fact that B maps $H^\perp$ into $H'^\perp$. Consider the sequence of contractions A_n mapping H_n into H'_n defined by $A_n^* = B^* \,|\, H'_n$. The definition of H_n and (3.3) show that A_{n+1} is a contractive lifting of A_n for all $n \geq 0$ where $A_0 = A$. From (3.2) it follows that

$$A_n^* T_n'^* h = B^* U'^* h = U^* B^* h = T_n^* A_n^* h \quad (h \in H'_n) .$$

Therefore A_n is in $I_1(T_n, T'_n)$. Finally, it is clear that $A_n P_{H_n} = P_{H'_n} B$ converges strongly to B. This completes the proof.

3.3 PROPOSITION. *If T^* or T' is an isometry, then there exists a unique contractive lifting B of A in $I_1(T, T')$.*

PROOF. To see this notice that if T' is isometric, then $T' = T'_n = U'$ for all n. Thus the factorization $T'_n \cdot A_n$ is regular for all $n \geq 0$. By Corollary 1.4 in Chapter V, A_{n+1} is uniquely determined by A_n. Thus all the A_n's are uniquely determined by $A = A_0$. This and the above remark show that B is unique.

If T^* is isometric, then T_n^* is isometric for all n. This follows from Remark 2.5. Thus $T_n^* \cdot A_n^*$ is a regular factorization for all n. Corollary V.2.3 implies that $A_n \cdot T_n$ is a regular factorization. Therefore by Corollary V.1.4 all of the A_n's are uniquely determined by $A = A_o$. Hence B is unique. This completes the proof.

4. CYCLIC SPACES AND UNIQUE CONTRACTIVE INTERTWINING LIFTINGS

In this section we will give some sufficient conditions for the uniqueness of a contractive intertwining lifting B of A. These conditions are easy to verify and will be applied to Hankel matrices with scalar entries.

This begins with the following result.

4.1 THEOREM *Let $\hat{U}$ on $\hat{K}$ be the minimal unitary dilation of* T *and* A *be in* $I_1(T, T')$. *If*

$$\hat{K} = \bigvee_{-\infty}^{\infty} \hat{U}^n(\ker D_A) \tag{4.1}$$

then there is a unique contractive intertwining lifting B *of* A. *Furthermore, this* B *is an isometry.*

PROOF. Let $\hat{U}'$ on $\hat{K}'$ be the minimal unitary dilation of T' and B a contractive intertwining lifting of A. By definition $\hat{U}'^*$ is the minimal isometric dilation of U'^*. Since B^* is in $I_1(U'^*, U^*)$ we infer by Proposition 3.3 that B^* has a unique contractive intertwining lifting $\hat{B}^*$, or equivalently, that B has a unique contractive extension $\hat{B}$ in $I_1(\hat{U}, \hat{U}')$. (This means that $\hat{B}$ is the only operator in $I_1(\hat{U}, \hat{U}')$ such that $\hat{B}K \subseteq K'$ and $B = \hat{B}|K$). For h in $\ker D_A$ we have

$$\|\hat{B}h\| = \|Bh\| \geq \|P_{H'}Bh\| = \|Ah\| = \|h\| \geq \|\hat{B}h\| .$$

and therefore

$$\hat{B}h = P_{H'}\hat{B}h = Ah \quad \text{(when } h \in \ker D_A) . \tag{4.2}$$

Since $\hat{B}$ is in $I_1(\hat{U}, \hat{U}')$ this gives $\hat{U}'^n Ah = \hat{B}\hat{U}^n h$ for all n. From (4.1) we readily conclude that $\hat{B}$ is uniquely determined by A and consequently so is $B = \hat{B}|K$.

Let h be in $\ker D_A$ and n be an integer in $(-\infty, \infty)$. Using (4.2) we have

$$\|\hat{B}\hat{U}^n h\| = \|\hat{U}'^n Ah\| = \|Ah\| = \|h\| = \|\hat{U}^n h\| .$$

Thus $D_{\hat{B}}\hat{U}^n h = 0$ for all n. Equation (4.1) implies that $D_{\hat{B}} = 0$ and $\hat{B}$ is an isometry. Clearly $B = \hat{B}\,|\,K$ is an isometry. This completes the proof.

4.2 COROLLARY. *Let A be a contraction in $I(T, T')$. If ker D_A is cyclic for U, then there exists a unique contractive intertwining lifting B of A. In this case, B is an isometry and A·T is a regular factorization.*

PROOF. If ker D_A is a cyclic for U, then obviously (4.1) holds. So the first part of the corollary follows from the previous theorem, which also shows that B is an isometry. To complete the proof it remains to show that A·T is a regular factorization. To this end, notice that $f \oplus g$ is in $G = G(A\cdot T)$ if and only if $f \oplus g$ is in $D_A \oplus D_T$ and $f \oplus g$ is orthogonal to $F = F(A\cdot T)$. So $f \oplus g$ is in G if and only if $T^* D_A f + D_T g = 0$. By the matrix form for the minimal isometric dilation of U in (VI.3.4), this implies that $D_A f \oplus g$ is in ker U^*. Thus $D_A f \oplus g$ is orthogonal to $U^n \ker D_A$ for all $n \geq 0$. Because ker D_A is cyclic, $D_A f \oplus g = 0$, or equivalently, $f \oplus g = 0$. Therefore $G = \{0\}$ and A·T is a regular factorization. This completes the proof.

4.3 REMARK. Actually the previous theorem and corollary can be obtained directly without using the commutant lifting theorem. We will illustrate this by providing a direct proof of the first two assertions in Corollary 4.2. To see this let x_i be in ker D_A, that is, $(I - A^* A)x_i = 0$ and set $y_i = Ax_i$. Using $A^* y_i = x_i$ for $0 \leq i \leq n$ we have for all $m \geq 0$ and $0 \leq i, j \leq n$

$$(U^m x_i, x_j) = (T^m x_i, A^* y_j) = (T'^m A x_i, y_j) = (U'^m y_i, y_j)$$

This implies that for all x_i in ker D_A

$$\left\|\sum_0^n U^i x_i\right\|^2 = \sum_{0,0}^{n,n}(U^i x_i, U^j x_j) = \sum_{i \geq j}(U^{i-j}x_i, x_j) + \sum_{j > i}(x_i, U^{j-i}x_j) =$$

$$\sum_{i \geq j}(U'^{i-j}y_i, y_j) + \sum_{j > i}(y_i, U'^{j-i}y_j) = \left\|\sum_0^n U'^i y_i\right\|^2 .$$

Because ker D_A is cyclic for U this implies that the operator B defined by

$$B \sum_0^n U^i x_i = \sum_0^n U'^i y_i \qquad (x_i \in \ker D_A \text{ and } Ax_i = y_i \text{ for } 0 \leq i \leq n, n = 0, 1, 2, \cdots)$$

is an isometry from K to K'. Obviously $U'B = BU$. Notice that

$$P_{H'}BU^i x_i = P_{H'}U'^i Bx_i = P_{H'}U'^i y_i = T'^i Ax_i = AT^i x_i = AP_H U^i x_i \,.$$

This and the fact that $\ker D_A$ is cyclic for U implies that $P_{H'}B = AP_H$, or equivalently, B is a lifting of A. Thus B is a contractive intertwining lifting of A. The uniqueness follows from the fact that if B' is any other contractive intertwining lifting of A, then

$$BU^i x_i = U'^i Bx_i = U'^i y_i = U'^i Ax_i = U'^i P_{H'}B' x_i = U'^i B' x_i = B'U^i x_i \,.$$

Since $\ker D_A$ is cyclic $B = B'$. Therefore B is the unique contractive intertwining lifting of A and B is an isometry.

Next we will show how one can use the previous results to construct the unique contractive intertwining lifting for Hankel matrices when $\ker D_A$ is nonzero. To this end, consider the contractive Hankel matrix A on l^2 defined by

$$A = \begin{bmatrix} a_0 & a_1 & a_2 & a_3 & \cdots \\ a_1 & a_2 & a_3 & a_4 & \cdots \\ a_2 & a_3 & a_4 & a_5 & \cdots \\ a_3 & a_4 & a_5 & a_6 & \cdots \\ \vdots & \vdots & \vdots & \vdots & \cdots \end{bmatrix} \tag{4.3}$$

A simple calculation shows that $U^* A = AU$ where U is now the unilateral shift on $l^2 (= l^2(C^1))$. Let $\hat U$ be the bilateral shift on $\hat l^2 (= \hat l^2(C^1))$, which moves elements in the two way infinite column vector one position down. It adjoint $\hat U^*$ is the minimal isometric dilation of U^*. By the commutant lifting theorem there exists a contractive intertwining lifting B (from l^2 to $\hat l^2$) of A. Using $\hat U^* B = BU$ we see that B admits a matrix representation of the form

$$B = \begin{bmatrix} \vdots & \vdots & \vdots & \vdots & \cdots \\ a_{-2} & a_{-1} & a_0 & a_1 & \cdots \\ a_{-1} & a_0 & a_1 & a_2 & \cdots \\ a_0 & a_1 & a_2 & a_3 & \cdots \\ a_1 & a_2 & a_3 & a_4 & \cdots \\ a_2 & a_3 & a_4 & a_5 & \cdots \\ \vdots & \vdots & \vdots & \vdots & \cdots \end{bmatrix} \tag{4.4}$$

This sets the stage for the following result.

4.4 COROLLARY. *Let A in (4.3) be a contractive Hankel matrix. If $\ker D_A$ is nonzero, then there is only one contraction B of the form (4.4).*

PROOF. Let $\ker D_A$ be nonzero, or equivalently, $\|Ax\| = \|x\|$ for some nonzero $x = [x_0, x_1, x_2, \cdots]^{tr}$ where tr denotes the transpose. We can always choose an x in $\ker D_A$ such that $x_0 = 1$. To verify this, let n be the first integer where x_n is nonzero. Then $x = U^n y$ where the first component of y is nonzero. Using $U^* A = AU$ we have

$$\|y\| = \|U^n y\| = \|x\| = \|Ax\| = \|AU^n y\| = \|U^{*n} Ay\| \leq \|Ay\| \leq \|y\| .$$

Thus $\|y\| = \|Ay\|$ or y is in $\ker D_A$. So by choosing $x = y$ and normalizing the first component to one we assume that $x_0 = 1$.

Using the fact that B is a contractive lifting of A and $\|Ax\| = \|x\|$ we see that $Bx = Ax$. This and (4.4) implies

$$[a_{-i}, a_{-i+1}, a_{-i+2}, \cdots]x = 0 \qquad \text{(for all } i > 0) .$$

Therefore the a_i for $i < 0$ in (4.4) can be computed recursively by

$$a_{-i} = -\sum_{j=1}^{\infty} a_{j-i} x_j \qquad (i > 0) . \tag{4.5}$$

All contractive intertwining liftings B of A are computed by (4.5). So there is only one contractive intertwining lifting B of A. It is given by (4.4) and (4.5). This completes the proof.

4.5 REMARK. One can also use Theorem 4.1 to obtain a proof of the previous corollary, and thus establish that the unique contraction B in (4.4) is an isometry. Since this argument involves ideas developed later in the text, we will only provide a hint of the proof. To this end associate to x the function $x(e^{it})$ in L^2 defined by

$$x(e^{it}) = \sum_{0}^{\infty} x_n e^{int} .$$

In this case, the space

$$\bigvee_{-\infty}^{\infty} \hat{U}^n x \ \text{ corresponds to } \ \bigvee_{-\infty}^{\infty} e^{int} x(e^{it}) = X .$$

by the Riesz theorem $x(e^{it})$ is nonzero a.e. This implies that $X = L^2$. Under this identification.

$$\bigvee_{-\infty}^{\infty} \hat{U}^n x = \hat{l}^2 .$$

Therefore (4.1) holds, there is a unique contractive intertwining lifting B of A, this B is an isometry, and it is given by (4.4) and (4.5).

The following results will play a basic role in the next section.

4.6 PROPOSITION. *There is a one to one correspondence between the set of all contractive intertwining liftings* B *of* A *in* $I_1(T, T')$, *and the set of all contractive intertwining lifting* B$_*$ *of* A* *in* $I_1(T'^*, T')$. *In particular, there is a unique contractive intertwining lifting of* A *in* $I_1(T, T')$ *if and only if there is a unique contractive intertwining lifting of* A* *in* $I_1(T'^*, T^*)$.

PROOF. As before $\hat{U}$ on $\hat{K}$ and $\hat{U}'$ on $\hat{K}'$ are the minimal unitary dilations of T and T', respectively. According to Corollary 2.4 in Chapter VI,

$$\hat{B} = \text{strong} \lim_{n \to \infty} \hat{U}'^{*n} B \hat{U}^n P_{-n} \tag{4.6}$$

is the only operator in $I(\hat{U}, \hat{U}')$ extending B. As before P_{-n} is the orthogonal projection onto $\hat{U}^{*n}K$. We claim that

$$\hat{B}^* K'_* \subseteq K_* \quad \text{and} \quad \hat{B}^* (K'_* \ominus H') \subseteq K_* \ominus H . \tag{4.7}$$

where K'_* is defined analogously to K_* in (VI.5.4). Using the fact that B is an intertwining lifting of A with $\hat{B} | K = B$ and (VI.5.10) we have

$$\hat{B}(\hat{K} \ominus K_*) = \hat{B}(K \ominus H) = B(K \ominus H) \subseteq K' \ominus H' = \hat{K}' \ominus K'_* .$$

This readily yields the first relation in (4.7). The second relation follows from $\hat{B}K \subseteq K'$ and $K^\perp = K_* \ominus H$ (see (VI.5.10)) with the corresponding result for $K'^\perp$.

Let B$_*$ mapping K'_* into K_* be the operator defined by $B_* = \hat{B}^* | K_*$. Recall that the isometric lifting of T^* and T'^* is $U_* = U^* | K_*$ and $U'_* = \hat{U}'^* | K'_*$ respectively. Furthermore, $\hat{U}^*$ and $\hat{U}'^*$ is the minimal unitary dilations of T^* and T'^*, respectively (see Section VI.5). We claim that B$_*$ is an intertwining lifting of A* in $I_1(T'^*, T^*)$. Using the above definitions and the first equation in (4.7) it is easy to verify that B$_*$ is in $I(U'_*, U_*)$. By the second equation in (4.7) we have

$$P_H B_* = P_H \hat{B}^* | H' + P_H \hat{B}^* | (K'_* \ominus H') = P_H B^* P_{H'} = B^* P_{H'} = A^* P_{H'} .$$

Therefore B$_*$ is an intertwining lifting of A*.

By construction $\hat{B}^*$ in $I(\hat{U}'^*, \hat{U}^*)$ is an extension of B$_*$. By our previous results applied to B$_*$ we see that $\hat{B}^*$ and B$_*$ uniquely determine each other and $\|B_*\| = \|\hat{B}\|$. Therefore any intertwining dilation B of A in $I_1(T, T')$ uniquely determines an intertwining lifting B$_*$ of A* in $I_1(T'^*, T^*)$ and $\|B_*\| = \|B\|$. By duality there is a one to one correspondence (through $\hat{B}$) between the set of all intertwining liftings B($=\hat{B} | K$) of

A in $I_1(T, T')$, and the set of all intertwining liftings $B_*(=\hat{B}^*|K_*')$ of A^* in $I_1(T'^*, T^*)$. In our correspondence $\|B\| = \|B_*\|$. From this Proposition 4.6 readily follows.

4.7 REMARK. The operator $\hat{B}^*$ can be obtained by

$$\hat{B}^* = \text{strong} \lim_{n \to \infty} \hat{U}^{*n} B^* \hat{U}'^n P'_{-n} \tag{4.8}$$

where P'_{-n} is the orthogonal projection onto $\hat{U}'^{*n} K'$. To prove this first notice that $P_K \hat{B}^* = B^* P_{K'}$. This is a consequence of the second equation in (4.7) and $\hat{K} = K \oplus (K_* \ominus H)$ with the corresponding result for $\hat{K}'$ (see (VI.5.10)). Now we are ready to prove (4.8). Let k be in K and $n > m$. Using $\hat{U}\hat{B}^* = \hat{B}^*\hat{U}'$ we have

$$\hat{U}^{*n} B^* \hat{U}'^n P'_{-n} \hat{U}'^{*m} k = \hat{U}^{*n} P_K \hat{B}^* \hat{U}'^{n-m} k =$$

$$\hat{U}^{*n} P_K \hat{U}^{(n-m)} \hat{B}^* k \to \hat{U}^{*m} \hat{B}^* k = \hat{B}^* \hat{U}'^{*m} k \quad (\text{as } n \to \infty) .$$

The limit follows from (VI.5.13). This and the fact that $\{\hat{U}'^{*n} K'\}$ is dense in $\hat{K}'$ proves (4.8).

The previous proposition with Theorem 4.1 readily yields the following result.

4.8 COROLLARY. *Let $\hat{U}'$ on $\hat{K}'$ be the minimal unitary dilation of T' and A be in $I_1(T, T')$. If*

$$\bigvee_{-\infty}^{\infty} \hat{U}'^n \ker D_{A^*} = \hat{K} \tag{4.9}$$

then there is a unique contractive intertwining lifting B of A.

It is emphasized that if (4.9) holds, then B is not necessarily a co-isometry. All we know from the proof of Proposition 4.6 is that $B = \hat{B}_*|K$ where $\hat{B}_*$ is an isometry. Here $\hat{B}_*$ is the unique isometric extension of B_* the unique contractive intertwining lifting of A^*. In fact choose

$$A = T = T' = \begin{bmatrix} 0 & 0 \\ 1 & 0 \end{bmatrix} \quad \text{on } C^2$$

then both (4.1) and (4.9) hold. In this case the unique contractive intertwining lifting B of A is $B = U$ the unilateral shift on l^2. Obviously B is not a co-isometry.

5. REGULAR FACTORIZATIONS AND UNIQUE
CONTRACTIVE INTERTWINING LIFTINGS

In this section we show that there exists a unique contractive intertwining lifting of A if and only if A·T or T$'$·A is a regular factorization.

We begin with the following geometric result.

5.1 THEOREM. *Let* A *be a contraction in* $I(T, T')$.

(i) If A $\cdot$ T *is a regular factorization, then there is only one operator* B *in* $I(U, U')$ *such that* B $|H$ *is a contraction and* B *is a lifting of* A. *This operator* B *is the contractive intertwining lifting of* A *given by the commutant lifting theorem.*

(ii) If T$'$ $\cdot$ A *is a regular factorization, then there is only one operator* B *in* $I(U, U')$ *such that*

$$\lim_{n \to \infty} \|B^* U'^n h\| \le \|h\| \qquad (\textit{for all } h \in H') \tag{5.1}$$

and B *is a lifting of* A. *This operator* B *is the contractive intertwining lifting of* A *given by the commutant lifting theorem.*

PROOF (i). Without loss of generality we assume that U on K is in matrix form, that is,

$$U = \begin{bmatrix} T & 0 & 0 & 0 & 0 & \ldots \\ D_T & 0 & 0 & 0 & 0 & \ldots \\ 0 & I & 0 & 0 & 0 & \ldots \\ 0 & 0 & I & 0 & 0 & \ldots \\ 0 & 0 & 0 & 1 & 0 & \ldots \\ 0 & 0 & 0 & 0 & 1 & \ldots \\ \vdots & \vdots & \vdots & \vdots & \vdots & \vdots \end{bmatrix} \tag{5.2}$$

where $K = H \oplus l^2(D_T)$. The isometry U$'$ on $K' = H' \oplus l^2(D_{T'})$ is defined in exactly the same way, but T$'$ replaces T in (5.2). Let B satisfy the hypothesis of (i). Matrix multiplication with (5.2) shows that if B $\in I(U, U')$ and $A^* = B^* |H'$, then B is of the form

$$
B = \begin{bmatrix}
A & 0 & 0 & 0 & 0 & \cdots \\
Z_1 & Y_1 & 0 & 0 & 0 & \cdots \\
Z_2 & Y_2 & Y_1 & 0 & 0 & \cdots \\
Z_3 & Y_3 & Y_2 & Y_1 & 0 & \cdots \\
Z_4 & Y_4 & Y_3 & Y_2 & Y_1 & \cdots \\
\cdot & \cdot & \cdot & \cdot & \cdot & \cdots \\
\cdot & \cdot & \cdot & \cdot & \cdot & \cdots \\
\cdot & \cdot & \cdot & \cdot & \cdot & \cdots
\end{bmatrix}
\tag{5.3}
$$

where $Z_i : H \to D_{T'}$ are contractions and $Y_i : D_T \to D_{T'}$ are operators for all $i \geq 1$.

Since $\|Bh\|^2 \leq \|h\|^2$ for h in H, equation (5.3) implies $\|Z_i h\| \leq \|D_A h\|$. Thus, $Z_i = X_i D_A$ for $i \geq 1$ where X_i is a contraction from D_A into $D_{T'}$. Finally, using $U'Bh = BUh$ for h in H along with (5.3) gives

$$
[X_1, \, Y_1] \begin{bmatrix} D_A Th \\ D_T h \end{bmatrix} = D_{T'} Ah \, ,
$$

$$
[X_{n+1}, \, Y_{n+1}] \begin{bmatrix} D_A Th \\ D_T h \end{bmatrix} = X_n D_A h \qquad (\text{if } n \geq 1) \, .
\tag{5.4}
$$

Since $A \cdot T$ is a regular factorization, the Z_i's and Y_i's in (5.3) are uniquely determined by (5.4). Hence B is unique. The commutant lifting theorem guarantees that there exists a contraction C in $I(U, U')$ which lifts A. The uniqueness gives $B = C$ and completes the proof of (i).

Now assume that the hypothesis of (ii) holds. By Corollary 2.3 in Chapter V the factorization $A^* \cdot T^{'*}$ is regular. Part (i) and Proposition 4.6 show that both A^* in $I_1(T^{'*}, T^*)$ and A in $I_1(T, T')$ admit a unique contractive intertwining lifting B_* and B respectively. Part (i) shows that B_* is the only intertwining lifting of A^* such that $B_* | H$ is a contraction. On the other hand, if B in $I(U, U')$ is a lifting of A and $\hat{B}$ is its unique intertwining extension in $I(\hat{U}, \hat{U}')$, then (4.8) holds for $\hat{B}^*$ and B^*. The relation (5.1) shows that $\hat{B}^* | H'$ is a contraction. Therefore $\hat{B}^* | K_*'$ is uniquely determined and equals B_*. By virtue of Proposition 4.6 and Remark 4.7, the operator B is uniquely determined. This completes the proof.

The following theorem is the main result of this section.

5.2 THEOREM. *The contraction* A *in* $I(T, T')$ *has a unique contractive intertwining lifting if and only if* $A \cdot T$ *or* $T' \cdot A$ *is a regular factorization.*

PROOF. If $A \cdot T$ or $T' \cdot A$ is a regular factorization, then by the previous theorem, A has a unique contractive intertwining lifting.

To prove the other half let T_1 and T'_1 be the one step dilations of T and T', respectively. Assume $T' \cdot A$ and $A \cdot T$ are not regular. Corollary 1.4 in Chapter V implies that there exists two different contractions A_1, A'_1 in $I_1(T_1, T'_1)$ of the form

$$A_1 = \begin{bmatrix} A & 0 \\ Z_1 & X_1 \end{bmatrix} \text{ and } A'_1 = \begin{bmatrix} A & 0 \\ Z'_1 & X'_1 \end{bmatrix}. \tag{5.5}$$

Notice that T and T_1 have the same minimal isometric dilation U. Likewise U' is the minimal isometric dilation for both T' and T'_1. Applying the commutant lifting theorem to A_1, A'_1 in $I_1(T_1, T'_1)$ gives two contractive intertwining liftings B and B' for A_1 and A'_1, respectively. Since obviously the lifting of a lifting is a lifting, B and B' are both contractive intertwining lifting of A. Obviously $B \neq B'$. Hence A does not have a unique contractive intertwining lifting B of A. This completes the proof.

5.3 REMARK. There exists a unique contractive intertwining lifting $B(=0)$ of $A = 0$ in $I_1(T, T')$ if and only if

$$D_{T'} = \{0\} \text{ or } D_{T^*} = \{0\}. \tag{5.6}$$

This follows from the previous theorem and the fact that $A \cdot T$ is regular if and only if $T^* \cdot A^*$ is regular; see Corollary 2.3 in Chapter V.

(ii) If (5.7) does not hold and there exists a unique contractive intertwining lifting B of $A \in I_1(T, T')$, then $\|A\| = 1$.

To see this assume that $\|A\| < 1$. Choose by the commutant lifting theorem a lifting B in $I_1(U, U')$ of A such that $\|B\| = \|A\| < 1$. Then choose (by the preceding remark) any lifting C in $I_1(T, T')$ of 0 such that $\|C\| = 1 - \|B\|$. Obviously $B \pm C$ are two distinct contractive liftings of A. This yields (ii).

5.4 COROLLARY. *Let H be finite dimensional and A in $I_1(T, T')$. Assume that the dimension of $D_{T'}$ is one and all the eigenvalues of T are contained in the open unit disc. Then A admits a unique contractive intertwining lifting if and only if the norm of A is one.*

PROOF. Corollaries V.1.4 and V.2.6 show that $\|A\| = 1$ if and only if $T' \cdot A$ or $A \cdot T$ is a regular factorization. This and Theorem 5.2 completes the proof.

If T or T' has an eigenvalue of modulus one, then the previous corollary does not hold. A counter example is given by T, T' on C^2 and A on C^2 defined by

$$T = T' = \begin{bmatrix} 1/\sqrt{2} & 0 \\ 0 & 1 \end{bmatrix}, \quad A = \begin{bmatrix} 1/\sqrt{2} & 0 \\ 0 & 1 \end{bmatrix}.$$

5.5 REMARK. It is emphasized that if $A \cdot T$ is a regular factorization, then one can use (5.4) with the fact that $[T^* D_A, D_T]^*$ is densely onto $D_A \oplus D_T$ to recursively compute the unique contractive intertwining lifting B of A in (5.3). If $\dim G' \geq \dim G$, then one can choose an isometry Γ in the contractive one step intertwining lifting A_1 of A in (V.1.16). By Corollary V.1.7 the factorization $A_1 \cdot T_1$ is regular. Now replacing A by A_1 and T by T_1 one can use (5.4) to recursively compute the unique contractive intertwining lifting B of A_1 in (5.3). Since B is also a contractive intertwining lifting of A, this provides us with the following recursive procedure to compute at least one contractive intertwining lifting B of A when $\dim G' \geq \dim G$: choose Γ in (V.1.16) to be an isometry. Then use (5.3) and (5.4) on A_1 and T_1 to recursively compute a contractive intertwining lifting B of A. Finally, it is noted that if $\dim G > \dim G'$, then one can use the above procedure to recursively compute a contractive intertwining lifting B_* of A^*. By the proof of Proposition 4.6 this readily produces a contractive intertwining lifting B of A.

6. COMMUTING ISOMETRIC LIFTINGS AND THE THIRD PROOF OF THE COMMUTANT LIFTING THEOREM

In this section we will use the concept of a commuting isometric lifting to obtain another proof of the commutant lifting theorem. We say that $\{V, W\}$ on N is a *commuting isometric lifting* of a pair of commuting contractions $\{T, A\}$ on H if V and W are commuting isometries on N lifting T and A, respectively. In this section we will show that any pair of commuting contractions admits a commuting isometric lifting. We will also show that this result is equivalent to the commutant lifting theorem.

We begin with the following result due to Ando.

6.1 THEOREM. *Every pair of commuting contractions $\{T, A\}$ on H admits a commuting isometric lifting $\{V, W\}$ on N.*

PROOF. Consider the isometries U_T and U_A on $l^2(H)$ defined by

$$U_T(h_0, h_1, h_2, ...) = (Th_0, D_T h_0, 0, h_1, h_2, h_3, ...)$$
$$U_A(h_0, h_1, h_2, ...) = (Ah_0, D_A h_0, 0, h_1, h_2, h_3, ...)$$

(6.1)

where h_i is in H. If we identify h in H with $(h, 0, 0, ...)$ in $l_2(H)$, then U_T and U_A are isometric liftings of T and A, respectively. A simple calculation gives

$$U_T U_A(h_0, h_1, h_2, ...) = (TAh_0, D_T Ah_0, 0, D_A h_0, 0, h_1, h_2, h_3, ...)$$
$$U_A U_T(h_0, h_1, h_2, ...) = (ATh_0, D_A Th_0, 0, D_T h_0, 0, h_1, h_2, h_3, ...) .$$

(6.2)

Clearly U_A and U_T do not commute but a minor modification of these operators will induce commutativity.

As in Section V.1 for the operator ω, the operator ω_o mapping F_o into F_o' defined by

$$\omega_o (D_A Th \oplus 0 \oplus D_T h \oplus 0) = D_T Ah \oplus 0 \oplus D_A h \oplus 0 \qquad (h \in H)$$
$$F_o = \{D_A Th \oplus 0 \oplus D_T h \oplus 0 : h \in H\}^- \text{ and } F_o' = \{D_T Ah \oplus 0 \oplus D_A h \oplus 0 : h \in H\}^-$$

(6.3)

is unitary. Let R be any unitary operator on $H \oplus H \oplus H \oplus H$ expanding ω_o, that is, $R | F_o = \omega_o$. Notice that if H is finite dimensional, then the orthogonal complement of F_o and F_o' are of the same dimension. So there exists a unitary expansion R of ω_o. On the other hand, if H is infinite dimensional, then the orthogonal complement of both F_o and F_o' are both infinite dimensional. Thus there exists a unitary expansion R of ω_o. Now let R also be the unitary operator on $l^2(H)$ defined by

$$R(h_0, h_1, h_2, h_3, ...) = (h_0, R(h_1, h_2, h_3, h_4), R(h_5, h_6, h_7, h_8), ...) .$$

Set $V = U_T R^{-1}$ and $W = RU_A$. Obviously V and W are contractive liftings of T and A, respectively. By (6.3) and (6.2) we have

$$WV(h_0, h_1, h_2, ...) = (ATh_0, R(D_A Th_0 \oplus 0 \oplus D_T h_0 \oplus 0), h_1, h_2, h_3, h_4, ...) =$$

$$(TAh_0, D_T Ah_0 \oplus 0 \oplus D_A h_0 \oplus 0, h_1, h_2, h_3, ...) = U_T U_A = VW .$$

Therefore $\{V, W\}$ on $l^2(H)$ is a commuting isometric lifting of $\{T, A\}$. This completes the proof.

Next we will use the commutant lifting theorem to obtain an alternate proof of the previous theorem.

ALTERNATE PROOF. Recall that TA = AT. Using the commutant lifting theorem there exists a contractive lifting B of A satisfying UB = BU where U on K is the minimal isometric dilation of T. Denoting by W the minimal isometric dilation of B and applying the commutant lifting theorem to U in $I_1(B, B)$ there exists a contractive lifting V in $I_1(W, W)$ of U. Since V is a contractive lifting of the isometry U we have $V | K = U$. This and VW = WV gives

$$\|VW^n k\| = \|W^n Vk\| = \|Vk\| = \|Uk\| = \|k\| = \|W^n k\| \quad (k \in K)$$

Therefore $D_V W^n k = 0$ for all $n \geq 0$. Because W is the minimal isometric dilation of B on K the minimality implies that $D_V = 0$, or equivalently, V is an isometry. (Theorem 4.1 with $A = U$ also shows that V is an isometry.) Using again the simple fact that a lifting of a lifting is a lifting we conclude that $\{V, W\}$ is also commuting isometric lifting of $\{T, A\}$. This completes the proof.

It is remarkable that the previous theorem can be used to prove the commutant lifting theorem. This is our next task.

THIRD PROOF OF THE COMMUTANT LIFTING THEOREM. Let A be in $I_1(T, T')$. To put us in the framework of Theorem 6.1 notice that

$$\begin{bmatrix} T & 0 \\ 0 & T' \end{bmatrix} \begin{bmatrix} 0 & 0 \\ A & 0 \end{bmatrix} = \begin{bmatrix} 0 & 0 \\ A & 0 \end{bmatrix} \begin{bmatrix} T & 0 \\ 0 & T' \end{bmatrix}. \tag{6.4}$$

According to Theorem 6.1 there exists a pair of commuting isometric operators $\{V, W\}$ lifting the pair of contractions

$$\left\{ \begin{bmatrix} T & 0 \\ 0 & T' \end{bmatrix}, \ \begin{bmatrix} 0 & 0 \\ A & 0 \end{bmatrix} \right\}.$$

Let K and K' be the spaces defined by

$$K = \bigvee_0^\infty V^n H \text{ and } K' = \bigvee_0^\infty V^n H' \tag{6.5}$$

where H and H' are identified with $H \oplus \{0\}$ and $\{0\} \oplus H'$, respectively. Since V is an isometric lifting of $T \oplus T'$ we see that $U = V | K$ and $U' = V | K'$ are the minimal isometric dilation of T and T', respectively. Let B be the contraction defined by $B = P_{K'} W | K$. Using the lifting property of W we have

$$P_{H'} B = P_{H'} W | K = P_{H'} P_{H \oplus H'} W | K = P_{H'} \begin{bmatrix} 0 & 0 \\ A & 0 \end{bmatrix} P_{H \oplus H'} | K = A P_H | K.$$

So B is a lifting of A. Remark VI.3.3 shows that both K and K' reduce V. Thus

$$U' B = V P_{K'} W | K = P_{K'} V W | K = P_{K'} W V | K = B U.$$

Therefore B is a lifting of A in $I_1(U, U')$, or equivalently, B is a contractive intertwining lifting of A. This completes the proof.

Summing up the previous analysis yields the following result.

6.2 REMARK. Let A be in $I_1(T,T')$ and $\{V,W\}$ be a commuting isometric lifting of

$$\left\{ \begin{bmatrix} T & 0 \\ 0 & T' \end{bmatrix}, \begin{bmatrix} 0 & 0 \\ A & 0 \end{bmatrix} \right\}.$$

If K and K' are the spaces defined in (6.5), then $B = P_{K'}W|K$ is a contractive intertwining lifting of A.

The commutant lifting theorem naturally leads to the following question: If A_1 and A_2 are two *commuting* contractions in $I(T,T')$, then does there exist two *commuting* contractive intertwining liftings B_1 of A_1 and B_2 of A_2? The following provides a counter example. In fact this counter example shows that in general, there does not exist a pair of commuting contractive liftings B_1 of A_1 and B_2 of A_2 satisfying $UB_1 = B_1U$ and $UB_2 = B_2U$ where U is any (even nonminimal) isometric lifting of T and $T = T'$.

6.3 EXAMPLE. Consider the following contractions on $H = C^2 \oplus C^2$

$$T = T' = \begin{bmatrix} 0 & I \\ 0 & 0 \end{bmatrix} \text{ and } A_i = \begin{bmatrix} 0 & C_i \\ 0 & 0 \end{bmatrix} \text{ on } C^2 \oplus C^2 \tag{6.6}$$

for $i = 1, 2$, where C_1 and C_2 are the unitary operators defined by

$$C_1 = \begin{bmatrix} 1 & 0 \\ 0 & -1 \end{bmatrix} \text{ and } C_2 = \begin{bmatrix} 0 & 1 \\ 1 & 0 \end{bmatrix}. \tag{6.7}$$

Obviously A_1 and A_2 are commuting contractions in $I_1(T,T)$. In fact the product of any pair of contractions in (6.6) is zero. Now assume that there exists a pair of commuting contractive intertwining liftings B_1 of A_1 and B_2 of A_2 satisfying $UB_1 = B_1U$ and $UB_2 = B_2U$ where U is a (possibly nonminimal) isometric lifting of T. By Corollary VI.2.4 there exists two contractions $\hat{B}_1$ extending B_1 and $\hat{B}_2$ extending B_2 commuting with $\hat{U}$, where $\hat{U}$ is the minimal unitary extension of U. Formula (VI.2.6) guarantees that $\hat{B}_1$ and $\hat{B}_2$ also commute. Since $\hat{U}$, $\hat{B}_1$ and $\hat{B}_2$ commute $\hat{U}^*$, $\hat{B}_1$, and $\hat{B}_2$ commute. Because I, C_1 and C_2 are all unitary operators we have

$$\hat{U}(0 \oplus h) = h \oplus 0 \text{ and } \hat{B}_i(0 \oplus h) = C_i h \oplus 0 \quad (\text{for } i=1,2) \tag{6.8}$$

when $0 \oplus h$ is in the subspace $\{0\} \oplus C^2$ of H. Using $\hat{U}^*(h \oplus 0) = 0 \oplus h$ we have

$$\hat{B}_1 \hat{U}^* \hat{B}_2 (0 \oplus h) = \hat{B}_1 \hat{U}^* (C_2 h \oplus 0) = C_1 C_2 h \oplus 0 \, ,$$

$$\hat{B}_2 \hat{U}^* \hat{B}_1 (0 \oplus h) = \hat{B}_2 \hat{U}^* (C_1 h \oplus 0) = C_2 C_1 h \oplus 0 \, .$$

Since $\hat{U}^*$, $\hat{B}_1$ and $\hat{B}_2$ commute this implies that $C_1 C_2 = C_2 C_1$. This is a contradiction. So there does not exist commuting contractive intertwining liftings of A_1 and A_2.

Since an isometric lifting is a special case of a contractive lifting, the previous example also shows that there does not in general exist a triple of commuting isometric liftings $\{V, W_1, W_2\}$ on N for a triple of commuting contractions $\{T, A_1, A_2\}$ on H.

7. THE COUPLING OF T, A AND T$'$

In this section we will introduce the concept of a coupling of $\{T, A, T'\}$ and present a complete classification of all one step couplings. In the next section we will use the idea of a coupling to obtain another proof of the commutant lifting theorem.

As before throughout this section, both T on H and T$'$ on H' are contractions and A is in $I_1(T, T')$. We say that C on $\hat{H}$ is a *coupling* of $\{T, A, T'\}$ if $\hat{H}$ is a space satisfying $\hat{H} \supseteq H \bigvee H'$ where $A = P_{H'}|H$, and C is a lifting of T$'$ and C is an extension of T. To simplify some of the calculations we will always assume that $T = U$ is an isometry. According to Lemma 1.1 this will also produce the set of all contractive intertwining lifting of A when T is a contraction.

Assume that $\hat{H}_1 = H \bigvee H'$ where H and H' satisfy $P_{H'}|H = A$. For h in H and h$'$ in H' this implies that

$$\|h'+h\|_1^2 = \|h'\|_1^2 + 2\mathrm{Re}(h, h')_1 + \|h\|_1^2 =$$
$$\|h'\|^2 + 2\mathrm{Re}(Ah, h') + \|Ah\|^2 + \|h\|^2 - \|Ah\|^2 = \|h' + Ah\|^2 + \|D_A h\|^2 \qquad (7.1)$$

where the subscript 1 refers to the norm or inner product in $\hat{H}_1$. By virtue of (7.1) the relation

$$\gamma(h'+h) = (h'+Ah) \oplus D_A h \qquad (h \in H \text{ and } h' \in H') \qquad (7.2)$$

defines a unitary operator γ from $\hat{H}_1$ onto $H' \oplus D_A$. Obviously h$'$ in H' is identified with $\gamma h' = h' \oplus 0$ and h in H with $\gamma h = Ah \oplus D_A h$. On the other hand, if (7.2) defines a unitary operator γ from $\hat{H}_1$ onto $H' \oplus D_A$ then clearly $\hat{H}_1 = H \bigvee H'$. Moreover, for h in H and h$'$ in H' we have

$$(P_{H'} h, h')_{H'} = (h, h')_1 = (\gamma h, \gamma h') = (Ah, h')_{H'} \, .$$

Therefore $P_{H'}|H = A$. Summing up the previous analysis yields the following result.

7.1 LEMMA. *Let A be a contraction mapping H into H'. A space $\hat{H}_1 = H \bigvee H'$ where H and H' satisfy $P_{H'}|H = A$ if and only if the operator γ defined in (7.2) is a unitary operator from $\hat{H}_1$ onto $H' \oplus D_A$.*

Given any contraction A from H to H' one can always embed H and H' in a Hilbert space $\hat{H}_1$ satisfying $\hat{H}_1 = H \bigvee H'$ where $P_{H'}|H = A$. To see this let ϕ be the isometry from H into $H' \oplus D_A$ defined by

$$\phi h = Ah \oplus D_A h \qquad (h \in H).\qquad (7.3)$$

(Lemma IV.1.1 guarantees that ϕ is an isometry.) Now we embed H in $H' \oplus D_A$ by identifying H with $\tilde{H} \stackrel{\Delta}{=} \phi H$. (The space H' is identified with $H' \oplus \{0\}$.) Under this identification the contraction A becomes the contraction $\tilde{A}$ from $\tilde{H}$ to H' defined by $\tilde{A}\phi h = Ah \oplus 0$ where h is in H. Equation (7.3) implies that $\tilde{A} = P_{H'}|\tilde{H}$. Obviously $H' \oplus D_A = \tilde{H} \bigvee H'$. Since H is identified with $\tilde{H}$ this proves our claim.

Let $\hat{H}$ be any space containing $H \bigvee H'$ satisfying $A = P_{H'}|H$ and γ be the unitary operator from $H \bigvee H'$ onto $H' \oplus D_A$ defined in (7.2). Let I_1 be the identity on $\hat{H} \ominus (H \bigvee H')$. Then using the definitions of γ and ϕ we see that C on $\hat{H}$ is a coupling of $\{U, A, T'\}$ if and only if

$$C = (I_1 \oplus \gamma^*)\tilde{C}(I_1 \oplus \gamma)\qquad (7.4)$$

where $\tilde{C}$ is an operator on $M = (\hat{H} \ominus (H \bigvee H')) \oplus (H' \oplus D_A)$ satisfying

$$P_{H'}\tilde{C} = T'P_{H'} \quad \text{and} \quad \tilde{C}\phi = \phi U.\qquad (7.5)$$

Consider the system $\{T_1, A_1, T_1'\}$ where T_1 is on H_1, A_1 maps H_1 onto H_1' and T_1' is on H_1'. We say that two systems $\{T, A, T'\}$ and $\{T_1, A_1\, T_1'\}$ are unitarily equivalent if there exists two unitary operator α from H onto H_1 and α' from H' onto H_1' satisfying $\alpha T = T_1\alpha$, $\alpha'A = A_1\alpha$ and $\alpha'T' = T_1'\alpha'$. Notice that if $\tilde{T}$ on $\tilde{H}$ is the contraction defined by $\tilde{T}\phi = \phi U$, then $\{\tilde{T}, \tilde{A}, T'\}$ is unitarily equivalent to $\{U, A, T'\}$. This follows because $\tilde{T}\tilde{\phi} = \tilde{\phi}U$ and $\tilde{A}\tilde{\phi} = A \oplus \{0\}$ which is identified with A. Here $\tilde{\phi}$ is the unitary from H onto $\tilde{H}$ defined by $\tilde{\phi} = \phi$. (As always we identify H' with $H' \oplus \{0\}$.) Moreover, the couplings C of $\{U, A, T'\}$ and $\tilde{C}$ of $\{\tilde{T}, \tilde{A}, T'\}$ are related by (7.4). The set of all couplings $\tilde{C}$ of $\{\tilde{T}, \tilde{A}, T'\}$ is given by the set of all $\tilde{C}$ on $M \supseteq H' \oplus D_A$ satisfying (7.5).

Our next task is to construct a matrix representation for all contractions satisfying (7.5). By (7.4) this produces the set of all couplings C of $\{U, A, T'\}$ up to a unitary operator. To begin recall that Corollary 2.3 in Chapter IV shows that $\tilde{C}$ on $M \supseteq H' \oplus D_A$ is a contractive lifting of T', or equivalently, the first condition in (7.5) holds if and only if $\tilde{C}$ admits a matrix representation of the form

$$\tilde{C} = \begin{bmatrix} T' & 0 \\ XD_{T'} & Y \end{bmatrix} \text{ on } H' \oplus H'^{\perp} \tag{7.6}$$

where $[X,Y]$ is a contraction from $D_{T'} \oplus H'^{\perp}$ to $H'^{\perp}$. Notice that $\tilde{C}\phi = \phi U$ if and only if

$$T'Ah \oplus (XD_{T'}Ah + YD_A h) = \tilde{C}\phi h = \phi Uh = AUh \oplus D_A Uh \quad (h \in H) .$$

Using $AU = T'A$ this implies that the second condition in (7.5) holds if and only if

$$[X,Y] \begin{bmatrix} D_{T'}A \\ D_A \end{bmatrix} = D_A U . \tag{7.7}$$

As in Section V.1, let ω^* be the unitary operator mapping F' onto F now defined by

$$\omega^*(D_{T'}Ah \oplus D_A h) = D_A Uh \qquad (h \in H)$$
$$F' = \{D_{T'}Ah \oplus D_A h : h \in H\}^- \text{ and } F = \overline{D_A UH} . \tag{7.8}$$

Then (7.7) is equivalent to

$$[X,Y]|F' = \omega^* . \tag{7.9}$$

Therefore, $\tilde{C}$ is a contraction satisfying (7.7) if and only if $\tilde{C}$ admits a matrix representation of the form (7.6), where $[X,Y]$ is a contraction satisfying (7.9). This readily yields the following result.

7.2 LEMMA. *An operator C is a contractive coupling of $\{U,A,T'\}$ if and only if C admits a matrix representation $\tilde{C}$ of the form (7.6), where $[X,Y]$ is a contraction from $D_{T'} \oplus H'^{\perp}$ to $H'^{\perp}$ satisfying (7.9), or equivalently, C is given by (7.4) where $\tilde{C}$ is a matrix of the form (7.6) and $[X,Y]$ is a contraction satisfying (7.7) or (7.9).*

We say that C on $\hat{H}$ is a *one step coupling* of $\{U,A,T'\}$ if C is a coupling of $\{U,A,T'\}$ and $\hat{H} = \hat{H}_1 = H \bigvee H'$. If we use the matrix representation for C in (7.6), this is equivalent to requiring that $M = \tilde{H}_1 = H' \oplus D_A$. To complete this section we will construct the set of all contractive one step couplings of $\{U,A,T'\}$. This analysis is similar to constructing the set of all contractive one step intertwining liftings in Section 1 of Chapter V. In fact up to a change in variables (7.6) and (7.7) are precisely (1.9) and

(1.10) in Chapter V. For completeness we will present this analysis. To this end let $M = \tilde{H}_1 = H' \oplus D_A$. Then the residual spaces G' and G are defined in the usual way:

$$G' = (D_{T'} \oplus D_A) \ominus F' \text{ and } G = D_A \ominus F . \tag{7.10}$$

Since ω is unitary Remark IV.1.7 shows that the set of all contractive [X,Y] satisfying (7.9), or equivalently, (7.7) is given by

$$[X,Y] = \omega^* P_{F'} + \Gamma^* P_{G'} \tag{7.11}$$

where Γ is an arbitrary contraction mapping G into G'. Hence the set of all contractive one step couplings C of $\{U,A,T'\}$ is unitarily equivalent to a matrix $\tilde{C}$ of the form (7.6) where [X,Y] is given by (7.11). Summing up the previous analysis with Lemma 7.2 yields the following result.

7.3 THEOREM. *Let A be in $I_1(U,T')$. Then C is a contractive one step coupling of $\{U,A,T'\}$ if and only if C admits a matrix representation $\tilde{C}$ of the form:*

$$\tilde{C} = \begin{bmatrix} T' & 0 \\ (\omega^* P_{F'} + \Gamma^* P_{G'}) D_{T'} & (\omega^* P_{F'} + \Gamma^* P_{G'}) P_{D_A} \end{bmatrix} \text{ on } H' \oplus D_A \tag{7.12}$$

where Γ is an arbitrary contraction from G into G'. In this case $C = \gamma^* \tilde{C} \gamma$ where γ is the unitary operator from $H \vee H'$ onto $H' \oplus D_A$ defined in (7.2).

7.4 COROLLARY. *There is a unique contractive one step coupling of $\{U,A,T'\}$ if and only if $G = \{0\}$ or $G' = \{0\}$. Moreover, if $G' = \{0\}$, then C is an isometry.*

PROOF. The first statement of the corollary is an obvious consequence of the previous theorem. If $G' = \{0\}$, then (7.11) implies that $[X,Y] = \omega^*$ is an isometry from $D_{T'} \oplus D_A$ into D_A. Using this in (7.6) gives

$$\|h \oplus g\|^2 - \|\tilde{C}(h \oplus g)\|^2 = \|h\|^2 + \|g\|^2 - \|T'h\|^2 - \|XD_{T'}h + Yg\|^2 =$$

$$\|D_{T'}h\|^2 + \|g\|^2 - \|XD_{T'}h + Yg\|^2 = \|D_{T'}h \oplus g\|^2 - \|D_{T'}h \oplus g\|^2 = 0$$

where h is in $D_{T'}$ and g is in D_A. This implies that the contractive coupling C of $\{U,A,T'\}$ is an isometry. The proof is now complete.

7.5 REMARK. Let A be a contraction in $I_1(U,T')$ and $\tilde{T}$ the isometry on $\phi H = \tilde{H}$ defined by $\tilde{T}\phi = \phi U$. A simple calculation shows that $\tilde{A} \overset{\Delta}{=} P_{H'} | \tilde{H}$ is in $I_1(\tilde{T},T')$. As noted earlier $\{U,A,T'\}$ is unitarily equivalent to $\{\tilde{T}, \tilde{A}, T'\}$. In particular, the couplings of $\{U,A,T'\}$ and $\{\tilde{T}, \tilde{A}, T'\}$ are unitarily equivalent. The set of all contractions $\tilde{C}$ in (7.12) is precisely the set of all contractive one step coupling of $\{\tilde{T}, \tilde{A}, T'\}$, which are unitarily equivalent to $(C = \gamma^* \tilde{C}\gamma)$ the set of all contractive one step couplings of $\{U,A,T'\}$. Finally, it is noted that due to unitary equivalence the contractive intertwining lifting of A and $\tilde{A}$ are equivalent up to a unitary operator on the right. These observations will lead to an algorithm for constructing the set of all contractive intertwining lifting B of A in the next section.

8. THE FOURTH PROOF OF THE COMMUTANT LIFTING THEOREM

In this section we will use the coupling of $\{T, A, T'\}$ to present another proof of the commutant lifting theorem. Then we will use the results on non-minimal isometric liftings in Section VI.7 to show that there is a one to one correspondence between the set of all contractive intertwining liftings B of A and a certain choice sequence. This will readily give another proof of fact that there exists a unique contractive intertwining lifting if and only if $A \cdot T$ or $T' \cdot A$ is a regular factorization.

Let us begin by presenting a coupling proof of the commutant lifting theorem. By Lemma 1.1 without loss of generality we assume that A is in $I_1(U,T')$ where $U = T$ is an isometry on H. Let V on N be any isometric coupling of $\{U,A,T'\}$. To establish the existence of at least one isometric coupling, let C be any contractive coupling and V the minimal isometric dilation of C. Because U is an isometry and $C|H = U$ we have $V|H = U$. Obviously $A = P_{H'}|H$. Since the lifting of a lifting is still a lifting V is a lifting of T'. By definition this V is an isometry coupling of $\{U,A,T'\}$.

As in the third proof of the commutant lifting theorem, let K' be the minimal space generated by H', that is,

$$K' = \bigvee_0^\infty V^n H' . \tag{8.1}$$

Since V is a coupling $P_{H'}V = T'P_{H'}$. Thus $U' \stackrel{\Delta}{=} V|K'$ is the minimal isometric dilation of T'. We claim that $B = P_{K'}|H$ is a contractive intertwining lifting of A. Using the fact that K' reduces V (see Remark VI.3.3) we have

$$U'B = VP_{K'}|H = P_{K'}V|H = P_{K'}U = BU . \tag{8.2}$$

The third equality follows from the coupling property $V|H = U$. The coupling property $A = P_{H'}|H$ gives

$$P_{H'}B = P_{H'}P_{K'} \,|\, H = P_{H'} \,|\, H = A \,. \tag{8.3}$$

Equations (8.2) and (8.3) verify that $B = P_{K'} \,|\, H$ is a contractive intertwining lifting of A. This completes the coupling proof of the commutant lifting theorem.

8.1 REMARK. Assume that A is a contraction in $I_1(U,T')$ where $U = T$ is an isometry on H and T' is a contraction on H'. The previous proof shows that there exists at least one isometric coupling V of $\{U,A,T'\}$. Moreover, if V is any isometric coupling of $\{U,A,T'\}$, then $B = P_{K'} \,|\, H$ is a contractive intertwining lifting of A where K' is the subspace generated by H' in (8.1).

To complete this section we show that there is a one to one correspondence between the set of all contractive intertwining liftings and a certain choice sequence. As before A is a contraction in $I_1(U, T')$. Let V on N be an isometric coupling of $\{U,A,T'\}$. The definition of K' in (8.1) shows that

$$\bigvee_0^\infty V^n(H \vee H') = H \vee K' \,.$$

We say that V on N is a *minimal isometric coupling* of $\{U,A,T'\}$ if V is an isometric coupling of $\{U,A,T'\}$ and

$$N = \bigvee_0^\infty V^n(H \vee H'), \text{ or equivalently, } N = H \vee K' \,. \tag{8.4}$$

If V is any isometric coupling of $\{U, A, T'\}$, then obviously $V \,|\, (H \vee K')$ is a minimal isometric coupling of $\{U, A, T'\}$. Minimal isometric couplings play a fundamental role in the characterization of all contractive intertwining liftings, which is demonstrated by the following result.

8.2 PROPOSITION. *Let A be a contraction in* $I_1(U,T')$. *There is a one to one correspondence between the set of all contractive intertwining liftings* B *of A and the set of all minimal isometric couplings* V *of* $\{U,A,T'\}$.

PROOF. Let B mapping H into K' be a contractive intertwining lifting of A where U' on K' is the minimal isometric dilation of T'. Because U' is an isometry $U' \cdot B$ is a regular factorization, or equivalently, $G(U' \cdot B) = \{0\}$. Applying Corollary 7.4 to $\{U,B,U'\}$ there exists a unique one step isometric coupling V on N of $\{U,B,U'\}$. We claim that V on N is also a minimal isometric coupling of $\{U,A,T'\}$. By the definition of a one step coupling $N = H \vee K'$. So the minimality condition in (8.4) is satisfied. Since

V is a lifting of U' and U' is a lifting of T' we see that V is a lifting of T'. Obviously $V | H = U$. Using $P_{K'} | H = B$ we have

$$P_{H'} | H = P_{H'} P_{K'} | H = P_{H'} B = A .$$

Thus V on N is a minimal isometric coupling of $\{U, A, T'\}$. Therefore every contractive intertwining lifting B of A uniquely determines a minimal isometric coupling V on N of $\{U, A, T'\}$. This coupling V is precisely the unique one step isometric coupling of $\{U, B, U'\}$.

On the other hand, let V on N be a minimal isometric coupling of $\{U, A, T'\}$. Remark 8.1 says that $B = P_{K'} | H$ is a contractive intertwining lifting of A where K' is the space generated by H' in (8.1). The minimality condition (8.4) implies that $N = H \bigvee K'$. Obviously V is a lifting of $U' = V | K'$ and $V | H = U$. So V is a one step isometric coupling of $\{U, B, U'\}$. Corollary 7.4 shows that V is the only contractive one step coupling of $\{U, B, U'\}$. Therefore each minimal isometric coupling V of $\{U, A, T'\}$ uniquely determines a contractive intertwining lifting B of A by $B = P_{K'} | H$. Moreover, this V is precisely the unique one step isometric coupling of $\{U, B, U'\}$. This establishes the one to one correspondence which completes the proof.

The proof of the previous proposition with Remark 8.1 gives the following result.

8.3 REMARK. Let A be a contraction in $I_1(U, T')$ where $U = T$ is an isometry on H. The set of all contractive intertwining liftings B of A is given by $B = P_{K'} | H$ where V on N is a minimal isometric coupling of $\{U, A, T'\}$ and K' is the space generated by H' in (8.1).

The following provides a complete characterization of all contractive intertwining liftings of A.

8.4 THEOREM. *Let A be a contraction in $I(U, T')$. There is a one to one correspondence between the set of all contractive intertwining liftings B of A, and the set of all choice sequences $\{\Gamma_i\}_1^\infty$ initiated from G' to G, where $G' - G'(T' \cdot A)$ is the residual space defined in (V.1.13) and $G = D_A \ominus \overline{D_A UH}$.*

PROOF. As in Section 7 we identify $H \bigvee H'$ with $H' \oplus D_A$; see (7.2). According to Lemma 7.2 the operator V on N is a minimal isometric coupling of $\{U, A, T'\}$ if and only if V is an isometry and admits a matrix representation of the form

$$\tilde{V} = \begin{bmatrix} T' & 0 & 0 \\ X_1 D_{T'} & Y_1 & * \\ * & * & * \end{bmatrix} \text{ on } H' \oplus D_A \oplus D \tag{8.5}$$

where $\Gamma_1 = [X_1, Y_1]$ is a contraction from $D_{T'} \oplus D_A$ to D_A satisfying $\Gamma_1 | F' = \omega^*$ and $H' \oplus D_A$ is cyclic for $\tilde{V}$. The contraction $[X_1, Y_1]$ in (8.5) equals $P_{D_A}[X, Y] | (D_{T'} \oplus D_A)$ where $[X, Y]$ is the contraction in Lemma 7.2 or (7.7), (7.9). (The unitary operator ω is defined in (7.8).) By Section VI.7 this is equivalent to $\tilde{V}$ being a D_A minimal isometric lifting of T' where its first choice contraction Γ_1 satisfies the constraint $\Gamma_1 | F' = \omega^*$. Therefore the set of all minimal isometric couplings V of $\{U, A, T'\}$ is unitarily equivalent to the matrix given in (VI.7.9), or equivalently, by (VI.7.12) where $\{\Gamma_i\}_1^\infty$ is a choice sequence subject to the constraint $\Gamma_1 | F' = \omega^*$. Since ω^* is unitary operator from F' to F the first choice contraction Γ_1 is only free from G' to G. So the set of all minimal isometric coupling are uniquely determined by a choice sequence initiated from G' to G. This with the previous proposition completes the proof.

The proof of the previous theorem along with the fact that $\{U, A, T'\}$ is unitarily equivalent to $\{\tilde{T}, \tilde{A}, T'\}$ (see Remark 7.5) readily yields the following result.

8.5 COROLLARY. *Let* A *be a contraction in* $I(U, T')$. *The set of all contractive intertwining liftings* B *of* A *is given by* $B = P_{K'} | \phi H$ *where* $\tilde{V}$ *is any* D_A *minimal isometric lifting of* T' *of the form* (VI.7.9), *whose choice sequence* $\{\Gamma_i\}_1^\infty$ *initiated from* $D_{T'} \oplus D_A$ *to* D_A *satisfies the constraint* $\Gamma_1 | F' = \omega^*$ *and* K' *is the space generated by* H' *in* (8.1). *In this case* $\tilde{V} | K'$ *is the minimal isometric dilation of* T'.

The previous theorem also gives us another proof of Theorem 5.2, that is, *a contraction* A *in* $I(T, T')$ *has a unique contractive intertwining lifting if and only if* $A \cdot T$ *or* $T' \cdot A$ *is a regular factorization*. To see this recall that Lemma 1.1 implies that $A P_H$ is in $I_1(U, T')$. The previous theorem shows that $A P_H$ admits a unique contractive intertwining lifting if and only if $G = \{0\}$ or $G' = \{0\}$, or equivalently, $A P_H \cdot U$ or $T' \cdot A P_H$ is a regular factorization. It is easy to verify that $A \cdot T$ or $T' \cdot A$ is a regular factorization if and only if $A P_H \cdot U$ or $T' \cdot A P_H$ is a regular factorization, respectively. This and Lemma 1.1 provides the uniqueness result for the commutant lifting theorem in Theorem 5.2.

To complete this section we will use minimal isometric couplings to obtain an explicit characterization of the set of all contractive intertwining lifting B of A. This is done by converting the previous corollary to a more useable form. To this end let $\tilde{V}$ be any D_A minimal isometric lifting of T' given by the matrix in (VI.7.9), where $\{\Gamma_i\}_1^\infty$ is a choice sequence initiated from $D_{T'} \oplus D_A$ to D_A satisfying $\Gamma_1 | F' = \omega^*$. By our previous analysis these $\tilde{V}$'s are precisely the set of all minimal isometric couplings of $\{\tilde{T}, \tilde{A}, T'\}$. Consulting (VI.7.9) we see that $\tilde{V}$ admits a matrix representation of the form

$$\tilde{V} = \begin{bmatrix} T' & 0 \\ \Psi D_{T'} & Z \end{bmatrix} \tag{8.6}$$

where Ψ is the isometry from $D_{T'}$ to $D_A \oplus D_1 \oplus D_2 \oplus \cdots$ defined by

$$\Psi = [\Gamma_1^*, D_1, 0, 0, 0, \cdots]^* | D_{T'} . \tag{8.7}$$

(Lemma 1.1 guarantees that Ψ is an isometry.) The operator Z on $D_A \oplus D_1 \oplus D_2 \oplus \cdots$ is defined by

$$Z = \begin{bmatrix} \Gamma_1 | D_A & D_{1*}\Gamma_2 & D_{1*}D_{2*}\Gamma_3 & \cdots \\ D_1 | D_A & -\Gamma_1^*\Gamma_2 & -\Gamma_1^*D_{2*}\Gamma_3 & \cdots \\ 0 & D_2 & -\Gamma_2^*\Gamma_3 & \cdots \\ 0 & 0 & D_3 & \cdots \\ 0 & 0 & 0 & \cdots \\ \vdots & \vdots & \vdots & \vdots \end{bmatrix} \tag{8.8}$$

Because $\tilde{V}$ is an isometry Ψ and Z are both isometries and the range of Ψ is orthogonal to the range of Z.

Notice that $\{0\} \oplus \Psi D_{T'} = \{(V-T')H'\}^-$. By the geometric structure of the minimal isometry dilation in Proposition VI.4.1, the space K' for the minimal isometric dilation $\tilde{V} | K'$ of T' admits a decomposition of the form

$$K' = H' \oplus (\overset{\infty}{\underset{0}{\oplus}} Z^n \Psi D_{T'}) \tag{8.9}$$

and $\{0\} \oplus \Psi D_{T'}$ is wandering for $\tilde{V}$. Now consider the unitary operator α from K' onto $H' \oplus l^2(D_{T'})$ defined by

$$\alpha(h' \oplus (\overset{\infty}{\underset{0}{\oplus}} Z^n \Psi d_n)) = h' \oplus [d_0, d_1, d_2, \cdots]^{tr} \quad (h' \in H' \text{ and } d_i \in D_{T'}) \tag{8.10}$$

where tr denotes the transpose. It is easy to verify that $\alpha(\tilde{V} | K') = U' \alpha$ where U' on $H' \oplus l^2(D_{T'})$ is the minimal unitary dilation of T' in matrix its form, that is, U' is given by (VI.3.4) where U' replaces U and T' replaces T.

Under the previous identification with Corollary 8.5, the set of all contractive intertwining liftings B of A is given by $B = \alpha P_{K'} \,|\, \phi H$ where $\tilde{V}$ is any D_A minimal isometric lifting of the form (8.6), (8.7), (8.8), whose choice sequence $\{\Gamma_i\}_1^\infty$ satisfies $\Gamma_1 \,|\, F' = \omega^*$ and K' is the space generated by H' in (8.1) corresponding to $\tilde{V}$. The term $\phi H (= \tilde{H})$ appears in B, because $\tilde{V}$ is a minimal isometric coupling for $\{\tilde{T}, \tilde{A}, T'\}$, which is unitarily equivalent to $\{U, A, T'\}$. (In particular, $\tilde{T}\phi = \phi U$ and $\tilde{A}\phi = A$.) To compute $\alpha P_{K'} \,|\, \phi H$ it is convenient to calculate $P_{L_n} \,|\, \phi H$ where $L_n = Z^n \Psi D_{T'}$ for all $n \geq 0$. To this end notice that for h in H and d in $D_{T'}$, the space $Z^n \Psi D_{T'}$ is orthogonal to $\phi h - Z^n \Psi d$ if and only if $d = \Psi^* Z^{*n} D_A h$. This follows because both Z and Ψ are isometries, and $Ah \in H'$ is orthogonal to $Z^n \Psi D_{T'}$ for all n. (Recall that $\phi h = Ah \oplus D_A h$). Thus

$$P_{L_n} \phi h = Z^n \Psi \Psi^* Z^{*n} D_A h \qquad (h \in H) . \tag{8.11}$$

Now consulting (8.9) and (8.10) we have

$$B = \alpha P_{K'} \,|\, \phi H = [A, \Psi^* D_A, \Psi^* Z^* D_A, \Psi^* Z^{*2} D_A, \cdots]^{\mathrm{tr}} . \tag{8.12}$$

This precisely describes the set of all contractive intertwining liftings B of A. In this case the minimal isometric dilation U' of T' is in its matrix form. Summing up the previous analysis produces the following result.

8.6 THEOREM. *Let A be a contraction in $I(U, T')$ and U' on $H' \oplus l^2(D_{T'})$ the minimal isometric dilation of T' given by the matrix in (VI.3.4) where U' replaces U and T' replaces T. The set of all contractive intertwining liftings B of A is given by (8.12) where Ψ and Z are the isometries defined in (8.7) and (8.8) and $\{\Gamma_i\}_1^\infty$ is a choice sequence initiated from $D_{T'} \oplus D_A$ to D_A satisfying the constraint $\Gamma_1 \,|\, F' = \omega^*$. In this case each contractive intertwining lifting B of A uniquely determines and is uniquely determined by a choice sequence $\{\Gamma_i\}_1^\infty$ satisfying $\Gamma_1 \,|\, F' = \omega^*$.*

Probably the easiest contractive intertwining lifting B to build by (8.12) is the unique contractive intertwining lifting B whose choice sequence $\{\Gamma_i\}_1^\infty$ is given by $\Gamma_1 = \omega^* P_{F'}$ and $\Gamma_i = 0$ for $i \geq 2$.

8.7 REMARK. In many applications one is only interested in computing

$$B_n = [A, \Psi^* D_A, \Psi^* Z^* D_A, \Psi^* Z^{*2} D_A, ..., \Psi^* Z^{*n-1} D_A]^{\mathrm{tr}} . \tag{8.13}$$

To this end let Z_n be the n by n matrix contained in the upper left hand corner of Z and Ψ_n be the column matrix of length n contained at the top of Ψ. Then using the upper triangular structure of Z one can show that B_n is also given by (8.13) where Ψ_n replaces Ψ and Z_n replaces Z. In other words there is a one to one correspondence between B_n and a choice sequence $\{\Gamma_i\}_1^n$ of length n initiated from $D_{T'} \oplus D_A$ to D_A satisfying $\Gamma_1 | F' = \omega^*$.

As hinted by the previous results we will see that the role played by choice sequences for the commutant lifting theorem is analogous to the role played by Schur numbers in the Carathéodory interpolation problem.

9. ANOTHER METHOD OF CONSTRUCTING ALL CONTRACTIVE ONE STEP COUPLINGS

In this section we will use the special form of all 2 by 2 contractive matrices in Theorem IV.3.1 to construct the set of all contractive one step couplings of $\{U, A, T'\}$. The analysis in this section is only sketched, because it is similar to the analysis in Section V.4 for constructing the set of all contractive one step liftings of A. The results in this section will not be used in the rest of the monograph.

To begin recall that the rotation matrix R_A defined in equation (IV.1.8), where A replaces T, is unitary. Then following ideas similar to those in Section V.4, $\tilde{C}$ on $H' \oplus D_A$ is a contraction of the form (7.6) satisfying (7.7) if and only if

$$\tilde{C}R_A = \begin{bmatrix} T'A & T'D_{A^*} \\ D_A U & D \end{bmatrix} \tag{9.1}$$

where D is an arbitrary operator provided that $\tilde{C}R_A$ is a contraction. Since $R_A^* = R_{A^*}$ Lemma 7.2 implies that $\tilde{C}$ is a contractive one step coupling of $\{\tilde{T}, \tilde{A}, T'\}$ if and only if $\tilde{C}$ admits a matrix representation of the form

$$\tilde{C} = \begin{bmatrix} T'A & T'D_{A^*} \\ D_A U & D \end{bmatrix} R_{A^*} \tag{9.2}$$

where D forces $\tilde{C}$ to be a contraction. Therefore the set of all contractive one step couplings of $\tilde{C}$ of $\{\tilde{T}, \tilde{A}, T'\}$ is given by (9.2) where D is chosen such that

$$\begin{bmatrix} T'A & T'D_{A^*} \\ D_A U & D \end{bmatrix} : H \oplus D_{A^*} \to H' \oplus D_A \tag{9.3}$$

is a contraction. This is precisely the set up in Theorem 3.1 in Chapter IV.

Using $T'A = AU$ we have

$$\|D_{AU}h\|^2 = \|h\|^2 - \|AUh\|^2 = \|Uh\|^2 - \|AUh\|^2 = \|D_A Uh\|^2 \qquad (h \in H) .$$

This implies that $\|D_{AU}h\| = \|D_A Uh\|$. So there exists an isometry Y mapping D_{AU} into D_A satisfying

$$D_A U = YD_{AU} . \qquad (9.4)$$

The dual of the previous calculations is given by

$$\|D_{A'T'^*}f\|^2 = \|f\|^2 - \|A^*T'^*f\|^2 =$$

$$\|f\| - \|T'^*f\|^2 + \|T'^*f\|^2 - \|A^*T'^*f\|^2 = \|D_{T'^*}f\|^2 + \|D_{A'}T'^*f\|^2 \ge \|D_{A'}T'^*f\|^2$$

where f is in H'. So there exists a contraction X_* mapping $D_{A'T'^*}$ into $D_{A'}$ satisfying $X_* D_{A'T'^*} = D_{A'}T'^*$, or equivalently, its adjoint X is a contraction satisfying

$$D_{(T'A)'}X = T'D_{A'} . \qquad (9.5)$$

Theorem 3.1 in Chapter IV with (9.1) to (9.5) yields the following result.

9.1 PROPOSITION. *Let* A *be in* $I_1(U,T')$. *Then* C *is a contractive one step coupling of* $\{U,A,T'\}$ *if and only if* C *admits a matrix representation* $\tilde{C}$ *of the form (9.2) where*

$$D = D_{Y^*}\Gamma D_X - YU^*A^*X \qquad (9.6)$$

and Γ *is a free contraction from* D_X *to* D_{Y^*}.

To complete this section we will follow the ideas in Section 4 of Chapter V to convert D to a more usable form. In fact we do not have to perform any calculations because (9.3) is minor modification of (V.4.4). To begin, let $F'_* = F(A^*{\cdot}T'^*)$ and $G'_* = G(A^*{\cdot}T'^*)$ where $F(\cdot)$ and $G(\cdot)$ are defined in (V.1.19) and (V.1.20). Following the calculations in (V.4.10) to (V.4.13) with T replaced by A and A replaced by T', there exists a unitary operator α mapping D_X onto G'_* satisfying (see (V.4.13)).

$$\alpha D_X = P_{G'_*} | D_{A'} . \qquad (9.7)$$

Recall that $F = F(A{\cdot}U)$ and $G = G(A{\cdot}U)$. Following the calculations in (V.4.14) to (V.4.18) with T' replaced by A and A replaced by U, there exists a unitary operator α_* mapping G onto D_{Y^*} satisfying

$$D_{Y^*}\alpha_* = P_{D_A} | G = I_G \qquad (9.8)$$

see (V.4.18). Following the calculation in (V.4.19) we have

$$-YA^*T'^*X = \omega^*P_{F'}\begin{bmatrix} D_{T'}D_{A^*} \\ -A^* \end{bmatrix}.\tag{9.9}$$

Finally, using (9.6) to (9.9) in Proposition 9.1 we obtain the following characterization of the set of all contractive one step couplings, which is analogous to the set of all one step liftings in Theorem V.4.4.

9.2 THEOREM. *Let* A *be in* $I_1(U,T')$. *Then* C *is a contractive one step coupling of* $\{U,A,T'\}$ *if and only if* C *admits a matrix representation* $\tilde{C}$ *of the form (9.2) where*

$$D = \Gamma_* P_{G'_*}|D_{A^*} + \omega^*P_{F'}\begin{bmatrix} D_{T'}D_{A^*} \\ -A^* \end{bmatrix}\tag{9.10}$$

and Γ_* *is an arbitrary contraction from* G'_* *to* G.

Following the proof of Proposition V.4.5 one can show that $\Gamma^*\beta = \Gamma_*$ where β is a unitary operator from G'_* onto G' and Γ is the contraction in Theorem 7.3. Therefore the representation for the set of all contractive one step couplings in Theorem 9.2 and Theorem 7.3 are equivalent.

VII.10. NOTES AND COMMENTS

One of the great mathematical discovery of the 1960's is certainly that made by Sarason, that is, a large class of classical interpolation problems reduces to one problem concerning the commutant of some adequate contractions [Sa 2]. Almost immediately it was discovered that the natural framework for Sarason's theorems was that of isometric liftings and unitary dilations, which lead to a purely geometric proof of his theorems [Sz.-NF 7]. This was the commutant lifting theorem (the name was coined later by Douglas, Muhly and Pearcy [DoMP]) and the first proof presented here is essentially the original one [Sz.-NF 7]. The second proof is taken from [DoMP]. Shortly after that it was also realized that the commutant lifting theorem yields the basic representation theorems for Hankel operators [Pag 2], [AAK 4]. The uniqueness Theorem 4.1 is from [AnCF] and is a generalization of some of the theorems in [Sa 2] and [AAK 1,4]. The uniqueness Theorem 5.1 is taken from [Frz 6]. It constitutes a partial strengthen of the uniqueness Theorem 5.2, obtain previously in [AnCF]. The connection between Ando's Theorem 6.1 [An] and the commutant lifting theorem was first published in [Parr 1]. For a further analysis of this connection see [CeS 1,2]. The coupling of two isometric operators was introduced in [AA 2, 3], in their context of the Lax-Phillips scattering theory [LaP 1,2]. The use of this concept to study the commutant lifting is due to

Arocena [Ar 1-5] and was inspired by the work of Cotlar on generalized Toeplitz operators. The proof of the commutant lifting theorem given in Section 8, which is based on the coupling of two contractions is essentially that of Arocena. Obtaining all contractive intertwining liftings through nonminimal isometric dilations is new, although it is inspired by [CeF 3]. Other coupling approaches for obtaining the set of all contractive intertwining liftings based on generalized Toeplitz kernels is given in [Con 4] and [Morán]. Section 9 is the coupling analogue of the characterization of the set of all one step contractive intertwining liftings given in Section V.4. Finally, it is clear that the set of all contractive intertwining liftings form a compact set in the weak operator topology. For a nice characterization of the extreme points of this set see [CeS 3]. In connection with this chapter see [BFHT], [BG], [Ber], [Br 5], [CeS 1, 2], [Con 2], [CoS], [Do 3, 4], [DoF], [DursSz.-N], [FoF 1], [Frz 3,4], [Helt 4], [Mu], [Ni], [Parr 2], [PtV 1,2], [Sar 1,2] and [Sz.-NF 11,12].

CHAPTER VIII

GEOMETRIC APPLICATIONS OF THE COMMUTANT LIFTING THEOREM

In this chapter we will show how one can use the commutant lifting theorem to solve several classical interpolation problems. Indeed we will treat the Carathéodory interpolation problem, the Nevanlinna-Pick interpolation problem, a Hankel matrix interpolation problem, an extension problem involving contractions intertwining isometries and a Toeplitz inversion problem. We will also use the commutant lifting theorem to obtain some Corona theorems for analytic Toeplitz operators.

1. OPERATOR VALUED CARATHEODORY INTERPOLATION

This section is devoted to the block Carathéodory interpolation problem in both its classical and tangential form.

To begin, let $l_n^2(E)$ be n orthogonal copies of E, that is, $l_n^2(E) = E \oplus E \oplus \cdots \oplus E$ where E appears n times. If n is infinite, then $l_\infty^2(E) = l^2(E)$ the usual Hilbert space of square summable sequences with values in E. Recall that $L(E,E')$ is the set of all operators from E to E'. Let $\{A_i\}_0^{n-1}$ be a sequence of operators in $L(E,E')$. The matrix $\hat{A}_n$ from $l_n^2(E)$ to $l_n^2(E')$ defined by

$$\hat{A}_n = \begin{bmatrix} A_0 & 0 & 0 & \ldots & 0 \\ A_1 & A_0 & 0 & \ldots & 0 \\ A_2 & A_1 & A_0 & \ldots & 0 \\ \vdots & \vdots & \vdots & \vdots\vdots\vdots & \vdots \\ A_{n-2} & A_{n-3} & A_{n-4} & \ldots & 0 \\ A_{n-1} & A_{n-2} & A_{n-3} & \ldots & A_0 \end{bmatrix} \tag{1.1}$$

is the n by n *analytic Toeplitz matrix generated* by $\{A_i\}_0^{n-1}$. Obviously $\hat{A}_n$ and $\{A_i\}_0^{n-1}$ uniquely determine each other. The *infinite analytic Toeplitz matrix* $\hat{A}_\infty$ generated by an infinite sequence of operators $\{A_i\}_0^\infty$ in $L(E,E')$ is the matrix from $l^2(E)$ to $l^2(E')$ given by

$$\hat{A}_\infty = \begin{bmatrix} A_0 & 0 & 0 & 0 & \cdots \\ A_1 & A_0 & 0 & 0 & \cdots \\ A_2 & A_1 & A_0 & 0 & \cdots \\ A_3 & A_2 & A_1 & A_0 & \cdots \\ \vdots & \vdots & \vdots & \vdots & \vdots \end{bmatrix} . \tag{1.2}$$

The adjective analytic is justified in Lemma 1.2 below. This sets the stage for the following block or operator version of the classical Carathéodory interpolation problem.

1.1 PROBLEM. Let $\{A_i\}_0^{n-1}$ be a specified sequence of operators in $L(E,E')$. Find necessary and sufficient conditions for the existence of a set of operators $\{A_i\}_n^\infty$ in $L(E,E')$, such that the infinite analytic Toeplitz matrix $\hat{A}_\infty$ generated by $\{A_i\}_0^\infty$ is a contraction.

To reformulate this problem in the form of Problem I.1.1 we introduce some supplementary remarks. First the set of all bounded analytic functions $A(z)$ in the open unit disc D with values in $L(E,E')$ is denoted by $H^\infty(E,E')$. The space $H^\infty(E, E')$ defines a Banach space under the norm $\|A\|_\infty \overset{\Delta}{=} \sup\,\{\|A(z)\| : z \in D\}$ $(A \in H^\infty(E, E'))$. So the closed unit ball, denoted by $H_1^\infty(E, E')$, is the set of all analytic functions $A(z)$ in the open unit disc with values in $L(E, E')$ satisfying $\|A(z)\| \le 1$ for all z in D. Also to $\{A_i\}_0^\infty$ a sequence of operators in $L(E,E')$ we associate $A(z)$ its formal power series defined by

$$A(z) = \sum_0^\infty A_i z^i .$$

If this series is convergent, then $A(z)$ also represents the sum of this series. In Lemma 1.2 below, we will show that $\hat{A}_\infty$ is a contraction if and only if $A(z)$ is in $H_1^\infty(E,E')$. Therefore Problem 1.1 is equivalent to: Given a sequence of operators $\{A_i\}_0^{n-1}$ in $L(E,E')$, find necessary and sufficient conditions for the existence of a function $A(z)$ in $H_1^\infty(E,E')$ of the form

$$A(z) = \sum_{i=0}^{n-1} A_i z^i + O(z^n) .$$

So if both E and E' are one dimensional, then Problem 1.1 reduces to the classical Carathéodory interpolation problem in Chapter I.

1.2 **LEMMA.** *Let $\{A_i\}_0^\infty$ be a sequence of operators in $L(E, E')$. Then $\hat{A}_\infty$ is a contraction if and only if $A(z)$ is in $H_1^\infty(E, E')$.*

PROOF. If $\hat{A}_\infty$ is a contraction, then A_n is a contraction for all n. This implies that the power series for $A(z)$ converges absolutely and $A(z)$ is analytic in D. Consider the analytic function $\tilde{A}(z)$ defined by

$$\tilde{A}(z) = (A(\overline{z}))^* = \sum_0^\infty A_i^* z^i \quad (z \in D) .$$

For all z in D and f in E' we have

$$(1-|z|^2)^{-1}\|\tilde{A}(z)f\|^2 = \sum_0^\infty \|z^i \tilde{A}(z)f\|^2 = \|\hat{A}_\infty^*(\bigoplus_0^\infty z^n f)\|^2 \leq \|\bigoplus_0^\infty z^n f\|^2 = (1-|z|^2)^{-1}\|f\|^2 .$$

The inequality follows from the fact that $\hat{A}_\infty$ is a contraction. Thus $\|\tilde{A}(z)f\| \leq \|f\|$ for all z in D and f in E'. This implies that $\tilde{A}(z)$ is in $H_1^\infty(E', E)$. Because $A(z) = (\tilde{A}(\overline{z}))^*$ the analytic function $A(z)$ is also in $H_1^\infty(E, E')$.

Now assume that $A(z)$ is in $H_1^\infty(E, E')$. Let

$$p = \sum_0^{n-1} z^i f_i \quad \text{and} \quad q = \sum_0^{n-1} z^i g_i$$

be two polynomials with values in E and E', respectively, that is, f_i is in E and g_i is in E' for all i. Using the power series expansion of $A(z)$ with the fact that the radial limits $A(e^{it})$ exists a.e. we have

$$|(\hat{A}_n(\bigoplus_0^{n-1} f_i), \bigoplus_0^{n-1} g_i)| = |\frac{1}{2\pi}\int_0^{2\pi}(A(e^{it})p(e^{it}), q(e^{it}))_{E'} dt| \leq$$

$$\|A(e^{it})p(e^{it})\|_{L^2}\|q(e^{it})\|_{L^2} \leq \|p(e^{it})\|_{L^2}\|q(e^{it})\|_{L^2} = \|\bigoplus_0^{n-1} f_i\|\|\bigoplus_0^{n-1} q_i\| .$$

The first inequality follows from the Cauchy-Schwartz inequality and the second from the fact that $A(e^{it})$ is a.e. a contraction. Thus $\hat{A}_n$ is a contraction for all n. An easy argument similar to the proof of Lemma VII.2.1 shows that the limit $\hat{A}_\infty$ is a contraction. This completes the proof.

If $\{A_i\}_n^\infty$ is a solution to Problem 1.1, then $\hat{A}_\infty$ is a contractive lifting of $\hat{A}_n$. So $\hat{A}_n$ being a contraction is necessary for the existence of a solution to Problem 1.1. On the other hand, if $\hat{A}_n$ is a contraction, then we claim that there exists a solution to Problem 1.1. To prove this let T_n be the contraction on $l_n^2(E)$ defined by

$$T_n = \begin{bmatrix} 0 & 0 & 0 & \ldots & 0 & 0 & 0 \\ I & 0 & 0 & \ldots & 0 & 0 & 0 \\ 0 & I & 0 & \ldots & 0 & 0 & 0 \\ 0 & 0 & I & \ldots & 0 & 0 & 0 \\ \vdots & \vdots & \vdots & \vdots\vdots\vdots & \vdots & \vdots & \vdots \\ 0 & 0 & 0 & \ldots & I & 0 & 0 \\ 0 & 0 & 0 & \ldots & 0 & I & 0 \end{bmatrix}. \tag{1.3}$$

The identity I appears immediately below the diagonal and zeros elsewhere. The operator T_n is the compression of the unilateral shift S on $l^2(E)$ to $l_n^2(E)$. Obviously $S^*|l_n^2(E) = T_n^*$. So S is an isometric dilation of T_n. It is easy to check that S is also minimal. (Moreover, consulting (VII.2.16) we see that T_n is the n−1-step dilation of 0 on E.) Let T_n' be the contraction on $l_n^2(E')$ defined by (1.3) where E' replaces E. Obviously the unilateral shift S' on $l^2(E')$ is the minimal isometric dilation of T_n' (and T_n' is the n−1-step dilation of 0 on E').

A simple calculation shows that an operator A is in $I(T_n, T_n')$ if and only if $A = \hat{A}_n$ where $\hat{A}_n$ is a n by n analytic Toeplitz matrix. Since $S = T_\infty$ and $S' = T_\infty'$ we see that an operator B is in $I(S, S')$ if and only if $B = \hat{A}_\infty$ where $\hat{A}_\infty$ is an infinite analytic Toeplitz matrix, that is, $\hat{A}_\infty$ is of the form (1.2). Therefore $B = \hat{A}_\infty$, or equivalently, its corresponding sequence $\{A_i\}_0^\infty$ is a solution to Problem 1.1 if and only if B is a contractive intertwining lifting of $\hat{A}_n$. If $\hat{A}_n$ is a contraction, or equivalently, if $\hat{A}_n$ is in $I_1(T_n, T_n')$, then the commutant lifting theorem shows that there exists a contraction B $(=\hat{A}_\infty)$ in $I_1(S, S')$ lifting $\hat{A}_n$. In this case there exists a solution to the Carathéodory interpolation problem. Summing up provides the following solution to the Carathéodory interpolation Problem 1.1.

1.3 THEOREM. *Let* $\{A_i\}_0^{n-1}$ *be a specified sequence of operators in* $L(E, E')$. *Then there exists a solution to the Carathéodory interpolation Problem 1.1 if and only if* $\hat{A}_n$ *is a contraction. In this case,* $\{A_i\}_0^\infty$ *is a solution to Problem 1.1 if and only if* B $(= \hat{A}_\infty)$ *is a contractive intertwining lifting of* $\hat{A}_n$ *(as considered in* $I_1(T_n, T_n')$*). In particular, there is a one to one correspondence between the set of all solutions* $\{A_i\}_0^\infty$ *to Problem 1.1 and the set of all contractive intertwining liftings* B$(=\hat{A}_\infty)$ *of* $\hat{A}_n$.

The previous analysis with Corollary VII.5.4 yields the following result.

1.4 COROLLARY. *Let* E *and* E' *both be one dimensional. Then there exists a unique solution to the Carathéodory interpolation problem if and only if the norm of* $\hat{A}_n$

is one.

When E and E' are both one dimensional, then Theorem 1.3 with Corollary 1.4 reduce to Theorem 6.7 in Chapter I. Here we have obtained these results by different techniques. To complete this section we will solve the following tangential Carathéodory interpolation problem.

1.5 PROBLEM. Let X from F to $l_n^2(E)$ and Y from F to $l_n^2(E')$ be two specified operators of the form:

$$X^* = [X_1^*, X_2^*, ..., X_n^*] \text{ and } Y^* = [Y_1^*, Y_2^*, ..., Y_n^*] \tag{1.4}$$

where X_i maps F to E and Y_i maps F to E' for $1 \leq i \leq n$. Find necessary and sufficient conditions for the existence of an infinite contractive analytic Toeplitz matrix $\hat{A}_\infty$ satisfying $Y = \hat{A}_n X$ where $\hat{A}_n$ is the n by n analytic Toeplitz matrix in the upper left hand corner of $\hat{A}_\infty$.

By choosing $F = E$

$$X^* = [I, 0, 0, ..., 0] \text{ and } Y^* = [A_0^*, A_1^*, ..., A_{n-1}^*]$$

where $\{A_i\}_0^{n-1}$ are now apriori given, Problem 1.5 reduces to the previous Carathéodory interpolation Problem 1.1.

To obtain a solution to Problem 1.5, let $\hat{X}$ from $l_n^2(F)$ to $l_n^2(E)$ be the n by n analytic Toeplitz matrix generated by $\{X_i\}_1^n$ and let $\hat{Y}$ from $l_n^2(F)$ to $l_n^2(E')$ be the n by n analytic Toeplitz matrix generated by $\{Y_i\}_1^n$. Notice that

$$\hat{X} = [X, T_n X, T_n^2 X, ..., T_n^{n-1} X] \text{ and } \hat{Y} = [Y, T_n' Y, T_n'^2 Y, ..., T_n'^{n-1} Y]. \tag{1.5}$$

If $\hat{A}_\infty$ is a solution to Problem 1.5, then $\hat{A}_\infty$ is a contractive intertwining lifting of $\hat{A}_n$ in $I_1(T_n, T_n')$. Using $T_n' \hat{A}_n = \hat{A}_n T_n$ and $Y = \hat{A}_n X$ we have $T_n'^i Y = \hat{A}_n T_n^i X$. By (1.4) we see that $\hat{Y} = \hat{A}_n \hat{X}$. Since $\hat{A}_n$ is a contraction one must have $\hat{Y}^* \hat{Y} \leq \hat{X}^* \hat{X}$. This proves the first half of the following result.

1.6 THEOREM. *There exists a solution to the tangential Carathéodory interpolation Problem 1.5 if and only if $\hat{Y}^* \hat{Y} \leq \hat{X}^* \hat{X}$ where $\hat{X}$ and $\hat{Y}$ are the n by n analytic Toeplitz matrices generated by $\{X_i\}_1^n$ and $\{Y_i\}_1^n$ respectively.*

PROOF. To complete the proof, let V_* be the co-isometry defined by

$$V_* = \begin{bmatrix} \vdots & \vdots & \vdots & \vdots & \vdots & \vdots & \vdots \\ \cdot & \cdot & \cdot & 0 & 0 & 0 & 0 \\ \cdot & \cdot & \cdot & I & 0 & 0 & 0 \\ \cdot & \cdot & \cdot & 0 & I & 0 & 0 \\ \cdot & \cdot & \cdot & 0 & 0 & I & 0 \end{bmatrix} \text{ on } \bigoplus_n^{-\infty} E \tag{1.6}$$

where the identity I on E appears immediately below the main diagonal and zeros appear elsewhere. Notice that V_* is a co-isometric extension of T_n, that is, $l_n^2(E)$ is invariant for V_* and $V_* | l_n^2(E) = T_n$. In fact V_*^* is the minimal isometric dilation of T_n^*. Moreover, the contraction T_n is the n by n matrix contained in the lower right hand corner of V_*. The co-isometry V_*' is the co-isometry defined by (1.6) where V_*' replaces V_* and E' replaces E. Obviously V_*' is a co-isometric extension of T_n'. Finally, notice that $(l_n^2(E), l_n^2(E'))$ is a hyperinvariant pair of subspaces for (V_*, V_*') (see section VII.1), since $l_n^2(E) = \ker V_*^n$ and $l_n^2(E') = \ker V_*'^n$.

Assume that $\hat{Y}^*\hat{Y} \leq \hat{X}^*\hat{X}$. This implies that there exists a contraction A from the closed range R of $\hat{X}$ to $l_n^2(E')$ satisfying $A\hat{X} = \hat{Y}$. Introducing the lower shift T on $l_n^2(F)$ defined as in (1.3) where F replaces E, we notice that $T_n\hat{X} = \hat{X}T$ and $T_n'\hat{Y} = \hat{Y}T$; thus

$$AT_n\hat{X} = A\hat{X}T = \hat{Y}T = T_n'\hat{Y} = T_n'A\hat{X}. \tag{1.7}$$

Since the range of $\hat{X}$ is invariant for T_n this implies that A is in $I_1(T_n | R, T_n')$. By Corollary VII.1.6 of the commutant lifting theorem, there exists a contraction $\hat{A}$ in $I(T_n, T_n')$ extending A, that is, $\hat{A}\hat{X} = A\hat{X} = \hat{Y}$. Obviously $\hat{A}(=\hat{A}_n)$ is an n by n analytic Toeplitz matrix. By Theorem 1.3 (or by the commutant lifting theorem), there exists a contraction B in $I(S,S')$ lifting $\hat{A}$. (Recall that the minimal isometric dilation of T_n and T_n' is S and S', respectively.) Clearly $B(=\hat{A}_\infty)$ is an infinite analytic Toeplitz matrix, and $\hat{A}_n = \hat{A}$ is the n by n analytic Toeplitz matrix contained in the upper left hand corner of $\hat{A}_\infty = B$. This and $\hat{A}_n\hat{X} = \hat{Y}$ completes the proof.

1.7 REMARK. The set of all solutions to Problem 1.5 is given by the set of all contractive intertwining liftings $B(=\hat{A}_\infty)$ of $\hat{A}(=\hat{A}_n)$ where $\hat{A}$ runs through the set of all contraction in $I(T_n, T_n')$ satisfying $\hat{A}\hat{X} = \hat{Y}$.

1.8 REMARK. If $R = l_n^2(E)$, then one can apply the commutant lifting theorem directly to obtain a proof of the previous theorem. However, if R is strictly contained in $l_n^2(E)$, then S is not necessarily an isometric lifting of $T_n | R$ and one cannot directly apply the commutant lifting theorem. This is why we used Corollary VII.1.6 (which is a corollary of the commutant lifting theorem) to prove Theorem 1.6.

1.9 REMARK. If $\hat{X}$ is invertible, then $\hat{A}_n$ is uniquely determined by $\hat{A}_n = \hat{Y}\hat{X}^{-1}$. In this case Problem 1.5 reduces to the usual Carathéodory interpolation Problem 1.1

1.10 REMARK. The tangential Carathéodory interpolation Problem 1.5 plays an important role in systems identification. In this case, one is given an input operator X and an output operator Y from experimental data. Then one must construct all (if any) contractive linear time invariant systems $\hat{A}_\infty$ satisfying $Y = \hat{A}_n X$. This is equivalent to constructing all solutions to Problem 1.5. Theorem 1.6 provides necessary and sufficient conditions for the existence of a solution. In Chapters XIII and XIV we will give several algorithms to compute the set of all contractive intertwining liftings in the commutant lifting theorem. This will readily provide all solutions to Problem 1.5, or equivalently, provide several algorithms to construct the set of all contractive linear time invariant systems satisfying the input-output constraint $Y = \hat{A}_n X$.

2. OPERATOR VALUED NEVANLINNA-PICK INTERPOLATION

In this section we will use the commutant lifting theorem to solve the following *tangential Nevanlinna-Pick interpolation problem.*

2.1 PROBLEM. Let $\{B_i\}_1^n$ and $\{C_i\}_1^n$ be a set of operators from F to E and from F to E', respectively. Let $\{\alpha_i\}_1^n$ be a set of n distinct complex numbers in D. Find necessary and sufficient conditions for the existence of an analytic function $A(z)$ in $H_1^\infty(E,E')$ satisfying

$$C_i = A(\alpha_i)B_i \quad \text{(for all } 1 \leq i \leq n) . \tag{2.1}$$

To obtain a solution to this problem, let $\{A_i\}_0^\infty$ be a sequence of operators from E to E'. The matrix $\hat{A}_\infty^{tr}$ from $l^2(E)$ to $l^2(E')$ defined by

$$\hat{A}_\infty^{tr} = \begin{bmatrix} A_0 & A_1 & A_2 & A_3 & \cdots \\ 0 & A_0 & A_1 & A_2 & \cdots \\ 0 & 0 & A_0 & A_1 & \cdots \\ 0 & 0 & 0 & A_0 & \cdots \\ \vdots & \vdots & \vdots & \vdots & \vdots \ddots \\ \vdots & \vdots & \vdots & \vdots & \vdots \ddots \end{bmatrix} \tag{2.2}$$

is the *co-analytic Toeplitz matrix generated* by $\{A_i\}_0^\infty$. It is easy to show that

$\|\hat{A}_n\| = \|\hat{A}_\infty^{tr} \mid l_n^2(E)\|$ where $\hat{A}_n$ is the n by n analytic Toeplitz matrix in (1.1). (In fact $\hat{A}_n$ and $\hat{A}_\infty^{tr} \mid l_n^2(E)$ are equivalent up to a unitary operator on the left and right. The unitary operators which reverses the order of the components in $l_n^2(E)$ and $l_n^2(E')$ are the unitary operators which intertwine $\hat{A}_n$ and $\hat{A}_\infty^{tr} \mid l_n^2(E)$.) Letting n approach infinity one easily obtains that $\|\hat{A}_\infty\| = \|\hat{A}_\infty^{tr}\|$. In particular, $\hat{A}_\infty$ is a contraction if and only if $\hat{A}_\infty^{tr}$ is a contraction. By Lemma 1.2 this implies that $\hat{A}_\infty^{tr}$ is a contraction if and only if its corresponding formal series

$$A(z) = \sum_0^\infty A_n z^n \tag{2.3}$$

is in $H_1^\infty(E,E')$. Let X be the operator from $l_n^2(F)$ to $l^2(E)$ defined by

$$X = \begin{bmatrix} B_1 & B_2 & \ldots & B_n \\ \alpha_1 B_1 & \alpha_2 B_2 & \ldots & \alpha_n B_n \\ \alpha_1^2 B_1 & \alpha_2^2 B_2 & \ldots & \alpha_n^2 B_n \\ \alpha_1^3 B_1 & \alpha_2^3 B_2 & \ldots & \alpha_n^3 B_n \\ \vdots & \vdots & \vdots\vdots & \vdots \end{bmatrix}. \tag{2.4}$$

Let Y be the operator from $l_n^2(F)$ to $l^2(E')$ defined in (2.4) where C replaces B and Y replaces X. Assume that A(z) is an analytic function in D with a power series expansion of the form (2.3). Then a simple calculation shows that (2.1) holds if and only if $Y = \hat{A}_\infty^{tr}X$. Therefore the Nevanlinna-Pick interpolation problem is equivalent to finding necessary and sufficient conditions for the existence of a contractive $\hat{A}_\infty^{tr}$ satisfying $Y = \hat{A}_\infty^{tr}X$.

Assume that there exists a solution to the Nevanlinna-Pick interpolation problem. This implies that there exists a co-analytic contractive Toeplitz matrix $\hat{A}_\infty^{tr}$ satisfying $Y = \hat{A}_\infty^{tr}X$. Therefore $Y^*Y \le X^*X$, or equivalently, $X^*X - Y^*Y$ is positive. Using (2.4) with $(1-\bar{\alpha}_i\alpha_j)^{-1} = \sum_{n=0}^\infty (\bar{\alpha}_i\alpha_j)^n$ we have

$$X^*X - Y^*Y = \begin{bmatrix} \dfrac{B_1^*B_1 - C_1^*C_1}{1-\bar{\alpha}_1\alpha_1} & \ldots & \dfrac{B_1^*B_n - C_1^*C_n}{1-\alpha_n\bar{\alpha}_1} \\ \vdots & \vdots\vdots & \vdots \\ \dfrac{B_n^*B_1 - C_n^*C_1}{1-\bar{\alpha}_n\alpha_1} & \ldots & \dfrac{B_n^*B_n - C_n^*C_n}{1-\alpha_n\bar{\alpha}_n} \end{bmatrix} \tag{2.5}$$

So if there exists a solution to the tangential Nevanlinna-Pick interpolation problem, then the matrix in (2.5) is positive. This proves the first part of the following result.

2.2 THEOREM. *There exists a solution to the tangential Nevanlinna-Pick interpolation problem if and only if the matrix* $X^*X - Y^*Y$ *on* $l_n^2(F)$ *in (2.5) is positive.*

PROOF. As before let S and S' be the unilateral shifts on $l^2(E)$ and $l^2(E')$, respectively. Using the matrix form for the unilateral shift in (VI.1.1) along with (2.4) we have

$$S^*X = X\Lambda \ \text{ and } \ S'^*Y = Y\Lambda \tag{2.6}$$

where Λ is the diagonal matrix on $l_n^2(F)$ defined by

$$\Lambda = \alpha_1 I \oplus \alpha_2 I \oplus \alpha_3 I \oplus \ \cdots \ \oplus \alpha_n I \,.$$

Now assume that $X^*X - Y^*Y$ is positive, or equivalently, $\|Yf\| \le \|Xf\|$ for all f in $l_n^2(F)$. This implies that there exists a contraction A_* from the closed range H of X to $l^2(E')$ satisfying $Y = A_*X$. Equation (2.6) gives

$$S'^*A_*X = S'^*Y = Y\Lambda = A_*X\Lambda = A_*S^*X \,. \tag{2.7}$$

Since the range of X is invariant for S^* we see that A_* is in $I_1(S^* \,|H,S'^*)$. Obviously S^* is a co-isometric extension of $S^*|H$. By Corollary VII.1.4 of the commutant lifting theorem, there exists a contraction B_* in $I(S^*,S'^*)$ extending A_*, that is, $A_* = B_*|H$. A simple calculation shows that B_* is in $I(S^*,S'^*)$ if and only if $B_* = \hat{A}_\infty^{tr}$ where $\hat{A}_\infty^{tr}$ is a matrix of the form (2.2). Therefore $B_* = \hat{A}_\infty^{tr}$ is a co-analytic contractive Toeplitz matrix extending A_* and thus satisfying $Y = \hat{A}_\infty^{tr}X$. This $B_*(= \hat{A}_\infty^{tr})$ produces an $A(z)$ by (2.3) which solves the Nevanlinna-Pick interpolation problem. The proof is now complete.

2.3 REMARK. The previous proof shows that the set of all solutions $A(z)$ to the Nevanlinna-Pick interpolation problem, is given by $A(z)$ in (2.3) where $B_*(= \hat{A}_\infty^{tr})$ is a contraction in $I(S^*,S'^*)$ extending A_*, or equivalently, satisfying $Y = B_*X$. In Chapters XIII and XIV we will give several algorithms to compute the set of all contractive intertwining liftings in the commutant lifting theorem. This will provide a computational method to obtain all solutions to the Nevanlinna-Pick interpolation problem.

If $E = E' = \mathbb{C}^1$, then $\{B_i\}_1^n$ and $\{C_i\}_1^n$ are complex numbers. If $B_i = 0$, then (2.1) shows that $C_i = 0$. If B_i is zero and C_i is not zero, then there is no solution to the tangential Nevanlinna-Pick interpolation problem. If B_i is nonzero, then we can divide both sides by B_i. So without loss of generality we can assume that $B_i = 1$ for all i. This observation shows that Problem 2.1 reduces to following classical Nevanlinna-Pick interpolation problem.

2.4 PROBLEM. Let $\{c_i\}_1^n$ be a set of n complex numbers, and $\{\alpha_i\}_1^n$ be a set of n distinct complex numbers in D. Find necessary and sufficient conditions for the existence of an analytic function A(z) in H_1^∞ satisfying

$$c_i = A(\alpha_i) \qquad \text{(for all } 1 \le i \le n) . \tag{2.8}$$

In this way Theorem 2.2 yields the following classical result.

2.5 COROLLARY. *There exists a solution to the classical Nevanlinna-Pick interpolation problem if and only if the matrix*

$$\begin{bmatrix} \dfrac{1-\bar{c}_1 c_1}{1-\bar{\alpha}_1 \alpha_1} & \cdots & \dfrac{1-\bar{c}_1 c_n}{1-\alpha_n \bar{\alpha}_1} \\ \vdots & \vdots\vdots\vdots & \vdots \\ \dfrac{1-\bar{c}_n c_1}{1-\bar{\alpha}_n \alpha_1} & \cdots & \dfrac{1-\bar{c}_n c_n}{1-\alpha_n \bar{\alpha}_n} \end{bmatrix} \tag{2.9}$$

on $\mathbb{C}^n$ *is positive.*

To complete this section we will use the commutant lifting theorem to prove the following uniqueness result.

2.6 THEOREM. *There exists a unique solution to the classical Nevanlinna-Pick interpolation problem if and only if the matrix in (2.9) is positive and singular.*

PROOF. We obtain a proof to this theorem by presenting a finer analysis of the proof of Theorem 2.2. Because $B_i = 1$ for all i the operator X in (2.4) becomes

$$X = \begin{bmatrix} 1 & 1 & \ldots & 1 \\ \alpha_1 & \alpha_2 & \ldots & \alpha_n \\ \alpha_1^2 & \alpha_2^2 & \ldots & \alpha_n^2 \\ \alpha_1^3 & \alpha_2^3 & \ldots & \alpha_n^3 \\ \vdots & \vdots & \vdots\vdots\vdots & \vdots \end{bmatrix} \tag{2.10}$$

Notice that X is a Vandermonde matrix from $\mathbb{C}^n$ to l^2. This implies that X is a one to one. The first equation in (2.6) shows that the range H of X is invariant for S. (Here $E = E' = C^1$ and $S = S'$ is the unilateral shift on l^2.) Let T be the contraction on H defined by $T^* = S^*|H$. Equation (2.6) gives $T^*X = X\Lambda$. Therefore T^* is similar to a diagonal matrix Λ and $\{\alpha_i\}_1^n$ are the eigenvalues for T^*.

Obviously S is an isometric lifting of T. We claim that S is the minimal isometric dilation of T. To prove this it is sufficient to show that the range H of X is cyclic for S. Assume that f in l^2 is orthogonal to $S^n x_j$ for all $n \geq 0$ where $x_j = [1, \alpha, \alpha^2, \alpha^3, ...]^{tr}$ is the j-th column of X, that is, $\alpha = \alpha_j$. This implies that f is orthogonal to the columns of the following matrix

$$M = \begin{bmatrix} 1 & 0 & 0 & 0 & \cdots \\ \alpha & 1 & 0 & 0 & \cdots \\ \alpha^2 & \alpha & 1 & 0 & \cdots \\ \alpha^3 & \alpha^2 & \alpha & 1 & \cdots \\ \vdots & \vdots & \vdots & \vdots & \vdots & \vdots \end{bmatrix}$$

These columns span l^2 because M is invertible. Indeed its inverse is given the infinite analytic Toeplitz matrix $\hat{A}_\infty$ generated by $\{1, -\alpha, 0, 0, 0, ...\}$. So $f = 0$ and $x_j \in H$ is cyclic for S. Therefore H is cyclic for S and S is the minimal isometric dilation of T.

We claim that the dimension δ_T of D_T is one. Proposition VI.4.2 along with the fact that the multiplicity of S is one show that $\delta_{T^*} = 1$. Recall that T is a unitary operator from $H \ominus D_T$ onto $H \ominus D_{T^*}$ (see Lemma V.2.1). Because H is finite dimensional this implies that $\delta_T = \delta_{T^*}$. Therefore both of the defect indices δ_T and δ_{T^*} for T are one.

Without loss of generality we assume that there exists a solution to the classical Nevanlinna-Pick interpolation problem. By Corollary 2.5 and (2.5) the matrix $X^*X - Y^*Y$ in (2.9) is positive. The range of Y is contained in H the range of X, since Y $= XC$ where C is a diagonal matrix with diagonal entries $c_1, c_2, ..., c_n$. Notice that there exists a unique contraction A_* on H satisfying $Y = A_*X$. Equation (2.7) and $T^* = S^*|H$ shows that $A_*T^* = T^*A_*$. Recall that A(z) is a solution to the Nevanlinna-Pick interpolation problem if and only if its corresponding co-analytic Toeplitz matrix $\hat{A}_\infty^{tr}$ is a contraction extending A_*. Therefore A(z) is a solution to the classical Nevanlinna-Pick interpolation problem if and only if A(z) corresponds to a contractive intertwining extension $B_*(= \hat{A}_\infty^{tr})$ of A_*. In other words, there is a one to one correspondence between the set of all solutions to the classical Nevanlinna-Pick interpolation problem, and the set of contractive intertwining extensions B_* of A_*, or equivalently, the set of all contractive intertwining liftings B of $A = (A_*)^*$.

Corollary VII.5.4 implies that there exists a unique contractive intertwining lifting of A if and only if the norm of A is one. Using $A_*X = Y$ we see that the norm of A_* is one if and only if the matrix $X^*X - Y^*Y$ is singular. This completes the proof.

It is worth emphasizing the principle facts in the previous proof.

2.7 REMARK. Assume that there exists a solution to the classical Nevanlinna-Pick interpolation problem, or equivalently, assume that the matrix $X^*X - Y^*Y$ in (2.9) is positive. Let $A = (A_*)^*$ be the contraction on the range H of X defined by $Y = A_*X$. Let T be the contraction on H defined by $T^* = S^* | H$ where S is the unilateral shift on l^2. Then S is the minimal isometric dilation of T. The eigenvalues of T are $\{\overline{\alpha_i}\}_1^n$, and both of the defect indices δ_T and δ_{T^*} are one. The set of all solutions to the classical Nevanlinna-Pick interpolation problem is given by $A(z)$ in (2.3) where $B = (\hat{A}_\infty^{tr})^*$ is a contractive intertwining lifting of A.

3. THE CARSWELL-SCHUBERT RESULT

In this section we will use the commutant lifting theorem to study contractive extensions of contractions intertwining the restrictions of isometries.

As noted earlier the commutant lifting theorem can be stated as follows: Let U on K and U' on K' be two isometries. Let H and H' be an invariant subspace for U^* and U'^* respectively. If A is a contraction from H to H' satisfying $U'^*A = AU^* | H$, then there exists a contraction B in $I(U^*, U'^*)$ extending A, that is, $A = B | H$. Therefore the commutant lifting theorem is a theorem concerning norm preserving extension of operators intertwining restrictions of co-isometries. A natural question is to decide what happens if the role of the co-isometries is replaced by isometries. First of all the exact analogue of the commutant lifting theorem does not hold. This is illustrated by the following instructive example.

3.1 EXAMPLE. Let $U = S \oplus S$ be the unilateral shift on $K = l^2 \oplus l^2$ where S is the unilateral shift on l^2. Obviously $M = l^2 \oplus Sl^2$ is invariant for U and $U_+ = U | M$ is an isometry on M. Let C be the contraction on M defined by

$$C(f \oplus g) = S^*g \oplus 0 \quad (f \oplus g \in M).$$

Using the fact that g is in Sl^2 we have

$$CU_+(f \oplus g) = C(Sf \oplus Sg) = (g \oplus 0) = (SS^*g \oplus 0) = U_+C(f \oplus g).$$

Therefore C is in the commutant of U_+. We claim that there does not exist an operator B in the commutant of U extending C. To prove this let h be the unit vector in l^2 where the first component is one and all the other components are zero. If there exists an operator B in the commutant of U extending C, then

$$1 = \|h \oplus 0\|^2 = (C(0 \oplus Sh), h \oplus 0) = (B(0 \oplus Sh), h \oplus 0) =$$

$$(BU(0 \oplus h), h \oplus 0) = (UB(0 \oplus h), h \oplus 0) = 0$$

which is absurd. Therefore C cannot be extended to an operator in the commutant of U.

Throughout U on K and U' on K' are two isometries. The subspaces M and M' are invariant for U and U', respectively. The operators U_+ on M and U'_+ on M' are defined by $U_+ = U | M$ and $U'_+ = U' | M'$, respectively. Let C be a contraction in $I(U_+, U'_+)$. We say that B is a *contractive intertwining extension* of C if B is a contraction in $I(U, U')$ extending C, that is, $B | M = C$. The previous example shows that C does not always admit a contractive intertwining extension. The following proposition provides a useful property of contractive intertwining extensions.

3.2 PROPOSITION. *Let C be a contraction in $I(U_+, U'_+)$. If C admits a contractive intertwining extension, then*

$$\|D_{U'^{*n}} Ch\| \le \|D_{U^{*n}} h\| \qquad (for\ all\ h \in M\ and\ n \ge 1). \tag{3.1}$$

PROOF. Let B be a contractive intertwining extension of C. For $n \ge 1$ we have $BU^n K \subseteq U'^n K'$; hence

$$U'^n U'^{*n} BU^n U^{*n} = BU^n U^{*n} .$$

From this it follows that

$$(I - U'^n U'^{*n})B = (I - U'^n U'^{*n})B(I - U^n U^{*n}) .$$

Using $C = B | M$ this implies that

$$\|D_{U'^{*n}} Ch\| = \|(I - U'^n U'^{*n})Bh\| \le \|(I - U^n U^{*n})h\| = \|D_{U^{*n}} h\| .$$

The equalities follow from the fact that if V is an isometry, then $I - VV^*$ is an orthogonal projection and $D_{V^*} = I - VV^*$. This completes the proof.

To complete this section we will use the commutant lifting theorem to show that the conditions in (3.1) are both necessary and sufficient for the existence of a contractive intertwining extension B of C. To this end, let V on $\hat{K}$ and V' on $\hat{K}'$ be the minimal unitary extensions of U and U', respectively. Let M_∞ be the reducing subspace of $\hat{K}$ defined by

$$M_\infty = \bigvee_{-\infty}^{\infty} V^n M .$$

Obviously $U_\infty = V|M_\infty$ on M_∞ is the minimal unitary extension of U_+. Corollary VI.2.4 shows that there exists a unique contraction C_∞ in $I(U_\infty, V')$ extending C. In fact this contraction is given by the following formula

$$C_\infty f = \lim_{n\to\infty} V'^{*n} C P_M V^n f \quad (f \in M_\infty) . \tag{3.2}$$

Finally, let Q and Q' be the orthogonal projection onto $K^\perp = \hat{K} \ominus K$ and $K'^\perp = \hat{K}' \ominus K'$, respectively. This sets the stage for the following result.

3.3 LEMMA. *Let C be a contraction in $I(U_+, V')$ and let C_∞ in $I_1(U_\infty, U'_\infty)$ be its unique contractive extension. Then the conditions in (3.1) are equivalent to*

$$\|Q' C_\infty f\| \le \|Qf\| \qquad (for\ all\ f \in M_\infty) . \tag{3.3}$$

PROOF. If (3.1) holds, then for h in M and $n \ge 0$ we have

$$\|Q' C_\infty V^{*n} h\|^2 = \|Q' V'^{*n} C_\infty h\|^2 = \|C_\infty h\|^2 - \|P_{K'} V'^{*n} C_\infty h\|^2 =$$

$$\|Ch\|^2 - \|U'^{*n} Ch\|^2 = \|D_{U'^n} Ch\|^2 \le \|D_{U^n} h\|^2 =$$

$$\|h\|^2 - \|U^{*n} h\|^2 = \|V^{*n} h\|^2 - \|P_K V^{*n} h\|^2 = \|Q V^{*n} h\|^2 .$$

Since $\{V^{*n} M\}$ for $n \ge 0$ forms a dense set in M_∞, the relation (3.3) follows. By rearranging the order in the above computations, we see that (3.3) implies that the conditions in (3.1) hold. This completes the proof.

Now assume that (3.3) holds. This implies that there exists a contraction A from $H \overset{\Delta}{=} \overline{QM_\infty}$ to $H' \overset{\Delta}{=} \overline{Q' C_\infty M_\infty}$ satisfying

$$AQf = Q' C_\infty f \qquad (for\ all\ f \in M_\infty) . \tag{3.4}$$

Obviously, $K^\perp$ and $K'^\perp$ are invariant subspaces for V^* and V'^*, respectively. So the operators $V_* \overset{\Delta}{=} QV|K^\perp$ on $K^\perp$ and $V'_* = Q'V'|K'^\perp$ on $K'^\perp$ are both co-isometries. Using $V'_* Q' = Q' V'$ we have

$$V'_* Q' C_\infty M_\infty = Q' V' C_\infty M_\infty = Q' C_\infty U_\infty M_\infty = Q' C_\infty M_\infty \subseteq H' ,$$

which implies that H' is an invariant subspace for V'_*. Using $V_* Q = QV$, a similar calculation shows that H is an invariant subspace for V_*.

If f is in M_∞, then

$$V'_* AQf = V'_* Q' C_\infty f = Q' V' C_\infty f = Q' C_\infty U_\infty f =$$

$$AQU_\infty f = AQVf = AV_* Qf .$$

Therefore A is a contraction in $I(V_* | H, V'_* | H')$. By applying Corollary VII.1.4 of the commutant lifting theorem, there exists a contraction B_* in $I(V_*, V'_*)$ extending A, that is, $B_* | H = A$. Notice that V and V' is an isometric lifting of V_* and V'_*, respectively. By Corollary VII.1.3 of the commutant lifting theorem there exists a contraction $\hat{B}$ in $I(V, V')$ lifting B_*. Using the fact that B_* is an extension of A and $\hat{B}$ is a lifting of B_* we have

$$Q' \hat{B} f = B_* Qf = AQf = Q' C_\infty f \qquad (f \in M_\infty) ,$$

which implies that

$$Q' (\hat{B} - C_\infty) | M_\infty = 0 . \tag{3.5}$$

Let $K' = K'_s \oplus K'_u$ be the Wold decomposition of U'. Obviously $\hat{K}' = \hat{K}'_s \oplus K'_u$ where $V' | \hat{K}'_s$ is the minimal unitary extension of the unilateral shift $V' | K'_s$. Notice that $V' | \hat{K}'_s$ is a bilateral shift and $K'^\perp = \hat{K}'_s \ominus K'_s$ is cyclic for $V' | \hat{K}'_s$, that is,

$$\bigvee_0^\infty V'^n K'^\perp = \hat{K}'_s . \tag{3.6}$$

If $n \geq 0$, then (3.5) and the fact that M_∞ is a reducing subspace for V gives

$$0 = (Q' (\hat{B} - C_\infty) V^{*n} f, g) = ((\hat{B} - C_\infty) f, V'^n g) \quad \text{(for all } f \in M_\infty \text{ and } g \in K'^\perp) .$$

This and (3.6) implies that $(\hat{B} - C_\infty) | M_\infty$ is in K_u, or equivalently,

$$P_s (\hat{B} - C_\infty) | M_\infty = 0 \tag{3.7}$$

where P_s is the orthogonal projection onto $\hat{K}'_s$. Since $\hat{B}$ is a contractive lifting of B_*

$$\hat{B} K \subseteq K' \quad \text{and} \quad B_1 \triangleq \hat{B} | K \text{ is in } I_1(U, U') . \tag{3.8}$$

Using this in (3.7) gives $P_s (B_1 - C) | M = 0$. In particular, if $K'_u = 0$, or equivalently, U' is a unilateral shift, then B_1 is a contractive intertwining extension of C. This with Proposition 3.2 and Lemma 3.3 yields the following result due to Carswell and Schubert.

3.4 PROPOSITION. *Let C be a contraction in $I(U_+, U'_+)$ and assume that U' is a unilateral shift. Then there exists a contractive intertwining extension B_1 of C (that is B_1*

is in $I_1(U, U')$) if and only if the conditions in (3.1), or equivalently, (3.3) hold. In this case the contraction B_1 in (3.8) is a contractive intertwining extension of C.

If U' is not a unilateral shift, then we will modify $\hat{B}$ to obtain an operator which is a contractive intertwining extension of C. To this end, set $B = P_s\hat{B}$ and let P_u be the orthogonal projection onto K'_u. Obviously B is a contraction in $I(V, V')$ and by (3.7) we have

$$\|P_u C_\infty f\|^2 = \|C_\infty f\|^2 - \|P_s C_\infty f\|^2 \le \|f\|^2 - \|Bf\|^2 = \|D_B f\|^2 \qquad (f \in M_\infty).$$

So there exists a contraction Γ from $G \overset{\Delta}{=} \overline{D_B M_\infty}$ to K'_u satisfying

$$P_u C_\infty = \Gamma D_B \mid M_\infty . \tag{3.9}$$

Since D_B is in the commutant of V, the space G is a reducing subspace for V. Equation (3.9) yields

$$V' \Gamma D_B f = V' P_u C_\infty f = P_u C_\infty V f = \Gamma D_B V f = \Gamma V D_B f \qquad (f \in M_\infty).$$

This implies that $V'\Gamma = \Gamma V \mid G$. Because G reduces V this shows that ΓP_G is a contraction in $I(V, V')$. Therefore $\Gamma P_G D_B$ is a contraction in $I(V, V')$.

Notice that

$$B_2 = B + \Gamma P_G D_B = P_s\hat{B} + \Gamma P_G D_B = P_s B + \Gamma P_G D_B . \tag{3.10}$$

is an operator in $I(V, V')$. The following calculation shows that B_2 is a contraction

$$\|B_2 k\|^2 = \|P_s B k\|^2 + \|\Gamma P_G D_B k\|^2 \le \|Bk\|^2 + \|D_B k\|^2 = \|k\|^2 \qquad (k \in \hat{K}).$$

Combining (3.7) with (3.9) we have

$$B_2 \mid M_\infty = (B + \Gamma P_G D_B) \mid M_\infty = (P_s + P_u) C_\infty = C_\infty . \tag{3.11}$$

So B_2 is an extension of C_∞. Equations (3.8) and (3.9) give

$$B_2 K \subseteq P_s K' \bigvee K'_u \subseteq K'_s \oplus K'_u = K' .$$

This implies that

$$B_2 K \subseteq K' \quad \text{and} \quad B_3 = B_2 \mid K \text{ is in } I_1(U, U') . \tag{3.12}$$

By (3.11) this B_3 is a contractive intertwining extension of C. This proves the following generalization of Proposition 3.4.

3.5 THEOREM. *Let* C *be a contraction in* $I(U_+, U'_+)$. *Then there exists a contractive intertwining extension* B_3 *of* C *(that is* B_3 *is in* $I_1(U, U')$*) if and only if the conditions in (3.1), or equivalently, (3.3) hold. In this case the contraction* B_3 *in (3.12) is a contractive intertwining extension of* C.

3.6 REMARK. The contraction B_3 is uniquely determined by the contraction $\hat{B}$.

4. TOEPLITZ AND HANKEL OPERATORS

In this section we will introduce and characterize Toeplitz and Hankel operators. We will also use the commutant lifting theorem to obtain necessary and sufficient conditions for the invertibility of a Toeplitz operator with unitary symbol.

Throughout this section V is a unitary operator on $\hat{K}$ and M is an invariant subspace for V. The operator V_+ on M is defined by $V_+ = V|M$ and P is the orthogonal projection onto M. We say that T_B on M is a *Toeplitz operator* if $T_B = PB|M$ where B is an operator on $\hat{K}$ commuting with V. The operator B is called the *symbol* of T_B. The following result plays a fundamental role in the theory of Toeplitz operators.

4.1 PROPOSITION. *An operator* A *on* M *is Toeplitz if and only if*

$$V_+^* A V_+ = A .\tag{4.1}$$

PROOF. Let $A = T_B$ be a Toeplitz operator. Using the fact that V^* is a lifting of V_+^*, or equivalently, $PV^* = V_+^* P$ we have

$$V_+^* A V_+ = V_+^* PBV_+ = PV^* BV|M = PB|M = T_B = A .$$

Therefore (4.1) holds.

Now assume that $V_+^* A V_+ = A$. Let M_n be the space defined by

$$M_n \overset{\Delta}{=} \bigvee_{-\infty}^{n} V^{*i} M = V^{*n} M \qquad (n \geq 0) .\tag{4.2}$$

The last equality follows from:

$$M_n = \bigvee_{i=-\infty}^{n} V^{*i} M = \bigvee_{i=-\infty}^{n} V^{*n} V^n V^{*i} M \subseteq \bigvee_{i=-\infty}^{n} V^{*n} M = V^{*n} M \subseteq M_n .$$

Obviously M_∞ is a reducing subspace for V and $V_\infty = V|M_\infty$ is the minimal unitary extension of V_+. We can assume without loss of generality that $\hat{K} = M_\infty$. If $A = P_\infty B_\infty|M_\infty$ where B_∞ is in the commutant of V_∞, then the reducing property of M_∞ implies that $B = B_\infty P_\infty$ is in the commutant of V and $A = PB|M$, or equivalently, A is a

Toeplitz operator with symbol B. (Here P_∞ is the orthogonal projection onto M_∞.)

Let A_n on $\hat{K}(=M_\infty)$ be the operators defined by

$$A_n = V^{*n} A P V^n \qquad \text{(for all } n \geq 0) .$$

Obviously $A_0 = AP$. We claim that

$$P_i A_n | M_i = A_i | M_i \qquad \text{(for all } 0 \leq i \leq n) \tag{4.3}$$

where P_n is the orthogonal projection onto M_n. Indeed using $V_+^* A V_+ = A$ we have for h and g in M and $0 \leq i \leq n$:

$$(A_n V^{*i}h, \ V^{*i}g) = (A V^{(n-i)}h, \ V^{(n-i)}g) = (V_+^{*(n-i)} A V_+^{(n-i)}h, \ g) =$$

$$(Ah, \ g) = (V^i V^{*i} A P V^i V^{*i}h, \ g) = (A_i V^{*i}h, \ V^{*i}g) ,$$

this with the last equality in (4.2) gives (4.3). Equation (4.3) with $\hat{K} = M_\infty$ and $\|A_n\| \leq \|A\|$ for all n implies that A_n converges weakly to an operator B on $\hat{K}$.

We claim that B is in the commutant of V. To prove this first notice that (4.3) yields

$$P_n B | M_n = A_n | M_n \qquad \text{(for all } n{\geq}0) . \tag{4.4}$$

For $n \geq i \geq 0$ and h, g in M this gives

$$(BV V^{*(i+1)}h, \ V^{*i}g) = (A_n V^{*i}h, \ V^{*i}g) =$$

$$(Ah, g) = (V^{i+1} V^{*(i+1)} A V^{(i+1)} V^{*(i+1)}h, \ g) =$$

$$(A_{i+1} V^{*(i+1)}h, \ V^{*(i+1)}g) = (B V^{*(i+1)}h, \ V^{*(i+1)}g) = (VB V^{*(i+1)}h, \ V^{*i}g) .$$

It follows that V commutes with B on a dense set, and B is in the commutant of V. Equation (4.4) implies that $A = A_0 | M = PB | M$ is a Toeplitz operator with symbol B. This completes the proof.

Let $Q = I - P$ be the orthogonal projection onto $\hat{K} \ominus M$. In this setting a *Hankel operator* H_B is an operator from M to $\hat{K} \ominus M$ of the form $H_B = QB | M$ where B is in the commutant of V. The operator B is the *symbol of* H_B. Obviously $\hat{K} \ominus M$ is an invariant subspace for V^*. So the operator V_* defined by $V_* = QV | \hat{K} \ominus M$ is a co-isometry on $\hat{K} \ominus M$. This sets the stage for the following result.

4.2 PROPOSITION. *An operator A from M to $\hat{K} \ominus M$ is a Hankel operator if and only if A is in $I(V_+, V_*)$.*

PROOF. If H_B is a Hankel operator, then $V_*Q = QV$ implies that

$$V_*H_B = V_*QB\,|\,M = QVB\,|\,M = QBV_+ = H_BV_+ \,.$$

So A is in $I(V_+, V_*)$. On the other hand assume that A is in $I(V_+, V_*)$. Notice that V is a unitary dilation of both V_+ and V_*. Corollary VII.1.5 of the commutant lifting theorem implies that there exists a B in the commutant of V satisfying $A = QB\,|\,M$, or equivalently, A is a Hankel operator with symbol B. This completes the proof.

4.3 THEOREM. *If* H_B *is a Hankel operator, then*

$$\|H_B\| = \inf\{\|B-F\|\colon F \text{ is in } I(V) \stackrel{\Delta}{=} I(V, V) \text{ and } FM \subseteq M\} \,. \tag{4.5}$$

Moreover, there exists an optimal F_* *in* $I(V)$ *which satisfies* $F_*M \subseteq M$ *and solves the minimization problem in (4.5), that is,*

$$\|H_B\| = \|B-F_*\| = \inf\{\|B-F\|\colon F \text{ is in } I(V) \text{ and } FM \subseteq M\} \,. \tag{4.6}$$

PROOF. If F is in $I(V)$ and $FM \subseteq M$, then $H_F = 0$ and

$$\|H_B\| = \|H_B - H_F\| = \|Q(B-F)\,|\,M\| \leq \|B-F\| \,. \tag{4.7}$$

This shows that $\|H_B\|$ is a lower bound for the right hand side of (4.5). Because H_B is in $I(V_+, V_*)$, Corollary VII.1.5 implies that there exists a B_1 in $I(V)$ satisfying $H_B = H_{B_1}$ and $\|H_B\| = \|B_1\|$. Using $F_* = B - B_1$ this gives

$$\|H_B\| = \|B_1\| = \|B-F_*\| \geq \inf\{\|B-F\|\colon F \in I(V) \text{ and } FM \subseteq M\} \,. \tag{4.8}$$

Since $H_B = H_{B_1}$ we see that $0 = Q(B-B_1)\,|\,M = QF_*\,|\,M$. Thus $F_*M \subseteq M$. Obviously F_* is in $I(V)$. This yields the inequality in (4.8) and shows that $\|H_B\|$ is a lower bound for the right hand side of (4.5). Therefore (4.5) holds. Equation (4.6) follows from (4.5) and (4.8). This completes the proof.

An operator T on M is *left invertible* if there exists an operator L on M satisfying $LT = I$. (Recall that an operator is by definition bounded.) It is well known that T is left invertible if and only if there exists a $\delta > 0$ such that $\|Th\| \geq \delta\|h\|$ for all h in M. The operator T is *right invertible* if there exists an operator R on M satisfying $TR = I$. It is easy to verify that T is left invertible if and only if T^* is right invertible. It is also obvious that an operator T is invertible if and only if T is both right and left invertible. Recall that if T is a strict contraction ($\|T\| < 1$), then $I - T$ is invertible. In this case

$$(I-T)^{-1} = \sum_0^\infty T^n \, .$$

In particular, if $I - T$ is a strict contraction, then T is invertible. The following result will play an important role in determining necessary and sufficient conditions for the invertibility of a Toeplitz operator.

4.4 LEMMA. *Let* W *be an isometry in the commutant of* V. *Then the Toeplitz operator* T_W *is left invertible if and only if the Hankel operator* H_W *is a strict contraction.*

PROOF. If T_W is left invertible, then there exists a $\delta > 0$ satisfying

$$\delta\|h\|^2 \le \|T_W h\|^2 = \|Wh\|^2 - \|H_W h\|^2 = \|h\|^2 - \|H_W h\|^2 \quad (h \in H) \, .$$

This implies that $\|H_W h\|^2 \le (1-\delta)\|h\|^2$, or equivalently, H_W is a strict contraction. On the other hand if H_W is a strict contraction, then

$$\|T_W h\|^2 = \|Wh\|^2 - \|H_W h\|^2 \ge (1-\|H_W\|^2)\|h\|^2 \ge \delta\|h\|^2 \quad (h \in H)$$

for some $\delta > 0$. This completes the proof.

4.5 THEOREM. *Let* W *be a unitary operator in the commutant of* V. *The following statements are equivalent*

(i) The Toeplitz operator T_W *is invertible.*

(ii) The Hankel operators H_W *and* H_{W*} *are both strict contractions.*

(iii) There exists an invertible operator F *in the commutant of* V *satisfying* $FM = M$ *and* $\|W - F\| < 1$.

PROOF. The previous lemma with $T_W^* = T_{W*}$ shows that the first two statements are equivalent. Now assume that (i), or equivalently, (ii) holds. Recall that H_W is a contraction in $I(V_+, V_*)$. Corollary VII.1.5 implies that there exists an operator G in the commutant of V satisfying

$$H_W = QG\,|\,M = H_G \quad \text{and} \quad \|H_W\| = \|G\| \, .$$

If we set $F = W - G$, then

$$1 > \|H_W\| = \|G\| = \|W - F\| = \|I - W^*F\| \, .$$

This implies that W^*F is invertible, or equivalently, F is invertible. Notice that $H_W = H_G$

gives

$$QF \,|\, M = Q(W-G) \,|\, M = 0 \,.$$

So F maps M into M. This implies that

$$1 > \|I - W^* F\| \geq \|I - PW^* F \,|\, M\| = \|I - T_W^* F \,|\, M\| \,.$$

Therefore $T_W^* F \,|\, M$ is invertible. Since T_W is invertible $F \,|\, M$ is an invertible operator on M. In particular, $FM = M$. This proves (iii).

If (iii) holds, then

$$1 > \|W-F\| = \|I-W^* F\| \geq \|I-T_W^* F \,|\, M\| \,.$$

So $T_W^* F \,|\, M$ is invertible. Because $F \,|\, M$ is invertible on M this implies that T_W^* is invertible. Therefore (i) holds. This completes the proof.

4.6 REMARK. It is emphasized that the norm of H_W does not necessarily equal the norm of H_{W*}. For a counter example choose V_+ to be a unilateral shift on M and let V be any unitary extension of V_+. If we set $W = V$, then $H_V = 0$ and

$$H_{V*} = V^* (I - V_+ V_+^*) \,.$$

Thus in this case $\|H_V\| = 0$ and $\|H_{V^.}\| = 1$.

5. INVERTIBLE TOEPLITZ OPERATORS

In this section we will factor the symbol B in the Toeplitz operator T_B to obtain necessary and sufficient conditions for T_B to be invertible when B is a general symbol.

To begin let B be an operator in the commutant of V. We say that B admits a *left invertible factorization* if $B = WR$ where both W and R are in the commutant of V, the operator R is invertible, $RM = M$ and W is an isometry. (In a left invertible factorization the isometry W appears on the left.) This sets the stage for the following result

5.1 THEOREM. *Let* B *be an operator in the commutant of* V. *Then* B *is left invertible if and only if* B *admits a left invertible factorization.*

PROOF. If B admits a left invertible factorization, then clearly B is left invertible. So we assume that B is left invertible. Obviously $BM (= \overline{BM})$ is invariant for V and $VB \,|\, M = BV_+$. Because B is left invertible this implies that $V \,|\, BM$ and V_+ are similar. By Proposition VI.1.3 the isometries V_+ and $V \,|\, BM$ are unitarily equivalent. Recall that the subspace M_∞ defined in (4.2) is a reducing subspace for V. Since B commutes with

V^* and B is left invertible this also implies that BM_∞ is a reducing subspace for V. Moreover, $V_\infty = V|M_\infty$ on M_∞ and $V'_\infty = V|BM_\infty$ on BM_∞ is the minimal unitary extension of V_+ and $V|BM$, respectively. By Corollary VI.2.4 there exists a unitary operator $\dot{W}_\infty$ from M_∞ onto BM_∞ satisfying

$$V'_\infty W_\infty = W_\infty V_\infty \quad \text{and} \quad W_\infty M = BM \ . \tag{5.1}$$

The operator B admits a matrix decomposition of the form

$$B = \begin{bmatrix} B_0 & * \\ 0 & B_1 \end{bmatrix} : M_\infty \oplus M_\infty^\perp \to (BM_\infty) \oplus (BM_\infty)^\perp \tag{5.2}$$

where $B_0 = B|M_\infty$. Because M_∞ and BM_∞ are reducing subspaces, V admits a decomposition of the form $V = V_\infty \oplus V_1$ on $M_\infty \oplus M_\infty^\perp$ and $V = V'_\infty \oplus V'_1$ on $BM_\infty \oplus (BM_\infty)^\perp$. The matrix decomposition of B in (5.2) with $(V'_\infty \oplus V'_1)B = B(V_\infty \oplus V_1)$ implies that $V'_1 B_1 = B_1 V_1$. Using the fact that B_0 is invertible and B is left invertible it is easy to check that B_1 is also left invertible. Therefore V_1 and $V'_1|B_1 M_\infty^\perp$ are similar. Proposition VI.1.3 shows that there exists a unitary operator W_1 from $M_\infty^\perp$ onto the range of B_1 satisfying $V'_1 W_1 = W_1 V_1$. This and (5.1) implies that $W = W_\infty \oplus W_1$ is an isometry on $\hat{K}$ in the commutant of V. Moreover, $WM = BM$ (see (5.1)) and $W\hat{K} = B\hat{K}$. The last equality follows from the fact $W\hat{K} = BM_\infty \oplus B_1 M_\infty^\perp$ which is precisely the range of B. Finally, since B is left invertible and $W\hat{K} = B\hat{K}$ we see that $R = W^*B$ is an invertible operator in the commutant of V. Therefore WR is a left invertible factorization of $WW^*B = B$. Finally, the condition $RM = M$ follows from the second relation in (5.1) and the definition of W. This completes the proof.

5.2 COROLLARY. *Let* B *be an operator in the commutant of* V. *Then* B *is invertible if and only if* B *admits a left invertible factorization* $B = WR$ *where* W *is a unitary operator on* $\hat{K}$.

PROOF. This follows from the previous theorem, and from the fact that an operator B on $\hat{K}$ is invertible if and only if B is left invertible and onto $\hat{K}$.

5.3 THEOREM. *Let* B *be a left invertible operator in the commutant of* V. *The Toeplitz operator* T_B *is left invertible if and only if* B *admits a left invertible factorization* $B = WR$ *where the Hankel operator* H_W *is a strict contraction.*

PROOF. The previous theorem guarantees that B admits a left invertible factorization B = WR. Because R is invertible and $RM = M$, the operator R_+ defined by $R_+ = R|M$ is an invertible operator on M. This and the factorization B = WR shows that $T_B = T_W R_+$. Therefore T_B is left invertible if and only if T_W is left invertible, or equivalently, by Lemma 4.4 if the Hankel operator H_W is a strict contraction. This completes the proof.

If the reducing subspace M_∞ defined in (4.2) equals $\hat{K}$, then the following result shows that we can eliminate the hypothesis that B is left invertible in Theorem 5.3.

5.4 COROLLARY. *Assume that* $M_\infty = \hat{K}$ *and* B *is in the commutant of* V. *Then the Toeplitz operator* T_B *is left invertible if and only if* B *admits a left invertible factorization* B = WR *where the Hankel operator* H_W *is a strict contraction. In this case* B *is also left invertible.*

PROOF. If T_B is left invertible, then

$$\|Bh\| \geq \|T_B h\| \geq \delta\|h\| \qquad (h \in M)$$

for some $\delta > 0$. Since V^* commutes with B this implies that $\|BV^{*n}h\| \geq \delta\|V^{*n}h\|$ for all $n \geq 0$. The hypothesis $M_\infty = \hat{K}$ implies that $\|Bh\| \geq \delta\|h\|$ for all h on a dense set in $\hat{K}$, or equivalently, B is left invertible. The previous theorem shows that B admits a left invertible factorization WR such that H_W is a strict contraction. On the other hand if B = WR is a left invertible factorization, then obviously B is left invertible. If moreover, H_W is a strict contraction the previous theorem shows that T_B is left invertible. This completes the proof.

Let B be an operator in the commutant of V. We say that B admits a *right invertible factorization* if B^* admits a left invertible factorization. This says that B admits a factorization of the form $B = L^* W_*$ where both L and W_* are in the commutant of V, the operator L is invertible, $LM = M$ and W_* is a co-isometry. Theorem 5.1 shows that B is right invertible if and only if B admits a right invertible factorization. By taking adjoints in Theorem 5.3 we arrive at the following result.

5.5 COROLLARY. *Let* B *be a right invertible operator in the commutant of* V. *The Toeplitz operator* T_B *is right invertible if and only if* B *admits a right invertible factorization* $B = L^* W_*$ *where the Hankel operator* H_{W_*} *is a strict contraction.*

The following is the dual of Corollary 5.4.

5.6 COROLLARY. *Assume that $M_\infty = \hat{K}$ and B is in the commutant of* V. *Then the Toeplitz operator T_B is right invertible if and only if B admits a right invertible factorization $B = L^* W_*$ where the Hankel matrix H_{W_*} is a strict contraction. In this case B is also right invertible.*

Let B be in the commutant of V. We say that B admits an *invertible factorization* if $B = L^* WR$ where R, L and W are all in the commutant of V, both R and L are invertible, $RM = M$ and $LM = M$ and W is unitary. This sets the stage for the following result.

5.7 THEOREM. *Let B be in the commutant of* V. *Then B is invertible if and only if B admits an invertible factorization.*

PROOF. If B is invertible, then Corollary 5.2 shows that B admits a left invertible factorization $B = \Omega R$ where Ω is unitary. Applying Corollary 5.2 to Ω^* we see Ω^* admits a left invertible factorization $\Omega^* = W^* L$ where W is unitary. By construction $B = L^* WR$ is an invertible factorization of B. If B admits an invertible factorization, then obviously B is invertible. This completes the proof.

The following uses invertible factorizations to determine the invertibility of a Toeplitz operator.

5.8 THEOREM. *Let B be an invertible operator in the commutant of* V. *The following statements are equivalent.*

(i) The Toeplitz operator T_B is invertible.

(ii) The operator B admits an invertible factorization $B = L^* WR$ where the Hankel operators H_W and H_{W_*} are both strict contractions.

(iii) The operator B admits an invertible factorization $B = L^* WR$ and there exists an invertible F in the commutant of V *satisfying $FM = M$ and $W - F$ is a strict contraction.*

PROOF. Theorem 5.7 shows that B admits an invertible factorization $B = L^* WR$. Notice that $T_B = T_{L^* W} R | M$. Since $R | M$ is invertible on M this implies that T_B is invertible if and only if $T_{L^* W}$ is invertible. However, $T_{L^* W}^* = T_{W^*} L | M$ and $L | M$ is invertible on M. Therefore T_B is invertible if and only if T_W is invertible. Now the theorem follows from Theorem 4.5. This completes the proof.

5.9 COROLLARY. *Assume that $M_\infty = \hat{K}$ and* B *is in the commutant of* V. *The following statements are equivalent*

(i) The Toeplitz operator T_B *is invertible.*

(ii) The operator B *admits an invertible factorization* $B = L^*WC$ *where the Hankel operators* H_W *and* H_{W^*} *are both strict contractions.*

(iii) The operator B *admits an invertible factorization* $B = L^*WR$ *and there exists an invertible* F *in the commutant of* V *satisfying* $FM = M$ *and* $W - F$ *is a strict contraction.*

PROOF. If T_B is invertible, then both T_B and $T_B^* = T_{B^*}$ are left invertible. The proof of Corollary 5.4 shows that B is invertible. On the other hand if B admits an invertible factorization, then B is invertible. In either case B is invertible. Now the corollary follows by applying the previous theorem. This completes the proof.

Recall that an operator A on H is *strictly positive* if $A \geq \delta I$ where δ is a positive constant. The following provides an alternative method for determining the invertibility of T_B.

5.10 THEOREM. *Let* B *be an invertible operator in the commutant of* V. *The Toeplitz operator* T_B *is invertible if and only if there exists an invertible operator* G *in the commutant of* V *such that* $GM = M$ *and* $\mathrm{Re}(B^*G)$ *is a strictly positive operator.*

PROOF. If T_B is invertible, then Theorem 5.8 shows that B has an invertible factorization $B = L^*WR$ and that there exists an invertible F in the commutant of V satisfying

$$\|(W-F)f\|^2 \leq (1-\delta)\|f\|^2 \qquad (f \in \hat{K}) \tag{5.3}$$

where $\delta > 0$ and $FM = M$. Expanding (5.3) gives $2\mathrm{Re}(W^*Ff,f) \geq \delta\|f\|^2$, or equivalently, $\mathrm{Re}(W^*F)$ is strictly positive. Using $B = L^*WR$ and the fact that R is invertible there exists an $\alpha > 0$ satisfying

$$0 < \alpha I \leq R^*(W^*F + F^*W)R = R^*W^*LL^{-1}FR + R^*F^*L^{*-1}L^*WR = B^*G + G^*B$$

where $G = L^{-1}FR$. Since $RM = FM = LM = M$ we see that $GM = M$. By construction G is an invertible operator in the commutant of V such that $\mathrm{Re}(B^*G)$ is strictly positive.

On the other hand, assume that G is an invertible operator in the commutant of V such that $GM = M$ and $\mathrm{Re}(B^*G)$ is strictly positive. If $C = B^*G$, then $GM = M$ implies that the Toeplitz operator $T_C = T_{B^*}G\,|\,M$ is invertible if and only if T_B is invertible. So to complete the proof it is sufficient to show that T_C is invertible. If h is a unit vector in

M, then $\mathrm{Re}(B^*G) \geq \alpha > 0$ gives

$$\|T_C h\| \geq |(T_C h, h)| = |(Ch, h)| \geq \mathrm{Re}(Ch, h) \geq \alpha \ .$$

Thus T_C is left invertible. A similar calculation shows that $T_C^* = T_{C^*}$ is left invertible. Therefore $T_C = T_B \cdot G \,|\, M$ is invertible. This completes the proof.

To complete this section let B be an operator in the commutant of V and H an invariant subspace for V_+^*. The Toeplitz operator obtained by compressing B to $M \ominus H$ is denoted by T_B'. The compression of B to H is denoted by C_B. A problem which naturally arises in some problems in control theory, is to find necessary and sufficient conditions on B for C_B to be invertible. The following provides a partial solution to this problem.

5.11 THEOREM. *Let H be an invariant subspace for V_+^* and B an invertible operator in the commutant of V satisfying $B(M \ominus H) \subseteq H^\perp$. If T_B' is invertible, then the following statements are equivalent.*

(i) The operator C_B on H is invertible

(ii) The operator B admits an invertible factorization $B = L^* W R$ where the Hankel operators H_W and H_{W^*} are both strict contractions.

(iii) The operator B admits an invertible factorization $B = L^* W R$ and there exists an invertible F in the commutant of V satisfying $FM = M$ and $W - F$ is a strict contraction.

PROOF. The hypothesis implies that T_B admits a matrix representation of the form:

$$T_B = \begin{bmatrix} T_B' & * \\ 0 & C_B \end{bmatrix} \text{ on } (M \ominus H) \oplus H \ . \tag{5.4}$$

Since T_B' is invertible, T_B is invertible if and only if C_B is invertible. Theorem 5.8 shows that (i), (ii) and (iii) are equivalent. This completes the proof.

Notice that if T_B' is invertible, then T_B in (5.4) is left invertible if and only if C_B is left invertible. This and Theorem 5.3 proves the following result.

5.12 THEOREM. *Let H be an invariant subspace of V_+^*. Let B be a left invertible operator in the commutant of V where $B(M \ominus H) \subseteq H^\perp$ and T_B' is invertible. Then C_B is left invertible if and only if B admits a left invertible factorization $B = W R$ where the Hankel operator H_W is a strict contraction.*

In many applications involving stochastic processes and filtering theory the operator B is strictly positive. In this case its Toeplitz matrix T_B is also strictly positive, and one does not need the commutant lifting theorem to obtain conditions on B for T_B to be invertible. The Toeplitz operator T_B is invertible. Moreover, the role of invertible factorizations can be replaced by outer factorizations.

We say that B admits an *outer factorization* if $B = R^*R$ where R is an invertible operator in the commutant of V and $RM = M$. Obviously this B is strictly positive. If we set $L = R^*$ and $W = I$, then we see that an outer factorization is a special case of an invertible factorization. If B is strictly positive, then its positive square root B_1 is strictly positive. Corollary 5.2 shows that $B_1 = WR$ where W is unitary. Thus $B = R^*R$ is an outer factorization of B. This proves the following result.

THEOREM 5.13. *Let* B *be an operator in the commutant of* V. *Then* B *is strictly positive if and only if* B *admits an outer factorization. In this case* T_B *is strictly positive.*

COROLLARY 5.14. *Let* $M_\infty = \hat{K}$ *and assume that* B *is an operator in the commutant of* V. *Then the following statements are equivalent*

(i) The operator B *is strictly positive.*

(ii) The operator B *admits an outer factorization.*

(iii) The Toeplitz operator T_B *is strictly positive.*

PROOF. The previous theorem shows that (i) and (ii) are equivalent and they imply (iii). If (iii) holds, then for all h in M and $n \geq 0$ we have

$$(BV^{*n}h, V^{*n}h) = (Bh, h) = (T_B h, h) \geq \delta\|h\|^2 = \delta\|V^{*n}h\|^2$$

for some $\delta > 0$. Therefore B is strictly positive on the dense set $\bigcup_0^\infty V^{*n}M$, or equivalently, B is strictly positive. This completes the proof.

6. CORONA THEOREMS FOR ANALYTIC TOEPLITZ OPERATORS

In this section we will use the commutant lifting theorem to obtain some Corona type theorems for analytic Toeplitz operators. Then we will apply these results to obtain necessary and sufficient conditions for n analytic Toeplitz operators to be coprime.

To begin we say that A is an *analytic Toeplitz operator* if A intertwines two isometries. For example the operator $\hat{A}_\infty$ in Section 1 is an analytic Toeplitz operator, since it intertwines the unilateral shifts S and S'. The operator L, W and R in the previous section are also examples of analytic Toeplitz operators.

Throughout this section U on K, U' on K' and U_i on K_i for $i = 1, 2, ..., n$ are all specified isometries. Now let A be in $I(U, U')$ and C in $I(U_1, U')$. Obviously A and C are both analytic Toeplitz operators. Assume that there exists a contractive analytic Toeplitz operator B in $I(U, U_1)$ satisfying A = CB. Then $\| A^* h \| \leq \| C^* h \|$ for all h in K', or equivalently, $CC^* - AA^*$ is positive. This proves the first part of the following fundamental result.

6.1 THEOREM. *Let* A *and* C *be analytic Toeplitz operators in* $I(U, U')$ *and* $I(U_1, U')$, *respectively. Then there exists a contractive analytic Toeplitz operator* B *in* $I(U, U_1)$ *satisfying* A = CB *if and only if* $CC^* - AA^*$ *is positive.*

PROOF. Assume that $CC^* - AA^*$ is positive. This implies that $\|A^* h\| \leq \|C^* h\|$ for all h in K'. So there exists a contraction A_* from $H = \overline{C^* K'}$ to K satisfying $A_* C^* = A^*$. Obviously H is an invariant subspace for U_1^*. Let T be the contraction on H defined by $T = U_1^* \,|\, H$. Since A^* is in $I(U'^*, U^*)$ and C^* is in $I(U'^*, U_1^*)$, it follows that A_* is in $I(T, U^*)$. By Corollary VII.1.4 of the commutant lifting theorem there exists a contraction B_* in $I(U_1^*, U^*)$ satisfying $B_* \,|\, H = A_*$. Therefore setting B equal to the adjoint of B_* we have A = CB. This completes the proof.

Recall that an operator C from K to K' admits a *right inverse* if there exists an operator B from K' to K satisfying $CB = I'$ where I' is the identity on K'. In this case B is called the *right inverse of* C. Obviously C admits a right inverse if and only if the range of C is onto K', or equivalently, $\|C^* h\| \geq \delta \|h\|$ for all h in K' and some $\delta > 0$. The following result shows that if an analytic Toeplitz operator admits a right inverse, then it has an analytic right inverse.

6.2 COROLLARY. *Let* C *be an analytic Toeplitz operator in* $I(U, U')$. *Then* C *admits a right inverse if and only if* C *admits a right analytic Toeplitz inverse, that is, there exists an analytic Toeplitz operator* B *in* $I(U', U)$ *satisfying* $CB = I'$. *Moreover, there exists a* $\delta > 0$ *satisfying*

$$\|C^* h\| \geq \delta \|h\| \quad \textit{(for all } h \in K') \tag{6.1}$$

if and only if there exists an analytic Toeplitz operator B *in* $I(U', U)$ *satisfying*

$$CB = I' \ and \ \|B\| \leq 1/\delta . \tag{6.2}$$

PROOF. If (6.1) holds, then we set $A = \delta I'$. Obviously A is in $I(U', U')$ and $CC^* - AA^*$ is positive. By the previous theorem there exists a contractive analytic

Toeplitz operator B_1 in $I(U', U)$ satisfying $\delta I' = A = CB_1$. Dividing by δ produces an operator $B = \delta^{-1} B_1$ in $I(U', U)$ satisfying (6.2). On the other hand if (6.2) holds, then

$$\|h\| = \|B^* C^* h\| \leq \delta^{-1} \|C^* h\| \qquad \text{(for all } h \in K').$$

This produces (6.1) and completes the proof.

Let C_i be an operator in $I(U_i, U')$ for $i = 1, 2, ..., n$. We say that the set of analytic Toeplitz operators $\{C_i\}_1^n$ is *left coprime* if there exists a sequence of analytic Toeplitz operators B_i in $I(U', U_i)$ for $i = 1, 2, ..., n$ satisfying

$$\sum_1^n C_i B_i = I' \overset{\Delta}{=} I_{K'} . \qquad (6.3)$$

Obviously $\{C_i\}_1^n$ is left coprime if and only if the analytic Toeplitz row operator

$$C = [\, C_1, C_2, ..., C_n \,] \text{ in } I(\overset{n}{\underset{1}{\oplus}} U_i, U') \qquad (6.4)$$

admits a right analytic Toeplitz inverse B of the form

$$B = [\, B_1, B_2, ..., B_n \,]^{\text{tr}} \text{ in } I(U', \overset{n}{\underset{1}{\oplus}} U_i) . \qquad (6.5)$$

that is, $C B = I'$. By consulting the previous corollary we readily obtain the following Toeplitz-Corona result.

6.3 COROLLARY. *Let C_i be an analytic Toeplitz operator in $I(U_i, U')$ for $i = 1, 2, ..., n$. Then $\{C_i\}_1^n$ is left coprime if and only if there exists a $\delta > 0$ satisfying*

$$\sum_1^n \| C_i^* h \|^2 \geq \delta \| h \|^2 \qquad \text{(for all } h \in K'). \qquad (6.6)$$

We notice that if $\{C_i\}_1^n$ is left coprime and $C_i = L\tilde{C}_i$ for $i = 1, 2, ..., n$ where L and all the $\tilde{C}_i$'s are analytic Toeplitz operators, then (6.3) implies that the range of L equals K'. In other words if $\{C_i\}_1^n$ is left coprime and L is a left common factor to all C_i, then the range of L is all of K'.

To complete this section we will use Theorem 3.5 to obtain a dual of Theorem 6.1, and present some results analogous to Corollary 6.3 concerning right coprime factorizations. Throughout the rest of this section V on $\hat{K}$ and V' on $\hat{K}'$ are the minimal unitary extensions of U and U', respectively; and V_i on $\hat{K}_i$ is the minimal unitary extension of U_i for $i = 1, 2, ..., n$. The orthogonal projection onto $\hat{K}' \ominus K'$ is denoted by Q' and Q_i is the orthogonal projection onto $\hat{K}_i \ominus K_i$ for $i = 1, 2, ..., n$.

Now let A and B be specified analytic Toeplitz operators in $I(U, U')$ and $I(U, U_1)$, respectively. Throughout A_∞ in $I(V, V')$ and B_∞ in $I(V, V_1)$ are the unique operators with norm $\|A\|$ and $\|B\|$ extending A and B, respectively provided by Corollary VI.2.4., that is,

$$A_\infty f = \lim_{n \to \infty} V'^{*n} A \, P_K \, V^n \, f \quad (f \in \hat{K})$$

$$B_\infty f = \lim_{n \to \infty} V_1^{*n} B \, P_K \, V^n \, f \quad (f \in \hat{K}) \tag{6.7}$$

The following useful result is similar to Lemma 3.3.

6.4 LEMMA. *Let A and B be analytic Toeplitz operators in* $I(U, U')$ *and* $I(U, U_1)$, *respectively. Then*

$$\|D_{U'^n} Ah\| \le \| D_{U_1^n} Bh\| \quad (\textit{for all } h \in K \textit{ and } n \ge 1) \tag{6.8}$$

if and only if

$$\|Q' A_\infty f\| \le \|Q_1 B_\infty f\| \quad (\textit{for all } f \in \hat{K} \ominus K) \tag{6.9}$$

PROOF. If (6.8) holds, then for all h in K and $n \ge 0$ we have

$$\|Q' A_\infty V^{*n} h\|^2 = \|Q' V'^{*n} A_\infty h\|^2 = \|A_\infty h\|^2 - \|P_{K'} V'^{*n} A_\infty h\|^2 =$$
$$\|Ah\|^2 - \|U'^{*n} Ah\|^2 = \|D_{U'^n} Ah\|^2 \le \|D_{U_1^n} Bh\|^2 =$$
$$\|Bh\|^2 - \|U_1^{*n} Bh\|^2 = \|V_1^{*n} Bh\|^2 - \|P_{K_1} V_1^{*n} Bh\|^2 = \|Q_1 V_1^{*n} Bh\|^2 = \|Q_1 B_\infty V^{*n} h\|^2 .$$

Since $\{V^{*n} K\}$ for $n \ge 0$ forms a dense set in $\hat{K}$ the relation (6.9) follows from the lifting properties

$$Q' A_\infty = Q' A_\infty Q \text{ and } Q_1 B_\infty = Q_1 B_\infty Q$$

where Q is the orthogonal projection onto $\hat{K} \ominus K$. Conversely if (6.9) holds, then using these lifting properties we obtain

$$\|Q' A_\infty g\| \le \|Q_1 B_\infty g\| \qquad (g \in \hat{K}) . \tag{6.9a}$$

Taking $g = V^{*n} h$ in (6.9a) and rearranging the order of the previous computations we obtain (6.8). This completes the proof.

The following which is the dual to Theorem 6.1 is the result we have been looking for.

6.5 THEOREM. *Let* A *and* B *be analytic Toeplitz operators in* $I(U, U')$ *and* $I(U, U_1)$, *respectively. Then there exists a contractive analytic Toeplitz operator* C *in* $I(U_1, U')$ *satisfying* A = CB *if and only if* $B^*B - A^*A$ *is positive and the conditions in (6.8), or equivalently, (6.9) hold.*

PROOF. Clearly $M = \overline{BK}$ is an invariant subspace for U_1. Let U_+ be the isometry on M defined by $U_+ = U_1 | M$. If A = CB where C is a contraction in $I(U_1, U')$, then obviously $B^*B - A^*A$ is positive. Moreover, the operator $C_1 = C | M$ is a contractive analytic Toeplitz operator in $I(U_+, U')$. Consulting Proposition 3.2

$$\|D_{U'^n} C_1 g\| \le \|D_{U_1^n} g\| \quad \text{(for all } g \in M \text{ and } n \ge 1) . \tag{6.10}$$

By setting g = Bh and using $A = C_1 B$ we readily obtain the conditions in (6.8).

Conversely assume that the conditions in (6.8) hold and $B^*B - A^*A$ is positive. This implies that there exists a contraction C_1 mapping M into K' satisfying $C_1 B = A$. By setting g = Bh and using $A = C_1 B$ in (6.8) we readily obtain the conditions in (6.10). Theorem 3.5 shows that there exists a contractive analytic Toeplitz operator C in $I(U_1, U')$ extending C_1. Obviously A = CB. This completes the proof.

6.6 REMARK. Let A and B be analytic Toeplitz operators in $I(U, U')$ and $I(U, U_1)$ respectively. Assume that $B^*B - A^*A$ is positive and (6.8), or equivalently, (6.9) holds. The proof of the previous theorem shows that the set of all contractive analytic Toeplitz operators C in $I(U_1, U')$ satisfying A = CB is given by the set of all contractive extensions C in $I(U_1, U')$ of C_1 in $I(U_+, U')$ defined by $A = C_1 B$. The operator B_3 in the proof of Theorem 3.5 with C replaced by C_1 provides at least one contractive analytic Toeplitz extension in $I(U_1, U')$ of C_1.

Recall that an operator B from K to K_1 admits a *left inverse* if there exists an operator C from K_1 into K satisfying CB = I where I is the identity in K. In this case C is called the *left inverse* of B. Obviously B admits a left inverse if and only if there exists a $\delta > 0$ such that $\|Bh\| \ge \delta\|h\|$ for all h in K, or equivalently, the range of B^* is onto. Clearly B admits a left inverse if and only if B^* admits a right inverse. By choosing $A = \delta I$ in the previous Theorem we obtain the following which is the dual of Corollary 6.2. It provides necessary and sufficient conditions for an analytic Toeplitz operator to admit an analytic Toeplitz left inverse.

6.7 COROLLARY. *Let* B *be in* $I(U, U')$. *The following are equivalent.*

(i)   *The analytic Toeplitz operator* B *admits an analytic Toeplitz left inverse* C *in* $I(U', U)$.

(ii)   *There exists a* $\delta > 0$ *satisfying*

$$\|Bh\| \geq \delta\|h\| \ \textit{and} \ \delta\|D_{U^{*n}}h\| \leq \|D_{U'^{*n}}Bh\| \quad \textit{(for all h in K and all n} \geq 1) \quad (6.11)$$

(iii)   *There exists a* $\delta > 0$ *satisfying*

$$\|Bh\| \geq \delta\|h\| \ \textit{and} \ \delta\|f\| \leq \|Q'B_\infty f\| \quad \textit{(for all h} \in K \textit{ and all f} \in \hat{K} \ominus K) . \quad (6.12)$$

In this case there exists and analytic Toeplitz left inverse C *of* B *whose norm* $\|C\| \leq 1/\delta$.

It is emphasized that the conditions in (6.12) is not automatically satisfied if B is left invertible. The unilateral shift S viewed as an analytic Toeplitz operator in $I(S, S)$ provides a counter example.

Let B_i be an operator in $I(U, U_i)$ for $i = 1, 2, ..., n$. We say that the set of analytic Toeplitz operator $\{B_i\}_1^n$ is *right coprime* if there exists a sequence of analytic Toeplitz operators C_i in $I(U_i, U)$ for $i = 1, 2, ..., n$ satisfying

$$\sum_1^n C_i B_i = I \ . \tag{6.13}$$

Obviously $\{B_i\}_1^n$ is right coprime if and only if the analytic Toeplitz column operator

$$B = [B_1, B_2, ..., B_n]^{tr} \text{ in } I(U, \bigoplus_1^n U_i) \tag{6.14}$$

admits a left analytic Toeplitz inverse C of the form

$$C = [C_1, C_2, ..., C_n] \text{ in } I(\bigoplus_1^n U_i, U), \tag{6.15}$$

that is, CB = I. By consulting the previous corollary we readily obtain the following Toeplitz-Corona result which is the dual of Corollary 6.3.

6.8 COROLLARY. *Let* B_i *be an operator in* $I(U, U_i)$ *for* $i = 1, 2, ..., n$. *The following are equivalent.*

(i)   *The set of analytic Toeplitz operators* $\{B_i\}_1^n$ *is right coprime.*

(ii)   *There exists a* $\delta > 0$ *satisfying*

$$\sum_1^n \|B_i h\|^2 \geq \delta\|h\|^2 \ \textit{and} \ \delta\|D_{U^{*n}}h\|^2 \leq \sum_1^n \|D_{U_i^{*n}}B_i h\|^2 \quad \textit{(for all h in K and n} \geq 1). \quad (6.16)$$

(iii)   *There exists a $\delta > 0$ satisfying*

$$\sum_1^n \|B_i h\|^2 \geq \delta \|h\|^2 \ \text{ and } \ \delta \|f\|^2 \leq \sum_1^n \|Q_i B_{i\infty} f\|^2 \quad \text{(for all } h \in K \text{ and all } f \in \hat{K} \ominus K) \quad (6.17)$$

where $B_{i\infty}$ *for* $i = 1, 2, ..., n$ *is the unique analytic Toeplitz extension in* $I(V, V_i)$ *of* B_i.

Finally, we notice that if $\{B_i\}_1^n$ is right coprime and $B_i = \tilde{B}_i R$ for $i = 1, 2,..., n$ where R and all the $\tilde{B}_i$'s are analytic Toeplitz operators, then (6.13) implies that the range of R^* is all of K. In other words if $\{B_i\}_1^n$ is right coprime and R is a common right factor to all B_i, then the range of R^* is all of K.

7. GENERALIZED HANKEL MATRICES AND CONTRACTIVE ONE STEP INTERTWINING LIFTING

In this section we will introduce generalized Hankel matrices. Then using the theory of contractive one step intertwining liftings in Chapter V, we will establish the existence of a solution to a generalized Hankel matrix interpolation problem. Finally, we will show that solving this generalized Hankel matrix interpolation problem is equivalent to obtaining a solution to the commutant lifting theorem.

To begin, let $k_n^2(E)$ be the Hilbert space formed by the set of all vectors

$$x = [..., x_{-2}, x_{-1}, x_0, x_1, x_2,..., x_n]^{tr}$$

where x_i is in the Hilbert space E for all i, and whose inner product

$$(x, x) = \sum_{i=-\infty}^n \|x_i\|^2$$

is finite. We set $k_\infty^2(E) = \hat{l}^2(E)$. A *generalized Hankel matrix* is an operator H_n mapping E_o into E_n' of the form

$$H_n = \begin{bmatrix} A_\infty & \cdots & R'^2 B_\infty & R' B_\infty & B_\infty \\ \cdot & \cdots & \cdot & \cdot & \cdot \\ \cdot & \cdots & \cdot & \cdot & \cdot \\ \cdot & \cdots & \cdot & \cdot & \cdot \\ C_\infty R^{*(n-2)} & \cdots & E_{n-4} & E_{n-3} & E_{n-2} \\ C_\infty R^{*(n-1)} & \cdots & E_{n-3} & E_{n-2} & E_{n-1} \\ C_\infty R^{*n} & \cdots & E_{n-2} & E_{n-1} & E_n \end{bmatrix} : E_o \to E_n' \qquad (7.1)$$

where

$$E_o = R \oplus k_o^2(E) \text{ and } E_n' = R' \oplus k_n^2(E') . \tag{7.1a}$$

Here R on R and R$'$ on R' are both unitary operators, A_∞ is an operator from R into R' satisfying

$$R'A_\infty = A_\infty R \qquad (\text{or equivalently, } A_\infty \in I(R,R')), \tag{7.2}$$

B_∞ is an operator from E into R' and C_∞ is an operator from R to E'. Finally, E_i is an operator from E into E' for all $i \le n$. If $R = R' = \{0\}$, then the generalized Hankel matrix in (7.1) reduces to the usual Hankel matrix

$$H_n = \begin{bmatrix} \cdots & \cdot & \cdot & \cdot \\ \cdots & \cdot & \cdot & \cdot \\ \cdots & \cdot & \cdot & \cdot \\ \cdots & E_{n-3} & E_{n-3} & E_{n-2} \\ \cdots & E_{n-1} & E_{n-2} & E_{n-1} \\ \cdots & E_{n-2} & E_{n-1} & E_n \end{bmatrix} \tag{7.3}$$

mapping $k_o^2(E)$ into $k_n^2(E')$. However, the standard Hankel matrix is in a slightly different form. By rearranging the rows and columns in (7.3) we obtain the following standard Hankel matrix from $l^2(E)$ to $l^2(E')$, which is often referred to as a block Hankel matrix (recall that block matrices are matrices with operator or matrix entries)

$$\begin{bmatrix} E_n & E_{n-1} & E_{n-2} & \cdots \\ E_{n-1} & E_{n-2} & E_{n-3} & \cdots \\ E_{n-2} & E_{n-3} & E_{n-4} & \cdots \\ \cdot & \cdot & \cdot & \cdots \\ \cdot & \cdot & \cdot & \cdots \\ \cdot & \cdot & \cdot & \cdots \end{bmatrix} .$$

If $n \ge 1$, then H_n in (7.1) is called a *n step generalized Hankel matrix lifting* of H_o. If $n = \infty$, then H_∞ is an *infinite generalized Hankel matrix lifting* of H_o. In this case the infinite generalized Hankel matrix H_∞ is a block matrix mapping E_o into $E_\infty' = R' \oplus \hat{l}^2(E')$ of the form

$$
\begin{bmatrix}
A_\infty & \ldots & R'^2 B_\infty & R' B_\infty & B_\infty \\
\cdot & \ldots & \cdot & \cdot & \cdot \\
\cdot & \ldots & \cdot & \cdot & \cdot \\
\cdot & \ldots & \cdot & \cdot & \cdot \\
C_\infty R & \ldots & E_{-3} & E_{-2} & E_{-1} \\
C_\infty & \ldots & E_{-2} & E_{-1} & E_0 \\
C_\infty R^* & \ldots & E_{-1} & E_0 & E_1 \\
\cdot & \ldots & \cdot & \cdot & \cdot \\
\cdot & \ldots & \cdot & \cdot & \cdot \\
\cdot & \ldots & \cdot & \cdot & \cdot
\end{bmatrix}
\tag{7.4}
$$

satisfying (7.2) and $P_{E'_0} H_\infty = H_0$. This leads to the following generalized Hankel matrix interpolation problem.

7.1 PROBLEM. Given a (fixed) contractive generalized Hankel matrix H_0 of the form (7.1) satisfying (7.2), find all contractive infinite generalized Hankel matrices H_∞ lifting H_0, that is, find all contractive infinite generalized Hankel matrices satisfying $P_{E'_0} H_\infty = H_0$. In other words find all sequences of operators $\{E_j\}_1^\infty$ such that H_∞ is a contractive generalized Hankel lifting of H_0.

Corollary 7.6 below shows that this interpolation problem is solvable. (As noted earlier this interpolation problem is equivalent to finding all contractive intertwining liftings in the commutant lifting theorem.) Obviously

$$
H_0 = \begin{bmatrix} 0 & 0 \\ 0 & A'_n \end{bmatrix}
$$

where A'_n is the Hankel matrix defined in (IV.7.3), is a generalized Hankel matrix. In this case the generalized Hankel matrix interpolation problem reduces to the Carathéodory interpolation problem.

An important special case of the previous problem is the following one step generalized Hankel matrix interpolation problem:

7.2 PROBLEM. *Given a contractive generalized Hankel matrix H_n find all operators E_{n+1} from E to E' such that H_{n+1} is a contractive one step generalized Hankel matrix lifting of H_n, that is, $P_{E'_n} H_{n+1} = H_n$.*

We will solve this problem by using the theory of contractive one step intertwining liftings in Section V.1.

To this end, let S on $k_o^2(E)$ and S_n' on $k_n^2(E')$ be the shift operators defined by

$$S = \begin{bmatrix} \cdots \cdots \cdots \cdots & \cdot \\ \cdots \cdots \cdots \cdot & \cdot \\ \cdots \cdots \cdots \cdot & \cdot \\ \cdots \cdots \cdots\ 0\ I\ 0\ 0 \\ \cdots \cdots\ 0\ 0\ I\ 0 \\ \cdots \cdots\ 0\ 0\ 0\ I \\ \cdots \cdots\ 0\ 0\ 0\ 0 \end{bmatrix} \text{ and } S_n' = \begin{bmatrix} \cdots \cdots \cdots \cdots \cdots \cdot \\ \cdots \cdots \cdots \cdots \cdot \\ \cdots \cdots \cdots \cdots \cdot \\ \cdots \cdots\ 0\ 0\ 0\ 0 \\ \cdots \cdots\ I\ 0\ 0\ 0 \\ \cdots \cdots\ 0\ I\ 0\ 0 \\ \cdots \cdots\ 0\ 0\ I\ 0 \end{bmatrix} . \tag{7.5}$$

Notice that S is a unilateral shift, while S_n' is the adjoint of a unilateral shift. Let T on E_o be the isometry, and T_n' on E_n' be the co-isometry defined by

$$T = R \oplus S \text{ and } T_n' = R' \oplus S_n' . \tag{7.6}$$

The following is basic to our approach.

7.3 LEMMA. *Let* A *be an operator mapping* E_o *into* E_n'. *Then* A *is in* $I(T, T_n')$ *if and only if* A = H_n *where* H_n *is a generalized Hankel matrix of the form (7.1).*

PROOF. If H_n is a generalized Hankel matrix, then a simple calculation shows that $T_n' H_n = H_n T$. To prove the other half assume that A is in $I(T, T_n')$. Then A admits a matrix representation of the form

$$A = \begin{bmatrix} A_\infty & B \\ C & D \end{bmatrix} : \begin{bmatrix} R \\ k_o^2(E) \end{bmatrix} \to \begin{bmatrix} R' \\ k_n^2(E') \end{bmatrix} . \tag{7.7}$$

Using $T_n' A = AT$ we have

$$\begin{bmatrix} R'A_\infty & R'B \\ S_n'C & S_n'D \end{bmatrix} = \begin{bmatrix} A_\infty R & BS \\ CR & DS \end{bmatrix} . \tag{7.8}$$

Obviously $R'A_\infty = A_\infty R$, or equivalently, (7.2) holds. Equation (7.7) shows that B admits a matrix representation of the form:

$$B = [\ldots, B_{-2}, B_{-1}, B_o] \tag{7.9}$$

where B_i is an operator from E into R' for all $i \le 0$. Equation (7.8) implies that $R'B = BS$. By (7.9)

$$[..., R'B_{-2}, R'B_{-1}, R'B_0] = [..., B_{-3}, B_{-2}, B_{-1}] .$$

Thus $B_{-1} = R'B_0$ and $B_{-2} = R'^2 B_0$. Continuing in this fashion $B_{-i} = R'^i B_0$. Setting $B_\infty = B_0$ in (7.9) gives

$$B = [..., R'^2 B_\infty, R'B_\infty, B_\infty] . \tag{7.10}$$

This is precisely part of the first row of H_n in (7.1).

Equation (7.7) also shows that C admits a matrix representation of the form

$$C = [..., C_{n-2}, C_{n-1}, C_n]^{tr} \tag{7.11}$$

where C_i is an operator from R into E' for all $i \le n$. Equation (7.8) implies that $S'_n C = CR$. By (7.11)

$$[..., C_{n-3}, C_{n-2}, C_{n-1}]^{tr} = [..., C_{n-2}R, C_{n-1}R, C_n R]^{tr} .$$

Thus $C_{n-1} = C_n R$ and $C_{n-2} = C_n R^2$. Continuing in this fashion $C_{n-i} = C_n R^i$. Setting $C_\infty = C_0$ with (7.11) yields

$$C = [..., C_\infty R^{*(n-2)}, C_\infty R^{*(n-1)}, C_\infty R^{*n}]^{tr} \tag{7.12}$$

This is precisely part of the first column of H_n in (7.1).

Equation (7.7) shows that D admits a matrix representation of the form

$$D = \begin{bmatrix} \cdots & & \cdot & & \cdot & & \cdot \\ \cdots & & \cdot & & \cdot & & \cdot \\ \cdots & & \cdot & & \cdot & & \cdot \\ \cdots & E_{2,n-2} & E_{1,n-2} & E_{0,n-2} \\ \cdots & E_{2,n-1} & E_{1,n-1} & E_{0,n-1} \\ \cdots & E_{2,n} & E_{1,n} & E_{0,n} \end{bmatrix} \tag{7.13}$$

where $E_{i,j}$ is an operator from E into E' for all $i \ge 0$ and $j \le n$. Equation (7.8) implies that $S'_n D = DS$. Hence

$$\begin{bmatrix} \cdot \\ \cdot \\ \cdot \\ E_{i,n-2} \\ E_{i,n-1} \\ E_{i,n} \end{bmatrix} = DS^i \begin{bmatrix} \cdot \\ \cdot \\ \cdot \\ 0 \\ 0 \\ I \end{bmatrix} = S'^i D \begin{bmatrix} \cdot \\ \cdot \\ \cdot \\ 0 \\ 0 \\ I \end{bmatrix} = \begin{bmatrix} \cdot \\ \cdot \\ \cdot \\ E_{0,n-i-2} \\ E_{0,n-i-1} \\ E_{0,n-i} \end{bmatrix} .$$

Thus $E_{i,n} = E_{0,n-i}$ for all $i \ge 0$. This and (7.13) yields

$$
D = \begin{bmatrix}
\cdots & \cdot & \cdot & \cdot \\
\cdots & \cdot & \cdot & \cdot \\
\cdots & \cdot & \cdot & \cdot \\
\cdots & E_{n-4} & E_{n-3} & E_{n-2} \\
\cdots & E_{n-3} & E_{n-2} & E_{n-1} \\
\cdots & E_{n-2} & E_{n-1} & E_n
\end{bmatrix}
\tag{7.14}
$$

where $E_j = E_{0,j}$ for all $j \leq n$. Substituting (7.9), (7.12), (7.14) into (7.7) along with $R'A_\infty = A_\infty R$ shows that $A = H_n$ is a generalized Hankel matrix. This completes the proof.

7.4 REMARK. The previous lemma shows that if A is in $I(U, U'^*)$ where U on K and U' in K' are two isometries, then A can be identified with a generalized Hankel matrix H_n. To see this let $R \oplus (U|M_+(E))$ on $R \oplus M_+(E)$ and $R'^* \oplus (U'|M_+(E'))$ on $R' \oplus M_+(E')$ be the Wold decompositions of U and U', where $E = K \ominus UK$ and $E' = K' \ominus U'K'$, respectively. Here R and R' are unitary, and $U|M_+(E)$ and $U'|M_+(E')$ are unilateral shifts. Now let W be the unitary operator from K onto $R \oplus k_o^2(E)$ and W' the unitary operator from K' onto $R' \oplus k_n^2(E')$ defined by

$$
W(r + \sum_o^\infty U^i x_i) = r \oplus [..., x_3, x_2, x_1, x_o]^{tr}
$$

$$
W'(r' + \sum_o^\infty U'^j y_{n-j}) = r' \oplus [..., y_{n-2}, y_{n-1}, y_n]^{tr}
$$

where r is in R, r' is in R', and $\{x_i\}$ is a sequence in E and $\{y_i\}$ is a sequence in E' with finite support. Obviously $WU = TW$ and $WU'^* = T'_n W$ where T is the isometry and T'_n is the co-isometry in (7.6). Therefore by Lemma 7.3 the operator $W'AW^* = H_n$ is a generalized Hankel matrix in $I(T, T'_n)$. In other words any operator intertwining an isometry with a co-isometry can be identified with a generalized Hankel matrix.

The following shows that the one step generalized Hankel matrix interpolation Problem 7.2 is solvable.

7.5 PROPOSITION. *Let H_n be a contractive generalized Hankel matrix of the form (7.1). Then there exists an operator E_{n+1} mapping E into E' such that H_{n+1} is a contractive one step generalized Hankel matrix lifting of H_n.*

PROOF. We will prove this result by using the theory of one step liftings in Section V.1. A simple calculation shows that $D_{T'_n}$ is the orthogonal projection onto $\{0\} \oplus (\dots \{0\} \oplus \{0\} \oplus E')$. So the one step lifting $(T'_n)_1$ of T'_n can be identified with T'_{n+1}. Without loss of generality we set $(T'_n)_1 = T_{n+1}$ on $E'_n \oplus E'$. Here $D_{T'_n}$ is identified with the last E' in E'_n. Since T is an isometry, its one step lifting T_1 is still T. By Theorem V.1.2 or Theorem V.4.4 with $A = H_o$ in $I_1(T, T'_n)$ there exists a contractive one step intertwining lifting A_1 of A. In particular, $T'_{n+1}A_1 = A_1T_1 = A_1T$. The previous lemma shows that $A_1 = H_{n+1}$ is a contractive generalized Hankel matrix. Moreover, the lifting property $P_{E'_n}H_{n+1} = AP_{E_o} = H_n$ implies that $A_1 = H_{n+1}$ is a contractive generalized Hankel matrix lifting of A. The form of H_{n+1} produces an appropriate E_{n+1} and completes the proof.

One can also use Theorem IV.3.1 or Corollary IV.3.6 to obtain a proof of the previous theorem. The details are left to the reader as a simple exercise.

Notice that if H_{n+i} is an i step generalized Hankel matrix lifting of H_n for all $i \geq 0$ and all $n \geq 0$, then H_n strongly converges to an infinite generalized Hankel matrix H_∞ which is also an lifting of H_o (see Lemma VII.2.1). Now by recursively applying the previous proposition we obtain the following corollary, which shows that our generalized Hankel matrix interpolation Problem 7.1 is also solvable.

7.6 COROLLARY. *Let* H_o *be a contractive generalized Hankel matrix of the form* *(7.1). Then there exists a sequence of operators* $\{E_j\}_1^\infty$ *mapping E into* E' *such that* H_∞ *is a contractive infinite generalized Hankel matrix lifting of* H_o.

Now we will use Theorem V.1.2 to give a characterization of the set of all contractive one step generalized Hankel matrix liftings H_1 of H_o. To begin let $H_o = A$ be a generalized contractive Hankel matrix mapping E_o into E'_o. As in the proof of the previous proposition we set the one step lifting $(T'_o)_1$ of T'_o equal to T'_1 and identify $D_{T'}$ with the last E' in E'_o. Using $T' = T'_o$ and the fact that T is an isometry, the spaces F, G, F' and G' in Section V.1 become

$$F = F(H_o \cdot T) = \overline{D_{H_o} TH}, \quad G = D_{H_o} \ominus F,$$

$$F' = F'(T' \cdot H_o) = \{D_{T'} H_o h \oplus D_{H_o} h : h \in E_o\}^- \tag{7.15}$$

$$\text{and } G' = (E' \oplus D_{H_o}) \ominus F'.$$

In this setting the unitary operator ω from F onto F' defined in (V.1.12) is

$$\omega D_{H_o} Th = D_{T'} H_o h \oplus D_{H_o} h \qquad (h \in E_o) \,. \qquad (7.16)$$

Recall that $D_{T'}$ is the orthogonal projection onto the last E' in E'_0. This is important in applications because in this case we do not have to compute $D_{T'}$. Finally, applying Theorem V.1.2 and Corollary V.1.7 proves the following result.

7.7 COROLLARY. *The set of all contractive one step generalized Hankel matrix liftings* H_1 *of* H_o *is given by* $H_1 = [H_o^*, Z_1^*]^*$ *from* E_o *to* $E'_o \oplus E'$ *where*

$$Z_1 = \Pi'(\omega P_F + \Gamma P_G)D_{H_o} \qquad (7.17)$$

and Γ *is an arbitrary contraction from* G *into* G'. *Moreover, the factorization* $H_1 \cdot T[T'_1 \cdot H_1]$ *is regular if and only if* Γ *is an isometry [co-isometry] respectively.*

7.8 REMARK. Let H_o be a contractive generalized Hankel matrix. The previous Corollary shows that the set of all E_1 such that H_1 is a contractive generalized Hankel matrix lifting of H_o is given by H_1 in (7.1) where n=1,

$$E_1 = \Pi'(\omega P_F + \Gamma P_G)D_{H_o} \mid (\{0\} \oplus (\cdots \oplus \{0\} \oplus \{0\} \oplus E)) \qquad (7.18)$$

and Γ is an arbitrary contraction from G to G'.

One can also use Theorem IV.3.1 or Theorem V.4.4 to obtain the set of all E_1 such that H_1 is a contractive one step generalized Hankel matrix lifting of H_o. The details are left to the reader as simple exercises.

To complete this section we will show that Corollary 7.6 can be used to prove the commutant lifting theorem and vice versa. To this end assume that H_o is a generalized Hankel matrix in $I(T, T'_o)$. Notice that $T'_\infty \triangleq R' \oplus S'_\infty$ on $R' \oplus \hat{l}^2(E')$ is the minimal isometric dilation of T'_0 where S'_∞ is the bilateral shift on $\hat{l}^2(E')$. (The matrix representation for S'_∞ has the identity I directly below the main diagonal and zero's elsewhere.) By the commutant lifting theorem there exists a contractive intertwining lifting B in $I(T, T'_\infty)$ of H_o. As in the proof of Lemma 7.3, it is easy to verify that B admits a matrix representation of the form (7.4). Therefore $B(= H_\infty)$ is a contractive infinite generalized Hankel matrix lifting of H_o. So the commutant lifting theorem provides another proof of Corollary 7.6.

Now assume that A is a contraction in $I(T, T')$ where T on H and T' on H' are both contractions. Let U_* on K_* be the minimal isometric dilation of T'^*. Using $U_*^* | H' = T'$ along with the lifting property $P_H U = TP_H$ we see that AP_H is a contraction in $I(U, U_*^*)$, where U is the minimal isometric dilation of T. By Remark 7.4 this AP_H can be identified with a generalized Hankel matrix H_n. In this case $U_*'^*$ is identified with T_n'. Moreover, the minimal isometric dilation $\hat{U}'$ on $\hat{K}'$ of $U_*'^*$ is identified with the minimal isometric dilation T_∞' of T_n'. By Corollary 7.6 this identification implies that there exists a contractive lifting B in $I(U, \hat{U}')$ of AP_H. Since $P_{K'} B = AP_H$ we have

$$P_{K' \ominus H'} B = 0 . \tag{7.19}$$

Because U_*' is the minimal isometric dilation of T'^* and $\hat{U}'$ is the minimal isometric dilation of $U_*'^*$, it follows that $\hat{U}'$ is the minimal unitary dilation of T'; see Section VI.5. According to (VI.5.10) adjusted to this setting, the operator $U' = \hat{U}' \,|\, K'$ where $K' = (\hat{K} \ominus K_*') \oplus H'$ is the minimal isometric dilation of T'. Equation (7.19) shows that B maps K into K'. Thus B is a contraction in $I(U, U')$. Obviously $P_{H'} B \,|\, H = A$. By Remark VII.1.7 this B is a contractive intertwining lifting of A. Therefore Corollary 7.6 provides another proof of the commutant lifting theorem. In other words obtaining a contractive intertwining lifting of A is equivalent to solving a generalized Hankel matrix interpolation problem. In Chapters XIII and XIV we will present several methods to compute the set of all contractive intertwining lifting in the commutant lifting theorem. This readily provides a solution to the generalized Hankel matrix interpolation Problem 7.1.

VIII.8. NOTES AND COMMENTS

Operator valued Nevalinna-Pick interpolation was first studied in [Sz.-NK 1]. This and other operator valued interpolation problems are also studied in [Kor], [Sz.-NK 2], [Er], [Fed 1], [DeGK 1,2], [RosR 2] and [BH]. Our approach through the commutant lifting theorem is that of Sarason [Sa 2]. Another approach through Hankel operators is due to [AAK 4] (see also Chapter IX). We recall that the commutant lifting theorem can also be stated as an extension property for contractive intertwining restrictions of co-isometries. The fact that one cannot replace co-isometries with isometries was noted by Page [Pag 3]. Page also proved Proposition 3.2 and conjectured Proposition 3.4; its nice proof in the text is due to Carswell-Schubert [CarS]. The general Theorem 3.5 was given in [Ce 2]. The characterization of Toeplitz operators in Proposition 4.1 is from [Do 2], which is another interpretation some of the results in [BroH]. The characterization of Hankel operators in Proposition 4.2 is from [AAK 4] and [Pag 2]. The general invertibility conditions as well as the proofs in 4.3 to 5.9 is due to Page [Pag 2]. However, Theorem 5.10 is from [DevS]. The formulation of the Corona Problem for Toeplitz operators is due to Arveson [Arv]. The exact solution to this problem was given in [Sz.-NF 14]. However, we have chosen the approach of Rosenblum [Ros] based on

Leech's result [Lee]. We mention that Rosenblum was able to show (by using a deep technique of T. Wolff) that the Corona Theorem for Toeplitz operators implies the classical Corona Theorem of Carleson, as well as some highly nontrivial generalizations of it (see also [Fu 2,3] and [Tol]). Coprime factorizations play an important role in many engineering applications [Fu 3], [Ka 1]. Finally, the equivalence between the commutant lifting theorem and the lifting of generalized Hankel operators given in Section 7 is an extension of the proof of a particular version of the commutant lifting theorem in [Ni]. For other applications of the commutant lifting theorem see [Ber], [Kri], [PtV] and [Sz.-NF 11,12].

CHAPTER IX

H$^\infty$ OPTIMIZATION AND FUNCTIONAL MODELS

In this chapter we will apply the commutant lifting theorem to Hankel operators to solve several H$^\infty$ optimization problems. Then we will use a functional model due to Sz.-Nagy-Foias along with the commutant lifting theorem to solve some other H$^\infty$ optimization problems.

1. MULTIPLICATION OPERATORS

In this section we present some classical results concerning multiplication operators.

Throughout this chapter all Hilbert spaces are separable. The Hardy space of analytic function in the unit disc with values in E is denoted by $H^2(E)$. By definition f is in $H^2(E)$ if and only if f admits a power series expansion of the form

$$f = \sum_{0}^{\infty} f_n z^n \qquad (z \in D) \qquad \text{and} \qquad \|f\|_{H^2} = \sum_{0}^{\infty} \|f_n\|^2 < \infty \;,$$

where f_n is in E for all $n \geq 0$. Recall that $L^2(E)$ is the Lebesgue space of all square intergable functions on $[0, 2\pi)$ with values in E. Since $\{f_n\}_0^\infty$ is square summable the series

$$f(e^{it}) = \sum_{0}^{\infty} f_n e^{int}$$

converges in $L^2(E)$ to a function denoted by $f(e^{it})$ a.e. Noting that

$$\int_{0}^{2\pi} \|f(e^{it}) - f(re^{it})\|^2 dt = \sum_{1}^{\infty} (1 - r^n)^2 \|f_n\|^2$$

converges to zero as $r \to 1-$ we see that $f(re^{it})$ converges to $f(e^{it})$ in $L^2(E)$. Moreover, the Fatou theorem also holds, that is, if $z \in D$ and $z \to e^{it}$ nontangentially, then $f(z) \to f(e^{it})$ at least a.e. The proof is omitted because it is analogous to the classical proof in [Ho]. Conversely let g(t) be in $L^2(E)$ and a_n its Fourier coefficient of e^{int}

defined by

$$a_n = \frac{1}{2\pi} \int_0^{2\pi} e^{-int} g(t) dt \qquad (\text{for } n = 0, \pm 1, \pm 2, \dots).$$

If $a_n = 0$ for all $n < 0$, then the function

$$f(z) = \sum_0^\infty a_n z^n \qquad (z \in D)$$

is in $H^2(E)$ and $f(e^{it}) = g(t)$ a.e., that is, they coincide as elements in $L^2(E)$. In this way by identifying $f(z)$ with its boundary value function $f(e^{it})$ we can view $H^2(E)$ as a subspace of $L^2(E)$ formed by the functions whose Fourier expansion is of the form

$$\sum_0^\infty a_n e^{int} \, .$$

Finally, $K_0^2(E) \triangleq L^2(E) \ominus H^2(E)$.

Let E and E' be Hilbert spaces. Throughout this chapter S, S', V, and V' are multiplication by e^{it} on $H^2(E)$, $H^2(E')$, $L^2(E)$ and $L^2(E')$, respectively. Clearly S, S' are unilateral shifts, and V, V' are bilateral shifts. Notice that V^* and V'^* are the minimal isometric dilations of S^* and S'^*, respectively. Let Ψ be in $L^\infty(E, E')$, that is, Ψ is a Lebesgue strongly measurable function over $[0, 2\pi)$ with values in $L(E, E')$ and there exists a finite constant M such that $\|\Psi(e^{it})\| \leq M$ a.e. Let us mention that a function Ψ with values in $L(E, E')$ is strongly measurable if the function $(\Psi h, g)_{E'}$ is Lebesgue measurable for all h in E and g in E'; see [DuS] for further details. We also recall that $L^\infty(E, E')$ defines a Banach space under the norm

$$\|\Psi\|_\infty \triangleq \operatorname{ess\,sup} \{\|\Psi(e^{it})\| : t \in [0, 2\pi)\} \, .$$

This means that $\|\Psi\|_\infty$ is the infimum of the values $\lambda \geq 0$ such that the set $\{t : \|\Psi(e^{it})\| > \lambda\}$ has Lebesgue measure zero. Obviously if $\|\Psi(e^{it})\| \leq M$ a.e., then $\|\Psi\|_\infty \leq M$. The multiplication operator mapping $L^2(E)$ into $L^2(E')$ is denoted by M_Ψ. To be precise $M_\Psi f = \Psi f$ for f in $L^2(E)$. It is easy to show that $\|M_\Psi\| \leq \|\Psi\|_\infty$. If Ψ is in $H^\infty(E, E')$, then the multiplication operator by Ψ from $H^2(E)$ to $H^2(E')$ is denoted by Ψ_+, that is, $\Psi_+ f = \Psi f$ for all f in $H^2(E)$. When there is no possible confusion we will simply represent the operator M_Ψ or Ψ_+ by Ψ. A function Ψ in $L^\infty(E, E')$ is *rigid* if $\Psi(e^{it})$ is a.e. an isometry. The function Ψ is *inner* if Ψ is rigid and in $H^\infty(E, E')$. For a function $\Psi(z)$ in $H^\infty(E, E')$ we can define its boundary values $\Psi(e^{it})$ by applying the Fatou theorem to $\Psi(z)h$ for h in E. Indeed if E_0 is a countable dense set in E which is linear with respect to rational scalars, then

$$\Psi_\partial(e^{it})h \overset{\Delta}{=} \lim_{r \to 1-} \Psi(re^{it})h$$

exists for all h in E_o and t in $[0, 2\pi)\backslash\sigma$ where σ has Lebesgue measure zero. It is clear that we can now define by continuity $\Psi_\partial(e^{it})$ on all E provided that t $\notin \sigma$ and that the $L(E, E')$ valued function Ψ_∂ obtained in this way is in $L^\infty(E, E')$. We will view Ψ_∂ as the extension to the boundary of Ψ in $H^\infty(E, E')$ and therefore drop the subscript ∂. In this way if Ψ is in $H^\infty(E, E')$ we can also consider besides Ψ_+ the operator M_Ψ from $L^2(E)$ to $L^2(E')$. The following result is classic.

1.1 THEOREM. *a) An operator* B *is in* $I(V, V')$ *if and only if* B *is a multiplication operator of the form* B $= M_\Psi$ *where* Ψ *is in* $L^\infty(E, E')$. *Furthermore,*

$$\|B\| = \|M_\Psi\| = \|\Psi\|_\infty . \tag{1.1}$$

The operator B $= M_\Psi$ *is an isometry if and only if* Ψ *is rigid.*

b) An operator A *is in* $I(S, V')$ *if and only if* A $= M_\Psi|H^2(E)$ *where* Ψ *is in* $L^\infty(E, E')$. *Furthermore,*

$$\|A\| = \|M_\Psi|H^2(E)\| = \|\Psi\|_\infty . \tag{1.2}$$

The operator A $= M_\Psi|H^2(E)$ *is an isometry if and only if* Ψ *is rigid.*

c) An operator A *is in* $I(S, S')$ *if and only if* A $= \Psi_+$ *where* Ψ *is in* $H^\infty(E, E')$. *Furthermore,*

$$\|\Psi\|_\infty = \|M_\Psi\| = \|A\| = \|\Psi\| \overset{\Delta}{=} \sup_{z \in D} \|\Psi(z)\| . \tag{1.3}$$

Finally, the operator A $= \Psi_+$ *is an isometry if and only if* Ψ *is inner.*

PROOF. It is obvious that the multiplication operators M_Ψ of the form indicated in the statements, enjoy the required intertwining properties. Part a) follows by a well-known measure theoretic argument which will only be sketched. Assume that B is in $I(V, V')$. Notice that for an h in E and a numerical trigonometric polynomial $\phi(e^{it}) \overset{\Delta}{=} \sum_{k=-N}^{N} \alpha_k e^{ikt}$ we have

$$\frac{1}{2\pi} \int_0^{2\pi} |\phi(e^{it})|^2 \|(Bh)(e^{it})\|^2 dt = \|\phi(V')Bh\|^2 = \|B\phi(V)h\|^2 \leq$$

$$\|B\|^2 \|\phi(V)h\|^2 = \|B\|^2 \|h\|^2 \frac{1}{2\pi} \int_0^{2\pi} |\phi(e^{it})|^2 dt$$

(with equality if B is an isometry). Using the fact that the polynomials are dense in the continuous function of period 2π, the previous inequality becomes

$$\frac{1}{2\pi} \int_0^{2\pi} \rho(t) \|(Bh)(e^{it})\|^2 \, dt \le \|B\|^2 \|h\|^2 \frac{1}{2\pi} \int_0^{2\pi} \rho(t) dt$$

where $\rho(t)$ is any positive $(\rho(t) \ge 0)$ continuous function. (There is equality if B is an isometry). It follows by a standard argument that

$$\|(Bh)(e^{it})\|^2 \le \|B\|^2 \|h\|^2 \quad \text{a.e. on } [0, 2\pi), \tag{1.4}$$

(with equality if B is an isometry). Let E_0 be a countable dense subset of E which is a vector subspace over the rational complex number field. There exists a set $\sigma \subset [0, 2\pi)$ of Lebesgue measure zero such that (1.4) holds for all $t \in [0, 2\pi)\backslash\sigma$ and all h in E_0. For these t and h there exists a map $\Psi(e^{it})$ from E_0 into E' such that $\Psi(e^{it})h = (Bh)(e^{it})$. Using (1.4) it is easy to see that $\Psi(e^{it})$ can be extented by continuity to an element in $L(E, E')$ which we still denote by $\Psi(e^{it})$. Furthermore,

$$\|\Psi(e^{it})h\| \le \|B\| \, \|h\| \quad \text{and} \quad (Bh)(\cdot) = \Psi(\cdot)h \text{ in } L^2(E') \tag{1.5}$$

first for all $h \in E_0$ and then by continuity, for all $h \in E$. From (1.5) we can also infer that $\Psi \in L^\infty(E, E')$, and therefore $\|\Psi\|_\infty \le \|B\|$. This implies that $(B - M_\Psi)h = 0$ for all $h \in E$. Since M_Ψ commutes with the bilateral shift we have

$$(B - M_\Psi)L^2(E) = \bigvee_{-\infty}^{\infty} (B - M_\Psi)V^n E = \bigvee_{-\infty}^{\infty} V'^n (B - M_\Psi)E = \{0\} \, .$$

Thus $B = M_\Psi$ and

$$\|\Psi\|_\infty \le \|B\| = \|M_\Psi\| \le \|\Psi\|_\infty \, .$$

So we have (1.1). If B is an isometry, then $\|B\| = 1$ and there is equality in (1.5). Therefore Ψ is rigid. This proves Part a.

Corollary VI.2.4 shows that an operator A in $I(S, V')$ has an unique extension B in $I(V, V')$ and $\|B\| = \|A\|$. Furthermore, A is an isometry if and only if B is an isometry. Therefore Part b follows from Part a with $A = B \,|\, H^2(E)$ and $B = M_\Psi$.

To prove Part c, let A be in $I(S, S')$. Since $I(S, S')$ is contained in $I(S, V')$, we deduce from Part b that $A = M_\Psi \,|\, H^2(E)$ for some Ψ in $L^\infty(E, E')$ satisfying (1.2). However,

$$\Psi H^2(E) = A H^2(E) \subseteq H^2(E') \ .$$

This implies that $\Psi(e^{it})h$ is in $H^2(E')$ for all h in E. Therefore

$$\int_0^{2\pi} e^{int}\Psi(e^{it})dt = 0 \qquad \text{(for all } n > 0) \ .$$

On the other hand we can define the Fourier coefficients Ψ_n in $L(E, E')$ by

$$\Psi_n h = \frac{1}{2\pi}\int_0^{2\pi} e^{-int}\Psi(e^{it})h dt \qquad\qquad (h \in E \text{ and } n \geq 0) \ .$$

Obviously $\|\Psi_n\| \leq \|\Psi\|_\infty$ for all $n \geq 0$ and thus

$$\Psi(z) \overset{\Delta}{=} \sum_0^\infty \Psi_n z^n$$

is analytic in D. Moreover, for all h in E by the Fatou Theorem the boundary function of $\Psi(z)h$ is $\Psi(e^{it})h$. These two functions are connected as in classical analysis by the Poisson Formula

$$\Psi(z)h = \frac{1}{2\pi}\int_0^{2\pi} \mathrm{Re}\left[\frac{e^{it} + z}{e^{it} - z}\right]\Psi(e^{it})h dt \ .$$

Taking the norms we obtain $\|\Psi(z)h\| \leq \|\Psi\|_\infty \|h\|$. Hence

$$\|\Psi\| \overset{\Delta}{=} \sup \ \{\|\Psi(z)\| : z \in \mathrm{D}\} \leq \|\Psi\|_\infty \ .$$

So Ψ is in $H^\infty(E, E')$ and $\|\Psi\| = \|\Psi\|_\infty$. Now (1.3) follows from (1.2). Finally, the last statement in Part c follows from the last statement in Part b. This completes the proof.

1.2   COROLLARY. *Let* A *be an operator in* $I(S, S')$. *Then* A *is unitary if and only if* $A = \Psi_+$ *where* Ψ *is a constant function in z equal to a unitary operator from E onto* E'.

PROOF. If A is unitary, then the previous theorem shows that $A = M_\Psi | H^2(E)$ where Ψ is inner. But A^* is in $I(S',S)$. By the previous theorem $A^* = \Psi_{*+}$ where Ψ_* is inner. Moreover, $\Psi_{*+}\Psi_+ = A^*A = I$. So $\Psi_*(z)\Psi(z) = I$ for all z in D. In particular, $\Psi_*(0)\Psi(0) = I$. Since both $\Psi_*(0)$ and $\Psi(0)$ are contractions $\Psi(0)$ is an isometry. A similar argument shows that $\Psi(0)\Psi_*(0) = I$. Therefore $\Psi(0)$ is onto and $\Psi(0)$ is unitary. Let

$$\Psi(z) = \sum_0^\infty \Psi_n z^n$$

be the power series expansion of $\Psi(z)$. For h in E we have

$$\|h\|^2 = \|\Psi_+ h\|^2 = \sum_0^\infty \|\Psi_n h\|^2 = \|\Psi(0)h\|^2 + \sum_1^\infty \|\Psi_n h\|^2 = \|h\|^2 + \sum_1^\infty \|\Psi_n h\|^2 .$$

It follows that $\Psi_n = 0$ for all $n \geq 1$. Thus $\Psi(z) \equiv \Psi(0)$ for all z in D. This completes the proof.

The multiplication operators Ψ_+ considered in Theorem 1.1 can be identified with infinite by infinite block analytic Toeplitz matrices. To see this let us identify S and S' with the unilateral shifts on $l^2(E)$ and on $l^2(E')$ defined as in (VI.1.1). For h in E and h' in E' we identify $z^n h$ in $H^2(E)$ and $z^n h'$ in $H^2(E')$ with the elements $[0,...,0, h, 0,...]^{\mathrm{tr}}$ in $l^2(E)$ and $[0,...,0, h', 0,...]^{\mathrm{tr}}$ in $l^2(E')$ where the n+1 components are h and h', respectively. Under this identification we see that an operator A is in $I(S, S')$ if and only if A admits a matrix representation of the form

$$\hat{A}_\infty = \begin{bmatrix} A_0 & 0 & 0 & \ldots \\ A_1 & A_0 & 0 & \ldots \\ A_2 & A_1 & A_0 & \ldots \\ A_3 & A_2 & A_1 & \ldots \\ \vdots & \vdots & \vdots & \vdots \end{bmatrix} \tag{1.6}$$

from $l^2(E)$ to $l^2(E')$ where the $\{A_i\}_0^\infty$ are operators form E to E'. The previous theorem shows that A is in $I_1(S, S')$ if and only if $A = \Psi_+$ where Ψ is in $H_1^\infty(E, E')$. If

$$\Psi(z) = \sum_0^\infty A_n z^n \quad \text{and} \quad f = \sum_0^\infty f_n z^n \qquad (z \in D) \tag{1.7}$$

is the power series extension of Ψ and an element f in $H^2(E)$, then $\Psi_+ f$ has a power series extension of the form

$$(\Psi_+ f)(z) = \sum_{n=0}^\infty z^n [\sum_{i=0}^n A_{n-i} f_i] . \tag{1.8}$$

The coefficient of z^n is given by the n+1 component of $\hat{A}_\infty(\bigoplus_0^\infty f_n)$. The coefficients A_i in the power series expansion of Ψ are precisely the entries in $\hat{A}_\infty$. Therefore the previous theorem provides us with another proof of Lemma VIII.1.2, that is, $\hat{A}_\infty$ is a contraction if and only if $\Psi(z)$ in (1.7) is in $H_1^\infty(E, E')$.

2. THE BEURLING-LAX-HALMOS THEOREM

This section is devoted to the Beurling-Lax-Halmos Theorem and some of its consequences concerning the factorization of functions in $H^\infty(E, E')$.

We begin with the following fundamental result known as the Beurling-Lax-Halmos Theorem.

2.1 THEOREM. *A subspace M of $H^2(E')$ is invariant for S' if and only if $M = \Theta H^2(E)$ where Θ is an inner function in $H^\infty(E, E')$, and E is an adequate Hilbert space with dimension less than or equal to the dimension of E'. Furthermore, Θ is unique up to a unitary constant right factor, that is, if $M = \Psi H^2(F)$ where Ψ is an inner function in $H^\infty(F, E)$, then $\Theta = \Psi W$ where W is a (constant in z) unitary operator mapping E onto F.*

PROOF. Let M be an invariant subspace for S'. Notice that $S'|M$ is an isometry on M and

$$\bigcap_0^\infty S'^n M \subseteq \bigcap_0^\infty S'^n H^2(E') = \{0\} \ .$$

Applying the Wold decomposition to $S'|M$ we have

$$M = \bigoplus_0^\infty S'^n E \tag{2.1}$$

where $E = M \ominus S'M$; see Theorem VI.1.1 and the formulas in (VI.1.2). Let B be the operator embedding $H^2(E)$ into $H^2(E')$ defined by

$$B\left(\sum_0^\infty f_n z^n\right) = \bigoplus_0^\infty S'^n f_n \quad \text{where} \quad f = \sum_0^\infty f_n z^n \text{ is in } H^2(E) \ .$$

Using the fact that E is wandering for S' it is easy to verify that B is an isometry. Obviously $S'B = BS$. Part c of Theorem 1.1 shows that $B = \Theta_+$ where Θ is an inner function. (In particular, Θ is in $H_1^\infty(E, E')$.) Because the range of B is M we have obtained that the space $M = \Theta H^2(E)$. Also $\Theta(e^{it})$ is an isometry from E to E' a.e. Therefore $\dim E \leq \dim E'$. On the other hand if Θ is an inner function, then obviously $\Theta H^2(E)$ is invariant for S'. This proves the first part of the theorem.

Now assume that Θ and Ψ are inner functions in $H^\infty(E, E')$ and $H^\infty(F, E')$, respectively such that

$$\Theta H^2(E) = \Psi H^2(F) \ . \tag{2.2}$$

Using this we have for every g in $H^2(E)$, that there exists a unique f_g in $H^2(F)$ such that $\Theta g = \Psi f_g$. The map W which takes g to f_g is a unitary operator from $H^2(E)$ onto $H^2(F)$ which intertwines the unilateral shifts. Corollary 1.2 implies that W is a unitary constant from E onto F. This completes the proof.

Let Θ and Ψ be in $H^\infty(E, E')$ and $H^\infty(F, E')$, respectively. We say that Ψ is a *left divisor* of Θ if there exists a Ω in $H^\infty(E, F)$ such that $\Theta = \Psi\Omega$. Notice that if Θ and Ψ are both inner, then Ω is also inner. A right divisor of an operator valued function is defined analogously. This definition is used in the following result.

2.2 COROLLARY. *Let Θ be in $H^\infty(E, E')$ and Ψ be an inner function in* $H^\infty(F, E')$. *Then $\Theta H^2(E) \subseteq \Psi H^2(F)$ if and only if Ψ is a left inner divisor of Θ.*

PROOF. If $\Theta H^2(E) \subseteq \Psi H^2(F)$, then given any h in $H^2(E)$ there exists a $f = Qh$ in $H^2(F)$ such that $\Theta h = \Psi f = \Psi Q h$. Since Ψ is inner, Q is an operator mapping $H^2(E)$ into $H^2(F)$. Clearly Q intertwines the unilateral shifts. By Part c of Theorem 1.1 the operator $Q = \Omega_+$ where Ω is in $H^\infty(E, F)$. Hence $\Theta = \Psi\Omega$. Since the converse is obvious the proof is complete

Let Θ and Ψ be inner functions. By the Beurling-Lax-Halmos Theorem there exists an inner function Θ_d in $H^\infty(F', E')$ satisfying

$$\Theta_d H^2(F') = \Theta H^2(E) \bigvee \Psi H^2(F).$$

Corollary 2.2 implies that Θ_d is a left inner divisor of both Θ and Ψ. If Ω is any other left inner divisor of both Θ and Ψ, then

$$\Theta_d H^2(F') = \Theta H^2(E) \bigvee \Psi H^2(F) \subseteq \Omega H^2(\cdot)$$

Applying Corollary 2.2 again shows that Ω is a left inner divisor of Θ_d. Therefore Θ_d is the *greatest common left inner divisor* of Θ and Ψ. Indeed by definition the greatest common left inner divisor of Θ and Ψ is a left inner divisor of both Θ and Ψ which moreover is divided from the left by any other left inner divisor of Θ and Ψ. Finally, the Beurling-Lax-Halmos Theorem also shows that the greatest common left inner divisor Θ_d is unique up to a constant unitary right factor.

The Beurling-Lax-Halmos Theorem also shows that there exists an inner function Θ_m in $H^\infty(F_1, E')$ such that

$$\Theta_m H^2(F_1) = \Theta H^2(E) \cap \Psi H^2(F).$$

Corollary 2.2 shows that both Θ and Ψ divide Θ_m on the left. If both Ψ and Θ are left inner divisors to any inner function Ω, then

$$\Theta_m H^2(F') = \Theta H^2(E) \cap \Psi H^2(F) \supsetneq \Omega H^2(\cdot) \ .$$

Corollary 2.2 implies that Θ_m is a left inner divisor of Ω. Therefore Θ_m is the *least common left inner multiple* of Θ and Ψ. By definition the least common left inner multiple of Θ and Ψ is an inner function which is divided by both Θ and Ψ on the left and moreover is a left inner divisor to any other inner function which is also divided by Θ and Ψ on the left. The Beurling-Lax-Halmos Theorem shows that the least common left inner multiple is unique up to a unitary constant right factor.

The previous definitions extend to a family of inner functions $\{\Theta_\alpha$ in $H^\infty(E_\alpha, E'){:}\alpha \in A\}$ in the obvious way. In this case, the greatest common left inner divisor Θ_d and least common left multiple inner multiple Θ_m of $\{\Theta_\alpha{:}\alpha \in A\}$ is the unique inner function defined by

$$\Theta_d H^2(F') = \bigvee_{\alpha \in A} \Theta_\alpha H^2(E_\alpha) \quad \text{and} \quad \Theta_m H^2(F') = \bigcap_{\alpha \in A} \Theta_\alpha H^2(E_\alpha) \ . \tag{2.3}$$

A function Θ_o is *outer* if Θ_o is in $H^\infty(E, F)$ and

$$\overline{\Theta_o H^2(E)} = H^2(F) \ . \tag{2.4}$$

The following theorem shows that any function in H^∞ admits a unique inner-outer factorization.

2.3 THEOREM. *Let Θ be in $H^\infty(E, E')$. Then Θ admits a unique factorization of the form $\Theta = \Theta_i \Theta_o$ where Θ_i is an inner function in $H^\infty(F, E')$ and Θ_o is an outer function in $H^\infty(E, F)$. By unique we mean that if $\Theta = \Theta_i' \Theta_o'$ is any other factorization of Θ where Θ_i' is an inner function in $H^\infty(F', E')$ and Θ_o is an outer function in $H^\infty(E, F')$, then there exists a constant (in z) unitary operator W mapping F onto F' satisfying $\Theta_i = \Theta_i' W$ and $\Theta_o = W^* \Theta_o'$.*

PROOF. Since $\overline{\Theta H^2(E)}$ is an invariant subspace for the unilateral shift the Beurling-Lax-Halmos Theorem shows that there exists an inner function Θ_i satisfying

$$\overline{\Theta H^2(E)} = \Theta_i H^2(F) \ . \tag{2.5}$$

Corollary 2.2 implies that there exists a Θ_o in $H^\infty(E, F)$ such that $\Theta = \Theta_i \Theta_o$. This and (2.5) gives:

$$\Theta_i H^2(F) = \overline{\Theta H^2(E)} = \Theta_i \overline{[\Theta_o H^2(E)]} \ . \tag{2.6}$$

Since the multiplication operator Θ_i is an isometry, this proves that (2.4) holds. Hence

Θ_o is outer.

Now let $\Theta = \Theta_i' \Theta_o'$ be any factorization where Θ_i' is inner and Θ_o' is outer. The definition of an outer function and (2.6) yield

$$\Theta_i H^2(F) = \overline{\Theta H^2(E)} = \Theta_i' H^2(F') .$$

By the uniqueness property in the Beurling-Lax-Halmos Theorem we have $\Theta_i = \Theta_i' W$ where W is a unitary operator from F onto F' which is constant in z. Thus

$$\Theta_i \Theta_o = \Theta = \Theta_i' \Theta_o' = \Theta_i' W W^* \Theta_o' = \Theta_i W^* \Theta_o' .$$

Since Θ_i is inner $\Theta_o = W^* \Theta_i'$. This completes the proof.

The factorization $\Theta = \Theta_i \Theta_o$ in Theorem 2.3 is called the *inner-outer factorization* of Θ. The function Θ_i and Θ_o is the *inner* and *outer part of* Θ, respectively. Finally, we say that two functions in Θ in $H^\infty(E, E')$ and Ψ in $H^\infty(F, E')$ are *left prime* if the only common left inner factor to both Θ and Ψ is a unitary constant. From the above analysis it is easy to show that Θ and Ψ are left prime if and only if

$$H^2(E') = \Theta H^2(E) \bigvee \Psi H^2(F) ,$$

or equivalently, if the greatest common left divisor of the inner parts of Θ and Ψ is a unitary constant.

If Θ is in $H^\infty(E, E')$, then $\tilde{\Theta}$ is the operator valued analytic function in $H^\infty(E', E)$ defined by $\tilde{\Theta}(e^{it}) = \Theta(e^{-it})^*$ a.e. in $[0, 2\pi)$. An analytic function $\Theta_{*i}[\Theta_{*o}]$ is called *∗-inner* [*∗-outer*] if $\tilde{\Theta}_{*i}$ is inner [$\tilde{\Theta}_{*o}$ is outer] respectively. By applying the above theorem to $\tilde{\Theta}$ we see that Θ admits a unique *∗-outer-∗-inner factorization* of the form $\Theta = \Theta_{*o} \Theta_{*i}$ where Θ_{*o} is ∗-outer and Θ_{*i} is ∗-inner. By unique we mean that if $\Theta = \Theta_{*o}' \Theta_{*i}'$ is a ∗-outer-∗-inner factorization, then there exists a constant in z unitary operator W such that $\Theta_{*i} = W \Theta_{*i}'$ and $\Theta_{*o} = \Theta_{*o}' W^*$. Finally, Θ and $\Psi \in H^\infty(E, F)$ are *right prime* if $\tilde{\Theta}$ and $\tilde{\Psi}$ are left prime.

3. HANKEL OPERATORS AND THE NEHARI PROBLEM

In this section we solve the Nehari H^∞ optimization problem by applying the commutant lifting theorem to Hankel operators.

Throughout $K_0^\infty(E, E')$ is the subspace of $L^\infty(E, E')$ whose Fourier expansion only contains strictly negative powers of e^{int}, that is, K is in $K_0^\infty(E, E')$ if K is in $L^\infty(E, E')$ and

$$K = \sum_{-\infty}^{-1} K_n e^{int} .$$

We begin by formulating the Nehari H^∞ optimization problem.

3.1 PROBLEM. Let Q be a fixed function in $L^\infty(E, E')$. Find μ where

$$\mu = \inf \|Q + K_o^\infty(E, E')\|_\infty \overset{\Delta}{=} \inf \{\|Q+K\|_\infty : K \in K_o^\infty(E, E')\} . \tag{3.1}$$

We will obtain an optimal K_* in $K_o^\infty(E, E')$, that is,

$$\|Q + K_*\| = \inf\|Q+K_o^\infty(E, E')\| = \mu \tag{3.2}$$

and will show that μ is the norm of the Hankel operator determined by Q. In order to solve this problem we will only need *Hankel operators* Γ in $I(S, S'^*)$ which are special cases of the Hankel operators considered in Sections VIII.4 and VIII.7. Therefore in this section by a Hankel operator we mean an operator in $I(S, S'^*)$. (Recall that S and S' are the unilateral shifts on $H^2(E)$ and $H^2(E')$, respectively. Here we use the symbol Γ and not H to represent a Hankel operator, since the symbol H is used in the notation for Hardy spaces.)

Let Γ be a Hankel operator as specified above, that is, Γ is in $I(S, S'^*)$ and $h \in E$. Then

$$\Gamma h = \sum_0^\infty A_n h z^n \qquad (z \in D) \tag{3.3}$$

where A_n is an operator mapping E into E' for all $n \geq 0$. We identify $e^{int}h$ and $e^{int}h'$ for h in E and h' in E' with the standard basis $[0, ..., 0, h, 0, ...]^{tr}$ and $[0, ..., 0, h', 0, ...]^{tr}$ in $l^2(E)$ and $l^2(E')$, respectively. (The nonzero element appears in the n+1 position and tr denotes the transpose). Using $S'^{*n}\Gamma h = \Gamma S^n h$ for all h in E with this standard basis and identifying S and S' with the unilateral shifts on $l^2(E)$ and $l^2(E')$, respectively it is easy to show that Γ admits a matrix representation of the form

$$\Gamma = \begin{bmatrix} A_o & A_1 & A_2 & A_3 & \cdots \\ A_1 & A_2 & A_3 & A_4 & \cdots \\ A_2 & A_3 & A_4 & A_5 & \cdots \\ A_3 & A_4 & A_5 & A_6 & \cdots \\ A_4 & A_5 & A_6 & A_7 & \cdots \\ \vdots & \vdots & \vdots & \vdots & \cdots \end{bmatrix} \tag{3.4}$$

Therefore a Hankel operator can be identified with a matrix of the form (3.4), which is a special case of a generalized Hankel matrix studied in Section VIII.7.

Let Q be a specified function in $L^\infty(E, E')$ and let J be the unitary operator on $L^2(\cdot)$ defined by

$$(Jf)(e^{it}) = f(e^{-it}) \qquad (f \in L^2(\cdot)) . \tag{3.5}$$

Consider the operator $\Gamma(Q)$ mapping $H^2(E)$ into $H^2(E')$ defined by

$$\Gamma(Q) = P_+ M_Q J \,|\, H^2(E) \tag{3.6}$$

where P_+ is the orthogonal projection from $L^2(E')$ onto $H^2(E')$. We will call the operator Q a *symbol* for the operator $\Gamma(Q)$. The operator $\Gamma(Q)$ does not uniquely determine the symbol Q. For example $\Gamma(Q) = \Gamma(Q+K)$ where K is any element in $K_0^\infty(E, E')$. The following calculation demonstrates that $\Gamma(Q)$ is a Hankel operator :

$$S'^* \Gamma(Q) = S'^* P_+ M_Q J \,|\, H^2(E) = P_+ V'^* M_Q J \,|\, H^2(E) = \Gamma(Q)S .$$

The next theorem will show that all Hankel operators (in $I(S, S'^*)$) are of the form $\Gamma(Q)$, and one can even choose a Q such that $\|Q\|_\infty = \|\Gamma(Q)\|$. We start with the following observation.

3.2 LEMMA. *Let Q and Q' be in* $L^\infty(E, E')$. *Then* $\Gamma(Q) = \Gamma(Q')$ *if and only if* $Q - Q'$ *is in* $K_0^\infty(E, E')$.

PROOF. Notice that $\Gamma(Q) = \Gamma(Q')$ if and only if the Hankel operator $\Gamma(Q-Q') = 0$. We claim that $\Gamma(Q-Q') = 0$ if and only if $\Gamma(Q-Q')E = 0$. This follows because $S'^{*n}\Gamma = \Gamma S^n$ and the span of $S^n E$ for $n \geq 0$ is dense in $H^2(E)$. Clearly

$$\Gamma(Q-Q')h = P_+(Q-Q')h \qquad \text{(for all } h \in E)$$

and $(Q-Q')h$ for all h in E is in $K_0^2(E')$, or equivalently, $Q - Q'$ is in $K_0^\infty(E, E')$. Therefore $\Gamma(Q) = \Gamma(Q')$ if and only if $Q-Q'$ is in $K_0^\infty(E, E')$. This proves the Lemma.

We now pass to the main result of this section due to Adamjan-Arov and Krein [AAK 4].

3.3 THEOREM. *The operator* Γ *mapping* $H^2(E)$ *into* $H^2(E')$ *is a Hankel operator if and only if* $\Gamma = \Gamma(Q)$ *for some Q in* $L^\infty(E, E')$. *In this case, one can always choose an optimal* Q_* *in* $L^\infty(E, E')$, *that is, such that*

$$\|\Gamma\| = \|Q_*\|_\infty \quad \text{and} \quad \Gamma = \Gamma(Q_*) . \tag{3.7}$$

Moreover, there exists at least one optimal K_* *in* $K_0^\infty(E, E')$ *solving the optimization*

Problem 3.1 and μ equals the norm of $\Gamma(Q)$, that is,

$$\|Q + K_*\|_\infty = \inf \|Q + K_0^\infty(E, E')\|_\infty = \|\Gamma(Q)\| . \tag{3.8}$$

The equation $Q_* = Q + K_*$ provides a one to one correspondence between the optimal solutions Q_* and K_*.

PROOF. Assume that Γ is a Hankel operator. Recall that V'^* is the minimal isometric dilation of S'^*. By virtue of the commutant lifting theorem there exists an operator B in $I(S, V'^*)$ such that

$$\Gamma = P_+ B \quad \text{and} \quad \|\Gamma\| = \|B\| . \tag{3.9}$$

Clearly $B_* = JB$ is in $I(S, V')$. Part b of Theorem 1.1 gives

$$B_* = M_\Psi \,|\, H^2(E) \quad \text{and} \quad \|B_*\| = \|\Psi\|_\infty \tag{3.10}$$

for some Ψ in $L^\infty(E, E')$. This and $B = JB_*$ implies that $B = M_{Q_*} J | H^2(E)$ where $Q_* = J\Psi$. Now (3.9) yields $\Gamma = \Gamma(Q_*)$. The identity $\|\Gamma\| = \|Q_*\|_\infty$ follows form the second equation in (3.9), (3.10) and $Q_* = J\Psi$, which finishes the proof of (3.7).

Now let $\Gamma(Q_*) = \Gamma(Q)$ and let Q_* be the above optimal choice for a symbol of Γ. Since $\Gamma(Q) = \Gamma(Q_*)$, by Lemma 3.2 we have a $Q + K_* = Q_*$ for some K_* in $K_0^\infty(E, E')$. Thus

$$\|\Gamma(Q)\| = \|\Gamma\| = \|Q + K_*\|_\infty . \tag{3.11}$$

If K is any element in $K_0^\infty(E, E')$, then $\Gamma(Q) = \Gamma(Q + K)$ and

$$\|\Gamma(Q)\| = \|\Gamma(Q + K)\| \leq \|Q + K\|_\infty .$$

Combining this with (3.11) gives

$$\|Q + K_*\|_\infty = \|\Gamma(Q)\| \leq \inf \|Q + K_0^\infty(E, E')\|_\infty .$$

This yields (3.8) and completes the proof.

3.4 REMARK. The proof of the previous theorem shows that K_* in $K_0^\infty(E, E')$ is an optimal solution to Problem 3.1 if and only if $M_{Q + K_*} J | H^2(E)$ is a intertwining lifting of Γ and $\|Q + K_*\| = \|\Gamma\|$. In other words, there is a one to one correspondence between the set of all intertwining liftings B of Γ of norm equal to $\|\Gamma\|$ and the set of all optimal solutions K_* to Problem 3.1.

Throughout the rest of this section we choose $E = C^1$; then S becomes the unilateral shift on H^2. We also assume that Γ is a contractive Hankel matrix and there exists a nonzero x in H^2 satisfying $\|\Gamma x\| = \|x\|$, or equivalently, $D_\Gamma x = 0$. Under this assumption we will use some of our uniqueness results to compute the optimal Q_* and K_* in the previous theorem. Recall that any nonzero x in H^2 is cyclic for $\hat{S} \equiv V$. By Theorem VII.4.1 there exists a unique contractive intertwining lifting B of Γ and this B is an isometry. Consulting the proof of the previous theorem and Remark 3.4 this implies that there exists a unique Q_* in $L^\infty(C^1, E')$ satisfying (3.7). Moreover, because $M_{Q_*} J \mid H^2$ is an isometry and $B = M_{Q_*} J \mid H^2$, Part b of Theorem 1.1 shows that Q_* is rigid. Setting $y = \Gamma x$ we have

$$y = \Gamma x = \Gamma(Q_*)x = P_+ Q_* Jx = Q_* Jx \ . \tag{3.12}$$

The second equality follows from the second equation in (3.7), and the last one from the fact that $P_+ Q_* Jx$ and $Q_* Jx$ have the same norm, namely $\|x\|$. Since any nonzero x in H^2 is nonzero a.e. equation (3.12) yields $Q_* = y/Jx$. Summing up the previous analysis with Theorem 3.3 produces the following result due to Adamjan-Arov and Krein [AAK 3].

3.5 COROLLARY. *Let $\Gamma = \Gamma(Q)$ be a Hankel operator mapping H^2 into $H^2(E')$ where Q is in $L^\infty(C^1, E')$. Assume x is a nonzero vector in H^2 satisfying $\|\Gamma x\| = \|\Gamma\| \|x\|$ and set $y = \Gamma x$. Then $\|\Gamma\|^{-1}(y/Jx)$ is rigid and $Q_* = y/Jx$ is the only function in $L^\infty(C^1, E')$ satisfying (3.7). Moreover, $K_* = (y/Jx) - Q$ is the only function in $K_0^\infty(C^1, E')$ satisfying (3.8), or equivalently, solving the optimization Problem 3.1.*

To complete this section we will present the original proof of this result due to Adamjan-Arov and Krein.

ALTERNATE PROOF OF COROLLARY 3.5. Without loss of generality we assume that $\|\Gamma\| = 1$. Using $\Gamma^* y = x$ with $S'^* \Gamma = \Gamma S$ we have for $n \geq 0$

$$\frac{1}{2\pi} \int_0^{2\pi} e^{-int} |Jx|^2 dt = \frac{1}{2\pi} \int_0^{2\pi} e^{int} x \bar{x} dt = (S^n x, x) =$$

$$(S^n x, \Gamma^* y) = (S'^{*n} y, y) = \frac{1}{2\pi} \int_0^{2\pi} e^{-int}(y, y) dt \ .$$

By taking the complex conjugate we see that the previous equation holds for all n. So $|Jx|^2$ and $(y, y)_{E'}$ have the same Fourier series. Therefore

$$|(Jx)(e^{it})|^2 = (y(e^{it}), y(e^{it}))_{E'} \quad \text{a.e.}$$

Because x is in H^2, x is nonzero a.e. This implies that y/Jx is a rigid function in $L^\infty(C^1, E')$. Therefore if x is any nonzero function in H^2 satisfying $\|\Gamma x\| = \|x\|$ and $\Gamma x = y$, then y/Jx is rigid.

Let $x_i x_o$ be the inner outer factorization of x. Using $S'^{*n}\Gamma = \Gamma S^n$ with our functional calculus in Theorem VI.6.5 we have

$$\|x_o\| = \|x\| = \|\Gamma x\| = \|\Gamma x_i(S)x_o\| = \|x_i(S'^{*})\Gamma x_o\| \le \|\Gamma x_o\| \le \|x_o\| . \tag{3.13}$$

Therefore $\|\Gamma x_o\| = \|x_o\|$ where x_o is an outer function.

From the above analysis for $y_o = \Gamma x_o$ we have that $Q_* \equiv y_o/Jx_o$ is rigid. Now consider the Hankel operator $\Gamma(Q_*)$. Obviously $\Gamma x_o = y_o = \Gamma(Q_*)x_o$. Using $S'^{*}\Gamma = \Gamma S$ this implies that $\Gamma S^n x_o = \Gamma(Q_*)S^n x_o$. Because x_o is outer $\Gamma = \Gamma(Q_*)$. Since Q_* is rigid $\|Q_*\| = 1 = \|\Gamma\|$. So Q_*J is a contractive intertwining lifting of Γ. Moreover, the operator Q_* is the only function in $L^\infty(C^1, E')$ satisfying (3.7). If Q_1 is another function in $L^\infty(C^1, E')$ satisfying (3.7), then $\|\Gamma x_o\| = \|x_o\|$ implies that

$$Q_*Jx_o = \Gamma(Q_*)x_o = \Gamma x_o = \Gamma(Q_1)x_o = Q_1 Jx_o .$$

Because x_o is in H^2 and functions in H^2 are nonzero a.e. $Q_* = Q_1$. Finally, using $\Gamma x = y$ and $\|\Gamma x\| = \|x\|$ we obtain $y = \Gamma(Q_*)x = Q_*Jx$ and consequently, $Q_* = y/Jx$. This proves the first part of the Corollary.

The second part, of the proof has a standard proof given here for completeness. To this end notice that $\Gamma(Q_*) = \Gamma = \Gamma(Q)$ implies $K_* = Q_* - Q = y/Jx - Q$ is in $K_0^\infty(C^1, E')$. For any K in $K_0^\infty(C^1, E')$ we have

$$\|Q + K_*\|_\infty = \|Q_*\|_\infty = \|\Gamma\| = \|\Gamma(Q+K)\| \le \inf\|Q + K_0^\infty(C^1, E')\| .$$

This, the uniqueness of Q_* and $K_* = Q_* - Q$ implies that K_* is the unique function in $K_0^\infty(C^1, E')$ solving the optimization problem in (3.8). This completes the proof.

4. H^∞ OPTIMIZATION

In this section we will use the commutant lifting theorem to solve several H^∞ optimization problems. In particular, we will give a solution to the four block problem which naturally arises in H^∞ control theory.

We begin with the following H^∞ optimization problem.

4.1 PROBLEM. Let Q be a fixed function in $L^\infty(E, E')$ and Θ_1 a fixed inner function $H^\infty(F, E')$. Find μ where

$$\mu = \inf \|Q + \Theta_1 H^\infty(E, F)\| \overset{\Delta}{=} \inf\{\|Q + \Theta_1 \Psi\|_\infty : \Psi \in H^\infty(E,F)\} . \tag{4.1}$$

In this section we will use the commutant lifting theorem to obtain an optimal Ψ_* in $H^\infty(E, F)$ solving Problem 4.1, that is,

$$\|Q + \Theta_1 \Psi_*\|_\infty = \inf\|Q + \Theta_1 H^\infty(E, F)\|_\infty \tag{4.2}$$

and show that μ is the norm of a generalized Hankel operator $\Gamma_1(Q)$.

Before solving Problem 4.1 we show that Problem 3.1 is a special case of Problem 4.1. As before S and S' are the unilateral shifts on $H^2(E)$ and $H^2(E')$, respectively. Also, V and V' are the bilateral shifts on $L^2(E)$ and $L^2(E')$, respectively. Recall that these operators are given by multiplication by e^{it} on the corresponding spaces. Using the fact that $V'^* J$ is unitary and $V'^* J K_o^\infty(E, E') = H^\infty(E, E')$ we have

$$\inf\|Q + K_o^\infty(E, E')\|_\infty = \inf\|V'^* JQ + H^\infty(E, E')\|_\infty .$$

Replacing Q in Problem 4.1 with $V'^* JQ$ and setting $F = E'$ and $\Theta_1 = I$ we see that Problem 3.1 is a special case of Problem 4.1. On the other hand if $\Theta_1(e^{it})$ is unitary a.e., then Problem 4.1 can also be converted into Problem 3.1. This follows from

$$\inf\|Q + \Theta_1 H^\infty(E, F)\|_\infty = \inf\|V'^* J\Theta_1^* Q + K_o^\infty(E, F)\|_\infty .$$

So replacing Q in Problem 3.1 by $V'^* J\Theta_1^* Q$ and setting $E' = F$ we can convert Problem 4.1 into Problem 3.1, when $\Theta_1(e^{it})$ is unitary a.e.

To obtain a solution to Problem 4.1 we set $H' = L^2(E') \ominus \Theta_1 H^2(F)$ and $T' = P_{H'} V' | H'$. Obviously $\Theta_1 H^2(F)$ is invariant for V'. So H' is invariant for V'^* and $T'^* = V'^* | H'$ is an isometry on H'. Therefore T' is a co-isometry on H' and V' is an isometric lifting of T'. Using

$$H' = \Theta_1 K_o^2(F) \oplus (L^2(E') \ominus \Theta_1 L^2(F))$$

it is easy to show that V' is the minimal isometric dilation of T'. For any Ψ in $L^\infty(E,E')$ we define the operator $\Gamma_1(\Psi)$ from $H^2(E)$ to H' by

$$\Gamma_1 = \Gamma_1(\Psi) \overset{\Delta}{=} P_{H'} M_\Psi | H^2(E) . \tag{4.3}$$

Let $\Gamma_1 = \Gamma_1(Q)$ where Q is the specified function in $L^\infty(E, E')$. Using the fact that V' is a lifting of T' we have

$$T' \Gamma_1 = P_{H'} V' P_{H'} M_Q | H^2(E) = P_{H'} V' M_Q | H^2(E) = \Gamma_1 S .$$

Hence Γ_1 is an operator in $I(S, T')$. Because Γ_1 intertwines the isometry S and co-isometry T' the operator Γ_1 is a Hankel operator in the sense of Section VIII.4. This is a generalized version of the Hankel operator considered in the previous section.

By the commutant lifting theorem there exists an intertwining lifting $B \in I(S, V')$ of Γ_1 satisfying $\|B\| = \|\Gamma_1\|$. Part b of Theorem 1.1 shows that $B = M_{Q_*} \,|\, H^2(E)$ where Q_* is in $L^\infty(E, E')$ and $\|B\| = \|Q_*\|_\infty$. Thus

$$\Gamma_1(Q) = \Gamma_1 = \Gamma_1(Q_*) \quad \text{and} \quad \|\Gamma_1\| = \|Q_*\|_\infty . \tag{4.4}$$

The first equation in (4.4) implies that $\Gamma_1(Q_* - Q) = 0$, or equivalently,

$$(Q_* - Q)H^2(E) \subseteq \Theta_1 H^2(F) .$$

By Corollary 2.2 of the Beurling-Lax-Halmos Theorem there exists an Ψ_* in $H^\infty(E, F)$ satisfying $Q_* - Q = \Theta_1 \Psi_*$. Thus $Q_* = Q + \Theta_1 \Psi_*$. If Ψ is any function in $H^\infty(E, F)$, then

$$\|Q + \Theta_1 \Psi_*\| = \|\Gamma_1\| = \|\Gamma_1(Q + \Theta_1 \Psi)\| \le \inf \|Q + \Theta_1 H^\infty(E, F)\|_\infty . \tag{4.5}$$

The first equality follows from (4.4) and the second from $\Gamma_1(\Theta_1 \Psi) = 0$. So there is equality in (4.5), that is, (4.2) holds and Ψ_* is an optimal solution to Problem 4.1. Moreover, $\mu = \|\Gamma_1(Q)\|$. Finally, it is noted that if Ψ_* is an optimal solution to Problem 4.1, that is, (4.2) holds, then $B = M_{Q_*} \,|\, H^2(E)$ where $Q_* = Q + \Theta_1 \Psi_*$ intertwining lifting of Γ_1 of norm equal to $\|\Gamma_1\|$. Summing up this analysis readily gives the following result.

4.2 THEOREM. *The set of all optimal solutions* Ψ_* *in* $H^\infty(E, F)$ *to Problem 4.1 is given by* $\Psi_* = \Theta_1^*(Q_* - Q)$ *where* $B = M(Q_*)\,|\,H^2(E)$ *is an intertwining lifting of* $\Gamma_1(Q)$ *with norm equal to* $\|\Gamma_1(Q)\|$ *and where* $Q_* = Q + \Theta_1 \Psi_*$ *is in* $L^\infty(E, E')$. *Moreover, in this case*

$$\mu = \|Q + \Theta_1 \Psi_*\| = \|\Gamma_1(Q)\| = \inf \|Q + \Theta_1 H^\infty(E, F)\|_\infty . \tag{4.6}$$

To complete this section we will use the commutant lifting theorem to solve the following H^∞ optimization problem, arising in H^∞ control theory, which is usually referred to as the four block problem. Obviously it is a generalization of Problem 4.1.

4.3 PROBLEM. Let Ω be a fixed function of the form

$$\Omega = \begin{bmatrix} A & B \\ C & D \end{bmatrix} \quad \text{in} \quad L^\infty(E_1 \oplus E_2, E_1' \oplus E_2') . \tag{4.7}$$

Let Θ_1 be a fixed inner function in $H^\infty(F, E_1')$. Then find δ where

$$\delta = \inf \left\| \begin{bmatrix} A+\Theta_1 H^\infty(E_1, F) & B \\ C & D \end{bmatrix} \right\|_\infty \triangleq \inf \left\{ \left\| \begin{bmatrix} A + \Theta_1 \Psi & B \\ C & D \end{bmatrix} \right\|_\infty : \Psi \in H^\infty(E_1, F) \right\}. \quad (4.8)$$

As before we will use the commutant lifting theorem to obtain an optimal Ψ_* in $H^\infty(E_1, F)$ solving Problem 4.3, that is,

$$\left\| \begin{bmatrix} A+\Theta_1 \Psi_* & B \\ C & D \end{bmatrix} \right\|_\infty = \delta = \inf \left\| \begin{bmatrix} A+\Theta_1 H^\infty(E_1, F) & B \\ C & D \end{bmatrix} \right\|_\infty \quad (4.9)$$

and show that δ equals the norm of a generalized Hankel operator $\Gamma_2(\Omega)$.

To this end let S_i and S_i' be the unilateral shifts on $H^2(E_i)$ and $H^2(E_i')$, respectively for $i = 1, 2$. The operators V_i and V_i' are the bilateral shifts on the L^2 appropriate spaces. To put this problem in the framework of the commutant lifting theorem we set

$$H = H^2(E_1) \oplus L^2(E_2) \quad \text{and} \quad H' = (L^2(E_1') \ominus \Theta_1 H^2(F)) \oplus L^2(E_2'). \quad (4.10)$$

The contractions T on H and T' on H' are defined by

$$T = S_1 \oplus V_2 \quad \text{and} \quad T' = P_{H'} W' | H' \quad (4.11)$$

where $W' = V_1' \oplus V_2'$. Obviously T is an isometry. Because H' is invariant for W^* it follows that T' is a co-isometry and $T'^* = W'^* | H'$. It is also clear that W' is an isometric lifting of T'. Moreover, using

$$H' = [\Theta_1 K_o^2(F) \oplus (L^2(E_1') \ominus \Theta_1 L^2(F))] \oplus L^2(E_2')$$

it is easy to verify that W' is the minimal isometric dilation of T'. Now we introduce the operator $\Gamma_2(\Psi)$ from H to H' defined by

$$\Gamma_2(\Psi) \triangleq P_{H'} M_\Psi | H \quad (4.12)$$

for any Ψ in $L^\infty(E_1 \oplus E_2, E_1' \oplus E_2')$ and we consider $\Gamma_2 = \Gamma(\Omega)$. Using the fact that W' is a lifting of T' it follows that Γ_2 is in $I(T, T')$. As before because Γ_2 intertwines the isometry T and co-isometry T', the operator Γ_2 is a Hankel operator in the sense of Section VIII.4. Moreover, Remark VIII.7.4 shows that Γ_2 can be identified with a generalized Hankel matrix of the form H_n in (VIII.7.1).

By the commutant lifting theorem there exists an intertwining lifting $B_2 \in I(T, W')$ of Γ_2 satisfying $\|B_2\| = \|\Gamma_2\|$. We notice that $W = V_1 \oplus V_2$ is the minimal unitary dilation of T. Applying Corollary VI.2.4 to B_2 in $I(T, W')$, we obtain a unique extension $\hat{B}_2$ in $I(W, W')$ of B_2 and $\|\hat{B}_2\| = \|B_2\|$. Theorem 1.1 show that $\hat{B}_2 = M_\Omega$ and $\|\hat{B}_2\| = \|\Omega_*\|_\infty$ where

$$\Omega_* = \begin{bmatrix} A_* & B_* \\ C_* & D_* \end{bmatrix} \text{ is in } L^\infty(E_1 \oplus E_2, E_1' \oplus E_2') . \tag{4.13}$$

Therefore

$$\Gamma_2(\Omega) = \Gamma_2 = \Gamma_2(\Omega_*) \quad \text{and} \quad \|\Gamma_2\| = \|B_2\| = \|\hat{B}_2\| = \|\Omega_*\|_\infty . \tag{4.14}$$

Obviously $\Gamma_2(\Omega_* - \Omega) = 0$. This with (4.13), (4.14) and the definition in (4.12) gives $C = C_*$ and $D = D_*$. Moreover,

$$[A_* - A, \ B_* - B] \begin{bmatrix} H^2(E_1) \\ L^2(E_2) \end{bmatrix} \subseteq \Theta_1 H^2(F) . \tag{4.15}$$

This implies that $(B_* - B)L^2(E_2)$ is contained in $\Theta_1 H^2(F)$. Thus

$$(B_* - B)L^2(E_2) = (B_* - B)(\bigcap_0^\infty V_2^n L^2(E_2)) \subseteq \bigcap_0^\infty V_1'^n (B_* - B)L^2(E_2)$$

$$\subseteq \bigcap_0^\infty S_1'^n \Theta_1 H^2(F) = \bigcap_0^\infty \Theta_1 S^n H^2(F) = \Theta_1 (\bigcap_0^\infty S^n H^2(F)) = \{0\}$$

where in the second to last equality we used the fact that Θ_1 is an isometry and S is the unilateral shift on $H^2(F)$. So $B_* = B$ and (4.15) implies that $(A_* - A)H^2(E_1)$ is contained in $\Theta_1 H^2(F)$. By Corollary 2.2 of the Beurling-Lax-Halmos Theorem there exists a Ψ_* in $H^\infty(E_1, F)$ satisfying $A_* - A = \Theta_1 \Psi_*$, or equivalently, $A_* = A + \Theta_1 \Psi_*$. This and (4.14) gives

$$\delta \le \left\| \begin{bmatrix} A + \Theta_1 \Psi_* & B \\ C & D \end{bmatrix} \right\|_\infty = \|\Gamma_2(\Omega)\| \le \delta \tag{4.16}$$

where δ is defined in (4.8). So there is equality in (4.16), that is, (4.9) holds and Ψ_* is an optimal solution to Problem 4.3. Moreover, $\|\Gamma_2(\Omega)\| = \delta$. Finally, it is noted that if Ψ_* is an optimal solution to Problem 4.3, then $B_2 = M_\Omega | H$ where Ω_* is the matrix in (4.13) or (4.16) is an intertwining lifting of $\Gamma_2(\Omega)$. Summing the preceding analysis readily proves the follow result.

4.4 THEOREM. *The set of all optimal solutions Ψ_* in $H^\infty(E_1, F)$ to Problem 4.3 is given by $\Psi_* = \Theta_1^*(A_* - A)$ where $B_2 = M_\Omega | H$ is any intertwining lifting of $\Gamma_2(\Omega)$ with norm equal to $\|\Gamma_2(\Omega)\|$ and Ω_* is of the form (4.13). Moreover, $A_* = A + \Theta_1 \Psi_*$, $B_* = B$, $C_* = C$, $D_* = D$ and*

$$\left\| \begin{bmatrix} A + \Theta_1 \Psi_* & B \\ C & D \end{bmatrix} \right\|_\infty = \|\Gamma_2(\Omega)\| = \delta$$

where δ is defined in (4.8).

5. THE BASIC FUNCTIONAL MODEL

In this section we will present our basic functional model which will be used in solving several interpolation and H^∞ optimization problems.

To begin recall that a function Θ in $H^\infty(E, E')$ is called *contractive* if $\|\Theta(e^{it})\| \leq 1$ a.e. in $[0, 2\pi)$, or equivalently, if $\|\Theta(z)\| \leq 1$ for all z in D. A contractive Θ is *purely contractive* if moreover, $\|\Theta(0)e\| < \|e\|$ for all nonzero e in E. For a contractive Θ in $H^\infty(E, E')$ we define the operator valued function Δ_Θ by

$$\Delta_\Theta(e^{it}) = (1 - \Theta(e^{it})^* \Theta(e^{it}))^{1/2} \qquad (0 \leq t < 2\pi) . \tag{5.1}$$

This is defined a.e. in $[0, 2\pi)$. For every contractive Θ in $H^\infty(E, E')$ we define the space $H(\Theta)$ by

$$H(\Theta) = [H^2(E') \oplus \overline{\Delta_\Theta L^2(E)}] \ominus \{\Theta h \oplus \Delta_\Theta h \colon h \in H^2(E)\} \tag{5.2}$$

and the operator $S(\Theta)$ on $H(\Theta)$ by

$$S(\Theta) = P_{H(\Theta)}(S' \oplus V) | H(\Theta) . \tag{5.3}$$

This isometry $U(\Theta)$ on $K(\Theta)$ and unitary operator $\hat{U}(\Theta)$ on $\hat{K}(\Theta)$ is defined by

$$U(\Theta) = (S' \oplus V) | K(\Theta) \quad \text{where} \quad K(\Theta) = H^2(E') \oplus \overline{\Delta_\Theta L^2(E)} ,$$

$$\tag{5.4}$$

$$\hat{U}(\Theta) = (V' \oplus V) | \hat{K}(\Theta) \quad \text{where} \quad \hat{K}(\Theta) = L^2(E') \oplus \overline{\Delta_\Theta L^2(E)} .$$

In many important applications Θ is inner. In this case

$$S(\Theta)^* = S'^* | H(\Theta) \quad \text{where} \quad H(\Theta) = H^2(E') \ominus \Theta H^2(E) .$$

Moreover, $U(\Theta)$ is the unilateral shift S' on $H^2(E')$ and $\hat{U}(\Theta)$ is the bilateral shift V' on $L^2(E')$. The operator $S(\Theta)$ is often referred to as the Sz.-Nagy-Foias functional model. The following theorem presents some important properties of $S(\Theta)$.

5.1 THEOREM. *Let Θ be a purely contractive analytic function in $H^\infty(E, E')$. Then the following hold:*

(i) *The minimal isometric (unitary) dilation of $S(\Theta)$ is $U(\Theta)(\hat{U}(\Theta)$ respectively).*

(ii) *The operator $S(\Theta)$ is completely non-unitary.*

(iii) *The operator $S(\Theta)$ is $*$-stable if and only if Θ is inner.*

(iv) *The operator $S(\Theta)$ is stable if and only if Θ is $*$-inner.*

PROOF. Let $M = \{\Theta h \oplus \Delta_\Theta h\colon h \in H^2(E)\}$. Obviously M is an invariant subspace for $U(\Theta)$. So $U(\Theta)$ is an isometric lifting of $S(\Theta)$. Using the fact that $[\Theta^*, \Delta_\Theta^*]^* \,|\, H^2(E)$ from $H^2(E)$ to $K(\Theta)$ is an isometry it is easy to verify that

$$P_M(f \oplus g) = \Theta w \oplus \Delta_\Theta w \quad \text{where} \quad w = P_+(\Theta^* f + \Delta_\Theta g) \tag{5.5}$$

and P_+ is the orthogonal projection of $L^2(E)$ onto $H^2(E)$. Clearly $M = K(\Theta) \ominus H(\Theta)$. To prove that $U(\Theta)$ is minimal we show that $H(\Theta)$ is cyclic for $U(\Theta)$ and thus $U(\Theta)$ is the minimal isometric dilation of $S(\Theta)$. To this end notice that by (5.5) for $u \oplus v$ in $K(\Theta)$ we have

$$(u \oplus v) \in H(\Theta) \quad \text{if and only if} \quad (\Theta^* u + \Delta_\Theta v) \in K_0^2(E). \tag{5.6}$$

Thus $u \oplus v$ is in $H(\Theta)$ if and only if the Fourier expansion of $\Theta^* u + \Delta_\Theta v$ is of the form

$$\Theta^* u + \Delta_\Theta v = w_1 e^{-it} + w_2 e^{-2it} + \cdots . \tag{5.7}$$

So for $u \oplus v$ in $H(\Theta)$ we have

$$e^{it}(\Theta^* u + \Delta_\Theta v) = w_1 + e^{-it} w_2 + e^{-2it} w_3 + \cdots$$

This and (5.3), (5.5), (5.7) gives

$$S(\Theta)(u \oplus v) = P_{H(\Theta)} U(\Theta)(u \oplus v) = S' u \oplus V v - \Theta w_1 \oplus \Delta_\Theta w_1 \tag{5.8}$$

where

$$w_1 = \frac{1}{2\pi} \int_0^{2\pi} e^{it}(\Theta^* u + \Delta_\Theta v)(e^{it}) dt = P_+ V(\Theta^* u + \Delta_\Theta v). \tag{5.9}$$

Let us denote by F the (obviously linear) space of the w_1's which occur in (5.9). From (5.8) and (5.9) we can easily deduce that

$$\bigvee_{n=0}^{\infty} U(\Theta)^n H(\Theta) = H(\Theta) \oplus \{\Theta w \oplus \Delta_\Theta w\colon w \in H^2(\bar{F})\}^-$$

where $\bar{F}$ denotes the closure of F. Therefore by virtue of (5.4) it remains to prove that $\bar{F} = E$. For this we consider elements of the form

$$u_a \oplus v_a = e^{-it}[\Theta - \Theta(0)]a \oplus e^{-it}\Delta_\Theta a \text{ in } K(\Theta),$$

where a is in E. Notice that

$$\Theta^* u_a + \Delta_\Theta v_a = e^{-it}(1 - \Theta^*\Theta(0))a = e^{-it}(1 - \Theta(0)^*\Theta(0))a + \cdots$$

belongs to $K_o^2(E)$. So $u_a \oplus v_a$ is in $H(\Theta)$ by (5.6). Therefore

$$(I - \Theta(0)^*\Theta(0))E \subseteq F \subseteq E. \tag{5.10}$$

The purely contractive property implies that $I - \Theta(0)^*\Theta(0)$ is injective and thus with range dense in E. Consequently $\bar{F} = E$ follows from (5.10). So $U(\Theta)$ is the minimal isometric dilation of $S(\Theta)$. The operator $\hat{U}(\Theta)$ is the minimal unitary dilation of $S(\Theta)$ because $\hat{U}(\Theta)$ is obviously the minimal unitary extension of $U(\Theta)$.

To verify (ii) notice that in the present case (5.8) along with $\bar{F} = E$ show that the wandering subspace L defined in Proposition VI.4.1 is given by

$$L = \overline{(U(\Theta) - S(\Theta))H} = \{\Theta a \oplus \Delta_\Theta a: \ a \in E\}. \tag{5.11}$$

Now let us compute the corresponding wandering subspace L_* in Proposition VI.4.2. If $u \oplus v$ is in $H(\Theta)$, then

$$(I - U(\Theta)S(\Theta)^*)(u \oplus v) = u \oplus v - (u - u(0)) \oplus v = u(0) \oplus 0. \tag{5.12}$$

We claim that $H_o \triangleq \{u(0): (u \oplus v) \in H(\Theta)\}^-$ is actually E'. Obviously $H_o \subseteq E'$. If f is in E' and f is orthogonal to H_o, then f is orthogonal to $H(\Theta)$. This implies that $f \oplus 0 = \Theta h \oplus \Delta_\Theta h$ for some h in $H^2(E)$. Since $[\Theta^*, \Delta_\Theta]^*$ is an isometry $\|f\| = \|h\|$. Because $\Delta_\Theta h = 0$ we have $\|h\| = \|\Theta h\|$. This and $f = \Theta(0)h(0)$ gives

$$\|h\| = \|\Theta h\| = \|f\| = \|\Theta(0)h(0)\| \le \|h(0)\| \le \|h\|.$$

Hence $\|\Theta(0)h(0)\| = \|h(0)\|$. Since Θ is purely contractive $h(0) = 0$. Hence the vector $f = 0$ and $H_o = E'$. This and (5.12) gives

$$L_* = \overline{(I - U(\Theta)S(\Theta)^*)H(\Theta)} = E'. \tag{5.13}$$

Obviously $M(L) \bigvee M(L_*) = \hat{K}(\Theta)$; see Section VI.2 for the definition of $M(L)$ and $M(L_*)$. Proposition VI.6.2 shows that $S(\Theta)$ is completely non-unitary. This proves (ii).

Theorem VI.5.3 implies that $S(\Theta)$ is $*$-stable if and only if $L^2(E') = M(L_*) = \hat{K}(\Theta)$, or equivalently, $\Delta_\Theta = 0$. Therefore $S(\Theta)$ is $*$-stable if and only if Θ is inner. This proves (iii). To verify (iv) notice that the "rotation" operator

$$\Omega = \begin{bmatrix} M_\Theta & \Delta_{\Theta^\cdot} \\ \Delta_\Theta & -M_{\Theta^\cdot} \end{bmatrix} : (L^2(E) \oplus \overline{\Delta_{\Theta^\cdot} L^2(E')}) \to (L^2(E') \oplus \overline{\Delta_\Theta L^2(E)}) = \hat{K}(\Theta)$$

is unitary. This fact is established by a calculation similar to that used in Corollary IV.1.4 to verify that the rotation matrix formed by a contraction is unitary. The range of the first column of Ω is $M(L)$. So $M(L) = \hat{K}(\Theta)$ if and only if the range of the second column of Ω is zero, or equivalently, $\Delta_{\Theta^\cdot} = 0$ Therefore $M(L) = \hat{K}(\Theta)$ if and only if $\Theta(e^{it})\Theta^*(e^{it}) = I.$ a.e., or equivalently, Θ is $*$-inner. This and Theorem VI.5.3 proves (iv). The proof is now complete.

5.2 REMARK. Let Θ be a purely contractive analytic function in $H^\infty(E, E')$ and set $T = S(\Theta)$. Propositions VI.4.1 and VI.4.2 with equations (5.11) and (5.13) show that the defect indices $\delta_T = \dim E$ and $\delta_{T^*} = \dim E'$.

The following shows that our assumption of purely contractive is not too restrictive.

5.3 THEOREM. *Let Θ be a contractive function in $H^\infty(E, E')$. Then Θ admits a unique decomposition of the form $\Theta = \Theta_0 \oplus \Theta_u$ where Θ_0 is purely contractive in $H^\infty(E_0, E_0')$, Θ_u is a unitary constant in $H^\infty(E_u, E_u')$, $E = E_0 \oplus E_\mu$ and $E' = E_0 \oplus E_u'$. Uniqueness means that if $\Theta = \Theta_1 \oplus \Theta_2$ is another decomposition of Θ where Θ_1 is purely contractive in $H^\infty(E_1, E_1')$ and Θ_2 is a unitary constant in $H^\infty(E_2, E_2')$ and $E = E_1 \oplus E_2$ and $E' = E_1' \oplus E_2'$, then $\Theta_0 = \Theta_1$ and $\Theta_u = \Theta_2$.*

PROOF. According to Lemma V.2.1 the operator $Z = \Theta(0)$ from E to E' admits a unique decomposition of the form $Z = Z_0 \oplus Z_u$ from $D_Z \oplus (E \ominus D_Z)$ to $D_{Z^\cdot} \oplus (E' \ominus D_{Z^\cdot})$ where Z_0 is pure and Z_u is unitary. Let $E_0 = D_Z$ and $E_u = (E \ominus D_Z)$ and $E_0' = D_{Z^\cdot}$ and $E_u' = (E' \ominus D_{Z^\cdot})$. Now let Θ_n be the coefficient in $L(E, E')$ of z^n in the power series expansion of $\Theta(z)$. Obviously $\Theta(0)|E_0 = \Theta_0$. Since $\Theta(0)|E_u$ is unitary we have

$$\|h\|^2 = \|\Theta(0)h\|^2 \le \sum_{0}^{\infty}\|\Theta_n h\|^2 = \|\Theta_+ h\|^2 \le \|h\|^2$$

when h is in E_u. So there is equality and $\Theta_n|E_u$ is zero for all $n \ge 1$. Therefore $\Theta(z)|E_u = \Theta(0)|E_u$ is a unitary constant in z mapping E_u onto E_u'. Because of this and because for all z in D the operator $\Theta(z)$ is a contraction from E to E', our $\Theta(z)$ must map E_0 into E_0'. (If $A = \Theta(z)|E_u$ in Theorem IV.3.1 is unitary, then the off diagonal terms in the matrix representation of the type (IV.3.1) of $\Theta(z)$ are zero.) Thus $\Theta_o(z) \stackrel{\Delta}{=} \Theta(z)|E_o$ is a purely contractive analytic function in $H^\infty(E_o, E_o')$. This shows that Θ admits a decomposition of the form $\Theta = \Theta_o \oplus \Theta_u$ where Θ_o is purely contractive and Θ_u is a unitary constant. To prove uniqueness let $\Theta = \Theta_1 \oplus \Theta_2$ be another decomposition of Θ where Θ_1 is purely contractive and Θ_2 is a unitary constant. For all f in E_2

$$\|f\| = \|\Theta_2(0)f\| = \|\Theta(0)f\| = \|Zf\| \ .$$

So f is in $(E \ominus D_Z) = E_u$ and $E_2 \subseteq E_u$. If g is a nonzero vector in E_u and g is orthogonal to E_2, then

$$\|g\| = \|\Theta_u(0)g\| = \|\Theta(0)g\| = \|\Theta_1(0)g\| < \|g\|$$

which is a contradiction. Therefore $g = 0$ and $E_u = E_2$. Since both Θ_u and Θ_2 are unitary constants $E_2' = \Theta_2 E_2 = \Theta(0)E_u = E_u'$ and $\Theta_2 = \Theta|E_u = \Theta_u$. Then obviously $\Theta_o = \Theta_1$. This completes the proof.

Let Θ be any contraction in $H^\infty(E, E')$. We call Θ_o and Θ_u in Theorem 5.3 the *purely contractive part* and *unitary part* of Θ, respectively. Using the notation in Theorem 5.3 we have

$$S(\Theta) = S(\Theta_o) \quad \text{and} \quad H(\Theta) = H(\Theta_o) \ . \tag{5.14}$$

This follows because $S(\Theta) = S(\Theta_o) \oplus S(\Theta_u)$ and $H(\Theta_u) = 0$. Therefore the contraction $S(\Theta)$ generated by Θ equals the contraction $S(\Theta_o)$ generated by the pure part Θ_o of Θ. It is easy to verify that $U(\Theta)$ is an isometric lifting of $S(\Theta)$. If Θ is not pure, then $U(\Theta) \neq U(\Theta_o)$. However, Theorem 5.1 shows that $U(\Theta_o)$ is the minimal isometric dilation of $S(\Theta) = S(\Theta_o)$.

Let Θ be the previous contractive function and Θ_1 be in $H^\infty(E_1, E_1')$. We say that Θ and Θ_1 *coincide* if there exists two constant unitary operators γ mapping E onto E_1 and γ' mapping E' onto E_1', satisfying $\Theta_1(z)\gamma = \gamma'\Theta(z)$ for all z in D. This definition is used in the following result.

5.4 PROPOSITION. *Let Θ and Θ_1 be contractive analytic functions in $H^\infty(E, E')$ and $H^\infty(E_1, E_1')$, respectively. If Θ and Θ_1 coincide, then $S(\Theta)$ is unitarily equivalent to $S(\Theta_1)$.*

PROOF. Notice that $\Delta_{\Theta_1}\gamma = \gamma\Delta_\Theta$. It follows that the operator $\Omega = (M\gamma' \oplus M\gamma)|K(\Theta)$ is a unitary operator mapping $K(\Theta)$ onto $K(\Theta_1)$. Clearly $\Omega U(\Theta) = U(\Theta_1)\Omega$. Moreover,

$$\Omega\{\Theta h \oplus \Delta_\Theta h : h \in H^2(E)\} = \{\Theta_1 h \oplus \Delta_{\Theta_1} h : h \in H^2(E_1)\} .$$

Hence $\Omega H(\Theta) = H(\Theta_1)$. This and $\Omega U(\Theta) = U(\Theta_1)\Omega$ yields

$$\Omega S(\Theta) = \Omega P_{H(\Theta)} U(\Theta) \,|\, H(\Theta) = P_{H(\Theta_1)} U(\Theta_1) \,|\, \Omega H(\Theta) = S(\Theta_1)\Omega \,|\, H(\Theta) .$$

Therefore $S(\Theta)$ is unitarily equivalent to $S(\Theta_1)$. This completes the proof.

5.5 REMARK. In [Sz.-NF 10] it is shown that $S(\Theta)$ is unitarily equivalent to $S(\Theta_1)$ if and only if the pure parts of Θ and Θ_1 coincide.

To complete this section we show that $S(\Theta)^*$ is unitarily equivalent to $S(\tilde\Theta)$ where $\tilde\Theta$ is defined by $\tilde\Theta(z) = \Theta(\bar z)^*$ for all z in D. To this end let $W(\Theta)$ be the (unitary) operator defined by

$$W(\Theta) = \begin{bmatrix} V'^* M_\Theta J & V'^* \Delta_{\Theta^\bullet} J \\ V^* \Delta_\Theta J & -V^* M_{\Theta^\bullet} J \end{bmatrix} \tag{5.15}$$

$$W(\Theta) : L^2(E) \oplus \overline{\Delta_{\tilde\Theta} L^2(E')} \to L^2(E') \oplus \overline{\Delta_\Theta L^2(E)} .$$

Using the fact that for any contraction T we have $T D_T = D_{T^\bullet} T$ where T is a contraction it is easy to verify as for a rotation matrix that $W(\Theta) = W'$ is unitary, that is, $W^* W = I = W W^*$.

5.6 PROPOSITION. *Let Θ be a purely contractive analytic function in $H^\infty(E, E')$ and $W = W(\Theta)$. The following properties hold;*
    (i) $W(K_o^2(E) \oplus \{0\}) = \{\Theta u \oplus \Delta_\Theta u : u \in H^2(E)\}$;
    (ii) $W\{\tilde\Theta u \oplus \Delta_{\tilde\Theta} u : u \in H^2(E')\} = K_o^2(E') \oplus \{0\}$;
    (iii) $W H(\tilde\Theta) = H(\Theta)$ *and* $W S(\tilde\Theta) = S(\Theta)^* W \,|\, H(\tilde\Theta)$. *In particular, $S(\tilde\Theta)^*$ and $S(\Theta)$ are unitarily equivalent.*

PROOF. Part (i) easily follows from the definition of $K_o^2(E)$. Part (ii) is obtained by using again the identity $D_T T^* = T^* D_{T^\bullet}$ for a contraction T. For Part (iii) notice that

$$L^2(E) \oplus \overline{\Delta_{\tilde\Theta} L^2(E')} = K_o^2(E) \oplus \{\tilde\Theta u \oplus \Delta_{\tilde\Theta} u : u \in H^2(E')\} \oplus H(\tilde\Theta) .$$

This with the analogous formula involving Θ and the properties (i), (ii) yield the first part of (iii). To verify the second identity in (iii) let $u \oplus v$ be in $H(\tilde\Theta)$. Then

$$WS(\tilde{\Theta})(u \oplus v) = WP_{H(\tilde{\Theta})}S \oplus V'(u \oplus v) =$$

$$P_{H(\Theta)}W(S \oplus V')(u \oplus v) = P_{H(\Theta)}(S' \oplus V)^* W(u \oplus v) = S(\Theta)^* W(u \oplus v) \ .$$

This completes the proof.

5.7 REMARK. In many important applications the contractive function Θ is inner and $*$-inner. In this case the unitary operator W in (5.15) reduces to

$$W(\Theta) = V'^* M_\Theta J \ : \ L^2(E) \to L^2(E') \ . \tag{5.16}$$

6. THE CHARACTERISTIC FUNCTION

Let T be a contraction on H. The *characteristic* function $\Theta_T(z)$ for T is defined by

$$\Theta_T(z) \overset{\Delta}{=} [-T + zD_{T^*}(I - zT^*)^{-1}D_T]\,|\,D_T \quad (z \in D) \ . \tag{6.1}$$

It is easy to check that $\Theta_{T^*}(z) = \tilde{\Theta}_T(z) \overset{\Delta}{=} \Theta_T(\overline{z})^*$ for all z in D. Recall that T maps D_T onto D_{T^*} and therefore it is obvious that Θ_T is an analytic function in D with values in $L(D_T, D_{T^*})$.

The following theorem is the basic result of this section.

6.1 THEOREM. *Let T be a completely non-unitary contraction on H. Then*

(i) The characteristic function $\Theta_T(z)$ is a purely contractive analytic function in
   $H^\infty(D_T, D_{T^*})$.

(ii) The contraction T is unitarily equivalent to $S(\Theta_T)$.

(iii) The contraction T is $*$-stable if and only if Θ_T is inner.

(iv) The contraction T is stable if and only if Θ_T is $*$-inner.

PROOF. Let $\hat{U}$ on $\hat{K}$ be the minimal unitary dilation of T. Consider the unitary operator Φ from $M(L)$ onto $L^2(L)$ and Φ_* from $M(L_*)$ onto $L^2(L_*)$ defined by

$$\Phi(\sum_n \hat{U}^n l_n) = \sum_n e^{int} l_n \quad \text{and} \quad \Phi_*(\sum_n \hat{U}^n l_n^*) = \sum_n e^{int} l_n^* \tag{6.2}$$

where l_n is in L and l_n^* is in L_* for all n. Here L and L_* are the wandering subspaces for $\hat{U}$ defined in Section VI.4. Obviously Φ maps $M_+(L)$ onto $H^2(L)$ and Φ_* maps $M_+(L_*)$ onto $H^2(L_*)$. Let P_* be the orthogonal projection of $\hat{K}$ onto $M(L_*)$ and B the contraction from $L^2(L)$ to $L^2(L_*)$ defined by $B = (\Phi_* P_* | M(L))\Phi^*$. Since $P_* M_+(L) \subseteq M_+(L_*)$ the operator B maps $H^2(L)$ into $H^2(L_*)$; see (VI.5.10). Using the fact that P_* commutes with $\hat{U}$ it follows that $B\,|\,H^2(L)$ intertwines the unilateral shift S on $H^2(L)$ with the unilateral shift S' on $H^2(L_*)$. Theorem 1.1 implies that $B = M_\Theta$ where Θ is a contractive analytic function in $H^\infty(L, L_*)$.

Since T is completely non-unitary, Proposition VI.6.2 and (VI.5.12) show that $\overline{(I-P_*)M(L)} = K_u$. If f is in $M(L)$, then

$$\|(I-P_*)f\|^2 = \|f\|^2 - \|P_*f\|^2 = \|\Phi f\|^2 - \|\Theta\Phi f\|^2 = \|\Delta_\Theta \Phi f\|^2 \ .$$

Thus there exists a unitary operator Φ_u from K_u onto $\overline{\Delta_\Theta L^2(L)}$ satisfying $\Phi_u(I-P_*)f = \Delta_\Theta \Phi f$. Now let V be multiplication by e^{it} on $L^2(L)$. Using the fact that P_* commutes with $\hat{U}$ we have

$$V\Phi_u(I-P_*)f = V\Delta_\Theta \Phi f = \Delta_\Theta \Phi \hat{U}f = \Phi_u \hat{U}(I-P_*)f \ .$$

Thus $V\Phi_u = \Phi_u \hat{U} \,|\, K_u$.

Now let Ω be the unitary operator mapping $K = M_+(L_*) \oplus K_u$ onto $K(\Theta) = H^2(L_*) \oplus \overline{\Delta_\Theta L^2(L)}$ defined by $\Omega = (\Phi_* | M_+(L_*)) \oplus \Phi_u$. By construction $\Omega U = U(\Theta)\Omega$ where U is the minimal isometric dilation of T and $U(\Theta)$ is the minimal isometric dilation of $S(\Theta)$ in (5.3). (However, here $E = L$ and $E' = L_*$). If f is in $M_+(L)$, then

$$\Omega f = \Phi_* P_* f + \Phi_u(I-P_*)f = \Theta\Phi f \oplus \Delta_\Theta \Phi f \ .$$

This readily implies that under the unitary operator Ω the space $M_+(L)$ is identified with

$$\Omega M_+(L) = \{\Theta h \oplus \Delta_\Theta h: \ h \in H^2(L)\} \ . \tag{6.3}$$

Applying (VI.5.5) and (5.2) we have

$$\Omega H = \Omega[(M_+(L_*) \oplus K_u) \ominus M(L)] =$$

$$[H^2(L_*) \oplus \overline{\Delta_\Theta L^2(E)}] \ominus \{\Theta h \oplus \Delta_\Theta h: \ h \in H^2(L)\} = H(\Theta) \ . \tag{6.4}$$

So we have shown that the minimal isometric dilation U of T is unitarily equivalent to $U(\Theta)$ and under this identification and that $\Omega H = H(\Theta)$. Therefore T is unitarily equivalent to $S(\Theta)$ the compression of $U(\Theta)$ to $H(\Theta)$. The unitary equivalence also shows that $U(\Theta)$ is the minimal isometric dilation of $S(\Theta)$.

Now we will show that Θ coincides with Θ_T the characteristic function for T. Recall that L_* is the closure of $(I-UT^*)H$ (see Proposition VI.4.2). A simple calculation gives

$$P_{U^n L_*} h = U^n(I-UT^*)T^{*n}h \qquad (\text{for all h in } H) \ .$$

Since P_*h is in $M_+(L_*)$ this implies that

$$P_*h = \sum_0^\infty U^n(I-UT^*)T^{*n}h \qquad \text{(for all h in } H) \ .$$

Using the fact that $\hat{U}$ commutes with P_* the previous equation yields

$$P_*(U-T)h = -(I-UT^*)Th + \sum_1^\infty U^n(I-UT^*)T^{*n-1}D_T^2 h \qquad \text{(for all h in } H) \ . \tag{6.5}$$

Recall that L is the closure of $(U-T)H$ (see Proposition VI.4.1). By (VI.4.4) an (VI.4.8) there exists unitary operators ϕ from D_T onto L and ϕ_* from D_{T^*} onto L_* such that

$$\phi D_T h = (U-T)h \quad \text{and} \quad \phi_* D_{T^*}h = (I-UT^*)h \qquad (h \in H) \ . \tag{6.6}$$

Let Θ_n be the Fourier coefficient of z^n in the power series expansion of $\Theta(z)$. Substituting (6.6) into (6.5) with the definition of Θ we have

$$\sum_0^\infty U^n\Theta_n\phi D_T h = P_*(U-T)h = -\phi_* D_{T^*}Th + \sum_1^\infty U^n\phi_* D_{T^*}T^{*n-1}D_T D_T h \qquad (h \in H) \ . \tag{6.7}$$

Comparing like coefficients of U^n with $D_{T^*}T = TD_T$ we have

$$\Theta_0\phi = -\phi_*T|D_T \quad \text{and} \quad \Theta_n\phi = \phi_* D_{T^*}T^{*n-1}D_T \qquad (n \geq 1) \ .$$

In other words $\Theta(z)\phi = \phi_*\Theta_T(z)$ for all z in D. This implies that $\|\Theta_T(z)\| = \|\Theta(z)\| \leq 1$ for all z in D, that is, Θ_T is a contractive analytic function in $H^\infty(D_T, D_{T^*})$. Actually it is also pure because $\Theta_T(0) = -T|D_T$ and $T|D_T$ is a pure contraction from D_T to D_{T^*} (see Lemma V.2.1). In turn this implies that Θ is a purely contractive analytic function in $H^\infty(L, L_*)$. Obviously $\Theta(z)$ coincides with $\Theta_T(z)$. By Proposition 5.4 the contractions $S(\Theta)$ and $S(\Theta_T)$ are unitarily equivalent. The proof of Parts (i) and (ii) are now complete. Parts (iii) and (iv) follow from Theorem 5.1 and the fact that Θ and Θ_T coincide. This completes the proof.

In Chapter XIV we will use some concepts from network theory to give a simple proof of the fact that the characteristic function Θ_T is a contractive analytic function.

If two contractions are unitarily equivalent, then it is easy to verify that their characteristic functions coincide. Combining this with Proposition 5.4 and the previous theorem readily proves the following result.

6.2 COROLLARY. *Two completely non-unitary contractions are unitarily equivalent if and only if their characteristic functions coincide.*

Finally, it is noted that if Θ is a purely contractive analytic function in $H^\infty(E, E')$, then the characteristic function of $S(\Theta)$ coincides with Θ. To see this notice that (5.11) and (5.13) yield a unitary identification of L with E and L_* with E'. Now the result follows from the fact that the characteristic function of $S(\Theta)$ coincides with the multiplication operator corresponding to $P_* | M_+(L)$ and that $P_*(\Theta a + \Delta_\Theta a) = \Theta a$ for all a in E. From the previous results we readily obtain the following corollary.

6.3 COROLLARY. *Let Θ and Θ_1 be two purely contractive analytic functions in $H^\infty(E, E')$ and $H^\infty(E_1, E'_1)$, respectively. Then $S(\Theta)$ is unitarily equivalent to $S(\Theta_1)$ if and only if Θ coincides with Θ_1.*

To complete this section let T on H be a contraction and U on $K = H \oplus H^2(D_T)$ be the minimal isometric dilation of T defined by

$$U(h \oplus f) = Th \oplus (D_T h + zf) \quad (h \oplus f \in H \oplus H^2(D_T)) . \tag{6.8}$$

(Notice that by identifying $l^2(D_T)$ with $H^2(D_T)$, the minimal isometric dilation U of T in (6.8) is the H^2 version of the matrix minimal isometric of T in (VI.3.4).) To complete this section we will present the following result, which will be used in solving some interpolation problems in Chapters XIII and XIV.

6.4 THEOREM. *Let T on H be a $*$-stable contraction and U on $K = H \oplus H^2(D_T)$ be its minimal isometric dilation in (6.8). Then the operator W_T from K to $H^2(D_{T^*})$ defined by*

$$W_T(h \oplus f) = D_{T^*}(I - zT^*)^{-1}h + \Theta_T f \quad (h \oplus f \in H \oplus H^2(D_T)) \tag{6.9}$$

is unitary, and $SW_T = W_T U$ where S is the unilateral shift on $H^2(D_{T^*})$. Moreover, Θ_T is inner, $W_T H = H(\Theta_T)$ and $W_T T = S(\Theta_T)W_T$, that is, T is unitarily equivalent to $S(\Theta_T)$. Finally, $U(\Theta_T) = S$ is the minimal isometric dilation of $S(\Theta_T)$.

PROOF. Part of this theorem follows from Theorem 6.1. To obtain some further insight, we will give a proof of this theorem independent of Theorem 6.1. To begin, recall that equations (VI.4.9) and (VI.4.10) show that the operator C from H to $H^2(D_{T^*})$ defined by

$$Ch = D_{T^*}(I - zT^*)^{-1}h \quad (h \in H) \tag{6.10}$$

is an isometry. It is easy to check that $S^*C = CT^*$. Obviously $H_* = CH$ is an invariant subspace for S^*. So if we define the contraction T_* on H_* by $T_*^* = S^* | H_*$, then S is an isometric lifting of T_* and T_* is unitarily equivalent to T. (Infact $T_* C_* = C_* T$ where C_* is the unitary operator from H onto H_* defined by $C_* = C$.) Notice that for h in H we have

$$(I - ST_*^*)Ch = (I - SS^*)Ch = (Ch)(0) = D_{T^*}h \; .$$

This implies that

$$H^2(D_{T^*}) \supseteq \bigvee_0^\infty S^n H_* \supseteq \bigvee_0^\infty S^n(I - ST_*^*)H_* = \bigvee_0^\infty S^n D_{T^*} = H^2(D_{T^*}) \; .$$

Therefore S is the minimal isometric dilation of T_*.

Because T and T_* are unitarily equivalent and all minimal isometric dilations are isomorphic, there exists a unitary operatory W_T from K onto $H^2(D_{T^*})$ satisfying $SW_T = W_T U$ and $W_T | H = C$. For h in H we have

$$W_T(0 \oplus D_T h) = W_T(U - T)h = SCh - CTh =$$

$$zD_{T^*}(I - zT^*)^{-1}h - D_{T^*}(I - zT^*)^{-1}Th =$$

$$- D_{T^*}Th + zD_{T^*}(I - zT^*)^{-1}(I - T^*T)h = \Theta_T(z)D_T h \; .$$

This and the fact that $W_T U = SW_T$ readily implies that $W_T(0 \oplus f) = \Theta_T f$ for all f in $H^2(D_T)$. Since W_T is unitary the operator $\Theta_T(= (\Theta_T)_+)$ is an isometry from $H^2(D_T)$ into $H^2(D_{T^*})$. By Theorem 1.1 the characteristic function $\Theta_T(z)$ is inner. The above analysis produces the unitary operator W_T in (6.9) satisfying $W_T U = SW_T$. Since $H = K \ominus H^2(D_T)$ it follows that $H_* = W_T H = H(\Theta_T)$. So T_* is the compression of S to $H(\Theta_T)$, or equivalently, $T_* = S(\Theta_T)$. Obviously $S = U(\Theta_T)$ is the minimal isometric dilation of T_*. This completes the proof.

7. THE OPERATORS A(F)

Let Θ and Θ_1 be two pure inner functions in $H^\infty(E, E')$ and $H^\infty(E_1, E_1')$, respectively. In this section we use the commutant lifting theorem to describe the set of operators in $I(S(\Theta), S(\Theta_1))$. This will lead to a special case of the H^∞ optimization Problem 4.1, with the emphasizes on a new geometric structure.

To begin we define the operator A(F) from $H(\Theta)$ to $H(\Theta_1)$ associated with a symbol F in $H^\infty(E', E_1')$ by

$$A(F) = P_{H(\Theta_1)}M_F | H(\Theta) \; . \tag{7.1}$$

The following result plays a basic role in describing $I(S(\Theta), S(\Theta_1))$.

7.1 LEMMA. *Let Θ and Θ_1 be two pure inner functions in $H^\infty(E, E^{'})$ and $H^\infty(E_1, E_1^{'})$, respectively. Let F be an analytic function in $H^\infty(E^{'}, E_1^{'})$. The operator A(F) is in $I(S(\Theta), S(\Theta_1))$ if and only if*

$$F\Theta = \Theta_1 G \quad \textit{(for some G in } H^\infty(E, E_1)) \ . \tag{7.2}$$

Moreover, in this case G is uniquely determined by F.

PROOF. If (7.2) holds, then

$$\overline{F\Theta H^2(E)} \subseteq \Theta_1 H^2(E_1) \ .$$

From this it readily follows that F_+ is a lifting of A(F), that is,

$$P_1 F_+ = A(F)P \ . \tag{7.3}$$

where P is the orthogonal projection onto $H(\Theta)$ and P_1 is the orthogonal projection onto $H(\Theta_1)$. Since $U(\Theta)$ and $U(\Theta_1)$ the minimal isometric dilations of $S(\Theta)$ and $S(\Theta_1)$ are unilateral shifts (see Theorem 5.1), we have $S(\Theta_1)A(F) = A(F)S(\Theta)$ by the lifting property (7.3). Hence A(F) is in $I(S(\Theta), S(\Theta_1))$. Conversely if A(F) is in $I(T, T_1)$ where $T = S(\Theta)$ and $T_1 = S(\Theta_1)$, then

$$P_1 M_F(I - P)S^{'}P = P_1 S_1 M_F P - P_1 M_F TP = (T_1 A(F) - A(F)T)P_, = 0$$

where S_1 is multiplication by e^{it} on $H^2(E_1^{'})$. Therefore

$$M_F(I - P)S^{'}PH^2(E^{'}) \subseteq \Theta_1 H^2(E_1) \ .$$

However, using (5.11) and the fact that $S^{'}$ is the minimal isometric dilation of T, this implies that

$$M_F\Theta E = M_F\overline{(S^{'} - T)H(\Theta)} = M_F\overline{(I - P)S^{'}PH^2(E^{'})} \subseteq \Theta_1 H^2(E_1) \ .$$

By applying S_1^n on the left we readily see that $F\Theta H^2(E)$ is contained in $\Theta_1 H^2(E_1)$. Corollary 2.2 shows that (7.2) holds. The uniqueness follows from the fact that Θ_{1+} is an isometry. This completes the proof.

Let $F^{'}$ in $H^\infty(E^{'}, E_1^{'})$ be another symbol. We claim that

$$A(F) = A(F^{'}) \quad \text{if and only if} \quad (F - F^{'}) \in \Theta_1 H^\infty(E^{'}, E_1) \ . \tag{7.4}$$

To prove this result it is sufficient to show that A(F) = 0 if and only if $F = \Theta_1\Psi$ where Ψ is in $H^\infty(E^{'}, E_1)$. Clearly $F = \Theta_1\Psi$ implies A(F) = 0. If A(F) = 0, then (7.1) places Fh in $\Theta_1 H^2(E_1)$ for all h in $H(\Theta)$. Since $H(\Theta)$ is cyclic for $S^{'}$ this yields

$\overline{FH^2(E')} \subseteq \Theta_1 H^2(E_1)$. By Corollary 2.2 of the Beurling-Lax-Halmos Theorem, $F = \Theta_1 \Psi$ where Ψ is in $H^\infty(E', E_1)$. Therefore (7.4) holds. Finally, we are ready for the main result of this section.

7.2 THEOREM. *Assume that Θ in $H^\infty(E, E')$ and Θ_1 in $H^\infty(E_1, E_1')$ are both purely inner. The operator A is in $I(S(\Theta), S(\Theta_1))$ if and only if $A = A(F)$ where F is a symbol in $H^\infty(E', E_1')$ satisfying (7.2). In this case one can always choose an optimal symbol F_* in $H^\infty(E', E_1')$ such that*

$$A = A(F_*) \quad and \quad \|A\| = \|F_*\|_\infty \quad and \quad F_* \Theta = \Theta_1 G_* \tag{7.5}$$

for some G_* is in $H^\infty(E, E_1)$. Moreover, there exists a Ψ_* in $H^\infty(E', E_1)$ solving the corresponding H^∞ optimization Problem 4.1 with $F = Q$ and μ is the norm of $A(F)$, that is,

$$\|A(F)\| = \|F + \Theta_1 \Psi_*\|_\infty = \inf\|F + \Theta_1 H^\infty(E', E_1)\|_\infty . \tag{7.6}$$

The equation $F_* = F + \Theta_1 \Psi_*$ provides a one to one correspondence between the optimal solutions F_* and Ψ_*.

PROOF. The first part follows from the previous lemma. So assume that A is in $I(S(\Theta), S(\Theta_1))$. By the commutant lifting theorem there exists an operator B in $I(U(\Theta), U(\Theta_1))$ such that

$$A = P_1 B \,|\, H(\Theta), \ \|A\| = \|B\| \quad and \quad \overline{B\Theta H^2(E)} \subseteq \Theta_1 H^2(E_1) . \tag{7.7}$$

Since Θ and Θ_1 are both inner, Theorem 5.1 implies that $U(\Theta)$ and $U(\Theta_1)$ are both unilateral shifts. Therefore B commutes with the unilateral shift. Theorem 1.1 shows that $B = M_{F_*} \,|\, H^2(E')$ where F_* is in $H^\infty(E', E_1')$ and $\|B\| = \|F_*\|_\infty$. This and (7.7) prove the first two equations in (7.5).

The last condition in (7.7) with $B = M_{F_*}$ implies that

$$\overline{F_* \Theta H^2(E)} \subseteq \Theta_1 H^2(E_1) .$$

Corollary 2.2 gives $F_* \Theta = \Theta_1 G_*$ for some G_* in $H^\infty(E, E_1)$. Therefore (7.5) holds.

Now assume that $A = A(F)$ for a specified symbol F. Equation (7.4) shows that $A(F) = A(F + \Theta_1 \Psi)$ where Ψ is arbitrary in $H^\infty(E', E_1)$. Thus

$$\|A\| = \|A(F + \Theta_1 \Psi)\| \le \|F + \Theta_1 \Psi\|_\infty . \tag{7.8}$$

Since $A(F) = A = A(F_*)$ for an adequate symbol F_* in $H^\infty(E', E_1')$, equation (7.4) also shows that $F_* = F + \Theta_1 \Psi_*$ for some Ψ_* in $H^\infty(E', E_1)$. This (7.8) and (7.5) easily leads to (7.6). The proof is now complete.

The previous proof also gives

7.3 COROLLARY. *Let* A *be a contraction in* $I(S(\Theta), S(\Theta_1))$ *where* Θ *and* Θ_1 *are purely inner. There is a one to one correspondence between the set of all symbols* F_* *for* A *satisfying* $\|F_*\|_\infty \leq 1$, *where* $F_*\Theta = \Theta_1 G_*$ *for an analytic* G_* *and the set of all contractive intertwining liftings* B *of* A; *this correspondence is given by* $B = M_{F_*} \,|\, H^2(E')$.

Obviously the optimization problem in (7.6) is a special case of the H^∞ optimization Problem 4.1. Moreover, Theorem 4.2 shows that $\|\Gamma_1(F)\| = \|A(F)\|$. In fact one can prove this directly. Using $F\Theta = \Theta_1 G$ we have

$$\Gamma_1(F) = P_1 \Gamma_1(F)P = A(F)P$$

where $\Gamma_1(F)$ is defined according to (4.3) with $H' = L^2(E_1') \ominus \Theta_1 H^2(E_1)$. This readily gives $\|\Gamma_1(F)\| = \|A(F)\|$. Now one can obtain (7.6) as a corollary of Theorem 4.2.

An inner and $*$-inner function is referred to as *inner from both sides* or *two sided inner*. An function Θ is inner from both sides if and only if Θ is analytic in D and $\Theta(e^{it})$ is a.e. unitary. The following allows us to determine when A(F) is one to one, or when the range of A(F) is dense in $H(\Theta_1)$.

7.4 PROPOSITION. *Let* Θ *in* $H^\infty(E, E')$ *and* Θ_1 *in* $H^\infty(E_1, E_1')$ *both be inner from both sides. Let* $A = A(F)$ *be in* $I(S(\Theta), S(\Theta_1))$ *with symbol* F *in* $H^\infty(E', E_1')$ *satisfying* $F\Theta = \Theta_1 G$ *where* G *is in* $H^\infty(E, E_1)$. *Then*
(i) The closed range of A(F) *equals* $H(\Theta_1)$ *if and only if* F *and* Θ_1 *are left prime.*
(ii) The operator A(F) *is one to one if and only if* $\tilde{\Theta}$ *and* $\tilde{G}$ *are left prime.*

PROOF. Using $F\Theta = \Theta_1 G$ we have

$$FH^2(E') \bigvee \Theta_1 H^2(E_1) = \overline{A(F)H(\Theta)} \oplus \Theta_1 H^2(E_1) \ .$$

Therefore $\overline{A(F)H(\Theta)} = H(\Theta_1)$ if and only if $FH^2(E') \bigvee \Theta_1 H^2(E_1) = H^2(E_1')$, or equivalently, F and Θ_1 are left prime (see the end of Section 2). This proves Part (i).

For (ii) we see that Proposition 5.6 with (5.16) and $\tilde{\Theta}\tilde{F} = \tilde{G}\tilde{\Theta}_1$ yield

$$W(\tilde{\Theta})A(F)^* = W(\tilde{\Theta})PM_{F^*}\,|\,H(\Theta_1) =$$

$$\tag{7.9}$$

$$\tilde{P}V^*M_{\tilde{\Theta}}M_{\tilde{F}}J\,|\,H(\Theta_1) = \tilde{P}M_{\tilde{G}}W(\tilde{\Theta}_1)\,|\,H(\Theta_1) = A(\tilde{G})W(\tilde{\Theta}_1)\,|\,H(\Theta_1)\,,$$

where $\tilde{P}$ is the orthogonal projection onto $H(\tilde{\Theta})$. Since the operator W is unitary the above shows that A(F) is one to one if and only if $A(\tilde{G})$ is onto. This and Part (i) proves Part (ii), which completes the proof.

By virtue of the commutativity of H^∞, Theorem 7.2 and Proposition 7.4 can be condensed in the following way.

7.5 PROPOSITION. *Let Θ be a nonconstant inner function in H^∞. The commutant $\{S(\Theta)\}'$ of $S(\Theta)$ (or equivalently, $I(S(\Theta)) = I(S(\Theta), S(\Theta))$) consists of operators of the form*

$$A = A(f) = P_{H(\Theta)}\, M_f\,|\,H(\Theta) \quad where \quad f \in H^\infty\,. \tag{7.10}$$

One can always choose an optimal f_* in H^∞ such that

$$A = A(f_*) \quad and \quad \|A\| = \|f_*\|_\infty\,. \tag{7.11}$$

For any function f in H^∞ such that $A = A(f)$, then there exists an optimal ψ_* in H^∞ solving the following optimization problem:

$$\|f + \Theta\psi_*\|_\infty = \inf\|f + \Theta H^\infty\|_\infty = \|A\|\,. \tag{7.12}$$

There is a one to one correspondence between the optimal solutions f_* and ψ_* by $f_* = f + \Theta\psi_*$. Finally, the following are equivalent.
(i) The operator A(f) is one to one.
(ii) The closed range of A(f) is $H(\Theta)$.
(iii) The functions Θ and f are prime.

7.6 REMARK. Let Θ be a nonconstant inner function in H^∞. Since Θ is pure Theorem 5.1 shows that $S(\Theta)$ is a completely non-unitary contraction and the bilateral shift $\hat{S}$ on L^2 is the minimal unitary dilation of $S(\Theta)$. Applying our functional calculus in Theorem VI.6.5 we see that $A(f) = f(S(\Theta))$. Therefore A is in $I(S(\Theta))$ if and only if $A = f(S(\Theta))$ where f is in H^∞. In other words $I(S(\Theta)) = \{f(S(\Theta)): f \in H^\infty\}$.

If Θ is an inner function, then the minimal isometric dilation of $S(\Theta)$ is a unilateral shift S. Since S is $*$-stable, $S(\Theta)$ is $*$-stable and $S(\Theta)^*$ has no eigenvalues of modulus one. If Θ is a nonconstant inner function in H^∞, then Remark 5.2 shows that the defect indices for $S(\Theta)$ are both one. In the next chapter we will show that if Θ is a Blaschke product of order n, then the dimension of $H(\Theta)$ is n. In this case $S(\Theta)$ is a finite dimensional operator. This with Theorem 7.2, Corollary 7.3 and the uniqueness of contractive intertwining liftings in Corollary VII.5.4 proves the following result.

7.7 COROLLARY. *Let* Θ *be a finite Blaschke produce in* H^∞. *Let* A *be a contraction commuting with* $S(\Theta)$. *There exists a unique* f_* *in* H_I^∞ *satisfying* $A = A(f_*)$ *if and only if the norm of* A *is one.*

To complete this section we present the following result which is the dual to Corollary 3.5 for Hankel matrices.

7.8 COROLLARY. *Let* Θ *be a nontrivial inner function in* H^∞, *and let* $A = A(f)$ *be a contraction commuting with* $S(\Theta)$ *where* f *is in* H^∞. *Assume that* x *is a nonzero vector in* $H(\Theta)$ *satisfying* $\|Ax\| = \|x\|$ *and set* $y = Ax$. *Then* y/x *is an inner function and* $f_* \equiv y/x$ *is the only function in* H^∞ *satisfying (7.11). Moreover,* $\psi_* = \overline{\Theta}(y/x - f)$ *a.e. in* $[0, 2\pi)$ *is the only function in* H^∞ *solving the optimization problem in (7.12).*

PROOF. Notice that x is in $\ker D_A$. Recall that any nonzero x in H^2 is cyclic for the bilateral shift $\hat{S} = V$ on L^2 and $\hat{S}$ is the minimal unitary dilation of $S(\Theta)$. By Theorem VII.4.1 there exists a unique contractive intertwining lifting B of $S(\Theta)$ and this B is an isometry. Since B is an isometry commuting with the unilateral shift, Theorem 1.1 shows that $B = f_{*+}$ where f_* is inner. Using $\|y\| = \|Ax\| = \|x\|$ we have $\|P_{H(\Theta)}f_*x\| = \|f_*x\|$. Hence

$$y = Ax = P_{H(\Theta)}f_*x = f_*x .$$

Because any nonzero x in H^2 is a.e. nonzero this implies that $f_* = y/x$. Now the last part of the corollary follows from Theorem 7.2 and Corollary 7.3. This completes the proof.

By dividing through by the appropriate constant one can use the previous corollary to compute the optimal f_* when $\|Ax\| = \|A\|\,\|x\|$ for some nonzero x. Finally, we note that one can modify the alternate proof of Corollary 3.5 to obtain a proof of Corollary 7.8 which does not use the commutant lifting theorem. The details are left to the reader.

8. THE DOUGLAS-SHAPIRO-SHIELDS FACTORIZATION

In this section we will present the Douglas-Shapiro-Shields factorization for certain Q in $L^\infty(E, E')$. Then using this factorization we will develop some further relationships between the Hankel operators $\Gamma(Q)$ and $A(F)$.

First we will present some supplementary properties concerning the Hankel operator $\Gamma(Q)$ in Section 3. By a slight abuse of notation below we will also use e^{it} to denote the function $t \to e^{it}$ for t in $[0, 2\pi)$ as well as the function $e^{it} \to e^{it}$ for $|e^{it}| = 1$.

8.1 LEMMA. *Let $\Gamma(Q)$ be a Hankel operator with symbol Q in $L^\infty(E, E')$ (see (3.6)) and let Θ be a two sided inner function in $H^\infty(F, E')$. Then*

$$\overline{\Gamma(Q)H^2(E)} \subseteq H(\Theta) \overset{\triangle}{=} H^2(E') \ominus \Theta H^2(F) \tag{8.1}$$

if and only if Q admits a factorization of the form

$$Q = e^{-it}\Theta G^* \quad \text{a.e. in } [0, 2\pi) \text{ (for some G in $H^\infty(F, E)$)} . \tag{8.2}$$

PROOF. It is easy to see that the following equivalences hold:

$$\overline{\Gamma(Q)H^2(E)} \subseteq H(\Theta) <=> \Theta H^2(F) \text{ is orthogonal to } \Gamma(Q)H^2(E) <=>$$

$$e^{int}Q^*\Theta F \text{ is orthogonal to } JH^2(E) \quad \text{(for all $n \geq 0$)} <=> \tag{8.3}$$

$$Q^*\Theta F \text{ is orthogonal to } JH^2(E) <=> Q^*\Theta \text{ is in } e^{it}H^\infty(F, E) <=>$$

$$Q = e^{-it}\Theta G^* \text{ a.e. in } [0, 2\pi) \quad \text{(for some $G \in H^\infty(F, E)$)} .$$

This proves the lemma.

The following plays an important role in applications. It explains what happens if there is equality in (8.1).

8.2 THEOREM. *Let Q be in $L^\infty(E, E')$ and Θ be a two sided inner function in $H^\infty(F, E')$. Then*

$$\overline{\Gamma(Q)H^2(E)} = H(\Theta) , \tag{8.4}$$

if and only if $Q = e^{-it}\Theta G^*$ a.e. in $[0, 2\pi)$, where G is in $H^\infty(F, E)$, and Θ and G are prime on the right. Furthermore, this factorization of Q is unique up to a constant unitary factor: If $Q = e^{-it}\Theta_1 G_1^*$ is a factorization, where Θ_1 is two sided inner in $H^\infty(F_1, E')$ and G_1 is in $H^\infty(F_1, E)$, and Θ_1 and G_1 are prime on the right, then $\Theta_1\phi = \Theta$ and $G_1\phi = G$ where ϕ is a unitary constant (in z) mapping F onto F_1.

PROOF. Assume that (8.4) holds. Lemma 8.1 implies that Q admits a factorization of the form (8.2). Let Θ_2 be a *-inner right factor to both Θ and G. By definition $\Theta = \Theta_1\Theta_2$ where Θ_2 is *-inner. Because Θ is a two sided inner, both Θ_1 and Θ_2 are two sided inner. Using $G = G_1\Theta_2$ and (8.2) we have $Q = e^{-it}\Theta_1 G_1^*$. By Lemma

8.1, equation (8.4) and Corollary 2.2

$$H(\Theta_1) \subseteq H(\Theta) = \overline{\Gamma(Q)H^2(E)} \subseteq H(\Theta_1) .$$

Thus $H(\Theta_1) = H(\Theta)$. This and the uniqueness of the Beurling-Lax-Halmos Theorem implies that $\Theta = \Theta_1$ up to a unitary constant on the right. So Θ_2 is a unitary constant. Therefore Θ and G are right prime.

Now assume that $Q = e^{-it}\Theta G^*$ where Θ and G are right prime. Notice that the closed range of $\Gamma(Q)$ is an invariant subspace for the backward shift S'^*. By the Beurling-Lax-Halmos Theorem there exists an inner function Θ_1 such that

$$H(\Theta_1) = \overline{\Gamma(Q)H^2(E)} .$$

Lemma 8.1 implies that

$$H(\Theta_1) = \overline{\Gamma(Q)H^2(E)} \subseteq H(\Theta) . \tag{8.5}$$

Corollary 2.2 of the Beurling-Lax-Halmos Theorem implies that $\Theta = \Theta_1\Theta_2$ where Θ_2 is inner. Since Θ is two sided inner Θ_1 and Θ_2 are two sided inner. Equation (8.5) and the part of this theorem we have just proven, shows that Q admits a factorization of the form $Q = e^{-it}\Theta_1 G_1^*$ where Θ_1 and G_1 are right prime and G_1 is in $H^\infty(F_1, E)$. Thus

$$e^{-it}\Theta G^* = Q = e^{-it}\Theta_1 G_1^* = e^{-it}\Theta_1\Theta_2\Theta_2^* G_1^* = e^{-it}\Theta\Theta_2^* G_1^* .$$

This implies that $G = G_1\Theta_2$. Recall that $\Theta = \Theta_1\Theta_2$ and G are right prime. Therefore Θ_2 is a unitary constant and there is equality in (8.5), that is (8.4) holds. This proves the first statement in the theorem.

To complete the proof assume that $Q = e^{-it}\Theta G^* = e^{-it}\Theta_1 G_1^*$ are two factorizations where Θ (respectively Θ_1) is an inner from both sides function in $H^\infty(F, E')$ (respectively $H^\infty(F_1, E')$) and G is in $H^\infty(F, E)$ (respectively G_1 is in $H^\infty(F_1, E)$). Moreover, we assume that Θ and G, as well as Θ_1 and G_1, are right prime. Then from the established part of this theorem we obtain

$$H(\Theta) = \overline{\Gamma(Q)H^2(E)} = H(\Theta_1) .$$

By the uniqueness property in the Beurling-Lax-Halmos Theorem, $\Theta = \Theta_1\phi$ where ϕ is a unitary constant. Hence $\Theta_1 G_1^* = \Theta_1\phi G^*$. So $G = G_1\phi$ which completes the proof.

The factorization $Q = e^{-it}\Theta G^*$ where Θ is two sided inner and G is in $H^\infty(F, E)$, and Θ and G are right prime is known as the *Douglas-Shapiro-Shields factorization of Q*. If Q is a rational function in $L^\infty(E, E')$, then it is easy to show that Q admits a factorization of the form $Q = e^{-it}m G^*$ where m is a Blaschke product and G is in

$H^\infty(E', E)$. From this it readily follows that Q admits a Douglas-Shapiro-Shields factorization. It is emphasized that not all functions Q in $L^\infty(E, E')$ admit a Douglas-Shapiro-Shields factorization. This follows from the following fact.

8.3 COROLLARY. *Assume that E' is finite dimensional and Q is in $L^\infty(E, E')$. Let H be the closed range of $\Gamma(Q)$ and T the contraction on H defined by $T^* = S'^* | H$. Then Q admits a Douglas-Shapiro-Shields factorization if and only if the defect indices δ_T and δ_{T^*} of T are equal. In particular, if $E = E' = \mathbb{C}^1$, then Q admits a Douglas-Shapiro-Shields factorization if and only if P_+Q is not cyclic for S'^*.*

PROOF. As noted in the proof of Theorem 8.2, H is an invariant subspace for S'^*. Therefore $H = H(\Theta)$ for some inner function Θ in $H^\infty(F, E')$. Now let Θ_0 in $H^\infty(F_0, E'_0)$ be the purely contractive part of Θ, and Θ_u in $H^\infty(F_u, E'_u)$ be its unitary part (see Theorem 5.3). This implies that $H(\Theta) = H(\Theta_0)$ and $T = S(\Theta_0)$. So if $\delta_T = \delta_{T^*}$, then Remark 5.2 shows that $\dim F_0 = \delta_T = \delta_{T^*} = \dim E'_0$. But $\Theta_u | (F \ominus F_0)$ is a unitary constant from $F \ominus F_0$ onto $E' \ominus E'_0$. It follows that $\dim F = \dim E'$. Thus Θ is two sided inner. On the other hand if Θ is two sided inner, then its pure part is also two sided inner. By Remark 5.2 again $\delta_T = \delta_{T^*}$. Therefore Θ is two sided inner if and only if $\delta_T = \delta_{T^*}$. Now the remaining part of the first statement of the corollary follows from Theorem 8.2. The last statement follows from the fact that H equals the closed linear space generated by $\{S'^{*n}P_+QE : n \geq 0\}$. This completes the proof.

Since $Q = e^z$ is cyclic for the backward shift on H^2 (see [Ha 5]), it follows that e^z does not admit a Douglas-Shapiro-Shields factorization. (For an indepth study of the Douglas-Shapiro-Shields factorization see [DoSS] and [Fu 3].) To complete this section we will use the Douglas-Shapiro-Shields factorization to relate the operator $A(F)$ in (7.1) whose symbol F satisfies (7.2) to a certain Hankel operator. This begins with the following result.

8.4 THEOREM. *Let Θ and Θ_1 be two inner from both sides functions in $H^\infty(E, E')$ and $H^\infty(E_1, E'_1)$, respectively. If $P = P_{H(\Theta)}$ and $\tilde{P}_1 = P_{H(\tilde{\Theta}_1)}$, then*

$$\Gamma(e^{-it}\Theta G^*) = \Gamma(e^{-it}F^*\Theta_1) = A(F)^* W(\Theta_1)\tilde{P}_1 , \qquad (8.6)$$

$$\Gamma(e^{-it}\tilde{G}^*\tilde{\Theta}) = \Gamma(e^{-it}\tilde{\Theta}_1\tilde{F}^*) = W(\tilde{\Theta}_1)A(F)P \qquad (8.7)$$

where $A(F)$ is the operator in (7.1) with symbol F in $H^\infty(E', E'_1)$ and G the analytic function in $H^\infty(E, E_1)$, satisfying $F\Theta = \Theta_1 G$.

PROOF. Using Lemma 8.1 along with $\Gamma(Q)^* = \Gamma(\tilde{Q})$ we see that the range of $\Gamma(e^{-it}F^*\Theta_1)^*$ is contained in $H(\tilde{\Theta}_1)$. (Recall that $\tilde{Q}(e^{it}) = Q(e^{-it})^*$.) This and the definition of $W(\Theta)$ in (5.16) yields

$$\Gamma(e^{-it}F^*\Theta_1) = \Gamma(e^{-it}F^*\Theta_1)\tilde{P}_1 = P_+F^*W(\Theta_1)\tilde{P}_1 = A(F)^*W(\Theta_1)\tilde{P}_1 \ . \tag{8.8}$$

The last equality follow from the fact that F_+ is a lifting of $A(F)$ in $I(S(\Theta), S(\Theta_1))$, that is, $A(F)^* = F_+^* \,|\, H(\Theta_1)$ (see the proof of Lemma 7.1). Equation (8.8) and $F^* = \Theta G^* \Theta_1^*$ gives (8.6). Equation (8.6) with $\Gamma(Q)^* = \Gamma(\tilde{Q})$ and the relation $W(\tilde{\Theta}_1)^* = W(\Theta_1)$ yields (8.7). This completes the proof.

Recall that (see Theorem (3.3)) the H^∞ optimization Problem 3.1 associated with a Hankel matrix $\Gamma(Q)$ (whose symbol Q is now in $L^\infty(E_1, E')$) is

$$\|\Gamma(Q)\| = \|Q_*\|_\infty = \|Q + K_*\|_\infty = \inf\|Q + K_0^\infty(E_1, E')\|_\infty \ . \tag{8.9}$$

Here $Q_* = Q + K_*$ where K_* in $K_0^\infty(E_1, E')$ is an optimal solution to (8.9). Recall that (see Theorem 7.2) the optimization problem associated with the operator $A(F)$ is

$$\|A(F)\| = \|F_*\|_\infty = \|F + \Theta_1 \Psi_*\|_\infty = \inf\|F + \Theta_1 H^\infty(E', E_1)\|_\infty \ . \tag{8.10}$$

Here $F_* = F + \Theta_1\Psi_*$ with Ψ_* in $H^\infty(E', E_1)$ is an optimal solution to (8.10).

Using $Q = e^{-it}F^*\Theta_1$ in (8.9) we have

$$\|\Gamma(Q)\| = \|e^{-it}F^*\Theta_1 + K_*\|_\infty = \|F^* + e^{it}K_*\Theta_1^*\|_\infty = \|F + e^{-it}\Theta_1 K_*^*\|_\infty \ . \tag{8.11}$$

By (8.10)

$$\|A(F)\| = \|F + \Theta_1\Psi_*\|_\infty = \|e^{-it}F^*\Theta_1 + e^{-it}\Psi_*^*\|_\infty \ . \tag{8.12}$$

Theorem 8.4 shows that $\|A(F)\| = \|\Gamma(Q)\|$. So comparing (8.11) with (8.12) we obtain the following result.

8.5 COROLLARY. *Let Θ and Θ_1 be two inner from both sides functions in $H^\infty(E, E')$ and $H^\infty(E_1, E_1')$, respectively. Let $A(F)$ be the operator in (7.1) with symbol F in $H^\infty(E', E_1')$ satisfying (7.2) and $Q = e^{-it}F^*\Theta_1$. Then there is a one to one correspondence between the set of all optimal solutions $\{Q_* = e^{-it}F^*\Theta_1 + K_*\}$ to the Hankel H^∞ optimization problem in (8.9), and the set of all optimal solutions $\{F_* = F + \Theta_1\Psi_*\}$ to the $A(F)$ optimization problem in (8.10). This correspondence is given by $\Psi_* = e^{-it}K_*^*$. Moreover, Q_* and F_* are related by*

$$Q_* = e^{-it}(F_*)^*\Theta_1 \ . \tag{8.13}$$

9. CHARACTERIZING $I(S(\Theta), S(\Theta_1))$

In this section we will use the commutant lifting theorem to describe the set of all operators in $I(S(\Theta), S(\Theta_1))$ when Θ and Θ_1 are purely contractive analytic functions.

The following is a generalization of Theorem 7.2. We proved Theorem 7.2 separately because this special form is very important in applications, and it provides another exercise in using the commutant lifting theorem.

9.1 THEOREM. *Let* $\Theta \in H^\infty(E, E')$ *and* $\Theta_1 \in H^\infty(E_1, E_1')$ *be purely contractive. Then any* $A \in I(S(\Theta), S(\Theta_1))$ *is of the form*

$$A(u \oplus v) = P_{H(\Theta_1)}[(Fu) \oplus (Bu + Cv)] \qquad (u \oplus v \in H(\Theta)), \tag{9.1}$$

where

$$F \in H^\infty(E', E_1'), \qquad B \in L^\infty(E', E_1), \qquad C \in L^\infty(E, E_1) \tag{9.2}$$

satisfy the conditions

$$B(e^{it}) = 0 \quad \text{if} \quad \Delta_{\Theta_1}(e^{it}) = 0 \tag{9.3a}$$

$$C(e^{it}) = 0 \quad \text{if} \quad \Delta_{\Theta_1}(e^{it}) = 0 \quad \text{or} \quad \Delta_{\Theta}(e^{it}) = 0 \tag{9.3b}$$

$$\left\| \begin{bmatrix} F(e^{it}) & 0 \\ B(e^{it}) & C(e^{it}) \end{bmatrix} \right\| \leq \|A\| \quad \text{a.e.} \quad \text{in} \quad [0, 2\pi), \tag{9.4}$$

and

$$F\Theta = \Theta_1 G \quad \text{and} \quad B\Theta + C\Delta_\Theta = \Delta_{\Theta_1} G \tag{9.5}$$

for some $G \in H^\infty(E, E_1)$.

PROOF. We let $K = H^2(E') \oplus L^2(E)$, $K_1 = H^2(E_1') \oplus L^2(E_1)$ and denote by S_1' and V_1 multiplication by e^{it} on $H^2(E_1')$ and $L^2(E_1)$, respectively. Then $S' \oplus V$ and $S_1' \oplus V_1$ are isometric liftings (but not necessarily minimal isometric dilations) of $S(\Theta)$ and $S(\Theta_1)$, respectively (see Section 5). Applying Corollary VII.1.3 of the commutant lifting theorem to A in $I(S(\Theta), S(\Theta_1))$, we obtain an operator B_o in $I(S' \oplus V, S_1' \oplus V_1)$ such that

$$P_{H(\Theta_1)}B_o = AP_{H(\Theta)} \quad \text{and} \quad \|B_o\| = \|A\|. \tag{9.6}$$

Obviously B_o has the matrix form

$$\begin{bmatrix} B_1 & B_2 \\ B_3 & B_4 \end{bmatrix} \quad \text{where} \quad \begin{aligned} B_1 &\in I(S', S_1'), & B_2 &\in I(V, S_1') \\ B_3 &\in I(S', V_1), & B_4 &\in I(V, V_1) \end{aligned} \tag{9.7}$$

But $V^* B_2^* = B_2^* S_1'^*$ and consequently, for all h in $H^2(E_1')$ we have

$$\| B_2^* h \| = \| V^{*n} B_2^* h \| = \| B_2^* S_1'^{*n} h \| \le \| B_2^* \| \, \| S_1'^{*n} h \| \to 0 \,,$$

as $n \to \infty$; so $B_2 = 0$. Applying Theorem 1.1 to B_1, B_3 and B_4 we obtain functions F, B$'$ and C$'$ satisfying the corresponding conditions in (9.2) and

$$B_1 = F_+ \,, \qquad B_3 = M_{B'} \,| H^2(E') \qquad \text{and} \qquad B_4 = M_{C'} \,.$$

We set

$$\hat{B} = P_{K(\Theta_1)} B_o \,| K(\Theta) \,, \tag{9.8}$$

and

$$B(e^{it}) = 0 \ \text{if} \ \Delta_{\Theta_1}(e^{it}) = 0 \,, \quad C(e^{it}) = 0 \ \text{if} \ \Delta_{\Theta_1}(e^{it}) = 0 \ \text{or} \ \Delta_{\Theta}(e^{it}) = 0$$

$$= B'(e^{it}) \ \text{otherwise} \,, \qquad\qquad = C'(e^{it}) \ \text{otherwise} \,. \tag{9.9}$$

Then B and C satisfy (9.3) and

$$\hat{B}(u \oplus v) = (Fu) \oplus (Bu + Cv) \qquad (u \oplus v \in K(\Theta)) \,. \tag{9.10}$$

Now (9.1) follows directly from (9.6) to (9.10).

Equation (9.9) implies that

$$\left\| \begin{bmatrix} F(e^{it}) & 0 \\ B(e^{it}) & C(e^{it}) \end{bmatrix} \right\| \le \left\| \begin{bmatrix} F(e^{it}) & 0 \\ B'(e^{it}) & C'(e^{it}) \end{bmatrix} \right\|$$

a.e. in $[0, 2\pi)$. This along with (9.7) and $\| B_o \| = \| A \|$ yields (9.4). Finally, the lifting property in (9.6), and (9.8) imply that

$$\hat{B}(K(\Theta) \ominus H(\Theta)) \subseteq K(\Theta_1) \ominus H(\Theta_1) \,, \quad \text{that is,}$$

$$\hat{B}(\Theta w \oplus \Delta_\Theta w) \in \{ \Theta_1 w_1 \oplus \Delta_{\Theta_1} w_1 : \ w_1 \in H^2(E_1) \} \tag{9.11}$$

for all w in $H^2(E)$. From (9.11) it follows that we can define an operator A_1 mapping $H^2(E)$ into $H^2(E_1)$ such that

$$\hat{B}(\Theta w \oplus \Delta_\Theta w) = \Theta_1 A_1 w \oplus \Delta_{\Theta_1} A_1 w \quad (w \in H^2(E)) \,. \tag{9.12}$$

Obviously A_1 is in $I(S, S_1)$ where S_1 is the unilateral shift on $H^2(E_1)$. By Part c of

Theorem 1.1, there exists a G in $H^\infty(E, E_1)$ such that $A_1 = G_+$. Using this in (9.12) along with (9.10) yields (9.5). This completes the proof.

IX.10. NOTES AND COMMENTS

The first two sections on multiplication operators and the Beurling-Lax-Halmos theorem are classical. We refer the reader to [DuS], [Fi], [Ha 5],[RosR 3] and [Sz.-NF 10] for further details and historical comments. We only mention that the Beurling-Lax-Halmos theorem is due to Beurling [Beu] in the scalar setting, to Lax [La 1,2] in the matrix case and to Halmos [Ha 2] in the infinite dimensional setting. The scalar Nehari H^∞ optimization problem was first considered and solved in [Ne]. The theory of its matrix and operator version is due to Adamjan, Arov and Krein [AAK 4]. Independently the connection between the Nehari problem, Hankel operators and the commutant lifting theorem was discovered by Page [Pag 2]. The connection between the four block problem and the commutant lifting theorem is due to Feintuch and Francis [FeFr 2]. Our presentation in Section 3 and 4 follows these last two papers. Sections 5 and 6 are basically borrowed from [Sz.-N 10] and [Sz.-NF 10]. Proposition 5.6 is taken from [BL]. The special form of Proposition 5.6 in Remark 5.7 has also been used in linear system theory [Fu 3]. Theorem 6.4 is an explicit identification which was used in [Fo 5]. For other useful model theories for contractions see [BK], [Br 2,4], [BrR 1,2] and [NiV]. The results in Sections 7 and 9 on the functional representation of intertwining operators follows from the papers [Sz.-NF 11,12]. The Douglas–Shapiro–Shields factorization is due to Douglas–Shapiro and Shields in the scalar case [DoSS] and Furhmann [Fu 3] in the operator setting. Our presentation of these results follows that in [Fu 3]. The book [Fu 3] also contains many important applications of the Douglas–Shapiro–Shields factorization to linear systems theory. In connection with this chapter see [BGR], [BK], [Ber], [Bu], [DursSz.-N], [Frz 3,4], [Ha 5], [Helt 1], [Ho], [LaP], [Ni], [RosR 3], [Rota], [Sz.-N 10], [Sz.-NF 10] and [Won].

CHAPTER X

SOME CLASSICAL INTERPOLATION PROBLEMS

In this chapter we will use the commutant lifting theorem with the functional model S(m) to solve the Carathéodory, Nevanlinna-Pick, and Hermite-Fejér interpolation problems. Besides this we will present the natural connection between Hermite-Fejér interpolation theory and contractive Hankel operators with rational symbols. In particular, we will solve the Nehari H^∞ optimization problem for rational symbols by Hermite-Fejér interpolation. Also we will show how the Hankel operator in the Nehari optimization problem can be used in Hermite-Fejér interpolation. Furthermore we will present state space realization methods for solving the Nehari H^∞ optimization problem with rational symbols. Finally, we present a Schur-Cohn test based on Hankel operators, and give a geometric interpretation of the norm of Hankel operators with rigid rational symbols.

1. THE EIGENSPACE FOR S(m)

In this section we will present several useful lemmas concerning S(m) operators when m is a finite Blaschke product.

To begin let m be a finite Blaschke product of the form

$$m = \prod_{i=1}^{n} B_i^{n_i} \quad \text{where} \quad B_i = B_{\alpha_i} = \frac{z-\alpha_i}{1-\overline{\alpha_i}z} \,. \tag{1.1}$$

The order of m is $d = \sum n_i$. It is assumed that d is finite. Throughout $\alpha_1, \alpha_2, ..., \alpha_n$ are n distinct complex numbers in D and $x_{i,j}$ are the functions in H^2 defined by

$$x_{i,j} = \frac{z^j}{(1-\overline{\alpha_i}z)^{j+1}} \quad (\text{for } 1 \le i \le n \text{ and } 0 \le j < n_i) \,. \tag{1.2}$$

In (1.1) and (1.2) z is viewed as the polynomial function $z \to z$ for z in $\overline{D}$. These functions and the following lemma will play a basic role in solving several interpolation problems.

1.1 LEMMA. *Let* m *be the finite Blaschke product of order* d *in (1.1). The functions*

$$\{x_{i,j}: \ 1 \le i \le n \text{ and } 0 \le j < n_i\} \tag{1.3}$$

form a (not necessarily orthonormal) basis for $H(m)$. *Furthermore, the functions* $x_{i,j}$ *in (1.3) form a generalized eigenvector chain for* $S(m)^*$, *that is,*

$$\begin{aligned}(S(m)^* - \bar{\alpha}_i)x_{i,j} &= x_{i,j-1} \ \text{ if } \ 0 < j < n_i \\ &= 0 \quad \text{ if } \ j = 0 .\end{aligned} \tag{1.4}$$

In particular, the zeros $\bar{\alpha}_1, \bar{\alpha}_2, \ldots, \bar{\alpha}_n$ *of* $\tilde{m}$ *are the eigenvalues of* $S(m)^*$.

PROOF. Let S be the unilateral shift on H^2 and the functions x_k in H^2 defined by

$$x_k = \frac{z^k}{(1 - \bar{\alpha}z)^{k+1}} \tag{1.5}$$

where α is in D and $k \ge 0$ is an arbitrary integer. Using

$$S^*f = \frac{f - f(0)}{z} \quad (f \in H^2) \tag{1.6}$$

it is easy to verify that $\{x_k\}_0^\infty$ forms a generalized eigenvector chain for the backward shift S^*, that is,

$$\begin{aligned}(S^* - \bar{\alpha})x_k &= x_{k-1} \ \text{ if } \ k > 0 \\ &= 0 \quad \text{ if } \ k = 0\end{aligned} \tag{1.7}$$

Next we will show that x_k is in $H(B_\alpha^{k+1})$ for all $k \ge 0$. To do this it is sufficient to show that $\bar{B}_\alpha^{k+1} x_k$ is orthogonal to H^2 for all $k \ge 0$. To this end assume that $|z| = 1$; then

$$\bar{B}_\alpha x_0 = \frac{\bar{z} - \bar{\alpha}}{1 - \alpha\bar{z}} \frac{1}{1 - \bar{\alpha}z} = \frac{\bar{z}}{1 - \alpha\bar{z}} \tag{1.8}$$

is orthogonal to H^2. Thus x_0 is in $H(B_\alpha)$. From (1.5) and (1.8)

$$\bar{B}_\alpha^{k+1} x_k = z^k(\bar{B}_\alpha x_0)^{k+1} = \frac{\bar{z}}{(1 - \alpha\bar{z})^{k+1}}$$

is orthogonal to H^2. Therefore x_k is in $H(B_\alpha^{k+1})$ for all $k \ge 0$. Clearly $B_i^{n_i}$ divides m. Corollary IX.2.2 of the Beurling-Lax-Halmos Theorem implies that $H(B_i^{n_i})$ is contained in $H(m)$. From this and the previous discussion it readily follows that

$$\{x_{i,j}: \ 1 \le i \le n \text{ and } 0 \le j < n_i\} \subsetneq H(m) . \tag{1.9}$$

Recall that $S(m)^* = S^* | H(m)$. Therefore the $\{x_{i,j}\}_j$ in (1.9) form a generalized eigenvector chain for $S(m)^*$. This proves (1.4). Clearly the set of vectors in (1.9) are linearly independent. To complete the proof we show that $\dim H(m) = d$ where $d = \sum n_i$. To this end let g be any polynomial in z. Then g can be put in the form

$$g = q[\prod_{i=1}^{n} (z-\alpha_i)^{n_i}] + r(z)$$

where q, r are polynomial in z and $\deg r < d$. Hence $g = mh + r$ for some h in H^2. This implies

$$H(m) = \{P_{H(m)}g: \ g \text{ is a polynomial}\}^- = \{P_{H(m)}r: \ r \text{ is a polynomial of } \deg r < d\} .$$

Therefore $\dim H(m) \le d$ and the proof is complete.

The eigenvectors for S^* are $(1-\bar{\alpha}z)^{-1}$ where $\alpha \in D$. Since in the following power series,

$$f = \sum_{0}^{\infty} f_n z^n \quad \text{and} \quad \frac{1}{1-\bar{\alpha}z} = \sum_{0}^{\infty} \bar{\alpha}^n z^n$$

the functions $\{z^i\}_0^\infty$ form an orthonormal basis for H^2, we have

$$f(\alpha) = (f, (1-\bar{\alpha}z)^{-1}) \quad (f \in H^2 \text{ and } \alpha \in D) \tag{1.10}$$

where the inner product is obviously taken in H^2. Because of property (1.10) the function $(1-\bar{\alpha}z)^{-1}$ of α and z in D is often referred to as the *reproducing kernel* of H^2.

The generalized eigenvectors $z^n(1-\bar{\alpha}z)^{-(n+1)}$ for the backward shift S^* can be used to reproduce the derivatives of a function in H^2, that is,

$$\frac{f^{(n)}(\alpha)}{n!} = (f, z^n(1-\bar{\alpha}z)^{-(n+1)}) \quad (f \in H^2 \text{ and } \alpha \in D \text{ and } n \ge 0) . \tag{1.11}$$

This follows as in (1.10) or by using the Cauchy integral formula:

$$(f, z^n(1-\bar{\alpha}z)^{n+1}) = \frac{1}{2\pi} \int_{0}^{2\pi} \frac{fe^{-int}dt}{(1-\alpha e^{-it})^{n+1}} = \frac{1}{2\pi i} \int \frac{fdz}{(z-\alpha)^{n+1}} = \frac{f^{(n)}(\alpha)}{n!} .$$

Proposition IX.5.6 shows that $WS(\tilde{m})^* = S(m)W$ where $W = W(m)$ is the unitary operator mapping $H(\tilde{m})$ into $H(m)$ defined by $W = V^* M_m J$ and V is the bilateral shift on L^2. Setting $y_{i,j} = W\tilde{x}_{i,j}$ where $x_{i,j}$ is defined in (1.2) yields

$$y_{i,j} = \left[\frac{(z-\alpha_i)^{n_i-j-1}}{(1-\bar{\alpha}_i z)^{n_i}}\right]\left[\prod_{\substack{k=1\\k\neq i}}^{n} B_k^{n_k}\right] \quad \text{for } (1 \leq i \leq n \quad \text{and} \quad 0 \leq j < n_i) \qquad (1.12)$$

where $\alpha_1, ..., \alpha_n$ are in D. Now Lemma 1.1 and Proposition IX.5.6 readily gives the following result.

1.2 COROLLARY. *Let* m *be the finite Blaschke product of order* d *in (1.1). The functions*

$$\{y_{i,j} \colon 1 \leq i \leq n \quad \text{and} \quad 0 \leq j < n_i\} \qquad (1.13)$$

form a (not necessarily orthogonal) basis for $H(m)$. *Furthermore, the functions* $y_{i,j}$ *in (1.12) form a generalized eigenvector chain for* $S(m)$, *that is,*

$$\begin{aligned}(S(m)-\alpha_i)y_{i,j} &= y_{i,j-1} \quad \text{if } 0 < j < n_i\\ &= 0 \qquad \text{if } j = 0\end{aligned} \qquad (1.14)$$

In particular, the zeros of m *are the eigenvalues of* $S(m)$.

1.3 COROLLARY. *Let* S *be the unilateral shift on* H^2. *Then* H *is a finite dimensional invariant subspace for the backward shift* S^* *if and only if* $H = H(m)$ *where* m *is a finite Blaschke product.*

PROOF. Let $\{x_k\}_0^n$ be a generalized eigenvector chain for S^*, that is, (1.7) holds for some α in D. Using $S^*x_0 = \bar{\alpha}x_0$ along with (1.6) we have

$$\bar{\alpha}x_0 = S^*x_0 = z^{-1}(x_0-x_0(0)).$$

Solving for x_0 this implies that

$$x_0 = \frac{x_0(0)}{(1-\bar{\alpha}z)}.$$

Since x_0 is a nontrivial eigenvector x_0 is nonzero. So $x_0(0)$ is nonzero. Without loss of generality we assume that $x_0(0) = 1$, or equivalently, $x_0 = (1-\bar{\alpha}z)^{-1}$. By (1.6) and (1.7)

$$x_{k-1} \overset{\Delta}{=} (S^*-\bar{\alpha})x_k = z^{-1}(x_k-x_k(0)) - \bar{\alpha}x_k.$$

Solving for x_k produces the following difference equation

$$x_k = \frac{zx_{k-1} + x_k(0)}{(1 - \bar{\alpha}z)} \quad \text{where} \quad x_0 = \frac{1}{1 - \bar{\alpha}z}.$$

This readily implies that

$$\bigvee_0^n \{x_i\} = \bigvee_0^n \left\{ \frac{z^i}{(1 - \bar{\alpha}z)^{i+1}} \right\}. \tag{1.15}$$

Now assume that H is a finite dimensional invariant subspace for S^*. This implies that $S^*|H$ is a finite dimensional operator on H. Therefore there exists a set of generalized eigenvector chains for S^* which span H. Equation (1.15) along with Lemma 1.1 show that $H = H(m)$ where m is a finite Blaschke product. The other half is an obvious consequence of Lemma 1.1. This completes the proof.

Using partial fraction expansion Corollary 1.3 with Lemma 1.1 easily gives the following result.

1.4 COROLLARY. *Let S be the unilateral shift on* H^2. *Then H is a finite dimensional invariant subspace for* S^* *if and only if*

$$H = \bigvee_0^n \{z^i/q(z)\} \tag{1.16}$$

where $q(z)$ *is a fixed polynomial of degree* k *with no zero in* $\bar{D}$ *and* $k \leq n + 1$. *In this case* $H = H(m)$ *where m is the Blaschke product given by*

$$m = (z^{(n-k+1)}q^{\#}(z))/q(z) \quad and \quad q^{\#} = J_{k+1}q$$

is the reverse polynomial of q *(see Section I.2).*

Let m be a finite Blaschke produce of order d written under the form

$$m = \prod_{j=1}^d B_j \quad (\text{where } B_j = B_{\alpha_j}) \tag{1.17}$$

and each zero of m is repeated according to its multiplicity. For example, the first Blaschke factor B_1 is repeated n_1 times and so on. Notice that this Blaschke product is precisely the same Blaschke product in (1.1). Consider the sequence of functions $\{\phi_j\}$ defined by

$$\phi_j \stackrel{\Delta}{=} d_j(1 - \bar{\alpha}_j z)^{-1} B_{j-1} B_{j-2} \cdots B_1 \quad (1 \leq j \leq d) \tag{1.18}$$

where $\phi_1 \stackrel{\Delta}{=} d_1(1 - \bar{\alpha}_1 z)^{-1}$ and d_j is the positive square root of $1 - |\alpha_j|^2$. Notice that $\bar{B}_j(1 - \bar{\alpha}_j z)^{-1}$ is orthogonal to H^2. This implies that ϕ_j is orthogonal to mH^2, or equivalently, ϕ_j is in $H(m)$. Using the reproducing property of $(1-\bar{\alpha}_j z)^{-1}$ in (1.10) we see that the norm of ϕ_j is one. Because $\bar{B}_j(1-\bar{\alpha}_j z)^{-1}$ is orthogonal to H^2 it also follows that $\{\phi_j\}_1^d$ is an orthonormal set in $H(m)$. By Lemma 1.1 the dimension of $H(m)$ is d. Therefore $\{\phi_j\}_1^d$ is an orthonormal basis for $H(m)$. This sets the stage for the following matrix representation of $S(m)$ due to Ptak and Young [PtY 2].

1.5 THEOREM. *Let* m *be the Blaschke product of order d in (1.17). Then* $\{\phi_j\}_1^d$ *is an orthonormal basis for* $H(m)$. *Moreover,* $S(m)$ *is unitarily equivalent to the lower triangular matrix* M *on* $\mathbf{C}^d$ *defined by*

$$M = \begin{bmatrix} \alpha_1 & 0 & 0 & 0 & 0 & 0 & \cdots \\ d_1 d_2 & \alpha_2 & 0 & 0 & 0 & 0 & \cdots \\ -d_1\bar{\alpha}_2 d_3 & d_2 d_3 & \alpha_3 & 0 & 0 & 0 & \cdots \\ d_1\bar{\alpha}_2\bar{\alpha}_3 d_4 & -d_2\bar{\alpha}_3 d_4 & d_3 d_4 & \alpha_4 & 0 & 0 & \cdots \\ -d_1\bar{\alpha}_2\bar{\alpha}_3\bar{\alpha}_4 d_5 & d_2\bar{\alpha}_3\bar{\alpha}_4 d_5 & -d_3\bar{\alpha}_4 d_5 & d_4 d_5 & \alpha_5 & 0 & \cdots \\ d_1\bar{\alpha}_2\bar{\alpha}_3\bar{\alpha}_4\bar{\alpha}_5 d_6 & -d_2\bar{\alpha}_3\bar{\alpha}_4\bar{\alpha}_5 d_6 & d_3\bar{\alpha}_4\bar{\alpha}_5 d_6 & -d_4\bar{\alpha}_5 d_6 & d_5 d_6 & \alpha_6 & \cdots \\ \vdots & \vdots & \vdots & \vdots & \vdots & \vdots & \ddots \end{bmatrix} \quad (1.19)$$

PROOF. To prove this result we obtain the matrix representation for $S(m)^* = S^* | H(m)$ with respect to $\{\phi_j\}_1^d$. Then M is the adjoint of this matrix. To this end notice that (1.6) gives

$$S^* B_j = \frac{d_j^2}{1-\bar{\alpha}_j z} \quad \text{and} \quad S^*(fg) = (S^* f)g + f(0)(S^* g) \qquad (f, g \in H^2). \quad (1.20)$$

By recursively applying (1.20) with (1.18) and using the fact that $(1-\bar{\alpha}z)^{-1}$ is an eigenvector for S^* we have

$$S^* \phi_j = \bar{\alpha}_j d_j (1-\bar{\alpha}_j z)^{-1} B_{j-1} B_{j-2} \cdots B_1 + d_j S^*(B_{j-1} B_{j-2} \cdots B_1) =$$

$$\bar{\alpha}_j \phi_j + d_j d_{j-1}^2 (1-\bar{\alpha}_{j-1} z)^{-1} B_{j-2} B_{j-3} \cdots B_1 - d_j \alpha_{j-1} S^*(B_{j-2} B_{j-3} \cdots B_1) =$$

$$\bar{\alpha}_j \phi_j + d_j d_{j-1} \phi_{j-1} - d_j \alpha_{j-1} d_{j-2} \phi_{j-2} + d_j \alpha_{j-1} \alpha_{j-2} S^*(B_{j-3} B_{j-4} \cdots B_1) = \cdots =$$

$$\bar{\alpha}_j \phi_j + d_j d_{j-1} \phi_{j-1} + d_j \sum_{i=1}^{j-2} (-1)^{j-1-i} \alpha_{j-1} \alpha_{j-2} \cdots \alpha_{i+1} d_i \phi_i .$$

Since $S^* \phi_j$ corresponds to the j-th column in the matrix representation M^* of $S^* | H(m)$, the previous equation shows that M^* is the adjoint of the matrix M in (1.19). This completes the proof.

Let us note that if $m = z^n$, then M in (1.19) reduces to the n by n lower shift T_n in (VIII.1.3) and (I.4.1).

We say that m in H^∞ is a *minimal function* for a completely non-unitary contraction T on H if m is the "smallest" inner function satisfying $m(T) = 0$. By smallest we mean that if m_1 is any other inner function satisfying $m_1(T) = 0$, then m divides m_1 in H^∞. By Section IX.2 two inner functions divide one another if and only if they are equal up to a multiplicative constant of modulus one. Therefore the minimal function is unique up to a constant of modulus one. If T has a minimal function, then this function is denoted by m_T. If p is the minimal polynomial of degree n for a finite dimensional completely non-unitary contraction T, then $m_T = p/p^\#$ where $p^\# = J_{n+1}p$ is the reverse polynomial for p (see Section I.2). We emphasize that not all completely non-unitary contractions have minimal functions; take for instance any unilateral shift. The following lemma provides us with a useful class of contractions with minimal functions.

1.6 THEOREM. *A contraction T on H is $*$-stable and $\delta_T = \delta_{T^*} = 1$ if and only if T is unitarily equivalent to S(m) where m is a nontrivial inner function in H^∞. In this case $m = m_T$ is the minimal function for T.*

PROOF. The first statement follows from Theorem IX.6.1. In fact m coincides with the characteristic function Θ_T of T. Also, if m is a nontrivial inner function, then Theorem IX.5.1 and Remark IX.5.2 show that S(m) is $*$-stable and both of the defect indices are one. For the converse we will present a direct proof by using the ideas in the last part of Sections VI.4 and IX.6. Now assume that T is $*$-stable and $\delta_T = \delta_{T^*} = 1$. For h in H notice that

$$\sum_{0}^{n} \|D_{T^*} T^{*i} h\|^2 = \sum_{0}^{n} (\|T^{*i}h\|^2 - \|T^{*(i+1)}h\|^2) = \|h\|^2 - \|T^{*(n+1)}h\|^2 .$$

Since T is $*$-stable this implies that

$$\|h\|^2 = \sum_{0}^{\infty} \|D_{T^*} T^{*i} h\|^2 \qquad (h \in H) .$$

Let α be any unitary operator mapping D_{T^*} onto $\mathbb{C}^1$. The previous equation shows that

$$Wh \overset{\Delta}{=} \alpha D_{T^*} (I - zT^*)^{-1} h \qquad (h \in H) \tag{1.21}$$

is an isometry from H into H^2. Using (1.6) we have $S^*W = WT^*$. Therefore WH is invariant for the backward shift S^* and $S^*|WH$ is unitarily equivalent to T^*. If $WH = H^2$, then T^* is unitary equivalent to the backward shift S^*. This contradicts the fact that $\delta_T = 1$. If WH is strictly contained in H^2, the Beurling-Lax-Halmos theorem shows that there exists an inner function m in H^∞ satisfying $WH = H(m)$. This implies that T^* is unitarily equivalent to $S(m)^*$, or equivalently, T is unitarily equivalent to $S(m)$.

To complete the proof we must show that m is the minimal function for $S(m)$. It is clear that $m(S(m)) = 0$. If m_1 is any inner function satisfying $m_1(S(m)) = 0$, then our functional calcalus in Section VI.6 gives

$$P_{H(m)} m_1 h = 0 \qquad (1.22)$$

for all h in $H(m)$. Since as multiplication operators m_1 commutes with m this implies that (1.22) holds for all h in H^2. Thus $m_1 H^2$ is contained in mH^2. By Corollary IX.2.2 of the Beurling-Lax-Halmos theorem m divides m_1. Therefore m is the minimal function for $S(m)$, or equivalently, for T. This completes the proof.

Let T be a $*$-stable contraction on a space H of dimension $d < \infty$, and assume that $\delta_T = 1$. On a finite dimensional space Lemma V.2.1 implies that $\delta_T = \delta_{T^*}$. In particular, $\delta_{T^*} = \delta_T = 1$. Let p be the minimal polynomial for T and assume that the degree of p is n. Because the eigenvalues of T are in D, all the zeros of p are also in D. If $p^\# = J_{n+1} p$ is the reverse polynomial for p, then $m_1 \overset{\Delta}{=} p/p^\#$ is a Blaschke product of order $n \le d$. Since $m_1(T) = 0$ the previous theorem shows that $m_1(S(m_T)) = 0$. By the definition of a minimal function m_T divides m_1. Recall that T is unitarily equivalent to $S(m_T)$. By Lemma 1.1 the order of m_T is d. However, m_T divides m_1 and the order of m_1 is $n \le d$. Therefore $n = d$ and $m = m_T$. Summing up this analysis and using the previous theorem we obtain the following result.

1.7 COROLLARY. *Let p be the minimal polynomial for a contraction T on a space H of dimension* $d < \infty$, *and let* $p^\# = J_{d+1} p$ *be the reverse polynomial for p. The eigenvalues of T are in D and* $\delta_T = 1$ *if and only if T is unitarily equivalent to* $S(m)$ *where m is a Blaschke product of order d. In this case the degree of p is d and* $m = m_T = p/p^\#$. *In particular, T is unitarily equivalent to* $S(m_T)$ *and the minimal function* $m_T = p/p^\#$.

The previous corollary shows that any finite dimensional contraction T on H whose eigenvalues are in D and $\delta_T = 1$ is unitarily equivalent to the matrix M in Theorem 1.5, where $m = p/p^\#$ and p is the minimal polynomial of T. In other words M in (1.19) provides an explicit matrix representation of all finite dimensional contractions whose eigenvalues are in D and $\delta_T = 1$. It is emphasized that these contractions played an important role in the Schur-Cohn test in Section V.5, and also in many uniqueness results concerning the commutant lifting theorem; see Corollary VII.5.4, Corollary VIII.1.4 and

Theorem VIII.2.6.

We complete this section with the following result which shows that there is a one to one correspondence between the invariant subspaces for S(m) and the inner divisors m_1 of m.

1.8 THEOREM. *Let* m *be a nontrivial inner function in* H^∞. *Then* H_1 *is an invariant subspace for* S(m) *if and only if* $H_1 = m_2 H(m_1)$ *where* $m = m_2 m_1$ *and* m_2 *and* m_1 *are two inner functions in* H^∞. *In this case* $H(m) = m_2 H(m_1) \oplus H(m_2)$ *and* $S(m)|H_1$ *on* H_1 *is unitarily equivalent to* $S(m_1)$. *Moreover, if* H_1 *is nonzero, then the minimal isometric dilation of* $S(m)|H_1$ *is* $S|m_2 H^2$ *on* $m_2 H^2$.

PROOF. Recall that $S^*|H(m) = S(m)^*$. So H_1 is an invariant subspace for S(m) if and only if $H_2 = H(m) \ominus H_1$ is an invariant subspace for the backward shift S^*. By the Beurling-Lax-Halmos theorem there exists an inner function m_2 in H^∞ satisfying $H_2 = H(m_2)$. Corollary IX.2.2 of the Beurling-Lax-Halmos Theorem shows that $m = m_2 m_1$ where m_1 is inner. Therefore H_2 is an invariant subspace for $S(m)^*$ if and only if $H_2 = H(m_2)$ where m_2 is an inner divisor of m. Using $m = m_2 m_1$ with $m_2 m_1 H^2 \subseteq m_2 H^2$ we have

$$H(m) = \mathrm{H}^2 \ominus m_2 m_1 \mathrm{H}^2 = (\mathrm{H}^2 \ominus m_2 \mathrm{H}^2) \oplus (m_2 \mathrm{H}^2 \ominus m_2 m_1 \mathrm{H}^2) = H(m_2) \oplus m_2 H(m_1) \,.$$

In other words H_1 is an invariant subspace for S(m) if and only if $H_1 = m_2 H(m_1)$ where $m = m_2 m_1$ and m_2 and m_1 are both inner.

Let H_1 be the previous invariant subspace for S(m). Using the decomposition $H(m) = m_2 H(m_1) \oplus H(m_2)$ with $h = m_2 h_1$ in $H_1 = m_2 H(m_1)$ we have

$$S(m)h = P_{H(m)} S m_2 h_1 = P_{H(m_2)} S m_2 h_1 + P_{m_2 H(m_1)} m_2 S h_1 = m_2 S(m_1) h_1 \,.$$

Let W be the unitary operator from $H(m_1)$ onto $m_2 H(m_1)$ defined by $W h_1 = m_2 h_1$ where h_1 is in $H(m_1)$. The previous equation shows that $S(m)|H_1 = W S(m_1) W^*$. In other words $S(m)|H_1$ is unitarily equivalent to $S(m_1)$. This proves the first two statements of the theorem.

To complete the proof assume that H_1 is a nonzero invariant subspace. This implies that m_1 is a nontrivial inner function in H^∞. Recall that the unilateral shift S is the minimal isometric dilation for $S(m_1)$; see Theorem IX.5.1. The fact that $H(m_1)$ is cyclic for S gives

$$H_{1+} \triangleq \bigvee_{o}^{\infty} S^n H_1 = \bigvee_{o}^{\infty} S^n m_2 H(m_1) = m_2 H^2 .$$

Since S is the minimal isometric dilation for S(m) and $H_1 = m_2 H(m_1)$ is an invariant subspace for S(m), it follows that $S \,|\, H_{1+} = S \,|\, m_2 H^2$ is the minimal isometric dilation for $S(m) \,|\, H_1$. This completes the proof.

If m is the Blaschke product of order d in (1.17), then the previous theorem shows that

$$H(m) = H(B_1) \oplus B_1 H(B_2) \oplus B_1 B_2 H(B_3) \oplus$$

$$(1.23)$$

$$\oplus B_1 B_2 B_3 H(B_4) \oplus \cdots \oplus (B_1 B_2 \cdots B_{d-1}) H(B_d) .$$

Since $d_j (1 - \overline{\alpha}_j z)^{-1}$ is a unit vector which spans $H(B_j)$, equation (1.23) also shows that $\{\phi_j\}_1^d$ in (1.18) is a complete orthonormal basis for $H(m)$ and the dimension of $H(m)$ is d.

The following is the operator version of the previous theorem.

1.9 THEOREM. *Let Θ be an inner function in $H^\infty(E, E')$. Then H_1 is an invariant subspace for $S(\Theta)$ if and only if $H_1 = \Theta_2 H(\Theta_1)$ where $\Theta = \Theta_2 \Theta_1$ and Θ_2 in $H^\infty(F, E')$ and Θ_1 in $H^\infty(E, F)$ are two inner functions. In this case $H(\Theta) = \Theta_2 H(\Theta_1) \oplus H(\Theta_2)$ and $S(\Theta) \,|\, H_1$ is unitarily equivalent to $S(\Theta_1)$. Moreover, if Θ_{1o} in $H^\infty(E_o, F_o)$ is the pure part of Θ_1, then the minimal isometric dilation of $S(\Theta) \,|\, H_1$ is $S' \,|\, \Theta_2 H^2(F_o)$. Finally, $S' \,|\, \Theta_2 H^2(F)$ is an isometric lifting of $S(\Theta) \,|\, H_1$. (Recall that S' is the unilateral shift on $H^2(E')$).*

PROOF. Except for the last statement the proof of this theorem is identical to the proof of the previous theorem. To complete the proof it remains to show that $S' \,|\, \Theta_2 H^2(F)$ is an isometric lifting of $S(\Theta) \,|\, H_1$. To this end notice that if $F_u \triangleq F \ominus F_o$ is the space corresponding to the unitary part of Θ_1, then $\Theta_2 H^2(F_u)$ is an invariant subspace for S'. Therefore $\Theta_2 H^2(F_o)$ is a reducing subspace for $S' \,|\, \Theta_2 H^2(F)$. Since $S' \,|\, \Theta_2 H^2(F_o)$ is the minimal isometric dilation of $S(\Theta) \,|\, H_1$ this implies that $S' \,|\, \Theta_2 H^2(F)$ is an isometric lifting of $S(\Theta) \,|\, H_1$. This completes the proof.

2. TANGENTIAL CARATHEODORY INTERPOLATION REVISITED

In this section we will use our functional models with the commutant lifting theorem to solve the tangential Carathéodory interpolation Problem VIII.1.5.

From now on in this chapter V and V' are the bilateral shifts on $L^2(E)$ and $L^2(E')$, respectively. Also as before S and S' are the unilateral shifts on $H^2(E)$ and $H^2(E')$, respectively. Finally, I is the identity on E and I' is the identity on E'. We begin with the following lemma which will play an important role in our approach for solving several classical interpolation problems.

　　2.1 LEMMA. *Let* m *be an inner function in* H^∞ *and R an invariant subspace for* $T = S(mI)$. *Then* A *is a contraction in* $I\,(T\,|\,R,\,S(mI'))$ *if and only if*

$$A = P_{H'}F_+\,|\,R \qquad (H' = H(mI')) \tag{2.1}$$

where F *is a contractive analytic function in* $H^\infty(E,\,E')$.

　　PROOF. Notice that $mH^2(E)$ and $mH^2(E')$ are invariant subspaces for V and V', respectively. The subspaces

$$M = L^2(E)\ominus mH^2(E) = K_0^2(E)\oplus H(mI)$$

$$\tag{2.2}$$

$$\text{and}\quad M' = L^2(E')\ominus mH^2(E') = K_0^2(E')\oplus H(mI')$$

are invariant for V^* and V'^*, respectively. So the operators V_* on M and V'_* on M' defined by compressing V to M and V' to M', that is, $V_* = P_M V\,|\,M$ and $V'_* = P_{M'}V'\,|\,M'$ are both co-isometries. Using (2.2)

$$V_* h = P_M Vh = P_{K_0^2}Vh + P_H Vh = S(m)h$$

for all h in $H = H(mI)$. Therefore H is an invariant subspace for V_* and V_* is a co-isometric extension of $S(m)$. By a similar argument it follows that H' is an invariant subspace for V'_* and V'_* is a co-isometric extension of $T' = S(mI')$. In fact it turns out that V_*^* and $V_*'^*$ are precisely the minimal isometric dilation for T^* and T'^*, respectively in Theorem VI.5.3 part (iv).

　　We claim that (H, H') is a hyperinvariant pair of subspaces for (V_*, V'_*). (A hypervariant pair of subspaces is defined in Section VII.1.) To prove this assume that C is in $I(V_*, V'_*)$. Notice that V and V' are the minimal isometric dilations of V_* and V'_* respectively. By the commutant lifting theorem there exists an operator B is $I(V, V')$ lifting C. (One can also obtain this B by using Corollary VI.2.4). By Theorem IX.1.1 there exists a G in $L^\infty(E, E')$ such that $B = M_G$. The lifting property gives $BmH^2(E) \subseteq mH^2(E')$. Since m commutes with G this implies that $BH^2(E) \subseteq H^2(E')$. Using this with (2.2) we have

$$CH = P_{M'}BH = P_{K_0^2}BH + P_{H'}BH = P_{H'}BH \subseteq H' .$$

Therefore CH is contained in H'. So by definition (H, H') is a hyperinvariant pair of subspaces for (V_*, V_*').

Now assume that A is a contraction in $I(T\,|\,R, T')$. By Corollary VII.1.6 of the commutant lifting theorem, there exists a contraction $\hat{A}$ in $I(T, T')$ extending A. In fact we can choose $\|\hat{A}\| = \|A\|$. Moreover, if B is any contractive intertwining lifting of $\hat{A}$, then $P_{H'}B\,|\,R = A$. Recall that the unilateral shifts S and S' are the minimal isometric dilations of T and T', respectively (see Theorem IX.5.1). Since $S'B = BS$ Theorem IX.1.1 shows that $B = F_+$ where F is a contractive analytic function in $H^\infty(E, E')$. Thus $A = P_{H'}F_+\,|\,R$ and (2.1) holds.

On the other hand if (2.1) holds for some contractive analytic F, then obviously $FmH^2(E)$ is contained in $mH^2(E')$. Thus

$$P_{H'}F_+ = P_{H'}F_+P_H . \tag{2.3}$$

Since S and S' are liftings of T and T' respectively, equation (2.3) implies that

$$T'P_{H'}F_+\,|\,H = P_{H'}S'F_+\,|\,H = P_{H'}F_+P_HS\,|\,H = P_{H'}F_+T .$$

Therefore $\hat{A} = P_{H'}F_+\,|\,H$ is a contraction in $I(T, T')$. Because R is invariant for T the contraction $A = P_{H'}F_+\,|\,R$ is in $I(T\,|\,R, T')$. This completes the proof.

The following lemma gives us a useful uniqueness result.

2.2 LEMMA. *Let $E = E' = C^1$ and m be a finite Blaschke product. Let R be an invariant subspace for $T = S(m)$ on $H = H(m)$ and A a contraction in $I(T\,|\,R, S(m))$. Then there is a unique contractive analytic function F in H^∞ satisfying*

$$A = P_H F_+\,|\,R \tag{2.4}$$

if and only if the norm of A is one. In this case the unique F in H_1^∞ satisfying (2.4) is given by $F = Ah/h$ where h is any nonzero vector in the kernel of D_A.

PROOF. According to Lemma 1.1 the space $H(= H')$ is finite dimensional. If the norm of A is one, there exists a nonzero vector h satisfying $\|h\| = \|Ah\|$, or equivalently, $D_A h = 0$. The previous lemma shows that there exists an F in H_1^∞ satisfying (2.4). Using $\|h\| = \|Ah\|$ we have

$$\|Ah\| = \|P_H Fh\| \le \|h\| = \|Ah\| .$$

So there is equality in the previous equation. Thus $Fh = Ah$. Since h is nonzero $h(e^{it}) \ne 0$

a.e. So we can divide by h to obtain $F = Ah/h$. Therefore there is only one contractive analytic F satisfying (2.4), namely $F = Ah/h$.

Assume that there is only one contractive analytic function F in H^∞ satisfying (2.4). We proceed by contradiction and assume that A is a strict contraction. Recall that H is a hyperinvariant subspace for V_* (see the proof of the previous lemma). Applying Corollary VII.1.6 there exists a contraction $\hat{A}$ in $I(T, T)$ extending A and satisfying $\|\hat{A}\| = \|A\| < 1$. By Corollary VII.5.4 there exists at least two different contractive intertwining liftings B_1 and B_2 of $\hat{A}$. Obviously $B_1 = F_{1+}$ and $B_2 = F_{2+}$ where F_1 and F_2 are two different functions in H_1^∞. Applying Corollary VII.1.6 again $P_H F_{1+} | R = A$ and $P_H F_{2+} | R = A$. Therefore (2.4) holds for two different F in H_1^∞. This conducts our assumptions of a unique solution. The proof is now complete.

Now we will use the commutant lifting theorem with the invariant subspace structure in Theorem 1.9 to obtain

ANOTHER PROOF OF LEMMA 2.1. To this end assume that A is a contraction in $I(T|R, T')$. By Theorem 1.9 the subspace $R = \Theta_2 H(\Theta_1)$ where $mI = \Theta_2 \Theta_1$ and Θ_1 is an inner function in $H^\infty(E, F)$ and Θ_2 is an inner function in $H^\infty(F, E)$. Theorem 1.9 also shows that $S | \Theta_2 H^2(F)$ is an isometric lifting of $T|R$. Recall that S' is the minimal isometric dilation of $S(mI)$; see Theorem IX.5.1. By virtue of Corollary VII.1.3 of the commutant lifting theorem, there exist a contractive intertwining lifting B in $I(S | \Theta_2 H^2(F), S')$ of A. The lifting property implies that

$$mH^2(E') \supseteq B[\Theta_2 H^2((F)) \ominus R] = B\Theta_2(H^2(F) \ominus H(\Theta_1)) = B\Theta_2 \Theta_1 H^2(E) = BmH^2(E),$$

or equivalently, $BmH^2(E)$ is contained in $mH^2(E')$.

Since $mH^2(E)$ is cyclic for V^* and $mH^2(E)$ is contained in $\Theta_2 H^2(F)$, the operator V is the minimal unitary extension of $S | \Theta_2 H^2(F)$. By Corollary VI.2.4 there exists a unique operator $\hat{B}$ in $I(V, V')$ extending B. Theorem IX.1.1 shows that $\hat{B} = M_F$ where F is a contraction in $L^\infty(E, E')$. Because $BmH^2(E) \subseteq mH^2(E')$ and m commutes with F, we see that $FH^2(E) \subseteq H^2(E')$. Therefore F is a contractive analytic function in $H^\infty(E, E')$. Moreover,

$$A = P_{H'} B | R = P_{H'} F_+ | R$$

which verifies (2.1). This proves half of Lemma 2.1. The other half follows by the relatively easy argument given in the first proof. This completes our alternate proof of Lemma 2.1.

As expected we will use Theorem 1.8 to obtain

ANOTHER PROOF OF LEMMA 2.2. Under the hypothesis of Lemma 2.2, Theorem 1.8 shows that $S | m_2 H^2$ is the minimal isometric dilation of $S(m) | R$ where

$R = m_2 H(m_1)$. (Here $m_1 = \Theta_1$ and $m_2 = \Theta_2$ are both inner functions in H^∞ satisfying $m = m_2 m_1$.) We claim that B is a contractive intertwining lifting of A if and only if $B = F_+ \,|\, m_2 H^2$ where F is a contractive analytic function in H^∞ satisfying (2.4). In this case B and F uniquely determine each other. If B is a contractive intertwining lifting of A, then our previous proof of Lemma 2.1 shows that B uniquely determines an F in H_1^∞ satisfying (2.4) and $B = M_F \,|\, m_2 H^2$. On the other hand if F is a contractive analytic function in H^∞ satisfying (2.4), then

$$F(m_2 H^2 \ominus R) = F m_1 m_2 H^2 \subseteq m H^2 \ .$$

This and (2.4) shows that $B = F_+ \,|\, m_2 H^2$ from $m_2 H^2$ to H^2 is a lifting of A. Obviously B is in $I(S \,|\, m_2 H^2, S)$. Thus B is a contractive intertwining lifting of A. This proves our claim. Now Lemma 2.2 directly follows from Corollary VII.5.4. This completes the proof.

2.3 REMARK. Let $E = E' = C^1$. The previous proof shows that there is a one to one correspondence through $B = F_+ \,|\, m_2 H^2$ between the set of all contractive intertwining liftings B of A and the set of all contractive analytic functions F in H^∞ satisfying (2.4).

To complete this section we will use Lemmas 2.1 and 2.2 to solve the tangential Carathéodory interpolation Problem VIII.1.5. For convenience this problem is restated in this section. To this end recall that $\hat{A}_n$ is the n by n analytic Toeplitz matrix from $l_n^2(E)$ to $l_n^2(E')$ defined by

$$\hat{A}_n = \begin{bmatrix} A_0 & 0 & 0 & \dots & 0 \\ A_1 & A_0 & 0 & \dots & 0 \\ A_2 & A_1 & A_0 & \dots & 0 \\ \vdots & \vdots & \vdots & \vdots & \vdots \\ A_{n-2} & A_{n-3} & A_{n-4} & \dots & 0 \\ A_{n-1} & A_{n-2} & A_{n-3} & \dots & A_0 \end{bmatrix} \tag{2.5}$$

where A_i is an operator from E to E' for all i. In our setting the tangential Carathéodory interpolation Problem VIII.1.5 becomes:

2.4 PROBLEM. Let $\{B_i\}_0^{n-1}$ and $\{C_i\}_0^{n-1}$ be a sequence of operators with values in $L(F, E)$ and $L(F, E')$, respectively. Find necessary and sufficient conditions for the existence of a contractive analytic function F in $H^\infty(E, E')$ satisfying

$$F(z) = A_0 + A_1 z + A_2 z^2 + \cdots + A_{n-1} z^{n-1} + 0(z^n)$$

and

$$[C_0, C_1, ..., C_{n-1}]^{tr} = \hat{A}_n [B_0, B_1, ..., B_{n-1}]^{tr} \tag{2.6}$$

where $\hat{A}_n$ is the n by n analytic Toeplitz matrix in (2.5) generated by $\{A_i\}_0^{n-1}$.

To obtain a solution to Problem 2.4 let Φ from $l_n^2(F)$ to $H^2(E)$ and Ψ from $l_n^2(F)$ to $H^2(E')$ be the operators defined by

$$\Phi = [I, zI, ..., z^{n-2}I, z^{n-1}I]\hat{B} \quad \text{and} \quad \Psi = [I', zI', ..., z^{n-2}I', z^{n-1}I']\hat{C}. \tag{2.7}$$

Here $\hat{B}(=\hat{B}_n)$ is the n by n analytic Toeplitz matrix from $l_n^2(F)$ to $l_n^2(E)$ generated by $\{B_i\}_0^{n-1}$ and $\hat{C}(=\hat{C}_n)$ is the n by n analytic Toeplitz matrix from $l_n^2(F)$ to $l_n^2(E')$ generated by $\{C_i\}_0^{n-1}$. Because $\{z^i E\}$ and $\{z^i E'\}$ are orthogonal systems in $H^2(E)$ and $H^2(E')$, respectively one can easily verify that $\Phi^*\Phi = \hat{B}^*\hat{B}$ and $\Psi^*\Psi = \hat{C}^*\hat{C}$. Let $H' = H(z^n I')$. Notice that H' equals the span of $\{z^i E' : 0 \le i < n\}$. Now assume that F is a solution to Problem 2.4. By using the power series expansion of F with (2.6) and (2.7) we have

$$P_{H'} F_+ \Phi = \Psi. \tag{2.8}$$

Since F is contractive this implies that $\|\Psi f\|^2 \le \|\Phi f\|^2$ for all f in $l_n^2(F)$, or equivalently, $\Phi^*\Phi - \Psi^*\Psi$ is positive. Recall that $\Phi^*\Phi = \hat{B}^*\hat{B}$ and $\Psi^*\Psi = \hat{C}^*\hat{C}$. This proves the first part of the following result which is precisely Theorem VIII.1.6.

2.5 THEOREM. *There exists a solution to the tangential Carathéodory interpolation Problem 2.4 if and only if* $\hat{B}^*\hat{B} - \hat{C}^*\hat{C}$ *is positive where* $\hat{B}$ *and* $\hat{C}$ *are the* n *by* n *analytic Toeplitz matrices generated by the data* $\{B_i\}_0^{n-1}$ *and* $\{C_i\}_0^{n-1}$ *respectively.*

PROOF. Notice that the closed range of Φ denoted by R is an invariant subspace for $T = S(z^n I)$. If $\hat{B}^*\hat{B} - \hat{C}^*\hat{C}$ is positive, then there exists a contraction A from R to H' satisfying $A\Phi = \Psi$. Using $T' = S(z^n I')$ we have

$$T'A\Phi = T'\Psi = [z^1 I', z^2 I', ..., z^{n-1} I', 0]\hat{C} =$$

$$[I', zI', ..., z^{n-2}I', z^{n-1}I']T_n'\hat{C} = \Psi T_n = A\Phi T_n = AT\Phi$$

where T_n' and T_n is the n by n lower shift on $l_n^2(E')$ and $l_n^2(F)$, respectively (see VIII.1.3). Thus A is a contraction in $I(T|R, T')$. By Lemma 2.1 there exists a contractive analytic function F in $H^\infty(E, E')$ satisfying $P_{H'}F_+|R = A$. Therefore (2.8) holds. By matching like coefficients of z^i in the power series expansion of (2.8) we obtain (2.6). In other words F is a solution to Problem 2.4. This completes the proof.

2.6 REMARK. The above analysis shows that there exists a solution to the tangential Carathéodory interpolation Problem 2.4 if and only if $A\Phi = \Psi$ defines a contraction from R to H'. Moreover, in this case A is in $I_1(T|R, T')$ and F is a solution to Problem 2.4 if and only if F is a contractive analytic function in $H^\infty(E, E')$ satisfying $P_{H'}F_+|R = A$.

2.7 COROLLARY. *Let* $E = E' = F = C^1$. *Then there exists a unique solution to Problem 2.4 if and only if the matrix* $\hat{B}^*\hat{B}^* - \hat{C}^*\hat{C}$ *is positive and there exists an f in* C^n *satisfying*

$$\hat{B}^*\hat{B}f = \hat{C}^*\hat{C}f \neq 0 . \tag{2.9}$$

In this case the unique solution to Problem 2.4 is given by $F = \Psi f/(\Phi f)$.

PROOF. The previous remark with Lemma 2.2 shows that there exists a unique solution to Problem 2.4 if and only if the norm of A is one. However, the norm of A is one if and only if $\hat{B}^*\hat{B} - \hat{C}^*\hat{C}$ is positive and there exists a f in C^n satisfying (2.9). This proves the first part of the Corollary. To verify the second part let f satisfy (2.9) and F be any solution to Problem 2.4, then

$$\|\Psi f\| = \|P_H F\Phi f\| \leq \|\Phi f\| = \|\Psi f\| .$$

So there is equality in the previous equation. Thus $F\Phi f = \Psi f$. Since Φf is a nonzero element in H^2 we can divide by Φf to obtain $F = \Psi f/(\Phi f)$. This completes the proof.

3. TANGENTIAL NEVANLINNA-PICK INTERPOLATION REVISITED

In this section we will use our functional model theory along with the commutant lifting theory to present another viewpoint of the tangential Nevanlinna-Pick interpolation Problem VIII.2.1.

We begin by restating this tangential Nevanlinna-Pick interpolation problem.

3.1 PROBLEM. Let $\{B_i\}_1^n$ and $\{C_i\}_1^n$ be a set of operators from F to E and from F to E', respectively. Let $\{\alpha_i\}_1^n$ be a set of n distinct complex numbers in D. Find necessary and sufficient conditions for the existence of a contractive analytic function $F(z)$ in $H^\infty(E, E')$ satisfying

$$C_i = F(\alpha_i)B_i \qquad \text{(for all } 1 \le i \le n). \tag{3.1}$$

To obtain a solution to this problem let $x_i = (1-\overline{\alpha_i}z)^{-1}$ be the eigenvector for the backward shift determined by its eigenvalue α_i. If F is any function in $H^\infty(E, E')$, we set $F_1 = JF$, that is $F_1(e^{it}) \triangleq F(e^{-it})$ for all t in $[0, 2\pi)$. We claim that

$$P_+F_1 x_i B_i f = F(\alpha_i)x_i B_i f \qquad \text{(for all } f \in F) \tag{3.2}$$

where P_+ is the orthogonal projection onto $H^2(E')$. This follows from the following calculation where we use the reproducing property (1.10) of x_i:

$$(h, P_+F_1 x_i B_i f) = (\tilde{F}h, x_i B_i f) = (\tilde{F}(\overline{\alpha_i})h_i(\overline{\alpha_i}), B_i f)_E =$$
$$(h(\overline{\alpha_i}), F(\alpha_i)B_i f)_{E'} = (h, F(\alpha_i)x_i B_i f) \qquad (f \in F \text{ and } h \in H^2(E'))$$

Now let Φ from $l_n^2(F)$ to $H^2(E)$ and Ψ from $l_n^2(F)$ to $H^2(E')$ be the operators defined by

$$\Phi \triangleq [x_1 B_1, x_2 B_2, ..., x_n B_n] \quad \text{and} \quad \Psi \triangleq [x_1 C_1, x_2 C_2, ..., x_n C_n]. \tag{3.3}$$

If F is a solution to Problem 3.1, then (3.2), (3.3) and $F(\alpha_i)B_i = C_i$ imply that $P_+F_1\Phi = \Psi$. Since F is a contractive analytic function $\|\Psi f\|^2 \le \|\Phi f\|^2$ for all f in F, or equivalently, $\Phi^*\Phi - \Psi^*\Psi$ is positive. Using $(x_i, x_j) = (1-\alpha_i\overline{\alpha_j})^{-1}$ which follows from the reproducing property (1.10), it is easy to verify that

$$\Phi^*\Phi - \Psi^*\Psi = \begin{bmatrix} \dfrac{B_1^*B_1 - C_1^*C_1}{1-\overline{\alpha}_1\alpha_1} & \cdots & \dfrac{B_1^*B_n - C_1^*C_n}{1-\alpha_n\overline{\alpha}_1} \\ \vdots & \vdots & \vdots \\ \dfrac{B_n^*B_1 - C_n^*C_1}{1-\overline{\alpha}_n\alpha_1} & \cdots & \dfrac{B_n^*B_n - C_n^*C_n}{1-\alpha_n\overline{\alpha}_n} \end{bmatrix} \tag{3.4}$$

Therefore if F is a solution to Problem 3.1, the matrix in (3.4) is positive. This proves the first part of Theorem VIII.2.2 restated here for convenience as

3.2 THEOREM. *There exists a solution to the tangential Nevanlinna-Pick interpolation Problem 3.1 if and only if the matrix* $\Phi^*\Phi - \Psi^*\Psi$ *on* $l_n^2(F)$ *in (3.4) is positive*

PROOF. Assume that $\Phi^*\Phi - \Psi^*\Psi$ is positive. There exists a contraction A from the range space $H \triangleq \overline{\Phi F}$ to $H^2(E')$ satisfying $A\Phi = \Psi$. Using the fact that the x_i's are eigenvectors for the backward shift we have

$$S^{'*}A\Phi = S^{'*}\Psi = \Psi \, \text{diag} \, [\alpha_1, \alpha_2, ..., \alpha_n] = A\Phi \, \text{diag}[\alpha_1, \alpha_2, ..., \alpha_n] = AS^{*}\Phi . \qquad (3.5)$$

Since the range of Φ is invariant for S^* this implies that $S^{'*}A = A(S^* | H)$. By Corollary VII.1.4 of the commutant lifting theorem, there exists a contraction B in $I(S^*, S^{'*})$ extending A, that is, $B | H = A$. Theorem IX.1.1 shows that $B^* = \tilde{F}_+ | H^2(E^{'})$ with a contractive analytic function F in $H^\infty(E, E^{'})$. Thus $B = P_+ M_{F_1} | H^2(E)$ where as before $F_1(e^{it}) = F(e^{-it})$ for all t in $[0, 2\pi)$. Now using (3.2) and $P_+ M_{F_1} | H = A$ along with $A\Phi = \Psi$ we have $F(\alpha_i)B_i = C_i$ for all i. This completes the proof.

3.3 COROLLARY. *Let $E = E^{'} = F = \mathbf{C}^1$. There exists a unique solution to Problem 3.1 if and only if the matrix $\Phi^* \Phi - \Psi^* \Psi$ on l_n^2 in (3.4) is positive and there exists an f in $\mathbf{C}^n$ satisfying*

$$\Phi^* \Phi f = \Psi^* \Psi f \neq 0 . \qquad (3.6)$$

In this case the unique solution F is given by $F(e^{it}) = (\Psi f/(\Phi f))(e^{-it})$ for t in $[0, 2\pi)$.

PROOF. Without loss of generality we assume that $\Phi^* \Phi - \Psi^* \Psi$ is positive. By Lemma 1.1 the range of Ψ is contained in $H(m^{'})$ where $m^{'}$ is the Blaschke product of order n whose zeros are $\bar{\alpha}_1, \bar{\alpha}_2, ..., \bar{\alpha}_n$. We also notice that the operator A defined by $A\Phi = \Psi$ as in the previous proof maps H into $H(m^{'})$. Obviously H is an invariant subspace for the backward shift. By the Beurling-Lax-Halmos Theorem $H = H(m)$ where m is an inner function in H^∞. Equation (3.5) shows that $S(m^{'})^* A = AS(m)^*$, or equivalently, A^* is a contraction $I(S(m^{'}), S(m))$.

We claim that $B^{'}$ is a contractive intertwining lifting of A^* if and only if $B^{'} = \tilde{F}_+$ where F is a solution to Problem 3.1. The proof of the previous theorem shows that a contractive intertwining lifting $B^{'}$ of A^* uniquely determines a solution F, by $B^{'} = \tilde{F}_+$ to Problem 3.1. On the other hand if F is a solution to Problem 3.1, then (3.2) implies that $P_+ F_1 x_i B_i = x_i C_i$ for $1 \leq i \leq n$. Thus $P_+ F_1 \Phi = \Psi$. If $B = P_+ M_{F_1} | H^2$, then $B | H = A$. Clearly $B^{'} = B^* = \tilde{F}_+$ is a contraction in $I(S, S)$. Therefore $B^{'}$ is a contractive intertwining lifting of A^*. This proves our claim.

The previous analysis shows that there is a one to one correspondence by $B^{'} = \tilde{F}_+$ between the set of all contractive intertwining lifting $B^{'}$ of A^* and the set of all solutions to Problem 1.1. By Corollary VII.5.4 there is a unique solution to Problem 3.1 if and only if the norm of A is one, or equivalently, $\Phi^* \Phi - \Psi^* \Psi$ is positive and there exists an f in $\mathbf{C}^n$ satisfying (3.6). This proves the first part of the corollary.

Let f be any nonzero vector satisfying (3.6) and F any solution to Problem 3.1. Then

$$\|\Psi f\| = \|P_+ F_1 \Phi f\| \le \|\Phi f\| = \|\Psi f\| .$$

So there is equality and $F_1 \Phi f = \Psi f$. Since Φf is a nonzero element in H^2 we can divide by Φf. This completes the proof.

3.4 REMARK. Let $E = E' = C^1$. The previous proof shows that there is a one to one correspondence through $B' = \tilde{F}_+$ between the set of all contractive intertwining lifting B' of A^* and the set of all solutions F to Problem 3.1.

To complete this section we will use Lemma 2.1 to obtain

ANOTHER PROOF OF THEOREM 3.2. To this end let y_i be the vectors in $H(m)$ defined by

$$y_i \overset{\Delta}{=} \frac{m}{z - \alpha_i} = (1 - \bar{\alpha}_i z)^{-1} \prod_{\substack{j=1 \\ j \ne i}}^{n} B_j \qquad (1 \le i \le n) \tag{3.7}$$

where m is the Blaschke product of order n whose zeros are $\alpha_1, \alpha_2, \cdots, \alpha_n$. Corollary 1.2 shows that $\{y_i\}_1^n$ span $H(m)$ and are the eigenvectors for $S(m)$. Notice that

$$(y_i, y_j) = (1 - \alpha_i \bar{\alpha}_j)^{-1} \tag{3.8}$$

This follows from $y_i = W(m)(1 - \alpha_i z)^{-1}$ (for $1 \le i \le n$) where $W(m)$ is the unitary operator defined in Proposition IX.5.6, or by using the reproducing property (1.10) in the following calculation:

$$(y_i, y_j) = (zy_i, zy_j) = ((1 - \bar{\alpha}_j z)^{-1}, (1 - \bar{\alpha}_i z)^{-1}) = (1 - \alpha_i \bar{\alpha}_j)^{-1} .$$

Consider the operators Φ_1 from $l_n^2(F)$ to $H^2(E)$ and Ψ_1 from $l_n^2(F)$ to $H^2(E')$ defined by

$$\Phi_1 \overset{\Delta}{=} [y_1 B_1, y_2 B_2, ..., y_n B_n] \quad \text{and} \quad \Psi_1 \overset{\Delta}{=} [y_1 C_1, y_2 C_2, ..., y_n C_n] . \tag{3.9}$$

Using $y_i = W(m)(I - \alpha_i z)^{-1}$ for $1 \le i \le n$ or (3.7) it is easy to verify that $\Phi_1^* \Phi_1 = \Phi^* \Phi$ and $\Psi_1^* \Psi_1 = \Psi^* \Psi$. In particular, $\Phi_1^* \Phi_1 - \Psi_1^* \Psi_1$ is precisely the matrix given in (3.4).

Corollary 1.2 shows that the range of Ψ_1 is contained in $H' \overset{\Delta}{=} H(mI')$. Let F be any function in $H^\infty(E, E')$. Using the fact that y_i is an eigenvector for $S(m)$ with eigenvalue α_i we have for all f in F

$$P_{H'} F y_i B_i f = \sum_0^\infty F_n (P_{H(m)} e^{int} y_i) B_i f = \sum_0^\infty F_n S(m)^n y_i B_i f = F(\alpha_i) y_i B_i f$$

where F_n is the coefficient of z^n in the power series expansion of F. Our computations make sense because the above series are convergent in $H^2(E')$. The previous equation gives

$$P_{H'} F y_i B_i f = F(\alpha_i) y_i B_i f \qquad (f \in F) . \tag{3.10}$$

If F is a solution to Problem 3.1, then (3.10) and $F(\alpha_i)B_i = C_i$ imply that $P_{H'} F \Phi_1 = \Psi_1$. Since F is a contractive analytic function $\|\Psi_1 f\|^2 \leq \|\Phi_1 f\|^2$, or equivalently, $\Phi_1^* \Phi_1 - \Psi_1^* \Psi_1$ is positive. This proves the first part of Theorem 3.2.

Assume that $\Phi_1^* \Phi_1^* - \Psi_1^* \Psi_1$ is positive. This implies that there exists a contraction A from $R \triangleq \overline{\Phi_1 l_n^2(F)}$ to $H(mI')$ satisfying $A\Phi_1 = \Psi_1$. Since the y_i's are eigenvectors for $S(m)$, the range space R is invariant for $S(mI)$ and

$$T' A\Phi_1 = T' \Psi_1 = \Psi_1 \, \text{diag} \, [\alpha_1, \alpha_2, ...,\alpha_n] = A\Phi_1 \, \text{diag}[\alpha_1, \alpha_2, ..., \alpha_n] = AT\Phi_1$$

where $T' = S(mI')$ and $T = S(mI)$. Thus A is a contraction in $I(T|R, T')$. By Lemma 2.1 there exists a contractive analytic function F in $H^\infty(E, E')$ satisfying $P_{H'} F_+ | R = A$, or equivalently, $P_{H'} F_+ \Phi_1 = \Psi_1$. This and (3.10) shows that $F(\alpha_i)B_i = C_i$ for all i. The alternative proof is now complete.

3.5 REMARK. The previous analysis shows that there exists a solution to the tangential Nevanlinna-Pick interpolation Problem 3.1 if and only if $A\Phi_1 = \Psi_1$ defines a contraction from R to $H(mI')$. Moreover, in this case A is in $I_1(T|R, T')$ and F is a solution to Problem 3.1 if and only if F is a contractive analytic function in $H^\infty(E, E')$ satisfying $P_{H'} F_+ | R = A$.

3.6 REMARK. Let $E = E' = F = C^1$. By using the previous remark and mimicking the proof of Corollary 2.7, there exists a unique solution to Problem 3.1 if and only if the matrix $\Phi^* \Phi - \Psi^* \Psi$ in (3.4) is positive and there exists an f satisfying (3.6). In this case the unique solution to Problem 3.1 is given by $F = \Psi_1 f/(\Phi_1 f)$.

4. THE TANGENTIAL HERMITE-FEJER INTERPOLATION PROBLEM

In this section we will use Lemma's 2.1 and 2.2 to solve the following tangential Hermite-Fejér interpolation problem which generalizes both the Carathéodory interpolation Problem 2.4 and the Nevanlinna-Pick interpolation Problem 3.1.

First we formulate the tangential Hermite-Fejér interpolation problem.

4.1 PROBLEM. Let $\{B_{i,j}: 1 \leq i \leq n$ and $0 \leq j < n_i\}$ and $\{C_{i,j}: 1 \leq i \leq n$ and $0 \leq j < n_i\}$ be a set of operators in $L(F, E)$ and $L(F, E')$, respectively; and let $\alpha_1, \alpha_2, ..., \alpha_n$ be n distinct complex numbers in D. Find necessary and sufficient conditions for the existence of a contractive analytic function F in $H^\infty(E, E')$ satisfying

$$
\begin{bmatrix} C_{i,0} \\ C_{i,1} \\ \vdots \\ C_{i,n_i-1} \end{bmatrix} = \begin{bmatrix} F(\alpha_i) & 0 & \cdots & 0 \\ \dfrac{F^{(1)}(\alpha_i)}{1!} & F(\alpha_i) & \cdots & 0 \\ \vdots & \vdots & \vdots & \vdots \\ \dfrac{F^{(n_i-1)}(\alpha_i)}{(n_i-1)!} & \dfrac{F^{(n_i-2)}(\alpha_i)}{(n_i-2)!} & \cdots & F(\alpha_i) \end{bmatrix} \begin{bmatrix} B_{i,0} \\ B_{i,1} \\ \vdots \\ B_{i,n_i-1} \end{bmatrix} \tag{4.1}
$$

for all $1 \le i \le n$ where $F^{(j)}(\alpha)$ is the j-th derivative of F evaluated at α.

Throughout this section m is the Blaschke product defined in (1.1). The zeros α_i of m are repeated n_i times and $d = \Sigma n_i$ is the order of m. The functions $x_{i,j}$ and $y_{i,j}$ are the generalized eigenvectors for $S(m)^*$ and $S(m)$ defined in (1.2) and (1.12), respectively. If F is in $H^\infty(E, E')$, then A(F) is the operator mapping $H = H(mI)$ into $H' = H(mI')$ defined by $A(F) = P_{H'} F_+ | H$. Since F commutes with m the function F is also the symbol of A(F) in the sense of Section IX.7, and (IX.7.2) holds for $\Theta = mI$ and $\Theta_i = mI'$. In particular, A(F) is in $I(S(mI), S(mI'))$. This sets the stage for the following basic result.

4.2 LEMMA. *Let F be a function in* $H^\infty(E, E')$ *and m the Blaschke product in (1.1). Then*

$$
A(F)y_{i,j}f = \sum_{k=0}^{j} \frac{F^{(k)}(\alpha_i)}{k!}\, y_{i,j-k}f \qquad (1 \le i \le n \text{ and } 0 \le j < n_i \text{ and } f \in E). \tag{4.2}
$$

PROOF. We claim that

$$
A(F)^* x_{i,j}g = \sum_{k=0}^{j} \frac{(F^{(k)}(\alpha_i))^*}{k!}\, x_{i,j-k}g \qquad (1 \le i \le n \text{ and } 0 \le j < n_i \text{ and } g \in E'). \tag{4.3}
$$

To verify this let h be in $H^2(E)$. The derivative reproducing property (1.11) of $x_{i,j}$ and the binomial expansion of $(Fh)^{(j)}$ yields:

$$
(h, A(F)^* x_{i,j}g) = (Fh, x_{i,j}g) = \frac{((Fh)^{(j)}(\alpha_i), g)}{j!}
$$

$$
= \sum_{k=0}^{j} \frac{(F^{(k)}(\alpha_i)h^{(j-k)}(\alpha_i), g)}{k!(j-k)!} = (h, \sum_{k=0}^{j} \frac{(F^{(k)}(\alpha_i))^*}{k!}\, x_{i,j-k}g).
$$

This gives (4.3).

A direct calculation, or by adjusting (IX.7.9) to this setting we have

$$A(F)W(mI) \,|\, H(\tilde{m}I) = W(mI')A(\tilde{F})^* \tag{4.4}$$

where the W's are the unitary operators defined in Proposition IX.5.6. Using this in (4.3) along with $y_{i,j} = W(m)\tilde{x}_{i,j}$ yields

$$A(F)y_{i,j}f = W(mI')A(\tilde{F})^*\tilde{x}_{i,j}f = W(mI')\sum_{k=0}^{j} \frac{(\tilde{F}^{(k)}(\bar{\alpha}_i))^*}{k!}\,\tilde{x}_{i,j-k}f = \sum_{k=0}^{j}\frac{F^{(k)}(\alpha_i)}{k!}\,W(m)\tilde{x}_{i,j-k}f.$$

This produces (4.2) and completes the proof.

To obtain a solution to Problem 4.1 let $\hat{B}_i$ be the n_i by n_i analytic Toeplitz matrix from $l^2_{n_i}(F)$ to $l^2_{n_i}(E)$ generated by $\{B_{i,j}: 0 \le j < n_i\}$. Let $\hat{B}$ be the d by d matrix from $l^2_d(F)$ to $l^2_d(E)$ defined by $\hat{B} = \text{diag}\,[\hat{B}_1, \hat{B}_2, ..., \hat{B}_n]$ where $d = \sum n_i$ is the order of m. In a similar manner let $\hat{C}_i$ be the n_i by n_i analytic Toeplitz matrix from $l^2_{n_i}(F)$ to $l^2_{n_i}(E')$ generated by $\{C_{i,j}: 0 \le j < n_i\}$, and $\hat{C}$ is the d by d matrix from $l^2_d(F)$ to $l^2_d(E')$ defined by $\hat{C} = \text{diag}\,[\hat{C}_1, \hat{C}_2, ..., \hat{C}_n]$. Now let Φ_2 be the operator from $l^2_d(E)$ to $H^2(E)$ defined by

$$\Phi_2 = [y(1, n_1{-}1)I, y(1, n_1{-}2)I, ..., y(1, 0)I, y(2, n_2{-}1)I,$$

$$\tag{4.5}$$

$$y(2, n_2{-}2)I, ..., y(2, 0)I, ..., y(n, n_n{-}1)I, y(n, n_n{-}2)I, ..., y(n, 0)I]$$

where $y(i,j) = y_{i,j}$. Let Ψ_2 be the operator from $l^2_d(E')$ to $H^2(E')$ defined by replacing I by I' in (4.5). Finally, let Φ from $l^2_d(F)$ to $H^2(E)$ and Ψ from $l^2_d(F)$ to $H^2(E')$ be the operators defined by $\Phi = \Phi_2\hat{B}$ and $\Psi = \Psi_2\hat{C}$. These operators reduce to the operators in (2.6) and (3.9) when $m = z^n$ and $m = \Pi B_j$ respectively. Corollary 1.2 shows that the closed range of Φ denoted by R is contained in $H = H(mI)$ and the closed range of Ψ is contained in $H' = H(mI')$.

Now assume that F is a solution to Problem 4.1. Using (4.1) and (4.2) it can be shown that $P_{H'}F_+\Phi = \Psi$. Because F is a contractive analytic function $\|\Psi f\|^2 \le \|\Phi f\|$ for all f in $l^2_d(F)$, or equivalently, $\Phi^*\Phi - \Psi^*\Psi$ is positive.

If M is a Hilbert space, then $J_{M,i}$ is the n_i by n_i lower Jordan back matrix on $l^2_{n_i}(M)$ defined by

$$J_{Mi} = \begin{bmatrix} \alpha_i I & 0 & 0 & . & . & . & . & . & 0 & 0 \\ I & \alpha_i I & 0 & . & . & . & . & . & 0 & 0 \\ 0 & I & \alpha_i I & . & . & . & . & . & 0 & 0 \\ 0 & 0 & I & . & . & . & . & . & 0 & 0 \\ 0 & 0 & 0 & . & . & . & . & . & 0 & 0 \\ \vdots & \vdots & \vdots & \vdots & \vdots & \vdots & \vdots & \vdots & \vdots & \vdots \\ 0 & 0 & 0 & . & . & . & . & . & I & \alpha_i I \end{bmatrix} \tag{4.6}$$

The d by d Jordan matrix J_M on $l_d^2(M)$ is defined by $J_M = \mathrm{diag}\,[J_{M,1}, J_{M,2}, ..., J_{M,n}]$. Using the fact that $y_{i,j}$ is a generalized eigenvector for $S(m)$ (see Corollary 1.2) with $J_E\hat{B} = \hat{B}J_F$ we have $T\Phi = \Phi_2 J_E \hat{B} = \Phi J_F$ where $T = S(mI)$. In particular, the closed range of Φ denoted by R is an invariant subspace for T.

If $\Phi^*\Phi - \Psi^*\Psi$ is positive, then there exists a contraction A from R into H' satisfying $A\Phi = \Psi$. Using again the fact that $y_{i,j}$ is a generalized eigenvector for $S(m)$ with $T' = S(mI')$ and $T = S(mI)$ we have

$$T'A\Phi = T'\Psi = \Psi_2 J_{E'}\hat{C} = \Psi J_F = A\Phi J_F = A\Phi_2 J_E \hat{B} = AT\Phi\,.$$

Thus A is a contraction in $I(T\,|\,R, T')$. By Lemma 2.1 there exists a contractive analytic function F in $H^\infty(E, E')$ satisfying $P_{H'}F_+\,|\,R = A$, or equivalently, $P_{H'}F_+\Phi = \Psi$. Using this and (4.2) and $P_{H'}F_+\Phi = \Psi$ it can be shown that (4.1) holds, that is, F is a solution to Problem 4.1.

The previous analysis shows that there is a solution to Problem 4.1 if and only if $\Phi^*\Phi - \Psi^*\Psi$ is positive. To obtain a matrix form for $\Phi^*\Phi - \Psi^*\Psi$ notice that the block Gram matrix $G = \Phi_2^*\Phi_2$ on $l_d^2(E)$ is given by

$$G = \begin{bmatrix} (y(1, n_1 - 1), y(1, n_1 - 1))I & ... & (y(1, n_1 - 1), y(n; 0))I \\ . & ... & . \\ . & ... & . \\ . & ... & . \\ (y(n, 0), y(1, n_1 - 1))I & ... & (y(n, 0), y(n, 0))I \end{bmatrix}. \tag{4.7}$$

The block Gram matrix $G' = \Psi_2^*\Psi_2$ on $l_d^2(E')$ is given by the matrix in (4.7) where I' replaces I. The derivative reproducing property of $x_{i,j}$ in (1.11) with $y_{i,j} = W\tilde{x}_{i,j}$ gives

$$(y_{i,j}, y_{k,r}) = (\tilde{x}_{i,j}, \tilde{x}_{k,r}) = \frac{\overline{x_{i,j}^{(r)}(\alpha_k)}}{r!}\,. \tag{4.8}$$

So the entries of G and G' are of the form

$$\frac{\overline{x_{i,j}^{(r)}(\alpha_k)}I}{r!} \quad \text{and} \quad \frac{\overline{x_{i,j}^{(r)}(\alpha_k)}I'}{r!} \, . \tag{4.9}$$

Therefore one can explicitly compute the entries of G and G'. Summing up our analysis proves the following main result of this section.

4.3 THEOREM. *There exists a solution to the tangential Hermite-Fejér interpolation Problem 4.1 if and only if the matrix* $\hat{B}^* G \hat{B} - \hat{C}^* G' \hat{C}$ *on* $l_d^2(F)$ *is positive.*

4.4 REMARK. The previous analysis shows that there exists a solution to Problem 4.1 if and only if $A\Phi = \Psi$ defines a contraction from R to $H' = H(mI')$. Moreover, in this case A is in $I_1(T \,|\, R, T')$ and F is a solution to Problem 4.1 if and only if F is a contractive analytic function in $H^\infty(E, E')$ satisfying $P_{H'}F_+ \,|\, R = A$.

By using the previous remark and mimicking the proof of Corollary 2.7 we obtain the following result.

4.5 COROLLARY. *Let* $E = E' = F = \mathbb{C}^1$. *Then there exists a unique solution to Problem 4.1 if and only if the matrix* $\hat{B}^* G \hat{B} - \hat{C}^* G' \hat{C}$ *on* l_d^2 *is positive and there exists an f in* $\mathbb{C}^d$ *satisfying*

$$\hat{B}^* G \hat{B} f = \hat{C}^* G \hat{C} f \neq 0 \, . \tag{4.10}$$

In this case the unique solution to Problem 4.1 is given by $F = \Psi f/(\Phi f)$.

Finally, it is noted that one can use the $\{x_{i,j}\}$ along with (4.3) to generalize the first proof of Theorem 3.2 to give another proof of Theorem 4.3. The details are left to the reader.

5. CLASSICAL HERMITE-FEJÉR INTERPOLATION

In this section we study the classical Hermite-Fejér interpolation problem for operator valued functions as a particular case of the commutant lifting theorem. We also connect this problem to the theory of Hankel operators with rational symbols.

We begin with a formulation of this classical interpolation problem.

5.1 PROBLEM. Given the sequence $\{A_{i,j}: 1 \leq i \leq n \text{ and } 0 \leq j < n_i\}$ of operator in $L(E, E')$ and a set of distinct complex numbers $\alpha_1, \alpha_2, ..., \alpha_n$ in D, find necessary and sufficient conditions for the existence of a contractive analytic function F in $H^\infty(E, E')$

satisfying

$$\frac{F^{(j)}(\alpha_i)}{j!} = A_{i,j} \qquad (\text{for } 1 \le i \le n \text{ and } 0 \le j < n_i) . \tag{5.1}$$

By setting $B_{i,0} = I$ and $B_{i,j} = 0$ for $1 \le i \le n$ and $1 \le j < n_i$ and by setting $C_{i,j} = A_{i,j}$ for all i,j we see that the classical Hermite-Fejér interpolation Problem 5.1 is a special case of the tangential Hermite-Fejér interpolation Problem 4.1. In this section we will again use the commutant lifting theorem along with some of the results in Section 1 to present a detailed study of the easier Problem 5.1.

To this end we recall the classical fact that one can always construct a polynomial $P(z)$ of degree at most $d-1$ where $d = \sum n_i$ satisfying (5.1). However, this P is not necessarily contractive, that is, P may not be in $H_1^\infty(E, E')$. To construct such a P let $p_i(z)$ be the polynomial of order $d-n_i$ defined by

$$p_i(z) \overset{\Delta}{=} \prod_{\substack{k=1 \\ k \ne i}}^{n} \left[\frac{z-\alpha_k}{\alpha_i-\alpha_k} \right]^{n_k} , \tag{5.2}$$

then $p_i(\alpha_k) = 0$ if $k \ne i$. Now consider the polynomial $P(z)$ of degree $d-1$ defined by

$$P(z) = \sum_{i=1}^{n} [A'_{i,0} + A'_{i,1}(z-\alpha_i) + A'_{i,2}(z-\alpha_i)^2 + \cdots + A'_{i,n_i-1}(z-\alpha_i)^{n_i-1}]p_i(z) \tag{5.3}$$

where the operators $A'_{i,j}$ are obtained by recursively solving the following equations:

$$A'_{i,j} = A_{i,j} - \sum_{k=0}^{j-1} \frac{A'_{i,k}p_i^{(j-k)}(\alpha_i)}{(j-k)!} \qquad (1 \le i \le n \text{ and } 0 \le j < n_i) . \tag{5.4}$$

subject to the initial condition $A_{i,0} = A'_{i,0}$ for $1 \le i \le n$.

We claim that the polynomial $P(z)$ defined by (5.3) and (5.4) satisfies (5.1), that is,

$$\frac{P^{(j)}(\alpha_i)}{j!} = A_{i,j} \qquad (\text{for } 1 \le i \le n \text{ and } 0 \le j < n_i) . \tag{5.5}$$

To verify this first notice that

$$p_i(\alpha_i) = 1 \quad \text{and} \quad p_i^{(j)}(\alpha_k) = 0 \quad (\text{for all } 0 \le j < n_k \text{ and } k \ne i) . \tag{5.6}$$

Using the second equation in (5.6) with the binomial expansion for the derivative of the product of two functions we have

$$P^{(j)}(\alpha_i) = (A_i(z)p_i(z))^{(j)}(\alpha_i) = \sum_{k=0}^{j} \frac{j!}{k!(j-k)!} A_i^{(k)}(\alpha_i) p_i^{(j-k)}(\alpha_i) =$$

$$\sum_{k=0}^{j} \frac{j!}{(j-k)!} A_{i,k}' p_i^{(j-k)}(\alpha_i) = j! A_{i,j}$$

where $A_i(z)$ is the polynomial of degree n_i-1 defined by

$$A_i(z) = \sum_{j=0}^{n_i-1} A_{i,j}'(z-\alpha_i)^j \ .$$

The last equality follows from (5.4). Thus (5.5) holds. In other words the polynomial P defined by (5.3) and (5.4) satisfies (5.5). Finally, it is emphasized that P may not be in $H_1^\infty(E, E')$.

To obtain a solution to Problem 5.1, let m be the Blaschke product of order d in (1.1). The zeros α_i of m are repeated n_i times. If F is any function in $H^\infty(E, E')$, then $A(F)$ is the operator from $H = H(mI)$ to $H' = H(mI')$ defined by $A(F) \triangleq P_{H'}F_+|H$. Since $Fm = mF$ the operator $A(F)$ associated with the symbol F in the sense of Section IX.7 is in $I(S(mI), S(mI'))$. Moreover, F_+ is a lifting of $A(F)$, that is, $P_{H'}F_+ = A(F)P_H$. The operator $A(F)$ plays a fundamental role in our solution to Problem 5.1, which begins with the following result.

5.2 LEMMA. *Let* F *and* G *be in* $H^\infty(E, E')$ *and* m *the Blaschke product of order* d *in (1.1). The following statements are equivalent*

    (i) F = G *modulo* $mH^\infty(E, E')$

(ii) $\dfrac{F^{(j)}(\alpha_i)}{j!} = \dfrac{G^{(j)}(\alpha_i)}{j!}$ *(for* $1 \le i \le n$ *and* $0 \le j < n_i$)   (5.7)

    (iii) $A(F) = A(G)$

In particular, the set of all F *in* $H^\infty(E, E')$ *satisfying (5.1) produces the same operator* $A = A(F)$.

PROOF. The equivalence of (i) and (ii) follows from the fact that the zero's α_i of m are repeated n_i times. If (i) holds, then obviously (iii) holds. If (iii) holds, then the lifting property gives $P_{H'}(F_+ - G_+) = 0$. Therefore $F = G$ modulo $mH^\infty(E, E')$. This completes the proof.

We can now say that A *is the operator uniquely determined by Problem 5.1.* In fact $A = A(F)$ where F is any function in $H^\infty(E, E')$ satisfying (5.1). Since one can always construct a polynomial $P(z)$ satisfying (5.5), the operator A is well defined. It is convenient to call this operator A *the Hermite-Fejér operator* determined by the data in

Problem 5.1.

5.3 LEMMA. *There exists a solution to Problem 5.1 if and only if the corresponding Hermite-Fejér operator* A *is a contraction.*

PROOF. Let F in $H^{\infty}(E, E^{'})$ be a solution to Problem 5.1 and A the corresponding, Hermite-Fejér operator. Since F is in $H_1^{\infty}(E, E^{'})$ and satisfies (5.1), we have $A = A(F)$ and A is a contraction.

Now let F be any function in $H^{\infty}(E, E^{'})$ satisfying (5.1) (for instance F can be chosen to be the polynomial constructed at the beginning of this section), and assume that the Hermite-Fejér operator $A = A(F)$ is a contraction. Recall that the unilateral shifts S and $S^{'}$ are the minimal isometric dilations of $T = S(mI)$ and $T^{'} = S(mI^{'})$; see Theorem IX.5.1. The lifting property $P_{H^{'}} F_+ = A(F) P_H$ gives

$$T^{'} A(F) = P_{H^{'}} S^{'} P_{H^{'}} F_+ \,|\, H = P_{H^{'}} F_+ S \,|\, H = A(F) P_H S \,|\, H = A(F) T\,.$$

So $A = A(F)$ is a contraction in $I(T, T^{'})$. By consulting Section IX.7 or by using the commutant lifting theorem, there exists a contractive intertwining lifting B in $I(S, S^{'})$ of $A(F)$. Moreover, $B = G_+$ where G is a contractive analytic function in $H^{\infty}(E, E^{'})$; see Theorem IX.1.1. This implies that $A(F) = A(G)$. Lemma 5.2 and (5.1) shows that G is a solution to Problem 5.1. This completes the proof.

5.4 REMARK. The previous proof shows that the Hermite-Fejér operator A uniquely determined by Problem 5.1 is in $I(S(mI), S(mI^{'}))$. Moreover, there is a one to one correspondence $F \leftrightarrow B = F_+$ between the set of all solutions F in $H_1^{\infty}(E, E^{'})$ to Problem 5.1 and the set of all contractive intertwining liftings B of A in $I(S(mI), S(mI^{'}))$. In particular, there is a unique contractive intertwining liftings B of A if and only if there is a unique solution to Problem 5.1.

Recall that the defect indices of $S(m)$ are one. By Lemma 1.1 the space $H(m)$ is finite dimensional. This with the uniqueness assertion of the previous remark and Corollary VII.5.4 proves the first part of the following result.

5.5 COROLLARY. *Let* $E = E^{'} = C^1$ *and* A *be the Hermite-Fejér operator uniquely determined by Problem 5.1. Then there exists a unique solution to Problem 5.1 if and only if the norm of* A *is one. In this case if* h *is any nonzero vector in* $H(m)$ *satisfying* $\|Ah\| = \|h\|$, *then the unique solution* F *to Problem 5.1 is given by* $F = (Ah)/h$.

PROOF. To complete the proof it only remains to verify the last statement. To this end if F is a solution to Problem 5.1 and $\|Ah\| = \|h\|$, then

$$\|h\| = \|Ah\| = \|P_{H'} Fh\| \leq \|h\| .$$

Therefore there is equality and $Ah = Fh$. Because h is a nonzero element in H^2, we can divide by h to obtain $F = (Ah)/h$. This completes the proof.

Let M be a matrix on C^d with entries $[m_{i,j}]$. If L is an operator in $L(E, E')$, then the matrix $L \otimes M$ is the operator from $l^2_d(E)$ to $l^2_d(E')$ defined by the block matrix

$$L \otimes M = \begin{bmatrix} Lm_{1,1} & Lm_{1,2} & ... & Lm_{1,d} \\ Lm_{2,1} & Lm_{2,2} & ... & Lm_{2,d} \\ \vdots & \vdots & \vdots & \vdots \\ Lm_{d,1} & Lm_{d,2} & ... & Lm_{d,d} \end{bmatrix} . \tag{5.8}$$

Now let P(z) be a polynomial of degree k with values in $L(E, E')$. The operator P(M) from $l^2_d(E)$ to $l^2_d(E')$ is defined by

$$P(M) = \sum_{i=0}^{k} P_i \otimes M^i \quad \text{where} \quad P(z) = \sum_{i=0}^{k} P_i z^i . \tag{5.9}$$

If $E = E' = C^1$ and P is scalar valued, then P(M) reduces to the usual definition of $P(M) = \sum P_i M^i$.

Let M be the matrix on C^d defined in (1.19) corresponding to the Blaschke product m of order d written under the form (1.17). Let P be the polynomial of degree at most $d - 1$ defined in (5.3), (5.4) and which satisfies (5.5). The operator P(M) from $l^2_d(E)$ to $l^2_d(E')$ is called *the Hermite-Fejér matrix determined by* Problem 5.1. This matrix is lower triangular and plays a fundamental role in the next solution to Problem 5.1. Recall that $\{\phi_i\}_1^d$ in (1.18) is an orthonormal basis for $H(m)$ where m is the Blaschke product in (1.17). Now let W_ϕ be the unitary operator from $l^2_d(E)$ onto $H(mI)$ defined by

$$W_\phi = [\phi_1 I, \ \phi_2 I, \ \phi_3 I, \ ..., \ \phi_d I],$$

and let W'_ϕ be the corresponding unitary operator from $l^2_d(E')$ onto $H(mI')$. These operators are used in the following basic result.

5.6 THEOREM. *Let* A *be the Hermite-Fejér operator determined by Problem 5.1 and* P(M) *be the Hermite-Fejér matrix also determined by Problem 5.1. Then* $AW_\phi = W'_\phi P(M)$, *that is,* P(M) *is a matrix representation for* A. *In particular, there exists a solution to Problem 5.1 if and only if* P(M) *is a contraction.*

PROOF. Recall that $\{\phi_j\}_1^d$ in (1.18) is an orthonormal basis for $H(m)$. Let $\{e_j\}_1^d$ be the standard unit vectors in C^d. By identifying ϕ_j with e_j for $1 \leq j \leq d$ the proof of Theorem 1.5 shows that $S(m)$ becomes M and consequently $S(m)^i$ becomes M^i for all $i \geq 0$. Let fe_j and ge_j be the vector in $l_d^2(E)$ and $l_d^2(E')$ whose j–th component is f in E and g in E', and all other components are zero. We identify fe_j and ge_j with $f\phi_j = W_\phi fe_j$ in H and $g\phi_j = W'_\phi ge_j$ in H', respectively for all $1 \leq j \leq d$ and f in E and g in E'. If P is the polynomial of degree at most d–1 satisfying (5.5), then obviously $A = A(P)$. Equation (5.9) and the fact that our identification yields by linearity a unitary operator gives

$$(Af\phi_j, g\phi_k) = (A(P)f\phi_j, g\phi_k) = \sum_{i=0}^{d-1}(P_i z^i f\phi_j, g\phi_k) = \sum_{i=0}^{d-1}(P_i fS(m)^i \phi_j, g\phi_k) =$$

$$\sum_{i=0}^{d-1}(P_i f, g)(M^i e_j, e_k) = (P(M)fe_j, ge_k) .$$

This implies that $AW_\phi = W'_\phi P(M)$. The last statement follows from Lemma 5.3. This completes the proof.

Corollary 5.5 readily proves the first part of the following result.

5.7 COROLLARY. *Let $E = E' = C^1$ and P(M) be the Hermite-Fejér matrix uniquely determined by Problem 5.1. Then there exists a unique solution to Problem 5.1 if and only if the norm of* P(M) *is one. In this case if* $x = [x_j]_1^d$ *is any nonzero vector in* C^d *satisfying* $\|P(M)x\| = \|x\|$, *then the unique solution* F *to Problem 5.1 is given by*

$$F = (\sum_{j=1}^d y_j\phi_j)(\sum_{j=1}^d x_j\phi_j)^{-1} \tag{5.10}$$

where $P(M)x = y = [y_j]_1^d$ *and* $\{\phi_j\}_1^d$ *is the orthonormal basis for $H(m)$ defined in (1.18).*

PROOF. To complete the proof it only remains to verify (5.10). The previous proof shows that $A = A(P)$ is unitarily equivalent to $P(M)$ under the identification of ϕ_j to e_j for $1 \leq j \leq d$. By this identification $h = \sum x_i\phi_i$ and $Ah = \sum y_i\phi_i$ satisfies $\|Ah\| = \|h\|$. Now (5.10) follows from $F = (Ah)/h$ in Corollary 5.5. The proof is now complete.

If $A_j = A_{1,j}$ for $0 \leq j < n_1$ and $n = 1$ and $\alpha_1 = 0$ are the data for Problem 1.5, then Problem 1.5 reduces to the classical Carathéodory interpolation Problem VIII.1.1. In this case $m = z^{n_1}$ and $P(M) = \hat{A}_{n_1}$ where $\hat{A}_{n_1}$ is the n_1 by n_1 analytic Toeplitz matrix generated by $\{A_j\}_0^{n_1-1}$. In particular, Theorem 5.6 and Corollary 5.7 reduce to Theorem VIII.1.3 and Corollary VIII.1.4, respectively.

To complete this section we will present a connection between the classical Hermite-Fejér interpolation problem and the Nehari H^∞ optimization problem for rational functions.

Now let F be any function in $H^\infty(E, E')$ satisfying (5.1) and set $\tilde{H}' = H(\tilde{m}I')$. According to Theorem IX.8.4 we have

$$\Gamma(e^{-it}mF^*) = A(F)^* W(mI')P_{\tilde{H}'} \tag{5.11}$$

where $\Gamma(Q)$ is the Hankel operator with symbol $Q = e^{-it}mF^*$ in (IX.3.6), $F = G$ and $W(mI')$ is the unitary operator in Proposition IX.5.6. Therefore there exists a solution to the classical Hermite-Fejér interpolation Problem 5.1 if and only if the Hankel operator $\Gamma(e^{-it}mF^*)$ is a contraction. (Here F is any function in $H^\infty(E, E')$ satisfying (5.1)). Moreover, if $E = E' = C^1$, then there exists a unique solution to Problem 5.1 if and only if the norm of $\Gamma(e^{-it}mF^*)$ is one.

One can also use (5.11) with our previous results to compute the norm of certain Hankel matrices. If $Q = Q(e^{it})$ is a rational function in $L^\infty(E', E)$, then Q admits a factorization of the form $Q = e^{-it}mF^*$ where m is a Blaschke product of the form (1.1) and F is in $H^\infty(E, E')$. (In fact m is a Blaschke product of the form $m = z^k m_1$ where the zeros $\alpha_1, \alpha_2, ..., \alpha_j$ of m_1 are given by all the poles including their multiplicity $q_i = 1/\bar{\alpha}_i$ of $Q(z)$ in $|z| > 1$.) According to (5.11) and Theorem 5.6

$$\inf \|Q + K_o^\infty(E', E)\|_\infty = \|\Gamma(Q)\| = \|A(F)^*\| = \|P(M)\| \tag{5.12}$$

where $P(M)$ is the Hermite-Fejér matrix corresponding to the data (5.1) obtained from F. The first equality in (5.12) follows from Theorem IX.3.3. Equation (5.12) provides the following procedure for computing the norm of $\Gamma(Q)$:

First factor Q into $e^{-it}mF^*$. Then compute all the $A_{i,j}$ in (5.1) from F and construct the Hermite-Fejér matrix $P(M)$ determined by Problem 5.1. Finally, to compute find the norm of $\Gamma(Q)$ compute the norm of $P(M)$.

If Q is rational and $E = E' = C^1$, then Corollary 5.7 can be used to compute the optimal K_* in solving the optimization problem in (5.12). To see this let $\{\psi_j\}_1^d$ be the orthonormal basis for $H(\tilde{m})$ defined by

$$\psi_j = d_j(1-\alpha_j z)^{-1} \tilde{B}_{j+1} \tilde{B}_{j+2} \cdots \tilde{B}_d \tag{5.13}$$

where $\psi_d = d_d(1-\alpha_d z)^{-1}$. Here m is the Blaschke product of order d defined in (1.17). Theorem 1.5 guarantees that $\{\psi_j\}_1^d$ is indeed an orthonormal basis for $H(\tilde{m})$. Indeed by the definition of $W(m)$ we have

$$\phi_j = W(m)\psi_j \qquad (1 \le j \le d) \qquad (5.14)$$

where ϕ_j for $1 \le j \le d$ is the orthonormal basis for $H(m)$ defined by (1.18). Now let $x = [x_j]_1^d$ be a nonzero vector in $\mathbb{C}^d$ satisfying $\|P(M)^*x\| = \|\Gamma(Q)\| \, \|x\|$ and set $P(M)^*x = y = [y_j]_1^d$. Using the identification $e_j \leftrightarrow \phi_j$ with (5.11) and (5.14) we have

$$\Gamma(Q) \sum_{j=1}^{d} x_j \psi_j = \sum_{j=1}^{d} y_j \phi_j \, . \qquad (5.15)$$

According to Corollary IX.3.5 the unique optimal solution K_* to (5.12) is given by

$$K_* = (\sum_{j=1}^{d} y_j \phi_j)(\sum_{j=1}^{d} x_j J\psi_j)^{-1} - Q \, . \qquad (5.16)$$

If $E = E^{'} = \mathbb{C}^1$ and Q is rational, then (5.16) provides a computational procedure for finding the unique optimal K_* in K_0^∞ satisfying

$$\inf \|Q + K_0^\infty\|_\infty = \|Q + K_*\|_\infty = \|\Gamma(Q)\| = \|P(M)\| \, . \qquad (5.17)$$

To compute K_* first factor Q into $Q = e^{-it}mF^*$ where m is the Blaschke product of order d in (1.1) and F is in H^∞. Then compute all the $A_{i,j}$ in (5.1) from F and find the Hermite-Fejér matrix $P(M)$ uniquely determined by Problem 5.1. Compute the nonzero vector x in $\mathbb{C}^d$ satisfying $\|P(M)^*x\| = \|P(M)\| \, \|x\|$ and find $y = P(M)^*x$. Finally, the unique optimal K_* in K_0^∞ solving the optimization problem (5.12) is given by (5.16).

6. MATRIX REPRESENTATIONS FOR FINITE RANK HANKEL MATRICES

The procedure presented at the end of the previous section solves the Nehari optimization problem for rational symbols, by using the solution of the classical Hermite-Fejér interpolation Problem 5.1. In this section we will present a direct method for solving H^∞ optimization problems involving Hankel operators with rational symbols. This is done by using the orthonormal basis $\{\phi_i\}_1^d$ for $H(m)$ to obtain a matrix representation for Hankel operators of the form $\Gamma(e^{-it}mF^*)$.

We begin with the following result.

6.1 LEMMA. *Let* m *be the Blaschke product of order* d *in (1.17) and* $\{\phi_j\}_1^d$ *the orthonormal basis for* $H(m)$ *defined by (1.18). Then the orthogonal projection of* h *onto* $H(mI)$ *is given by*

$$P_{H(mI)}h = \sum_{k=1}^{d} h_k(\alpha_k)d_k\phi_k \qquad (h \in H^2(E)) \qquad (6.1)$$

where h_k *are the functions in* $H^2(E)$ *recursively defined by*

$$h_{k+1}(z) = \frac{h_k(z)(1-\bar{\alpha}_k z) - h_k(\alpha_k)d_k^2}{z-\alpha_k} \qquad (for \ 1 \le k \le d \ and \ h_1 = h) . \tag{6.2}$$

PROOF. The proof of the lemma will show that h_k is indeed a function in $H^2(E)$. (The point α_k is a removable singularity for h_{k+1}. So h_{k+1} is obviously analytic in D.). Notice that if $x_k = (1-\bar{\alpha}_k z)^{-1}$, then $x_k E = H(B_k I)$ for $1 \le k \le d$. We claim that

$$P_{H(B_k I)}f = f(\alpha_k)d_k^2 x_k \qquad (f \in H^2(E)) . \tag{6.3}$$

Clearly the right hand side of (6.3) is in $H(B_k I)$. Using the reproducing property (1.10) of $(1-\bar{\alpha}_k z)^{-1}$ it is easy to verify that $f - f(\alpha_k)d_k^2 x_k$ is orthogonal to $x_k E = H(B_k I)$. Thus (6.3) holds. Equation (6.3) implies that $h = h_1$ admits an orthogonal decomposition of the form

$$h_1 = h = h_1(\alpha_1)d_1^2 x_1 + B_1 h_2 \tag{6.4}$$

where h_2 is some function in $H^2(E)$. Repeating the previous argument on h_2 we see that h_2 admits an orthogonal decomposition of the form

$$h_2 = h_2(\alpha_2)d_2^2 x_2 + B_2 h_3$$

where h_3 is some function in $H^2(E)$. Substituting h_2 back in (6.4) and using $\phi_1 = d_1 x_1$ and $\phi_2 = B_1 d_2 x_2$ gives

$$h = h_1(\alpha_1)d_1\phi_1 + h_2(\alpha_2)d_2\phi_2 + B_1 B_2 h_3 .$$

Continuing in this fashion produces (6.1). To obtain the recursion in (6.2) notice that

$$h_k(z) = h_k(\alpha_k)d_k^2 x_k + B_k h_{k+1}(z) .$$

Solving for h_{k+1} yields (6.2). This completes the proof.

It is emphasized that if h is rational, then one can use the Euclid-Horner algorithm to recursively compute each h_{k+1}.

Let m be the Blaschke product in (1.17) and F any function in $H^\infty(E, E')$. We can define $A_{i,j}$ for $1 \le i \le n$ and $0 \le j < n_i$ in Problem 5.1 by (5.1) in terms of F. The Hermite-Fejér matrix P(M) viewed as a block matrix whose entries are operators from E to E' is the matrix representation of the Hermite-Fejér operator A = A(F) with respect to the splitting

$$H(mI) = \overset{d}{\underset{1}{\oplus}} \phi_i E \quad and \quad H(mI') = \overset{d}{\underset{1}{\oplus}} \phi_i E' .$$

This observation and Lemma 6.1 leads to another way of computing the operator entries of $P(M) = [C_{i,j}]$ for $1 \le i,\ j \le d$. Thus for f in E, g in E' and $1 \le i,\ j \le d$ we have

$$(C_{i,j}f,\ g) = (A(F)\phi_j f,\ \phi_i g) = (F\phi_j f,\ \phi_i g) = (d_j x_j B_{j-1}...B_1 Ff,\ d_i x_i B_{i-1}...B_1 g) \ .$$

Hence if $j > i$

$$(C_{i,j}f,\ g) = (d_j x_j B_{j-1}...B_i Ff,\ d_i x_i g) = 0$$

because of the reproducing property (1.10) and $B_i(\alpha_i) = 0$. On the other hand if $j \le i$, then

$$(C_{i,j}f,g) = (d_j x_j Ff,\ d_i x_i B_{i-1}...B_j g) \ . \tag{6.5}$$

Now in this last case we can apply Lemma 6.1 with $h = d_j x_j Ff$ and $m = m_j = B_j B_{j+1}...B_d$ to obtain

$$P_{H(m_j I')} d_j x_j Ff = \sum_{i=j}^{d} G_{i,j}(\alpha_i) f d_i^2 x_i B_{i-1}...B_j \qquad (f \in E)$$

where for every $j = 1,\ 2,\ ...,\ d$

$$G_{j,j} = d_j x_j F$$
$$G_{i+1,\,j}(z) = \frac{G_{i,j}(z)(1-\bar{\alpha_i}z) - G_{i,j}(\alpha_i)d_i^2}{z-\alpha_i} \qquad (\text{for } j \le i < d) \ . \tag{6.6}$$

Obviously $G_{i,j}$ is in $H^\infty(E, E')$ for all $1 \le j \le i \le d$. So we can rewrite (6.5) as $(C_{i,j}f,\ g) = (d_i G_{i,j}(\alpha_i)f,\ g)$. Therefore

$$C_{i,j} = 0 \qquad\qquad \text{if} \quad j > i \tag{6.7}$$
$$C_{i,j} = d_i G_{i,j}(\alpha_i) \quad \text{if} \quad j \le i \ .$$

As before $A(F)$ is the operator from $H = H(mI)$ to $H' = H(mI')$ defined by $A(F) = P_{H'} F_+ \,|\, H$, and W_ϕ is the unitary operator mapping $l_d^2(E)$ onto $H(mI)$ defined by

$$W_\phi = [\phi_1 I,\ \phi_2 I,\ \phi_3 I,\ ...,\ \phi_d I] \ . \tag{6.8}$$

Let W'_ϕ be the corresponding unitary operator mapping $l_d^2(E')$ onto $H(mI')$ defined by (6.8) with I' replacing I. We can now conclude with the following result.

     6.2 PROPOSITION. *Let* m *be the Blaschke product of order* d *in (1.17) and let* P(M) *be the Hermite-Fejér matrix corresponding to the data (5.1) provided by* F. *Then* $A(F)W_\phi = W'_\phi P(M)$ *and the entries* $C_{i,j}$ *of* P(M) *can be recursively computed by the formulas (6.6) and (6.7).*

If F is rational, then one can use the Euclid-Horner algorithm to easily compute the entries $C_{i,j}$ of the lower triangular matrix P(M). Moreover, if F is any function in $H^\infty(E, E')$ satisfying (5.1), then by Theorem 5.6 there exists a solution to the Hermite-Fejér interpolation Problem 5.1 if and only if Hermite-Fejér the matrix P(M) generated by m and F is a contraction. However, in certain applications it may be easier to compute the matrix P(M) by using Proposition 6.2 instead of the formula (5.9) and (1.19) defining P(M).

The previous proposition also provides another proof of the fact that S(m) is unitarily equivalent to M in (1.19) (see Theorem 1.5). To see this notice that if $E = E' = C^1$, then S(m) = A(z). Proposition 6.2 implies that S(m) is unitarily equivalent to the lower triangular matrix $[C_{i,j}]$ on C^d where the entries are computed by (6.6) and (6.7). In this case, simple calculations show that the relations in (6.7) yield

$$C_{j,j} = \alpha_j \ (1 \le j \le d), \ C_{j+1,\,j} = d_j d_{j+1} \ (1 \le j < d)$$
$$C_{i,j} = (-1)^{i-j+1} d_j \bar{\alpha}_{j+1} \bar{\alpha}_{j+2} \bar{\alpha}_{j+3} \cdots \bar{\alpha}_{i-1} d_i \qquad (2 \le j+1 < i \le d) \tag{6.9}$$

These are precisely the entries of the matrix M in (1.19). Therefore $S(m)W_\phi = W_\phi M$ and S(m) is unitarily equivalent to M.

Let W_ψ be the unitary operator mapping $l_d^2(E)$ onto $H(\tilde{m}I)$ defined by (6.8) where the $\psi_j's$ defined by (5.13) replaces the $\phi_j's$. According to Theorem 1.5 and the form of ψ_i in (5.13)

$$S(\tilde{m})[\psi_d, \ \psi_{d-1}, \ ..., \ \psi_1] = [\psi_d, \ \psi_{d-1}, \ ..., \ \psi_1]M_r$$

where M_r is the matrix in (1.19) with $\alpha_1, \ \alpha_2, \ ..., \ \alpha_d$ replaced by $\bar{\alpha}_d, \ \bar{\alpha}_{d-1}, \ ..., \ \bar{\alpha}_1$. Now let $\tilde{M}$ be the upper triangular matrix on C^d defined by $\tilde{M} = \gamma M_r \gamma$ where γ is the unitary operator on C^d whose off diagonal contains 1 and all other entries are zero. Using the fact that $\gamma^2 = I$ we readily obtain that

$$S(\tilde{m}I)W_\psi = W_\psi(I \otimes \tilde{M}) .$$

If G is in $H^\infty(E', E)$, then Lemma IX.8.1 implies that range of $\Gamma = \Gamma(e^{-it}mG^*)$ is contained $H(mI')$. Since $\Gamma^* = \Gamma(e^{-it}\tilde{m}\tilde{G}^*)$ the range of Γ^* is contained in $H(\tilde{m}I)$. This sets the stage for the following result.

6.3 PROPOSITION. *Let* $\Gamma = \Gamma(e^{-it}mF^*)$ *be a Hankel operator where* m *is the Blaschke product of order* d *in (1.17) and* F *is now in* $H^\infty(E', E)$. *Then* $\Gamma W_\psi = W_\phi' C^*$ *where* $C(= P(M))$ *is the Hermite-Fejér matrix corresponding to the data provided by* m *and* F. *Moreover,*

$$\Gamma = W_\phi' C^* W_\psi^* \ and \ S'^* \Gamma = \Gamma S = W_\phi' C^* (I \otimes \tilde{M}) W_\psi^* . \tag{6.10}$$

where S and S$'$ are the unilateral shifts on $H^2(E)$ *and* $H^2(E')$, *respectively.*

PROOF. The first part follows from Proposition 6.2 and (5.11), (5.14) where we switched the roles of E and E'. Here we will give a direct proof of this result. By following the proof of Proposition 6.2 the i, jth entry of the block matrix representation of $\Gamma(e^{-it}mF^*)$ corresponding to the orthogonal splitting $\bigoplus_1^d \psi_j E = H(\tilde{m}I)$ and $\bigoplus_1^d \phi_j E' = H(mI')$ is given by

$$(\Gamma(e^{-it}mF^*)\psi_j f, \phi_i g) = (B_1 B_2 \cdots B_{j-1} x_j d_j f, F\phi_i g) = (\phi_j f, F\phi_i g) = (f, C_{j,i} g) = (C_{j,i}^* f, g)$$

where $C_{i,j}$ is computed according to (6.6) and (6.7). Therefore $\Gamma W_\psi = W_\phi' C^*$. Since W_ψ is an isometry whose range is $H(\tilde{m}I) = \tilde{H}$ the orthogonal projection $P_{\tilde{H}}$ equals $W_\psi W_\psi^*$. This and the fact that the range of Γ^* is contained in $\tilde{H}$ produces the first equality in (6.10). The second equation follows from the fact that $W_\psi^* S(\tilde{m}I) P_{\tilde{H}} = I \otimes \tilde{M} W_\psi^*$ and S is a lifting of $S(\tilde{m}I)$. This completes the proof.

6.4 REMARK. Assume that Q in L^∞ admits a factorization of the form $Q = e^{-it}mF^*$ where m is the Blaschke product of order d in (1.17) and F is in H^∞. Let $C(= P(M))$ be the matrix on C^d generated by m and F according to (6.6) and (6.7). The previous proposition with Corollary IX.3.5 shows that if x is any nonzero vector in C^d satisfying $\|C^* x\| = \|C\| \|x\|$, then $K_* = ((W_\phi C^* x)/(J W_\psi x) - Q)$ is the unique element in K_0^∞ solving the following H^∞ optimization problem:

$$\inf \|Q + K_0^\infty\|_\infty = \|Q + K_*\|_\infty = \|\Gamma(Q)\| = \|C\| . \tag{6.11}$$

As before if F is rational, then we can use the Euclid-Horner algorithm to compute C and, then find the optimal solution K_* to (6.11).

7. STATE SPACE REALIZATIONS

In this section we will introduce state space realizations of a function in $H^\infty(E, E')$ and use these realizations to compute the norm of a finite rank Hankel operator.

Throughout G is a rational function in $H^\infty(E, E')$ where both E and E' are finite dimensional. We say that $\Sigma = \{A, B, C, X\}$ is a *realization* of G(z) if

$$G(z) = C(I - zA)^{-1} B \qquad (z \in D) \tag{7.1}$$

where A is an operator on a finite dimensional space X whose spectrum $\sigma(A) \subseteq D$, B is an operator from E to X and C is an operator from X to E'. The spectral condition guarantees that the inverse in (7.1) is well defined in a neighborhood of $\bar{D}$. Let $G(z) = \Sigma G_n z^n$ for z in D be the power series expansion for G. If $\sigma(A) \subseteq D$, then $(I-zA)^{-1} = \Sigma A^n z^n$ for z in D. So (7.1) is equivalent to

$$G_n = CA^n B \qquad (n \geq 0) \,. \tag{7.2}$$

Therefore $\Sigma = \{A, B, C, X\}$ is a realization of G if $\sigma(A) \subseteq D$ and (7.2) holds. The space X is called the *state space* for Σ. We say that Σ is a *minimal realization* of G if $\dim X \leq \dim X_1$ where $\Sigma_1 = \{A_1, B_1, C_1, X_1\}$ is any other realization of G. Finally, Σ and Σ_1 are *similar [unitarily equivalent] realizations* if there exists a similarity transformation [unitary operator] W from X onto X_1 satisfying

$$WA = A_1 W \quad \text{and} \quad WB = B_1 \quad \text{and} \quad C_1 W = C \,. \tag{7.3}$$

To construct a realization of G consider the invariant subspace H' for the backward shift S'^* defined by

$$H' \triangleq \bigvee_0^\infty S'^{*n} GE = \overline{\Gamma(G)H^2(E)} \,. \tag{7.4}$$

The second equality in (7.4) follows from the fact that $S'^*\Gamma = \Gamma S$ and E is cyclic for the unilateral shift S. Since G is rational the range of $\Gamma(G) = \Gamma$ equals H' which is finite dimensional. This also follows from the fact that G admits a factorization of the form $G = e^{-it} mF^*$ where m is a Blaschke product of the form (1.17). Then a direct calculation or Lemma IX.8.1 implies that $H' = \operatorname{ran}\Gamma(G) \subseteq H(mI')$. Thus $\dim H' \leq d \cdot \dim E'$ (see Lemma 1.1). Therefore the closure of the last term in (7.4) is superfluous. Because S'^* is stable and H' is finite dimensional, the spectrum of $S'^* | H'$ is contained in D.

Let B_G be the operator from E to H' defined by $B_G = G_+ | E$ and C_o be the operator from H' to E' defined by $C_o = P_{E'} | H'$. We claim that $\{S'^* | H', B_G, C_o, H'\}$ is a realization of G. Using the fact that S'^{*n} shifts the coefficient of z^n to the coefficient of $z^o = 1$ we have

$$C_o(S'^* | H')^n B_G e = P_{E'} S'^{*n} Ge = G_n e \qquad (e \in E) \,.$$

Thus (7.2) holds for $A = S'^* | H', B = B_G$ and $C = C_o$. So $\{S'^* | H', B_G, C_o, H'\}$ is a realization of G. The realization $\{S'^* | H', B_G, C_o, H'\}$ is called the *backward shift realization* of G. The following shows that the backward shift realization is the "only" minimal realization of G.

7.1 PROPOSITION. *Let* G *be a rational function in* $H^\infty(E, E')$ *where both* E *and* E' *are finite dimensional. The backward shift realization of* G *is a minimal realization of* G. *Moreover, all minimal realizations of* G *are similar.*

PROOF. Let $\Sigma = \{A, B, C, X\}$ be any realization of G. Consider the *observability operator* Θ_Σ from X into $H^2(E')$ defined by

$$\Theta_\Sigma x = C(I-zA)^{-1}x \qquad\qquad (x \in X) . \tag{7.5}$$

This operator is well defined because $\sigma(A) \subseteq D$. Using (1.6) where S' replaces S we have $S'^*\Theta_\Sigma = C(I-zA)^{-1}A = \Theta_\Sigma A$. By (7.1) and (7.4)

$$H' = \bigvee_0^\infty S'^{*n}GE = \bigvee_0^\infty S'^{*n}\Theta_\Sigma BE = \bigvee_0^\infty \Theta_\Sigma A^n BE \subseteq \Theta_\Sigma X . \tag{7.6}$$

Thus $H' \subseteq \Theta_\Sigma X$ and $\dim H' \leq \dim X$. Therefore the backward shift realization is minimal.

If Σ is a minimal realization, then $H' \subseteq \Theta_\Sigma X$ implies that Θ_Σ is one to one and onto H'. Now let W be the linear invertible operator from X onto H' provided by $W = \Theta_\Sigma$. The equation $S'^*\Theta_\Sigma = \Theta_\Sigma A$ shows that $(S'^* \,|\, H')W = WA$. The definition of C_0 along with (7.5) gives $C_0 W = C_0 \Theta_\Sigma = C$. Finally, (7.1) yields $B_G = G_+ \,|\, E = \Theta_\Sigma B = WB$. Therefore the backward shift realization is similar to Σ. In turn this implies that all minimal realizations are similar. This completes the proof.

The realization $\Sigma = \{A, B, C, X\}$ is *observable* if $CA^n x = 0$ for all $n \geq 0$, then $x = 0$. Notice that Σ is observable if and only if its observability operator Θ_Σ defined in (7.5) is one to one. The *observability Grammian* O_Σ is the positive operator defined by

$$O_\Sigma \triangleq \Theta_\Sigma^* \Theta_\Sigma = \sum_0^\infty A^{*n}C^* CA^n . \tag{7.7}$$

Since $\sigma(A) \subseteq D$ the sum in (7.7) is well defined. Notice that Σ is observable if and only if its observability Grammian O_Σ is strictly positive. The second equation in (7.7) implies that O_Σ satisfies the following equation

$$O_\Sigma = A^* O_\Sigma A + C^* C , \tag{7.8}$$

which is usually referred to as a Lyapunov equation.

If $\Sigma_b = \{S'^* \,|\, H', B_G, C_0, H'\}$ is the backward shift realization, then its observability Grammian $O_{\Sigma_b} = I$. To see this it is sufficient to prove that Θ_{Σ_b} is an isometry. This follows from the following calculation where $h = \Sigma h_n z^n$ is the power series expansion of h in H'

$$\|\Theta_{\Sigma_b} h\|^2 = \|C_o(I-zS'^*|H')^{-1}h\|^2 = \|\sum_o^\infty z^n P_{E'} S'^{*n} h\|^2 = \sum_o^\infty \|P_{E'} S'^{*n} h\|^2 = \sum_o^\infty \|h_n\|^2 = \|h\|^2 .$$

Therefore Θ_{Σ_b} is an isometry, thus indeed the backward shift realization is observable and $O_{\Sigma_b} = I$.

The realization $\Sigma = \{A, B, C, X\}$ is *controllable* if the linear span of $\{A^n BE: n \geq 0\}$ equals X. Therefore by virtue of the definition of H' the backward shift realization is controllable. Notice that Σ is controllable if and only if $B^* A^{*n} x = 0$ for all $n \geq 0$, then $x = 0$. The *controllability operator* Ω_Σ from X to $H^2(E)$ is defined by

$$\Omega_\Sigma x = B^*(I-zA^*)^{-1}x \qquad (x \in X) . \tag{7.9}$$

The controllability operator Ω_Σ is the dual of the observability operator Θ_Σ in (7.5). Equation (1.6) implies that $S^* \Omega_\Sigma = \Omega_\Sigma A^*$. Obviously Σ is controllable if and only if Ω_Σ is one to one. The *controllability Grammian* C_Σ for Σ is the positive operator defined by

$$C_\Sigma \overset{\Delta}{=} \Omega_\Sigma^* \Omega_\Sigma = \sum_o^\infty A^n BB^* A^{*n} \tag{7.10}$$

which is the dual of the observability Grammian in (7.7). The realization Σ is controllable if and only if C_Σ is strictly positive. The second equation in (7.10) implies that C_Σ satisfies the following Lyapunov equation

$$C_\Sigma = AC_\Sigma A^* + BB^* . \tag{7.11}$$

The realization $\Sigma = \{A, B, C, X\}$ is controllable if and only if $\Sigma_* = \{A^*, C^*, B^*, X\}$ is observable. Moreover,

$$\Theta_\Sigma = \Omega_{\Sigma_*} \quad \text{and} \quad \Omega_\Sigma = \Theta_{\Sigma_*} \quad \text{and} \quad O_\Sigma = C_{\Sigma_*} \quad \text{and} \quad C_\Sigma = O_{\Sigma_*} . \tag{7.12}$$

We recall that the backward shift realization is both controllable and observable.

7.2 PROPOSITION. *Let* G *be a rational function in* $H^\infty(E, E')$ *where both E and* E' *are finite dimensional. Then* Σ *is a minimal realization of* G *if and only if* Σ *is a controllable and observable realization of* G. *In particular, all controllable and observable realizations of* G *are similar.*

PROOF. If Σ is a minimal realization of G, then Proposition 7.1 shows that Σ is similar to the backward shift realization of G. The backward shift realization is controllable and observable. Because similarity preserves controllability and observability, Σ is controllable and observable.

If $\Sigma = \{A, B, C, X\}$ is a controllable realization of G, then (7.6) implies that $H^{'} = \Theta_{\Sigma} X$. If moreover Σ is observable, then Θ_{Σ} is one to one. Therefore Θ_{Σ} viewed as a map from X into $H^{'}$ defines an invertible operator W from X onto $H^{'}$ by $W = \Theta_{\Sigma}$. The proof of Proposition 7.1 shows that

$$(S^{'*} | H^{'})W = WA \quad \text{and} \quad C_o W = C \quad \text{and} \quad B_G = WB .$$

Therefore Σ is similar to the backward shift realization. By Proposition 7.1 the backward shift realization is minimal. Hence Σ is minimal. This completes the proof.

To compute a matrix representation for the backward shift realization, recall that any rational G in $H^{\infty}(E, E^{'})$ admits a factorization of the form $G = e^{-it} m F^{*}$, where m is the Blaschke product of order d in (1.17) and F is a rational function in $H^{\infty}(E^{'}, E)$. By Lemma IX.8.1 and (7.4) the space $H^{'} \subseteq H(mI^{'})$. In particular, $S^{'*} | H^{'} = S(mI^{'})^{*} | H^{'}$. According to Theorem 1.5 the operator $S(mI^{'})^{*}$ is unitarily equivalent to $I^{'} \otimes M^{*}$ on $l_d^2(E^{'})$ under the identification $\phi_j g$ to $e_j g$ where g is in $E^{'}$ and M is the lower triangular matrix in (1.19). (The tensor product $\otimes$ is defined in (5.8) and the orthonormal basis $\{\phi_j\}_1^d$ for $H(m)$ in (1.18).) Under this identification Lemma 6.1 implies that the operator $G_+ | E$ becomes

$$B^{'} = [G_1(\alpha_1)d_1, G_2(\alpha_2)d_2, ..., G_d(\alpha_d)d_d]^{tr} \tag{7.13}$$

where $G_k(z)$ is computed recursively using (6.2) with the $h_k^{'s}$ replaced by the $G_k^{'s}$. The operator $P_{E^{'}} | H(mI^{'})$ is identified with

$$C^{'} = [\phi_1(0)I^{'}, \phi_2(0)I^{'}, ..., \phi_d(0)I^{'}] . \tag{7.14}$$

If we define H_b as the subspace of $l_d^2(E^{'})$ by

$$H_b = \bigvee_o^{\infty} (I^{'} \otimes M^{*})^n B^{'} E , \tag{7.15}$$

then our identification shows that the backward shift realization is unitarily equivalent to

$$\Sigma_b^{'} = \{(I^{'} \otimes M)^{*} | H_h, B_b, C^{'} | H_b, H_b\}$$

where B_b is the operator from E to H_b defined by $B_b = B^{'}$. Let $\{x_i\}_1^n$ be an orthonormal basis for H_b and consider the isometry W from $\mathbb{C}^n$ to $l_d^2(E^{'})$ defined by

$$W = [x_1, x_2, ..., x_n] . \tag{7.16}$$

The range of W is H_b. It is now obvious that $\Sigma_b^{'}$ is unitarily equivalent to the following system:

$$\Sigma = \{ W^*(I' \otimes M^*)W, \; W^*B', \; C'W, \; C^n \} . \tag{7.17}$$

The previous analysis provides a procedure for computing a minimal realization Σ of G, which is unitarily equivalent to the backward shift realization of G. This is done first by factoring G into $G = e^{it}mF^*$ and then by computing M, B$'$ and C$'$. Next find a complete orthonormal basis $\{x_i\}_1^n$ for H_b. Finally, Σ is given by (7.17). Since Σ is unitarily equivalent to the backward shift realization its observability Grammian $O_\Sigma = I$.

If $E = E' = C^1$ and in the factorization $G = e^{-it}mF^*$ the functions m and F are prime, then Theorem IX.8.2 and (7.4) show that $H' = H(m)$. In this case the backward shift realization is unitarily equivalent to $\Sigma = \{ M^*, B', C', C^d \}$. Thus in this case one does not have to calculate the space H_b and the operator W, since $H_b = C^d$.

Let $\Sigma = \{A, B, C, X\}$ be a realization of G. Using $S'^* \Theta_\Sigma = \Theta_\Sigma A$ and $S^* \Omega_\Sigma = \Omega_\Sigma A^*$ it follows that $\Theta_\Sigma \Omega_\Sigma^*$ is a Hankel operator in $I(S, S'^*)$. Equation (7.1) gives $Gf = \Theta_\Sigma Bf$ for all f in E. Also equation (7.9) gives $\Omega_\Sigma^* f = Bf$. We conclude that $\Gamma(G)f = Gf = \Theta_\Sigma \Omega_\Sigma^* f$. Because E is cyclic for S, and $\Gamma(G)$ and $\Theta_\Sigma \Omega_\Sigma^*$ are both Hankel operators, the equation $\Gamma(G)f = \Theta_\Sigma \Omega_\Sigma^* f$ produces

$$\Gamma = \Gamma(G) = \Theta_\Sigma \Omega_\Sigma^* . \tag{7.18}$$

Therefore H' which is the range of Γ is contained in the range of Θ_Σ; also the range of Γ^* which we denote by H is contained in Ω_Σ. Moreover, if Σ is minimal, then Θ_Σ is one to one and Ω_Σ^* is onto. In this case H' equals the range of Θ_Σ and H equals the range of Ω_Σ. Consulting (7.7), (7.10) and (7.18) we have

$$\Gamma^* \Gamma \Omega_\Sigma = \Omega_\Sigma O_\Sigma C_\Sigma \quad \text{and} \quad \Gamma \Gamma^* \Theta_\Sigma = \Theta_\Sigma C_\Sigma O_\Sigma . \tag{7.19}$$

In particular, if Σ is minimal (7.19) shows that $\Gamma^* \Gamma | H$ is similar to $O_\Sigma C_\Sigma$ and $\Gamma \Gamma^* | H'$ is similar to $C_\Sigma O_\Sigma$.

If $\Sigma = \Sigma_b$ is the backward shift realization, then Θ_Σ is an isometry $\Gamma \Gamma^* | H'$ is unitarily equivalent to C_{Σ_b}. In this case the second equation in (7.19) becomes $\Gamma \Gamma^* \Theta_{\Sigma_b} = \Theta_{\Sigma_b} C_{\Sigma_b}$. It is emphasized that C_{Σ_b} can be computed. This is done first by computing the backward shift realization. Then C_{Σ_b} is obtained by solving the corresponding Lyapunov equation in (7.11). By summing up the previous analysis we obtain the following result.

7.3 PROPOSITION. *Let Σ_b be the backward shift realization of G and set* $\Gamma = \Gamma(G)$. *Then Θ_{Σ_b} is an isometry whose range is H' and*

$$\Gamma\Gamma^*\Theta_{\Sigma_b} = \Theta_{\Sigma_b}C_{\Sigma_b} \ . \tag{7.20}$$

In particular, the nonzero eigenvalues of $\Gamma\Gamma^*$ *equal the eigenvalues of* C_{Σ_b} *and* $\|\Gamma\|^2 = \|C_{\Sigma_b}\|$.

7.4   REMARK.   Using $S(mI')^*|H' = S^{'*}|H'$ it is easy to verify that $\Sigma' = \{S^*(mI'), G_+|E, P_{E'}|H(mI'), H(mI')\}$ is a not necessarily minimal realization of G. (Recall that $G = e^{-it}mF^*$.) Moreover, $\Theta_{\Sigma'}$ is an isometry and $O_{\Sigma'} = I$. Furthermore Σ' is unitarily equivalent to $\Sigma_m = \{I' \otimes M^*, B', C'\}$. Equation (7.19) and $O_{\Sigma_m} = I$ implies that $\Gamma\Gamma^*\Theta_{\Sigma_m} = \Theta_{\Sigma_m}C_{\Sigma_m}$. Since Θ_{Σ_m} is an isometry the nonzero eigenvalues of $\Gamma\Gamma^*$ equal the nonzero eigenvalues of C_{Σ_m}. In particular, $\|\Gamma\|^2 = \|C_{\Sigma_m}\|$. So one can compute the norm of Γ or even the eigenvalues of $\Gamma\Gamma^*$, by computing the norm of C_{Σ_m} or the eigenvalues of C_{Σ_m}. The advantage in using Σ_m over the backward shift realization Σ_b to compute the eigenvalues of $\Gamma\Gamma^*$ is that one does not have to compute H_b. The disadvantage in using Σ_m is that C_{Σ_m} lives on a space whose dimension can possibly be much larger than that of H_b. So one must compute C_{Σ_m} and its eigenvalues on a larger space than H_b.

7.5   COROLLARY. *Let* $E = E' = C^1$ *and* Σ_b *be the backward shift realization of* G. *Let* x *be any nonzero eigenvector satisfying* $C_{\Sigma_b}x = \lambda^2 x$ *where* λ^2 *is the largest eigenvalue for* C_{Σ_b}. *Then* $K_* = (\lambda^2\Theta_{\Sigma_b}x)/(J\Omega_{\Sigma_b}x) - G$ *is the unique element in* K_0^∞ *solving the following* H^∞ *optimization problem:*

$$\inf\|G + K_0^\infty\|_\infty = \|G + K_*\|_\infty = \|\Gamma(G)\| = \lambda \ . \tag{7.21}$$

PROOF. By using $\Gamma = \Theta_{\Sigma_b}\Omega_{\Sigma_b}^*$ with the fact that Θ_{Σ_b} is an isometry we have

$$\Gamma\Omega_{\Sigma_b}x = \Theta_{\Sigma_b}C_{\Sigma_b}x = \lambda^2\Theta_{\Sigma_b}x \ . \tag{7.22}$$

Notice that $\|\Omega_{\Sigma_b}x\|^2 = (C_{\Sigma_b}x, x) = \lambda^2\|x\|^2 = \|\Gamma\|^2\|x\|^2$ (see Proposition 7.3). Therefore (7.22) yields

$$\|\Gamma\Omega_{\Sigma_b}x\| = \lambda^2\|\Theta_{\Sigma_b}x\| = \|\Gamma\|\,\lambda\|x\| = \|\Gamma\|\,\|\Omega_{\Sigma_b}x\| \ . \tag{7.23}$$

Since the backward shift realization is controllable Ω_{Σ_b} is one to one. Therefore $\Omega_{\Sigma_b}x$ is a nonzero vector on which Γ attains its norm (see (7.23)). Now (7.2.1) follows from (7.22) and Corollary IX.3.5. This completes the proof.

We point out that the previous corollary allows one to compute the optimal K_* in K_0^∞ solving the Nehari optimization problem (7.21). This is done by first computing the backward shift realization $\Sigma_b = \{M^*, B', C', C^d\}$ where $G = e^{-it}mF^*$ is a factorization such that m and F are prime (see Section IX.8). Then one must solve the Lyapunov equation (7.11) with $\Sigma = \Sigma_b$ and find the largest eigenvalue λ^2 and corresponding eigenvector x for C_{Σ_b}. Finally, K_* is given by $K_* = ((\lambda^2 \Theta_{\Sigma_b} x)/(J\Omega_{\Sigma_b} x) - G)$. The vector $\Theta_{\Sigma_b} x$ and $\Omega_{\Sigma_b} x$ can be computed by standard procedures for calculating the resolvent of a matrix or by using the fact that M^* is upper triangular.

Let $\Sigma = \{A, B, C, H\}$ be the backward shift realization of $\tilde{G}$. Then $\Sigma_f = \{A^*, C^*, B^*, H\}$ is called the *forward shift realization* of G. Obviously Σ_f is a minimal, controllable and observable realization of G. The system Σ_f is called the forward shift realization because the state space operator $A^* = P_H S \,|\, H$ is the compression of the unilateral, or forward shift S to H. Since $\Theta_\Sigma (= \Omega_{\Sigma_f})$ is an isometry $C_{\Sigma_f} = I$. To complete this section we present the following results which are the dual of Proposition 7.3 and Corollary 7.5. The proofs are similar to our earlier proofs and therefore omitted.

7.6 PROPOSITION. *Let Σ_f be the forward shift realization of G and set $\Gamma = \Gamma(G)$. Then Ω_{Σ_f} is an isometry whose range is H and*

$$\Gamma^* \Gamma \Omega_{\Sigma_f} = \Omega_{\Sigma_f} O_{\Sigma_f} \; . \tag{7.24}$$

In particular, the nonzero eigenvalues of $\Gamma^* \Gamma$ equal the eigenvalues of O_{Σ_f} and $\|\Gamma\|^2 = \|O_{\Sigma_f}\|$.

7.7 COROLLARY. *Let $E = E' = C^1$ and Σ_f be the forward shift realization of G. Let x be any nonzero eigenvector satisfying $O_{\Sigma_f} x = \lambda^2 x$ where λ^2 is the largest eigenvalue for O_{Σ_f}. Then $K_* = (\Theta_{\Sigma_f} x)/(J\Omega_{\Sigma_f} x) - G$ is the unique element in K_0^∞ solving the following H^∞ optimization problem:*

$$\inf \|G + K_0^\infty\|_\infty = \|G + K_*\|_\infty = \|\Gamma(G)\| = \lambda \; . \tag{7.25}$$

We conclude this section with some comments on the actual implementation of the computational procedures presented in the last part of Section 6 as well as this section. In particular, a basic role in these procedures is played by the factorization $G = e^{-it} m F^*$ from which we obtained a matrix representation for the Hankel operator $\Gamma(G)$. To compute this factorization one must compute the poles of G. Computing the zeros of a polynomial can introduce some numerical errors. In many problems such as the Hermite-Fejér interpolation problems the poles of G are known. However, in other problem the poles are not known. We will show how one can use the backward shift realization to compute a minimal realization of G without computing the poles of G. This method is demonstrated in the following example.

7.8 EXAMPLE. In this example we will use the backward shift realization to compute a minimal realization of

$$G(z) = \frac{z+2}{z^3 + 2z^2 - 4z - 8} \; . \tag{7.26}$$

First we compute the space H' in (7.4). To this end we set $G(z) = \gamma_1$ and use (1.6) to obtain

$$\gamma_2 = S^* \gamma_1 = \frac{.25(z^2 + 2z)}{z^3 + 2z^2 - 4z - 8} \; . \tag{7.27}$$

Obviously γ_1 and γ_2 are linearly independent in H^2. Another calculation using (1.6) shows that $S^* \gamma_2 = .25\gamma_1$. This implies that the space H' is spanned by γ_1 and γ_2. Now we identify γ_1 and γ_2 with e_1 and e_2 the standard orthonormal basis for C^2. Under this identification $S^* | H'$, B_G and C_o become the matrices A on C^2, B and C defined by

$$A = \begin{bmatrix} 0 & .25 \\ 1 & 0 \end{bmatrix} \quad \text{and} \quad B = \begin{bmatrix} 1 \\ 0 \end{bmatrix}$$

$$C = [-.25, \, 0] = [\gamma_1(0), \, \gamma_2(0)] \; . \tag{7.28}$$

Proposition 7.1 guarantees that $\Sigma = \{A, B, C, C^2\}$ is a minimal realization of G. Moreover, since γ_1 and γ_2 are not orthogonal Σ is similar and not unitarily equivalent to the backward shift realization of G.

We note that in the minimal realization (7.28) of G the state space C^2 is of dimension 2 although the denominator of G is of degree 3. The obvious explanation is that the denominator and numerator have a common factor $z+2$ and therefore the function G is actually $(z^2 - 4)^{-1}$. In engineering this fact is expressed by saying that there is a pole zero cancellation in G. We emphasize that the method presented in this section and illustrated by this example, allows us to obtain a minimal realization without cancelling all the common poles and zeros.

The previous example demonstrates how one can compute a minimal realization of G without computing the poles of G. This is done by using the rational form $G = N/q$ to compute a basis $\{x_i\}_1^n$ for H'. Then a minimal realization $\{A, B, C, C^n\}$ of G is obtained by identifying $\{S^* | H', B_B, C_o\}$ with $\{A, B, C\}$. Now assume that $\Sigma = \{A, B, C, X\}$ is any minimal realization of G. Let W be the positive square root of O_Σ. Since Σ is minimal it is observable. So O_Σ is strictly positive and W is invertible. Consider the minimal realization Σ_1 of G defined by

$$\Sigma_1 = \{WAW^{-1},\ WB,\ CW^{-1},\ C^n\} \tag{7.29}$$

Obviously Σ_1 is similar to Σ. Consulting (7.8) it is easy to verify that the observability Grammian $O_{\Sigma_1} = I$. This implies that Θ_{Σ_1} is an isometry. The proof of Proposition 7.1 shows that Σ_1 is unitarily equivalent to the backward shift realization of G. Therefore one can obtain a realization Σ_1 which is unitarily equivalent to the backward shift realization without computing the zeros of G. This is done by first using the method in Example 7.8 to compute a minimal realization of G. Then solving the Lyapunov equation (7.8) to obtain O_Σ. Finally, Σ_1 is given by (7.29) where W is the positive square root of O_Σ. By applying the above method to $\tilde{G}$ one can also obtain a minimal realization which is unitarily equivalent to the forward shift realization of G without computing the poles of G.

Due to the previous observations we say that Σ is a *backward shift realization of* G if Σ is unitarily equivalent to the backward shift realization of G, or equivalently, Σ is a minimal realization and $O_\Sigma = I$. Analogously we say that Σ is a *forward shift realization of* G is Σ is unitarily equivalent to the forward shift realization of G, or equivalently, Σ is a minimal realization and $C_\Sigma = I$. Now let $\Sigma_b = \{A_b, B_b, C_b, X_b\}$ be a backward shift realization of G and $\Sigma_f = \{A_f, B_f, C_f, X_f\}$ be a forward shift realization of G. Since both of these realizations are minimal there exists an invertible transformation W from X_f onto X_b intertwining Σ_f with Σ_b, that is, the corresponding conditions in (7.3) hold $(\Sigma = \Sigma_f$ and $\Sigma_1 = \Sigma_b)$. This readily implies that $W\Omega_{\Sigma_f}^* = \Omega_{\Sigma_b}^*$. Using this in $\Gamma(G) = \Theta_{\Sigma_b}\Omega_{\Sigma_b}^*$ we have

$$\Gamma(G) = \Theta_{\Sigma_b} W\, \Omega_{\Sigma_f}^* \quad \text{and} \quad \Gamma(G)S = \Theta_{\Sigma_b} WA_f\Omega_{\Sigma_f}^* \tag{7.30}$$

where the second equation follows from $S^*\Omega_{\Sigma_f} = \Omega_{\Sigma_f}A_f^*$. The operators Θ_{Σ_b} and Ω_{Σ_f} are both isometries, because Σ_b is the backward shift realization and Σ_f is the forward shift realization. Therefore $\Gamma(G)$ is (unitarily) identified with $W \oplus 0$. In particular, the nonzero singular values of $\Gamma(G)$ are precisely the singular values of W. Summing up the previous analysis produces the following remark.

7.9 REMARK. Let Σ_b be the backward shift realization of G and Σ_f be its forward shift realization. Let W from X_f to X_b be the invertible transformation intertwining Σ_f with Σ_b. Then Θ_{Σ_b} and Ω_{Σ_f} are both isometries and (7.30) holds. In other words the Hankel operator $\Gamma(G)$ is (unitarily) identified with the operator $W \oplus 0$ where W is the invertible transformation which intertwines Σ_f with Σ_b. Moreover, if $E = E' = C^1$, then the unique solution to the H^∞ optimization problem in (7.25) is given by $K_* = (\Theta_{\Sigma_b}Wx)/(J\Omega_{\Sigma_f}x) - G$ where x is any nonzero vector in X_f satisfying

$\| Wx \| = \| W \| \| x \|$. Finally, it is emphasized that one can obtain both Σ_b and Σ_f without computing the poles of G.

One can easily compute W from the forward and backward shift realizations with the identity $W \Omega_{\Sigma_f}^* = \Omega_{\Sigma_b}^*$. To see this notice that $C_{\Sigma_f} = I$ gives $W = \Omega_{\Sigma_b}^* \Omega_{\Sigma_f}$. Using the fact that

$$X = \Omega_{\Sigma_b}^* \Omega_{\Sigma_f} = \sum_0^\infty A_b^n B_b B_f^* A_f^{*n}$$

it follows that X satisfies the following Lyapunov equation

$$X = A_b X A_f^* + B_b B_f^* . \tag{7.31}$$

Therefore the intertwining operator W is given by $W = X$ where X is the solution the Lyapunov equation in (7.31). Finally, it is noted that if $G = \tilde{G}$, then one can always choose $A_f = A_b^*$, $B_f = C_b^*$ and $C_f = B_b^*$. This means that one only has to compute a forward or backward shift realization which simplifies some of the above computations.

If W is the positive square root of C_{Σ_b}, then it is easy to verify that

$$A_f = W^{-1} A_b W, \ B_f = W^{-1} B_b \text{ and } C_f = C_b W$$

is a forward realization of G, or equivalently, $C_{\Sigma_f} = I$. In this case the operator W which intertwines the forward shift realization with the backward shift realization is precisely the positive square root of C_{Σ_b} and of course (7.30) holds. However, if both forward and backward shift realizations are known, then (7.31) provides us with a method of computing W without taking the square root. This can be advantageous in certain situations for instance when $G = \tilde{G}$.

8. A HANKEL VERSION OF THE SCHUR-COHN TEST REVISITED

In this section we will present a Hankel version of the Schur-Cohn test in Section V.5, and the interpretation of the norm of certain Hankel operators as the distance between adequate $H(m)$ spaces.

To begin recall that if f is in H^∞ and m is an inner function, then A(f) on $H(m)$ is the compression of f_+ to $H(m)$. The functional calculus in Section VI.6 shows that $A(f) = f(S(m))$. As before if Q is in H^∞, then the Hankel operator $\Gamma(Q)$ on H^2 is $\Gamma(Q) = P_+ M_Q J | H^2$ where P_+ is the orthogonal projection of L^2 onto H^2. The following lemma plays a fundamental role in this section.

8.1 LEMMA. *If* b *and* m *are two finite Blaschke products, then*

$$\|(I-P_{H(b)})\,|\,H(m)\| = \|b_+^*\,|\,H(m)\| = \|A(b)\| = \|\Gamma(e^{-it}m\overline{b})\| \,. \tag{8.1}$$

Moreover, the norm of $\Gamma = \Gamma(e^{-it}m\overline{b})$ *is one if and only if the order of* b *is strictly less than the order of* m.

PROOF. Since the range of the isometry b_+ is bH^2, the kernel of the co-isometry b_+^* is $H(b)$. The first equation in (8.1) follows from $b_+^* = b_+^*(I-P_{H(b)})$ and the fact that b_+^* is an isometry on bH^2. The second equality in (8.1) follows from the fact that b_+ is a lifting of $A(b)$, that is

$$P_H b_+ = P_H b_+ P_H = A(b)P_H \quad \text{or} \quad A(b)^* = b_+^* \,|\, H$$

where $H = H(m)$ and P_H denotes the orthogonal projection of H^2 onto H.

To prove the last equality let P_- be the orthogonal projection onto K_o^2. We claim that

$$P_H = M_m P_- M_{\overline{m}} \,|\, H^2 \,. \tag{8.2}$$

Let P be the operator on the right hand side of (8.2). Obviously $P^2 = P = P^*$. So P is an orthogonal projection. The range of P is

$$PH^2 = M_m P_- M_{\overline{m}}(H(m) \oplus mH^2) = M_m P_- M_{\overline{m}} H(m) = M_m M_{\overline{m}} H(m) = H(m) \,.$$

Thus P is the orthogonal projection onto $H(m)$ which proves our claim. Using (8.2) and $VP_+JV = JP_-$ where V is the bilateral shift on L^2 we have

$$\|\Gamma(e^{-it}\widetilde{m}\overline{\widetilde{b}})h\| = \|P_+ e^{-it}\widetilde{m}\overline{\widetilde{b}}Jh\| = \|P_+ Je^{it}\overline{m}bh\| =$$
$$\|P_- M_{\overline{m}}bh\| = \|P_H bh\| = \|A(b)P_H h\| \quad (h \in H^2) \tag{8.3}$$

which shows that $\|\Gamma(e^{-it}\widetilde{m}\overline{\widetilde{b}})\| = \|A(b)P_H\| = \|A(b)\|$. Because $\Gamma(e^{-it}\widetilde{m}\overline{\widetilde{b}})$ is the adjoint of Γ this yields the last equality in (8.1). To complete the proof assume that the norm of Γ is one. According to (8.1) the norm of $(I - P_{H(b)})\,|\,H$ is one. This implies that there exists a nonzero h in $H(m) \cap bH^2$. Now let d be the order of m and d' the order of b. Because h is in $H(m)$, Lemma 1.1 shows that $h = p/q$ where p is a polynomial of degree less than or equal to $d-1$ and q is a polynomial of degree at most d with no zeros in $\overline{D}$. In fact q is the denominator of m. Since $p/q = h = bg$ for some g in H^2, it follows that the number of zeros counted according to their multiplicity in D of b, that is, d' is less than or equal to the number of zeros in D of p which is less than or equal to $d-1$. Therefore $d' < d$.

On the other hand if $d' < d$, then Lemma 1.1 shows that $\dim H(b) < \dim H(m)$. This readily implies that $H(m)$ contains a nonzero vector x which is orthogonal to $H(b)$. Therefore $(I - P_{H(b)})x = x$. Hence the norm of $(I-P_{H(b)})|H$ one, or equivalently, the norm of Γ is one. This completes the proof.

An alternate proof of the last statement of the previous lemma can be obtained by applying Corollary V.5.4 to $A(b) = b(S(m))$. This is left to the reader as an instructive exercise. We also note that the proof of the relations in (8.1) is valid even in the general case when m and b are inner functions in H^∞. However, in the proof of the last statement m has to be a finite Blaschke product and b can be any inner function. The following is a version of the Schur-Cohn test in Section V.5 based on Hankel operators. It is probably useful to recall first that a polynomial is stable if all its zeros are in D.

8.2 THEOREM. *Let* p *be a polynomial of degree* d *and set* $f = p/p^{\#}$ *where* $p^{\#} = J_{d+1}p$ *is the reverse polynomial of* p. *Let* m *be any Blaschke product of order* d. *Then* p *is a stable polynomial if and only if the Hankel operator* $\Gamma = \Gamma(e^{it}m\bar{f})$ *is a strict contraction. In this case f is in* H^∞, *and* $\|\Gamma\| = \|A(f)\|$.

PROOF. If p is stable, then f is a Blaschke product of order d. Thus f is in H^∞. Lemma 8.1 shows that Γ is a strict contraction and that $\|\Gamma\| = \|A(f)\|$. On the other hand, if p is unstable and α_1, α_2, ..., α_d are the zeros of p, then

$$f = \gamma \prod_{1}^{d} \frac{z-\alpha_i}{1-\bar{\alpha}_i z}$$

where γ is a constant of modulus one. Because $|\alpha_i| \geq 1$ for at least one i the previous equation shows that on the unit circle $f = f_c \bar{f}_a$ where f_c is a Blaschke product of order $d_c < d$ and f_a is a Blaschke product of $d_a < d$. (Obviously $d_c + d_a \leq d$. However, we can have a pole zero cancellation. In that case $d_c + d_a < d$.) Now Γ becomes $\Gamma = \Gamma(e^{-it}mf_a\bar{f}_c)$. Lemma 8.1 with $b = f_c$ and m replaced by mf_a shows that the norm of Γ is one. This completes the proof.

One can also use Rouche's Theorem to obtain an alternate proof of part of the previous theorem, that is, if Γ is a strict contraction, then p is stable. To this end recall that Rouche's theorem says that if f and g are rational functions with no poles or zeros on the unit circle and

$$|f(e^{it}) + g(e^{it})| < |f(e^{it})| \qquad \text{(for all } t \in [0, 2\pi)) , \qquad (8.4)$$

then $Z_f - P_f = Z_g - P_g$ where Z_h (respectively P_h) is the number of zeros (respectively poles) of h in D counted according to their multiplicity. Now assume that the Hankel matrix $\Gamma(e^{-it}m\bar{f})$ is a strict contraction. Then Remark 6.4, or Corollary IX.3.5 shows that there exists a rational function k in K_o^∞ satisfying

$$1 > \|\Gamma\| = \|e^{-it}m\bar{f} + k\|_\infty = \|\bar{f} + e^{it}\bar{m}k\|_\infty = \|f + mh\|_\infty$$

where h is the function in H^∞ defined by $h = e^{-it}\bar{k}$. Since $|f| = 1$ on the unit circle (8.4) is satisfied for f and $g = mh$. By Rouche's Theorem $Z_f - P_f = Z_{mh} \geq d$. However, f can have at most d zeros. So all the zeros of f are in D. Therefore p is stable and this is precisely what we wanted to prove. Finally, let us make a supplementary remark, namely that $P_f = 0$ and the previous inequality becomes $d = Z_f = Z_m + Z_h = d + Z_h$. Thus $Z_h = 0$ and the rational function $h = e^{-it}\bar{k}$ in H^∞ is outer.

According to Theorem 1.6 the Schur-Cohn test given in Theorem V.5.3 is equivalent to: The polynomial p of degree d is stable if and only if

$$N = p^\#(T)^* p^\#(T) - p(T)^* p(T) \tag{8.5}$$

is strictly positive where $T = S(m)$ and $p^\# = J_{d+1}p$. Theorem 8.2 shows that N is strictly positive if and only if the Hankel operator $\Gamma = \Gamma(e^{-it}m\bar{f})$ is a strict contraction where $f = p/p^\#$. To establish a direct link between Γ and N we assume that p is stable and we consider the following generalized eigenvalue problem:

$$\rho = \inf\{\lambda \geq 0: \ \lambda^2 p^\#(T)^* p^\#(T) - p(T)^* p(T) \geq 0\} \ . \tag{8.6}$$

Obviously since N is strictly positive $\rho < 1$, equation (8.6) implies that for all x

$$\{\rho\|p^\#(T)x\| - \|p(T)x\|\} \geq 0$$

with equality for at least one nonzero x. Since $p^\#(T)$ is invertible ρ equals the norm of $p(T)p^\#(T)^{-1} = f(T)$. Because $A(f) = f(T)$, Lemma 8.1 shows that $\|f(T)\| = \|\Gamma\|$. Therefore $\rho = \|\Gamma\|$. In other words if p is stable, or equivalently, $\rho < 1$, then $\|\Gamma\| = \rho$. We note that in general ρ can be any positive number. To see this choose $T = 0$ on C^1 and set $p = z + \alpha$. Then $\rho = |\alpha|$.

To complete this section we will show that the norm of $\Gamma = \Gamma(e^{-it}m\bar{f})$ is precisely the distance between the spaces mH^2 and fH^2. For this we begin by recalling the following definition: If H_1 and H_2 are two subspaces in a Hilbert space K, then the *distance between H_1 and H_2* is defined by $\operatorname{dist}(H_1, H_2) = \|P_1 - P_2\|$ where P_i is the orthogonal projection of K onto H_i for $i = 1, 2$.

Obviously the distance between H_1 and H_2 equals the distance between $K \ominus H_1$ and $K \ominus H_2$. Notice that

$$\|(P_1-P_2)x\|^2 = \|(I-P_2)P_1x - P_2(I-P_1)x\|^2 = \|(I-P_2)P_1x\|^2 + \|P_2(I-P_1)x\|^2 \le \|x\|^2$$

This implies that $\mathrm{dist}\,(H_1, H_2) \le 1$. If $\mathrm{dist}\,(H_1, H_2) < 1$, then $\dim H_1 = \dim H_2$. To prove this we can assume that $\dim H_2 > \dim H_1$. Since $\dim P_2 H_1 \le \dim H_1 < \dim H_2$ there exists a nonzero vector h in H_2 orthogonal to $P_2 H_1$. This h is also orthogonal to H_1. Thus $\|(P_1 - P_2)h\| = \|h\|$, which proves the claim.

The following classical result plays an important role in developing a connection between the norm of Γ and the distance between mH^2 and fH^2.

8.3 THEOREM. *If H_1 and H_2 are two subspaces in K, then*

$$\mathrm{dist}\,(H_1, H_2) = \max\{\|(I-P_1)|H_2\|, \|(I-P_2)|H_1\|\} \ . \tag{8.7}$$

Moreover, if H_1 and H_2 are two finite dimensional subspaces with the same dimension, then the norm of $(I-P_2)|H_1$ equals the norm of $(I-P_1)|H_2$. In this case

$$\mathrm{dist}\,(H_1, H_2) = \|(I-P_1)|H_2\| = \|(I-P_2)|H_1\| \ . \tag{8.8}$$

PROOF. Let $Q_i = I - P_i$ for $i = 1, 2$ and denote by δ the right hand side of (8.7). Notice that $(P_1 - P_2)|H_1 = Q_2|H_1$ and thus $\|P_1 - P_2\| \ge \|Q_2|H_1\|$. By symmetry we obtain $\|P_1 - P_2\| \ge \delta$. On the other hand for any x in K we have

$$\|(P_1 - P_2)x\|^2 = \|Q_2 P_1 x - P_2 Q_1 x\|^2 = \|Q_2 P_1 x\|^2 + \|P_2 Q_1 x\|^2 \le$$
$$\delta^2 \|P_1 x\|^2 + \|(Q_1|H_2)^* Q_1 x\|^2 \le \delta^2 \|P_1 x\|^2 + \delta^2 \|Q_1 x\|^2 = \delta^2 \|x\|^2 \ .$$

It follows that $\|P_1 - P_2\| \le \delta$. This establishes (8.7).

To complete the proof assume that $\dim H_1 = \dim H_2 < \infty$. If the norm of $(I-P_2)|H_1$ is one, then the kernel of $P_2|H_1$ is nonzero. Since $\dim H_1 = \dim H_2$ there exists a nonzero h_2 in H_2 orthogonal to $P_2 H_1$. So h_2 is orthogonal to H_1. This implies that $\|(I-P_1)h_2\| = \|h_2\|$ and the norm of $\|(I-P_1)|H_2\|$ is one. Reversing the roles of H_1 and H_2 shows that the norm of $(I-P_2)|H_1$ is one if and only if the norm of $(I-P_1)|H_2$ is one, or equivalently, by (8.7) the distance between H_1 and H_2 is one.

Now assume that $\mathrm{dist}\,(H_1, H_2) < 1$. If λ is the norm of $A = (I-P_2)|H_1$ from H_1 to K, then there exists a nonzero h_1 in H_1 satisfying

$$P_1(I-P_2)h_1 = A^* A h_1 = \lambda^2 h_1 \ .$$

Multiplying this by P_2 gives

$$P_2(I-P_1)P_2 h_1 = B^* B P_2 h_1 = \lambda^2 P_2 h_1 \tag{8.9}$$

where B is the operator from H_2 to K defined by $B = (I-P_1)|H_2$. Notice that $P_2 h_1$ is not zero. Because if $P_2 h_1 = 0$, then the norm of $(I-P_2)|H_1$ is one, and this contradicts the assumption that $\operatorname{dist}(H_1, H_2) < 1$. Because $P_2 h_1$ is nonzero (8.9) implies that the norm of $(I-P_2)|H_1$ is less than or equal to the norm of $(I-P_1)|H_2$. Reversing the roles of H_1 and H_2 shows that there is equality, that is, the norm of $(I-P_2)|H_1$ equals the norm of $(I-P_1)|H_2$. This completes the proof.

Now we are ready for the second main result of this section.

8.4 THEOREM. *Let* p *be a polynomial of degree* d *and set* $f = p/p^{\#}$ *where* $p^{\#} = J_{d+1} p$ *is the reverse polynomial of p. Let* m *be a Blaschke product of order* d *and set* $\Gamma = \Gamma(e^{-it} m \bar{f})$. *Then*

$$\|\Gamma\| = \operatorname{dist}(mH^2, fH^2) . \tag{8.10}$$

In particular, p *is stable if and only if the distance between* mH^2 *and* fH^2 *is strictly less than one. In this case*

$$\|\Gamma\| = \|A(f)\| = \operatorname{dist}(H(m), H(f)) = \|\Gamma(e^{-it} f \bar{m})\| . \tag{8.11}$$

PROOF. If Γ is a strict contraction, then p is stable and f is a Blaschke product of order d. In this case (8.1) shows that the norm of Γ equals the norm of $(I-P_{H(f)})|H(m)$ and that of $A(f)$. Equation (8.8) in Theorem 8.3 produces (8.11). So (8.10) is valid when Γ is a strict contraction. If the norm of Γ is one, then p is unstable and $f = f_c \bar{f_a}$ where f_c is a Blaschke product of order $d_c < d$ and f_a is a Blaschke product of order $d_a \le d$. Since $\dim H(mf_a) > \dim H(f_c)$ the distance between $H(mf_a)$ and $H(f_c)$ is one. So the distance between $mf_a H^2$ and $f_c H^2$ is also one. If b is any rigid function, then $M_b P_+ M_{\bar{b}}$ is the orthogonal projection onto bH^2. This implies that

$$1 = \operatorname{dist}(mf_a H^2, f_c H^2) = \|P_{mf_a H^2} - P_{f_c H^2}\| =$$
$$\|M_{mf_a} P_+ M_{\overline{mf_a}} - M_{f_c} P_+ M_{\overline{f_c}}\| = \|P_{mH^2} - P_{fH^2}\|$$

and thus (8.10) is satisfied in this case too. The proof is now complete.

Finally, it is noted that the previous theorem along with Proposition IX.7.5 and (8.1) show that if m and b are two finite Blaschke products of the same order, then

$$\inf \| b + mH^\infty \|_\infty = \inf \| m + bH^\infty \|_\infty . \tag{8.12}$$

X.9. NOTES AND COMMENTS

Recall that classical interpolation theory began around 1907 with Carathéodory [Ca 1], Carathéodory-Fejér in 1911 [CaFe] and Nevanlinna-Pick in 1916 [Nev 1], [Pi1]. The form of the Nevanlinna-Pick Theorem 3.2 for infinitely many data points was obtained in 1938 [KrR]. For a historical perspective on classical interpolation see [Heins] and [Szegö]. After Sarason's paper [Sa 2] in 1967 it became quite clear that operator versions of classical interpolation problems are special cases of the commutant lifting theorem. This approach was already exploited in the monographs, [Ni] and [RosR3]. (A treatment of some of the results in Rosenblum and Rovnyak's book is given in the next chapter.) The geometric study of S(m) operators was initiated in [Sz.-NF 4], but their connection with classical interpolation was discovered by Sarason in [Sa 2]. Our presentation in Section 1 follows classical lines as in [Ni]. The matrix representation M in (1.19) for a finite dimensional S(m) operator is taken from [PtY 2]. Triangular representation of S(m) operators is also presented in [Ni]. Our approach of using the geometric structure of S(m) operators along with the commutant lifting theorem to solve the operator versions of Carathéodory, Nevanlinna-Pick and Hermite-Fejér interpolation problems in Sections 2 to 4 was inspired by the work of Sarason [Sa 2]. (These results are also presented in the next chapter by using a momentum theorem of Rosenblum and Rovnyak's.) Here we have tried to pay a special attention to uniqueness results, and in this case provide a formula to actually compute the contractive intertwining lifting. The tangential form of the Hermite-Fejér interpolation problem in Section 4 is not an artificial generalization of the classical Hermite-Fejér interpolation problem. It was first considered by Fedcina [Fed]. Problem 4.1 naturally occurs in H^∞ control theory (see the monographs [Fran 2] and [Helt 5]). The idea of using the lower triangular matrix M in (1.19) to solve the classical Hermite-Fejér interpolation Problem 5.1 came from [AllY] and [Y 3]. The connection between Hankel operators and Hermite-Fejér interpolation was established in [AAK 1,3]. Our connection relating Hankel operators to A(F) operators by using the Douglas–Shapiro–Shields factorization is taken from [AAK 2] and [Fu 3]. We also notice that besides using the commutant lifting theorem there are at least two other approaches to solving these interpolation problems. One approach developed in [AAK 1,2,3,4] is to reduce these problems to Nehari interpolation problems and then solve these Nehari problems. Another approach developed by Ball and Helton [BH] is based on their extension of the Beurling-Lax-Halmos theorem from Hilbert spaces to Krein space (that is, a space with an indefinite metric induced by a self adjoint unitary operator not equal to the identity.) We refer the reader to the interesting monograph [Helt 5] for further details. The matrix representation for finite rank Hankel operators involving the Euclid-Horner algorithm is taken from [FrzL]. Shift realizations for linear systems are due to Helton [Helt 1] and Furhmann [Fu 3]. For further results on shift realizations and historical comments see [Fu 3]. Realization theory is also discussed in [Ka 1], [Rugh] and [Sk]. For shift realizations of certain nonlinear systems see [Frz 1,2], [Rugh] and [Won]. Our matrix representation for a Hankel operator in (7.30) is a slight modification of balanced realizations discussed in [Gl 1], [Sk] and elsewhere. In fact the main purpose of Section 7 was to present just enough state space results to obtain the matrix representation for a Hankel operator in (7.30). For further results on how state space techniques can be used to solve the Nehari H^∞ optimization problem see [BGR], [BR 1,2,3], [Gl 1,2] and [Fran 2]. Finally, Theorem 8.3 concerning the distance between two subspaces is classical; see [AkG] and [RieszSz.-N]. The Schur-Cohn test in Theorem 8.2 is taken from [FrzL]. However, this result is just a Hankel version of the Schur- Cohn tests in [PtY 1,2]. In fact many of the results in this section follow from the results in [PtY 1,2] and [Y 1,2]. For other results concerning the angle between subspaces and Hankel operators see [Ni]. In connection with this chapter see [Ak], [AlpD], [BGR], [Br 3], [Ca], [CaFe], [DeGK], [Fed], [Fejér], [FoT 2], [FoTZ], [Helt 2,3], [KrN], [Nev], [Nu], [Sz.-NK] and [Wa].

CHAPTER XI

INTERPOLATION AS A MOMENTUM PROBLEM

In the scalar setting the Carathéodory interpolation Problem 2.4 in Chapter X is equivalent to the following: Given b and c in C^n find necessary and sufficient conditions for the existence of a contractive analytic f in H^∞ satisfying $f(T)b = c$ where T is the lower shift matrix on C^n. In this chapter we will present a fundamental momentum theorem due to Rosenblum and Rovnyak, to obtain necessary and sufficient conditions for the existence of contractive analytic F in $H^\infty(E, E')$ satisfying $F(T)b = c$. Here b and c are arbitrary operators and F(T) defined below is generalization of the function of an operator T. In the first section the commutant lifting theorem will be used to prove the basic momentum result. In the subsequent sections we will follow Rosenblum and Rovnyak's ideas to solve the Carathéodory, Nevanlinna-Pick, Hermite-Fejér and Loewner interpolation problems.

1. THE BASIC MOMENTUM RESULT

This section is devoted to the Rosenblum and Rovnyak momentum theorem, along with some simple sufficient conditions for the uniqueness condition provided by this theorem.

First we establish some notation. As before E and E' are separable Hilbert spaces. Recall that $L(E, E')$ denotes the Banach space of all operators mapping E into E'. Let T be a linear map on a complex linear space X. (A linear map from X to Y is a linear function from X to Y.) The space X is not necessarily a Hilbert space and a topological structure on X is not needed. The set of all linear maps from X into E is denoted by $L_0(X, E)$. The *dual of* T *with respect to* $L_0(X, E)$ is the linear map T^d on $L_0(X, E)$ defined by:

$$x'(Tx) = <Tx, x'> = <x, T^d x'> = T^d x'(x) \qquad (x \in X \quad \text{and} \quad x' \in L_0(X, E)). \quad (1.1)$$

Notice that we have used the notation $<x, x'> = x'(x)$ to represent the linear map x' acting on x. This reduces to the usual notation for a linear functional when $E = C^1$. In this case T^d is the usual dual of T.

To present the basic momentum result referred to above, we establish the following conventions. Throughout T and T_1 are linear maps on the complex linear (but not necessarily Hilbert) spaces X and X_1 respectively. The dual of T_1 with respect to $L_o(X_1, E')$ is denoted by T_1^d. We say that Y is an *invariant subspace for* $\{T^d, T_1^d\}$ *with respect to* $\{E, E_1\}$, if E from Y into $L_o(X, E)$ and E_1 from Y into $L_o(X_1, E')$ are two injective linear maps such that the space $\{[Ey, E_1y]^{tr} : y \in Y\}$ in $[L_o(X, E), L_o(X_1, E')]^{tr} = L_o(X, E) \times L_o(X_1, E')$ is invariant for

$$\begin{bmatrix} T^d & 0 \\ 0 & T_1^d \end{bmatrix} .$$

Notice that in particular this implies that Ey is invariant for T^d and $E_1 Y$ is invariant for T_1^d. We formally define *the resolvent* $R_T(z)$ *of* T by the following formal series

$$R_T(z) \stackrel{\Delta}{=} (I - zT)^{-1} = \sum_0^\infty T^n z^n .$$

The notation $R_T(z)$ for the resolvent of T should not be confused with the notation R_T for the rotation matrix formed by a contraction T. We will always insert the argument z when referring to the resolvent $R_T(z)$ of T, otherwise R_T is the rotation matrix generated by T. For x' in $L_o(X, E)$ we say that $<R_T(z)x, x'>$ is in $H^2(E)$ if

$$<R_T(z)x, x'> \stackrel{\Delta}{=} \sum_0^\infty <T^n x, x'> z^n \quad \text{is in } H^2(E) . \tag{1.2}$$

Clearly $<R_T(z)x, x'>$ is in $H^2(E)$ if and only if

$$\|<R_T(z)x, x'>\|^2 \stackrel{\Delta}{=} \sum_0^\infty \|<T^n x, x'>\|^2 < \infty .$$

Let F be in $H^\infty(E, E')$ and c in X_1 and b in X. We say that $c = F(T)b$ *on* $\{Y, E, E_1\}$ if Y is an invariant for $\{T^d, T_1^d\}$ with respect to $\{E, E_1\}$ and

$$<c, E_1 y> = \sum_0^\infty F_n <T^n b, Ey> \qquad \text{(for all } y \in Y)$$

$$\|<R_T(z)b, Ey>\|^2 < \infty \qquad \text{(for all } y \in Y) \tag{1.3}$$

where F_n is the coefficient of z^n in the power series expansion of F(z) for z in D. If $E = E_1$, then $\{Y, E, E_1\}$ is denoted by $\{Y, E\}$. Because $\{<T^n b, Ey>\}$ is square summable the sum in (1.3) is finite and $<c, E_1 y>$ is well defined. The following result is the basic momentum theorem of this Section.

1.1 THEOREM. *Let Y be invariant for $\{T^d, T_1^d\}$ with respect to $\{E, E_1\}$. Let b be in X and c be in X_1. Then $c = F(T)b$ on $\{Y, E, E_1\}$ for some F in $H_1^\infty(E, E')$ if and only if*

$$\|<R_{T_1}(z)c, E_1 y>\|^2 \leq \|<R_T(z)b, Ey>\|^2 < \infty \quad (\textit{for all } y \in Y). \tag{1.4}$$

For clarification let us notice that (1.4) is equivalent to

$$\sum_0^\infty \| < T_1^n c, E_1 y > \|_{E'}^2 \leq \sum_0^\infty \| < T^n b, Ey > \|_E^2 \quad (\text{for all } y \in Y)$$

where the subscript is used to indicate the space where the norm is taking place.

PROOF. Recall that S and S' are the unilateral shifts on $H^2(E)$ and $H^2(E')$, respectively. Assume that (1.4) holds. Consider the linear maps L from EY into $H^2(E)$ and L_1 from $E_1 Y$ into $H^2(E')$ defined by

$$LEy = <R_T(z)b, Ey> \quad \text{and} \quad L_1 E_1 y = <R_{T_1}(z)c, E_1 y> \quad (y \in Y). \tag{1.5}$$

Equation (1.4) guarantees that these linear maps are well defined. Furthermore by (1.2) and the shifting property (X.1.6) of the backward shift

$$S^* L = L T^d | EY \quad \text{and} \quad S'^* L_1 = L_1 T_1^d | E_1 Y. \tag{1.6}$$

Since EY and $E_1 Y$ are invariant for T^d and T_1^d, respectively, this implies that $\overline{LEY} = H$ and $\overline{L_1 E_1 Y} = H'$ are invariant for S^* and S'^*, respectively. Now consider the operators T_2 on H and T_2' on H' defined by

$$T_2 = S^* | H \quad \text{and} \quad T_2' = S'^* | H'. \tag{1.7}$$

Clearly S^* and S'^* are co-isometric extensions of T_2 and T_2', respectively.

By the equations (1.4) and (1.5):

$$\|L_1 E_1 y\|^2 \leq \|LEy\|^2 \quad (\text{for all } y \in Y). \tag{1.8}$$

So there exists a contraction A mapping H into H' such that $AL = L_1$. Using (1.6) and (1.7) with the fact that Y is invariant for $\{T^d, T_1^d\}$ with respect to $\{E, E_1\}$, it is easy to verify that A is in $I(T_2, T_2')$. By Corollary VII.1.4 of the commutant lifting theorem there exists a contraction B in $I(S^*, S'^*)$ such that $A = B|H$. Since S and S' are unilateral shifts $B^* = G_+$ where G is in $H_1^\infty(E', E)$; see Theorem IX.1.1. Thus,

$$<R_{T_1}(z)c, E_1 y> = L_1 E_1 y = BLEy = P_+ M_{G^*} <R_T(z)b, Ey> \tag{1.9}$$

where y is in Y and P_+ is the orthogonal projection from $L^2(E')$ onto $H^2(E')$. Let $F = \tilde{G}$ and recall that $F_n = G_n^*$ for all $n \geq 0$ where F_n and G_n are the coefficients of z^n in the power series expansion of F and G, respectively. Using this in (1.9) and (1.2) gives

$$<c, E_1 y> = P_{E'} <R_{T_1}(z)c, E_1 y> = P_{E'} M_{G^\cdot} <R_T(z)b, Ey> =$$

$$P_{E'}(\sum_0^\infty G_n^* e^{-int}) \cdot (\sum_0^\infty <T^n b, Ey> e^{int}) = \sum_0^\infty F_n <T^n b, Ey> .$$

This proves (1.3), or equivalently, $c = F(T)b$ on $\{Y, E, E_1\}$ for some F in $H_1^\infty(E, E')$.

For the other half assume that (1.3) holds, or equivalently, $c = F(T)b$ on $\{Y, E, E_1\}$ for some F in $H_1^\infty(E, E')$. Using $\tilde{G} = F$ and (1.3) we have

$$P_+ M_{G^\cdot} LEy = P_+ \sum_0^\infty F_n e^{-int} \cdot \sum_0^\infty <T^n b, Ey> e^{+int} = \sum_{n=0}^\infty z^n (\sum_{j=0}^\infty F_j <T^{n+j} b, Ey>) =$$

$$\sum_{n=0}^\infty z^n (\sum_{j=0}^\infty F_j <T^j b, T^{dn} Ey>) = \sum_{n=0}^\infty z^n <c, T_1^{dn} E_1 y> = L_1 E_1 y . \tag{1.10}$$

Since $P_+ M_{G^\cdot} | H$ is a contraction (1.10) and (1.5) yield (1.4). This completes the proof.

In many applications $E = E' = C^1$ and $T = T_1$. In this case EY is an invariant subspace for T^d the dual of T with respect to $L_o(X, E)$. In this set up we obtain the following uniqueness result.

1.2 COROLLARY. *Let $E = E' = C^1$ and Y be a finite dimensional space such that EY is invariant for T^d. Let b and c be in X and $T = T_1$ and $E = E_1$. Then there exists a unique F in H_1^∞ satisfying $c = F(T)b$ on $\{Y, E\}$ if and only if (1.4) holds and*

$$\|<R_T(z)b, Ey>\| = \|<R_T(z)c, Ey>\| \neq 0 \tag{1.11}$$

for some y in Y. In this case if y is any vector in Y satisfying (1.11), the unique solution F is given by

$$F(e^{it}) = \frac{<(I - e^{-it}T)^{-1}c, Ey>}{<(I - e^{-it}T)^{-1}b, Ey>} . \tag{1.12}$$

PROOF. We first notice that obviously Y is invariant for $\{T^d, T_1^d\}$ with respect to $\{E, E_1\}$. Thus we are within the framework provided by Theorem 1.1. Since Y is finite dimensional the equations (1.6) and (1.7) show that H and H' are finite dimensional invariant subspaces for the backward shift. Corollary X.1.3 shows that $T_2 = S(m)^*$ and $T_2' = S(m_2)^*$ where m and m_2 are (rational) inner functions in H^∞ satisfying $H = H(m)$ and $H' = H(m_2)$. By Corollary X.1.6 all the defect indices of T_2 and T_2' are one, and all the eigenvalues of T_2 and T_2' are in D. The proof of the previous theorem shows that if $B^* (=G_+)$ is a contractive intertwining lifting of A^*, then $F = \tilde{G}$ is function in H_1^∞

satisfying $c = F(T)b$ on $\{Y, E\}$. If $F = \tilde{G}$ is a function in H_1^∞ satisfying $c = F(T)b$ on $\{Y, E\}$, then (1.10) shows that G_+^* maps $H(m)$ into $H(m_2)$. Thus G_+ is a contractive intertwining lifting of A^*. In other words there is a one to one correspondence between the set of all contractive intertwining liftings B^* $(=G_+)$ of A^* and the set of all F $(=\tilde{G})$ in H_1^∞ satisfying $c = F(T)b$ on $\{Y, E\}$. By Corollary VII.5.4 there is a unique F satisfying $c = F(T)b$ if and only if $\|A\| = 1$, or equivalently, (1.11) holds.

Now assume that (1.4) and (1.11) hold for some y. If $F = \tilde{G}$ is any function in H_1^∞ satisfying $c = F(T)b$ on $\{Y, E\}$, then (1.5) and (1.10), (1.11) give

$$\|L_1 Ey\| = \|P_+ M_G \cdot LEy\| \le \|LEy\| = \|L_1 Ey\|. \tag{1.13}$$

Thus $M_G \cdot LEy = L_1 Ey$. This yields (1.12). The proof is now complete.

By using the last part of the above proof we readily obtain the following result.

1.3 COROLLARY. *Let $E = E' = C^1$ and assume that (1.4) holds. If (1.11) holds for some y in Y, then there exists a unique F in H_1^∞ satisfying $F(T)b = c$ on $\{Y, E\}$. This F is given by (1.12).*

2. TANGENTIAL CARATHEODORY INTERPOLATION
AS A MOMENTUM PROBLEM

In this section we will use the Rosenblum and Rovnyak Theorem to obtain another approach for solving the tangential Carathéodory interpolation Problem X.2.4, or equivalently, Problem VIII.1.5. Precisely we will give another proof based on Theorem 1.1 of Theorem VIII.1.6, or equivalently, Theorem X.2.5, restated here for convenience as Theorem 2.1. (Throughout this section $\hat{B} = \hat{B}_n$ and $\hat{C} = \hat{C}_n$ are the n by n analytic Toeplitz matrices generated by $\{B_i\}_0^{n-1}$ and $\{C_i\}_0^{n-1}$; see Section VIII.1 for the definition of a n by n analytic Toeplitz matrix.)

2.1 THEOREM. *Let $\{B_i\}_0^{n-1}$ and $\{C_i\}_0^{n-1}$ be a set of operators in $L(F, E)$ and $L(F, E')$, respectively where E, E' and F are fixed Hilbert spaces. There exists a solution to the tangential Carathéodory interpolation Problem X.2.4 if and only if $\hat{B}^* \hat{B} - \hat{C}^* \hat{C}$ is positive where $\hat{B}$ and $\hat{C}$ are the n by n analytic Toeplitz matrices generated by $\{B_i\}_0^{n-1}$ and $\{C_i\}_0^{n-1}$ respectively.*

PROOF. Consider the spaces X, X_1 and Y defined by

$$X = [L(F, E)]_0^{n-1}, \qquad X_1 = [L(F, E')]_0^{n-1} \qquad \text{and} \qquad Y = [F]_0^{n-1}. \tag{2.1}$$

(Recall that $[G]_0^{n-1}$ is the linear space formed by n copies of the linear space G, that is,

$[g_i]_0^{n-1}$ is in $[G]_0^{n-1}$ if each g_i is in G. Moreover, $[g_i]_0^{n-1}$ is identified with $[g_0, g_1, ..., g_{n-1}]^{tr}$. The zero element in $[G]_0^{n-1}$ is $[0]_0^{n-1}$ and addition is defined by $\alpha[f_i]_0^{n-1} + \beta[g_i]_0^{n-1} = [\alpha f_i + \beta g_i]_0^{n-1}$ where α and β are scalars.) Let E from Y into $L_0(X, E)$ and E_1 from Y into $L_0(X_1, E')$ be the linear maps defined by

$$<[X_i]_0^{n-1}, E[y_i]_0^{n-1}> = \sum_0^{n-1} X_i y_i \qquad ([X_i]_0^{n-1} \in X)$$

$$<[X_i]_0^{n-1}, E_1[y_i]_0^{n-1}> = \sum_0^{n-1} X_i y_i \qquad ([X_i]_0^{n-1} \in X_1) \tag{2.2}$$

where $[y_i]_0^{n-1}$ is in Y. The linear map T on X is the n by n lower shift defined by

$$T = \begin{bmatrix} 0 & 0 & \cdots & 0 & 0 \\ I & 0 & \cdots & 0 & 0 \\ 0 & I & \cdots & 0 & 0 \\ \vdots & \vdots & \cdots & \vdots & \vdots \\ 0 & 0 & \cdots & I & 0 \end{bmatrix}. \tag{2.3}$$

The identity I on E appears immediately below the diagonal and all the other entries are zero. The linear map T_1 on X_1 is defined analogously, that is, by the matrix in (2.3) where I is replaced by I' the identity on E'. Notice that $T^d E = E T_F^{tr}$ and $T_1^d E_1 = E_1 T_F^{tr}$ where T_F^{tr} on Y is the transpose of the matrix in (2.3) with I replaced by the identity on F. Taking this into account it is easy to show that Y is invariant for $\{T^d, T_1^d\}$ with respect to $\{E, E_1\}$.

Let F be in $H^\infty(E, E')$ and F_n the coefficient of z^n in the power series expansion of F, and let $\hat{F}_n$ be the n by n analytic Toeplitz matrix generated by $\{F_i\}_0^{n-1}$ (see (VIII.1.1) with A replaced by F). For $b = [B_i]_0^{n-1}$ in X, $c = [C_i]_0^{n-1}$ in X_1 and $y = [y_i]_0^{n-1}$ in Y we have

$$\sum_0^\infty F_j <T^j b, Ey> = <\hat{F}_n [B_i]_0^{n-1}, Ey>,$$

$$C_0 y_0 + C_1 y_1 + \cdots + C_{n-1} y_{n-1} = <c, E_1 y>. \tag{2.4}$$

Since the y_i's are arbitrary this implies that $c = F(T)b$ on $\{Y, E, E_1\}$ if and only if $\hat{F}_n b = c$ holds. In other words there exists contractive analytic function F in $H^\infty(E, E')$ satisfying $c = F(T)b$ on $\{Y, E, E_1\}$ if and only if F is a solution to the tangential Carathéodory interpolation Problem X.2.4. By Theorem 1.1, there exists a solution to this tangential Carathéodory interpolation problem if and only if (1.4) holds. To complete this proof it is sufficient to show that (1.4) is equivalent to $\hat{B}^* \hat{B} - \hat{C}^* \hat{C}$ being positive.

A simple calculation shows that

$$R_T(z) = (I - zT)^{-1} = \sum_0^\infty z^i T^i = \begin{bmatrix} I & 0 & \ldots & 0 \\ zI & I & \ldots & 0 \\ \vdots & \vdots & \ldots & \vdots \\ z^{n-1}I & z^{n-2}I & \ldots & I \end{bmatrix}.$$

Using this we have

$$\langle R_T(z)b, Ey \rangle = [z^{n-1}I, z^{n-2}I, \ldots, zI, I]\hat{B}[y_{n-1}, y_{n-2}, \ldots, y_0]^{tr} \tag{2.5}$$

where $y = [y_i]_0^{n-1} = [y_0, \ldots, y_{n-1}]^{tr}$. The row operator on the left of $\hat{B}$ maps $l_n^2(E)$ into $H^2(E)$. A similar calculation gives

$$\langle R_{T_1}(z)c, E_1 y \rangle = [z^{n-1}I', z^{n-2}I', \ldots, zI', I']\hat{C}[y_{n-1}, y_{n-2}, \ldots, y_0]^{tr} \tag{2.6}$$

where I' is the identity on E'. The row operator on the left of $\hat{C}$ maps $l_n^2(E')$ into $H^2(E')$. Since $\{z^i\}_0^{n-1}$ is an orthonormal set, (2.5) and (2.6) show that (1.4) is equivalent to $\hat{B}^*\hat{B} - \hat{C}^*\hat{C}$ being positive. This completes the proof.

2.2 COROLLARY. *Let $E = E' = C^1$. The Carathéodory interpolation Problem X.2.4 has a unique solution if and only if $\hat{B}^*\hat{B} - \hat{C}^*\hat{C}$ is positive and*

$$\|\hat{B}y\| = \|\hat{C}y\| \neq 0 \tag{2.7}$$

for some y in C^n. In this case if $y = [y_0, y_1, \ldots, y_{n-1}]^{tr}$ is any vector in C^n satisfying (2.7), then the unique solution F in H_1^∞ is given by

$$F(e^{it}) = \frac{[e^{i(1-n)t}, e^{i(2-n)t}, \ldots, 1]\hat{C}[y_0, y_1, \ldots, y_{n-1}]^{tr}}{[e^{i(1-n)t}, e^{i(2-n)t}, \ldots, 1]\hat{B}[y_0, y_1, \ldots, y_{n-1}]^{tr}}. \tag{2.8}$$

PROOF. The first part of the corollary follows from the previous theorem along with (2.5), (2.6) and Corollary 1.2. Equations (2.5) and (2.6) show that (1.11) is equivalent to (2.7). Equation (2.8) follows by substituting (2.5) and (2.6) into (1.12), and then by rearranging the indices. This completes the proof.

3. TANGENTIAL NEVANLINNA-PICK INTERPOLATION
AS A MOMENTUM PROBLEM

In this section we will also use another approach based on Theorem 1.1 for solving the tangential Nevanlinna-Pick interpolation Problem X.3.1.

We start by recalling that the solution to Problem X.3.1 was given in Theorem X.3.2 (see also Theorem VIII.2.2) which is restated here for convenience as

3.1 THEOREM. *Let $\{B_i\}_1^n$ and $\{C_i\}_1^n$ be a sequence of operators in $L(F, E)$ and $L(F, E')$, respectively. The tangential Nevanlinna-Pick interpolation Problem X.3.1 has a solution if and only if the matrix*

$$\begin{bmatrix} \dfrac{B_1^* B_1 - C_1^* C_1}{1 - \bar{\alpha}_1 \alpha_1} & \cdots & \dfrac{B_1^* B_n - C_1^* C_n}{1 - \bar{\alpha}_1 \alpha_n} \\ \vdots & \vdots & \vdots \\ \dfrac{B_n^* B_1 - C_n^* C_1}{1 - \bar{\alpha}_n \alpha_1} & \cdots & \dfrac{B_n^* B_n - C_n^* C_n}{1 - \bar{\alpha}_n \alpha_n} \end{bmatrix} \tag{3.1}$$

on $l_n^2(F)$ is positive.

PROOF. Consider the spaces X, X_1 and Y defined by

$$X = [L(F, E)]_1^n, \qquad X_1 = [L(F, E')]_1^n \qquad \text{and} \qquad Y = [F]_1^n . \tag{3.2}$$

The linear maps E from Y into $L(X, E)$ and E_1 from Y into $L(X_1, E')$ are defined analogously to (2.2), that is,

$$<[X_i]_1^n, E[y_i]_1^n> = \sum_1^n X_i y_i \qquad ([X_i]_1^n \in X)$$

$$\tag{3.3}$$

$$<[X_{1,i}]_1^n, E_1[y_i]_1^n> = \sum_1^n X_{1,i} y_i \qquad ([X_{1,i}]_1^n \in X_1)$$

where $[y_i]_1^n$ is in Y. In this case the linear maps T on X and T_1 on X_1 are the diagonal maps defined by

$$T[X_i]_1^n = [\alpha_i X_i]_1^n \qquad \text{and} \qquad T_1[X_{1,i}]_1^n = [\alpha_i X_{1,i}]_1^n . \tag{3.4}$$

It is easy to verify that Y is invariant for $\{T^d, T_1^d\}$ with respect to $\{E, E_1\}$.

Let F be in $H^\infty(E, E')$ and F_j the coefficient of z^j in the power series expansion of F. Using $b = [B_i]_1^n$ in X and $c = [C_i]_1^n$ in X_1 with $[y_i]_1^n$ in Y we have

$$\sum_{0}^{\infty} F_j <T^j b, \, E[y_i]_1^n> = \sum_{j=0}^{\infty} F_j(\alpha_1^j B_1 y_1 + \alpha_2^j B_2 y_2 + \cdots + \alpha_n^j B_n y_n) =$$

$$F(\alpha_1) B_1 y_1 + F(\alpha_2) B_2 y_2 + \cdots + F(\alpha_n) B_n y_n \quad \text{and} \tag{3.5}$$

$$C_1 y_1 + C_2 y_2 + \cdots + C_n y_n = <c, \, [y_i]_1^n> \, .$$

Since the y_i's are arbitrary this implies that $c = F(T)b$ on $\{Y, E, E_1\}$ if and only if $C_i = F(\alpha_i)B_i$ for $1 \le i \le n$. By Theorem 1.1 there exists a solution to this tangential Nevanlinna-Pick interpolation problem if and only (1.4) holds. To complete this proof it is sufficient to show that (1.4) is equivalent to the matrix in (3.1) being positive.

The definition of T and T_1 give

$$<R_T(z)b, \, Ey> = \sum_{1}^{n} \frac{B_j y_j}{1 - \alpha_j z}$$

$$\tag{3.6}$$

$$<R_{T_1}(z)c, \, E_1 y> = \sum_{1}^{n} \frac{C_j y_j}{1 - \alpha_j z}$$

where $y = [y_i]_1^n$ is in $l_n^2(F)$. This and the reproducing property (X.1.10) of $(1 - \alpha z)^{-1}$ shows that

$$\|<R_T(z)b, \, Ey>\|^2 - \|<R_{T_1}(z)c, \, E_1 y>\|^2 = \sum_{0,0}^{n,n} \frac{((B_j^* B_k - C_j^* C_k)y_k, \, y_j)}{1 - \alpha_k \bar{\alpha}_j} \tag{3.7}$$

for all y_i in E. Therefore (1.4) is equivalent to the matrix in (3.1) being positive. This completes the proof.

By following the proof of Corollary 2.2 we readily obtain the following result.

3.2 COROLLARY. *Let* $E = E' = \mathbf{C}^1$. *There exists a unique solution to the tangential Nevanlinna-Pick interpolation Problem X.3.1 if and only if the matrix in (3.1) is positive and*

$$\sum_{0,0}^{n,n} \frac{\bar{B}_j B_k y_k \bar{y}_j}{1 - \alpha_k \bar{\alpha}_j} = \sum_{0,0}^{n,n} \frac{\bar{C}_j C_k y_k \bar{y}_j}{1 - \alpha_k \bar{\alpha}_j} \ne 0 \tag{3.8}$$

for some y *in* $\mathbf{C}^n$. *In this case if* $y = [y_i]_1^n$ *is any vector in* $\mathbf{C}^n$ *satisfying (3.8), then the unique* F *is given by*

$$F(e^{it}) = [\sum_{j=1}^{n} (1 - \alpha_j e^{-it})^{-1} C_j y_j][\sum_{j=1}^{n} (1 - \alpha_j e^{-it})^{-1} B_j y_j]^{-1} \, . \tag{3.9}$$

4. TANGENTIAL HERMITE-FEJER INTERPOLATION
AS A MOMENTUM PROBLEM

In this section we will use Theorem 1.1 to solve the tangential Hermite-Fejér interpolation Problem X.4.1.

Throughout this section $\{B_{i,j}: 1 \leq i \leq n$ and $0 \leq j < d_i\}$ and $\{C_{i,j}: 1 \leq i \leq n$ and $0 \leq j < d_i\}$ is a sequence of operators in $L(F, E)$ and $L(F, E')$, respectively. To begin let X (X_1 and Y) be the linear spaces formed by

$$[X_{i,j}] \triangleq [X_{i,j} : 1 \leq i \leq n \text{ and } 0 \leq j \leq d_i - 1] \tag{4.1}$$

with the entries $X_{i,j}$ in $L(F, E)$ ($L(F, E')$ and F, respectively). Naturally the vector operations on these spaces are component wise. The linear maps E from Y into $L(X, E)$ and E_1 from Y into $L(X_1, E')$ are defined analogously to (2.2) and (3.3), that is,

$$\begin{aligned}
<x, Ey> &= \sum_{i=1}^{n} \sum_{j=0}^{d_i-1} X_{i,j} f_{i,j} \qquad (x = [X_{i,j}] \in X) \\
<x, E_1 y> &= \sum_{i=1}^{n} \sum_{j=0}^{d_i-1} X_{i,j} f_{i,j} \qquad (x = [X_{i,j}] \in X_1)
\end{aligned} \tag{4.2}$$

where $y = [f_{i,j}] \in Y$.

Consider the Jordan matrix J_i on $[L(F, E)]_o^{d_i-1}$ defined by

$$J_i = \begin{bmatrix}
\alpha_i I & 0 & \ldots & 0 \\
I & \alpha_i I & \ldots & 0 \\
0 & I & \ldots & 0 \\
\vdots & \vdots & \vdots\vdots\vdots & \vdots \\
0 & 0 & \ldots & \alpha_i I
\end{bmatrix} \tag{4.3}$$

where I is the identity operator on E. The α_i's are distinct complex numbers in D. The operator $\alpha_i I$ appears on the diagonal, I is immediately below the diagonal and the rest of the matrix is filled in with zeros. The Jordan matrix J_i' on $[L(F, E')]_o^{d_i-1}$ is defined by (4.3) where the identity operator I is replaced by the identity operator I' on E'. The linear maps T on X and T_1 on X_1 are defined by

$$T = \text{diag}[J_1, J_2, ..., J_n] \tag{4.4}$$

and

$$T_1 = \text{diag}[J_1', J_2', ..., J_n'] .$$

If $x = [X_{i,j}]$ is in X, then

$$Tx = [J_1[X_{1,j}]_{j=0}^{d_1-1}, \, ..., \, J_n[X_{n,j}]_{j=0}^{d_n-1}]^{tr} \,.$$

The linear map T_1 on X_1 is defined analogously.

Let b and c be the elements in X and X_1 defined by $b = [B_{i,j}]$ and $c = [C_{i,j}]$. Notice that $b = [b_i]_1^n$ and $c = [c_i]_1^n$ where $b_i = [B_{i,j}]_{j=0}^{d_i-1}$ and $c_i = [C_{i,j}]_{j=0}^{d_i-1}$. Let $[y_i]_1^n$ be in Y where $y_i = [f_{i,j}]_{j=0}^{d_i-1}$. Assume that F_n is the coefficient of z^n in the power series expansion of F in $H^\infty(E, E')$. Notice that $J_i = \alpha_i I + S_i$ where S_i is the d_i by d_i lower shift matrix in (2.3), that is, the identity I appears immediately below the diagonal and zero's elsewhere. Obviously $S_i^k = 0$ if $k \geq d_i$. Clearly $\alpha_i I$ and S_i commute. This with the binomial expansion of $(\alpha_i I + S_i)^k$ yields:

$$\sum_{o}^{\infty} F_k <T^k b, Ey> = \sum_{i=1}^{n} \sum_{k=0}^{\infty} \sum_{m=0}^{k} \binom{k}{m} F_k \alpha_i^{(k-m)} <S_i^m b_i, Ey_i> =$$

$$\sum_{i=1}^{n} \sum_{m=0}^{d_i-1} \left(\sum_{k=m}^{\infty} \binom{k}{m} F_k \alpha_i^{k-m} \right) <S_i^m b_i, Ey_i> = \sum_{i=1}^{n} <\hat{F}_{i,d_i} b_i, Ey_i> \tag{4.5}$$

where $\hat{F}_{i,d_i}$ is the d_i by d_i matrix (X.4.1). Another calculation yields

$$\sum_{i=1}^{n} \sum_{j=0}^{d_i-1} C_{i,j} f_{i,j} = <c, E_1[f_{i,j}]> \,. \tag{4.6}$$

Since the $f_{i,j}$'s are arbitrary, (4.5) and (4.6) show that (X.4.1) holds if and only if $c = F(T)b$ on $\{Y, E, E_1\}$. By Theorem 1.1 there exists a solution to the tangential Hermite-Fejér interpolation Problem X.4.1 if and only if (1.4) holds.

Now we will show that (1.4) is equivalent to a certain matrix being positive; hence there exists a solution to the Hermite-Fejér interpolation Problem X.4.1 if and only if this matrix is positive. To this end notice that by using the standard functional calculus for a Jordan matrix or by direct calculation we have for $1 \leq i \leq n$

$$(I - zJ_i)^{-1} = \begin{bmatrix} x(i,0)I & 0 & ... & 0 \\ x(i,1)I & x(i,0)I & ... & 0 \\ \vdots & \vdots & \vdots & \vdots \\ x(i,d_i-1)I & x(i,d_i-2)I & ... & x(i,0)I \end{bmatrix} \tag{4.7}$$

where $x(i, j) = z^j(1 - \alpha_i z)^{-(j+1)}$ for $0 \leq j \leq d_i - 1$. In other words $(I - zJ_i)^{-1}$ is the d_i by d_i analytic Toeplitz matrix on $l_n^2(E)$ generated by $\{x(i,j): 0 \leq j < d_i\}$. It is worth noting that $x(i, j) = \tilde{x}_{i,j}$ where $x_{i,j}$ is the generalized eigenvector for the backward shift defined in (X.1.2).

Now let $\hat{B}_i$ and $\hat{C}_i$ be the d_i by d_i analytic Toeplitz matrix generated by $\{B_{i,j}: 0 \leq j < d_i\}$ and $\{C_{i,j}: 0 \leq j < d_i\}$, respectively. As in Section X.4 the operator $\hat{B}$

from $l_d^2(F)$ to $l_d^2(E)$ and $\hat{C}$ from $l_d^2(F)$ to $l_d^2(E')$ is defined by $\hat{B} = \hat{B}_1 \oplus \hat{B}_2 \oplus \cdots \oplus \hat{B}_n$ and $\hat{C} = \hat{C}_1 \oplus \hat{C}_2 \oplus \cdots \oplus \hat{C}_n$ where $d = \Sigma d_i$. We also introduce the operators $x(i)$ from $l_{d_i}^2(E)$ into $H^2(E)$ and the column vector $y(i)$ in $l_{d_i}^2(F)$ defined by

$$x(i) = [x(i, d_i-1)I, x(i, d_i - 2)I, ..., x(i, 0)I]$$
$$y(i) = [f_{i,d_i-1}, f_{i,d_i-2}, ..., f_{i,o}]^{tr} . \tag{4.8}$$

Using (4.2), (4.3), (4.4) and (4.7) one can show that for $y = [f_{i,j}] = [f_{1,o}, f_{1,1}, ..., f_{n,d_n-1}]^{tr}$

$$<R_T(z)b, Ey> = [x(1), x(2), ..., x(n)]\hat{B}[y(1), y(2), ..., y(n)]^{tr} . \tag{4.9}$$

The row operator on the left of $\hat{B}$ maps $l_d^2(E)$ into $H^2(E)$. The vector on the right of $\hat{B}$ in (4.9) is in $l_d^2(F)$. A calculation analogous to (4.9) gives

$$<R_{T_1}(z)c, E_1y> = [x'(1), x'(2), ..., x'(n)]\hat{C}[y(1), y(2), ..., y(n)]^{tr} \tag{4.10}$$

where $x'(i)$ is the operator mapping $l_{d_i}^2(E')$ into $H^2(E')$ defined by the first operator in (4.8) where I' replaces I. As in Section X.4 the block Gram matrix G on $l_d^2(E)$ and G' on $l_d^2(E')$ are defined by

$$G = [x(1), x(2), ..., x(n)]^*[x(1), x(2), ..., x(n)] \tag{4.11}$$
$$G' = [x'(1), x'(2), ..., x'(n)]^*[x'(1), x'(2), ..., x'(n)] .$$

In other words G is given by the following matrix

$$G = \begin{bmatrix} (x(1, n_1 - 1), x(1, n_1 - 1))I & \cdots & (x(1, n_1 - 1), x(n, 0))I \\ \cdot & \cdots & \cdot \\ \cdot & \cdots & \cdot \\ \cdot & \cdots & \cdot \\ (x(n, 0), x(1, n_1 - 1))I & \cdots & (x(n, 0), x(n, 0))I \end{bmatrix} . \tag{4.12}$$

The block Gram matrix G' is given by (4.12) where I replaces I'. According to the derivative reproducing property of $x_{i,j}$ in (X.1.11) we obtain

$$(x(i, j), x(k, r)) = (\tilde{x}_{i,j}, \tilde{x}_{k,r}) = \frac{\overline{x_{i,j}^{(r)}(\alpha_k)}}{r!} . \tag{4.13}$$

So the entries of G and G' are of the form

$$\overline{\frac{x_{i,j}^{(r)}(\alpha_k)I}{r!}} \quad \text{and} \quad \overline{\frac{x_{i,j}^{(r)}(\alpha_k)}{r!}}I' \,. \tag{4.14}$$

So one can explicitly compute the entries of G and G'. (By (X.4.8) equations (X.4.7) and (4.12) produce the same G and G'.) By (4.9) and (4.10) we see that (1.4) holds if and only if $\hat{B}^* G\hat{B} - \hat{C}^* G'\hat{C}$ is positive. In other words there exists a solution to the tangential Hermite-Fejér interpolation Problem X.4.1 if and only if this matrix is positive. Summing up the previous analysis provides another proof of Theorem X.4.3, restated as

4.1 THEOREM. *There exists a solution to the tangential Hermite-Fejér interpolation Problem X.4.1 if and only if the matrix* $\hat{B}^* G\hat{B} - \hat{C}^* G'\hat{C}$ *on* $l_d^2(F)$ *is positive.*

By following the proof of Corollary 2.2 we also obtain the following result.

4.2 COROLLARY. *Let* $E = E' = C^1$. *The tangential Hermite-Fejér interpolation Problem X.4.1 has a unique solution if and only if the matrix* $\hat{B}^* G\hat{B} - \hat{C}^* G'\hat{C}$ *is positive and*

$$(\hat{B}^* GBy, y) = (\hat{C}^* G'\hat{C}y, y) \neq 0 \tag{4.15}$$

for some y in C^d. *In this case if y is any vector in* C^d *satisfying (4.15), then the unique solution F in* H_1^∞ *is given by*

$$F(e^{it}) = \frac{[x(1), x(2), ..., x(n)]\hat{C}y}{[x(1), x(2), ..., x(n)]\hat{B}y}(e^{-it})\,. \tag{4.16}$$

By using $y_{i,j} = W(m)\tilde{x}_{i,j}$ where $W(m)$ is the unitary operator in Remark IX.5.7, it is easy to show that the previous expression for F in (4.16) is also given by $F = \Psi y/(\Phi y)$, where Ψ and Φ are the operators used in the corresponding uniqueness Corollary X.4.5.

5. LOEWNER INTERPOLATION

In this section we will use the basic momentum interpolation theorem of Rosenblum and Rovnyak to solve the following Loewner interpolation problem.

5.1 PROBLEM. Let Δ be a Borel subset of $[0, 2\pi)$, and let $B(t)$ and $C(t)$ be weakly measurable functions with values in $L(F, E)$ and $L(F, E')$, respectively. Find necessary and sufficient conditions for the existence of a contractive analytic function F in $H^\infty(E, E')$ satisfying $F(e^{it})B(t) = C(t)$ a.e. on Δ.

To obtain a solution to this problem recall that $L^2(\Delta, M)$ is the set of all square integrable weakly Lebesgue measurable function on Δ with values in M. This sets the stage for the following result.

5.2 THEOREM. *There exists a solution to the Loewner interpolation Problem 5.1 if and only if*

$$\lim_{r\uparrow 1} \iint_{\Delta\Delta} \frac{((B(s)^*B(t) - C(s)^*C(t))y(t),\, y(s))}{1 - r^2 e^{it}e^{-is}}\, dt\, ds \geq 0 \tag{5.1}$$

for all weakly measurable functions y *on* Δ *with values in* F *such that* By *is in* $L^2(\Delta, E)$ *and* Cy *is in* $L^2(\Delta, E')$.

The proof of this theorem shows that the limit in (5.1) exists as a finite real number for all weakly measurable functions y on Δ with values in F such that By is in $L^2(\Delta, E)$ and Cy is in $L^2(\Delta, E')$.

PROOF. Let X and X_1 be the linear spaces generated by $e^{int}B$ and $e^{int}C$ for all $n \geq 0$, respectively. The linear maps T on X and T_1 on X_1 are defined by $Tx = e^{it}x$ and $T_1 x_1 = e^{it}x_1$ where x is in X and x_1 is in X_1. Let Y be the linear space of all weakly measurable functions y on Δ with values in F such that By is in $L^2(\Delta, E)$ and Cy is in $L^2(\Delta, E')$. The linear maps E from Y into $L_o(X, E)$ and E_1 from Y into $L_o(X_1, E')$ are defined by

$$<x, Ey> = \int_{\Delta} x(t)y(t)dt \qquad (x \in X)$$

$$<x_1, E_1 y> = \int_{\Delta} x_1(t)y(t)dt \qquad (x_1 \in X_1). \tag{5.2}$$

It is easy to show that Y is an invariant subspace for $\{T^d, T_1^d\}$ with respect to $\{E, E_1\}$.

Let F be in $H^\infty(E, E')$ and F_n be the coefficient of z^n in the power series expansion of F. For $b \triangleq B$ in X and $c \triangleq C$ in X_1 and y in Y we have

$$\sum_o^\infty F_n <T^n b,\, Ey> = \lim_{r\uparrow 1} \sum_o^\infty r^n F_n \int_{\Delta} e^{int}B(t)y(t)dt = \lim_{r\uparrow 1} \int_{\Delta} F(re^{it})B(t)y(t)dt = \int_{\Delta} F(e^{it})B(t)y(t)dt,$$

$$\tag{5.3}$$

$$\int_{\Delta} C(t)y(t)dt = <c, E_1 y>.$$

Because By is in $L^2(\Delta, E)$ its Fourier coefficients

$$\int_\Delta e^{int} B(t) y(t) dt \qquad (n \ge 0) \qquad\qquad (5.4)$$

are square summable. This implies that the sum in (5.3) is finite and the limit exists. Since y is an arbitrary element in Y equation (5.3) shows that $F(T)b = c$ on $\{Y, E, E_1\}$ if and only if $F(e^{it})B(t) = C(t)$ a.e. on Δ.

To complete the proof it is sufficient to show that (1.4) is equivalent to (5.1). This follows by using Fubini's Theorem, and the reproducing property (X.1.10) of $(1 - \alpha z)^{-1}$ in the following calculation:

$$\| <R_T(z)b, Ey> \|^2 = \sum_o^\infty \| \int_\Delta e^{int} B(t) y(t) dt \|^2 =$$

$$\lim_{r\uparrow 1} \frac{1}{2\pi} \int_0^{2\pi} \| \sum_o^\infty r^n e^{inv} \int_\Delta e^{int} B(t) y(t) dt \|^2 dv = \lim_{r\uparrow 1} \frac{1}{2\pi} \int_0^{2\pi} \| \int_\Delta \frac{B(t)y(t)dt}{1 - re^{it}e^{iv}} \|^2 dv = \qquad (5.5)$$

$$\lim_{r\uparrow 1} \int\!\!\int_{\Delta\Delta} ((1-re^{it}z)^{-1}B(t)y(t), (1-re^{is}z)^{-1}B(s)y(s))_{H^2}\, dtds = \lim_{r\uparrow 1} \int\!\!\int_{\Delta\Delta} \frac{(B(s)^* B(t)y(t), y(s))}{1 - r^2 e^{it}e^{-is}}\, dt\, ds$$

Since the Fourier coefficients of By in (5.4) are square summable, (5.5) is finite for all y in Y. Replacing b by c and B by C in the previous calculation we obtain

$$\| <R_{T_1}(z)c, E_1 y> \|^2 = \lim_{r\uparrow 1} \int\!\!\int_{\Delta\Delta} \frac{(C(s)^* C(t)y(t), y(s))}{1 - r^2 e^{it}e^{-is}}\, dt\, ds \qquad (5.6)$$

which is finite for all y in Y. Equations (5.5) and (5.6) show that (1.4) is equivalent to (5.1). By Theorem 1.1 there exists a solution to the Loewner interpolation problem if and only if (5.1) holds. This completes the proof.

The previous proof along with Corollary 1.3 gives the following result.

5.3 COROLLARY. *If $E = E' = C^1$ and (5.1) holds for all y in Y and there exists a y_o in Y satisfying*

$$\lim_{r\uparrow 1} \int\!\!\int_{\Delta\Delta} \frac{\overline{B(s)}B(t)y_o(t)\overline{y_o(s)}}{1 - r^2 e^{it}e^{-is}}\, dt\, ds = \lim_{r\uparrow 1} \int\!\!\int_{\Delta\Delta} \frac{\overline{C(s)}C(t)y_o(t)\overline{y_o(s)}}{1 - r^2 e^{it}e^{-is}}\, dt\, ds \ne 0, \qquad (5.7)$$

then there exists a unique solution to the Loewner interpolation problem. In this case the unique solution F in H_1^∞ is given by

$$F(e^{it}) = [\sum_o^\infty e^{-int} \int_\Delta e^{int} C(t)y_o(t)\, dt] \cdot [\sum_o^\infty e^{-int} \int_\Delta e^{int} B(t)y_o(t)dt]^{-1} \qquad (\text{a.e. on } [0, 2\pi)) \qquad (5.8)$$

XI.6. NOTES AND COMMENTS

This chapter is devoted to an approach to general interpolation problems due to Rosenblum and Rovnyak. Their approach is based on their elegant momentum theorem (namely Theorem 1.1). This theorem provides an optimal solution to one of the momentum problems first considered by Nudelman [Nu 1]. Most of the results and proofs in this chapter are taken from the works of Rosenblum and Rovnyak [RosR 1,2,3]. Our supplement consists in the analysis of the uniqueness cases in their work.

CHAPTER XII

NUMERICAL ALGORITHMS FOR H^∞ OPTIMIZATION IN CONTROL THEORY

In this short chapter we will present one of the connections between operator theory and control theory. In Section 1 we will show how some of the previous H^∞ optimization problems introduced in Chapter IX naturally arise in robust control theory. One of the simplest of these problems will be solved in Section 3, by another method inspired by the functional model theory presented in Section IX.5. For this purpose some useful operator theoretical preparations are given in Section 2.

1. AN INTRODUCTION TO H^∞ CONTROL THEORY

This section presents a bird's eye view of the relationship between robust control theory and H^∞ optimization problems.

To begin, let P be an analytic function in some neighborhood of the origin with values in $L(F, E')$. Let $\{u_n\}_0^\infty$ be any sequence with values in F and u(z) its z-transform formally defined by $u(z) = \sum u_n z^n$. If $\{u_n\}_0^\infty$ is in $l^2(F)$, then u(z) is in $H^2(F)$. We define the sequence $\{y_n\}_0^\infty$ in E' by convolution

$$y_n = \sum_{i=0}^{n} P_{n-i}\, u_i \, . \tag{1.1}$$

Obviously y_n is well defined. Its z-transform y(z) is formally defined by $y(z) = \sum y_n z^n$. Notice that y(z) can be obtained by formally multiplying P(z) by u(z) and matching like coefficients of z^n, that is, $y(z) = P(z)u(z)$ where P_n is the coefficient of z^n in the power series expansion of P(z). Equation (1.1) is represented by the black box in Figure 1. The *input* is u(z), the *output is* y(z) and n is the time index. Finally, P is known as the *plant*. Notice that if for some n the inputs $u_0 = u_1 = \cdots = u_n = 0$, then $y_0 = y_1 = \cdots = y_n = 0$. This is usually referred to as the *causality* property of the plant. So these plants are linear causal and time invariant; see also Chapter III.

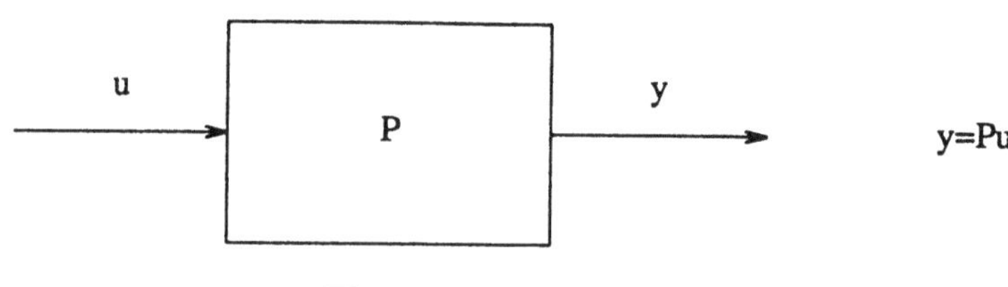

Figure 1

The *energy* of a signal u is the square of its l^2 norm, that is, $\|\{u_n\}_0^\infty\|^2 = \sum \|u_n\|^2$. Obviously u has finite energy if and only if u(z) is in $H^2(F)$, or equivalently, $\|u(z)\|$ is finite. We say that P is *stable* if P is in $H^\infty(F, E')$. This means that $M_P | H^2(F)$ is an operator from $H^2(F)$ to $H^2(E')$, or equivalently, $M_P | H^2(F)$ maps signals u with finite energy into signals y with finite energy. In most applications P(z) is rational. So in this case P defines a stable plant if and only if P admits a decomposition of the form $P(z) = N(z)/d(z)$ where d(z) is a scalar-valued polynomial whose zeroes are in $\{z: |z| > 1\}$.

In many applications the output signal y is perturbed by a noise $W_+ v$ where W is in $H^\infty(E, E')$ and v is a noise process in $H^2(E)$, that is, a function in $H^2(E)$. This is represented by the block diagram in Figure 2. In control engineering one wants to design a feedback control system of the form given by Figure 3 to minimize the effect of the noise v on the output y. The compensator C to be designed by the engineer is an analytic function in a neighborhood of the origin with values in $L(E', F)$ and C is not necessarily stable, that is, C may not be in $H^\infty(E', F)$, but C is almost always in $L^\infty(E', F)$.

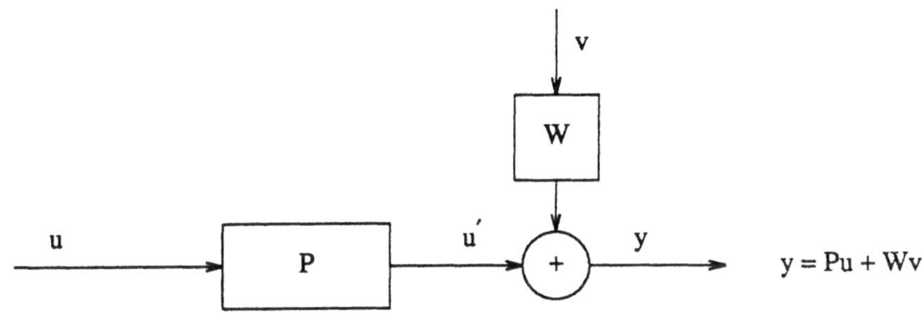

Figure 2

Consulting Figure 3 we see that the equations for the error signal e is given by

$$e = r - y = r - PCe - Wv \, .$$

Formally solving for e produces

$$e = (I + PC)^{-1}r - (I + PC)^{-1}Wv$$

$$\tag{1.2}$$

$$u = C(I + PC)^{-1}r - C(I + PC)^{-1}Wv \, .$$

Figure 3 also shows that the output signal y is given by $y = PC(r - y) + Wv$. By formally solving for y this produces

$$y = (I + PC)^{-1}PCr + (I + PC)^{-1}Wv \, . \tag{1.3}$$

In control systems one wants to minimize the effect of the noise v in $H^2(E)$ on the output y. By (1.3) this is equivalent to designing a compensator C to minimize the energy of $(I + PC)^{-1}Wv$ for all v in $H^2(E)$, that is, to design a compensator C to minimize the effect of all noise signals v whose energy (norm) is one. In other words we want to design a compensator C to minimize the L^∞ norm of $(I + PC)^{-1}W$; see Theorem IX.1.1.

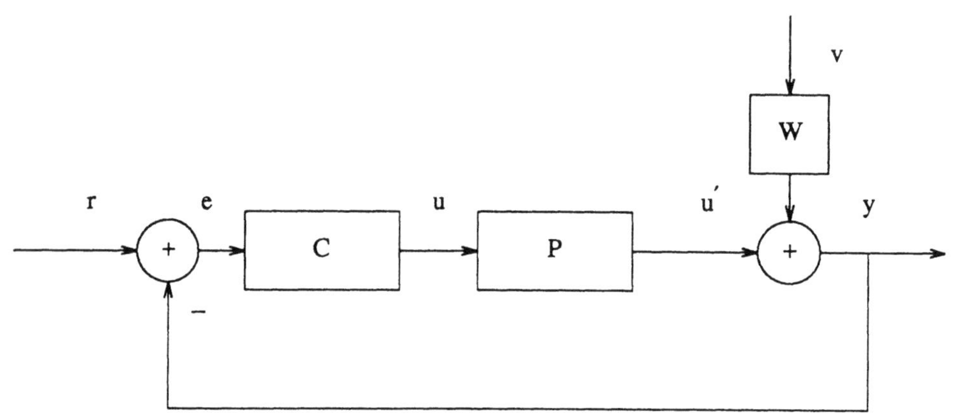

Figure 3

Choosing $C = +\infty$ obviously minimizes the L^∞ norm of $(I + PC)^{-1}W$. However, this choice can introduce some inherent instabilities which can lead to problems. To eliminate these instabilities we must add some additional constraints on our optimization problem. First of all we will assume that both P and W are stable, that is, P is in $H^\infty(F, E')$ and W is in $H^\infty(E, E')$. (We have chosen P to be stable only to simplify the presentation. In practice P can be unstable; see [Fran 2]). Next we want the system in Figure 3 to be *internally stable*. This means that the internal signals e, u and u' are in $H^2(E')$, $H^2(F)$ and $H^2(E')$ for all inputs r in $H^2(E')$ and all noise v in $H^2(E)$. Internal stability says that all the internal signals e, u and u' have finite energy when the external inputs r and v have finite energy. If the system was not internally stable, then one of the internal signals could become unbounded and cause the circuitry to malfunction. Consulting (1.2) and using the fact that both P and W are stable we see that the system in Figure 3 is internally stable if and only if

$$(I + PC)^{-1} \text{ is in } H^\infty(E', E') \quad \text{and} \quad C(I + PC)^{-1} \text{ is in } H^\infty(E', F). \qquad (1.4)$$

Notice that formally

$$(I + PC)^{-1} = I - PC(I + PC)^{-1}. \qquad (1.5)$$

This implies that (1.4) is redundant, that is, if $C(I + PC)^{-1}$ is in $H^2(E', F)$, then $(I + PC)^{-1}$

is in $H^2(E', E')$. Therefore internal stability is equivalent to the second condition in (1.4).

The above analysis shows that minimizing the effect of the noise v on the output y subject to the system in Figure 3 being internally stable is equivalent to

$$\inf_C \|(I+PC)^{-1}W\|_\infty \text{ subject to } C(I+PC)^{-1} \in H^\infty(E', F). \tag{1.6}$$

The quantity $(I+PC)^{-1}$ is known as the *sensitivity* of the closed loop system. To obtain a solution to (1.6). Let $F = C(I+PC)^{-1}$. Solving C in terms of F gives

$$C = (I-FP)^{-1}F. \tag{1.7}$$

Substituting $F = C(I+PC)^{-1}$ into (1.5) the optimization problem in (1.6) becomes

$$\mu = \inf\|W - PFW\|_\infty \text{ subject to } F \in H^\infty(E', F) \tag{1.8}$$

where P and W are specified. This is precisely the H^∞ optimization problem we have been looking for. In other words if F in $H^\infty(E', F)$ solves the H^∞ optimization problem in (1.8), then a compensator C which minimizes the effect of the noise v on y and maintains internal stability is given by $C = (I - FP)^{-1}F$.

We will obtain a solution to the H^∞ optimization problem in (1.8) for two different cases. For the first case assume that P admits an inner outer factorization (see Section IX.2) of the form $P = P_i P_o$ where P_i is an inner function in $H^\infty(F, E')$ and P_o is an invertible outer function in $H^\infty(F, F)$. By invertible we mean that $P_o(z)^{-1}$ is an outer function in $H^\infty(F, F)$. We also assume that W admits a $*$-outer $*$-inner factorization of the form $W = W_{*_o} W_{*_i}$ where W_{o*} is an invertible outer function in $H^\infty(E', E')$ and W_{*_i} is a two sided inner function in $H^\infty(E, E')$. In this case the optimization problem in (1.8) becomes

$$\mu = \inf\|W + PH^\infty(E', F)W\|_\infty = \inf\|W_{o*} + P_i P_o H^\infty(E', F)W_{o*}\|_\infty = \tag{1.9}$$

$$\inf\|W_{o*} + P_i H^\infty(E', F)\|_\infty = \|\Gamma_1(W_{o*})\|$$

where $\Gamma_1(W_{o*})$ is the Hankel operator defined in Section IX.4, see Theorem IX.4.2 for $Q = W_{o*}$ and $\Theta_1 = P_i$. In other words the H^∞ optimization problem in (1.8) is equivalent to the H^∞ optimization Problem IX.4.1. If F_* in $H^\infty(E', F)$ is a solution to the last H^∞ optimization problem in (1.9), that is,

$$\|W_{o*} + P_i F_*\|_\infty = \mu = \|\Gamma_1(W_{o*})\|,$$

then a compensator C satisfying (1.6), or equivalently, minimizing the effect of v on y is given by $C = (I - FP)^{-1}F$ where $F = P_0^{-1}F_*W_{0*}^{-1}$. If P_i is also inner from both sides, then (1.6) can be converted to the H^∞ Nehari optimization Problem 3.1. The details are left to the reader as a simple exercise. (In the rational case an explicit solution to this problem is given in Sections XIII.9 and XIV.9.) Finally, it is noted that if all the functions W, P and C are scalar valued, then the H^∞ optimization problem in (1.8) reduces to $\inf \|W_0 + P_iH^\infty\|_\infty$ which is precisely the H^∞ optimization problem discussed in Section IX.7. (Here $W_0 = W_{0*}$ is the outer part of W.) In this case μ in (1.8) is the norm of $W_0(S(P_i))$. Moreover, if $\mu = \|W_0 + P_iF_*\|_\infty$ for some F_* in H^∞, then the optimal controller C is given by $(1 - FP)^{-1}F$ where $F = P_0^{-1}F_*W_0^{-1}$.

The H^∞ optimization problem in (1.8) is a special case of the following H^∞ optimization problem

$$\inf\|Q+PH^\infty(E_1,F)W\|_\infty \tag{1.10}$$

where Q is in $H^\infty(E, E')$, W is in $H^\infty(E, E_1)$ and P is in $H^\infty(F, E')$. (In the previous control problem $E' = E_1$.) Now we will follow Doyle's reduction [Doy] to convert the H^∞ optimization problem in (1.10) to the four block H^∞ optimization Problem IX.4.3. To this end as before let us assume that $P = P_iP_0$ where P_i is an inner function in $H^\infty(F, E')$ and P_0 is an invertible outer function in $H^\infty(F, F)$. We also assume that $W = W_{0*}W_{i*}$ where W_{0*} is an invertible outer function in $H^\infty(E_1, E_1)$ and W_{i*} is a $*$-inner function in $H^\infty(E, E_1)$. Now as a side remark let M be a contractive function in $L^\infty(E, E')$. By definition $\Delta_M(t)$ is a.e. the positive square root of $I - M(e^{it})^*M(e^{it})$ and $\Delta_{M*}(t)$ is a. e. the positive square root of $I - M(e^{it})M(e^{it})^*$. A simple calculation shows that the "rotation matrix"

$$\begin{bmatrix} M(e^{it}) & \Delta_{M*}(t) \\ \Delta_M(t) & -M(e^{it})^* \end{bmatrix} \text{ in } L^\infty(E \oplus E', E' \oplus E)$$

is a.e. unitary (see Corollary IV.1.4). In particular, using the fact that $\Delta_{P_i} = 0$ and $\Delta_{W_{i*}^*} = 0$ we see that

$$L(t) \triangleq \begin{bmatrix} P_i(e^{it}) & \Delta_{P_i^*}(t) \\ 0 & -P_i(e^{it})^* \end{bmatrix} \text{ and } R(t) \triangleq \begin{bmatrix} W_{i*}(e^{it}) & 0 \\ \Delta_{W_{i*}(t)} & -W_{i*}(e^{it})^* \end{bmatrix} \tag{1.11}$$

are both a.e. unitary function in $L^\infty(F \oplus E', E' \oplus F)$ and $L^\infty(E \oplus E_1, E_1 \oplus E)$. Finally, let $\hat{Q}$ be the function in $L^\infty(E \oplus E_1, E' \oplus F)$ defined by

$$\hat{Q} = \begin{bmatrix} Q & 0 \\ 0 & 0 \end{bmatrix}. \tag{1.12}$$

Using the unitary operators in (1.11) the H^∞ optimization problem (1.10) becomes

$$\inf\|Q + PH^\infty(E_1, F)W\|_\infty = \inf\|\hat{Q} + L\begin{bmatrix} P_oH^\infty(E_1, F)W_{o*} & 0 \\ 0 & 0 \end{bmatrix} R\|_\infty =$$

$$\inf\|\begin{bmatrix} A_{1,1} + H^\infty(E_1, F) & A_{1,2} \\ A_{2,1} & A_{2,2} \end{bmatrix}\|_\infty \tag{1.13}$$

where $A_{i,j}$ are the $L^\infty(\cdot, \cdot)$ entries of the matrix $L^*\hat{Q}R^*$. Therefore the H^∞ optimization problem in (1.13) is a special case of the four block H^∞ optimization Problem IX.3.

There is a special case of the H^∞ optimization problem in (1.10) worth mentioning, that is, when $F = C^1 = E_1$ and $E = C^2 = E'$. In this case P admits an inner-outer factorization of the form $P = P_iP_o$ where $P_i = [p_1, p_2]^{tr}$ and $W = W_{o*}[w_1, w_2]$ where $W_{i*} = [w_1, w_2]$. Here p_1, p_2, w_1, w_2 are all functions in H^∞ satisfying

$$|p_1(e^{it})|^2 + |p_2(e^{it})|^2 = 1 \text{ and } |w_1(e^{it})|^2 + |w_2(e^{it})|^2 = 1 \quad \text{a.e.}$$

In this case one can choose the a.e., unitary matrices $L(t)$ and $R(t)$ to be

$$L(t) = \begin{bmatrix} p_1(e^{it}) & \overline{p_2(e^{it})} \\ p_2(e^{it}) & -\overline{p_1(e^{it})} \end{bmatrix} \text{ and } R(t) = \begin{bmatrix} w_1(e^{it}) & w_2(e^{it}) \\ \overline{w_2(e^{it})} & -\overline{w_1(e^{it})} \end{bmatrix}. \tag{1.14}$$

Then the H^∞ optimization problem in (1.10) becomes

$$\inf\|Q + PH^\infty W\|_\infty = \inf\|Q + L\begin{bmatrix} P_oH^\infty W_{o*} & 0 \\ 0 & 0 \end{bmatrix} R\|_\infty =$$

$$\inf\|\begin{bmatrix} a + H^\infty & b \\ c & d \end{bmatrix}\|_\infty = \delta \tag{1.15}$$

where a, b, c, d are the L^∞ entries in L^*QR^*. As before this is a special case of the four block H^∞ optimization Problem IX.3. Moreover, if f_* in H^∞ is an optimal solution to the last H^∞ optimization problem in (1.15), then $P_o^{-1}f_*W_{o*}^{-1}$ is the optimal solution to the first H^∞ optimization problem in (1.15), that is

$$\|Q + PP_o^{-1}f_*W_{o*}^{-1}W\|_\infty = \inf\|Q + PH^\infty W\|_\infty. \tag{1.16}$$

The advantages of using the unitary matrices L and R in (1.14) over the ones in (1.11) is that in the scalar case we do not have to compute square roots.

The H^∞ optimization problem in (1.10) can be regarded as a model matching problem. In this case one is given a plant Q in $H^\infty(E, E')$ to be matched. One is also given two other fixed plants P in $H^\infty(F, E')$ and W in $H^\infty(E, E_1)$; see Figure 4. In model matching, one must design a (stable) plant F in $H^\infty(E_1, F)$ such that PFW matches the given plant Q as close as possible. In other words one must find a stable plant F which solves the H^∞ optimization problem in (1.10). By minimizing the H^∞ norm in (1.10) the engineer is designing a plant F such that the error $\|(Q-PFW)u\|$ between Q and PFW is as small as possible for all inputs u whose energy equals one ($\|u\| = 1$).

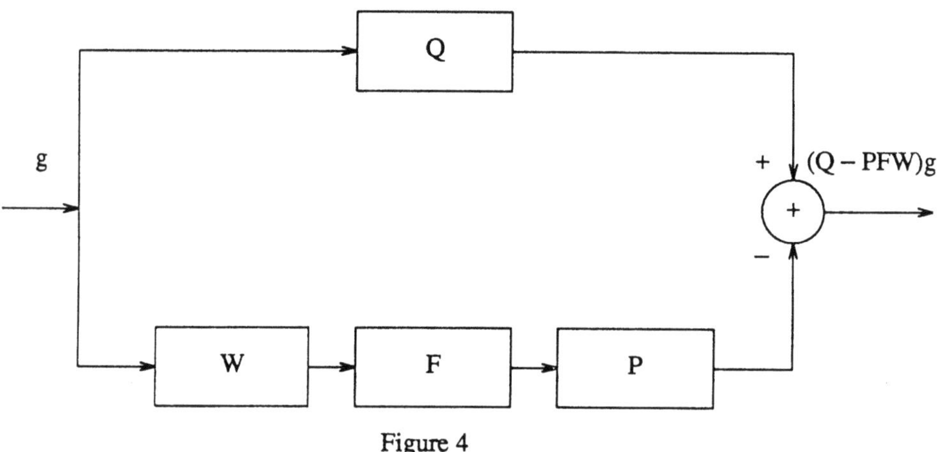

Figure 4

To complete this section we will briefly sketch the standard problem in H^∞ control theory. To this end consider the following system given by Figure 5.

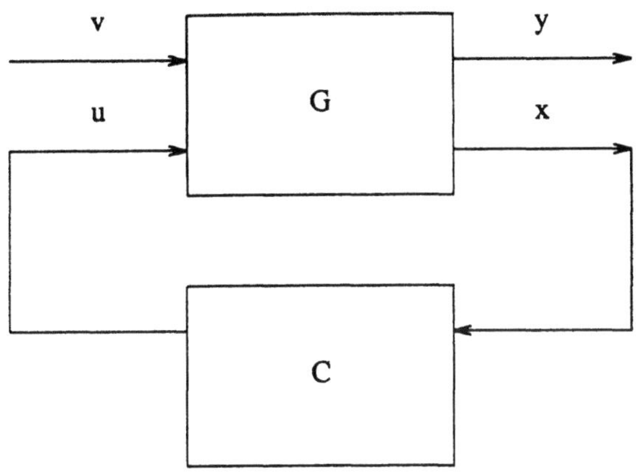

Figure 5

The equations for the block diagram in Figure 5 are

$$\begin{bmatrix} y \\ x \end{bmatrix} = \begin{bmatrix} G_{1,1} & G_{1,2} \\ G_{2,1} & G_{2,2} \end{bmatrix} \begin{bmatrix} v \\ u \end{bmatrix}$$

$$u = Cx$$

(1.17)

where $G_{i,j}$ are given analytic functions in some neighborhood of the origin with values in the appropriate spaces. The engineer must design a compensator C to minimize the effect of the noise v on y and also maintain internal stability, that is, all internal nodes must be stable for all inputs with finite energy. Solving for y in terms of v gives

$$y = [G_{1,1} + G_{1,2}C(I - G_{2,2}C)^{-1}G_{2,1}]v .$$

(1.18)

One of the conditions on internal stability imply that the system $F = C(I - G_{2,2}C)^{-1}$ must be stable, that is, in $H^\infty(E_1, F)$. In many applications as in (1.4) it turns out that some of the conditions on internal stability are redundant. In other words in many cases $F = C(I - G_{2,2}C)^{-1}$ in $H^\infty(E_1, F)$ is necessary and sufficient for internal stability (see Theorem 2 in Chapter IV, Section 3 of [Fran 2]). So to minimize the effect of v on y and maintain internal stability is equivalent to the following H^∞ optimization problem.

$$\inf \|G_{1,1} + G_{1,2}H^\infty(E_1, F)G_{2,1}\|$$

(1.19)

which is precisely the H^∞ optimization problem in (1.10).

Finally, it is noted that the model matching problem is a special case of the standard problem. The model matching problem is obtained by choosing

$$G = \begin{bmatrix} Q & P \\ W & 0 \end{bmatrix} \quad \text{and} \quad C = F.$$

The standard problem can also be used to solve a tracking problem and a robust stability problem [Fran 2]. For further results on H^∞ control theory see [DoyFT], [Fran 2], [Helt 5] and [V].

2. MINIMAL FUNCTIONS AND SPECTRUM

In this section we will present some basic results concerning minimal functions and the spectrum of C_0 contractions, which will be used to solve a numerical H^∞ optimization problem arising in control theory.

To this end recall that m is an inner function in H^∞ if and only if

$$m(z) = \gamma B(z) \, S(z) \qquad (z \in D) \tag{2.1}$$

where γ is a constant of modulus one, $B(z)$ is a Blaschke product of the form

$$B(z) = \prod_1^\infty \frac{\bar{\alpha}_i(\alpha_i - z)}{|\alpha_i|(1 - \bar{\alpha}_i z)} \quad (|\alpha_i| < 1 \quad \text{and} \quad \sum_1^\infty (1 - |\alpha_i|) < \infty) \tag{2.2}$$

and $S(z)$ is a singular function

$$S(z) = \exp\left[-\int_0^{2\pi} \frac{e^{it} + z}{e^{it} - z} \, d\mu\right] \tag{2.3}$$

where μ is positive finite measure which is singular with respect to the Lebesgue measure. The α_i's in (2.2) can be repeated, if $\alpha_i = 0$, then $\bar{\alpha}_i / |\alpha_i|$ is set equal to one. In this case the corresponding Blaschke factor becomes z (see [Ho], [Sz.-NF10] for further details). The $B(z)$ or $S(z)$ term can be one. In this case the inner function $m(z) = \gamma B(z)$ or $m(z) = \gamma S(z)$ or $m(z)$ is the constant function γ.

We say that T on H is a C_0 *contraction*, if T is a completely non-unitary contraction and T admits a minimal function m_T. Recall that a minimal function m_T is the "smallest" inner function in H^∞ satisfying $m_T(T) = 0$, that is, if m is any inner function in H^∞ satisfying $m(T) = 0$, then m_T divides m, see Section X.1. (The operator $m(T)$ is defined by the functional calculus in Section VI.6.) In fact one can show that if f is any function in H^∞ satisfying $f(T) = 0$ where T is a completely non-unitary contraction, then T is a C_0 contraction and there exists a unique minimal function for T (see [Ber], [Sz.-NF10] for further details). However, this fact is not needed here. We only use Theorem X.1.6 which shows S(m) is a C_0 contraction and its minimal function is m. Now let T be a C_0 contraction with minimal function m_T. Let S_T be the set $\{\alpha_i\}_{i \geq 1}$ of all the zeros of m_T in D and of all points on the unit circle C through which m_T can not be continued analytically into the exterior of D. If $m = m_T$ is given by (2.1), then (2.2) and (2.3) imply that S_T is the union of the closure of $\{\alpha_i\}_{i \geq 1}$ and the support of the singular measure μ.

In matrix algebra the eigenvalues which form the spectrum of a n by n matrix are precisely the zeros of its minimal polynomial. The following shows that an analogous result holds for C_0 contractions. For this we denote by $\sigma(T)$ the spectrum of an operator T, that is, the set of all complex numbers λ for which $\lambda I - T$ is not invertible.

2.1 THEOREM. *If T is a C_0 contraction, then* $\sigma(T) = S_T$.

PROOF. Let α be a complex number in $\overline{D}$ not in S_T. We claim that $m_T(\alpha)$ is nonzero. This obviously holds if α is in D. If α is in C, then the fact that m_T is analytic in a neighborhood of α and $|m_T(e^{it})| = 1$ a.e. shows that $m_T(\alpha)$ is nonzero which proves our claim. Consider the function m in H^∞ defined by

$$m(z) = \frac{m_T(\alpha) - m_T(z)}{\alpha - z} \qquad (z \in D) .$$

By applying our function calculus in Section VI.6 we have

$$m(T)(\alpha - T) = (\alpha - T)m(T) = m_T(\alpha)I - m_T(T) = m_T(\alpha)I .$$

Since $m_T(\alpha)$ is nonzero this implies that $\alpha - T$ is invertible. In particular,

$$\sigma(T) \subseteq S_T . \tag{2.4}$$

Let α be a point in D satisfying $m_T(\alpha) = 0$, that is, α is in $D \cap S_T$. Consulting (2.1) and (2.2) we see that there exists an inner function n in H^∞ satisfying

$$m_T(z) = \frac{z - \alpha}{1 - \overline{\alpha}z} \, n(z) \qquad (z \in D) .$$

By applying our functional calculus

$$(T - \alpha I)n(T) = m_T(T)(I - \overline{\alpha}T) = 0 .$$

Because n is a strict divisor of m_T the definition of a minimal function shows that $n(T)$ is nonzero. Thus $T - \alpha I$ is not invertible. In particular,

$$D \cap S_T \subseteq \sigma(T) . \tag{2.5}$$

Combining (2.4) and (2.5) it remains to prove that $C \cap S_T \subseteq C \cap \sigma(T)$, or equivalently, if ω is an open arc of C contained in the resolvent set $\rho(T)$ of T (which is the complement in C^1 of $\sigma(T)$), then m_T is analytic on ω.

To verify this let $m = m_T$ be given by the decomposition in (2.1). If α is in $\omega(\subseteq \rho(T))$, then the distance between α and $\sigma(T)$ is nonzero. This with (2.5) and (2.2) implies that α is not in the closure of the zeros of the Blaschke product B(z) for m_T. Therefore B(z) is analytic on ω.

To complete the proof we must show that the singular function S(z) is analytic on ω. We will accomplish this by showing that $\mu(\omega) = 0$. To this end we use contradiction and assume that $\mu(\beta) > 0$ for some closed subarc β of ω. Consider the inner function m_1 defined by

$$m_1(z) \triangleq \exp[-\int_\beta \frac{e^{it}+z}{e^{it}-z} \, d\mu]$$

Since $|m_1(0)| = e^{-\mu(\beta)} < 1$ the function m_1 is not a constant and an inner divisor of m_T. Consider the invariant subspace H_1 for T defined by $H_1 = \overline{m_2(T)H}$ where m_2 is the inner function given by $m_2 = m_T/m_1$. Because $m_T = m_1 m_2$ is the minimal function for T and m_1 is not a constant, H_1 is nonzero.

We claim that $T_1 \triangleq T|H_1$ on H_1 is a C_o contraction and m_1 is the minimal function for T_1. Since H_1 is an invariant subspace for T the contraction T_1 is completely non-unitary. If u is an inner function satisfying $u(T_1) = 0$, then $(um_2)(T) = u(T_1)m_2(T) = 0$. This implies that $m_T = m_1 m_2$ divides um_2. Therefore m_1 divides u. In other words m_1 is the minimal function for T_1, which proves our claim. Moreover, applying (2.4) to T_1 yields

$$\sigma(T_1) \subseteq S_{T_1} \subseteq \beta . \tag{2.6}$$

To complete the proof notice that for $|\lambda| > 1$ the following equations hold $(\lambda I - T)f = g$ and $f = \sum_o^\infty \lambda^{-(i+1)} T^i g$. This shows that $\lambda I - T$ maps the invariant subspace H_1 onto H_1. In particular,

$$(\lambda I - T_1)^{-1} = (\lambda I - T)^{-1} | H_1 .$$

By passing limits on λ to the unit circle $(\beta \subseteq)$ $C \cap \rho(T) \subseteq \rho(T_1)$. This with (2.6) proves $\sigma(T_1) \subseteq \rho(T_1)$ which is a contradiction, since the spectrum of an operator acting on a Hilbert space $(\neq \{0\})$ cannot be empty (see [Ha 5]). The contradiction arose from $\mu(\omega) > 0$. Therefore $\mu(\omega) = 0$. The proof is now complete.

Consulting Theorem X.1.6 we readily obtain the following result.

2.2 COROLLARY. *Let* m *be an inner function in* H^∞ *and* $T = S(m)$. *Then* T *is a* C_0 *contraction with minimal function* $m_T = m$. *Moreover,* $\sigma(T) = S_T$. *In particular, the eigenvalues of* T *are the zeros of* m *in* D.

Let m be a nontrivial inner function in H^∞ and set $H = H(m)$. A simple calculation also shows that

$$\phi_* \triangleq P_H 1 = 1 - \overline{m(0)}m , \tag{2.7}$$

where 1 is also viewed as the constant function one in H^2. This sets the stage for the following useful result.

2.3 LEMMA. *Let* $T = S(m)$ *and* h *be an element in* $H = H(m)$ *where* m *is an inner function in* H^∞. *Then*

$$h(z) = ((I - zT^*)^{-1}h, \phi_*) \qquad (\textit{for all } z \in D) . \tag{2.8}$$

Thus h(z) *can be extented analytically to*

$$\{z \in C^1 : I - zT^* \textit{ is invertible}\} = \{z \in C^1 : 1/\overline{z} \notin \sigma(T)\} . \tag{2.9}$$

PROOF. The last statement follows from (2.8), since the right hand side of (2.8) is analytic in the set described by (2.9). To obtain (2.8), recall that the unilateral shift S on H^2 is the minimal isometric dilation of T. In particular, $S^* | H = T^*$. This and (2.7) gives

$$((I - zT^*)^{-1}h, \phi_*) = ((I - zT^*)^{-1}h, 1) = ((I - zS^*)^{-1}h, 1) = (h, (I - \overline{z}S)^{-1}1) =$$

$$\frac{1}{2\pi} \int_0^{2\pi} \frac{h(e^{it})}{1 - ze^{-it}} dt = h(z) \qquad (\text{for } z \in D)$$

by virtue of the Cauchy formula. This completes the proof.

We say that an operator A on H *attains its norm,* if there exists a nonzero h in H, satisfying $\|Ah\| = \|A\| \|h\|$. We claim that A attains its norm if and only if there exists a nonzero h in H satisfying

$$(\|A\|^2 I - A^* A)h = 0 . \tag{2.10}$$

If (2.10) holds, then obviously A attains its norm. On the other hand, if A attains its norm, then

$$\| \, \|A\|^2 h - A^* A h \|^2 =$$

$$\|A\|^4 \|h\|^2 - 2\|A\|^2 \text{Re}(A^* A h, \, h) + \|A^* A h\|^2 =$$

$$\|A^* A h\|^2 - \|A\|^4 \|h\|^2 \leq 0 \, .$$

Therefore (2.10) holds, which proves our claim. If (2.10) holds for some nonzero h, then we say that h *is a maximal vector for* A. An operator in a finite dimensional vector space always attains its norm. In fact, a maximal vector for such an operator A is any eigenvector corresponding to the largest eigenvalue of $A^* A$.

Recall that $T = S(m)$ is a completely non-unitary contraction. So if f is in H^∞, then $A = f(T)$ is defined according the function calculus in Section VI.6. If m is rational (that is, a finite Blaschke product), then $H = H(m)$ is finite dimensional (see Section X.1). So A attains its norm. However, if m is nonrational, then $H(m)$ is infinite dimensional and A may not attain its norm. To see this let m be the inner function defined by

$$m(z) = \exp\left[\frac{z+1}{z-1}\right] \text{ and } f(z) = \frac{1}{2-z} \, .$$

Obviously $\|f\|_\infty \leq 1$ and thus $\|A\| \leq 1$. By the commutant lifting theorem there exists a g in H^∞ satisfying $A = g(T)$ and $\|A\| = \|g\|_\infty$. Since $f(T) = g(T)$ we see that there exists a ϕ in H^∞ satisfying $f - g = m\phi$. So as $z \to 1$ for z in D we have

$$\lim_{z \to 1} (f(z) - g(z)) = \lim_{z \to 1} m(z)\phi(z) = 0$$

because $\phi(z)$ is bounded. Using $f(1) = 1$ the previous equation shows that $\|g\|_\infty \geq 1$. This with $\|A\| \leq 1$ implies that $\|g\|_\infty = 1$ and $\|A\| = 1$. We claim that A does not attain its norm. If $\|Ah\| = \|h\|$ for some h in $H(m)$, then

$$\|h\|^2 = \|f(T)h\|^2 \leq \|f(S)h\|^2 \leq \|h\|^2 \, .$$

Therefore f h is in $H(m)$. Hence

$$\frac{1}{2\pi} \int_0^{2\pi} |f(e^{it})|^2 \, |h(e^{it})|^2 dt = \frac{1}{2\pi} \int_0^{2\pi} |h(e^{it})|^2 dt \, .$$

In particular, this implies that

$$\frac{1}{2\pi} \int_0^{2\pi} (1 - |f(e^{it})|^2) \, |h(e^{it})|^2 dt = 0 \, .$$

Since $|f(e^{it})| < 1$ a.e., the vector h must be equal to 0. Therefore, A does not attain its norm. This explains the intrinsic interest of the next result, which will play an important role in obtaining a numerical algorithm for solving a certain nonrational numerical H^∞ optimization problem in the next section.

2.4 COROLLARY. *Let* m *be a nonrational inner function in* H^∞. *Let* $A = f(T)$ *where* $T = S(m)$ *and* f *is a function in* H^∞ *continuous on* $\overline{D}$. *If*

$$\|A\| > \max \{ \, |f(z)| : z \in S_T \text{ and } |z| = 1 \} \,, \tag{2.11}$$

then A *attains its norm.*

PROOF. Let us observe, for the readers sake, that because m is not rational (not a finite Blaschke product) the maximum in (2.11) is not taken over an empty set. By applying the Cauchy-Schwartz inequality for the semi-positive form

$$< h, g > = (h, (\|A\|^2 I - A^* A)g)$$

we have for g and h in H

$$|(h, (\|A\|^2 I - A^* A)g)|^2 \le (h, (\|A\|^2 I - A^* A)h)(g, (\|A\|^2 I - A^* A)g) \,. \tag{2.12}$$

By the definition of the norm of A, there exists a sequence $\{h_j\}_1^\infty$ of unit vectors in H such that $\|Ah_j\| \to \|A\|$ as $j \to \infty$. From (2.12) we deduce that

$$\lim_{j \to \infty} (h_j, (\|A\|^2 I - A^* A)g) = 0 \qquad \text{(for all } g \in H) \,. \tag{2.13}$$

Since any sequence of unit vectors is weakly compact (see [Ha 5], or [RieszSz.-N]), there exists a subsequence, which we will still denote by $\{h_j\}_1^\infty$ and an element h in H such that

$$\lim_{j \to \infty} (h_j, x) = (h, x) \qquad \text{(for all } x \in H) \,. \tag{2.14}$$

Obviously (2.13) and (2.14) imply that (2.10) holds. So if $h \ne 0$, then the corollary is proven.

In the contrary case, when $h = 0$ we have by Lemma 2.3

$$h_j(z) = (h_j, (I - \overline{z} T)^{-1} \phi_*) \to 0 \qquad \text{(for all } z \notin S_T \text{ and } |z| = 1) \,. \tag{2.15}$$

Therefore, if (2.10) holds and β is strictly between the two sides of (2.11), and if we set

$$J_1 = \{t \in [0, 2\pi) : |f(e^{it})| \ge \beta\} \text{ and } J_2 = \{t \in [0, 2\pi) : |f(e^{it})| < \beta\} \,,$$

then we have

$$\|Ah_j\|^2 = \|f(T)h_j\|^2 \le \|f(S)h_j\|^2 =$$

$$\frac{1}{2\pi}\int_0^{2\pi} |f(e^{it})|^2 |h_j(e^{it})|^2 dt \le \frac{\beta^2}{2\pi}\int_{J_2} |h_j(e^{it})|^2 dt + \frac{1}{2\pi}\int_{J_1} |f(e^{it})|^2 |h_j(e^{it})|^2 dt \le$$

$$\beta^2 + \frac{1}{2\pi}\int_{J_1} |f(e^{it})|^2 |h_j(e^{it})|^2 dt .$$

According to Lemma 2.3 the functions in the last integral are dominated by

$$\|f\|_\infty^2 \|(I - e^{-it}T)^{-1}\phi_*\|^2$$

which is bounded on J_1 (due to the assumption that f is continuous on $\overline{D}$). Thus by (2.15) and the Lebesgue dominated convergence theorem (see [Riesz Sz.-N]) we obtain

$$\|A\|^2 = \lim_{j \to \infty} \|Ah_j\| \le \beta^2 < \|A\|^2$$

which is a contradiction. Therefore, if (2.11) holds, the last case $h = 0$ does not occur. This completes the proof.

The following result shows that the maximum in (2.11) is always bounded by the norm of A.

2.5 LEMMA. *Let* m *be a nonrational inner function in* H^∞. *Let* $A = f(T)$ *where* $T = S(m)$ *and* f *is a function in* H^∞ *continuous on* $\overline{D}$. *Then*

$$\|A\| \ge \max \{ |f(z)| : z \in S_T \text{ and } |z| = 1\} . \tag{2.16}$$

PROOF. By the commutant lifting theorem there exists a u in H^∞ satisfying $\|u\|_\infty = \|A\|$ and $A = u(T)$ (see Section IX.7). This implies that

$$f - u = mh \qquad \text{(for some h in } H^\infty) . \tag{2.17}$$

On the other hand, a classical result on inner functions (see [Ho]) states that for every z in $S_T \cap \{z : |z| = 1\}$ there exists a sequence $\{z_j\}_1^\infty$ in D such that $z_j \to z$ and $m(z_j) \to 0$ as $j \to \infty$. Therefore, (2.17) implies that

$$|f(z)| = \lim_{j \to \infty} |f(z_j)| \le \|u\|_\infty + \lim_{j \to \infty} |m(z_j)| \|h\|_\infty = \|A\|$$

This completes the proof.

3. A NUMERICAL ALGORITHM FOR H^∞ OPTIMIZATION

In this section we will use some particular skew Toeplitz operators and the properties of the operator $S(m)$ to solve a nonrational H^∞ optimization problem.

Consider the following numerical H^∞ optimization problem: Find the optimal h_* in H^∞ and ν satisfying

$$\nu = \inf\|f + mH^\infty\| = \|f + mh_*\| \tag{3.1}$$

where $f = p/q$ is a specified rational function in H^∞, the polynomials p and q have no common zeros, and m is a specified nontrivial inner function in H^∞. Let $T = S(m)$ on $H = H(m)$. For the moment, let us assume that $A = P_H f_+|H = f(T)$ attains its norm. Consulting Section IX.7, the optimal h_* satisfying (3.1) is given by

$$h_* = \frac{(A - f)x}{mx}, \tag{3.2}$$

where x is any maximal vector for A. Moreover, $\nu = \|A\|$ and $x = A^*y$, where y is any vector in the kernel of $\nu^2 I - AA^*$. Using the fact that $f = p/q$ is a rational function in H^∞, this implies that the norm of A is the largest positive number ρ such that the kernel of

$$T_\rho = \rho^2 q(T)q(T)^* - p(T)p(T)^* = \sum_{j=0,\,k=0}^{n,\,n} C_{j,\,k} T^j T^{*k} \tag{3.3}$$

is nonzero, where

$$C_{j,\,k} = \rho^2 b_j \overline{b}_k - a_j \overline{a}_k . \tag{3.4}$$

Here p and q are polynomials of degree less than or equal to n, and a_j, respectively b_j, is the coefficient for z^j in p, respectively q.

3.1 REMARK. Equation (3.3) shows that the norm of A is given by the largest positive number $\rho(= \nu)$ such that the kernel of T_ρ is nonzero. In other words, the ν solving the H^∞ optimization problem in (3.1) is precisely the largest positive number ρ such that the kernel of T_ρ is nonzero. Moreover, in this case, if y is any nonzero vector satisfying $T_\rho y = 0$, then $x = A^*y$ attains the norm of A. If $\rho = \nu$, then h_* in (3.2) solves the H^∞ optimization problem in (3.1).

At this point it is appropriate to mention that an operator $T(w)$ of the form

$$T(w) = \sum_{i=o,\, j=o}^{n,\, n} C_{i,\,j} T^i T^{*j} \tag{3.4a}$$

is called a *skew Toeplitz operator*. The *symbol* w of T(w) is the function

$$w = \sum_{i=o,\, j=o}^{n,\, n} C_{i,\,j} z^i \bar{z}^j \qquad (\text{for } z \in C). \tag{3.4b}$$

Obviously T_ρ in (3.3), when $C_{i,\,j}$ is defined by (3.4) and ρ is real, is a self-adjoint skew Toeplitz operator; notice that in (3.4) $C_{i,\,j} = \bar{C}_{j,\,i}$ for all $1 \le i \le n$ and $1 \le j \le n$, that is, w is real.

To complete this section we will use the above remark to compute the optimal h_* and v in (3.1). To this end, recall that the unilateral shift S on H^2 is the minimal isometric dilation of $T = S(m)$. This implies that

$$T_\rho = P_H \sum_{0,\,0}^{n,\,n} C_{j,\,k} S^j S^{*k} |H. \tag{3.5}$$

In particular, $T_\rho y = 0$ for some y in *H* if and only if

$$\sum_{0,\,0}^{n,\,n} C_{j,\,k} S^j S^{*k} y = mg \tag{3.6}$$

for some g in H^2. If y_j is the coefficient of z^j in the power series expansion of y, then

$$S^j S^{*k} y = S^j [z^{-k}(y - y_o - zy_1 - \cdots - z^{k-1} y_{k-1})] =$$

$$\tag{3.6a}$$

$$z^{j-k}(y - y_o - zy_1 - \cdots - z^{k-1} y_{k-1}).$$

Now let C(z), respectively $C_i(z)$ (for $0 \le i < n$) be the polynomial of degree less than or equal to 2n, respectively $2n - 1$ defined by

$$C(z) = \sum_{j=0,\, k=0}^{n,\, n} C_{j,\,k} z^{n+j-k} \quad \text{and} \quad C_i(z) = \sum_{i<k\le n,\, 0\le j\le n} C_{j,\,k} z^{n+j-k+i} \quad (\text{for } 0 \le i < n), \tag{3.7}$$

where $C_{j,\,k}$ is given by (3.4). Finally, substituting (3.6a) into (3.6) and using the definitions of C and of the C_i's we conclude that $T_\rho y = 0$ if and only if

$$C(z)y(z) - \sum_{i=0}^{n-1} C_i(z)y_i = z^n m(z)g(z) \tag{3.8}$$

for some g in H^2. We emphasize that equation (3.8) holds for all z in D and for almost all t in $[0, 2\pi)$ when $z = e^{it}$.

In order to solve the H^∞ optimization problem in (3.1) we will impose two further conditions. Our first assumption is that

$$C(z) \neq 0 \text{ for all } z \in \sigma(T) \cup \{0\} . \tag{A1}$$

Since

$$C(T) = \beta\Pi(T - z_o I) ,$$

where $\beta \neq 0$ and z_o runs over the zeros of $C(z)$ (and thus z_o is not in $\sigma(T)$ so each $T - z_o I$ is invertible) we deduce that $C(T)$ is invertible. Obviously $C(T)^{-1}$ commutes with T. By Proposition IX.7.4 there exists $C^{(-1)}$ in H^∞ such that $C^{(-1)}(T) = C(T)^{-1}$. Moreover, using the fact that $(C^{(-1)}C)(T) = I$ we see that there exists an h_1 in H^∞ satisfying $C^{(-1)}C = 1 + mh_1$. By multiplying (3.8) by $C^{(-1)}$ we obtain

$$y(z) - \sum_{i=0}^{n-1} (C^{(-1)}C_i)(z)y_i = m(z)[z^n C^{(-1)}g(z) - h_1(z)y(z)] = m(z)h(z) \qquad (\text{where } h \in H^2) .$$

Applying P_H to the previous equation gives:

$$y = \sum_{i=0}^{n-1} y_i P_H C^{(-1)}C_i = \sum_{i=0}^{n-1} y_i \psi_i , \tag{3.9}$$

where (using $P_H C_+^{(-1)} = C(T)^{-1}P_H$) the

$$\psi_i = P_H C^{(-1)}C_i = C(T)^{-1}P_H C_i \qquad (\text{for } 0 \le i < n) , \tag{3.9a}$$

do not depend on the particular choice of $C^{(-1)}$ and are obviously in H. Substituting (3.9) into (3.8) produces

$$\sum_{i=0}^{n-1} y_i(CP_H C^{(-1)}C_i - C_i)(z) = z^n m(z)g(z) \qquad (z \in D) . \tag{3.10}$$

Using in addition the lifting property $P_H C_+ P_H = P_H C_+$ we have

$$P_H CP_H C^{(-1)}C_i - C_i = P_H CC^{(-1)}C_i - C_i = (P_H - I)C_i \text{ is in } mH^2 \qquad (\text{for } 0 < i < n) .$$

So by (3.9a) and (3.10) there exists a sequence of functions g_i in H^2 such that

$$C\psi_i - C_i = mg_i \qquad (\text{for } 0 \le i < n) \tag{3.11}$$

$$\sum_{i=0}^{n-1} y_i g_i(z) = z^n g(z) \qquad (z \in D) \tag{3.11a}$$

Finally, if

$$g_i(z) = g_{i,0} + zg_{i,1} + z^2 g_{i,2} + \cdots \tag{3.12}$$

is the power series expansion of g_i, then equation (3.11a) yields:

$$\sum_{i=0}^{n-1} y_i g_{i,j} = 0 \quad \text{(for } 0 \le j < n\text{)} . \tag{3.12a}$$

Notice that by (3.9) if y is nonzero, then $[y_0, y_1, ..., y_{n-1}]$ is also nonzero. Summing up, the previous analysis proves the first part of the following result.

3.2 PROPOSITION. *Assume that condition (A1) holds. If* $T_\rho y = 0$ *for some nonzero* y *in H, then* y *is given by (3.9), where* $[y_0, y_1, ..., y_{n-1}]$ *is a nonzero row vector satisfying (3.12a). Conversely, if* $[\gamma_0, \gamma_1, ..., \gamma_{n-1}]$ *is a nonzero row vector satisfying*

$$[\gamma_0, \gamma_1, ..., \gamma_{n-1}] \, [g_{i,j}]_{0,0}^{n-1,\,n-1} = 0 , \tag{3.13}$$

then y *given by (3.9), where* $y_i = \gamma_i$ *for all* $0 \le i \le n-1$ *is a nonzero vector in H satisfying* $T_\rho y = 0$ *and* $y_i = \gamma_i$ *for* $0 \le i \le n-1$.

PROOF. The converse is the only part of the proposition left to prove. Obviously

$$y = \sum_{0}^{n-1} \gamma_i \, \psi_i \tag{3.13a}$$

is in *H*. Using (3.11), (3.12) and (3.13) we have

$$Cy - \sum_{0}^{n-1} \gamma_i C_i = \sum_{0}^{n-1} \gamma_i (C\psi_i - C_i) = \sum_{0}^{n-1} \gamma_i m g_i = z^n m g . \tag{3.14}$$

for some g in H^2. Since (3.8) holds if and only if $T_\rho y = 0$, this implies that $T_\rho y = 0$. Now observe that the polynomials in (3.8) have the following form:

$$C(z) = \sum_{j=0}^{2n} a_j z^j \text{ and } C_i(z) = \sum_{j=i}^{2n-1} a_{i,j} z^j \quad (0 \le i < n)$$

where for $i \le j < n$ we have

$$a_{i,j} = \sum_{\substack{i < k \le n \\ 0 \le r \le n \\ n+r-k+i=j}} C_{r,k} = \sum_{\substack{0 \le k \le n \\ 0 \le r \le n \\ n+r-k+i=j}} C_{r,k} = a_{j-i} .$$

Equating the coefficients of z^j (for $0 \le j < n$) in (3.14) we obtain

$$a_0(y_j - \gamma_j) + a_1(y_{j-1} - \gamma_{j-1}) + \cdots + a_j(y_0 - \gamma_0) =$$

$$a_0 y_j + a_1 y_{j-1} + \cdots + a_j y_0 - (\gamma_0 a_{0,j} + \gamma_1 a_{1,j} + \cdots + \gamma_j a_{j,j}) = 0 \,,$$

or equivalently,

$$\begin{bmatrix} a_0 & 0 & \ldots & 0 & 0 \\ a_1 & a_0 & \ldots & 0 & 0 \\ \vdots & \vdots & \ldots & \vdots & \vdots \\ a_{n-1} & a_{n-2} & \ldots & a_1 & a_0 \end{bmatrix} \begin{bmatrix} y_0 - \gamma_0 \\ y_1 - \gamma_1 \\ \vdots \\ y_{n-1} - \gamma_{n-1} \end{bmatrix} = 0 \,.$$

By our assumption (A1) the coefficient a_0 is not zero. Thus $y_i = \gamma_i$ for $0 \le i \le n - 1$. This shows that $y = 0$ if and only if $y_i = 0$ for $0 \le i \le n - 1$. Therefore, y is nonzero. This completes the proof.

Notice that in the previous proposition we did not use the particular form of the symbol (3.4b) of the Skew Toeplitz operator. So Proposition 3.2 is also valid for any skew Toeplitz operator $T(w)$ whose symbol is real. It is natural to call the system (3.13) the *singular system* of the skew Toeplitz operator $T(w)$.

In order to explicitly compute this system, we will make the following simplifying technical assumption:

$$\text{The zeros of } C(z) \text{ are simple and their modulus are not equal to one .} \qquad \text{(A2)}$$

We need the following result:

3.3 LEMMA. *The polynomial $C(z)$ has n zeros inside D and n zeros outside $\overline{D}$. Moreover, if $z_1, z_2, \cdots, z_n$ are the zeros of $C(z)$ inside D, then $1/\overline{z}_1, 1/\overline{z}_2, \cdots, 1/\overline{z}_n$ are the zeros of $C(z)$ outside $\overline{D}$; also $m(z_j) \ne 0$ for $1 \le j \le n$. Finally, $g_0, g_1, \cdots, g_{n-1}$ are polynomials of degree less than or equal to $2n - 1$.*

PROOF. Recall the notation $C^{\#} = J_{2n+1} C$ used in Section I.2, that is,

$$C^{\#}(z) = J_{2n+1} C(z) = z^{2n} \overline{C(\tfrac{1}{\overline{z}})} \,.$$

Since the symbol (3.4b) is real, a simple computation shows that $C = C^{\#}$. So that if z is a zero of C, then $1/\overline{z}$ is a zero of C. Since $C(0) \ne 0$, and $C^{\#} = C$ we see that C is a polynomial of degree $2n$. On the other hand, for $z = e^{it}$ we deduce from (3.11) that

$$g_j = C\overline{m}\psi_j - \overline{m}C_j \qquad (0 \le j < n) \qquad (3.15)$$

where $\overline{m}\psi_j$ is in K_o^2, (because ψ_j is in H). Therefore since C_j is a polynomial of degree at most $2n - 1$ and C is a polynomial of degree n, the Fourier coefficients of e^{imt} in the power series expansion of g_j are all zero for $m \ge 2n$. Finally, $z_j (1 \le j \le n)$ is not in $\sigma(T)$ (see (A1)). By virtue of Theorem 2.1, z_j is not a zero of m. This completes the proof.

Now we will give the explicit computation of the singular system (3.13).

3.4 PROPOSITION. *Assume that (A1) and (A2) hold. The polynomials* $g_1, g_2, \cdots, g_n$ *in (3.11) (determining, via (3.12), the singular system of* $T(w)$*) are uniquely determined by the interpolating conditions:*

$$g_j(z_k) = - \frac{C_j(z_k)}{m(z_k)} \qquad (3.16a)$$

$$g_j(1/\overline{z}_k) = - \overline{m(z_k)}C_j(1/\overline{z}_k) \qquad (3.16b)$$

for all $0 \le j < n$ *and* $1 \le k \le n$.

PROOF. To obtain (3.16a) just take $z = z_k$ in (3.11). Taking the conjugate values in (3.15) and, then multiplying by $e^{(2n-1)it}$, we obtain $J_{2n}g_j = CQ_j - mJ_{2n}C_j$ in H^2, where Q_j is the function in H^2 with boundary values $e^{-it}m(e^{it})(J_{2n}\psi_j)(e^{it})$ on the unit circle. Therefore, we also have

$$(J_{2n}g_j)(z) = C(z)Q_j(z) - m(z)(J_{2n}C_j)(z) \qquad (z \in D) . \qquad (3.17)$$

Taking $z = z_k$ in (3.17), we obtain (3.16b). Finally, the polynomial g_j $(0 \le j < n)$ is of degree $2n - 1$ at most. So that it is uniquely determined by its 2n interpolating conditions (3.16). This completes the proof.

By combining Corollary 2.4, Lemma 2.5, Propositions 3.2 and 3.4 we can now formulate the following computational procedure for solving the H^∞ optimization problem in 3.1.

3.5 PROCEDURE. Assume that

$$v = \|A\| > v_o = \max \{f(z) : z \in \sigma(T) \text{ and } |z| = 1\} \qquad (3.18)$$

and (A1) and (A2) hold for $\rho = v$. For any $\rho > v_o$ compute the zeros of the polynomial $C(z)$ and check, if the assumptions (A1), (A2) hold. If so, then compute the Lagrange polynomials $g_0, g_1, \cdots, g_{n-1}$ determined by the conditions (3.16) (see Section X.5). Take their power series expansion as in (3.12) and define (3.13) accordingly. Retain ρ if the (3.13) has a nonzero solution, that is, if the n by n matrix $[g_{i,j}]$ (for $0 \leq i < n$ and $0 \leq j < n$) is singular. Then the largest such ρ is ν. Now compute the function y according to the formulas (3.9), (3.11) and (3.12a), (3.13a) where $[\gamma_0, \gamma_1, ..., \gamma_{n-1}]$ is a nonzero solution of (3.13). Finally, compute $x = A^* y$. Then the solution h_* to the H^∞ optimization problem in (3.1) is given by (3.2).

XII.4. NOTES AND COMMENTS

The brief presentation in Section 1 was inspired by the monographs [Fran 2], [V] as well as the lecture notes of Doyle, Frances and Tannenbaum [DoyFT]. For an in-depth study of robust control we recommend these works as well as [Helt 5]. The operator theoretic preliminaries in Section 2 are quite well known [Sz.-NF 10], [Ber]. However, the proofs of the last three results in Section 2 are adjusted to the level of this monograph. Section 3 is a rather faithful adaptation of [FoT 2]. Skew Toeplitz operators were explicitly introduced in [BerFT] although the singular system was originally introduced in [FoTZ 2]. This paper inspired [BerFT], and [FoT 2] is the scalar version of [BerFT]. The algorithm presented in Procedure 3.5 seems to be quite effective in the case when the (McMillan) degree of the rational function f is small and m is either nonrational or a rational function of high order. If this is not the case, then it may be better to use the algorithms presented in Sections X.6 and X.7.

CHAPTER XIII

INVERSE SCATTERING ALGORITHMS FOR THE COMMUTANT LIFTING THEOREM

In this chapter we will obtain a complete characterization of all contractive intertwining liftings B of A in the commutant lifting theorem. As in Section VII.8 we show that there is a one to one correspondence between the set of all contractive intertwining liftings B of A and the set of all choice sequence initiated from G to G'. Then we present an inverse scattering algorithm to compute the choice sequence associated with B. By using an approach based on choice sequences we give a Schur type representation of B. This Schur representation shows that the set of all contraction intertwining liftings B of A is given by a "linear fractional transformation" of a contractive analytic function R in $H^\infty(G,G')$. Next by recursively using Theorem 3.1 in Chapter IV we present four more scattering algorithms for the commutant lifting theorem. We will give another proof of the fact that there is a one to one correspondence between the set of all contractive intertwining liftings of A and the set of all choice sequences initiated from G to G'. If A is a strict contraction, then by choosing the appropriate operators T and T' two of these algorithms reduce to the Schur algorithm, namely to Procedure 4.3 and Procedure 5.6 used in Chapter I to compute the reflection coefficients from the Carathéodory data. To complete this chapter we will use our Schur representation to obtain the Adamjan, Arov and Krein formula for characterizing all contractive intertwining liftings of a strictly contractive Hankel operator. This naturally leads to a computational procedure for computing all contractive intertwining liftings for a strictly contractive Hankel operator whose symbol is rational. As a demonstration of this procedure we provide another derivation of the Schur representation (I.3.8) used in solving the Carathéodory interpolation problem.

1. A PARAMETERIZATION OF BLOCK MATRIX CONTRACTIONS

In this section we will show that any n by m block matrix contraction can be parameterized by adequate systems of $n \cdot m$ block contractions.

We begin by studying column matrix contractions. To this end let Λ_n be the "column" operator defined by

$$\Lambda_n = [A, Z_1, Z_2, ..., Z_n]^{tr} : H \to \overset{n}{\underset{o}{\oplus}} H_i \tag{1.1}$$

where A is an operator from H to H_o and Z_i is an operator from H to H_i for $1 \le i \le n$.

1.1 LEMMA. *The operator Λ_n in (1.1) is a contraction if and only if $\Psi_o \overset{\Delta}{=} A$ is a contraction and there exists a set of contractions Ψ_i mapping $D_{\Psi_{i-1}}$ into H_i for $1 \le i \le n$, satisfying*

$$Z_1 = \Psi_1 D_{\Psi_o}, \; Z_2 = \Psi_2 D_{\Psi_1} D_{\Psi_o}, \; ...,$$
$$Z_n = \Psi_n D_{\Psi_{n-1}} D_{\Psi_{n-2}} \cdots D_{\Psi_o}. \tag{1.2}$$

Moreover, in this case the formula

$$\gamma_n D_{\Lambda_n} = D_{\Psi_n} D_{\Psi_{n-1}} \cdots D_{\Psi_o} \tag{1.3}$$

defines a unitary operator γ_n from D_{Λ_n} onto D_{Ψ_n}.

PROOF. This lemma reduces to Lemma 1.1 in Chapter IV when $n = 1$. So we use induction and assume that (1.2) and (1.3) holds for $n - 1$, or equivalently, Λ_{n-1} is a contraction. Then Λ_n is a contraction if and only if for all h in H

$$0 \le \|h\|^2 - \|\Lambda_{n-1}h\|^2 - \|Z_n h\|^2 , \tag{1.4}$$

or equivalently, by the induction hypothesis on (1.3)

$$\|Z_n h\|^2 \le \|D_{\Psi_{n-1}} D_{\Psi_{n-2}} \cdots D_{\Psi_o} h\|^2 .$$

Therefore Λ_n is a contraction if and only if there exists a contraction Ψ_n mapping $D_{\Psi_{n-1}}$ into H_n such that the last equation in (1.2) holds. To complete the proof notice that (1.2) and (1.4) give:

$$\|D_{\Lambda_n} h\|^2 = \|D_{\Lambda_{n-1}} h\|^2 - \|\Psi_n D_{\Psi_{n-1}} \cdots D_{\Psi_o} h\|^2 = \|D_{\Psi_n} D_{\Psi_{n-1}} \cdots D_{\Psi_o} h\|^2 .$$

This proves (1.3) for n and completes the proof.

1.2 COROLLARY. *Let A be a strict contraction. Then Λ_n in (1.1) is a strict contraction if and only if the operators Ψ_i in (1.2) are strict contractions for all $1 \le i \le n$.*

PROOF. If the Ψ_i's are all strict contractions, then D_{Ψ_i} is invertible for all $0 \le i \le n$. Equation (1.3) implies that D_{Λ_n} is invertible and thus Λ_n is a strict contraction. On the other hand if Λ_n is a strict contraction, then Λ_1 is a strict contraction. By Corollary 1.2 in Chapter IV and by the hypothesis on $A = \Psi_0$, both D_{Ψ_1} and D_{Ψ_0} are invertible. Now we use induction and assume that D_{Ψ_i} is invertible for all $0 \le i < n$. Since D_{Λ_n} is invertible (1.3) shows that D_{Ψ_n} is invertible, or equivalently, Ψ_n is a strict contraction. This completes the proof.

Let Λ_n' be the "row" operator defined by

$$\Lambda_n' = [\Gamma_0, Z_1, Z_2, ..., Z_n]: \overset{n}{\underset{0}{\oplus}} H_i \to H \tag{1.5}$$

where Γ is a operator from H_0 to H and Z_i is an operator mapping H_i into H for $1 \le i \le n$. Applying Lemma 1.1 to the adjoint of Λ_n' we obtain the following result.

1.3 COROLLARY. *The operator Λ_n' in (1.5) is a contraction if and only if Γ_0 is a contraction and*

$$\Lambda_n' = [\Gamma_0, D_{\Gamma_0^*}\Gamma_1, D_{\Gamma_0^*}D_{\Gamma_1^*}\Gamma_2, ...,D_{\Gamma_0^*}D_{\Gamma_1^*} \cdots D_{\Gamma_{n-1}^*}\Gamma_n]: \overset{n}{\underset{0}{\oplus}} H_i \to H \tag{1.6}$$

where Γ_i is a contraction mapping H_i into $D_{\Gamma_{i-1}^*}$ for $1 \le i \le n$.

Let Γ_i mapping H_i into $D_{\Gamma_{i-1}^*}$ be a set of contractions for $1 \le i \le n$ and Γ_0 a contraction from H_0 into H. Throughout

$$D_i = D_{\Gamma_i} , \quad D_{i*} = D_{\Gamma_i^*} , \quad D_i = D_{\Gamma_i} , \quad D_{i*} = D_{\Gamma_i^*} \quad \text{(for all i)} . \tag{1.7}$$

Also let α be a unitary operator from D_0 onto some D_0'. The upper triangular matrix $\mathfrak{T}_n$ mapping $\overset{n}{\underset{0}{\oplus}} H_i$ into $D_0' \oplus (\overset{n}{\underset{1}{\oplus}} D_i)$ is defined by

$$\Phi_n(\alpha) = \begin{bmatrix} \alpha D_o & -\alpha\Gamma_o^*\Gamma_1 & -\alpha\Gamma_o^*D_{1*}\Gamma_2 & \ldots & -\alpha\Gamma_o^*D_{1*}\cdots D_{n-1*}\Gamma_n \\ 0 & D_1 & -\Gamma_1^*\Gamma_2 & \ldots & -\Gamma_1^*D_{2*}\cdots D_{n-1*}\Gamma_n \\ 0 & 0 & D_2 & \ldots & -\Gamma_2^*D_{3*}\cdots D_{n-1*}\Gamma_n \\ 0 & 0 & 0 & \ldots & -\Gamma_3^*D_{4*}\cdots D_{n-1*}\Gamma_n \\ 0 & 0 & 0 & \ldots & \cdot \\ \cdot & \cdot & \cdot & \ldots & \cdot \\ \cdot & \cdot & \cdot & \ldots & \cdot \\ \cdot & \cdot & \cdot & \ldots & \cdot \\ 0 & 0 & 0 & \ldots & D_n \end{bmatrix} \qquad (1.8)$$

This operator will play an important role in many interpolation problems. In fact it is used in the following result.

1.4 COROLLARY. *Let Λ_n' be a contraction of the form (1.5) and Γ_i for $1 \leq i \leq n$ the contractions defined by (1.6). Then*

$$\beta_n D_{\Lambda_n'} = \Phi_n(\alpha) \qquad (1.9)$$

defines a unitary operator β_n from $D_{\Lambda_n'}$ onto $D_o' \oplus (\overset{n}{\underset{1}{\oplus}} D_i)$.

PROOF. It is clear that we only have to consider the case when $\alpha = I$. Let U_i be the unitary operator mapping $(\overset{i-1}{\underset{o}{\oplus}} H_j) \oplus (H_i \oplus D_{i*}) \oplus (\overset{n}{\underset{i+1}{\oplus}} D_j)$ onto $(\overset{i-1}{\underset{o}{\oplus}} H_j) \oplus (D_{i-1*} \oplus D_i) \oplus (\overset{n}{\underset{i+1}{\oplus}} D_j)$ (with the obvious notational conventions if $i = 0$ or $i = n$) defined by

$$U_o = R_{\Gamma_o} \oplus I \oplus I \oplus \cdots \oplus I \, ,$$

$$U_1 = I \oplus R_{\Gamma_1} \oplus I \oplus I \oplus \cdots \oplus I, \ldots,$$

$$U_{n-1} = I \oplus I \oplus \cdots \oplus I \oplus R_{\Gamma_{n-1}} \oplus I \, ,$$

where R_{Γ_i} is the rotation matrix

$$\begin{bmatrix} \Gamma_i & D_{i*} \\ D_i & -\Gamma_i^* \end{bmatrix} : H_i \oplus D_{i*} \to D_{i-1*} \oplus D_i$$

of Γ_i for $0 \leq i \leq n-1$. The isometry U_n mapping $\overset{n}{\underset{o}{\oplus}} H_i$ into $\overset{n-1}{\underset{o}{\oplus}} H_i \oplus (D_{n-1*} \oplus D_n)$ is defined by

$$U_n = I \oplus I \oplus \cdots \oplus I \oplus \begin{bmatrix} \Gamma_n \\ D_n \end{bmatrix} .$$

A simple calculation shows that

$$[\Lambda_n'^*, \Phi_n^*]^* = U_0 U_1 U_2 \cdots U_n \tag{1.10}$$

where Λ_n' and Φ_n are defined in (1.5), (1.6) and (1.8), respectively (and $\alpha = I$). Because all the U_i's are isometric, $[\Lambda_n'^*, \Phi_n^*]^*$ is an isometry. In particular, for h in $\overset{n}{\underset{0}{\oplus}} H_i$

$$\|D_{\Lambda_n'} h\|^2 = \|h\|^2 - \|\Lambda_n' h\|^2 = \|\Phi_n h\|^2 .$$

Thus (1.9) defines an isometry β_n mapping $D_{\Lambda_n'}$ into $\overset{n}{\underset{0}{\oplus}} D_i$. The upper triangular structure of Φ_n guarantees that β_n is onto. This completes the proof.

To complete this section we will show that any m by n contractive block matrix uniquely determines and is uniquely determined by $m \cdot n$ contractions. To this end let Λ_m be the contractive m+1 row by n+1 column matrix defined by

$$\Lambda_m = \begin{bmatrix} A \\ Z_1 \\ \cdot \\ \cdot \\ \cdot \\ Z_m \end{bmatrix} = \begin{bmatrix} A_{0,0} & A_{0,1} & \cdots & A_{0,n} \\ A_{1,0} & A_{1,1} & \cdots & A_{1,n} \\ \cdot & \cdot & \cdots & \cdot \\ \cdot & \cdot & \cdots & \cdot \\ \cdot & \cdot & \cdots & \cdot \\ A_{m,0} & A_{m,1} & \cdots & A_{m,n} \end{bmatrix} : \overset{n}{\underset{0}{\oplus}} H_j \to \overset{m}{\underset{0}{\oplus}} H_i' \tag{1.11}$$

where $A_{i,j}$ is a contraction mapping H_j into H_i' and Z_i mapping $\overset{n}{\underset{0}{\oplus}} H_j$ into H_i' is the i+1 row of Λ_m. Throughout this section $\{\Gamma_{i,j}\}_{0,0}^{m,n}$ is a set of contractions satisfying:

$$\Gamma_{0,0} = A_{0,0} \text{ and } \Gamma_{0,j+1} : H_{j+1} \to D_{\Gamma_{0,j}^*} \quad (\text{for } 0 \leq j < n)$$

$$\tag{1.12}$$

$$\Gamma_{i+1,0} : D_{\Gamma_{i,0}} \to H_{i+1}' \text{ and } \Gamma_{i+1,j+1} : D_{\Gamma_{i,j+1}} \to D_{\Gamma_{i+1,j}^*} \quad (\text{for } 0 \leq i < m \text{ and } 0 \leq j < n)$$

Let $\Lambda_{i,n}'$ be the row matrix Λ_n' defined in (1.6) where we replace $\{\Gamma_j\}_0^n$ by $\{\Gamma_{i,j}\}_{j=0}^n$. Let $\Phi_{i,n}$ be the matrix $\Phi_n(I)$ defined by (1.8) and we replace $\{\Gamma_j\}_0^n$ by $\{\Gamma_{i,j}\}_{j=0}^n$. Finally, we are ready for the following result.

1.5 THEOREM. *The* m+1 *by* n+1 *block matrix* Λ_m *defined in (1.11) is a contraction if and only if*

$$A = \Lambda'_{0,n}, \; Z_1 = \Lambda'_{1,n}\Phi_{0,n}, \; Z_2 = \Lambda'_{2,n}\Phi_{1,n}\Phi_{0,n}, \; ...,$$

$$\tag{1.13}$$

$$Z_m = \Lambda'_{m,n}\Phi_{m-1,n}\Phi_{m-2,n} \cdots \Phi_{0,n}$$

and $\{\Gamma_{i,j}\}_{0,0}^{m,n}$ is a set of contractions satisfying (1.12). In particular, there is a one to one correspondence between the set of all m+1 by n+1 contractive block matrices Λ_m of the form (1.11) and the set of all contractions $\{\Gamma_{i,j}\}_{0,0}^{m,n}$ satisfying (1.12). Moreover, if Λ_m is a contraction, then the formula

$$\alpha_m D_{\Lambda_m} = \Phi_{m,n}\Phi_{m-1,n} \cdots \Phi_{0,n} \tag{1.14}$$

defines a unitary operator α_m from D_{Λ_m} onto $\displaystyle\bigoplus_{j=0}^{n} D_{\Gamma^*_{m,j}}$.

PROOF. By Corollary 1.3 the row matrix $\Lambda_0 = A$ is a contraction if and only if $A = \Lambda'_{0,n}$ where $\Lambda'_{0,n}$ is the contraction in (1.6) with $\{\Gamma_j\}_0^n$ replaced by a set of contractions $\{\Gamma_{0,j}\}_0^n$ satisfying (1.12) for $m = 0$. In this case Corollary 1.4 shows that the equation $\alpha_0 D_A = \Phi_{0,n}$ defines a unitary operator α_0 from D_A onto $\displaystyle\bigoplus_{0}^{n} D_{\Gamma_{0,j}}$.

Now we proceed by induction. Assume that the theorem is true for $m = i$. Then Λ_{i+1} is a contraction if and only if for all h in $\displaystyle\bigoplus_{0}^{n} H_i$

$$0 \leq \|h\|^2 - \|\Lambda_i h\|^2 - \|Z_{i+1} h\|^2$$

or equivalently, by the induction hypothesis on (1.14)

$$\|Z_{i+1} h\|^2 \leq \|\Phi_{i,n}\Phi_{i-1,n} \cdots \Phi_{0,n} h\|^2 .$$

Since $\Phi_{i,n}$ is upper triangular the closed range of $\Phi_{i,n}$ is $\displaystyle\bigoplus_{j=0}^{n} D_{\Gamma_{i,j}}$. Therefore Λ_{i+1} is a contraction if and only if there exists a contraction $\Lambda'_{i+1,n}$ mapping $\displaystyle\bigoplus_{j=0}^{n} D_{\Gamma_{i,j}}$ into H'_{i+1} satisfying the last equation in (1.13) for $m = i + 1$. In this case, Corollary 1.3 guarantees that $\Lambda'_{i+1,n}$ is a contraction of the form (1.6) where $\{\Gamma_j\}_0^n$ is replaced by a set of contractions $\{\Gamma_{i+1,j}\}_{j=0}^n$ satisfying (1.12).

To complete the induction we can now assume that Λ_{i+1} is a contraction. Then Corollary 1.4 with (1.13) and (1.14) give

$$\|D_{\Lambda_{i+1}} h\|^2 = \|D_{\Lambda_i} h\|^2 - \|\Lambda'_{i+1,n}\Phi_{i,n}\Phi_{i-1,n} \cdots \Phi_{0,n} h\|^2 =$$

$$\|D_{\Lambda'_{i+1,n}}\Phi_{i,n}\Phi_{i-1,n}\ \cdots\ \Phi_{0,n}h\|^2 = \|\Phi_{i+1,n}\Phi_{i,n}\ \cdots\ \Phi_{0,n}h\|^2\ .$$

It follows that (1.14) with $m = i+1$ defines an isometry α_{i+1} mapping $D_{\Lambda_{i+1}}$ into $\bigoplus\limits_{j=0}^{n} D_{\Gamma_{i+1,j}}$. The upper triangular structure of $\Phi_{k,n}$ guarantees that α_{i+1} is onto, or equivalently, α_{i+1} is unitary. This completes the proof.

1.6 COROLLARY. *The* $m+1$ *by* $n+1$ *block matrix* Λ_m *defined in (1.11) is a strict contraction if and only if its corresponding contractions* $\{\Gamma_{i,j} : 0 \le i \le m$ *and* $0 \le j \le n\}$ *are all strict.*

PROOF. This follows from the proof of the previous theorem with Corollary 1.2 and its row matrix analogue.

1.7 REMARK. Let Λ_m be the contractive matrix in (1.11). The previous theorem can be used to obtain a lower - upper triangular factorization of $D_{\Lambda_m}^2$, that is, a factorization of the form LU where L is a lower triangular block matrix and U is an upper triangular block matrix. To see this let Θ be the upper triangular matrix defined by

$$\Theta = \Phi_{m,n}\Phi_{m-1,n}\ \cdots\ \Phi_{0,n}\ . \tag{1.15}$$

Equation (1.14) implies that $(D_{\Lambda_m}^2 h,h) = (\Theta^*\Theta h,h)$ for all h in $\bigoplus\limits_{o}^{n} H_i$. Therefore $\Theta^*\Theta = D_{\Lambda_m}^2$ is a lower-upper triangular factorization of $D_{\Lambda_m}^2$.

1.8 REMARK. Let Λ_m be the contractive matrix in (1.11). Let $\{\Psi_j\}_0^n$ be the contractions uniquely determined by Λ_m in Lemma 1.1. By Lemma 1.1 and the previous theorem, the equation

$$\alpha_i D_{\Psi_i} D_{\Psi_{i-1}}\ \cdots\ D_{\Psi_o} = \Phi_{i,n}\Phi_{i-1,n}\ \cdots\ \Phi_{0,n}\ . \tag{1.16}$$

defines a unitary operator α_i mapping D_{Ψ_i} onto $\bigoplus\limits_{j=0}^{n} D_{\Gamma_{i,j}}$.

The previous theorem shows that a n by n matrix is uniquely determined by n^2 contractions. Later we will see that certain n by n matrices, such as Toeplitz matrices, are uniquely determined by n contractions. These special matrices will play an important role in interpolation problems.

2. CHOICE SEQUENCES AND THE COMMUTANT LIFTING THEOREM

In this section we will present a refined analysis of the original proof of the commutant lifting theorem to show that there exists a one to one correspondence between the set of all contractive intertwining liftings B of A, and the set of all choice sequences initiated from G to G'. Then we will give an inverse scattering algorithm to compute the choice sequence $\{\Gamma_i\}$ associated with B. It turns out that this algorithm is a generalization of Procedure 3.3 in Chapter III.

Throughout this section $U = T$ on $K = H$ is an isometry, T' on H' is a contraction and A is a contraction in $I(U, T')$. Recall that the contraction T'_n on $H_n = H' \oplus (l^2_n (D_{T'}))$ defined by

$$
T'_n = \begin{bmatrix}
T' & 0 & 0 & \ldots & 0 & 0 \\
D_{T'} & 0 & 0 & \ldots & 0 & 0 \\
0 & I & 0 & \ldots & 0 & 0 \\
0 & 0 & I & \ldots & 0 & 0 \\
\cdot & \cdot & \cdot & \cdot & \cdot & \cdot \\
\cdot & \cdot & \cdot & \cdot & \cdot & \cdot \\
0 & 0 & 0 & \ldots & I & 0
\end{bmatrix}
\tag{2.1}
$$

is called the *n step dilation of* T' (see Section VII.2). Notice that $D_{T'_n}$ is the orthogonal projection onto $\{0\} \oplus \{0\} \oplus \cdots \oplus \{0\} \oplus D_{T'}$. Therefore the one step dilation $(T'_n)_1$ of T'_n is unitarily equivalent to T'_{n+1}. So without loss of generality we set $(T'_n)_1 = T'_{n+1}$. The n step dilation T'_n of T' is obtained by performing n one step dilations of T'; see Remark 2.4 in Chapter VII for further details.

We say that Λ_n is a *n step contractive intertwining lifting of* A if Λ_n is a contraction in $I(U, T'_n)$ lifting A, that is, $A = P_{H'}\Lambda_n$. In this case if $m \leq n$, then $\Lambda_m = P_{H_m}\Lambda_n$ is a m step contractive intertwining lifting of A. Moreover, if $\{\Lambda_n\}$ is a sequence of n step contractive intertwining liftings of A satisfying $\Lambda_m = P_{H_m}\Lambda_n$ for $m \leq n$, then Lemma 2.1 in Chapter VII shows that $\Lambda_n \to \Lambda_\infty$ strongly. Using this and the fact that $T'_n \to U'$ strongly, where U' is the minimal isometric dilation of T', it follows that Λ_∞ is a contractive intertwining lifting of A; for the details see the second proof of the commutant lifting theorem in Section VII.3. On the other hand if B is a contractive intertwining lifting of A, then $\{\Lambda_n \overset{\Delta}{=} P_{H_n}B\}_n$ defines a sequence of n step contractive intertwining liftings of A. Therefore there is a one to one correspondence between the set of all contractive intertwining liftings B of A, and the set of all sequences $\{\Lambda_n\}$ of n step contractive intertwining liftings of A satisfying $\Lambda_m = P_{H_m}\Lambda_n$ for $m \leq n$. So we obtain the set of all contractive intertwining liftings B of A by studying the set of all n step contractive intertwining liftings Λ_n of A.

Using $T_n'\Lambda_n = \Lambda_n U$ we see that Λ_n is a contractive n step intertwining lifting of A if and only if Λ_n is a contraction and Λ_n admits a matrix representation of the form:

$$\Lambda_n = [A, Z_1, Z_2, ..., Z_n]^{\mathrm{tr}} : K \to H' \oplus (l_n^2 (D_{T'})) \tag{2.2}$$

satisfying

$$Z_1 U = D_{T'} A \text{ and } Z_{i+1} U = Z_i \quad (\text{for } 1 \le i < n) . \tag{2.3}$$

Lemma 1.1 and (2.2), (2.3) yield the following result.

2.1 LEMMA. *Let* A *be a contraction in* $I(U, T')$ *and set* $\Psi_0 = A$. *The operator* Λ_n *in (2.2) is a contractive n step intertwining lifting of* A *if and only if there exists a set of contractions* Ψ_i *from* $D_{\Psi_{i-1}}$ *to* $D_{T'}$ *(for* $1 \le i \le n$) *satisfying*

$$Z_1 = \Psi_1 D_{\Psi_0}, \qquad Z_2 = \Psi_2 D_{\Psi_1} D_{\Psi_0}, ..., \tag{2.4}$$
$$Z_n = \Psi_n D_{\Psi_{n-1}} D_{\Psi_{n-2}} \cdots D_{\Psi_0}$$

and

$$\Psi_1 D_A U = D_{T'} A, \qquad \Psi_2 D_{\Psi_1} D_{\Psi_0} U = \Psi_1 D_{\Psi_0} ,$$
$$\Psi_3 D_{\Psi_2} D_{\Psi_1} D_{\Psi_0} U = \Psi_2 D_{\Psi_1} D_{\Psi_0}, ..., \tag{2.5}$$
$$\Psi_n D_{\Psi_{n-1}} D_{\Psi_{n-2}} \cdots D_{\Psi_0} U = \Psi_{n-1} D_{\Psi_{n-2}} D_{\Psi_{n-3}} \cdots D_{\Psi_0} .$$

We will find the set of all contractive intertwining liftings $B(=\Lambda_\infty)$ of A by recursively finding the set of all contractions Ψ_i for $i \ge 1$ satisfying (2.5) where $A = \Psi_0$. Then the set of all contractive intertwining liftings $B = \Lambda_\infty$ of A is given by inserting the Z_i''s in (2.4) into the infinite column matrix Λ_∞ from K to $H' \oplus l^2(D_{T'})$ in (2.2). Lemma 2.1 and our previous discussion guarantees that Λ_∞ is a contractive intertwining lifting of A. Notice that $A = \Psi_0$ is known and the constraints on Ψ_{i+1} depends only on the previous Ψ_i. So we begin our search by looking for the set of all contractions Ψ_1 satisfying (2.5), that is, $\Psi_1 D_A U = D_{T'} A$. This problem was solved in Section V.1. To present that solution recall that the formula

$$\Gamma_0 D_A U h = D_{T'} A h \qquad (h \in H) \tag{2.6}$$

defines a contraction Γ_0 from $\overline{D_A U H}$ to $D_{T'}$; compare (2.6) to (V.1.5) and notice that now T is the isometry U. Also in this case the spaces F and G defined in (V.1.4) become $F = \overline{D_A U H}$ and $G = D_A \ominus F$. Finally, let j_0 be the unitary operator from D_A onto $F \oplus G$ defined by

$$j_o h = \begin{bmatrix} P_F h \\ P_G h \end{bmatrix} \qquad (h \in D_A).$$

We can now state the following result.

2.2 LEMMA. *The set of all contractive one step intertwining liftings Λ_1 of A is given by $\Lambda_1 = [A^*, Z_1^*]^*$ where*

$$Z_1 = \Psi_1 D_A = [\Gamma_o, D_{\Gamma_o^*}\Gamma_1] j_o D_A \qquad (2.7)$$

and Γ_1 is an arbitrary contraction mapping G into $D_{\Gamma_o^*}$. Furthermore, there exists a unitary operator α mapping D_{Γ_o} onto $F \oplus G$ such that

$$\alpha D_{\Gamma_o} D_A U = j_o D_A. \qquad (2.8)$$

PROOF. This follows from Lemma 1.1 in Chapter V where T is the isometry U.

Recall that a *choice sequence* $\{\Gamma_j\}_1^\infty$ *initiated from G to $D_{\Gamma_o^*}$*, is a sequence of contractions such that Γ_1 is a contraction from G into $D_{\Gamma_o^*}$ and Γ_{j+1} is a contraction from D_j into D_{j*} for all $j \geq 1$. Here as before the operators D_j, D_{j*} and spaces D_j, D_{j*} are defined in (1.7). Throughout this section Γ_o is the contraction from $F = \overline{D_A UK}$ to $D_{T'}$ defined in (2.6) and α is the unitary operator from D_{Γ_o} onto $F \oplus G$ defined in (2.8). The following shows that the set of all contractive intertwining liftings $B = \Lambda_\infty$ of A is parameterized by the choice sequences initiated from G to $D_{o*} = D_{\Gamma_o^*}$.

2.3 THEOREM. *Let A be a contraction in $I(U, T')$. The set of all contractive n step intertwining liftings Λ_n of A is given by (2.2) where*

$$Z_1 = \Lambda_1' j_o D_A \quad and \quad Z_i = \Lambda_i' \Phi_{i-1} \Phi_{i-2} \cdots \Phi_1 j_o D_A \qquad (2 \leq i \leq n) \qquad (2.9)$$

Λ_i' and $\Phi_i = \Phi_i(\alpha)$ are the operators defined in (1.6) and (1.8) with the same choice sequence $\{\Gamma_i\}_1^n$ initiated from G to $D_{\Gamma_o^*}$. (Here Γ_o is the contraction and α is the unitary operator in (2.6) and (2.8), respectively.) In this case the formula

$$\delta_i D_{\Lambda_i} = \Phi_i \Phi_{i-1} \cdots \Phi_1 j_o D_A \qquad (1 \leq i \leq n) \qquad (2.10)$$

defines a unitary operator δ_i from D_{Λ_i} onto $D_A \oplus D_1 \oplus D_2 \oplus \cdots \oplus D_i$. In particular, the operators $Z_i (1 \leq i \leq n)$ corresponding to a contractive n step intertwining lifting Λ_n of A uniquely determine and are uniquely determined by a finite choice sequence $\{\Gamma_i\}_1^n$ initiated form G to D_{o*}. Finally, this gives a one to one correspondence between the set of all contractive intertwining liftings Λ_∞ of A and the set of all choice sequences $\{\Gamma_i\}_1^\infty$

initiated from G to D_{o*}.

PROOF. We prove this result by recursively applying Corollary 1.3 and Corollary 1.4 to find the set of all contractions Ψ_i satisfying (2.5). Then by Lemma 2.1 the set of all contractive n step intertwining liftings Λ_n of A is given by substituting (2.4) into (2.2). Lemma 2.2 with $\Lambda'_1 = \Psi_1 j_0^* = [\Gamma_o, D_{o*}\Gamma_1]$ shows that the set of all contractive one step intertwining liftings $\Lambda_1 = [A^*, Z_1^*]^*$ of A is given by (2.9) for n = 1. Here Γ_o is the contraction defined by (2.6) and Γ_1 is a contraction from G to D_{o*}. Obviously Ψ_1 and Γ_1 uniquely determine each other.

Our next task is to find all contractive Ψ_2 satisfying (2.5). By Corollary 1.4 applied to $\Lambda'_1 = \Psi_1 j_0^*$ the formula $\alpha_1 D_{\Psi_1} D_A = \Phi_1 j_0 D_A$ defines a unitary operator α_1 from D_{Ψ_1} onto $F \oplus G \oplus D_1$. Lemma 1.1 shows that $\delta_1 = \alpha_1 \gamma_1$ is a unitary operator satisfying (2.10) for i = 1. Therefore finding all contractive Ψ_2 satisfying (2.5) is equivalent to finding all contractions $\Lambda'_2 = \Psi_2 \alpha_1^*$ satisfying the constraint

$$\Lambda'_2 \Phi_1 j_0 D_A U = \Psi_1 D_A . \tag{2.11}$$

We claim that the constraint on Λ'_2 in (2.11) is equivalent to the fact that the operator Λ'_2 has the following matrix form

$$\Lambda'_2 = [\Lambda'_1, *] : (F \oplus G) \oplus D_1 \to D_{T'} . \tag{2.12}$$

To see this first notice that the definition (1.8) for Φ_1 and the equation (2.8) give

$$\Phi_1 j_0 D_A U = \begin{bmatrix} \alpha D_o & -\alpha\Gamma_o^*\Gamma_1 \\ 0 & D_1 \end{bmatrix} \begin{bmatrix} D_A U \\ 0 \end{bmatrix} = \begin{bmatrix} \alpha D_o D_A U \\ 0 \end{bmatrix} = \begin{bmatrix} j_0 D_A \\ 0 \end{bmatrix} .$$

Using this in (2.11) yields $\Lambda'_2 [j_0 D_A, 0]^{tr} = \Psi_1 D_A = \Lambda'_1 j_0 D_A$, which proves (2.12). Corollary 1.3 and $\Lambda'_1 = [\Gamma_o, D_{o*}\Gamma_1]$ show that the set of all contractive Λ'_2 satisfying (2.12) is given by $\Lambda'_2 = [\Gamma_o, D_{o*}\Gamma_1, D_{o*}D_{1*}\Gamma_2]$ where Γ_2 is a contraction mapping D_1 into D_{1*}. Hence there exists a one to one correspondence between the set of all contractions Ψ_i for i = 1, 2 satisfying (2.5), and the set of all finite choice sequences $\{\Gamma_i\}_1^2$ initiated from G to D_{o*}. Equivalently by Lemma 2.1 there is a one to one correspondence between the set of all contractive two step intertwining liftings Λ_2 of A, and the set of all choice sequences $\{\Gamma_i\}_1^2$ initiated from G to D_{o*}.

Corollary 1.4 for $\Psi_1 j_0^* = \Lambda'_1$ and $\Lambda'_2 = \Psi_2 \alpha_1^*$ give:

$$\|D_{\Psi_2} D_{\Psi_1} D_A h\|^2 = \|D_{\Psi_1} D_A h\|^2 - \|\Psi_2 D_{\Psi_1} D_A h\|^2 =$$

$$\|\Phi_1 j_o D_A h\|^2 - \|\Lambda_2' \Phi_1 j_o D_A h\|^2 = \|D_{\Lambda_2'} \Phi_1 j_o D_A h\|^2 = \|\Phi_2 \Phi_1 j_o D_A h\|^2 \ .$$

Therefore there exists an isometry α_2 mapping D_{Ψ_2} into $F \oplus G \oplus D_1 \oplus D_2$ such that

$$\alpha_2 D_{\Psi_2} D_{\Psi_1} D_A = \Phi_2 \Phi_1 j_o D_A \ .$$

The upper triangular structure of the $\Phi_i's$ implies that α_2 is onto, that is α_2 is unitary. This and Lemma 1.1 proves (2.9) and (2.10) with $\delta_2 = \alpha_2 \gamma_2$. Thus the theorem is true for $n = 2$.

Notice that for $i \geq 1$ the operator Φ_{i+1} in (1.8) admits an upper triangular decomposition of the form

$$\Phi_{i+1} = \begin{bmatrix} \Phi_i & * \\ 0 & D_{i+1} \end{bmatrix} \tag{2.13}$$

Using (2.8) and (2.13) we have

$$\Phi_2 \Phi_1 j_o D_A U = \Phi_2 \begin{bmatrix} \alpha D_o & -\alpha \Gamma_o^* \Gamma_1 \\ 0 & D_1 \end{bmatrix} \begin{bmatrix} D_A U \\ 0 \end{bmatrix} = \begin{bmatrix} \Phi_1 & * \\ 0 & D_2 \end{bmatrix} \begin{bmatrix} j_o D_A \\ 0 \end{bmatrix} = \begin{bmatrix} \Phi_1 j_o D_A \\ 0 \end{bmatrix}$$

The rest of the proof follows by induction. To this end assume that the theorem is true for some $n \geq 2$ and that

$$\Phi_n \Phi_{n-1} \cdots \Phi_1 j_o D_A U = [\Phi_{n-1} \Phi_{n-2} \cdots \Phi_1 j_o D_A, 0]^{\alpha} \ . \tag{2.14}$$

Then the operators $\alpha_i = \delta_i \gamma_i^*$ for $1 \leq i \leq n$ are unitary; here the operators γ_i are defined by (1.3). With these definitions the operators $\Psi_1 = \Lambda_1' j_o$ and $\Psi_i = \Lambda_i' \alpha_{i-1}^*$ for $2 \leq i \leq n$ satisfy (2.2), (2.4), (2.5) and

$$\alpha_i D_{\Psi_i} D_{\Psi_{i-1}} \cdots D_{\Psi_1} D_A = \Phi_i \Phi_{i-1} \cdots \Phi_1 j_o D_A \qquad (1 \leq i \leq n) \ . \tag{2.15}$$

This equation along with (2.9) shows that the set of all contractions Ψ_{n+1} satisfying (2.5) is given by the set of all contractions $\Lambda_{n+1}' = \Psi_{n+1} \alpha_n^*$ satisfying the constraint

$$\Lambda_{n+1}' \Phi_n \Phi_{n-1} \cdots \Phi_1 j_o D_A U = \Lambda_n' \Phi_{n-1} \Phi_{n-2} \cdots \Phi_1 j_o D_A \ ,$$

or equivalently, by (2.14)

$$\Lambda_{n+1}' \begin{bmatrix} \Phi_{n-1} \Phi_{n-2} \cdots \Phi_1 j_o D_A \\ 0 \end{bmatrix} = \Lambda_n' \Phi_{n-1} \Phi_{n-2} \cdots \Phi_1 j_o D_A \ .$$

But the range of $\Phi_{n-1} \Phi_{n-2} \cdots \Phi_1 j_o D_A$ is dense in $F \oplus G \oplus D_1 \oplus \cdots \oplus D_{n-1}$. This follows at once by using induction on the upper triangular form of Φ_{i+1} in (2.13). Therefore the constraint on Λ_{n+1}' means that the operator Λ_{n+1}' viewed as a map from

$$(F \oplus G \oplus D_1 \oplus \cdots \oplus D_{n-1}) \oplus D_n$$

into $D_{T'}$ has the matrix form

$$\Lambda'_{n+1} = [\Lambda'_n, *] \tag{2.16}$$

where Λ'_n is defined in (1.6) by the choice $\{\Gamma_i\}_1^n$. Corollary 1.3 implies that the set of all contractive Λ'_{n+1} satisfying (2.16) is given by (1.6) where n is replaced by n+1 and Γ_{n+1} is an arbitrary contraction from D_n to D_{n*}. So if we define now $\Psi_{n+1} = \Lambda'_{n+1}\alpha_n$, then by the conditions (2.14), (2.15) and according to (2.4) (with n+1 instead of n) we obtain

$$Z_{n+1} = \Psi_{n+1} D_{\Psi_n} \cdots D_{\Psi_1} D_A = \Lambda'_{n+1} \Phi_n \cdots \Phi_1 j_o D_A$$

which proves that (2.9) also holds for n+1 instead of n.

The equation (2.15) and $\Psi_{n+1}\alpha_n^* = \Lambda'_{n+1}$ with Corollary 1.4 give

$$\|D_{\Psi_{n+1}} D_{\Psi_n} \cdots D_{\Psi_1} D_A h\|^2 = \|\Phi_n \cdots \Phi_1 j_o D_A h\|^2 - \|\Lambda'_{n+1}\Phi_n \cdots \Phi_1 j_o D_A h\|^2 =$$

$$\|D_{\Lambda'_{n+1}}\Phi_n \cdots \Phi_1 j_o D_A h\|^2 = \|\Phi_{n+1}\Phi_n \cdots \Phi_1 j_o D_A h\|^2$$

where h is in H. This and the upper triangular structure of Φ_{n+1} (see (2.13) with i replaced by n) implies that $\alpha_{n+1} D_{\Psi_{n+1}} \cdots D_{\Psi_1} D_A = \Phi_{n+1} \cdots \Phi_1 j_o D_A$ defines an unitary operator α_{n+1} that is (2.15) holds for n+1 instead of n. By Lemma 1.1 this shows that (2.10) holds for n+1 with $\delta_{n+1} = \alpha_{n+1}\gamma_{n+1}$. Equations (2.14) and (2.13) for n instead of i yield

$$\Phi_{n+1} \cdots \Phi_1 j_o D_A U = \Phi_{n+1} \begin{bmatrix} \Phi_{n-1} \cdots \Phi_1 j_o D_A \\ 0 \end{bmatrix} =$$

$$\begin{bmatrix} \Phi_n & * \\ 0 & D_{n+1} \end{bmatrix} \begin{bmatrix} \Phi_{n-1} \cdots \Phi_1 j_o D_A \\ 0 \end{bmatrix} = \begin{bmatrix} \Phi_n \cdots \Phi_1 j_o D_A \\ 0 \end{bmatrix}$$

This proves that (2.14) also holds for n+1 instead of n and completes the induction argument. Finally, the proof is complete.

Due to the previous theorem we say that $\{\Gamma_i\}_1^n$ is the choice sequence *associated with the contractive n step intertwining lifting* Λ_n of A if $\Lambda_n = [A, Z_1, ..., Z_n]^{tr}$ where Z_i is given by (2.9) and Φ_i and Λ'_i are the matrices associated with the same finite choice sequence $\{\Gamma_i\}_1^n$. In a similar manner we also say that $\{\Gamma_i\}_1^\infty$ is the choice sequence *associated with the contractive intertwining lifting* B of A if $B = [A, Z_1, Z_2, Z_3, ...]^{tr}$ where Z_i is given by (2.9) for all i and Φ_i and Λ'_i are associated with the same choice sequence $\{\Gamma_i\}_1^\infty$.

Recall that $\{\Gamma_i\}_1^n$ is a *strictly contractive choice sequence* if $\{\Gamma_i\}_1^n$ is a choice sequence and Γ_i is a strict contraction for all $1 \le i \le n$.

2.4 COROLLARY. *Let* A *be a strict contraction in* $I(U,T')$. *Let* Λ_n *be a contractive n step intertwining lifting of* A *and* $\{\Gamma_i\}_1^n$ *be its choice sequence. Then* Λ_n *is a strict contraction if and only if* $\{\Gamma_i\}_1^n$ *is a strictly contractive choice sequence.*

PROOF. We claim that Γ_0 in (2.6) is a strict contraction, or equivalently, D_0 is invertible. If D_0 is not invertible there exists a sequence of unit vectors $d_n = D_A U h_n$ with h_n in H such that $D_0 d_n \to 0$ as $n \to \infty$. Equation (2.8) shows that $D_A h_n \to 0$ as $n \to \infty$. Since D_A is invertible this implies that $h_n \to 0$ as $n \to \infty$. Therefore the sequence of unit vectors $D_A U h_n$ must approach zero as $n \to \infty$. This is a contradiction and Γ_0 is a strict contraction. According to Corollary 1.2 and the previous theorem, the contractive n step intertwining lifting Λ_n of A is a strict contraction if and only if $\Psi_i \alpha_{i-1}^* = \Lambda_i'$ is a strict contraction for all $1 \le i \le n$. Applying again Corollary 1.2 to $\Lambda_i'^*$ we see that Λ_i' is a strict contraction for all $1 \le i \le n$ if and only if Γ_i is a strict contraction for all $1 \le i \le n$. Therefore Λ_n is a strict contraction if and only if its choice operators Γ_i are strict contractions for $1 \le i \le n$. This completes the proof.

As before let A be a contraction in $I(U,T')$. To complete this section we will use the unitary operator ω defined by the equation (V.1.12), to obtain an explicit characterization of the set of all contractive intertwining liftings of A. Using the fact that $T = U$ is an isometry, the spaces F, G, F' and G' defined in equations (V.1.4) and (V.1.13) become

$$F = \overline{D_A UK} \text{ and } G = D_A \ominus \overline{D_A UK} ,$$
$$F' = \{D_{T'} Ak \oplus D_A k{:}k \in K\}^- \text{ and } G' = (D_{T'} \oplus D_A) \ominus F' . \tag{2.17}$$

In this case the unitary operator ω from F onto F' defined in (V.1.12) becomes

$$\omega D_A Uk = D_{T'} Ak \oplus D_A k \qquad (k \in K) . \tag{2.18}$$

Throughout this section Π' respectively Π_A are the operators which pick out the $D_{T'}$ respectively D_A component from any space of the form $D_{T'} \oplus D_A \oplus X$ where X is an arbitrary Hilbert space, that is, if $y = d' \oplus d_A \oplus x$ is in $D_{T'} \oplus D_A \oplus X$, then $\Pi' y = d'$ and $\Pi_A y = d_A$. (The space X can be infinite dimensional, or the trivial space zero in which case we simply write $D_{T'} \oplus D_A$.) Now let W be the block matrix from $F \oplus G \oplus (\overset{\infty}{\underset{1}{\oplus}} D_i)$ to $D_{T'} \oplus (F \oplus G) \oplus (\overset{\infty}{\underset{1}{\oplus}} D_i)$ defined by

$$
W = \begin{bmatrix}
\Pi'\omega & \Pi'\Gamma_1 & \Pi'D_{1*}\Gamma_2 & \Pi'D_{1*}D_{2*}\Gamma_3 & \cdots \\
j_o\Pi_A\omega & j_o\Pi_A\Gamma_1 & j_o\Pi_A D_{1*}\Gamma_2 & j_o\Pi_A D_{1*}D_{2*}\Gamma_3 & \cdots \\
0 & D_1 & -\Gamma_1^*\Gamma_2 & -\Gamma_1^* D_{2*}\Gamma_3 & \cdots \\
0 & 0 & D_2 & -\Gamma_2^*\Gamma_3 & \cdots \\
0 & 0 & 0 & D_3 & \cdots \\
0 & 0 & 0 & 0 & \cdots \\
\cdot & \cdot & \cdot & \cdot & \cdots \\
\cdot & \cdot & \cdot & \cdot & \cdots \\
\cdot & \cdot & \cdot & \cdot & \cdots
\end{bmatrix}
\tag{2.19}
$$

where $\{\Gamma_i\}_1^\infty$ is a choice sequence initiated from G to G'. (As before the defect operators D_i and D_{i*} are defined in (1.7).) A simple calculation shows that if $h = f \oplus g \oplus d_1 \oplus d_2 \oplus \cdots$ is an element in $F \oplus G \oplus D_1 \oplus D_2 \oplus \cdots$ with only a finite number of nonzero entries, then $Wh = U_1 U_2 \cdots U_n h$ for all n large enough where the U_i's are the unitary operators defined by

$$
U_1 = \begin{bmatrix} [\Pi', j_o\Pi_A]^{tr} & 0 \\ 0 & I \end{bmatrix} \begin{bmatrix} \omega \oplus \Gamma_1 & D_{1*} \\ D_1 & -\Gamma_1^* \end{bmatrix} \oplus I \oplus I \oplus I \oplus \cdots
$$

$$
U_2 = I \oplus R_{\Gamma_2} \oplus I \oplus I \oplus I \oplus \cdots
\tag{2.20}
$$

$$
U_3 = I \oplus I \oplus R_{\Gamma_3} \oplus I \oplus I \oplus \cdots
$$

The rotation matrix R_{Γ_i} defined in (IV.1.8) appears in the i-th position and all the other entries are the identity. Therefore since these h's are dense we can define

$$
W = \text{strong} \lim_{n \to \infty} U_1 U_2 \cdots U_n \overset{\Delta}{=} U_1 U_2 U_3 U_4 \cdots
\tag{2.21}
$$

Because all the U_i's are unitary the operator W is an isometry.

We now introduce some elementary results concerning choice sequences.

2.5 LEMMA. *Let γ be a unitary operator mapping G' onto H'. Then $\{\Gamma_i\}_1^\infty$ is a choice sequence initiated from G to G' if and only if $\{\gamma\Gamma_i\}_1^\infty$ is a choice sequence initiated from G to H'.*

PROOF. Let $\{\Gamma_i\}_1^\infty$ be a choice sequence initiated from G to G'. Obviously $\{\Gamma_i' \overset{\Delta}{=} \gamma\Gamma_i\}_1^\infty$ is a sequence of contractions and $\gamma\Gamma_1$ is a contraction from G to H'. Clearly $D_{\Gamma_i'} = D_i$ for all i. A simple calculation shows that $D_{\Gamma_i'}^2 \gamma = \gamma D_{i*}^2$. Since the positive square root of a positive operator A is the norm limit of polynomials in A this implies that

$$D_{\Gamma_i^*}\gamma = \gamma D_{i*} \qquad (1 \le i) .$$

It follows that the operator $\gamma | D_{i*}$ is a unitary operator mapping D_{i*} onto $D_{\Gamma_i^*}$ for all i. Hence $\Gamma_{i+1}' = \gamma \Gamma_{i+1}$ is a contraction mapping D_i into $D_{\Gamma_i^*}$ for all i, that is, $\{\gamma \Gamma_i\}_1^\infty$ is a choice sequence. This completes the proof.

2.6 COROLLARY. *Let γ be a unitary operator mapping G' onto H' and α be a unitary operator mapping H onto G. Then $\{\Gamma_i\}_1^\infty$ is a choice sequence initiated from G to G' if and only if $\{\gamma \Gamma_i \alpha | \alpha^* D_{i-1}\}_1^\infty$ is a choice sequence initiated from H to H'.*

PROOF. Lemma 2.5 shows that $\{\Gamma_i\}_1^\infty$ is a choice sequence if and only if $\{\Gamma_i^* \gamma^*\}_1^\infty$ is a choice sequence. Applying Lemma 2.5 again proves the corollary.

At this point it is convenient to introduce the following definition. Two choice sequences finite or infinite $\{\Gamma_i\}$ initiated from G to G' and $\{\Gamma_i'\}$ initiated from H to H' are said to be *unitarily equivalent* if $\Gamma_i' = \gamma \Gamma_i \alpha | D_{\Gamma_{i-1}'}$ for all i where α is a unitary operator from H onto G and γ is a unitary operator from G' onto H'.

Let Q' be the operator from $D_{T'} \oplus F \oplus G \oplus D_1 \oplus D_2 \oplus \cdots$ onto $F \oplus G \oplus D_1 \oplus D_2 \oplus \cdots$ which eliminates the $D_{T'}$ component, that is,

$$Q'(d_{T'} \oplus f \oplus g \oplus d_1 \oplus d_2 \oplus \cdots) = f \oplus g \oplus d_1 \oplus d_2 \oplus \cdots$$

Notice that $Q'W$ is the upper triangular matrix on $F \oplus G \oplus D_1 \oplus D_2 \oplus \cdots$ defined by

$$Q'W = \begin{bmatrix} j_o\Pi_A\omega & j_o\Pi_A\Gamma_1 & j_o\Pi_A D_1 \cdot \Gamma_2 & j_o\Pi_A D_1 \cdot D_2 \cdot \Gamma_3 & \cdots \\ 0 & D_1 & -\Gamma_1^*\Gamma_2 & -\Gamma_1^* D_2 \cdot \Gamma_3 & \cdots \\ 0 & 0 & D_2 & -\Gamma_2^*\Gamma_3 & \cdots \\ 0 & 0 & 0 & D_3 & \cdots \\ 0 & 0 & 0 & 0 & \cdots \\ \cdot & \cdot & \cdot & \cdot & \cdots \\ \cdot & \cdot & \cdot & \cdot & \cdots \\ \cdot & \cdot & \cdot & \cdot & \cdots \end{bmatrix}$$

This sets the stage for the following basic result.

2.7 THEOREM. *Let A be a contraction in $I(U,T')$. The set of all contractive intertwining liftings Λ_∞ of A is given by the infinite column matrix Λ_∞ from K into $H' \oplus l^2(D_{T'})$ in (2.2) where*

$$Z_{i+1} = \Pi'W(Q'W)^i[j_0D_A, 0, 0, 0, \cdots]^{tr} \quad \textit{(for all } i \geq 0) \tag{2.22}$$

and W *is the isometry in (2.19) defined by a choice sequence* $\{\Gamma_i\}_1^\infty$ *initiated from* G *to* G'. *In particular, this gives a one to one correspondence between the set of all contractive intertwining liftings* $B(= \Lambda_\infty)$ *of* A *and the set of all choice sequences* $\{\Gamma_i\}_1^\infty$ *initiated from* G *to* G'.

PROOF. Let Φ_∞ be the block matrix on $F \oplus G \oplus D_1 \oplus D_2 \oplus \cdots$ defined by

$$\Phi_\infty = \begin{bmatrix} \alpha D_0 & -\alpha\Gamma_0^*\Gamma_1 & -\alpha\Gamma_0^*D_{1*}\Gamma_2 & \cdots \\ 0 & D_1 & -\Gamma_1^*\Gamma_2 & \cdots \\ 0 & 0 & D_2 & \cdots \\ 0 & 0 & 0 & \cdots \\ \vdots & \vdots & \vdots & \cdots \end{bmatrix}$$

where $\{\Gamma_i\}$ is a choice sequence from G to D_{0*}. Also let Λ'_∞ be the row matrix from $(F \oplus G) \oplus D_1 \oplus D_2 \oplus \cdots$ to $D_{T'}$ defined by

$$\Lambda'_\infty = [[\Gamma_0, D_{0*}\Gamma_1], D_{0*}D_{1*}\Gamma_2, D_{0*}D_{1*}D_{2*}\Gamma_3, \dots] \quad .$$

Notice that Φ_{n-1} is precisely the n by n matrix contained in the upper left hand corner of Φ_∞ and Λ'_{n-1} is the row matrix of length n contained in the left of Λ'_∞. Using the upper triangular structure of Φ_∞ Theorem 2.3 shows that the set of all contractive intertwining liftings Λ_∞ of A is given by $\Lambda_\infty = [A, Z_1, Z_2, \dots]^{tr}$ where

$$Z_{i+1} = \Lambda'_\infty\Phi_\infty^i[j_0D_A, 0, 0, \dots]^{tr} \quad (i \geq 0) \tag{2.23}$$

and $\{\Gamma_i\}_1^\infty$ is a choice sequence from G to D_{0*}. By (V.1.17), (V.1.18) and (V.1.22)

$$\Gamma_0 = \Pi'\omega, \quad D_{0*}\gamma = \Pi'|G', \quad \alpha D_0 = j_0\Pi_A\omega \text{ and } -\alpha\Gamma_0^*\gamma = j_0\Pi_A|G'$$

where γ is a unitary operator from G' onto D_{0*}. (Comparing (2.8) with (V.1.22) our unitary operator α in (2.8) is $j_0\alpha$ where the last α is the unitary operator in (V.1.8).) Substituting this into Φ_∞ and Λ'_∞ we see that Φ_∞ becomes $Q'W$ and Λ'_∞ becomes $\Pi'W$, with the choice sequence $\{\Gamma_i\}_1^\infty$ in the definition of W replaced by the choice sequence $\{\gamma^*\Gamma_i\}_1^\infty$ determined by Φ_∞. Lemma 2.5 guaranteed that $\{\gamma^*\Gamma_i\}_1^\infty$ is indeed a choice sequence initiated from G to G'. Now the previous theorem with (2.23) readily produces (2.22) and concludes the proof.

As a trivial consequence of the previous theorem we recover the main result of Section VII.5.

2.8 COROLLARY. *Let* A *be a contraction in* $I(U,T')$. *Then* A *admits a unique contractive intertwining lifting if and only if* $A \cdot U$ *or* $T' \cdot A$ *is a regular factorization.*

Indeed there is only one choice sequence initiated from G to G' if and only if G or G' is zero.

As before let A be a contraction in $I(U,T')$. In this context by *forward scattering* we mean given a choice sequence $\{\Gamma_i\}_1^n$ initiated from G to G' find its corresponding contractive intertwining lifting Λ_n of A. Equation (2.22) in Theorem 2.7 provides a recursive forward scattering algorithm. *Inverse scattering* is just the opposite, that is, given a contractive n step intertwining lifting Λ_n of A find its corresponding choice sequence $\{\Gamma_i\}_1^n$. To obtain an inverse scattering algorithm notice that the upper triangular structure of W in (2.19) and (2.22) give for $i \geq 1$

$$Z_{i+1} - \Pi'[\omega, \Gamma_1, D_{1*}\Gamma_2, ..., D_{1*}D_{2*} \cdots D_{i-1*}\Gamma_i, 0, 0, ...](Q'W)^i[j_0 D_A, 0, 0, ...]^{tr}$$

$$\tag{2.24}$$

$$= \Pi' D_{1*}D_{2*} \cdots D_{i*}\Gamma_{i+1}D_i D_{i-1} \cdots D_1 P_G D_A \ .$$

The $i+1$ choice operator Γ_{i+1} appears in the last term. Using the upper triangular structure of W it follows that one only needs $\{\Gamma_j\}_1^i$ to compute $(Q'W)^i[j_0 D_A, 0, 0, \cdots]^{tr}$. Therefore the second term in (2.24) is uniquely determined by the choice sequence $\{\Gamma_j\}_1^i$, or equivalently, by $\{Z_j\}_1^i$. This observation produces the following recursive inverse scattering algorithm.

2.9 PROCEDURE. Let A be a contraction in $I(U,T')$ and Λ_n in (2.2) a contractive n step intertwining lifting of A. Assume that one has computed the choice sequence $\{\Gamma_j\}_1^i$ from the data $\{Z_j\}_1^i$ where $i < n$. Since Z_{i+1} is known and the second term in (2.24) is computed from $\{\Gamma_j\}_1^i$ the next choice contraction Γ_{i+1} is computed by inverting the operators $\Pi'|G'$, D_{j*}, D_j and $P_G D_A$ in (2.24) for $1 \leq j \leq i$. (Obviously $\overline{P_G D_A} = G$. Equation (1.14) in Chapter V guarantees that $\Pi'|G'$ is one to one.) In almost all applications the spaces G and G' are finite dimensional. This means that one does not have to invert an operator whose range is not necessarily closed in order to find the choice sequence $\{\Gamma_i\}_1^n$ from the data $\{Z_i\}_1^n$. Finally, it is noted that this algorithm reduces to the geophysics algorithm Procedure 3.3 in Chapter III, when $A = A_n$ and $T = T' = T_n$ are the operators in (I.4.4) and (I.4.1), respectively.

3. SCHUR REPRESENTATIONS FOR THE COMMUTANT
LIFTING THEOREM

In this section we will obtain Schur representations for the commutant lifting theorem, that is, we will show that the set of all contractive intertwining liftings B of A is parameterized by the set of all contractive analytic functions in $H_1^\infty(G,G')$. We will give two explicit formulas involving $H_1^\infty(G,G')$ for the set of all contractive intertwining liftings B of A. The results in this section generalize the results in Chapter I, which showed that the solution to the Carathéodory interpolation problem is parameterized by the analytic functions in H_1^∞.

To begin let V be the isometry form $H \oplus D_1 \oplus D_2 \oplus \cdots$ to $H' \oplus D_1 \oplus D_2 \oplus \cdots$ defined by

$$V = \begin{bmatrix} \Gamma_1 & D_{1*}\Gamma_2 & D_{1*}D_{2*}\Gamma_3 & \cdots \\ D_1 & -\Gamma_1^*\Gamma_2 & -\Gamma_1^*D_{2*}\Gamma_3 & \cdots \\ 0 & D_2 & -\Gamma_2^*\Gamma_3 & \cdots \\ 0 & 0 & D_3 & \cdots \\ 0 & 0 & 0 & \cdots \\ \vdots & \vdots & \vdots & \cdots \end{bmatrix} \tag{3.1}$$

where $\{\Gamma_1\}_1^\infty$ is a choice sequence initiated from H to H'. As in (2.21) it is easy to verify that

$$V = V_1 V_2 V_3 V_4 \cdots \overset{\Delta}{=} \text{strong} \lim_{n \to \infty} V_1 V_2 \cdots V_n \tag{3.2}$$

where the V_i's are the unitary operators defined by

$$\begin{aligned} V_1 &= R_{\Gamma_1} \oplus I \oplus I \oplus I \oplus \cdots \\ V_2 &= I \oplus R_{\Gamma_2} \oplus I \oplus I \oplus \cdots \\ V_3 &= I \oplus I \oplus R_{\Gamma_3} \oplus I \oplus \cdots , \cdots \end{aligned} \tag{3.3}$$

and R_{Γ_i} is the rotation matrix corresponding to Γ_i. By mimicking the proof that W in (2.21) is an isometry equations (3.2) and (3.3) imply that V is an isometry. Let $Q_{H'}$ be the operator from $H' \oplus D_1 \oplus D_2 \oplus \cdots$ into $H \oplus D_1 \oplus D_2 \oplus \cdots$ which replaces the H' component by zero in H, that is,

$$Q_{H'}(h' \oplus d_1 \oplus d_2 \oplus d_3 \oplus \cdots) = 0 \oplus d_1 \oplus d_2 \oplus d_3 \oplus \cdots$$

(If $H = H'$, then $Q_{H'}$ is the orthogonal projection onto $D_1 \oplus D_2 \oplus \cdots$.) Notice that $Q_{H'}V$ is the matrix on $H \oplus D_1 \oplus D_2 \oplus \cdots$ defined by

$$
Q_{H'}V = \begin{bmatrix}
0 & 0 & 0 & \cdots \\
D_1 & -\Gamma_1^*\Gamma_2 & -\Gamma_1^*D_{2*}\Gamma_3 & \cdots \\
0 & D_2 & -\Gamma_2^*\Gamma_3 & \cdots \\
0 & 0 & D_3 & \cdots \\
0 & 0 & 0 & \cdots \\
\cdot & \cdot & \cdot & \cdots \\
\cdot & \cdot & \cdot & \cdots \\
\cdot & \cdot & \cdot & \cdots
\end{bmatrix}.
$$

Finally, let $\Pi_{H'}$ be the operator which picks out the H' component from $H' \oplus D_1 \oplus D_2 \oplus \cdots$, that is, $h' = \Pi_{H'}(h' \oplus d_1 \oplus d_2 \oplus \cdots)$. This sets the stage for the following result.

3.1 LEMMA. *A function* $F(z)$ *is a contractive analytic function in* $H^\infty(H,H')$ *if and only if*

$$F(z) = \Pi_{H'}V(I - z\,Q_{H'}V)^{-1}\,|\,H \qquad (for\ all\ z \in D) \tag{3.4}$$

where V *is the isometry in (3.1) corresponding to the choice sequence* $\{\Gamma_i\}_1^\infty$ *initiated from H to* H'. *In particular, (3.4) provides a one to one correspondence between the set of all contractive analytic functions* $F(z)$ *in* $H^\infty(H,H')$ *and the set of all choice sequences* $\{\Gamma_i\}_1^\infty$ *initiated from H to* H'.

PROOF. Let $U = S$ be the unilateral shift on $K = H^2(H)$ and $T' = 0$ on H'. Let $A(=0)$ be the zero operator mapping K into H'. Obviously the unilateral shift on S' on $K' = H^2(H')$ is the minimal isometric dilation of T'. Recall the $S'B = BS$ if and only if B is a multiplication operator. Therefore the set of all contractive intertwining liftings B of A is given by $Bk = zF(z)k$ where F is a contractive analytic function in $H^\infty(H,H')$. To complete the proof we will use Theorem 2.7 to obtain the representation of F given by (3.4). To this end notice that $D_A = I$ and $D_{T'} = I$. Thus $D_A U = U$ and $D_{T'}Ak \oplus D_A k = 0 \oplus k$ for k in K implies that

$$F = SK, \ \ G = H, \ \ F' = \{0\} \oplus K \ \ and \ \ G' = H' \oplus \{0\}$$

$$\omega Sk = 0 \oplus k \quad (k \in K) \tag{3.5}$$

Using this, the isometry W in (2.19) becomes

$$
W = \begin{bmatrix}
0 & \Gamma_1 & D_{1*}\Gamma_2 & D_{1*}D_{2*}\Gamma_3 & \cdots \\
j_o\Pi_A\omega & 0 & 0 & 0 & \cdots \\
0 & D_1 & -\Gamma_1^*\Gamma_2 & -\Gamma_1^*D_{2*}\Gamma_3 & \cdots \\
0 & 0 & D_2 & -\Gamma_2^*\Gamma_3 & \cdots \\
0 & 0 & 0 & D_3 & \cdots \\
0 & 0 & 0 & 0 & \cdots \\
\vdots & \vdots & \vdots & \vdots & \cdots
\end{bmatrix}
\tag{3.6}
$$

from $SK \oplus H \oplus D_1 \oplus D_2 \oplus \cdots$ to $H' \oplus K \oplus D_1 \oplus D_2 \oplus \cdots$. Comparing (3.6) to (3.1) we see that (2.22) becomes

$$
Z_{i+1}h = \Pi' W(Q'W)^i j_o D_A h = \Pi_{H'} V(Q_{H'}V)^i h \quad (h \in H \text{ and for all } i \geq 0) \tag{3.7}
$$

where $Z_i k$ is the orthogonal projection of Bk onto $S'^i D_{T'} = S'^i H'$. In (3.7) we have identified $j_o D_A h$ with $[j_o D_A h, 0, 0, \cdots]^{tr}$. Let F be the unique contractive analytic function in $H^\infty(H,H')$ corresponding to B $(= (zF)_+)$ and let $\sum F_i z^i$ be its power series expansion of F, then obviously $Z_{i+1} = F_i$ for all $i \geq 0$. This and (3.7) gives

$$
F(z)h = \sum_o^\infty F_i z^i = \Pi_{H'} V(I-zQ_{H'}V)^{-1}h \quad (h \in H \text{ and } z \in D).
$$

This proves (3.4). Now the lemma follows from Theorem 2.7 and from the fact that the set of all contractive intertwining liftings B of A is given by $B = (zF)_+$ where F is in $H_1^\infty(H,H')$. This completes the proof.

Recall that ω is the unitary operator defined in (2.18) and W is the isometry in (2.19), and $D_A = F \oplus G$. Let Q_1 be the operator from $D_{T'} \oplus D_A \oplus D_1 \oplus D_2 \oplus \cdots$ into $D_A \oplus D_1 \oplus D_2 \oplus \cdots$ which eliminates the $D_{T'} \oplus D_A$ component in the following way

$$
Q_1(d_{T'} \oplus d_A \oplus d_1 \oplus d_2 \oplus d_3 \oplus \cdots) = 0 \oplus d_1 \oplus d_2 \oplus d_3 \oplus \cdots
$$

where 0 is in D_A. Notice that $Q_1 = (I - P_A)Q'$ where P_A is the orthogonal projection onto D_A. Finally, let Π_1 be the operator which picks out the $D_{T'} \oplus D_A$ component from $D_{T'} \oplus D_A \oplus D_1 \oplus D_2 \oplus \cdots$, that is, $d_{T'} \oplus d_A = \Pi_1(d_{T'} \oplus d_A \oplus d_1 \oplus d_2 \oplus \cdots)$. This notation is used in the following result.

3.2   COROLLARY. *The set of all contractive analytic functions* F *in* $H^\infty(D_A, D_{T'} \oplus D_A)$ *satisfying* $F(0)|F = \omega$ *is given by*

$$
F(z) = \Pi_1 W(I - zQ_1 W)^{-1}|j_o D_A \quad (z \in D) \tag{3.8}
$$

where W *is the isometry in (2.19) corresponding to the choice sequence* $\{\Gamma_i\}_1^\infty$ *initiated from G to* G'*. In particular, (3.8) provides a one to one correspondence between the set of all contractive analytic functions* F *in* $H^\infty(D_A, D_{T'} \oplus D_A)$ *satisfying the constraint* $F(0)F = \omega$*, and the set of all choice sequences* $\{\Gamma_i\}_1^\infty$ *initiated from G to* G'*.*

PROOF. The previous lemma shows that the set of all contractive analytic functions F in $H^\infty(D_A, D_{T'} \oplus D_A)$ is given by (3.4) where $\{\Gamma_i\}_1^\infty$ is a choice sequence from $D_A (= H)$ to $D_{T'} \oplus D_A (= H')$. The constraint $F(0)|F = \omega$ shows that $\Gamma_1|F = \omega$ is a unitary operator from F onto F'. Therefore in this case the isometry V in (3.1) equals the isometry W in (2.19)(up to the unitary operator j_o). This and (3.4) yields (3.8). The rest of the proof follows from Lemma 3.1. The proof is now complete.

3.3 LEMMA. *Let* C *be a strict contraction on* H *and* D *an operator on* H *and* C+D *is a strict contraction. Then* $I - C$ *and* $I - (C+D)$ *and* $I - D(I-C)^{-1}$ *are all invertible. Moreover, in this case*

$$(I-(C+D))^{-1} = (I-C)^{-1}(I-D(I-C)^{-1})^{-1} . \tag{3.9}$$

PROOF. Recall that if T is a strict contraction on H, then I–T is invertible. Using this we have

$$I-(C+D) = I - C - D(I-C)^{-1}(I-C) = (I-D(I-C)^{-1})(I-C) \tag{3.10}$$

Since $I - (C+D)$ and I–C are both invertible (3.10) shows that $I - D(I-C)^{-1}$ is invertible. Now (3.9) follows by taking inverses in (3.10). This completes the proof.

To obtain the Schur representation for a contractive intertwining lifting B of A we identify $k' = h' \oplus f_1 \oplus f_2 \oplus f_3 \oplus \cdots$ in $K' = H' \oplus l^2(D_{T'})$ with

$$\hat{k}'(z) = h' \oplus \sum_1^\infty f_i z^{i-1} \quad \text{in} \quad H' \oplus H^2(D_{T'}) . \tag{3.11}$$

Since $f_1 \oplus f_2 \oplus f_3 \oplus \cdots$ is in $l^2(D_{T'})$ we see that $\hat{k}'(z)$ is defined for all z in D and

$$\|k'\|^2 = \|\hat{k}'(z)\|^2 = \|h'\|^2 + \sum_1^\infty \|f_i\|^2 .$$

Recall that B is a contractive intertwining lifting of A if and only if B is an operator of the form

$$B = [A, Z_1, Z_2, Z_3, \cdots]^{\text{tr}} : \ K \to H' \oplus l^2(D_{T'}) \tag{3.12}$$

where Z_{i+1} is given by (2.22) in Theorem 2.7. Using (3.11) and Theorem 2.7 we identify

B with the operator $B(z)$ from K to $H' \oplus H^2(D_{T'})$ by:

$$B(z) = A \oplus \sum_1^\infty z^{i-1} Z_i = A \oplus \sum_0^\infty z^i \Pi' W(Q'W)^i j_0 D_A \, .$$

This readily implies that

$$B(z) = A \oplus \Pi' W(I - zQ'W)^{-1} j_0 D_A \quad \text{(for all } z \in D) \tag{3.13}$$

where W is the isometry in (2.19) and $\{\Gamma_i\}_1^\infty$ is a choice sequence initiated from G to G'.

Let P_1 be the orthogonal projection onto $D_{T'} \oplus D_A$ and $P_A = P_{D_A}$. Using this in (3.13) along with Lemma 3.3 and Corollary 3.2 we have

$$B(z) = A \oplus \Pi' \Pi_1 W(I - z(Q_1 W + P_A P_1 W))^{-1} j_0 D_A =$$

$$A \oplus \Pi' \Pi_1 W(I - zQ_1 W)^{-1}(I - zP_A P_1 W(I - zQ_1 W)^{-1})^{-1} j_0 D_A = \tag{3.14}$$

$$A \oplus \Pi' F(z)(I - z\Pi_A F(z))^{-1} D_A$$

where $F(z)$ is defined by (3.8). Combining Corollary 3.2 along with Theorem 2.7 we see that $B(z)$ uniquely determines and is uniquely determined by a contractive analytic function $F(z)$ in $H^\infty(D_A, D_{T'} \oplus D_A)$ satisfying the constraint $F(0)|F = \omega$. Recall that ω is unitary. So the set of all contractive analytic functions F in $H^\infty(D_A, D_{T'} \oplus D_A)$ satisfying $F(0)|F = \omega$ is given by $F(z) = \omega \oplus R(z)$ where $R(z)$ is any contractive analytic function in $H_1^\infty(G, G')$. Summing up the above analysis proves the following result.

3.4 THEOREM. *Let A be a contraction in $I(U, T')$. The set of all contractive intertwining liftings $B(z)$ of A is given by*

$$B = A \oplus \Pi' F(z)(I - z\Pi_A F(z))^{-1} D_A \quad (z \in D) \tag{3.15}$$

where $F(z)$ is a contractive analytic function in $H^\infty(D_A, D_{T'} \oplus D_A)$ satisfying $F(0)|F = \omega$, or equivalently, $F(z) = \omega \oplus R(z)$ and $R(z)$ is a contractive analytic function in $H^\infty(G, G')$. Moreover, (3.15) provides a one to one correspondence between the set of all contractive intertwining liftings B of A and the set of all contractive analytic functions $R(z)$ in $H^\infty(G, G')$. In particular, A admits a unique contractive intertwining lifting B if and only if $A \cdot U$ or $T' \cdot A$ is a regular factorization, that is, $G = \{0\}$ or $G' = \{0\}$.

Let A be a contraction in $I(T, T')$ and U on $H \oplus l^2(D_T)$ be the minimal isometric dilation of T in its matrix form (VI.3.4). Recall that AP_H is a contraction in $I(U, T')$, and B is a contractive intertwining lifting of A if and only if B is a contractive intertwining lifting of AP_H (see Lemma 1.1 in Chapter VII). Now let ω_T be the unitary operator from $F(A{\cdot}T)$ onto $F(T'{\cdot}A)$ defined by

$$\omega_T(D_A Th \oplus D_T h) = D_{T'} Ah \oplus D_A h \quad (h \in H) \;. \tag{3.16}$$

By using (3.16), replacing A by AP_H in the previous theorem, and using the fact that $D_A \oplus D_T$ is invariant for $(I - P_{D_T})(\omega \oplus R(z))$ we deduce the following result.

3.5 COROLLARY. *Let A be a contraction in $I(T,T')$ and $P_A = P_{D_A}$. The set of all contractive intertwining liftings B of A are determined by*

$$B \,|\, (H \oplus D_T) = AP_H \oplus \Pi' F(z)(I-zP_A F(z))^{-1}(D_A P_H + I - P_H) \quad (z \in D) \tag{3.17}$$

where $F(z) = \omega_T \oplus R(z)$ and $R(z)$ is a contractive analytic function in $H^\infty(G,G')$. Moreover, (3.17) provides a one to one correspondence between the set of all contractive intertwining liftings B of A and the set of all contractive analytic functions $R(z)$ in $H^\infty(G,G')$. In particular, A admits a unique contractive intertwining lifting B if and only if $A{\cdot}T$ or $T'{\cdot}A$ is a regular factorization.

Since H is cyclic for U and $U'B = BU$, it is clear that the right hand side of (3.17) uniquely determines B. It is also important to emphasize that

$$G = G(AP_H{\cdot}U) = G(A{\cdot}T) \quad \text{and} \quad F(A{\cdot}T) \oplus G = D_A \oplus D_T \;.$$

Moreover, in this case the unitary operator ω becomes

$$\omega(D_A Th \oplus (D_T h + Uf)) = \omega_T(D_A Th \oplus D_T h) \oplus f$$

where $h \oplus f$ is in $H \oplus (K \ominus H)$. In the previous corollary $F(z)$ is any contractive analytic function in $H^\infty(D_A \oplus D_T, D_{T'} \oplus D_A)$ satisfying $F(0)\,|\,F(A{\cdot}T) = \omega_T$, or equivalently, $F(z) = \omega_T \oplus R(z)$ for z in D where $R(z)$ is a contractive analytic function in $H^\infty(G,G')$. Finally, we notice that we have abused the standard notation for the orthogonal projection P_A in (3.17). In (3.17) P_A is actually the operator from $D_{T'} \oplus D_A$ to $D_A \oplus D_T$ defined by $P_A(d_{T'} \oplus d_A) = d_A \oplus 0$.

Let T' be the co-isometry on $H' = K_0^2(E')$ defined by $T' = P_{H'} V' \,|\, H'$ where V' is the bilateral shift on $L^2(E')$. Let A be a contractive Hankel matrix in $I(S, T')$ where S is the unilateral shift on $H^2(E)$. (Recall that A is called a Hankel operator if A intertwines a co-isometry with an isometry; see Section XIII.4. In fact $A = V'^* J\Gamma$ where Γ is the Hankel operator in $I(S, S')$ defined in Section IX.3.) Obviously V' is the minimal isometric dilation of T' and the space $L' = \{(V' - T')H'\}$ defined in Section VI.4 equals

E'. In particular, $M_+(L') = H^2(E')$. Clearly any contractive intertwining lifting B of A is in $I(S, V')$. Theorem IX.1.1 implies that $B = M_C$ where C is in $L^\infty(E, E')$. By Theorem 3.4, the set of all contractive intertwinings liftings B of A is given by $B = M_C$ where C is the function in $L^\infty(E, E')$ defined by

$$C = [A + \Pi' F(I - z\Pi_A F)^{-1} D_A] | E \qquad (3.18)$$

$F = \omega \oplus R$ and R is a contractive analytic function in $H_1^\infty(G, G')$. Therefore (3.18) provides us with an explicit formula for computing all contractive intertwining liftings B $= M_C$ of A. (In Section 9 and in Chapter XIV we will use (3.18) to compute C when the rank of the Hankel operator A is finite.)

Let us assume that $A = P_- M_Q | H^2(E)$ where Q is in $L^\infty(E, E')$ and P_- denotes the orthogonal projection into $K_0^2(E')$. It is easy to verify that A is a Hankel operator, that is, A is in $I(S, T')$; see also Section VIII.4 and Section IX.3. Without loss of generality let us assume that the norm of A is one. This can always be done by dividing through by the appropriate constant. By consulting Theorem VIII.4.3 (or Section IX.3 adjusted to this setting) we see that the set of all optimal solutions H_* to the following H^∞ optimization problem

$$\inf \| Q + H^\infty(E, E') \|_\infty = \| Q + H_* \|_\infty \qquad (3.19)$$

is given by $Q + H_* = C$ where $B = M_C$ is a contractive intertwining lifting of A. Therefore by (3.18) the set of all optimal solutions H_* to the H^∞ optimization problem in (3.19) is given by

$$H_* = \Pi' F(I - z\Pi_A F)^{-1} D_A | E - P_+ M_Q | E \qquad (3.20)$$

where P_+ is the orthogonal projection onto $H^2(E')$ and $F = \omega \oplus R$ with R in $H_1^\infty(G, G')$. Equation (3.20) provides an explicit formula for computing all optimal solutions to (3.19).

The previous analysis show that the Schur type formula in (3.15) is a useful formula for computing all contractive intertwining liftings for a Hankel matrix. However, the formula for B in (3.15) or (3.17) is not very useful for Carathéodory or Nevanlinna-Pick interpolation. The reason for this is that the range of B is decomposed into $H' \oplus H^2(D_{T'})$ and for Carathéodory and Nevanlinna-Pick interpolation we want to express B in terms of a multiplication operator, $B = M_C$ where C is a contractive analytic function in $H^\infty(L_*, L_*')$. Recall that $L_* = \{(I - UT^*)H\}^-$ and $L_*' = \{(I - U'T'^*)H'\}^-$ are the wandering subspaces obtained from the Wold decomposition of U and U'; see Section VI.4. Let ϕ_* from D_{T^*} onto L_* and ϕ_*' from $D_{T'^*}$ onto L_*' be the unitary operators defined in Section VI.4 by $\phi_* D_{T^*} h = (I - UT^*)h$ and $\phi_*' D_{T'^*} h' = (I - U'T'^*)h'$

for all h in H and h' in H'. Now assume that both T and T' are $*$-stable. By Proposition VI.4.3 this implies that $K = M_+(L_*)$ and $K' = M_+(L'_*)$. Let ϕ_* also be the unitary operator from $H^2(D_{T^*})$ onto $M_+(L_*)$ defined by $\phi_* h = \sum_0^\infty U^n \phi_* h_n$ where $h(z) = \sum_0^\infty h_n z^n \in H^2(D_{T^*})$; also let ϕ'_* be its corresponding unitary operator from $H^2(D_{T'^*})$ onto $M_+(L'_*)$. Obviously $\phi_* S = U \phi_*$ and $\phi'_* S' = U' \phi'_*$ where S is the unilateral shift on $H^2(D_{T^*})$ and S' is the unilateral shift on $H^2(D_{T'^*})$. So according to Theorem IX.1.1 an operator B is in $I(U, U')$ if and only if $B = \phi'_* M_C \phi_*^*$ where C is in $H^\infty(D_{T^*}, D_{T'^*})$. Moreover, B and C uniquely determine each other. We are now ready to state our second Schur representation for the set of all contractive intertwining liftings of A.

3.6 THEOREM. *Let* A *be a contraction in* $I(T, T')$ *where both* T *and* T' *are* $*$-stable. *Let* $\Theta' = \Theta_{T'}$ *be the characteristic function for* T'. *Then the set of all contractive intertwining liftings* B *of* A *is given by* $B = \phi'_* M_C \phi_*^*$ *where* C *is a contractive analytic function in* $H^\infty(D_{T^*}, D_{T'^*})$ *defined by*

$$C = D_{T'^*}(I - zT'^*)^{-1} A + \Theta'(z) \Pi' F(z)(I - zP_A F(z))^{-1}(D_A P_H + I - P_H) \,|\, \phi_* D_{T^*} \quad (3.21)$$

$F(z) = \omega_T \oplus R(z)$ *and* $R(z)$ *is a contractive analytic function in* $H^\infty(G, G')$. *Moreover, (3.21) provides a one to one correspondence by* $B = \phi'_* M_C \phi_*^*$ *between the set of all contractive intertwining liftings* B *of* A *and the set of all contractive analytic functions* $R(z)$ *in* $H^\infty(G, G')$.

PROOF. According to Theorem IX.6.4 the operator $W' = W_{T'}$ from $H' \oplus H^2(D_{T'})$ onto $H^2(D_{T'^*})$ defined by

$$W'(h' \oplus f) = D_{T'^*}(I - zT'^*)^{-1} h + \Theta' f \qquad (h \oplus f \in H' \oplus H^2(D_{T'}))$$

is unitary. Moreover, $S'W' = W'U'$ where U' is the minimal isometric dilation of T'. In this setting

$$U'(h \oplus f) = T'h \oplus (D_{T'} h + zf) .$$

and S' is unitarily equivalent to the minimal isometric dilation of T'. Notice that for h in H' we have

$$W'(I - U'T'^*)h = W'h - S'W'T'^*h =$$
$$W'h - S'S'^*W'h = D_{T'^*}h = \phi'^*_*(I - U'T'^*)h$$

which readily implies that

$$\phi'_* W' (I - U' T'^*) h = (I - U' T'^*) h \qquad (h \in H') \ .$$

Since T' is $*$-stable $M_+(L'_*) = K'$ (see Proposition VI.4.3). Therefore $\phi'_* W'$ is a unitary operator mapping the decomposition $H' \oplus H^2(D_{T'})$ of K' onto the decomposition $M_+(L'_*)$ of K' commuting with U'. Because T is $*$-stable $M_+(L_*) = K$. Thus both of the minimal isometric dilations U and U' can be viewed as unilateral shifts on $H^2(D_{T^*})$ and $H^2(D_{T'^*})$, respectively. Therefore any contractive intertwining lifting B of A can be viewed as multiplication by C where C is in $H_1^\infty(D_{T^*}, D_{T'^*})$. So applying $\phi'_* W'$ to the left of B in (3.17) we see that B is given by $\phi'_* M_C \phi_*^*$ where C is determined by (3.21). Finally, an application of Corollary 3.5 finishes the proof.

The following example uses the previous theorem to solve the Carathéodory interpolation problem with one data point.

3.7 EXAMPLE. Let A be a contraction from H to H'. The set of all contractive analytic functions C(z) in $H_1^\infty(H,H')$ satisfying C(0) = A is given by

$$C(z) = A + zD_{A^*} F_1 (I + zA^* F_1(z))^{-1} D_A \qquad (3.22)$$

where F_1 is a contractive analytic function in $H^\infty(D_A, D_{A^*})$.

To obtain this result, let T = 0 on H and $T' = 0$ on H'. Obviously A is a contraction in $I(T,T')$. The minimal isometric dilations S' on $H^2(H')$ and S on $H^2(H)$ of T' and T are respectively both unilateral shifts. Therefore B is a contractive intertwining lifting of A if and only if B is a multiplication operator (B = C_+) where C(z) is a contractive analytic function in $H^\infty(H,H')$ and C(0) = A. (The constraint $P_{H'} B = A P_H$ is equivalent to C(0) = A). Formula (3.21) provides the set of all contractive intertwining liftings of A. In this case $L_* = H, L'_* = H', \Theta' = zI'$ and $\phi'_* = I'$ the identity on H'. Therefore the set of all C(z) in $H_1^\infty(H,H')$ satisfying C(0) = A is given by

$$C(z) = A + zII' F(z)(I - zP_A F(z))^{-1} D_A \quad (z \in D) \ . \qquad (3.23)$$

To obtain (3.22) from (3.23) first notice that since T = 0 and $T' = 0$

$$F = F(A \cdot T) = \{0\} \oplus D_T \quad \text{and} \quad F' = \{Ah \oplus D_A h : h \in H\}^-$$

$$G = D_A \oplus \{0\} \quad \text{and} \quad G' = \{D_{A^*} h \oplus -A^* h : h \in D_{A^*}\}^-$$

Obviously $F \oplus G = D_A \oplus D_T$ and using the rotation matrix R_A for A it is easy to verify that $F' \oplus G' = H' \oplus D_A$. Since $[D_{A^*}, -A]^*$ is a unitary operator from D_{A^*} onto G' and

$G = D_A \oplus \{0\}$ we see that the contractive analytic functions R(z) in Theorem 3.6 is given by

$$R(z)d = D_A \cdot F_1 d \oplus -A^* F_1 d \quad (d \in D_A) \qquad (3.24)$$

where $F_1(z)$ is a contractive analytic function in $H^\infty(D_A, D_{A^*})$. In other words (3.24) identifies the R in $H_1^\infty(G, G')$ with a F_1 in $H_1^\infty(D_A, D_{A^*})$ and visa versa. Using (3.23) (3.24) and the fact that $G = D_A \oplus \{0\}$ we have

$$C(z) = A + z\Pi' F(z)(I - zP_A F(z))^{-1} D_A = A + z\Pi' F(z)(I_{D_A} - z\Pi_A F(z)P_A)^{-1} D_A$$

$$= A + z\Pi' F(z)(I_{D_A} - z\Pi_A F(z)P_G)^{-1} D_A = A + zD_A \cdot F_1(I + zA^* F_1)^{-1} D_A .$$

This produces (3.22) by Theorem 3.6. So by the above discussion C(z) in (3.22) gives all contractive analytic functions satisfying C(0) = A when F_1 is in $H_1^\infty(D_A, D_{A^*})$.

3.8   REMARK. If $H = H' = C^1$, then formula (3.22) in the previous example reduces to

$$C(z) = (A + zF_1)(1 + z\overline{A}F_1)^{-1} . \qquad (3.25)$$

This is precisely the Schur solution to the one step Carathéodory interpolation problem in Chapter I.

4. OBTAINING THE CHOICE SEQUENCE FROM 2x2 MATRIX EXTENSIONS

In this section we will recursively use Theorem IV.3.1 to compute the set of all contractive n step intertwining liftings Λ_n of A. As a consequence of this approach we will also show that there is a one to one correspondence between the set of all contractive n step intertwining liftings Λ_n and the set of all choice sequences $\{\Gamma_i\}_1^n$ initiated from G to G'.

As before A is a contraction in $I(U, T')$ where U is an isometry on K and T' is a contraction on H'. Throughout $U' = T'_\infty$ is the minimal isometric dilation of T' in its matrix form. Let Λ_i be a contractive i-th step intertwining lifting of A. Obviously Λ_i admits a matrix decomposition of the form:

$$\Lambda_i = \begin{bmatrix} A_i & B_i \\ C_i & D_i' \end{bmatrix} : UK \oplus L_* \to H_{i-1}' \oplus D_{T'} \qquad (4.1)$$

where $L_* = K \ominus UK = D_{U^*}$ and $H_j' = H' \oplus (l_j^2(D_{T'}))$ for all $j \geq 1$. By (2.2), (2.3) we see that

$$A_i = \Lambda_{i-1} \mid UK, \quad B_i = \Lambda_{i-1} \mid L_* \text{ and } Z_i = [C_i, D_i'] \tag{4.2}$$

where Λ_{i-1} is a contractive i–1 step intertwining lifting of A. In Section 2 we used the form of a contractive column matrix in Lemma 1.1 to recursively construct the set of all contractive i+1 step intertwining liftings Λ_{i+1} of A lifting Λ_i. In this section we will use Theorem 3.1 in Chapter IV to construct, also recursively, the set of all contractive i+1 step intertwining liftings Λ_{i+1} of A lifting Λ_i. This will readily give all contractive n step intertwining liftings of A. We begin with the following result.

4.1 LEMMA. *Let Λ_i for $i \geq 1$ be the contractive i step intertwining lifting of A in (4.1). Then Λ_{i+1} is a contractive i+1 step intertwining lifting of A lifting Λ_i if and only if Λ_{i+1} is a contraction and Λ_{i+1} admits a matrix decomposition of the form:*

$$\Lambda_{i+1} = \begin{bmatrix} A_{i+1} & B_{i+1} \\ C_{i+1} & D_{i+1}' \end{bmatrix} : UK \oplus L_* \to H_i' \oplus D_{T'} \tag{4.3}$$

where

$$A_{i+1} = \begin{bmatrix} A_i \\ C_i \end{bmatrix}, \quad B_{i+1} = \begin{bmatrix} B_i \\ D_i' \end{bmatrix}$$

$$\tag{4.4}$$

$$A_{i+1} = U'[A_i, B_i]U_* \text{ and } C_{i+1} = [C_i, D_i']U_*$$

and U_* is the unitary operator mapping UK onto K defined by $U_* = U^* \mid UK$.

PROOF. Equations (2.2) and (2.3) imply that Λ_{i+1} is a contractive i+1 step intertwining lifting of A lifting Λ_i if and only if Λ_{i+1} is a contraction mapping K into $H_{i+1} = H_i \oplus D_{T'}$ of the form:

$$\Lambda_{i+1} = [\Lambda_i, Z_{i+1}]^{\text{tr}} \text{ satisfying } Z_{i+1}U = Z_i \quad \text{(for } i \geq 1\text{)} . \tag{4.5}$$

In this case Λ_{i+1} admits a matrix decomposition of the form (4.3). Moreover, the definitions of A_{i+1} and B_{i+1} in (4.3) along with (4.1), (4.5) give

$$A_{i+1} = \Lambda_i \mid UK = \begin{bmatrix} A_i \\ C_i \end{bmatrix} \text{ and } B_{i+1} = \Lambda_i \mid L_* = \begin{bmatrix} B_i \\ D_i' \end{bmatrix} .$$

This verifies the first two equations in (4.4). Notice that (2.2), (2.3) along with $\Lambda_i = [\Lambda_{i-1}, Z_i]^{\text{tr}}$ imply that $\Lambda_i U = U' \Lambda_{i-1}$. Equations (4.1) and (4.2) show that $\Lambda_{i-1} = [A_i, B_i]$. Thus

$$A_{i+1} U_*^* = A_{i+1} U = \Lambda_i U = U' \Lambda_{i-1} = U' [A_i, B_i] .$$

This proves the third equation in (4.4). Finally, by using the last equation in (4.2) with (4.5) we have

$$[C_i, D_i'] = Z_i = Z_{i+1} U = [C_{i+1}, D_{i+1}']U = C_{i+1} U = C_{i+1} U_*^*$$

This verifies the last equation in (4.4).

Assume that (4.3) and (4.4) hold. The first two equations in (4.4) with (4.1) give

$$\Lambda_{i+1} = \begin{bmatrix} A_i & B_i \\ C_i & D_i' \\ C_{i+1} & D_{i+1}' \end{bmatrix} = \begin{bmatrix} \Lambda_i \\ Z_{i+1} \end{bmatrix}$$

where $Z_{i+1} = [C_{i+1}, D_{i+1}']$. The last equation in (4.2) and (4.4) imply that

$$Z_{i+1} U = [C_i, D_i']U_* U = [C_i, D_i'] = Z_i .$$

By (4.5) the operator Λ_{i+1} is a contractive i+1 step intertwining lifting of A lifting Λ_i. This completes the proof.

Let Λ_i in (4.1) be a contractive i step intertwining lifting of A. Throughout X_i from L_* to $D_{A_i^*}$ and Y_i from D_{A_i} to $D_{T'}$ and $\Gamma = \Gamma_i'$ from D_{X_i} to $D_{Y_i^*}$ and $D = D_i'$ are the contractions uniquely determined by Theorem IV.3.1 for the contraction Λ_i in (4.1). As before G and G' are the residuals spaces defined in (2.17), the operators D_i and D_{i*} are the defect operators defined in (1.7) for the choice sequence $\{\Gamma_i\}_1^n$. This sets the stage for the main result in this section.

4.2 THEOREM. *Let* A *be a contraction in* $I(U,T')$. *The set of all contractive n step intertwining liftings* Λ_n *of* A *is given by*

$$\Lambda_n = \begin{bmatrix} A_n & B_n \\ C_n & D_n' \end{bmatrix} \tag{4.6}$$

where A_n, B_n, C_n *and* D_n' *are recursively obtained by solving the following difference equations:*

$$A_{i+1} = \begin{bmatrix} A_i \\ C_i \end{bmatrix}, \; B_{i+1} = \begin{bmatrix} B_i \\ D_i' \end{bmatrix}, \; C_{i+1} = [C_i, D_i']U_* \tag{4.7a}$$

$$D_{i+1}' = \Pi' D_{1*} D_{2*} \cdots D_{i*} \Gamma_{i+1} D_i D_{i-1} \cdots D_1 P_G D_A | L_* - Y_{i+1} A_{i+1}^* X_{i+1} \tag{4.7b}$$

for any choice sequence $\{\Gamma_i\}_1^n$ *initiated from* G *to* G'. *The initial condition for the* *difference equations in (4.7) is given by the following characterization of the set of all contractive one step intertwining liftings Λ_1 of A:*

$$\Lambda_1 = \begin{bmatrix} A\,|\,UK & A\,|\,L_* \\ D_{T'}AU^* & \Pi'\Gamma_1 P_G D_A\,|\,E-Y_1 U^* A^* X_1 \end{bmatrix} = \begin{bmatrix} A_1 & B_1 \\ C_1 & D_1' \end{bmatrix} \tag{4.8}$$

where Γ_1 from G to G' is the first choice contraction. Moreover, this gives a one to one correspondence between the set of all contractive n step intertwining liftings Λ_n of A and the set of all choice sequences initiated from G to G'.

PROOF. Proposition 4.1 in Chapter V adjusted to this setting along with $T = U$ shows that the set of all contractive one step intertwining liftings Λ_1 of A is given by

$$\Lambda_1 = \begin{bmatrix} A\,|\,UK & A\,|\,L_* \\ D_{T'}AU^* & D_{Y_1^*}\Gamma_1' D_{X_1} - Y_1 U^* A^* X_1 \end{bmatrix} = \begin{bmatrix} A_1 & B_1 \\ C_1 & D_1' \end{bmatrix} \tag{4.9}$$

where Γ_1' is a contraction mapping D_{X_1} into $D_{Y_1^*}$. From Section V.4 we infer that the equations

$$D_{Y_1^*}\alpha_* = \Pi'\,|\,G' \quad \text{and} \quad D_{X_1} = \gamma P_G D_A\,|\,L_* \tag{4.10}$$

define unitary operators α_* from G' onto $D_{Y_1^*}$ and γ from G onto D_{X_1}, respectively, for α_* refer to (4.18) in Chapter V with $Y = Y_1$. Notice that since $T = U$ is an isometry $T^* D_{T^*} = 0$. For γ the argument is a bit more involved. Namely by using the unitary operators α and β defined in (4.13) and (4.22) in Chapter V we infer that for $X = X_1$

$$D_{X_1} = \alpha^* P_{G_*}\,|\,L_* = \alpha^* \beta^* P_G \begin{bmatrix} D_A D_{T^*} \\ -T^* \end{bmatrix} |\,L_* = (\beta\alpha)^* P_G D_A\,|\,L_* \ .$$

Since for the isometry $T = U$ we have $G = D_A \ominus \overline{D_A UK}$ and $D_{T^*}h = h$ for h in L_* the operator γ in (4.10) is $(\beta\alpha)^*$. By setting $\Gamma_1 = \alpha_*^* \Gamma_1' \gamma$ and using (4.9), (4.10) we see that (4.8) describes all contractive one step intertwining lifting Λ_1 of A.

Notice that a contractive i+1 step intertwining lifting Λ_{i+1} of A lifts a contractive i step intertwining lifting Λ_i of A (see (4.5)). So we construct the set of all contractive n step intertwining liftings Λ_n of A by recursively using Sections IV.1 to IV.3, to construct the set of all contractive i+1 step intertwining liftings of A lifting Λ_i with the initial condition Λ_1 in (4.8). To this end let Λ_i be the contractive i step intertwining lifting of A in (4.1). By applying Corollary IV.1.5 to $A_{i+1} = U'[A_i, B_i]U_*$ defined in (4.4) the equation

$$W_i D_{A_{i+1}} = \begin{bmatrix} D_{A_i} & -A_i^* X_i \\ 0 & D_{X_i} \end{bmatrix} U_* \tag{4.11}$$

defines a unitary operator W_i from $D_{A_{i+1}}$ onto $D_{A_i} \oplus D_{X_i}$. Using also (IV.3.2) and (IV.3.4) adjusted to the present case (4.4):

$$[Y_i, D_{Y_i^*} \Gamma_i'] \begin{bmatrix} D_{A_i} & -A_i^* X_i \\ 0 & D_{X_i} \end{bmatrix} U_* = [C_i, D_i'] U_* = C_{i+1} = Y_{i+1} D_{A_{i+1}} . \tag{4.12}$$

This with (4.11) gives:

$$Y_{i+1} W_i^* = [Y_i, D_{Y_i^*} \Gamma_i'] \qquad (1 \le i \le n) . \tag{4.13}$$

By applying Corollary IV.1.5 to $A_{i+1}^* = [A_i^*, C_i^*]$ as defined in (4.4) and taking the adjoint of the present version of (IV.1.10) we obtain that the equation

$$D_{A_{i+1}^*} W_{i*} = \begin{bmatrix} D_{A_i^*} & 0 \\ -Y_i A_i^* & D_{Y_i^*} \end{bmatrix} \tag{4.14}$$

defines a unitary operator W_{i*} mapping $D_{A_i^*} \oplus D_{Y_i^*}$ onto $D_{A_{i+1}^*}$. Using (IV.3.2) adjusted to the present case and (4.4):

$$\begin{bmatrix} D_{A_i^*} & 0 \\ -Y_i A_i^* & D_{Y_i^*} \end{bmatrix} \begin{bmatrix} X_i \\ \Gamma_i' D_{X_i} \end{bmatrix} = \begin{bmatrix} B_i \\ D_i' \end{bmatrix} = B_{i+1} = D_{A_{i+1}^*} X_{i+1} . \tag{4.15}$$

This with (4.14) gives

$$W_{i*}^* X_{i+1} = \begin{bmatrix} X_i \\ \Gamma_i' D_{X_i} \end{bmatrix} \quad (1 \le i \le n) . \tag{4.16}$$

By applying repeatedly Lemma IV.1.1 we can now infer from (4.16) and (IV.1.5)(with A, Y and W replaced by X_i, Γ_i' and W_i' respectively)

$$D_{X_{i+1}} = D_{W_{i*}^* X_{i+1}} = W_i' D_{\Gamma_i'} D_{X_i} = \ldots =$$
$$W_i' D_{\Gamma_i'} W_{i-1}' D_{\Gamma_{i-1}'} W_{i-2}' \ldots W_1' D_{\Gamma_1'} D_{X_1} =$$
$$W_i' D_{\Gamma_i'} W_{i-1}' D_{\Gamma_{i-1}'} W_{i-2}' \ldots W_1' D_{\Gamma_1'} \gamma P_G D_A | L_* ;$$

here W_j' is a unitary operator from $D_{\Gamma_j'}$ onto $D_{X_{j+1}}$ which is in turn the domain of definition of Γ_{j+1}' for $j = 1, 2, \ldots, i$. Now we apply the same argument to the operators

$$W_i Y^*_{i+1} = \begin{bmatrix} Y^*_i \\ \Gamma'^*_i D_{Y^*_i} \end{bmatrix} \qquad (\text{see}(4.13))$$

We deduce as above that

$$D_{Y^*_{i+1}} = W'_{i*} D_{\Gamma^*_i} W'_{i-1*} \ldots W'_{1*} D_{\Gamma^*_1} D_{Y^*_1} =$$
$$W'_{i*} D_{\Gamma^*_i} \ldots W'_{1*} D_{\Gamma^*_1} \alpha_* P_G \Pi'^*$$

where W'_{j*} is a unitary operator from $D_{\Gamma^*_j}$ onto $D_{Y^*_{j+1}}$ which is the domain of definition of Γ'^*_{j+1} for $1 \le j \le i$. Using the isometric operators

$$\alpha_j = \gamma^* W'^*_1 W'^*_2 \ldots W'^*_{j-1} \text{ from } D_{X_j} \text{ to } G$$
$$\beta_j = \alpha^*_* W'^*_{1*} W'^*_{2*} \ldots W'^*_{j-1*} \text{ from } D_{Y^*_j} \text{ to } G'$$

we define

$$\Gamma_j = \beta_j \Gamma'_j \alpha^*_j \,|\, \alpha_j D_{X_j} \text{ for } 1 \le j \le i \ . \tag{4.17}$$

This produces a choice sequence $\{\Gamma_j\}^i_1$ initiated from G to G' and

$$D_{X_{i+1}} = \alpha^*_{i+1} D_i D_{i-1} \ldots D_1 P_G D_A \,|\, L_*$$
$$D_{Y^*_{i+1}} = \Pi' D_{1*} D_{2*} \ldots D_{i*} \beta_{i+1} \ . \tag{4.18}$$

Thus

$$D'_{i+1} = D_{Y^*_{i+1}} \Gamma'_{i+1} D_{X_{i+1}} - Y_{i+1} A^*_{i+1} X_{i+1} =$$
$$\Pi' D_{1*} D_{2*} \ldots D_{i*} \beta_{i+1} \Gamma'_{i+1} \alpha^*_{i+1} D_i D_{i-1} \ldots D_1 P_G D_A \,|\, L_* - Y_{i+1} A^*_{i+1} X_{i+1} = \tag{4.19}$$
$$\Pi' D_{1*} D_{2*} \ldots D_{i*} \Gamma_{i+1} D_i D_{i-1} \ldots D_1 P_G D_A \,|\, L_* - Y_{i+1} A^*_{i+1} X_{i+1} \ .$$

This finishes the induction argument and completes the proof.

The previous theorem readily yields another proof of Corollary 2.8, that is,

4.3 COROLLARY. *Let A be a contraction in $I(U, T')$. There is a one to one correspondence between the set of all contractive intertwining lifting Λ_∞ of A and the set of all choice sequences $\{\Gamma_i\}^\infty_1$ initiated from G to G'. In particular, A admits a unique contractive intertwining lifting if and only if the $A \cdot U$ or $T' \cdot A$ is a regular factorization.*

4.4 COROLLARY. *Let A be a strict contraction in $I(U, T')$. Let Λ_n be a contractive n step intertwining lifting of A and $\{\Gamma_i\}^n_1$ be its choice sequence initiated from G to G' defined in (4.7b). Then Λ_n is a strict contraction if and only if $\{\Gamma_i\}^n_1$ is a strictly contractive. In particular, this gives a one to one correspondence between the set* *of all strictly contractive n step intertwining lifting Λ_n of A and the set of all strictly contractive choice sequences initiated from G to G'.*

PROOF. Corollary 4.3 in Chapter V shows that Λ_1 is a strict contraction if and only if Γ_1' is a strict contraction. Let Λ_i be the contractive i step intertwining lifting of A defined in the previous Theorem. We proceed by induction and assume that Λ_i is a strict contraction if and only if Γ_j' is a strict contraction for all $1 \le j \le i$. Recall that Λ_{i+1} is a contractive one step intertwining lifting of Λ_i in $I(U,T_i)$. Applying Corollary 4.3 in Chapter V again Λ_{i+1} is a strict contraction if and only if Γ_{i+1}' is a strict contraction. This completes the induction, that is, Λ_n is a strict contraction if and only if Γ_i' is a strict contraction for all $1 \le i \le n$. Since $\{\Gamma_i\}$ and $\{\Gamma_i'\}$ are related by unitarily operators (see (4.17)) this completes the proof.

4.5 REMARK. Let A be a contraction in $I(U,T')$. Let Λ_n be a contractive n step intertwining lifting of A and $\{\Gamma_i\}_1^n$ its choice sequence initiated from G to G' defined in (4.7b). By recursively using (4.13), (4.16) and (4.17) we see that

$$Y_{i+1}\gamma_{i*} = [Y_1, \Pi'\Gamma_1, \Pi'D_{1*}\Gamma_2, ..., \Pi'D_{1*}D_{2*} \cdots D_{i-1*}\Gamma_i]$$

$$(4.20)$$

$$\gamma_i X_{i+1} = [X_1, \Gamma_1 P_G D_A |L_*, ..., \Gamma_i D_{i-1} D_{i-2} \cdots D_1 P_G D_A |L_*]^{tr}$$

where γ_{i*} mapping $D_{A_1} \oplus G \oplus (\overset{i-1}{\underset{1}{\oplus}} D_i)$ onto $D_{A_{i+1}}$ and γ_i mapping $D_{A_{i+1}^\bullet}$ onto $D_{A_1^\bullet} \oplus G' \oplus (\overset{i-1}{\underset{1}{\oplus}} D_{i*})$, are unitary.

5. TWO INVERSE SCATTERING ALGORITHMS FOR THE COMMUTANT LIFTING THEOREM

In this section we will use the results in Sections 4 and 5 in Chapter IV to obtain two inverse scattering algorithms for the commutant lifting theorem.

Throughout this section it is assumed that A is a strict contraction in $I(U,T')$ and Λ_n is a strict contractive n step intertwining lifting of A. By Corollary 4.4 this implies that the operators A_i in (4.1) and Γ_i in (4.7b) are strict contractions for all $1 \le i \le n$. Throughout $\hat{X}_i$ and $\hat{Y}_i$ are the operators defined by the formula (IV.4.4) corresponding to the representation (4.1) of Λ_i. The following lemma plays an important role in developing several inverse scattering algorithms. It expresses $\hat{X}_{i+1}$ and $\hat{Y}_{i+1}$ as a linear combination of $\hat{X}_i$ and $\hat{Y}_i$.

5.1 LEMMA. *Let* A *be a strict contraction in* $I(U,T')$. *Let* Λ_n *be a strictly contractive n step intertwining lifting of* A *with strictly contractive choice sequence* $\{\Gamma_i\}_1^n$ *initiated from G to* G'. *Then for* $1 \le i \le n$

$$\hat{X}_{i+1} = \hat{X}_i + \hat{Y}_i L_i^{*-1} \Gamma_i R_i$$

$$\hat{Y}_{i+1} = U \begin{bmatrix} \hat{X}_i \\ I \end{bmatrix} R_i^{-1} \Gamma_i^* L_i^* + U \begin{bmatrix} \hat{Y}_i \\ 0 \end{bmatrix} \tag{5.1}$$

$$R_{i+1} = D_i R_i \text{ and } L_{i+1} = L_i D_{i*} .$$

The initial conditions for the difference equations in (5.1) are

$$\hat{X}_1 = D_{A_1}^{-2} A_1^* B_1 \text{ and } \hat{Y}_1 = D_{A_1}^{-2} C_1^*$$

$$R_1 = P_G D_A \,|L_* \text{ and } L_1 = \Pi' \,|G' . \tag{5.2}$$

PROOF. The first two initial conditions in (5.2) are precisely the adequate definitions in (IV.4.4) for $\hat{X}_1$ and $\hat{Y}_1$ while the last two are just notation which make the first relation in (5.1) valid for i=1. By Lemma 4.2 in Chapter V both D_{X_1} and $D_{Y_1^*}$ are invertible. Corollary 4.4 along with (4.1), (4.18) guarantees that both D_{X_i} and $D_{Y_i^*}$ are invertible. The definition (IV.4.4) for $\hat{X}_{i+1}$ and (4.4) for A_{i+1} together with (4.16) yield

$$\hat{X}_{i+1} = D_{A_{i+1}}^{-1} A_{i+1}^* X_{i+1} = A_{i+1}^* D_{A_{i+1}^*}^{-1} X_{i+1} =$$

$$A_{i+1}^* (D_{A_{i+1}^*}^{-1})^* W_{i*} W_{i*}^* X_{i+1} = [A_i^*, C_i^*] D_{A_{i+1}^*}^{-1} W_{i*} \begin{bmatrix} X_i \\ \Gamma_i' D_{X_i} \end{bmatrix}$$

But by taking the adjoint and then the inverse in (4.14)

$$D_{A_{i+1}^*}^{-1} W_{i*} = \begin{bmatrix} D_{A_i}^{-1} & D_{A_i}^{-1} A_i Y_i^* D_{Y_i^*}^{-1} \\ 0 & D_{Y_i^*}^{-1} \end{bmatrix}$$

Therefore using the formula (IV.4.4) for $\hat{X}_i$ and $\hat{Y}_i$ we finally obtain

$$\hat{X}_{i+1} = \hat{X}_i + [A_i^* A_i D_{A_i}^{-1} Y_i^* + C_i^*] D_{Y_i^*}^{-1} \Gamma_i' D_{X_i} =$$

$$\hat{X}_i + [A_i^* A_i + D_{A_i}^2] \hat{Y}_i D_{Y_i^*}^{-1} \Gamma_i' D_{X_i} = \hat{X}_i + \hat{Y}_i D_{Y_i^*}^{-1} \Gamma_i' D_{X_i} .$$

This with (4.17) and (4.18) gives the first equation in (5.1).

In a similar way by using the definition (IV.4.4) for $\hat{Y}_{i+1}$, $\hat{Y}_i$ and $\hat{X}_{i+1}$, $\hat{X}_i$ as well as the formulas (4.11) and (4.13) we obtain

$$\hat{Y}_{i+1} = D_{A_{i+1}}^{-1} Y_{i+1}^* =$$

$$U \begin{bmatrix} D_{A_i}^{-1} & D_{A_i}^{-1} A_i^* X_i D_{X_i}^{-1} \\ 0 & D_{X_i}^{-1} \end{bmatrix} \begin{bmatrix} Y_i^* \\ \Gamma_i'^* D_{Y_i^*} \end{bmatrix} = U \begin{bmatrix} \hat{Y}_i \\ 0 \end{bmatrix} + U \begin{bmatrix} \hat{X}_i \\ I \end{bmatrix} D_{X_i}^{-1} \Gamma_i'^* D_{Y_i^*} .$$

This with (4.17) and (4.18) gives the second equation in (5.1). The proof is now complete.

ALTERNATIVE PROOF. Here we present an alternate proof of the previous lemma by using the matrix results in Section 5 in Chapter IV. Equation (IV.4.4) implies that $D_{A_{i+1}}^2 \hat{X}_{i+1} = A_{i+1}^* B_{i+1}$, or equivalently, using $A_{i+1}^* = [A_i^*, C_i^*]$ the operator $\hat{X}_{i+1}$ is the unique solution to

$$(D_{A_i}^2 - C_i^* C_i)\hat{X}_{i+1} = A_{i+1}^* B_{i+1} . \tag{5.3}$$

On the other hand, the equations (IV.4.4), (IV.4.5) for $\hat{X}_i$ and $\hat{Y}_i$ and (IV.4.6) for $\hat{Y}_i$ give

$$(D_{A_i}^2 - C_i^* C_i)(\hat{X}_i + \hat{Y}_i D_{Y_i}^{-1} \cdot \Gamma_i' D_{X_i}) =$$

$$A_i^* B_i - C_i^* C_i \hat{X}_i + (C_i^* - C_i^* C_i \hat{Y}_i) D_{Y_i}^{-1} \cdot \Gamma_i' D_{X_i} =$$

$$A_i^* B_i - C_i^* Y_i A_i^* X_i + C_i^* (I - C_i \hat{Y}_i) D_{Y_i}^{-1} \cdot \Gamma_i' D_{X_i} =$$

$$A_i^* B_i - C_i^* Y_i A_i^* X_i + C_i^* D_{Y_i^*} \Gamma_i' D_{X_i} .$$

The last operator equals $A_i^* B_i - C_i^* D_i'$ by virtue of the relation (IV.3.2) (for the matrix in (4.1)), which in turn equals $A_{i+1}^* B_{i+1}$ by (4.4). This with (4.17), (4.18) and (5.3) proves the first equation in (5.1).

To prove the second equation in (5.1) notice that $\hat{Y}_{i+1}$ is the unique solution to

$$D_{A_{i+1}}^2 \hat{Y}_{i+1} = C_{i+1}^* . \tag{5.4}$$

Using $A_{i+1} = U'[A_i, B_i]U_*$ along with Corollary IV.5.2 and Corollary IV.5.3 in the present setting we have

$$D_{A_{i+1}}^2 U \left\{ \begin{bmatrix} \hat{X}_i \\ I \end{bmatrix} D_{X_i}^{-1} \Gamma_i'^* D_{Y_i^*} - \begin{bmatrix} \hat{Y}_i \\ 0 \end{bmatrix} \right\} =$$

$$U(I-[A_i,B_i]^*[A_i,B_i])\left\{\begin{bmatrix}\hat{X}_i\\I\end{bmatrix}D_{X_i}^{-1}\Gamma_i^{'*}D_{Y_i^*}-\begin{bmatrix}\hat{Y}_i\\0\end{bmatrix}\right\}=$$

$$U\begin{bmatrix}0\\D_{X_i}\Gamma_i^{'*}D_{Y_i^*}\end{bmatrix}-U\begin{bmatrix}C_i^*\\-X_i^*A_iY_i^*\end{bmatrix}.$$

Using the above (IV.3.2) and (4.4) we see that the previous operator is equal to

$$U\begin{bmatrix}C_i^*\\D_i^{'*}\end{bmatrix}=C_{i+1}^*.$$

This with (4.17), (4.18) and (5.4) proves the second equation in (5.1), which completes the alternate proof of Lemma 5.1.

Lemma 5.1 with (IV.4.5) adjusted to this setting and Theorem 4.2 readily yields the following result.

5.2 THEOREM. *Let* A *be a strict contraction in* $I(U,T')$. *Let* Λ_n *be a strictly contractive n step intertwining lifting of* A *and* $\{\Gamma_i\}_1^n$ *be its strictly contractive choice sequence initiated from G to* G'. *Then* Λ_n *is given by (4.6) where* A_n, B_n, C_n *and* D_n' *are recursively obtained by solving the following difference equations:*

$$A_{i+1}=\begin{bmatrix}A_i\\C_i\end{bmatrix},\quad B_{i+1}=\begin{bmatrix}B_i\\D_i'\end{bmatrix},\quad C_{i+1}=[C_i,D_i']U_*$$

$$\text{(5.5)}$$

$$D_{i+1}'=L_{i+1}\Gamma_{i+1}R_{i+1}-C_{i+1}\hat{X}_{i+1}=L_{i+1}\Gamma_{i+1}R_{i+1}-\hat{Y}_{i+1}^*A_{i+1}^*B_{i+1}$$

where $\hat{X}_i$, $\hat{Y}_i$, L_i *and* R_i *are recursively obtained by solving the difference equations in (5.1) subject to the initial conditions in (5.2).*

5.3 PROCEDURE. Let A be a strict contraction in $I(U,T')$. Let Λ_n be a strictly contractive n step intertwining lifting of A and $\{\Gamma_i\}_1^n$ be its strictly contractive choice sequence initiated from G to G'. Equation (5.5) gives us two different methods to compute D_{i+1}'. Therefore Theorem 5.2 provides us with two different recursive algorithms to find Λ_n given its strictly contractive choice sequence $\{\Gamma_i\}_1^n$.

On the other hand, Theorem 5.2 also provides us with two different recursive inverse scattering algorithms to extract the strictly contractive choice sequence $\{\Gamma_i\}_1^n$ from Λ_n. To describe these algorithms notice that (5.5) yields:

$$\Gamma_{i+1} = L_{i+1}^{-1}(D_{i+1}' + C_{i+1}\hat{X}_{i+1})R_{i+1}^{-1} = L_{i+1}^{-1}(D_{i+1}' + \hat{Y}_{i+1}^{*}A_{i+1}^{*}B_{i+1})R_{i+1}^{-1} . \qquad (5.6)$$

The operators $\hat{X}_1$ and $\hat{Y}_1$ are defined in (5.2). Lemma V.4.2 and (4.10) guarantees that R_1 and L_1 are both invertible. So (5.6) gives us two different methods to compute Γ_1 from Λ_1. For the next step assume that we have computed Γ_i, $\hat{X}_i$, $\hat{Y}_i$, L_i and R_i. The difference equation in (5.1) yields $\hat{X}_{i+1}$, $\hat{Y}_{i+1}$, L_{i+1} and R_{i+1}. Since D_{i+1}' is known we can use (5.6) to compute Γ_{i+1} from Λ_{i+1} by two different methods. Corollary 4.4 guarantees that both L_{i+1} and R_{i+1} are invertible. This establishes the recursion, and yields two inverse scattering algorithms for the commutant lifting theorem.

5.4 REMARK. Let A be a strict contraction in $I(U,T')$. Let Λ_n be a strictly contractive n step intertwining lifting of A and $\{\Gamma_i\}_1^n$ be its choice sequence initiated from G to G'. Equations (IV.4.10) in the present setting and (4.18) show that

$$D_{X_{i+1}} = (I - B_{i+1}^{*}B_{i+1} - \hat{X}_{i+1}^{*}A_{i+1}^{*}B_{i+1})^{1/2} = \alpha_{i+1}^{*}D_iD_{i-1} \cdots D_1 P_G D_A | L_* = \alpha_{i+1}^{*}R_{i+1}$$
$$D_{Y_{i+1}^{*}} = (I - \hat{Y}_{i+1}^{*}C_{i+1}^{*})^{1/2} = \Pi' D_{1*}D_{2*} \cdots D_{i*}\beta_{i+1} = L_{i+1}\beta_{i+1} . \qquad (5.7)$$

In the previous procedure one may replace R_i, L_i and Γ_i by the operators D_{X_i}, $D_{Y_i^{*}}$ and Γ_i' respectively. The operators $D_{X_{i+1}}$ and $D_{Y_{i+1}^{*}}$ are computed by (5.7) where $\hat{X}_{i+1}$ and $\hat{Y}_{i+1}$ are recursively obtained by (5.1). In this case the forward scattering algorithm in Procedure 5.3 uses the contractions Γ_i'; and the inverse scattering algorithm in Procedure 5.3 produces the contractions Γ_{i+1}' from $D_{X_{i+1}}$ to $D_{Y_{i+1}^{*}}$. Finally, it is noted that the contractions Γ_i and Γ_i' are related by (4.17).

5.5 REMARK. Finally, it is noted that the previous procedure reduces to Procedure I.4.3 and I.5.6 when $T = T' = T_n$ the lower shift on C^n in (I.4.1) and $A = A_n$ the n by n Toeplitz matrix on C^n in (I.4.4).

6. TWO MORE INVERSE SCATTERING ALGORITHMS

In this section we will use the results in Sections 4 and 5 in Chapter IV to obtain two more inverse scattering algorithms for the commutant lifting theorem. These algorithms are the "dual" of the algorithms in Section 5.

As before A is a strict contraction in $I(U,T')$ and Λ_n is a strictly contractive n step intertwining lifting of A. Throughout $\hat{X}_{i*}$ and $\hat{Y}_{i*}$ are the operators defined in (IV.4.11) corresponding to the strictly contractive i step intertwining lifting Λ_i of A in (4.1). The following lemma is the "dual" of Lemma 5.1. It expresses $\hat{X}_{i+1*}$ and $\hat{Y}_{i+1*}$ as a recursive linear combination of $\hat{X}_{i*}$ and $\hat{Y}_{i*}$.

6.1 LEMMA. *Let A be a strict contraction in $I(U,T')$. Let Λ_n be a strictly contractive n step intertwining lifting of A with strictly contractive choice sequence $\{\Gamma_i\}_1^n$ initiated from G to G'. Then for $1 \le i \le n$*

$$\hat{X}_{i+1*} = \begin{bmatrix} \hat{X}_{i*} \\ 0 \end{bmatrix} + \begin{bmatrix} \hat{Y}_{i*} \\ I \end{bmatrix} L_i^{*-1} \Gamma_i R_i$$

$$\hat{Y}_{i+1*} = U'[\hat{X}_{i*} R_i^{-1} \Gamma_i^* L_i^* + \hat{Y}_{i*}] \tag{6.1}$$

$$R_{i+1} = D_i R_i \text{ and } L_{i+1} = L_i D_{i*} .$$

The initial conditions for the difference equations in (6.1) are

$$\hat{X}_{1*} = D_{A_1}^{-2} B_1 \text{ and } \hat{Y}_{i*} = D_{A_1}^{-2} A_1 C_1^*$$

$$R_1 = P_G D_A \,|L_* \text{ and } L_1 = \Pi' \,|G' . \tag{6.2}$$

PROOF. By Lemma V.4.2 both D_{X_1} and $D_{Y_1^*}$ are invertible. Corollary 4.4 along with (4.10) and (4.18) guarantees that both R_i and L_i are invertible. The definition of $\hat{X}_{i+1*}$ and (IV.3.2), (IV.4.2), (IV.4.11) adjusted to this setting with the adjoint of (4.14) and (4.16) yield

$$\hat{X}_{i+1*} = D_{A_{i+1}}^{-1} X_{i+1} = (D_{A_{i+1}}^{-1})^* W_{i*} W_{i*}^* X_{i+1} =$$

$$\begin{bmatrix} D_{A_i}^{-1} & D_{A_i}^{-1} A_i Y_i^* D_{Y_i}^{-1} \\ 0 & D_{Y_i}^{-1} \end{bmatrix} \begin{bmatrix} X_i \\ \Gamma_i' D_{X_i} \end{bmatrix} = \begin{bmatrix} \hat{X}_{i*} \\ 0 \end{bmatrix} + \begin{bmatrix} \hat{Y}_{i*} \\ I \end{bmatrix} D_{Y_i}^{-1} \Gamma_i' D_{X_i} .$$

This with (4.17) and (4.18) gives the first equation in (6.1).

Using the definition of $\hat{Y}_{i+1*}$ with (4.4), (4.11) and (4.13):

$$\hat{Y}_{i+1*} = A_{i+1} D_{A_{i+1}}^{-1} Y_{i+1}^* =$$

$$U'[A_i, B_i] \begin{bmatrix} D_{A_i}^{-1} & D_{A_i}^{-1} A_i^* X_i D_{X_i}^{-1} \\ 0 & D_{X_i}^{-1} \end{bmatrix} \begin{bmatrix} Y_i^* \\ \Gamma_i'^* D_{Y_i} \end{bmatrix} =$$

$$U'[\hat{Y}_{i*} + (A_i A_i^* \hat{X}_{i*} + D_{A_i}^2 \hat{X}_{i*}) D_{X_i}^{-1} \Gamma_i'^* D_{Y_i}] =$$

$$U'[\hat{Y}_{i*} + \hat{X}_{i*} D_{X_i}^{-1} \Gamma_i'^* D_{Y_i}] .$$

This with (4.17) and (4.18) gives this second equation in (6.1). The proof is now complete.

ALTERNATE PROOF. Here we present an alternate proof of the previous lemma by using the matrix results in Section IV.5. Equation (4.11) in Chapter IV implies that $\hat{X}_{i+1*}$ is the unique solution to

$$D^2_{A^*_{i+1}} \hat{X}_{i+1*} = B_{i+1} . \tag{6.3}$$

Using $A^*_{i+1} = [A^*_i, C^*_i]$ along with Corollary IV.5.5, Corollary IV.5.4 and equation (IV.3.2) adjusted to this setting and (4.4)

$$D_{A^*_{i+1}} \left\{ \begin{bmatrix} \hat{X}_{i*} \\ 0 \end{bmatrix} + \begin{bmatrix} \hat{Y}_{i*} \\ I \end{bmatrix} D^{-1}_{Y_i} \cdot \Gamma'_i D_{X_i} \right\} =$$

$$\begin{bmatrix} B_i \\ -Y_i A^*_i X_i \end{bmatrix} + \begin{bmatrix} 0 \\ D_{Y_i^*} \Gamma'_i D_{X_i} \end{bmatrix} = \begin{bmatrix} B_i \\ D'_i \end{bmatrix} = B_{i+1} .$$

This with (4.17), (4.18) and (6.3) proves the first equation in (6.1).

Equation (IV.4.11) implies that $D^2_{A^*_{i+1}} \hat{Y}_{i+1*} = A_{i+1} C^*_{i+1}$, or equivalently, using $A_{i+1} = U'[A_i, B_i]U_*$ the operator $\hat{Y}_{i+1*}$ is the unique solution to

$$(I - U'(A_i A^*_i + B_i B^*_i)U'^*) \hat{Y}_{i+1*} = A_{i+1} C^*_{i+1} . \tag{6.4}$$

Equations (IV.4.11), (IV.4.12), (IV.3.2) and (4.4) give

$$(I - U'(A_i A^*_i + B_i B^*_i)U'^*)U'(\hat{X}_{i*} D^{-1}_{X_i} \Gamma'^*_i D_{Y_i^*} + \hat{Y}_{i*}) =$$

$$U'(B_i - B_i B^*_i \hat{X}_{i*})D^{-1}_{X_i} \Gamma'^*_i D_{Y_i^*} + U'A_i C^*_i - U'B_i B^*_i \hat{Y}_{i*} =$$

$$U'B_i(I - B^*_i \hat{X}_{i*})D^{-1}_{X_i} \Gamma'^*_i D_{Y_i^*} + U'A_i C^*_i - U'B_i X^*_i A_i Y^*_i =$$

$$U'B_i D_{X_i} \Gamma'^*_i D_{Y_i^*} + U'A_i C^*_i - U'B_i X^*_i A_i Y^*_i =$$

$$U'(A_i C^*_i + B_i D'^*_i) = A_{i+1} C^*_{i+1} .$$

This with (4.17), (4.18) and (6.4) proves the second equation in (6.1). This completes the alternate proof of the lemma.

The previous lemma along with (IV.4.15) adjusted to this setting, and Theorem 4.2 applied to D'_{i+1} readily yields the following result.

6.2 THEOREM. *Let* A *be a strict contraction in* $I(U, T')$. *Let* Λ_n *be a strictly contractive n step intertwining lifting of* A *and* $\{\Gamma_i\}^n_1$ *be its strictly contractive choice sequence initiated from G to* G'. *Then* Λ_n *is given by (4.6) where* A_n, B_n, C_n *and* D'_n *are recursively obtained by solving the following difference equation:*

$$A_{i+1} = \begin{bmatrix} A_i \\ C_i \end{bmatrix}, \ B_{i+1} = \begin{bmatrix} B_i \\ D_i' \end{bmatrix}, \ C_{i+1} = [C_i, \ D_i']U_*$$

$$(6.5)$$

$$D_{i+1}' = L_{i+1}\Gamma_{i+1}R_{i+1} - \hat{Y}_{i+1*}^*B_{i+1} = L_{i+1}\Gamma_{i+1}R_{i+1} - C_{i+1}A_{i+1}^*\hat{X}_{i+1*}$$

where $\hat{X}_{i*}$, $\hat{Y}_{i*}$, L_i and R_i are recursively obtained by solving the difference equations in (6.1) subject to the initial conditions in (6.2).

6.3 PROCEDURE. Let A be a strict contraction in $I(U,T')$. Let Λ_n be a strictly contractive n step intertwining lifting of A and $\{\Gamma_i\}_1^n$ be its strictly contractive choice sequence initiated from G to G'. Equation (6.5) gives us two different methods to compute D_{i+1}'. Therefore Theorem 6.2 provides us with two different recursive algorithms to find Λ_n given its strictly contractive choice sequence $\{\Gamma_i\}_1^n$.

On the other hand, Theorem 6.2 also provides us with two different recursive inverse scattering algorithms to extract the strictly contractive choice sequence $\{\Gamma_i\}_1^n$ from Λ_n. To describe these algorithms notice that (6.5) yields:

$$\Gamma_{i+1} = L_{i+1}^{-1}(D_{i+1}'+\hat{Y}_{i+1*}^*B_{i+1})R_{i+1}^{-1} = L_{i+1}^{-1}(D_{i+1}'+C_{i+1}A_{i+1}^*\hat{X}_{i+1*})R_{i+1}^{-1} . \qquad (6.6)$$

The operators $\hat{X}_{1*}$ and $\hat{Y}_{1*}$ are defined in (6.2). Lemma V.4.2 and (4.10) guarantees that both R_1 and L_1 are invertible. So (6.6) gives us two different methods to compute Γ_1 from Λ_1. For the next step assume that we have computed Γ_i, $\hat{X}_{i*}$, $\hat{Y}_{i*}$, L_i and R_i. The difference equation in (6.1) yields $\hat{X}_{i+1*}$, $\hat{Y}_{i+1*}$, L_{i+1} and R_{i+1}. Since D_{i+1}' is known (6.6) gives us two different methods to compute Γ_{i+1} from Λ_{i+1}. Corollary 4.4 and (4.10) guarantees that both L_{i+1} and R_{i+1} are invertible. This establishes the recursion and yields two more inverse scattering algorithms for the commutant lifting theorem. As in Remark 5.5 it is noted that these two inverse scattering algorithms also reduce to Procedure 4.3 and Procedure 5.6 used in Chapter I, to compute the Schur numbers from the Carathéodory data when $T = T' = T_n$ and $A = A_n$.

6.4 REMARK. Let A be a strict contraction in $I(U,T')$. Let Λ_n be a strictly contractive n step intertwining lifting of A and $\{\Gamma_i\}_1^n$ be its choice sequence initiated from G to G'. Equations (IV.4.12) and (4.18) show that

$$(I-B_{i+1}^*\hat{X}_{i+1*})^{1/2} = \alpha_{i+1}^*D_iD_{i-1} \ \cdots \ D_1P_GD_A |L_* = \alpha_{i+1}^*R_{i+1}$$

$$(I-C_{i+1}C_{i+1}^*-C_{i+1}A_{i+1}^*\hat{Y}_{i+1*})^{1/2} = \Pi'D_{1*}D_{2*} \ \cdots \ D_{i*}\beta_{i+1} = L_{i+1}\beta_{i+1} .$$

$$(6.7)$$

In the previous procedure one may replace R_i, L_i and Γ_i by the operators D_{X_i}, $D_{Y_i^*}$ and Γ_i' respectively. The operators $D_{X_{i+1}}$ and $D_{Y_{i+1}^*}$ are computed by (6.7) where $\hat{X}_{i+1*}$ and $\hat{Y}_{i+1*}$ are recursively obtained by (6.1). In this case the forward scattering algorithm in Procedure 6.3 uses the contractions Γ_i' and the inverse scattering algorithm in Procedure 6.3 produces the contractions Γ_{i+1}' from $D_{X_{i+1}}$ to $D_{Y_{i+1}^*}$. The contractions Γ_i and Γ_i' are related by (4.17). Finally, it is noted that it is an interesting exercise to apply the above procedures to Hankel matrices.

7. OPERATOR QUASI-BALLS

Our previous results give us several different methods to compute a choice sequence corresponding to a contractive intertwining lifting B of A. By studying quasi-balls we will show that all of these methods produce the same choice sequence up to a unitary operator.

To begin we say that $B(C, L, R)$ is an *operator quasi-ball in* $L(E, E')$ if

$$B = B(C, L, R) = \{C + L\Gamma R: \ \Gamma \text{ is a contraction in } L(H_1, H_2)\} \tag{7.1}$$

By definition *the center* C is an operator from E to E', *the left radius* L is an injective operator from $H_2 (= \overline{L^*E'})$ to E' and the *right radius* R is an operator from E to H_1 satisfying $\overline{RE} = H_1$. The contraction Γ in (7.1) is a *choice contraction for* $B(C, L, R)$. Two operator quasi-balls B_1 and B_2 are *equal* if they are equal as sets $B_1 = B_2$. A quasi-ball B is *trivial* if its left radius L or right radius is zero. A quasi-ball is trivial if and only if B consists of a single point C, its center. Moreover, if two quasi-balls B_1 and B_2 are equal, then B_1 is trivial if and only if B_2 is trivial.

We say that two quasi-balls $B(C_1, L_1, R_1)$ and $B(C_2, L_2, R_2)$ *are* ρ *equivalent* if they have the same centers $C_1 = C_2$ and there exists a constant $\rho > 0$ satisfying

$$L_1 W_l = \rho L_2 \quad \text{and} \quad W_r R_1 = \rho^{-1} R_2 \tag{7.2}$$

where W_l from $\overline{L_2^* E'}$ onto $\overline{L_1^* E'}$ and W_r from $\overline{R_1 E}$ onto $\overline{R_2 E}$ are both unitary. Obviously two ρ equivalent quasi balls are equal. The following theorem due to Smuljian shows that the converse is also true.

7.1 THEOREM. *Two nontrivial quasi-balls are equal if and only they are* ρ *equivalent.*

PROOF. Assume that the two operator quasi-balls

$$B_1 = B(C_1, L_1, R_1) \quad \text{and} \quad B_2 = B(C_2, L_2, R_2)$$

are equal and x is in E and y is in E'. If Γ_1 is any choice contraction for B_1, then there exists a choice contraction Γ_2 for B_2 satisfying

$$|((C_1 - C_2)x, y) + (\Gamma_1 R_1 x, L_1^* y)| \leq |(\Gamma_2 R_2 x, L_2^* y)| \ . \tag{7.3}$$

If $R_1 x$ and $L_1^* y$ are both nonzero, then we can pick a choice contraction Γ_1 preserving the norm of $R_1 x$ such that $\Gamma_1 R_1 x = re^{it} L_1^* y$ for some $r > 0$; thus $(\Gamma_1 R_1 x, L_1^* y) = e^{it} \|R_1 x\| \, \|L_1^* y\|$. So by choosing the appropriate value of t and using Schwartz inequality in (7.3) we obtain

$$|((C_1 - C_2)x, y)| + \|R_1 x\| \, \|L_1^* y\| \leq \|R_2 x\| \, \|L_2^* y\| \ . \tag{7.4}$$

Equation (7.3) shows that (7.4) holds also if $R_1 x = 0$ or $L_1^* y = 0$. Therefore (7.4) holds for all x in E and y in E'. Obviously (7.4) is still valid if we interchange R_1 with R_2 and L_1 with L_2. This readily implies that $C_1 = C_2$ and

$$\|R_1 x\| \, \|L_1^* y\| = \|R_2 x\| \, \|L_2^* y\| \quad \text{(for all x in } E \text{ and y in } E') \ . \tag{7.5}$$

By the hypothesis both of the quasi-balls are nontrivial. So there exists an x in E such that $R_1 x$ is nonzero. Setting $\rho = \|R_2 x\|/\|R_1 x\|$ the equation (7.5) implies that $\rho > 0$ and there exists a unitary operator W_l from $\overline{L_2^* E'}$ onto $\overline{L_1^* E'}$ satisfying $\rho \, W_l L_2^* = L_1^*$, or equivalently, $L_1 W_l = \rho L_2$. Using this in (7.5) gives $\rho \|R_1 x\| = \|R_2 x\|$ for all x in E. This implies that there exists a unitary operator W_r from $\overline{R_1 E}$ onto $\overline{R_2 E}$ satisfying $W_r R_1 = \rho^{-1} R_2$. Therefore (7.2) holds. The proof is now complete.

Let $B_1 = B(C_1, L_1, R_1)$ and $B_2 = B(C_2, L_2, R_2)$ be two nontrivial equal quasi-balls; so $C_1 = C_2$. If Γ_1 is a choice contraction for B_1, then there exists a choice contraction Γ_2 for B_2 satisfying $L_1 \Gamma_1 R_1 = L_2 \Gamma_2 R_2$. By the previous theorem and (7.2) with the fact that both L_1 and R_1^* are injective, we obtain $\Gamma_1 = W_l \Gamma_2 W_r$. Therefore the choice contractions for nontrivial equal quasi-balls are equal up to a unitary operator on the right and left.

Thus far we have presented several different methods to show that there is a one to one correspondence between the set of all contractive intertwining liftings and a certain choice sequence. No matter what method we choose all of the choice sequences are unitarily equivalent. To see this notice that all of our methods parameterize all the i-th step contractive intertwining liftings Λ_i of A by expression Z_i as a quasi-ball of the form

$$Z_i = C_i + L_o D_{1*} \, \cdots \, D_{i-1*} \Gamma_i D_{i-1} \, \cdots \, D_1 R_o \tag{7.6}$$

where $\{\Gamma_i\}_1^\infty$ is a choice sequence initiated from H_1 to H_2. Now consider another quasi-ball representation of Z_i of the form

$$Z_i = C_i' + L_o' D_{\Gamma_1^*} \cdots D_{\Gamma_{i-1}^*} \Gamma_i' D_{\Gamma_{i-1}'} \cdots D_{\Gamma_1'} R_o' \tag{7.7}$$

where $\{\Gamma_i\}$ is a choice sequences from H_1' to H_2'. Obviously (7.6) and (7.7) provide two different quasi-ball representations of Z_1. By the previous theorem this implies that $C_1 = C_1'$ and $\Gamma_1 = W_l \, \Gamma_1' W_r$. Now repeated applications of the previous theorem with the special form of the left and right radii in (7.6) and (7.7) show that $C_i = C_i'$ and $\Gamma_i = W_l \Gamma_i' W_r | D_{i-1}$ for all i. In particular, the choice sequences $\{\Gamma_i\}$ and $\{\Gamma_i'\}$ are unitarily equivalent. Summing up these observation we conclude with the following observation.

7.2 REMARK. All of our choice sequences corresponding to a contractive intertwining lifting B of A are unitarily equivalent.

8. A PARAMETERIZATION OF ALL CONTRACTIVE INTERTWINING LIFTINGS FOR A STRICTLY CONTRACTIVE HANKEL OPERATOR.

In this section we will use the Schur representation in Theorem 3.4 to obtain the Adamjan, Arov and Krein parameterization of all contractive intertwining liftings B of a strictly contractive Hankel operator. It turns out that the operators $\hat{X}_1$ and $\hat{Y}_1$ in Section 5 naturally occur in their formula. Since the classical Hermite-Ferjér interpolation problem is a special case of the H^∞ Nehari optimization problem (see Section X.5) their formula can also be used to provide a complete characterization of all solutions to the classical Hermite-Ferjér interpolation problem. In the next section we will show that the Adamjan, Arov and Krein representation reduces to the Schur representation (I.3.8) used in parameterizing all solutions to the classical Carathéodory interpolation problem. ·

We begin with the following elementary result.

8.1 LEMMA. *Assume that* D *is an invertible operator from* U *to* Y. *Then the matrix*

$$L = \begin{bmatrix} A' & B \\ C & D \end{bmatrix} : H \oplus U \to H' \oplus Y \tag{8.1}$$

is invertible if and only if $A' - BD^{-1}C$ *is invertible. Moreover, in this case if we set* $E = (A' - BD^{-1}C)^{-1}$, *then*

$$L^{-1} = \begin{bmatrix} E & -EBD^{-1} \\ -D^{-1}CE & D^{-1} + D^{-1}CEBD^{-1} \end{bmatrix} . \tag{8.2}$$

PROOF. If the matrix L is invertible, then there exists a matrix satisfying

$$\begin{bmatrix} A' & B \\ C & D \end{bmatrix} \begin{bmatrix} E & F \\ G & H \end{bmatrix} = \begin{bmatrix} I & 0 \\ 0 & I \end{bmatrix} . \tag{8.3}$$

This implies that $CE + DG = 0$. Substituting $G = -D^{-1}CE$ into $A'E + BG = I$ produces

$$(A' - BD^{-1}C)E = I . \tag{8.4}$$

Now consider

$$\begin{bmatrix} E & F \\ G & H \end{bmatrix} \begin{bmatrix} A' & B \\ C & D \end{bmatrix} = \begin{bmatrix} I & 0 \\ 0 & I \end{bmatrix} \tag{8.5}$$

which readily implies that $EB + FD = 0$. So using $F = -EBD^{-1}$ in $EA' + FC = I$ we have

$$E(A' - BD^{-1}C) = I . \tag{8.6}$$

Equations (8.4) and (8.6) imply that $A' - BD^{-1}C$ is invertible. In fact its inverse is E.

Now assume that $A' - BD^{-1}C$ is invertible. In this case we have all ready seen that $E = (A' - BD^{-1}C)^{-1}$ and $F = -EBD^{-1}$ and $G = -D^{-1}CE$. So substituting this F in $CF + DH = I$ yields

$$H = D^{-1} + D^{-1}C(A' - BD^{-1}C)^{-1}BD^{-1} .$$

Using these definitions for E, F, G and H simple calculations show that (8.3) and (8.5) hold. Therefore L is invertible. Equation (8.2) readily follows from the definitions of E, F, G and H. This completes the proof.

Throughout this section A is a strictly contractive Hankel operator in $I(U, T')$, that is, $T = U$ is an isometry on $H = K$ and T' is a co-isometry. In this case $D_{U^*} = P_{L_*}$ the orthogonal projection onto the wandering subspace $L_* = K \ominus UK$ obtained in the Wold decomposition of U and $D_{T'} = P_{L'_*}$ the orthogonal projection onto $L'_* = H' \ominus T'^*H'$. By a slight abuse of notation we let Π' be the operator from H' onto $D_{T'}$ and Π_* the operator from K onto $D_{T'^*}$ defined by $\Pi' = D_{T'}$ and $\Pi_* = D_{T^*}$. The operators Π' and Π_* pick out the $D_{T'} = L'_*$ and $D_{U^*} = L_*$ "component" from H' and K, respectively.

As before F, G, F' and G' are the spaces defined in (2.17) and ω is the unitary operator from F onto F' in (2.18) defined by the following column matrix

$$\omega D_A U = \begin{bmatrix} D_A \\ \Pi' A \end{bmatrix} . \tag{8.7}$$

Now let us notice that if C is an operator from X to Y and C^*C is invertible, then $C(C^*C)^{-\frac{1}{2}}$ is an isometry from X to Y. Moreover, C and $C(C^*C)^{-\frac{1}{2}}$ have the same range. (Since C^*C is invertible the range of C is closed.) Choosing $C = D_A U$ we see that $C^*C = D_{AU}^2$. Because A is a strict contraction D_{AU} is invertible. Applying the inverse of D_{AU} to both sides of (8.7) gives

$$\omega D_A U D_{AU}^{-1} = \begin{bmatrix} D_A \\ \Pi' A \end{bmatrix} D_{AU}^{-1} .$$

However, since $C(C^*C)^{-\frac{1}{2}} = D_A U D_{AU}^{-1}$ is an isometry whose range is F, we can apply the adjoint of $D_A U D_{AU}^{-1}$ to the right hand side of the previous equation involving ω to obtain

$$\omega P_F = \begin{bmatrix} D_A \\ \Pi' A \end{bmatrix} D_{AU}^{-2} U^* D_A . \tag{8.8}$$

Since A is a strict contraction, D_A is invertible and $D_A = K$. This implies that g is in G if and only if g is in K and g is orthogonal to $D_A UK$, or equivalently, $U^* D_A g = 0$. Because D_A is invertible and $U^* h = 0$ if and only if h is in L_* we see that $U^* D_A g = 0$ if and only if g is in $D_A^{-1} L_*$. Thus

$$G = D_A^{-1} D_{T^*} = D_A^{-1} \Pi_*^* L_* . \tag{8.9}$$

On the other hand $f \oplus g$ in $D_A \oplus D_{T'}$ is in G', (which equals the orthogonal complement in $D_A \oplus D_{T'}$ of $\{D_A h \oplus D_{T'} Ah : h \in H\}$ where we reversed the order of D_A and $D_{T'}$ in (2.17)) if and only if $f \oplus g$ is in $K \oplus D_{T'}$ and $D_A f + A^* D_{T'} g = 0$. Since D_A is invertible $f = -D_A^{-1} A^* D_{T'} g$. Thus

$$G' = \begin{bmatrix} -D_A^{-1} A^* \Pi'^* \\ I \end{bmatrix} L'_* . \tag{8.10}$$

Now we will express G and G' as the range of two isometries. To this end let C be the operator from L_* to D_A with range G defined by $C = D_A^{-1} \Pi_*^*$. Then $C^*C = \Pi_* D_A^{-2} \Pi_*^* \triangleq N$ on L_*. Since A is a strict contraction D_A^{-2} is strictly positive. Thus $P_{L_*} D_A^{-2} | L_*$ is also a strictly positive operator on L_*. So C^*C is invertible. (If $L_* = \{0\}$, then $C^*C = 0$ on $\{0\}$ is also invertible.) Therefore using the normalizing operator N in (8.9) we see that

$$D_A^{-1} \Pi_*^* N^{-\frac{1}{2}} \text{ where } N = \Pi_* D_A^{-2} \Pi_*^* \tag{8.11}$$

is an isometry from L_* to D_A whose range is G. Now let C be the operator from L_* to $D_A \oplus L'_*$ defined by $C = [-D_A^{-1} A^* \Pi'^*,\ I]^{\text{tr}}$. Then using $A D_A^{-2} = D_{A^*}^{-2} A$ we have

$$C^* C = \Pi' A D_A^{-2} A^* \Pi'^* + I = \Pi'(A D_A^{-2} A^* + I)\Pi'^* =$$
$$\Pi'(D_{A^*}^{-2} A A^* + I)\Pi'^* = \Pi' D_{A^*}^{-2}(A A^* + D_{A^*}^2)\Pi'^* = \Pi' D_{A^*}^{-2} \Pi'^* \triangleq N_1 \ .$$

As before since A is a strict contraction the operator $\Pi' D_{A^*}^{-2} \cdot \Pi'^*$ on L'_* is invertible. Therefore

$$\begin{bmatrix} -D_A^{-1} A^* \Pi'^* \\ I \end{bmatrix} N_1^{-\frac{1}{2}} \quad \text{where } N_1 = \Pi' D_{A^*}^{-2} \cdot \Pi'^* \tag{8.12}$$

is an isometry from L'_* to $D_A \oplus L'_*$ whose range is G'. Since the operators in (8.11) and (8.12) are both isometries we readily see that G is identified with D_{U^*} and G' is identified with $D_{T'}$. In particular, $\dim G = \dim L_*$ and $\dim G' = \dim L'_*$.

Because the operators in (8.11) and (8.12) are isometries whose ranges are G and G', respectively, we see that the set of all contractive analytic functions R(z) in $H^\infty(G, G')$ are given by

$$R(z) = R(z) P_G = \begin{bmatrix} -D_A^{-1} A^* \Pi'^* \\ I \end{bmatrix} N_1^{-\frac{1}{2}} F_1(z) N^{-\frac{1}{2}} \Pi_* D_A^{-1}$$

where F_1 is a contractive analytic function in $H^\infty(L_*, L'_*)$. (Recall that $L_* = D_{T^*}$ and $L'_* = D_{T'}$.) Applying Theorem IX.5.3 to this R(z) with ωP_F in (8.8) readily gives the following useful result.

8.2 LEMMA. *Let A be a strictly contractive Hankel operator in $I(U, T')$. Then the set of all contractive analytic functions F(z) in $H^\infty(D_A, D_A \oplus L'_*)$ satisfying the constraint $F(0)|F = \omega$ is given by*

$$F(z) = \begin{bmatrix} D_A D_{AU}^{-2} U^* D_A - D_A^{-1} A^* \Pi'^* N_1^{-\frac{1}{2}} F_1(z) N^{-\frac{1}{2}} \Pi_* D_A^{-1} \\ \Pi' A D_{AU}^{-2} U^* D_A + N_1^{-\frac{1}{2}} F_1(z) N^{-\frac{1}{2}} \Pi_* D_A^{-1} \end{bmatrix} \tag{8.13}$$

where $F_1(z)$ is a contractive analytic function in $H^\infty(L_*, L'_*)$.

In order to compute the Schur representation in (3.15) we must calculate the inverse of $I - z \Pi_A F(z)$ where F(z) is given by (8.13) in the previous lemma. To this end notice that (8.13) yields

$$I - z\Pi_A F(z) = D_A(I - zD_{AU}^{-2}U^*D_A^2 + zD_A^{-2}A^*\Pi'^*N_1^{-\frac{1}{2}}F_1(z)N^{-\frac{1}{2}}\Pi_*)D_A^{-1} \quad . \quad (8.14)$$

The first nontrivial term in (8.14) becomes

$$D_{AU}^{-2}U^*D_A^2 = D_{AU}^{-2}U^*(I - A^*A(D_{U^*}^2 + UU^*)) =$$
$$- D_{AU}^{-2}(AU)^*AP_{L_*} + D_{AU}^{-2}(I - U^*A^*AU)U^* = - D_{AU}^{-2}(AU)^*AP_{L_*} + U^* \quad . \quad (8.15)$$

In order to compute $D_A^{-2}A^*\Pi'^*$ consider the unitary operator γ from H' onto $L_* \oplus T'^*H'$ defined by $\gamma = [\Pi', T']^{\text{tr}}$. Notice that since T' is a co-isometry γ is the column matrix obtained by interchanging T' and $D_{T'}$ in the rotation matrix $R_{T'}$ which is also a column matrix. Using this γ we have

$$D_A^{-2}A^*\Pi'^* = A^*D_A^{-2}\cdot\Pi'^* = A^*\gamma^*\gamma D_A^{-2}\cdot\gamma^*\gamma\Pi'^* = A^*\gamma^*(I - \gamma AA^*\gamma^*)^{-1}\gamma\Pi'^* =$$
$$[A^*\Pi'^*, A^*T'^*]\begin{bmatrix} \Pi'D_A^2\cdot\Pi'^* & -\Pi'AA^*T'^* \\ -T'AA^*\Pi'^* & D_{(AU)^\cdot}^2 \end{bmatrix}^{-1}\begin{bmatrix} I \\ 0 \end{bmatrix}$$

Notice that $L = \gamma D_A^2\cdot\gamma^*$ is a matrix on $L_* \oplus T'^*H'$. So the operator E in Lemma 8.1, that is, the entry in the upper left hand corner of L^{-1} becomes $E = \Pi'D_A^{-2}\cdot\Pi'^*$ which is precisely N_1 in (8.12). Now applying Lemma 8.1 to the above equation we have

$$D_A^{-2}A^*\Pi'^* = [A^*\Pi'^*, (AU)^*]\begin{bmatrix} N_1 \\ D_{(AU)^\cdot}^{-2}AUA^*\Pi'^*N_1 \end{bmatrix} =$$
$$(I + (AU)^*D_{(AU)^\cdot}^{-2}AU)A^*\Pi'^*N_1 = D_{AU}^{-2}A^*\Pi'^*N_1 \quad .$$

The last equation follows from the fact that for any strict contraction Γ

$$I + \Gamma^*D_\Gamma^{-2}\cdot\Gamma = I + D_\Gamma^{-2}\Gamma^*\Gamma = D_\Gamma^{-2}(D_\Gamma^2 + \Gamma^*\Gamma) = D_\Gamma^{-2} \quad . \quad (8.16)$$

Therefore

$$D_A^{-2}A^*\Pi'^*N_1^{-\frac{1}{2}}F_1(z)N^{-\frac{1}{2}} = D_{AU}^{-2}A^*\Pi'^*N_1^{\frac{1}{2}}F_1(z)N^{-\frac{1}{2}} \quad . \quad (8.17)$$

Finally, we will eventually need the following identity

$$N_1 = (I - \Pi'AD_{AU}^{-2}A^*\Pi'^*)^{-1} \quad . \quad (8.18)$$

This identity follows by applying (8.16) and Lemma 8.1 to $L = \gamma D_A^2\cdot\gamma^*$

$$N_1^{-1} = A' - BD^{-1}C = \Pi'(D_A^2\cdot - AU^*A^*D_{(AU)^\cdot}^{-2}T'AA^*)\Pi'^* =$$
$$\Pi'(I - A(I + U^*A^*D_{(AU)^\cdot}^{-2}AU)A^*)\Pi'^* = \Pi'(I - AD_{AU}^{-2}A^*)\Pi'^* \quad .$$

As in Section 5 let $\hat{X}_1$ from L_* to K and $\hat{Y}_1$ from L'_* to K be the operators defined by

$$\hat{X}_1 = D_{A_1}^{-2} A_1^* A \Pi_*^* \text{ and } \hat{Y}_1 = D_{A_1}^{-2} A^* \Pi'^* \tag{8.19}$$

where $A_1 = AU$. Notice that $\hat{X}_1$ and $\hat{Y}_1$ are precisely the operators in (5.2). Also let α be the unitary operator from K onto $L_* \oplus H$ defined by $\alpha = [\Pi_*, U^*]^{tr}$. Inserting (8.19) into (8.15), (8.17) and, then substituting these equations into (8.14) yields

$$\alpha D_A^{-1} (I - z\Pi_A F(z)) D_A \alpha^* = \begin{bmatrix} I + z\Pi_*(\hat{X}_1 + \hat{Y}_1 N_1^{1/2} F_1(z) N^{-1/2}) & -z\Pi_* \\ zU^*(\hat{X}_1 + \hat{Y}_1 N_1^{1/2} F_1(z) N^{-1/2}) & (I - zU^*) \end{bmatrix} \tag{8.20}$$

Consulting (8.13) and (8.15), (8.19) we have

$$\Pi' F(z) D_A \alpha^* = \Pi' A D_{AU}^{-2} U^* D_A^2 \alpha^* + N_1^{-1/2} F_1(z) N^{-1/2} \Pi_* \alpha^* =$$
$$[N_1^{-1/2} F_1(z) N^{-1/2} - \Pi' A \hat{X}_1, \ \Pi' A] \ . \tag{8.21}$$

Since F is a contractive analytic function $(I - z\Pi_A F)$ is invertible for all z in D. So adjusting Lemma 8.1 to this setting we see that the E(z) corresponding to the inverse of the matrix in (8.20) is an invertible analytic function in D with values in $L(L_*, L_*)$ and is given by

$$(E(z))^{-1} = I + z\Pi_*(I - zU^*)^{-1}(\hat{X}_1 + \hat{Y}_1 N_1^{1/2} F_1(z) N^{-1/2}) \ .$$

If we set

$$\hat{X}(z) = \Pi_*(I - zU^*)^{-1}\hat{X}_1 \text{ and } \hat{Y}(z) = \Pi_*(I - zU^*)^{-1}\hat{Y}_1 \ , \tag{8.22}$$

then E(z) on L_* becomes

$$E(z) = (I + z\hat{X}(z) + z\hat{Y}(z) N_1^{1/2} F_1(z) N^{-1/2})^{-1} \quad (z \in D) \tag{8.23}$$

Therefore using (8.20), (8.21) along with $\alpha | L_* = [I, 0]^{tr}$ and Lemma 8.1 we have

$$\Pi' F(z)(I - z\Pi_A F(z))^{-1} D_A | L_* =$$

$$\tag{8.24}$$

$$[\Pi' A \hat{Y}_1 N_1^{1/2} F_1(z) N^{-1/2} + N_1^{-1/2} F_1(z) N^{-1/2} - \Pi' A(I - zU^*)^{-1}(\hat{X}_1 + \hat{Y}_1 N_1^{1/2} F_1(z) N^{-1/2})] E(z) \ .$$

However, using (8.16), (8.18) and (8.19) we obtain

$$\Pi' A \hat{Y}_1 N_1^{1/2} + N_1^{-1/2} = (\Pi' A \hat{Y}_1 + N_1^{-1}) N_1^{1/2} =$$

$$\Pi' (A D_{A_1}^{-2} A^* + I - A D_{A_1}^{-2} A^*) \Pi'^* N_1^{1/2} = N_1^{1/2} \ .$$

So (8.24) becomes

$$\Pi' F(z)(I - z\Pi_A F(z))^{-1} D_A \,|\, L_* =$$
$$[N_1^{\frac{1}{2}} F_1(z) N^{-\frac{1}{2}} - \Pi' A (I - zU^*)^{-1} (\hat{X}_1 + \hat{Y}_1 N_1^{\frac{1}{2}} F_1(z) N^{-\frac{1}{2}})] \, E(z) \quad . \tag{8.25}$$

Equation (8.25) provides us with a very general formula for computing the set of all contractive intertwining liftings for generalized Hankel matrices in Section VIII.7 and for solving the four block H^∞ optimization problem in Section IX.4. Notice that for a generalized Hankel matrix the knowledge of a contractive intertwining lifting $B \,|\, L_*$ readily produces a contractive intertwining lifting B of A (see (VIII.7.4)). So Theorem 3.4 along with formula (8.25) can be used to construct the set of all contractive intertwining liftings for generalized strictly contractive Hankel matrices. Moreover, this formula can be simplified further when A is a Hankel matrix which intertwines a backward and forward shift. To see this as in Section 3 let A be a strictly contractive Hankel operator of the form $A = P_- M_{Q_o} \,|\, H^2(E)$ where Q_o is in $K_o^\infty(E, E')$ and $U = S$ the unilateral shift on $H^2(E)$ and T' is the compression of V' to $H' = K_o^2(E')$. (Recall that P_- is the orthogonal projection onto $K_o^2(E')$ and V' is the bilateral shift on $L^2(E')$.) In this case L_* is identified with E and L'_* with E'. If h is in $H^2(E)$, then $\Pi_* h = h(0)$ and $\Pi' h'$ is the Fourier coefficient of e^{-it} in the Fourier series expansion of h' in $K_o^2(E)$. This sets the stage for the main result of this section due to Adamjan, Arov and Krein.

8.3 THEOREM. *Let* $A = P_- M_{Q_o} \,|\, H^2(E)$ *be a strictly contractive Hankel operator in* $I(S, T')$ *where* Q_o *is in* $K_o^\infty(E, E')$. *Then the set of all contractive intertwining liftings* B *of* A *is given by* $B = M_C \,|\, H^2(E)$ *where*

$$C = [Q_o + (AS\hat{X}_1)(e^{it}) + (I + (AS\hat{Y}_1)(e^{it})) N_1^{\frac{1}{2}} F_1(e^{it}) N^{-\frac{1}{2}}] E(e^{it}) \tag{8.26}$$

F_1 *is a contractive analytic function in* $H^\infty(E, E')$ *and* $E(z)$ *is the analytic function in* D *with values in* $L(E, E)$ *defined in (8.23).*

PROOF. Let

$$Q_o = \sum_1^\infty Q_n e^{-int} \quad \text{and} \quad h = \sum_o^\infty h_n e^{int}$$

be the power series expansion of Q_o and h in $H^2(E)$, respectively. A simple calculation shows that

$$P_+ e^{it} Q_o h = \sum_{n=0}^\infty z^{int} \sum_{m=1}^\infty Q_m h_{n+m-1}$$

where P_+ is the orthogonal projection onto $H^2(E')$. Another calculation shows that

$$\Pi' A (I - z S^*)^{-1} h = \sum_{n=0}^{\infty} z^n \sum_{m=1}^{\infty} Q_m h_{n+m-1} = P_+ e^{it} Q_o h \ . \tag{8.27}$$

Moreover, for an h in $H^2(E)$ we have

$$\Pi_* (I - z S^*)^{-1} h = \sum_{0}^{\infty} z^n \Pi_* S^{*n} h = \sum_{0}^{\infty} z^n h_n = h(z) \ . \tag{8.27a}$$

This and (8.22) readily implies that

$$P_- e^{it} Q_o \hat{X} = A S \hat{X}_1 \quad \text{and} \quad P_- e^{it} Q_o \hat{Y} = A S \hat{Y}_1 \ . \tag{8.28}$$

Now using (8.27) and (8.28) we have

$$Q_o (I + e^{it} \hat{X}(e^{it}) + e^{it} \hat{Y}(e^{it}) N_1^{1/2} F_1 N^{-1/2}) - \Pi' A (I - z S^*)^{-1} (\hat{X}_1 + \hat{Y}_1 N_1^{1/2} F_1 N^{-1/2})(e^{it}) =$$

$$\tag{8.29}$$

$$Q_o + P_- e^{it} Q_o \hat{X} + (P_- e^{it} Q_o \hat{Y}) N_1^{1/2} F_1 N^{-1/2} = Q_o + A S \hat{X}_1 + (A S \hat{Y}_1) N_1^{1/2} F_1 N^{-1/2} \ .$$

So substituting (8.25) into (3.15) and using (8.29) and $A | L_* = Q_o E^{-1} E$ we obtain that $B | L_*$ is given by the right-hand side of (8.26) where B is the contractive intertwining lifting in (3.15) of A. Since B is in $I(S, V')$ Theorem IX.1.1 shows that $B = M_C | H^2(E)$ and obviously $C = B | L_*$. Therefore by Theorem 3.4 equation (8.26) describes the set of all contractive intertwining liftings B of A. This completes the proof.

By (8.27a) the operators $\hat{X}(z) = \hat{X}_1(z)$ and $\hat{Y}(z) = \hat{Y}_1(z)$ for all z in D. This observation may simplify the calculation of $E(z)$ in (8.23). It is emphasized that the previous theorem shows that one can construct the set of all contractive intertwining liftings by only using the operators $\hat{X}_1$ and $\hat{Y}_1$ which also naturally occur in Section 5 for obtaining two inverse scattering algorithms. In fact the results in Section 5 show that one can recursively compute $\hat{X}_n$ and $\hat{Y}_n$ where $\hat{X}_n$ and $\hat{Y}_n$ correspond to contractive n step liftings Λ_n of A. Then applying the previous theorem to $A = \Lambda_n$ with $\hat{X}$ and $\hat{Y}$ determined from $\hat{X}_n$ and $\hat{Y}_n$ formula (8.26) provides the set of all contractive intertwining liftings B of Λ_n. (The results in Section 4 show that if $\{\Gamma_i\}_1^n$ is a strictly contractive choice sequence, then it corresponding intertwining lifting Λ_n is also strictly contractive. So we can apply Theorem 8.3 to Λ_n.)

Using Lemma 8.1 it is an easy exercise to verfiy that

$$Q_o + AS\hat{X}_1 = AD_A^{-2}\Pi_*^* N^{-1}$$

$$I + AS\hat{Y}_1(e^{it}) = e^{it}D_A^{-2} \cdot \Pi'^* N_1^{-1}$$

$$I + z\hat{X}(z) = D_A^{-2}\Pi_*^* N^{-1}$$

$$\hat{Y}(z) = A^* D_A^{-2} \cdot \Pi'^* N_1^{-1} \,.$$

(8.30)

Substituting this into (8.26) gives

$$C(e^{it}) =$$

$$[AD_A^{-2}\Pi_*^* N^{-\frac{1}{2}} + zD_A^{-2} \cdot \Pi'^* N_1^{-\frac{1}{2}} F_1][D_A^{-2}\Pi_*^* N^{-\frac{1}{2}} + zA^* D_A^{-2} \cdot \Pi'^* N_1^{-\frac{1}{2}} F_1]^{-1}(e^{it}) \,.$$

(8.31)

Equation (8.31) (which is precisely the Schur representation in [GoKW 1]) provides us with another formula for computing the set of all contractive intertwining lifting of A. (As before F_1 runs through the set of all functions in $H_1^\infty(E, E')$.) Formula (8.26) depends on the inverses of D_{AS}^2 and $D_{(AS)}^2$ (see also (8.18)) while formula (8.31) depends on the inverses of D_A^2 and $D_{A^*}^2$. Each formula has advantages and disadvantages. For example (8.31) depends on fewer terms. However, formula (8.26) may be easier to use in conjunction with the scattering algorithms in Section 5. The following provides a further simplification of the formulas (8.26) and (8.31) when E and E' equal C^1.

8.4 COROLLARY. *Let $E = E' = C^1$ and assume that $A = P_- M_{Q_o} | H^2$ is a strictly contractive Hankel operator where Q_o is in K_o^∞, then $N_1 = N$ and the set of all contractive intertwining liftings B of A is given by $B = M_C | H^2$ where*

$$C = [Q_o + AS\hat{X}_1 + (I + AS\hat{Y}_1)f_1]\,[I + z\hat{X} + z\hat{Y}f_1]^{-1}(e^{it})$$

$$= [AD_A^{-2}\Pi_*^* + zD_A^{-2} \cdot \Pi'^* f_1][D_A^{-2}\Pi_*^* + zA^* D_A^{-2} \cdot \Pi'^* f_1]^{-1}(e^{it})$$

(8.32)

and f_1 is a contractive analytic function in H^∞.

PROOF. Recall that J is the unitary operator on $L^2(E')$ defined by $(Jf)(e^{it}) = f(e^{-it})$. Notice that $\beta = V'^* J | K_o^2(E')$ is a unitary operator from $K_o^2(E')$ onto $H^2(E')$ satisfying $S'^* \beta = \beta T'$ where S' is the unilateral shift on $H^2(E')$. Therefore βA is a strictly contractive Hankel operator in $I(S, S'^*)$. By consulting Section IX.3 we see that βA admits a matrix representation M on l^2 of the form

$$M = \begin{bmatrix} a_0 & a_1 & a_2 & a_3 & \cdots \\ a_1 & a_2 & a_3 & a_4 & \cdots \\ a_2 & a_3 & a_4 & a_5 & \cdots \\ a_3 & a_4 & a_5 & a_6 & \cdots \\ \vdots & \vdots & \vdots & \vdots & \vdots \end{bmatrix}$$

Obviously $M^* = \overline{M}$ the complex conjugate of M. This and the fact that the entries of $I - M^*M$ are real readily imply that

$$N = \Pi_o(I - M^*M)^{-1}\Pi_o^* = \Pi_o(I - \overline{M}M)^{-1}\Pi_o^* = \Pi_o(I - M\overline{M})^{-1}\Pi_o^* = N_1$$

where $\Pi_o[f_o, f_1, f_2, \ldots]^{\mathrm{tr}} = f_o$. (In the first equality Π_o is identified with Π_* while in the last equality Π_o is identified with Π'.) Now the Corollary follows from the previous theorem.

9. COMPUTING ALL CONTRACTIVE INTERTWINING LIFTINGS FOR RATIONAL HANKEL OPERATORS.

In this section we will use the Adamjan, Arov and Krein parameterization in Theorem 8.3, along with our matrix representations of rational Hankel operators in Sections X.6 and X.7 to compute the set of all contractive intertwining liftings B of a strictly contractive rational Hankel operator. As a demonstration of this we will show how Theorem 8.3 reduces to the Schur representation (I.3.8) for solving the Carathéodory interpolation problem.

To begin let Q_o be a rational function in $K_o^\infty(E, E')$ and $A = P_- M_{Q_o}|H^2(E)$ be a strictly contractive Hankel operator in $I(S, T')$. As before let β be the unitary operator from $K_o^2(E')$ onto $H^2(E')$ defined by $\beta = V'^*J|K_o^2(E')$. Using $V'^*JP_- = P_+V'^*J$ we have $\beta A = \Gamma(e^{-it}Q_o(e^{-it}))$ where Γ is the Hankel operator in $I(S, S'^*)$ with symbol $G = e^{-it}Q_o(e^{-it})$ defined in Section IX.3. According to Proposition X.6.3 or (X.7.30) the Hankel operator $\Gamma = \Gamma(G)$ admits a factorization of the form $\Gamma = W_1 M W^*$ where W from C^n to $H^2(E)$ and W_1 from C^n to $H^2(E')$ are two (rational) isometric operators which we can explicately obtain and M is a matrix on C^n. Moreover, $W^*S = TW^*$ where T is also a matrix on C^n. So if W' is the isometry from C^n to $K_o^2(E')$ defined by $W' = \beta^*W_1 = JV'W_1$, then

$$A = W'MW^* \quad \text{and} \quad AS = W'MTW^*,$$

$$(9.1)$$

$$D_A^{-2} = WD_M^{-2}W^* + I - WW^* \quad \text{and} \quad D_{AS}^{-2} = WD_{MT}^{-2}W^* + I - WW^*.$$

Since W is an isometry, $I - WW^*$ is the orthogonal projection onto the kernel of W^*. It is emphasized that we can compute W, W', M and T by using the orthogonal basis method in Section X.6 or the state space techniques in Section X.7.

Notice that $\Pi_* W = W(0)$ is the row operator from $\mathbb{C}^n$ to E defined by evaluating $W(z)$ at $z = 0$. This and $S^* W = WT^*$ readily implies that

$$W(z)x = \Pi_*(I - zS^*)^{-1}Wx = W(0)(I - zT^*)^{-1}x \qquad (x \in \mathbb{C}^n) . \tag{9.2}$$

(One may prefer to use the second equality in (9.2) to compute $W(z)x$ when one uses a state space realization to obtain $A = W'MW^*$.)

Using (9.1) and (9.2) in (8.19), (8.22) and (8.30) we obtain

$$\hat{X}(z) = W(z)D_{MT}^{-2}T^*M^*MW(0)^*$$

$$(I + z\hat{X})N^{\frac{1}{2}} \stackrel{\Delta}{=} \Phi_{2,1} = [W(z)D_M^{-2}W(0)^* + (I - W(z)W(0)^*)]N^{-\frac{1}{2}} . \tag{9.3}$$

Now notice that

$$W_1(0) = \Pi'W' = \frac{1}{2\pi}\int_0^{2\pi} e^{+it}W'(e^{it})dt$$

where $\beta W' = W_1$, or equivalently, $W_1 = (1/z)W'(1/z)$ for z in D. Using this a calculation similar to (9.3) along with (8.30) yields

$$\hat{Y}(z) = W(z)D_{MT}^{-2}M^*W_1(0)^*$$

$$\hat{Y}(z)N_1^{\frac{1}{2}} \stackrel{\Delta}{=} \Phi_{2,2} = W(z)M^*D_M^{-2}\cdot W_1(0)^*N_1^{-\frac{1}{2}} . \tag{9.4}$$

Equation (9.1) and the definition of $\hat{X}_1$ in (8.19) give

$$AS\hat{X}_1 = W'MTD_{MT}^{-2}T^*M^*MW(0)^* . \tag{9.5}$$

In particular, by (8.16), (8.30) and $Q_o = A\Pi_*^* = W'MW(0)^*$ we have

$$Q_o + AS\hat{X}_1 = W'D_{(MT)}^{-2}\cdot MW(0)^*$$

$$(Q_o + AS\hat{X}_1)N^{\frac{1}{2}} \stackrel{\Delta}{=} \Phi_{1,1} = W'MD_M^{-2}W(0)^*N^{-\frac{1}{2}} . \tag{9.6}$$

Equation (8.30), (9.1) and the definition of $\hat{Y}_1$ in (8.19) also give

$$AS\hat{Y}_1 = W'MTD_{MT}^{-2}M^*W_1(0)^* = W'D_{(MT)}^{-2}\cdot MTM^*W_1(0)^*$$

$$e^{-it}(I + AS\hat{Y}_1)N_1^{\frac{1}{2}} \stackrel{\Delta}{=} \Phi_{1,2} = [W'D_M^{-2}\cdot W_1(0)^* + (e^{-it}I - W'W_1(0))]N_1^{-\frac{1}{2}} . \tag{9.7}$$

Finally, by using (9.1) and the definitions for N in (8.11) and N_1 in (8.12) we have

$$N = W(0)D_M^{-2}W(0)^* + I - W(0)W(0)^* = (I - W(0)M^*D_{(MT)}^{-2}\cdot MW(0)^*)^{-1} \,,$$

$$(9.8)$$

$$N_1 = W_1(0)D_{M^*}^{-2}W_1(0)^* + I - W_1(0)W_1(0)^* = (I - W_1(0)MD_{MT}^{-2}M^*W_1(0)^*)^{-1} \;.$$

The second and fourth equality follows from (8.18) and (9.1). Summing up the previous analysis produces the following computational procedure for obtaining all contractive intertwining liftings for a rational Hankel operator.

9.1 PROCEDURE. Let $A = P_- M_{Q_o}\,|H^2(E)$ be a strictly contractive Hankel operator in $I(S, T')$ whose symbol Q_o is a rational function in $K_o^\infty(E, E')$. Using the results in Section X.6, or the state space techniques in Section X.7, compute a factorization for $\Gamma(e^{-it}Q_o(e^{-it})) = \Gamma$ of the form $W_1 MW^*$ where W is an isometry from $\mathbb{C}^n$ to $H^2(E)$ and W_1 is an isometry from $\mathbb{C}^n$ to $H^2(E')$ and M is a matrix on $\mathbb{C}^n$. (Note one can even choose M to be a diagonal matrix.) Set $W' = V^*JW_1$. Also compute the contraction T on $\mathbb{C}^n$ satisfying $TW^* = W^*S$. By Theorem 8.3 the set of all contractive intertwining lifting B of A is given by $B = M_C\,|H^2(E)$ where

$$C = [Q_o + (A S\hat{X}_1)(e^{it}) + (I + (A S\hat{Y}_1)(e^{it}))N_1^{\frac{1}{2}}F_1 N^{-\frac{1}{2}}]\,[I + e^{it}\hat{X} + e^{it}\hat{Y}N_1^{\frac{1}{2}}F_1 N^{-\frac{1}{2}}]^{-1}(e^{it})$$

$$(9.9)$$

$$= [\Phi_{1,1} + z\Phi_{1,2}F_1][\Phi_{2,1} + z\Phi_{2,2}F_1]^{-1}\,(e^{it})$$

and F_1 is a contractive analytic function in $H^\infty(E, E')$. Here $\hat{X}$, $\hat{Y}$, N and N_1 are computed by (9.3), (9.4) and (9.8). The functions $(A S\hat{X}_1)(e^{it})$ can be computed by (9.5) or $Q_o + (A S\hat{X}_1)(e^{it})$ is also given by (9.6). The function $(A S\hat{Y}_1)(e^{it})$ is given by (9.7). Notice that one can use either the inverse of D_{MT}^2 or the inverse of $D_{(MT)}^2\cdot$ to compute $A S\hat{X}_1$ or $A S\hat{Y}_1$, or the appropriate N or N_1. The functions $\Phi_{i,j}$ are computed by (9.3), (9.4), (9.6) and (9.7). If $E = E' = \mathbb{C}^1$, then Corollary 8.4 shows that $N_1 = N$. In this case one does not have to compute N_1 and N and these terms vanish from (9.9). Since Q_o is rational W and W' are rational, all the terms in (9.9) are rational, except perhaps F_1. So if F_1 is rational, C can be expressed as a rational function in z. Finally, it is noted that the easiest C to compute is the one where $F_1 = 0$.

If one obtains the factorization $A = W'MW$ according to the methods in Section X.6, then one does not have to compute $(I - zT^*)^{-1}$. However, if one choses to use the state space techniques in Section X.7, then $W(0)(I - zT^*)^{-1}$ turns out to be the controllability operator corresponding to the state space realization. As noted earlier one can use Cayley-Hamilton techniques to compute $(I - zT^*)^{-1}$. It is also noted that the previous procedure presents two different methods to compute the set of all contractive intertwining liftings B of A. The first equation in (9.9) involves the inverses of D^2_{MT} and $D^2_{(MT)^*}$ while the second uses the inverses of D^2_M and $D^2_{M^*}$. Each method has an advantage. The second formula in (9.9) involving $\Phi_{i,j}$ contains fewer terms, and does not use T. (However, one must compute T if one choses to use state space techniques to find $W'MW^*$.) So the second formula is easier to use and more computationally efficient in many situations. However, if the norm of A is close to one it may be numerically more robust to use the first equation in (9.9), because $\|MT\| \leq \|M\|$ and so D^2_{MT} and $D^2_{(MT)^*}$ may be numerically easier to invert. Infact one can construct examples where the norm of A is arbitrarily close to one and MT is zero. The first equation in (9.9) is compatible with the scattering algorithms in Section 5. This may be useful in conjunction with Procedure 5.3. Finally, we leave it to the reader, as a simple exercise, to express the first equation in (9.9), in terms of a matrix representation of the Hankel operator $A_1 = AS$ of the form $A_1 = W'_1 M_1 W^*_1$.

To complete this section we will use the previous procedure to solve the following Carathéodory interpolation problem: Let Q_o be a specified polynomial of the form

$$Q_o = a_0 e^{-int} + a_1 e^{-i(n-1)t} + a_2 e^{-i(n-2)t} + \cdots + a_{n-1} e^{-it} \tag{9.10}$$

where the a'_is are complex numbers. Then find the set of all h in H^∞ such that $\|Q_o + h\|_\infty \leq 1$. Since $\|Q_o + h\|_\infty = \|e^{int}Q_o + e^{int}h\|_\infty$ this is equivalent to the Carathéodory interpolation Problem I.1.2. So without loss of generality we will assume that the n by n analytic Toeplitz matrix A_n in (I.4.4) with $j = n$ generated by the data $(a_0, a_1, \ldots, a_{n-1})$ is strictly contractive. Moreover, if f is obtained by the Schur representation (I.3.8) for solving the Carathéodory interpolation Problem I.1.2 (with f_n in H^∞_1), then $h = e^{-int}f - Q_o$ provides all h in H^∞ satisfying $\|Q_o + h\|_\infty \leq 1$.

To obtain the Schur representation (I.3.8) from Procedure 9.1, let A be the Hankel matrix from H^2 to K^2_o defined by $A = P_- M_{Q_o} | H^2$. By consulting the end of Section 3 or Section VIII.4 or IX.3 we see that

$$\inf\{\|Q_o + h\|_\infty : h \in H^\infty\} = \|A\| \ . \tag{9.11}$$

In particular, there exists an h in H^∞ satisfying $\|Q_o + h\|_\infty \leq 1$ if and only if A is a contraction. Moreover, in this case the set of all contractive intertwining liftings $B(= M_C)$ of A provide the set of all h in H^∞ satisfying $\|Q_o + h\|_\infty \leq 1$ (by $C = Q_o + h$). So without loss of generality we will assume that A is strictly contractive. Now let W from C^n to H^2 and W' from C^n to K^2_o be the isometries defined by the row vectors.

$$W = [1, z, z^2, \cdots, z^{n-1}] \quad \text{and} \quad W' = [e^{-it}, e^{-2it}, \cdots, e^{-int}] \ . \tag{9.12}$$

Using these isometries a simple exercise shows that

$$A = W' R_n A_n W^* \tag{9.13}$$

where R_j is the unitary matrix on C^j whose entries along the off diagonal are one and zero elsewhere. (The matrix R_j reverse the order of the components of C^j.) In this case the matrix M equals $R_n A_n$. Since A is strictly contractive A_n is also strictly contractive. In this setting the operator T satisfying $S^* W = W T^*$ is precisely the lower shift on C^n, that is, $T = T_n$ where T_n is defined in (I.4.1) for $j = n$. In particular, this implies that

$$MT = R_n A_n T = \begin{bmatrix} R_{n-1} A_{n-1} & 0 \\ 0 & 0 \end{bmatrix} \text{ on } C^{n-1} \oplus C^1 \tag{9.14}$$

where A_{n-1} is the $n-1$ by $n-1$ analytic Toeplitz matrix (in (I.4.4) for $j = n-1$) generated by $(a_o, a_1, \ldots, a_{n-2})$. Finally, as expected by (9.2)

$$W(z) = W(0)(I - zT^*)^{-1} = [1, z, z^2, \cdots, z^{n-1}] \ . \tag{9.15}$$

To compute $\hat{X}(z)$ in (9.3) let us first notice that $R_j \overline{A_j} R_j = A_j^*$ where $-$ denotes the complex conjugate. This readily implies that

$$R_j \overline{(I - A_j A_j^*)}^{-1} R_j = (I - A_j^* A_j)^{-1} \triangleq D_{A_j}^{-2} \ . \tag{9.16}$$

So using this identity in (I.5.9) for $j = n-1$ gives

$$[\overline{y}_{n,n-2}, \overline{y}_{n,n-3}, \cdots, \overline{y}_{n,0}]^{tr} = R_{n-1}[\overline{y}_{n,0}, \overline{y}_{n,1}, \cdots, \overline{y}_{n,n-2}]^{tr} = \tag{9.17}$$

$$D_{A_{n-1}}^{-2} R_{n-1} \overline{A}_{n-1} R_{n-1} R_{n-1} [a_{n-1}, a_{n-2}, \cdots, a_1]^{tr} = D_{A_{n-1}}^{-2} A_{n-1}^* R_{n-1} [a_{n-1}, a_{n-2}, \cdots, a_1]^{tr} \ .$$

By using this (9.14) and $M = R_n A_n$ in (9.3) gives

$$\hat{X}(z) = W(z) \begin{bmatrix} D_{A_{n-1}}^{-2} & 0 \\ 0 & 1 \end{bmatrix} \begin{bmatrix} A_{n-1}^* R_{n-1} & 0 \\ 0 & 0 \end{bmatrix} R_n A_n W(0)^* =$$

$$[1, z, z^2, \cdots, z^{n-1}] [\overline{y}_{n,n-2}, \overline{y}_{n,n-3}, \cdots, \overline{y}_{n,0}, 0]^{tr} \ .$$

Recalling from (I.3.14) that $y_{n,n-1} = 1$ we see that $1 + z\hat{X}(z) = y_n^{\#}(z)$ where $y_n^{\#}(z) = J_n y_n(z)$ is the reverse polynomial of $y_n(z)$ defined in Section I.3.

To compute $\hat{Y}(z)$ notice that inserting (9.16) in (I.5.1) implies that

$$[\overline{x}_{n,n-1}, \overline{x}_{n,n-2}, \cdots, \overline{x}_{n,1}] = R_{n-1}[\overline{x}_{n,1}, \overline{x}_{n,2}, \cdots, \overline{x}_{n,n-1}] =$$

$$D_{A_{n-1}}^{-2} R_{n-1} [\overline{a}_1, \overline{a}_2, \cdots, \overline{a}_{n-1}] = D_{A_{n-1}}^{-2} [\overline{a}_{n-1}, \overline{a}_{n-2}, \cdots, \overline{a}_1] \ .$$

This equality along with (9.4), (9.14) and $W_1(0)^* = [1, 0, 0, \cdots, 0]^{tr}$ gives

$$\hat{Y}(z) = W(z) \begin{bmatrix} D_{A_{n-1}}^{-2} & 0 \\ 0 & 1 \end{bmatrix} A_n^* R_n W_1(0)^* =$$

$$(9.18)$$

$$[1, z, z^2, \cdots, z^{n-1}] \, [\bar{x}_{n,n-1}, \bar{x}_{n,n-2}, \cdots \bar{x}_{n,1}, \bar{a}_o]^{tr} = x_n^\#(z)$$

where $x_n^\#(z) = J_n x_n(z)$ is the reverse polynomial for $x_n(z)$ defined in Section I.3. (The last equality follows from the fact that $x_{n,o} = r_o = a_o$; see (I.3.14).) Corollary 8.4 shows that $N_1 = N$. Therefore

$$1 + z\hat{X}(z) + z\hat{Y}(z)N_1^{\frac{1}{2}} f_n N^{-\frac{1}{2}} = y_n^\#(z) + zx_n^\#(z)f_n \tag{9.19}$$

where $f_n = F_1$ is an arbitrary function in H_1^∞. This precisely the denominator in the Schur representation (I.3.8).

By using (I.5.1) for $j = n-1$ and $x_{n,o} = a_o$ in (9.6) we have

$$Q_o + AS\hat{X}_1 = W' \begin{bmatrix} R_{n-1} D_{A_{n-1}}^{-2} R_{n-1} & 0 \\ 0 & 1 \end{bmatrix} R_n A_n W(0)^* =$$

$$(9.20)$$

$$[e^{-i(n-1)t}, e^{-i(n-2)t}, \cdots, e^{-it}, e^{-int}] \, [x_{n,1}, x_{n,2}, \cdots, x_{n,n-1}, x_{n,0}]^{tr} = e^{-int}x_n(e^{it})$$

where $x_n(z)$ is the Schur polynomial in Section I.3. Finally, using (I.5.9) for $j = n-1$ in (9.7) and $y_{n,n-1} = 1$ we have

$$1 + AS\hat{Y}_1 = 1 + W' \begin{bmatrix} R_{n-1} D_{A_{n-1}}^{-2} R_{n-1} & 0 \\ 0 & 1 \end{bmatrix} \begin{bmatrix} R_{n-1} A_{n-1} & 0 \\ 0 & 0 \end{bmatrix} A_n^* R_n W_1(0)^* =$$

$$(9.21)$$

$$1 + W'(R_{n-1} \oplus 0) \, [y_{n,0}, y_{n,1}, \cdots, y_{n,n-2}, 0]^{tr} = e^{-int}e^{it}y_n(e^{it}) \ .$$

Summing up the previous analysis by inserting (9.19), (9.20) and (9.21) into (9.9) we readily obtain the following result which is just another way of stating part of Theorem I.3.2.

9.2 THEOREM. *Let* $A = P_- M_{Q_o} | H^2$ *be a strictly contractive Hankel operator whose symbol* Q_o *is a polynomial of the form (9.10). Then the set of all contractive interwining liftings* B *of* A *is given by* $B = M_C$ *where*

$$C = e^{-int} \frac{x_n + e^{it}y_n f_n}{y_n^\# + e^{it}x_n^\# f_n} \, (e^{it}) \tag{9.22}$$

and f_n *is an arbitrary analytic function in* H_1^∞.

The above analysis shows that Procedure 9.1 reduces to the Schur representation (I.3.8) for Carathéodory interpolation. (It is an interesting exercise to use (I.6.8) and (I.6.11) in the second equation in (9.9) to compute (9.22).) Obviously it is much easier to obtain Theorem 9.2 by using the Techniques in Chapter I. However, it is an amazing fact that one also obtains the Schur representation (I.3.8) as a special case of Procedure 9.1, which came from a very general theory of commutant lifting. Finally, it is also noted that Section I.3 provides a recursive algorithm for computing x_n and y_n. As expected these algorithms are special cases of the recursive algorithms used in Sections 5 and 6 to update $\hat{X}_n$, $\hat{Y}_n$, $\hat{X}_{n*}$ and $\hat{Y}_{n*}$.

XIII.10. NOTES AND COMMENTS

The characterization of all n by m block contractive matrices in Section 1 is due to Davis [Da 2]. In connection with this see also [Da 3] and [JoR]. The characterization of all contractive intertwining liftings given in Theorem 2.7 is from [ArsCF 2]. This characterization is based on the concept of choice sequences, which is the natural generalization of the sequence of Schur numbers arising in the Carathéodory interpolation problem. The proof is taken from [Frz 8] which is essentially a combination of the results in [CeF 3] and [ArsCF 2]. An original treatment of a theory of choice sequences for a large class of Hankel operators was done by Dym and Gohberg [DyG 3]. This theory and [GoKW 2] provides an elegant generalization of the maximal entropy approached sketched in Section II.5 (see also [Con 4]). The Schur representation in Theorem 3.4, Corollary 3.5 and Procedure 2.9 is also taken from [ArsCF 2]. However, the Schur representation, which is a more intimate generalization of the Schur representation in Chapter I, is given in Theorem 3.6. (The results in Section 9 show that the Schur representation for the commutant lifting theorem reduces to the classical Schur representation in Chapter I.) In Section 8 we also used the Schur representation for the commutant lifting theorem, to obtain the famous result (Theorem 8.3) of Adamjan, Avov and Krein, which parameterizes all solutions of the block Nehari H^∞ optimization problem [AAK 4]. Formula (8.31) is due to Gohberg, Kaashoek and Woerdeman [GoKW 1]. For a nice derivation of this formula based on the band extension method see [GoKW 1]. (Here we have tried to express (8.26) in terms of $\hat{X}_1$ and $\hat{Y}_1$ because these are the initial conditions to Procedure 5.3. So formula (8.26) can be used recursively along with the inverse scattering algorithms in Procedure 5.3.) Formula (8.25) is a generalization of their result, which may be useful in solving the four block problem discussed in Section IX.4. The direct and inverse scattering algorithms in Sections 5 and 6 are improved versions of the algorithms in [Frz 8]. All of the inverse scattering algorithms can obviously be modified to direct algorithms. (With quite a bit of work one can also use these algorithms to obtain Theorem 8.3 and Formula (8.25).) The geophysical algorithms in Chapter III motivated us in putting the emphasis on inverse scattering algorithms. Finally, it is noted that one can also use the coupling set up in Section VII.8, to obtain inverse scattering algorithms and a Schur representation for the commutant lifting theorem (see [Ar 5], [Con 4] and [Morán] for further details).

CHAPTER XIV

THE SCHUR REPRESENTATION

In this chapter we will use Redheffer products to obtain another Schur type representation for the commutant lifting theorem. We begin by presenting some elementary properties of Redheffer products. Then we will use Redheffer products to obtain the characterization of all 2 by 2 matrix contractions in Section IV.3, solve an operator version of the classical Carathéodory interpolation problem, and obtain a layer peeling algorithm to compute the corresponding choice sequence. Then we will use Redheffer products to obtain the Schur representation for the commutant lifting theorem in Corollary XIII.3.5. The advantage of this approach is that it bypasses the use of choice sequences. From this we will easily obtain another Schur representation for the commutant lifting theorem. This will readily lead to a computational procedure for solving the operator version of the classical Hermite-Fejér interpolation problem and other interpolation problems involving $*$-stable contractions. A network interpretation of these results will also be given. We will also give another computational procedure for computing the set of all contractive intertwining liftings for a finite rank Hankel operator.

1. REDHEFFER PRODUCTS

In this section, we will present a short introduction under a form suitable for our purposes to the theory of Redheffer cascading systems.

First we introduce the block diagram representation used in systems theory for a linear operator. If D_1 is a linear map from some linear space X' to another linear space X, then the block diagram representation of $x - D_1 x'$ is given by Figure 1.

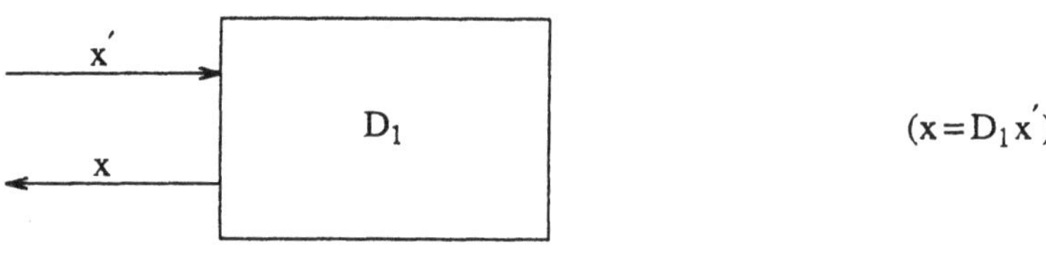

$$(x = D_1 x')$$

Figure 1

Also if L is a block matrix of the form

$$L = \begin{bmatrix} A & B \\ C & D \end{bmatrix} : X \oplus U \to X' \oplus Y$$

where X, X', U and Y are linear spaces, then its block diagram representation is given by Figure 2. (If X and U are linear spaces, then $X \oplus U$ is the direct sum of X and U, that is, $x \oplus u$ in $X \oplus U$ can be identified with the column vector $[x, u]^{tr}$. However, if X and U are both Hilbert spaces, then $X \oplus U$ always means the orthogonal direct sum of X and U.)

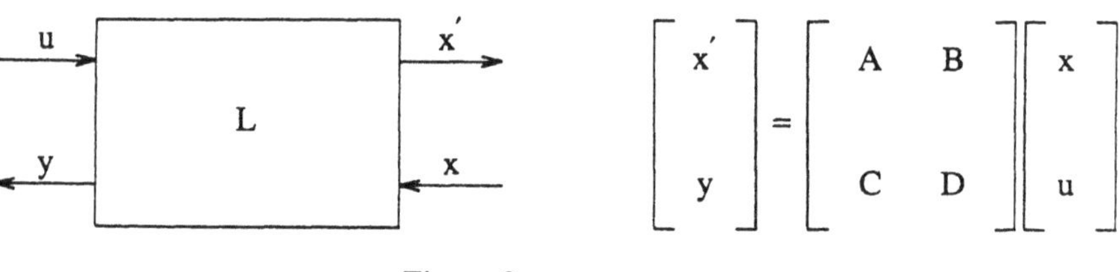

Figure 2

The operator L is referred to as the *scattering matrix* of the block diagram in Figure 2. It transfers the inputs u and x into the outputs x' and y. For example, if

$$L = \begin{bmatrix} r_j & 1 - r_j \\ 1 + r_j & -r_j \end{bmatrix} \text{ on } \mathbb{C}^2$$

is the scattering matrix formed by the j-th interface with reflection coefficient r_j in Section III.6, then L can also be viewed as the block diagram in Figure 2. In other words the scattering representation in Figure 2 can be identified with the scattering at the j-th interface, where the input and output immediately above (below) the j-th interface is given by x and x' (u and y respectively).

Now consider the linear maps L from $X \oplus U$ to $X' \oplus Y$ and L_1 from $X_1 \oplus U_1$ to $X_1' \oplus Y_1$ defined by

$$L = \begin{bmatrix} A & B \\ C & D \end{bmatrix} \text{ and } L_1 = \begin{bmatrix} A_1 & B_1 \\ C_1 & D_1 \end{bmatrix} \tag{1.1}$$

where all the spaces are linear spaces $X' \subseteq U_1$ and $Y_1 \subseteq X$. Analogous to the stacking of layered mediums consider the cascading of the scattering matrices L and L_1 given in the block diagram in Figure 3.

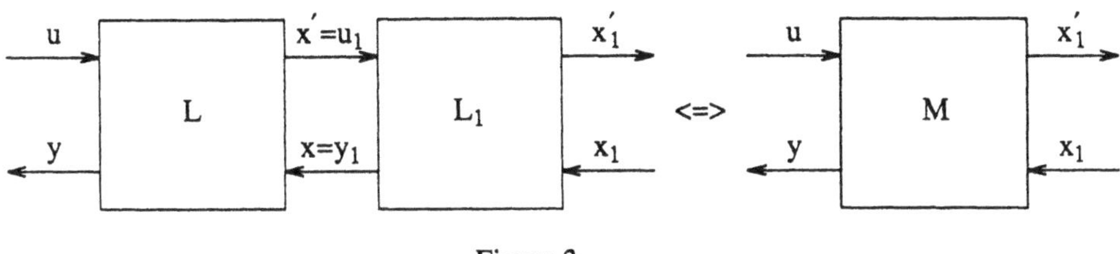

Figure 3

The equations for the block diagram corresponding to Figure 3 are

$$\begin{bmatrix} x' \\ y \end{bmatrix} = \begin{bmatrix} A & B \\ C & D \end{bmatrix} \begin{bmatrix} x \\ u \end{bmatrix} \text{ and } \begin{bmatrix} x_1' \\ y_1 \end{bmatrix} = \begin{bmatrix} A_1 & B_1 \\ C_1 & D_1 \end{bmatrix} \begin{bmatrix} x_1 \\ u_1 \end{bmatrix}$$
$$\tag{1.2}$$
$$\text{subject to } x' = u_1 \text{ and } x = y_1$$

We say that a linear map M from $X_1 \oplus U$ into $X_1' \oplus Y$ satisfies the *Redheffer cascading equation in the subspace* $\tilde{M} \subseteq (X_1 \oplus U)$ if for every $x_1 \oplus u$ in $\tilde{M}$ there exists a solution x, x', y, u_1, x_1', y_1 to (1.2) satisfying

$$\begin{bmatrix} x_1' \\ y \end{bmatrix} = M \begin{bmatrix} x_1 \\ u \end{bmatrix} = [L \circ L_1] \begin{bmatrix} x_1 \\ u \end{bmatrix}. \tag{1.3}$$

If M is uniquely determined by this condition and $\tilde{M} = X_1 \oplus U$, then M is called the *Redheffer product* $L \circ L_1$ *of the matrices* L and L_1. If all the spaces are Hilbert spaces and L and L_1 are operators, we only consider continuous maps M. (By definition an operator is a bounded linear map between the appropriate Hilbert spaces.) In this setting we say that M is *the Redheffer product* of L and L_1 if M is the only operator satisfying the Redheffer cascading equation on any dense set $\tilde{M}$ in $X_1 \oplus U$. This M can be extended by continuity to all of $X_1 \oplus U$. The resulting operator is also referred to as the Redheffer product of L and L_1. Finally, we notice that if $I - D_1 A$ is an invertible map or an invertible operator in the Hilbert space setting, then a simple calculation involving (1.2) shows that the Redheffer product $M = L \circ L_1$ exists and is given by

$$M = \begin{bmatrix} A_1 + B_1 A(I - D_1 A)^{-1} C_1 & B_1 A(I - D_1 A)^{-1} D_1 B + B_1 B \\ C(I - D_1 A)^{-1} C_1 & C(I - D_1 A)^{-1} D_1 B + D \end{bmatrix} . \tag{1.4}$$

As an application of (1.4) consider the operator J defined by

$$J = \begin{bmatrix} 0 & I \\ I & 0 \end{bmatrix} \tag{1.5}$$

where I is the identity operator on the appropriate spaces. The operator J acts like the identity operator for the Redheffer product, that is, a simple calculation involving (1.2) shows that $L = L \circ J = J \circ L$.

To demonstrate the possible nonuniqueness of the Redheffer cascading system in (1.2) let all the spaces be C^2 and set

$$A = C = D_1 = \begin{bmatrix} 0 & 0 \\ 0 & 1 \end{bmatrix} , \quad C_1 = \begin{bmatrix} 1 & 0 \\ 0 & 0 \end{bmatrix}$$

with B, D, A_1, and B_1 all zero. Then a simple calculation shows that the set of all M on C^4 satisfying the Redheffer cascading system (1.2) is given by

$$M = \begin{bmatrix} 0 & 0 & 0 & 0 \\ 0 & 0 & 0 & 0 \\ 0 & 0 & 0 & 0 \\ a & b & c & d \end{bmatrix}$$

where a, b, c and d are arbitrary complex numbers. Obviously M is not unique.

We have defined the Redheffer product on linear spaces rather than on Hilbert spaces, because in applications one can have unbounded inputs (that is, inputs not in a Hilbert space) to a network. However, when restricting these inputs to a Hilbert space there are networks which produce a contraction mapping these inputs into the outputs. Moreover, this is precisely the set up we need for developing our cascading interpretation of the commutant lifting theorem.

To obtain a more explicit form for M in the Hilbert space setting we introduce the psuedo inverse $T^{(-1)}$ of an operator. By definition if T is an operator from a Hilbert space X to a Hilbert space Y, then $T^{(-1)} y \triangleq x$ if $Tx = y$ and x is orthogonal to the kernel of

T. Obviously $T^{(-1)} = T^{-1}$ if T is invertible. Now assume that L and L_1 are operators acting between the appropriate Hilbert spaces. Using $x' = u_1$ and substituting $x' = Ax + Bu$ into the y_1 equation in (1.2) we have

$$x = y_1 = C_1 x_1 + D_1 x' = C_1 x_1 + D_1 Ax + D_1 Bu .$$

This readily implies that

$$(I - D_1 A)x = C_1 x_1 + D_1 Bu . \tag{1.6}$$

So if we define the space $\tilde{M}$ by

$$\tilde{M} = \{x_1 \oplus u \in X_1 \oplus U : C_1 x_1 + D_1 Bu \in (I - D_1 A)X\} , \tag{1.7}$$

then for $x_1 \oplus u$ in $\tilde{M}$ we have

$$x = (I - D_1 A)^{(-1)}(C_1 x_1 + D_1 Bu) + q$$

where q is an element in the kernel of $I - D_1 A$. Substituting this x into the equation for x_1' gives

$$\begin{aligned} x_1' &= A_1 x_1 + B_1 u_1 = A_1 x_1 + B_1 x' = A_1 x_1 + B_1(Ax + Bu) = \\ & A_1 x_1 + B_1 Bu + B_1 A(I - D_1 A)^{(-1)}(C_1 x_1 + D_1 Bu) + B_1 Aq . \end{aligned} \tag{1.8}$$

Another calculation involving (1.2) and the above equation for x yields

$$y = Cx + Du = C(I - D_1 A)^{(-1)}(C_1 x_1 + D_1 Bu) + Du + Cq . \tag{1.9}$$

Notice that (1.8) and (1.9) readily produces (1.4) when $I - D_1 A$ is invertible. Moreover, these equations are used in the following useful result.

1.1 LEMMA. *Let L and L_1 be linear operators of the form (1.1) acting between the appropriate Hilbert spaces. Assume that*

(i) $B_1 Aq$ *and* $Cq = 0$ *for all q in the kernel of* $I - D_1 A$;

(ii) *the space $\tilde{M}$ defined in (1.7) is dense in $X_1 \oplus U$;*

(iii) *there exists operators A_2, B_2, C_2, D_2 such that for all $x_1 \oplus u$ in $\tilde{M}$ and x_1' and y given by (1.8), (1.9) we have*

$$x_1' = A_2 x_1 + B_2 u \; and \; y = C_2 x_1 + D_2 u . \tag{1.10}$$

Then the Redheffer product $M = L \circ L_1$ exists and is given by

$$M = \begin{bmatrix} A_2 & B_2 \\ C_2 & D_2 \end{bmatrix} . \tag{1.11}$$

PROOF. By virtue of (i) the solution of the Redheffer cascading equations (1.2) is necessarily given by (1.8) and (1.9) with $q = 0$ for $x_1 \oplus u$ in $\tilde{M}$. Since the values of M are uniquely given by (1.8), (1.9) with $q = 0$ and $\tilde{M}$ is dense, M is the Redheffer product of $L \circ L_1$. This completes the proof.

Now consider the spaces $\tilde{X}_1$ and $\tilde{U}$ defined by

$$\tilde{X}_1 = \{x_1 \in X_1 : C_1 x_1 \in (I - D_1 A)X\} \text{ and } \tilde{U} = \{u \in U : D_1 B u \in (I - D_1 A)X\} \ .$$

So for all x_1 in $\tilde{X}_1$ and u in $\tilde{U}$ equations (1.8) and (1.9) become

$$M \begin{bmatrix} x_1 \\ u \end{bmatrix} = \begin{bmatrix} A_1 + B_1 A(I - D_1 A)^{(-1)}C_1 & B_1 A(I - D_1 A)^{(-1)}D_1 B + B_1 B \\ C(I - D_1 A)^{(-1)}C_1 & C(I - D_1 A)^{(-1)}D_1 B + D \end{bmatrix} \begin{bmatrix} x_1 \\ u \end{bmatrix} + \begin{bmatrix} B_1 Aq \\ Cq \end{bmatrix} \tag{1.12}$$

where q is in the kernel of $I - D_1 A$. In particular, if $B_1 Aq = 0$ and $Cq = 0$ for all q in the kernel of $I - D_1 A$ and $\tilde{X}_1 \oplus \tilde{U}$ is a dense set in $X_1 \oplus U$ and all the entries in the 2 by 2 matrix in (1.12) are operators defined on the appropriate $\tilde{X}_1$ and $\tilde{U}$ spaces, then the Redheffer product is given by the closure of the operator M determined by (1.12). It is noted that if $I - D_1 A$ is invertible, then $\tilde{X}_1 = X_1$, $\tilde{U} = U$ and the Redheffer product exists. Moreover, in this case (1.12) is precisely the Redheffer product M in (1.4).

To demonstrate how (1.12) can be used in practice let A be a contraction from X to X'. Let R_A from $X \oplus D_{A^*}$ to $X' \oplus D_A$ and R'_A from $D_A \oplus X'$ to $D_{A^*} \oplus X$ be the operators defined by

$$R_A = \begin{bmatrix} A & D_{A^*} \\ D_A & -A^* \end{bmatrix} \text{ and } R'_A = \begin{bmatrix} -A & D_{A^*} \\ D_A & A^* \end{bmatrix} = JR_{A^*}J . \tag{1.13}$$

Obviously R_A is the rotation matrix associated with A and R'_A is unitarily equivalent to the rotation matrix R_{A^*} associated with A^*. In particular, both R_A and R'_A are unitary. We claim that R'_A is the Redheffer inverse of R_A, that is,

$$J = R_A \circ R'_A \text{ and } J = R'_A \circ R_A . \tag{1.14}$$

To establish (1.14) we will use (1.12). To this end notice that if we set $L = R_A$ and $L_1 = R'_A$, then $U = D_{A^*}$, $Y = D_A$, $X_1 = D_A$ and $X'_1 = D_{A^*}$. In this case $I - D_1 A = I - A^* A$ and the kernel of $I - D_1 A$ equals $X \ominus D_A$. So using $C = D_A$ and $B_1 A = D_{A^*} A = A D_A$ we see that $Cq = 0$ and $B_1 Aq = 0$ for all q in the kernel of $I - D_1 A$.

Thus the last column vector in (1.12) vanishes. It is easy to check that $D_A X \subseteq \tilde{X}_1$, $D_{A^{\boldsymbol{\cdot}}} X^{'} \subseteq \tilde{U}$ and that $D_A(I - A^* A)^{(-1)} D_A x = x$ for all x in $D_A X$. Therefore

$$B_1 A(I - D_1 A)^{(-1)} C_1 \,|\, D_A X = A D_A (I - A^* A)^{(-1)} D_A \,|\, D_A X = A \,|\, D_A X$$

$$C(I - D_1 A)^{(-1)} C_1 \,|\, D_A X = D_A(I - A^* A)^{(-1)} D_A \,|\, D_A X = I \,|\, D_A X$$

$$B_1 A(I - D_1 A)^{(-1)} D_1 B \,|\, D_{A^{\boldsymbol{\cdot}}} X^{'} = A D_A(I - A^* A)^{(-1)} D_A A^* \,|\, D_{A^{\boldsymbol{\cdot}}} X^{'} = A A^* \,|\, D_{A^{\boldsymbol{\cdot}}} X^{'}$$

$$C(I - D_1 A)^{(-1)} D_1 B \,|\, D_{A^{\boldsymbol{\cdot}}} X^{'} = D_A(I - A^* A)^{(-1)} D_A A^* \,|\, D_{A^{\boldsymbol{\cdot}}} X^{'} = A^* \,|\, D_{A^{\boldsymbol{\cdot}}} X^{'} \,.$$

So M in (1.12) becomes $M = J$ on $D_A X \oplus D_{A^{\boldsymbol{\cdot}}} X^{'}$. Obviously $D_A X \oplus D_{A^{\boldsymbol{\cdot}}} X^{'}$ is dense in $D_A \oplus D_{A^{\boldsymbol{\cdot}}}$. Therefore the Redheffer product $M = R_A \circ R_A^{'}$ is uniquely determined and equal to J. A similar calculation shows that $R_A^{'} \circ R$ is also uniquely determined and equal to J.

Consider the contractions L and L_1 defined by

$$L = \begin{bmatrix} A & D_{A^{\boldsymbol{\cdot}}} X \\ 0 & 0 \end{bmatrix} \text{ and } L_1 = \begin{bmatrix} 0 & D_{A^{\boldsymbol{\cdot}}} \\ 0 & A^* \end{bmatrix}$$

where X is a rank one contraction from U to $D_{A^{\boldsymbol{\cdot}}}$ and A is a contraction such that $A^* X$ is not in the range of D_A. (Corollary IV.1.3 and Lemma IV.1.1 guarantees that both L and L_1 are contractions.) By consulting (1.6) which comes from the Redheffer cascading equation (1.2) we have $(I - A^* A)x = A^* D_{A^{\boldsymbol{\cdot}}} X u$. This and $A^* D_{A^{\boldsymbol{\cdot}}} = D_A A^*$ readily implies that $D_A x = A^* X u$. However, the only solution to this equation is $u = 0$ and x orthogonal to D_A. So in this case the Redheffer product does not exist. In other words the Redheffer product may not exist even if L and L_1 are both contractions.

The *energy* of a signal x is $\|x\|^2$. If L is an isometry, then the energy of the output $\|x^{'} \oplus y\|^2$ of L equals the energy of the input $\|x \oplus u\|^2$ of L. If L is a contraction, then the energy of the output is less than or equal to the energy of the input, that is, $\|x^{'} \oplus y\|^2 \leq \|x \oplus u\|^2$. If both L and L_1 are contractions, then it makes sense that the energy of the output $\|x_1^{'} \oplus y\|^2$ of the Redheffer product $M = L \circ L_1$ is less than the energy of the input $\|x_1 \oplus u\|^2$ of M, that is, M is a contraction. This suggests that the following result is true.

1.2 LEMMA. *Let L and L_1 be two contractions of the form (1.1). If the space $\tilde{M}$ defined in (1.7) is dense in $X_1 \oplus U$, then the Redheffer product $M = L \circ L_1$ exists and is a contraction. Moreover, in this case if $x_1 \oplus u$ is in $\tilde{M}$, then*

$$\|D_M(x_1 \oplus u)\|^2 = \|D_{L_1}(x_1 \oplus u_1)\|^2 + \|D_L(x \oplus u)\|^2 \qquad (1.15)$$

where x *and* u_1 *are obtained from*

$$x = (I - D_1 A)^{(-1)}(C_1 x_1 + D_1 Bu)$$
$$u_1 = x' = A(I - D_1 A)^{(-1)}(C_1 x_1 + D_1 Bu) + Bu \tag{1.16}$$

In particular, if $I - D_1 A$ *is invertible, then the Redheffer product exists is a contraction and is given by (1.4).*

PROOF. If q is in the kernel of $I - D_1 A$, then

$$\|q\| = \|D_1 Aq\| \le \|Aq\| \le \|q\| \ .$$

So there is equality. Thus $D_A q = 0$ and $D_{D_1} Aq = 0$. Since $[A^*, C^*]^*$ is a contraction Lemma IV.1.1. shows that $C = Y D_A$ where Y is a contraction. This implies that $Cq = Y D_A q = 0$. On the other hand because $[B_1^*, D_1^*]^*$ is a contraction $B_1 = Y_1 D_{D_1}$ for some contraction Y_1. Hence $B_1 Aq = Y_1 D_{D_1} Aq = 0$. Therefore part (i) of Lemma 1.1 is satisfied.

By inverting $I - D_1 A$ in (1.6) and substituting this into the Redheffer cascading equation (1.2) we obtain (1.16). In other words x, u_1, x_1' and y in (1.8), (1.9) (with $B_1 Aq = 0$ and $Cq = 0$) and (1.16) satisfy (1.2) for $x_1 \oplus u$ in $\tilde{M}$. By using the M defined in (1.3) for $x_1 \oplus u$ in $\tilde{M}$ we have

$$\|x_1 \oplus u\|^2 - \|M(x_1 \oplus u)\|^2 = \|x_1\|^2 + \|u\|^2 - (\|x_1'\|^2 + \|y\|^2) =$$
$$\|x_1\|^2 + \|u\|^2 + \|u_1\|^2 + \|x\|^2 - (\|x_1'\|^2 + \|y\|^2 + \|x'\|^2 + \|y_1\|^2) =$$
$$\|x_1 \oplus u_1\|^2 - \|x_1' \oplus y_1\|^2 + \|x \oplus u\|^2 - \|x' \oplus y\|^2 =$$
$$\|D_{L_1}(x_1 \oplus u_1)\|^2 + \|D_L(x \oplus u)\|^2 \ge 0 \ .$$

Therefore M is a contraction on the dense set $\tilde{M}$. Let M also denote the unique contractive extension of M to all of $X_1 \oplus U$. Equation (1.15) follows from the previous equation and the fact that $\|D_M f\|^2 = \|f\|^2 - \|Mf\|$ for all f in $X_1 \oplus U$. By construction this contraction M is precisely $x_1' \oplus y = M(x_1 \oplus u)$ where x_1', y is obtained from (1.8) and (1.9) for $x_1 \oplus u$ in $\tilde{M}$. So the third hypothesis in Lemma 1.1 holds. Now we can apply Lemma 1.1 to conclude that the Redheffer product $M = L \circ L_1$ exists and is a contraction. This completes the proof.

1.3 COROLLARY. *Let L and L_1 be two finite dimensional contractions of the form (1.1). Then the Redheffer product $M = L \circ L_1$ exists and is a contraction. Moreover, if $I - D_1 A$ is invertible, then the Redheffer product $M = L \circ L_1$ is given by (1.4); otherwise $L \circ L_1$ is given by*

$$
L \circ L_1 = \begin{bmatrix} A_1 + B_1 A(I - D_1 A)^{(-1)} C_1 & B_1 A(I - D_1 A)^{(-1)} D_1 B + B_1 B \\ C(I - D_1 A)^{(-1)} C_1 & C(I - D_1 A)^{(-1)} D_1 B + D \end{bmatrix} . \tag{1.17}
$$

PROOF. To begin notice that if T is a contraction on X and q is a vector in X satisfying $Tq = q$, then $T^* q = q$. This follows from

$$
\|q - T^* q\|^2 = \|q\|^2 - 2\,\mathrm{Re}(q, T^* q) + \|T^* q\|^2 \le 2\|q\|^2 - 2\,\mathrm{Re}(Tq, q) = 0 ,
$$

which readily implies that $T^* q = q$. By setting $T = D_1 A$ we see that the kernel of $I - D_1 A$ equals the kernel of $I - A^* D_1^*$. In particular, for q in the kernel of $I - D_1 A$ we have

$$
\|q\| = \|A^* D_1^* q\| \le \|D_1^* q\| \le \|q\|
$$

so there is equality $D_{D_1^*} q = 0$ and $D_{A} \cdot D_1^* q = 0$. In particular, the kernel of $I - A^* D_1^*$ is contained in the kernel of $D_{D_1^*}$. Thus $D_{D_1^*}$ is contained in $X \ominus (\ker(I - A^* D_1^*))$, or equivalently, bypassing to the adjoints $D_{D_1^*} \subseteq (I - D_1 A)X$. Now since $[C_1, D_1]$ is a contraction, Corollary IV.1.3 adjusted to this setting, shows that $C_1 = D_{D_1^*} Z$ where Z is a contraction. In particular, because X is finite dimensional

$$
C_1 X_1 \subseteq (I - D_1 A)X , \tag{1.18}
$$

also we see that the kernel of $I - A^* D_1^*$ is contained in the kernel of $D_A \cdot D_1^*$. By taking the orthogonal complement and passing to the adjoints, this implies that $D_1 D_A \cdot X' \subseteq (I - D_1 A)X$. Because $[A, B]$ is a contraction $B = D_A \cdot X$ (see Corollary IV.1.3). Thus

$$
D_1 B U \subseteq (I - D_1 A)X . \tag{1.19}
$$

By virtue of (1.18) and (1.19) the space $\tilde{M} = X_1 \oplus U$. The previous Lemma readily implies that the Redheffer product exists and is a contraction. Equation (1.17) follows from either (1.8), (1.9) or (1.12). This completes the proof.

The system L in Figure 2 or $M = L \circ L_1$ in Figure 3 is called a *two port system* because it has two inputs and two outputs. A *one port system* is for instance that described by Figure 1 it has only one input and one output. By cascading the two port system L in Figure 2 with the one port system D_1 in Figure 1, we obtain the one port system in Figure 4.

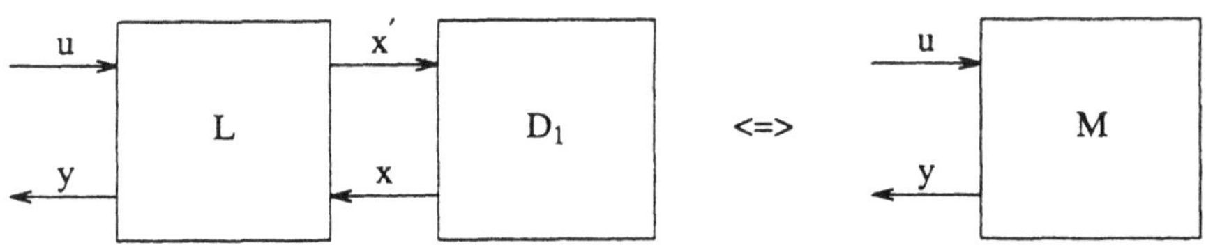

Figure 4

The equations for the block diagram corresponding to Figure 4 are

$$\begin{bmatrix} x' \\ y \end{bmatrix} = \begin{bmatrix} A & B \\ C & D \end{bmatrix} \begin{bmatrix} x \\ u \end{bmatrix} \tag{1.20}$$

subject to $x = D_1 x'$.

In systems theory M is known as the *closed loop system for the scattering map* L *with respect to the feedback* D_1. The operator D_1 is called the *feedback* because it operators on x' from L and feedsback the vector $x = D_1 x'$ to L. Here we say that an operator M is the *closed loop system defined by (1.20)* if there exists a unique solution u and y to (1.20) for all u in U and y is computed according to the formula $y = Mu$.

As before if all the spaces are Hilbert spaces and L and D_1 are operators we only consider continuous maps M. In this setting we say that M *is the closed loop system for the scattering operator* L *with respect to the feedback* D_1 if M is the only operator satisfying $y = Mu$ where u and y are computed according to (1.20) and u runs through any dense set $\tilde{U}$ in U. Finally, if $I - D_1 A$ is an invertible map or an invertible operator in the Hilbert space setting, then the closed loop system M determined by Figure 4, or equivalently, (1.20) is given by

$$M = C(I - D_1 A)^{-1} D_1 B + D \ . \tag{1.21}$$

Actually the closed loop system M corresponding to (1.20) is a particular case of Redheffer cascading. Indeed if L_1 is the operator defined by

$$L_1 = \begin{bmatrix} 0 & 0 \\ 0 & D_1 \end{bmatrix}$$

where $X_1 = X_1' = \{0\}$, then the Redheffer scattering matrix $M = L \circ L_1$ defined by (1.2) reduces the closed loop system M determined by (1.20). Moreover, if all the spaces are Hilbert spaces and L and D_1 are operators, then we set

$$\tilde{U} = \{u \in U : D_1 Bu \in (I - D_1 A)X\} \quad .$$

Then a solution to (1.20) is given by

$$Mu = y = C(I - D_1 A)^{(-1)} D_1 Bu + Du + Cq \qquad (u \in \tilde{U}) \tag{1.22}$$

where q is in the kernel of $(I - D_1 A)$. As before if $Cq = 0$ for all q in the kernel of $I - D_1 A$ and $\tilde{U}$ is dense in U and $C(I - D_1 A)^{(-1)} D_1 B \,|\, \tilde{U}$ is an operator, then there is a unique closed loop system M defined by (1.20). Moreover, this M is given by the closure of the operator M determined by (1.22) where $Cq = 0$.

Finally, we need to discuss the cascading of a three port system with a one port system given by Figure 5; of course the cascaded system is a two port system. This system can be viewed as a particular case of cascading two port systems.

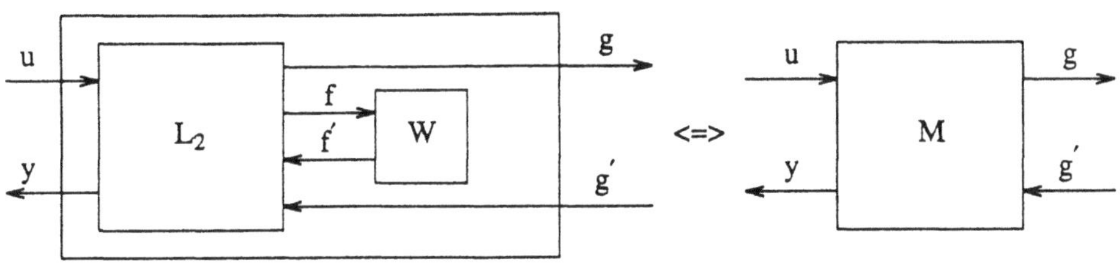

Figure 5

Indeed the cascading in Figure 5 is expressed mathematically as

$$\begin{bmatrix} f \\ g \\ y \end{bmatrix} = \begin{bmatrix} A_{1,1} & A_{1,2} & A_{1,3} \\ A_{2,1} & A_{2,2} & A_{2,3} \\ A_{3,1} & A_{3,2} & A_{3,3} \end{bmatrix} \begin{bmatrix} f' \\ g' \\ u \end{bmatrix} \tag{1.23}$$

subject to $f' = Wf$

where $f' \oplus g' \oplus u$ is in $F' \oplus G' \oplus U$ and $f \oplus g \oplus y$ is in $F \oplus G \oplus Y$. Here L_2 is the 3 by 3 matrix from $F' \oplus G' \oplus U$ to $F \oplus G \oplus Y$ in (1.23) and W is a fixed operator from F to F'. Now if we define

$$A = \begin{bmatrix} A_{1,1} & A_{1,2} \\ A_{2,1} & A_{2,2} \end{bmatrix}, \quad B = \begin{bmatrix} A_{1,3} \\ A_{2,3} \end{bmatrix}, \quad C = [A_{3,1} A_{3,2}], \quad D = A_{3,3}$$

$$A_1 = 0, \quad B_1 = [0, I], \quad C_1 = \begin{bmatrix} 0 \\ I \end{bmatrix}, \quad D_1 = \begin{bmatrix} W & 0 \\ 0 & 0 \end{bmatrix}$$

that is

$$\begin{bmatrix} \dfrac{g}{f'} \\ g' \end{bmatrix} = \begin{bmatrix} A_1 & B_1 \\ C_1 & D_1 \end{bmatrix} \begin{bmatrix} \dfrac{g'}{f} \\ g \end{bmatrix},$$

then the cascaded system described by Figure 5 or equation (1.23) is precisely the system obtained by the Redheffer product $L \circ L_1$.

In this setting $I - D_1 A$ becomes

$$I - D_1 A = \begin{bmatrix} I - WA_{1,1} & -WA_{1,2} \\ 0 & I \end{bmatrix}$$

Therefore if $I - WA_{1,1}$ is invertible, then the Redheffer product $M = L \circ L_1$ exists and M is given by

$$M = \begin{bmatrix} A_{2,1}(I - WA_{1,1})^{-1}WA_{1,2} + A_{2,2} & A_{2,1}(I - WA_{1,1})^{-1}WA_{1,3} + A_{2,3} \\ A_{3,1}(I - WA_{1,1})^{-1}WA_{1,2} + A_{3,2} & A_{3,1}(I - WA_{1,1})^{-1}WA_{1,3} + A_{3,3} \end{bmatrix} \tag{1.24}$$

Finally, it is noted that one can develop the Redheffer cascading theory by using the inverse of $I - AD_1$ or $I - A_{1,1}W$ instead of the inverses of $I - D_1 A$ or $I - WA_{1,1}$. The details are left to the reader.

2. REDHEFFER PRODUCTS AND THE STRUCTURE
OF 2 by 2 MATRIX CONTRACTIONS

In this section we will use Redheffer products to reprove the characterization of all 2 by 2 matrix contractions in Theorem IV.3.1. For the convenience of the reader we will restate the principle results in Section 1 to 3 of Chapter IV where they are needed.

We begin by restating the following result (see Corollaries IV.1.3 and IV.1.5).

2.1 LEMMA. *Let A be a contraction mapping X into X'. The matrix $T = [A, B]$ from $X \oplus U$ to X' is a contraction if and only if $B = D_{A^*} X$ where X is a contraction mapping U into D_{A^*}. In this case the equation*

$$WD_T = \begin{bmatrix} D_A & -A^*X \\ 0 & D_X \end{bmatrix} \tag{2.1}$$

defines a unitary operator W *from* D_T *onto* $D_A \oplus D_X$.

PROOF. The first part of the proof is identical to the proof of Corollary 1.3 in Chapter IV, namely if T is a contraction, then T^* is a contraction, or equivalently,

$$\|D_A \cdot x\|^2 = \|x\|^2 - \|A^*x\|^2 \geq \|B^*x\|^2 \qquad \text{(for all } x \in X').$$

This implies that there exists a contraction X mapping U into $D_A\cdot$ satisfying $X^*D_A\cdot = B^*$, or equivalently, $B = D_A\cdot X$. So without loss of generality we assume that $B = D_A\cdot X$ where X is an operator from U to $D_A\cdot$. To complete the proof by using Redheffer products, notice that T is a contraction if and only if

$$L = \begin{bmatrix} A & B \\ 0 & 0 \end{bmatrix} : X \oplus U \to X' \oplus \{0\}$$

is a contraction. We want to consider the Redheffer product $L \circ R'_A$. For this we notice that by using the notation in the previous section $I - D_1 A = D_A^2$. So for all q in the kernel of $I - D_1 A$ we have $B_1 Aq = D_A\cdot Aq = AD_A q = 0$ and $Cq = 0q = 0$. Thus the first hypothesis in Lemma 1.1 holds. Now consider the space $\tilde{M}$ defined by

$$\tilde{M} = \{x_1 \oplus u = (D_A x - A^*Xu) \oplus u : x \in D_A \text{ and } u \in U\}. \tag{2.2}$$

Obviously $\tilde{M}$ is dense in $X_1 \oplus U = D_A \oplus U$. By applying (1.8) and (1.9) to these $x_1 \oplus u$ in $\tilde{M}$ and using $D_A\cdot A = AD_A$ we obtain

$$x_1' = -AD_A x + AA^*Xu + D_A^2\cdot Xu + D_A\cdot A(I - A^*A)^{(-1)}D_A^2 x = Xu$$

$$y = 0x_1 + 0u .$$

Now according to Lemma 1.1 the Redheffer product $M = L \circ L_1$ exists and is given by

$$M = L \circ R'_A = \begin{bmatrix} 0 & X \\ 0 & 0 \end{bmatrix} : D_A \oplus U \to D_A\cdot \oplus \{0\} .$$

A trivial argument involving (1.4) also shows that $L = M \circ R_A$. This and Lemma 1.2 implies that M is a contraction if and only if L is a contraction. Therefore T is a contraction if and only if X is a contraction, or equivalently, $B = D_A\cdot X$ where X is a contraction.

To complete the proof it remains to verify (2.1). Solving for x and u_1 in (1.2) with L replaced by M and L_1 by R_A we obtain

$$x = D_A x_1 - A^* X u . \tag{2.3}$$

A simple calculation shows that $D_M = I \oplus D_X$. Using this along with (2.3) and Lemma 1.2 for $L = M \circ R_A$ we have

$$\|D_L(x_1 \oplus u)\|^2 = \|D_M(x \oplus u)\|^2 = \|(D_A x_1 - A^* X u) \oplus D_X u\|^2$$

which produces (2.1). This completes the proof.

It is clear that the previous proof also yields the following result.

2.2 COROLLARY. *The matrix*

$$\begin{bmatrix} A & B \\ 0 & 0 \end{bmatrix} : X \oplus U \to X' \oplus Y$$

is a contraction if and only if it can be represented as the Redheffer product

$$\begin{bmatrix} 0 & X \\ 0 & 0 \end{bmatrix} \circ R_A$$

where R_A is the rotation matrix associated with A *and* X *is a contraction.*

In a similar way one can also obtain the following supplementary result

2.3 COROLLARY. *The matrix*

$$\begin{bmatrix} 0 & X \\ 0 & D \end{bmatrix} : X \oplus U \to X' \oplus Y$$

is a contraction if and only if it can be represented as the Redheffer product

$$R_{D^*}' \circ \begin{bmatrix} 0 & \Gamma \\ 0 & 0 \end{bmatrix}$$

where D *and* Γ *are both contractions acting between the appropriate spaces. In this case* $X = \Gamma \Delta$ *where* Δ *is the positive square root of* $I - D^* D$.

The proof of the previous lemma shows that the Redheffer product can be used to convert a contractive row matrix [A, B] to [0, X]. Next we will obtain an alternate proof of Lemma IV.2.1, by using the Redheffer product to convert a 2 by 2 matrix contraction with three nonzero entries to a 2 by 2 matrix contraction with two nonzero entries.

2.4 LEMMA. *Let* A *from* X *to* X' *and* D *from* U *to* Y *be contractions. The matrix*

$$L = \begin{bmatrix} A & B \\ 0 & D \end{bmatrix} : X \oplus U \to X' \oplus Y$$

is a contraction if and only if $B = D_A \cdot \Gamma \Delta$ *where* Γ *is a contraction from* D_D *into* D_{A^*} *and* Δ *is the positive square root of* $I - D^* D$. *In this case the equation*

$$WD_L = \begin{bmatrix} D_A & -A^* \Gamma \Delta \\ 0 & D_\Gamma \Delta \end{bmatrix} \tag{2.4}$$

defines a unitary operator W *from* D_L *onto* $D_A \oplus D_\Gamma$.

PROOF. If L is a contraction, then [A, B] is a contraction. By the previous lemma this implies that $B = D_A \cdot X$ where X is a contraction. So without loss of generality we assume that $B = D_A \cdot X$ where X is a contraction from U to D_{A^*}. As before we can consider the Redheffer product $M = L \circ R_A'$. By using again the space $\tilde{M}$ in (2.2) and $D_A \cdot A = AD_A$ along with Lemma 1.1, we see that the Redheffer product $L \circ R_A'$ exists and is given by

$$M = L \circ R_A' = \begin{bmatrix} 0 & X \\ 0 & D \end{bmatrix} .$$

It is easy to check that conversely $L = M \circ R_A$. Lemma 1.2 shows that L is a contraction if and only if $M = L \circ R_A'$ is a contraction. Applying Lemma 2.1 or Corollary 2.3 we see that M is a contraction if and only if $X = \Gamma \Delta$, or equivalently, $B = D_A \cdot \Gamma \Delta$ where Γ is a contraction. This completes the proof of the first part.

To prove (2.4) notice that $X = \Gamma \Delta$ gives

$$\begin{aligned}
\|D_M(x \oplus u)\|^2 &= \|x \oplus u\|^2 - \|M(x \oplus u)\|^2 = \\
&\|x\|^2 + \|u\|^2 - \|\Gamma \Delta u\|^2 - \|Du\|^2 = \\
&\|x\|^2 + \|\Delta u\|^2 - \|\Gamma \Delta u\|^2 = \|x \oplus D_\Gamma \Delta u\|^2 .
\end{aligned} \tag{2.5}$$

Because C and D were not used in calculating (2.3), equation (2.3) is also valid in this setting. (As in the proof of Lemma 2.1, equation (2.3) is obtained from (1.2) corresponding to $L = M \circ R_A$.) Using (2.3), (2.5) and Lemma 1.2 applied to $L = M \circ R_A$ we have

$$\|D_L(x_1 \oplus u)\|^2 = \|D_M(x \oplus u)\|^2 = \|x \oplus D_\Gamma \Delta u\|^2 = \|(D_A x_1 - A^* X u) \oplus D_\Gamma \Delta u\|^2 .$$

This produces (2.4) which completes the proof.

From the previous proof and Corollary 2.3 we readily obtain the following result.

2.5 COROLLARY. *The matrix*

$$\begin{bmatrix} A & B \\ 0 & D \end{bmatrix} : X \oplus U \to X' \oplus Y$$

is a contraction if and only if it can be represented as a Redheffer product of the form

$$R'_{D^*} \circ \begin{bmatrix} 0 & \Gamma \\ 0 & 0 \end{bmatrix} \circ R_A$$

where A, D and Γ are contractions acting between the appropriate spaces.

To complete this section we use the previous results with Redheffer products to give another proof of Theorem 3.1 in Chapter IV, that is,

2.6 THEOREM. *The matrix L in (1.1) is a contraction if and only if A is a contraction*

$$B = D_{A^*} X \quad \text{and} \quad C = Y D_A \quad \text{and} \quad D = D_{Y^*} \Gamma D_X - Y A^* X \tag{2.6}$$

where X from U to D_{A^*} and Y from D_A to Y and Γ from D_X to D_{Y^*} are all contractions. In this case the equation

$$W D_L = \begin{bmatrix} D_Y D_A & -(D_Y A^* X + Y^* \Gamma D_X) \\ 0 & D_\Gamma D_X \end{bmatrix} \tag{2.7}$$

defines a unitary operator from D_L onto $D_Y \oplus D_\Gamma$.

PROOF. The proof proceeds along the lines of the proofs of the previous two lemmas. If L is a contraction, then [A, B] and $[A^*, C^*]^*$ are both contractions. According to Lemma 2.1 and Corollary 2.3 there exists contractions X from U to D_{A^*} and Y from D_A to Y satisfying $B = D_{A^*} X$ and $C = Y D_A$. So without loss of generality we assume that the first two equations in (2.6) hold. As before we define M by $M = L \circ R'_A$. In this case M is well defined and becomes

$$M = \begin{bmatrix} 0 & X \\ Y & YA^*X + D \end{bmatrix}$$

Moreover, $L = M \circ R_A$. Lemma 1.2 shows that M is a contraction if and only if L is a contraction. According to the previous lemma, M is a contraction if and only if X and Y are both contractions and $YA^*X + D = D_{Y^*} \Gamma D_X$ where Γ is a contraction. This verifies (2.6), which proves the first part of the theorem.

To complete the proof it remains to establish (2.7). Using the last equation in (2.6) we have

$$JM = \begin{bmatrix} Y & D_{Y^*} \Gamma D_X \\ 0 & X \end{bmatrix}.$$

This is precisely the form of Lemma 2.4. By Lemma 2.4 there exists a unitary operatory γ satisfying

$$\gamma D_M = \begin{bmatrix} D_Y & -Y^* \Gamma D_X \\ 0 & D_\Gamma D_X \end{bmatrix}. \tag{2.8}$$

Notice that (2.3) is also valid in this setting. Now (2.7) follows from the definition of J, (2.3), (2.8) and again by applying Lemma 1.2 to $L = M \circ R_A$. This completes the proof.

2.7 REMARK. It is emphasized that M and L in the previous proof do not necessarily have the same norm, even though R_A is unitary. To see this choose $A = 1/2$, $X = Y = \Gamma = 0$. Then $\|L\| \neq \|M\|$.

From the first part of the proof of Theorem 2.6 and Corollary 2.5 we easily deduce the following result.

2.8 COROLLARY. *The matrix*

$$\begin{bmatrix} A & B \\ C & D \end{bmatrix} : X \oplus U \to X' \oplus Y$$

is a contraction if and only if it can be represented as a Redheffer product of the form

$$(J(R'_{X^*} \circ \begin{bmatrix} 0 & \Gamma \\ 0 & 0 \end{bmatrix} \circ R_Y)) \circ R_A \tag{2.9}$$

where A, X, Y and Γ are contractions acting between the appropriate spaces and J is an operator of the form (1.5).

At this point it is appropriate to introduce the block diagram in Figure 6 for $JL \circ JL_1$.

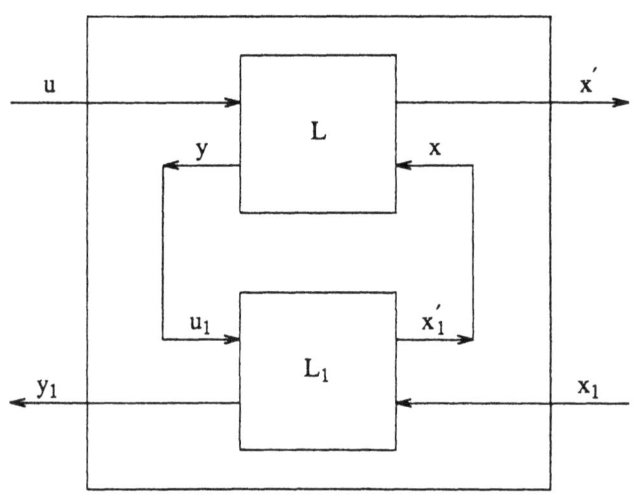

Figure 6

Therefore we can conclude with the following observation.

2.9 REMARK. The block diagram for the Redheffer product in (2.9) is given by Figure 7. In other words any 2 by 2 matrix contraction of the form

$$\begin{bmatrix} A & B \\ C & D \end{bmatrix}$$

admits a scattering matrix interpretation whose block diagram is given in Figure 7.

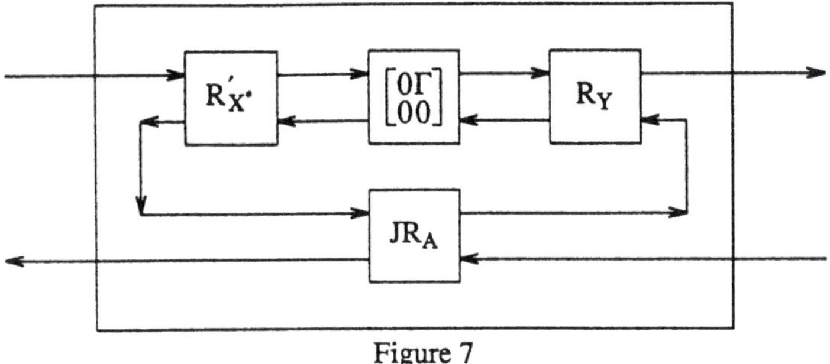

Figure 7

3. REDHEFFER PRODUCTS AND CARATHEODORY INTERPOLATION

In this section we will use Redheffer products to solve the operator version of the classical Carathéodory interpolation problem. The solution to this problem is similar to the solution of the scalar valued Carathéodory interpolation problem in Chapter I. Finally, it is noted that this section is not used in the rest of this chapter, that is, for obtaining the Schur representation for the commutant lifting theorem.

We begin by restating the operator version of the classical Carathéodory interpolation problem in Chapter I.

3.1 PROBLEM. Given a set of operators $\{A_i\}_0^{n-1}$ in $L(E,E')$ find necessary and sufficient conditions for the existence of a contractive analytic function F in $H^\infty(E,E')$ satisfying

$$F(z) = \sum_{i=0}^{n-1} A_i z^i + 0(z^n) \qquad (z \in D) . \tag{3.1}$$

3.2 PROBLEM. Give an explicit description of all solutions (if there are any) to Problem 3.1.

To eliminate some technical problems with inverses throughout this section only we will assume that both E and E' are finite dimensional. The following result which is an operator version of Lemma I.1.4 plays a basic role in this section.

3.3 LEMMA. *Let E and E' be finite dimensional Hilbert spaces. A function F is a contractive analytic function in* $H^\infty(E,E')$ *if and only if $F(0) \equiv A$ is a contraction and there exists a contractive analytic function F_1 in* $H^\infty(D_A, D_{A^*})$ *satisfying*

$$F(z) = A + zD_{A^*}F_1(z)(I + zA^*F_1)^{-1}D_A \quad (z \in D) . \tag{3.2}$$

In this case F *and* F_1 *uniquely determine each other, and* F_1 *is given by*

$$F_1(z) = z^{-1}[D_{A^*}F(z)(I - A^*F(z))^{(-1)}D_A - A] \,|\, D_A \quad (z \in D) . \tag{3.3}$$

PROOF. Assume that A is a contraction from E to $E^{'}$ and F_1 is a contractive analytic function in $H^\infty(D_A, D_{A^*})$. Consider the system given by

$$\begin{bmatrix} x^{'} \\ y \end{bmatrix} = \begin{bmatrix} -A^* & D_A \\ D_{A^*} & A \end{bmatrix} \begin{bmatrix} x \\ u \end{bmatrix} \tag{3.4}$$

$$\text{subject to } x = zF_1(z)x^{'}$$

where $x \oplus u$ is in $D_{A^*} \oplus E$ and $x^{'} \oplus y$ is in $D_A \oplus E^{'}$. The previous 2 by 2 matrix is $R_{A^*}^{'}$ which is a unitary operator. Solving for y in terms of u we have y = Fu where F is given by (3.2). Obviously F(0) = A and F is an analytic function in D. We claim that F is a contractive analytic function in $H^\infty(E, E^{'})$. To see this notice that

$$\|u\|^2 = \|u\|^2 + \|x\|^2 - \|x\|^2 = \|y\|^2 + \|x^{'}\|^2 - \|x\|^2 \geq \|y\|^2 .$$

The second equality follows from the fact that the 2 by 2 matrix in (3.4) is unitary, and the inequality follows from the fact that $F_1(z)$ is a contraction for all z in D, that is, $\|x^{'}\| \geq \|x\|$. Therefore $\|y\| = \|F(z)u\| \leq \|u\|$ for all u in H and z in D. So F in (3.2) is a contractive analytic function in $H^\infty(E, E^{'})$. (This also follows from Lemma 1.2 or Corollary 1.3 by noticing that (3.4) is the cascading of a unitary operator $R_{A^*}^{'}$ with the contraction $0 \oplus zF_1(z)$; see (1.21).)

Now assume that F(z) is a contractive analytic function. By the maximum principle $A \equiv F(0)$ is a contraction. Consider the system

$$\begin{bmatrix} x \\ u \end{bmatrix} = \begin{bmatrix} -A & D_{A^*} \\ D_A & A^* \end{bmatrix} \begin{bmatrix} x^{'} \\ y \end{bmatrix} \tag{3.5}$$

$$\text{subject to } y = F(z)u$$

where as before $x \oplus u$ is in $D_{A^*} \oplus E$ and $x^{'} \oplus y$ is in $D_A \oplus E^{'}$. However, (3.5) is precisely the Redheffer cascading corresponding to $(F(z) \oplus 0) \circ R_A^{'}$. By Corollary 1.3 adjusted to this setting, the solution $x = F_1^{'}(z)x^{'}$ to (3.5) for all z in D is given by

$$F_1^{'}(z) = D_{A^*}F(z)(I - A^*F(z))^{(-1)}D_A - A \quad (z \in D) . \tag{3.6}$$

According to the proof of Corollary 1.3, D_A is contained in the range of $I - A^*F(z)$. So

F_1' is analytic in D. (This also follows from Theorem IX.5.3. To see this notice that by this theorem $A^*F(z)$ is a strict contraction from D_A to D_A, because D_A is finite dimensional. Thus $I - A^*F(z)$ is an invertible operator on D_A and F_1' is analytic in D.) Since $\|F_1'(z)\| \leq 1$ for all z in D this implies that $F_1'(z)$ is a contractive analytic function in $H^\infty(D_A, D_{A^*})$. Using $F(0) = A$ in (3.6) we see that $F_1'(0) = 0$. Thus $F_1'(z) = zF_1(z)$ where $F_1(z)$ is a contractive analytic function in $H^\infty(D_A, D_{A^*})$. This and (3.6) produces (3.3).

To complete the proof it remains to show that the F_1 in (3.3) produces F by (3.2). To see this notice that by moving the unitary matrix in (3.5) to the other side and using $x = zF_1 x'$ we have

$$\begin{bmatrix} x' \\ y \end{bmatrix} = \begin{bmatrix} -A^* & D_A \\ D_{A^*} & A \end{bmatrix} \begin{bmatrix} x \\ u \end{bmatrix} \tag{3.7}$$

$$y = F(z)u \quad \text{and} \quad x = zF_1(z)x' .$$

However, this is precisely (3.4). Therefore F is given by (3.2). This completes the proof.

Notice that (3.2) is precisely (XIII.3.22) which was obtained by using the Schur representation for the commutant lifting theorem. The previous lemma readily produces the following solution to the one step Carathéodory interpolation problem.

3.4 COROLLARY. *If* n = 1, *then the classical Carathéodory interpolation Problem 3.1 has a solution if and only if* A_o *is a contraction. In this case the set of all solutions to this Carathéodory interpolation problem is given by (3.2) where* $A = A_o$ *and* F_1 *is a contractive analytic function in* $H^\infty(D_A, D_{A^*})$. *Moreover, there exists a unique solution to this Carathéodory interpolation problem if and only if* A_o *or* A_o^* *is an isometry.*

The previous corollary shows that the set of all solutions to the one step Carathéodory interpolation problem is parameterized by $H_1^\infty(D_A, D_{A^*})$ where $A = A_o$. In particular, there exists a unique solution to the one step Carathéodory interpolation problem if and only if $H_1^\infty(D_A, D_{A^*}) = \{0\}$, or equivalently, A or A^* is an isometry.

If $E = E' = C^1$ and $A = a$ is a complex number in the open unit disc, then (3.2) and (3.3) reduce to

$$F(z) = \frac{a + zF_1(z)}{1 + z\bar{a}F_1(z)} \quad \text{and} \quad F_1(z) = \frac{1}{z}\left[\frac{F(z) - a}{1 - \bar{a}F(z)}\right] \tag{3.8}$$

respectively. (If $|a| = 1$, then $D_a = D_{a^*} = \{0\}$ and (3.2), (3.3) give $F(z) = a$ and $F_1(z) = 0$.) The equations in (3.8) are precisely the Mobius transformations used in solving the Carathéodory interpolation problem in Chapter I. In Chapter I we recursively used the equations in (3.8) to solve the Carathéodory interpolation problem. By following the methods in Chapter I one can recursively use Lemma 3.3 and (3.2), (3.3) to solve the operator version of the Carathéodory interpolation problem. Rather than repeat these arguments here we will just present the key ideas.

To obtain a solution to the operator version of the Carathéodory interpolation problem, consider the following system of equations

$$F_j(z) = \Gamma_j + zD_{j*}F_{j+1}(z)(I + z\Gamma_j^* F_{j+1}(z))^{-1}D_j = \omega_j(zF_{j+1})$$

$$(3.9)$$

$$F_{j+1}(z) = z^{-1}[D_{j*}F_j(z)(I - \Gamma_j^* F_j(z))^{(-1)}D_j - \Gamma_j]|D_{j-1} \qquad (z \in D)$$

where $\Gamma_j = F_j(0)$ and $j \geq 0$. (As expected $D_{\Gamma_j} = D_j$ and $D_{\Gamma_j^*} = D_{j*}$ and $D_j = D_{\Gamma_j}$ and $D_{j*} = D_{\Gamma_j^*}$). According to Lemma 3.3 we see that F_0 is a contractive analytic function in $H^\infty(E,E')$ if and only if $F_0 = \omega_0(zF_1)$ where $\Gamma_0 \equiv F_0(0)$ is a contraction and F_1 is a contractive analytic function in $H^\infty(D_0, D_{0*})$. In this case F_0 uniquely determines and is uniquely determined by Γ_0 and F_1. Now applying Lemma 3.3 to F_1 we see that F_0 is a contractive analytic function in $H^\infty(E,E')$ if and only if $F_0 = \omega_0(z\omega_1(zF_2))$ where $\Gamma_0(=F_0(0))$ from E to E' and Γ_1 from D_0 to D_{0*} are both contractions and F_2 is a contractive analytic function in $H^\infty(D_1, D_{1*})$. In this case F_0 uniquely determines and is uniquely determined by Γ_0, Γ_1 and F_2. Proceeding by induction F_0 is a contractive analytic function in $H^\infty(E,E')$ if and only if

$$F_0 = \omega_0(z\omega_1(\cdots (z\omega_{n-1}(zF_n))\cdots)) \qquad (z \in D) \qquad (3.10)$$

where $\{\Gamma_i\}_0^{n-1}$ is a choice sequence initiated from E to E' and F_n is a contractive analytic function in $H^\infty(D_{n-1}, D_{n-1*})$. In this case F_0 uniquely determines, and is uniquely determined by the choice sequence $\{\Gamma_i\}_0^{n-1}$ initiated from E to E' and F_n a contractive analytic function in $H^\infty(D_{n-1}, D_{n-1*})$. Moreover, $\Gamma_i = F_i(0)$ for all i.

One can express (3.10) in terms of a Redheffer product. To see this notice that (3.2), (3.3), (3.7) and (3.9) imply that the system in (3.10) is equivalent to

$$\begin{bmatrix} x_j' \\ y_j \end{bmatrix} = \begin{bmatrix} -z\Gamma_j^* & D_j \\ zD_{j*} & \Gamma_j \end{bmatrix} \begin{bmatrix} x_j \\ u_j \end{bmatrix} \qquad (3.11)$$

$$y_j = F_j(z)u_j \quad \text{and} \quad x_j = F_{j+1}(z)x_j' \qquad (j \geq 0)$$

where $x_j \oplus u_j$ is in $D_{j*} \oplus D_{j-1}$ and $x_j' \oplus y_j$ is in $D_j \oplus D_{j-1*}$. (Here $D_{-1} = E$ and $D_{-1*} = E'$.) The second two equations in (3.11) imply that $x_j' = u_{j+1}$ and $x_j = y_{j+1}$ for all $j \geq 0$. This with the definition of the Redheffer product in (1.3) and (1.4) yield

$$\begin{bmatrix} x'_{n-1} \\ y_o \end{bmatrix} = [R'_o \circ R'_1 \circ R'_2 \circ \cdots \circ R'_{n-1}] \begin{bmatrix} x_{n-1} \\ u_o \end{bmatrix} \tag{3.12}$$

$$y_o = F_o(z)u_o \quad \text{and} \quad x_{n-1} = F_n(z)x'_{n-1} \quad (z \in D)$$

where $R'_i(=R_i(z))$ is the 2 by 2 matrix $R'_{\Gamma_i}(zI \oplus I)$ in (3.11). Because (3.9) and (3.11) are equivalent, (3.10) and (3.12) produce the same F_o. Summing up this analysis proves the following result.

3.5 LEMMA. *A function F_o is a contractive analytic function in $H^\infty(E,E')$ if and only if F_o is given by (3.12) where $\{\Gamma_i\}_0^{n-1}$ is a choice sequence initiated from E to E' and F_n is a contractive analytic function in $H^\infty(D_{n-1},D_{n-1*})$. In this case F_o uniquely determines and is uniquely determined by the choice sequence $\{\Gamma_i\}_0^{n-1}$ and F_n.*

The previous lemma with Figures 3 and 4 in Section 1 adjusted to this setting, shows that a contractive analytic function F_o admits a network interpretation shown by Figure 8 below.

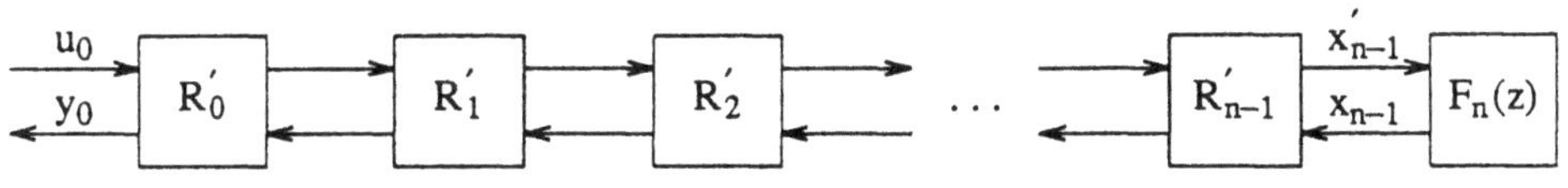

Figure 8

Therefore a contractive analytic function can be obtained by cascading R'_o through R'_{n-1} and feeding back F_n at the R'_{n-1} terminal.

Let F_j be the contractive analytic function in (3.9) and

$$F_j(z) = \sum_{i=0}^{\infty} A_{j,i} z^i \tag{3.13}$$

be its power series expansion. By expanding the first equation in (3.9) into a power series and using $\Gamma_j = F_j(0)$ we see that $\{A_{j,i}: 0 \leq i \leq n\}$ uniquely determines and is uniquely determined by $\{\Gamma_j, A_{j+1,i}: 0 \leq i \leq n-1\}$. So proceeding by induction we see that $\{A_{o,i}\}_0^{n-1}$ and $\{\Gamma_i\}_0^{n-1}$ uniquely determine each other. Therefore if $\{A_i = A_{o,i}\}_0^{n-1}$ is the data for the Carathéodory interpolation problem, then one can determine $\{\Gamma_i\}_0^{n-1}$ recursively. This is done setting $F_o(z) = A_o + A_1 z + \cdots A_{n-1} z^{n-1}$, then using this F_o in (3.9) to compute $F_j(0) = \Gamma_j$, recursively. The set of operators $\{\Gamma_i\}_0^{n-1}$ is called the *Schur operators determined by* $\{A_i\}_0^{n-1}$. (In the scalar setting this Schur sequence is precisely the Schur numbers determined by the data $\{A_i\}_0^{n-1}$.) The above analysis shows that $\{A_i\}_0^{n-1}$ and $\{\Gamma_i\}_0^{n-1}$ uniquely determine each other. If at any stage the Schur operators $\{\Gamma_i\}_0^m$ is not a choice sequence, then the calculation stops. In this case Lemma 3.5 shows that there is no solution to the Carathéodory interpolation problem. On the other hand if $\{\Gamma_i\}_0^{n-1}$ is a choice sequence, then Lemma 3.5 shows that there exists a solution to the Carathéodory interpolation problem, and (3.12) gives all solutions to this problem. Summing up the above analysis yields the following result.

3.6 THEOREM. *There exists a solution to the Carathéodory interpolation problem with data* $\{A_i\}_0^{n-1}$ *if and only if the Schur operators* $\{\Gamma_i\}_0^{n-1}$ *determined by* $\{A_i\}_0^{n-1}$ *is a choice sequence. In this case the set of all solutions* F_0 *to the Carathéodory interpolation problem is given by (3.12) where* F_n *is a contractive analytic function in* $H^\infty(D_{n-1}, D_{n-1*})$. *Moreover, there exists a unique solution to the Carathéodory interpolation problem if and only if* $\{\Gamma_i\}_0^{n-1}$ *is a choice sequence and* Γ_j *or* Γ_j^* *is an isometry for some* $0 \le j < n$.

To complete this section we will show how (3.11) and (3.12) can be used to obtain an operator version of the layer peeling algorithm in Chapter I. Recall that the layer peeling algorithm is a recursive inverse scattering algorithm to compute the choice sequence $\{\Gamma_i\}_0^{n-1}$ from the data $\{A_i\}_0^{n-1}$. To simplify some of the calculations throughout the rest of this chapter we assume that $\{\Gamma_i\}_0^{n-1}$ is a choice sequence of strict contractions initiated from E to E'. We begin by inverting the 2 by 2 matrix $R_i' \overset{\Delta}{=} R_{\Gamma_i'}(zI \oplus I)$ in (3.11) to obtain

$$\begin{bmatrix} x_j \\ u_j \end{bmatrix} = \begin{bmatrix} -z^{-1}\Gamma_j & z^{-1}D_{j*} \\ D_j & \Gamma_j^* \end{bmatrix} \begin{bmatrix} x_j' \\ y_j \end{bmatrix} \qquad (j \ge 0) \ . \qquad (3.14)$$

By using (3.11) and (3.14) to solve for $x_j' \oplus x_j$ in terms of $u_j \oplus y_j$ we obtain

$$\begin{bmatrix} x_j' \\ x_j \end{bmatrix} = \begin{bmatrix} D_j^{-1} & -D_j^{-1}\Gamma_j^* \\ -z^{-1}D_{j*}^{-1}\Gamma_j & z^{-1}D_{j*}^{-1} \end{bmatrix} \begin{bmatrix} u_j \\ y_j \end{bmatrix} \qquad (j \ge 0) \ . \qquad (3.15)$$

The 2 by 2 matrix in (3.15) denoted by $T_j(=T_j(z))$ is the *transmission matrix* for the reflection operator Γ_j. Using $x_j' = u_{j+1}$ and $x_j = y_{j+1}$ for all j in (3.15) we have the following difference equation:

$$x_j' \oplus x_j = T_j(x_{j-1}' \oplus x_{j-1}) \quad \text{and} \quad x_{-1}' \oplus x_{-1} = u_o \oplus y_o \,. \tag{3.16}$$

This difference equation with $\Gamma_i = F_i(0)$ and $x_i = F_{i+1}x_i'$ produces the following operator version of the layer peeling inverse scattering algorithm in Procedure I.1.7 for computing the choice sequence corresponding to a contractive analytic function.

3.7 PROCEDURE. Let $u(z)$ be an analytic function in $H^\infty(E, E)$ such that $u(0)$ is invertible. Let F be a contractive analytic function in $H^\infty(E,E')$, and let y be the analytic function defined by $y = Fu$. The choice sequence $\{\Gamma_i\}_0^\infty$ for F is given by

$$\Gamma_{i+1} = x_i(0)(x_i'(0))^{-1} \tag{3.17}$$

where $x_i' \oplus x_i$ is recursively computed by interlacing (3.16) with (3.17) subject to the initial condition $u_o \oplus y_o = u(z) \oplus y(z)$. Moreover, if the sequence $\{\Gamma_i\}_0^\infty$ computed by (3.17) is not strictly contractive, then the procedure must stop. In this case F may not be in $H_1^\infty(E,E')$.

3.8 REMARK. It is easy to show that $x_i'(0)$ in (3.17) is invertible. To see this we use induction. By the hypothesis $u_o(0)$ is invertible. Now assume that $u_j(0) = x_{j-1}'(0)$ is invertible. Evaluating (3.9) at $z = 0$ and using

$$\Gamma_j u_j(0) = F_j(0)x_{j-1}'(0) = x_{j-1}(0) = y_j(0)$$

in (3.15) we have $u_{j+1}(0) = x_j'(0) = D_j u_j(0)$. Therefore $x_j'(0)$ is invertible and (3.17) makes sense.

By choosing $u(z) \equiv I$ and $F(z)$ the polynomial

$$F(z) = \sum_{i=0}^{n-1} A_i z^i \,,$$

then one can use the previous procedure to compute the reflection operators associated with the data $\{A_i\}_0^{n-1}$. If the corresponding $\{\Gamma_i\}_0^{n-1}$ is a choice sequence, then Theorem 3.6 provides the set of all solutions to the Carathéodory interpolation Problem 3.2. The network interpretation of these solutions is given in Figure 8.

By recursively solving (3.16) we have

$$x_n' \oplus x_n = T_n T_{n-1} T_{n-2} \cdots T_1 T_0(u_o \oplus y_o) \,. \tag{3.18}$$

where $\Gamma_{n+1}x_n'(0) = x_n(0)$. Using this along with the form of T_j in (3.15) and $F = F_o$ in (3.1), one can easily show that there is a one to one correspondence between the data

$\{A_i\}_0^{n-1}$ and its choice sequence $\{\Gamma_i\}_0^{n-1}$. Finally, it is noted that one of the advantages in using the transmission operators T_j to recursively compute the choice sequence $\{\Gamma_i\}_0^{\infty}$ is that the algorithm only involves matrix multiplication. This is easier than cascading the scattering matrices R_i' which would require one to compute a Redheffer product at each stage.

4. A REDHEFFER PRODUCT APPROACH
TO THE SCHUR REPRESENTATION

In this section we will use the concept of a Redheffer product to present another approach for obtaining the Schur representation for the commutant lifting theorem presented in Section XIII.3. The advantage of this approach is that it bypasses the use of choice sequences.

Throughout T on H and T' on H' are both contractions and A is a contraction in $I(T, T')$. For convenience we introduce, in this section, the space $Z = H^2(D_A \oplus D_T \oplus D_{T'})$ which will be identified with $H^2(D_A) \oplus H^2(D_T) \oplus H^2(D_{T'})$. Moreover, we identify $H^2(D_A)$, $H^2(D_T)$ and $H^2(D_{T'})$ with the appropriate subspace of Z. For instance $H^2(D_T)$ is identified with both $H^2(\{0\} \oplus D_T \oplus \{0\})$ and $\{0\} \oplus H^2(D_T) \oplus \{0\}$. The space $H^2(D_A \oplus D_{T'})$ is identified with both $H^2(D_A \oplus \{0\} \oplus D_{T'})$ and $H^2(D_A) \oplus \{0\} \oplus H^2(D_{T'})$. The shift operator on Z, that is, multiplication by z on Z is denoted by S. Recall that the minimal isometric dilation U on K of T is given by

$$U(h \oplus f) = Th \oplus (D_T h + Sf) = Th \oplus (D_T h + zf) \quad (h \oplus f \in H \oplus H^2(D_T) \overset{\Delta}{=} K) \quad (4.1)$$

(see (3.4) in Chapter VI). The minimal isometric dilation U' on $K' = H \oplus H^2(D_{T'})$ is given by (4.1) where T' replaces T. As before ω is the unitary operator from

$$F = \{D_A Th \oplus D_T h : h \in H\}^- \quad \text{onto} \quad F' = \{D_A h \oplus D_{T'} Ah : h \in H\}^- \quad (4.2)$$

defined by

$$\omega(D_A Th \oplus D_T h) = D_A h \oplus D_{T'} Ah \quad (h \in H) . \quad (4.3)$$

Obviously ω is an inner and $*$-inner constant in z function in $H^\infty(F, F')$ and ω_+ is a unitary operator from $H^2(F)$ onto $H^2(F')$.

Throughout $X = H^2(D_A \oplus D_T)$ and $X' = H^2(D_A \oplus D_{T'})$. Obviously X and X' are both subspaces in Z. The defect spaces G and G' are defined by

$$G = (D_A \oplus D_T) \ominus F \quad \text{and} \quad G' = (D_A \oplus D_{T'}) \ominus F' . \quad (4.4)$$

We say that F is a *Schur contraction* if F is a contractive analytic function in

$H^\infty(D_A \oplus D_T, D_A \oplus D_{T'})$ satisfying $F(0)|F = \omega$. Since ω is unitary, from Theorem IX.5.3 we can infer that F is a Schur contraction if and only if $F(z) = \omega \oplus R(z)$ where R is a contractive analytic function in $H^\infty(G, G')$. By using $U'B = BU$ and the fact that $U|H^2(D_T)$ is a unilateral shift, Corollary XIII.3.5 can be formally reformulated as follows; the set of all contractive intertwining liftings B of A is given by

$$B = B_F = AP_H \oplus P'F_+(I - SP_A F_+)^{-1}[D_A P_H \oplus (I - P_H)] =$$
$$AP_H \oplus P'(I - F_+ SP_A)^{-1} F_+(D_A P_H \oplus I - P_H) . \tag{4.5}$$

where F is an adequate Schur contraction, and P' and P_A is the orthogonal projection onto $H^2(D_{T'})$ and $H^2(D_A)$, respectively. (In (4.5) we have slightly abused the standard defininition for the orthogonal projection. Actually P_A in (4.5) is the operator from $H^2(D_A \oplus D_{T'})$ into $H^2(D_A \oplus D_T)$ defined by $P_A(d_A \oplus d_{T'}) = d_A \oplus 0$.) We emphasize that B maps $H \oplus H^2(D_T)$ into $H \oplus H^2(D_{T'})$. Equation (4.5) is a *Schur type representation* of the set of all contractive intertwining liftings B of A.

Let us demonstrate how one can formally obtain the Schur representation in (4.5) without using choice sequences. To this end first notice that $P_{H'}B = AP_H$ is equivalent to the fact that B admits a representation of the form

$$B = AP_H \oplus Z \tag{4.6}$$

where Z is an operator mapping K into $H^2(D_{T'})$. The operator B in (4.6) is a contraction if and only if all k in K

$$0 \le \|k\|^2 - \|Bk\|^2 = \|k\|^2 - \|AP_H k\|^2 - \|Zk\|^2 = \|D_A P_H k + (I - P_H)k\|^2 - \|Zk\|^2 ,$$

or equivalently,

$$\|D_A P_H k + (I - P_H)k\|^2 \ge \|Zk\|^2 .$$

Thus B in (4.6) is a contraction if and only if there exists a contraction Y mapping $D_A \oplus H^2(D_T)$ into $H^2(D_{T'})$ such that $Z = Y(D_A P_H \oplus I - P_H)$. Therefore the set of all contractive B satisfying $P_{H'}B = AP_H$ is given by

$$B = AP_H + Y(D_A P_H \oplus (I - P_H)) \quad \text{with} \quad \|Y\| \le 1 . \tag{4.7}$$

Equation (4.7) with the definitions of U, U' and ω give:

$$BU = ATP_H + Y(D_A TP_H + D_T P_H + S(I-P_H)) \quad \text{and}$$

$$U'B = T'AP_H + D_{T'}AP_H + SY(D_A P_H + I - P_H) =$$

$$T'AP_H + [P'\omega + SYP_A\omega](D_A TP_H + D_T P_H) + SY(I-P_H) \, .$$

Using $T'A = AT$ this implies that $U'B = BU$ if and only if

$$Y|F = P'\omega + SYP_A\omega \quad \text{and} \quad YS(I-P_H) = SY(I-P_H) \, . \tag{4.8}$$

Therefore the set of all contractive intertwining liftings B of A are given by (4.7) where Y is any contraction mapping $D_A \oplus H^2(D_T)$ into $H^2(D_{T'})$ satisfying (4.8).

The formulas (4.7) and (4.8) do not yet yield the representation (4.5). However, to find at least one contractive Y satisfying (4.8) choose $Y|G = 0$. We set $Y|G = 0$ because (4.8) is an operator equation on $F \oplus SH^2(D_T)$ and we are free to choose $Y|G$ as long as Y is a contraction. So the simplest choice of $Y|G$ is zero. Formally solving the first equation in (4.8) gives

$$Y = \sum_0^\infty S^n P'\omega_+ (P_A\omega_+)^n = \sum_0^\infty P'\omega_+(SP_A\omega_+)^n = P'\omega_+(I - SP_A\omega_+)^{-1} \, . \tag{4.9}$$

We obtain the second equality in (4.9), because we also interpret ω_+ as a partial isometry $\omega_+ \oplus 0$ mapping $H^2(D_A \oplus D_T)$ into $H^2(D_A \oplus D_{T'})$ whose initial and final space is $H^2(F)$ and $H^2(F')$, respectively, and therefore $P_A\omega_+$ can be viewed as an operator on $H^2(D_A \oplus D_T)$. Moreover, since $S\omega_+ = \omega_+ S | H^2(D_A \oplus D_T)$ we have satisfied the second constraint in (4.8). We emphasize that in this point of our presentation the formulas (4.5) and (4.9) have no rigorous meaning, since we have neither proven that the series in (4.9) converges nor that the inverses in (4.5) and (4.9) exists. However, substituting this Y into (4.7) yields

$$B = AP_H + P'\omega_+(I - SP_A\omega_+)^{-1}(D_A P_H \oplus (I - P_H)) \, . \tag{4.10}$$

This is precisely the Schur representation of B for $F = \omega_+$, that is, $F(z) \equiv \omega \oplus 0$. From (4.10) one could probably guess the general representation in (4.5). However, this is no guarantee that these are all the contractive intertwining liftings of A. In the next section we will prove that they are.

Let us now formally verify that B given by (4.5) is an intertwining lifting of A. Clearly $P_{H'}B = AP_H$ is a lifting of A. The intertwining property $U'B = BU$ follows from the definition of ω with $T'A = AT$ and the following calculation

$$BU = AP_H U \oplus \{P'(I - F_+ SP_A)^{-1} F_+[D_A TP_H \oplus (D_T P_H + S(I - P_H))]\} =$$

$$ATP_H \oplus \{P'(I - F_+ SP_A)^{-1}[(D_A P_H) \oplus D_{T'} AP_H) + F_+ S(0 \oplus (I - P_H))]\} = T' AP_H \oplus$$

$$\{D_{T'} AP_H + P'(I - F_+ SP_A)^{-1} F_+ S(D_A P_H \oplus 0) + SP'(I - F_+ SP_A)^{-1} F_+(0 \oplus (1 - P_H))\} =$$

$$T' AP_H \oplus \{D_{T'} AP_H + SP'(I - F_+ SP_A)^{-1} F_+(D_A P_H \oplus (I - P_H))\} = \tag{4.11}$$

$$U' AP_H + SP'(I - F_+ SP_A)^{-1} F_+(D_A P_H \oplus (I - P_H)) =$$

$$U'[AP_H + P'(I - F_+ SP_A)^{-1} F_+(D_A P_H \oplus (I - P_H)) = U' B .$$

Therefore any B of the form (4.5) is an intertwining lifting of A.

To formally prove that this B is a contraction we use the following Redheffer cascading interpretation of (4.5). To this end we consider the following system

$$\begin{bmatrix} x \\ k' \end{bmatrix} = \begin{bmatrix} SP_A & (D_A P_H) \oplus (I - P_H) \\ P' & AP_H \end{bmatrix} \begin{bmatrix} x' \\ k \end{bmatrix} \tag{4.12}$$

$$\text{subject to } x' = F_+ x$$

where F is a Schur contraction, k is in $K (= H \oplus H^2(D_T))$ and x is in $X \triangleq H^2(D_A \oplus D_T)$ and x' is in $X' \triangleq H^2(D_A \oplus D_{T'})$ and k' is in $K' (= H \oplus H^2(D_{T'}))$. Solving (4.12) for k' in terms of k yields B_F in the Schur representation (4.5). To be precise $k' = B_F k$. Equation (4.12) is the Redheffer cascading system corresponding to the representation (4.5) , that is, the contractive intertwining lifting $B = B_F$ in (4.5) is the closed loop system for the 2 by 2 scattering matrix in (4.12) with respect to the feedback F (see Figure 4 and equation (1.20) in Section 1).

Equation (4.12) can be used to formally prove that B_F is a contraction. To see this first notice that the 2 by 2 matrix in (4.12) is an isometry. This and the fact that F_+ is a contraction yields

$$\|k'\|^2 + \|x\|^2 = \|x'\|^2 + \|k\|^2 = \|F_+ x\|^2 + \|k\|^2 \le \|x\|^2 + \|k\|^2 .$$

Hence $\|k'\| = \|B_F k\| \le \|k\|$. This proves that B_F in (4.5) is a contraction. With this we conclude our heuristic proof of the fact that B_F is a contractive intertwining lifting of A. (It is not yet a rigorous proof because the operator $F_+(I - SP_A F_+)^{-1}$ may be unbounded.) To complete this section we will present a rigorous proof of this fact inspired by the above analysis. Finally, it is emphasized that so far we have shown that any operator B_F

given by (4.5) is formally a contractive intertwining lifting of A.

To begin we present a geometric interpretation of the cascading system (4.12). The linear space of the set of all formal series with values in a Hilbert space M is denoted by $l(M)$. For example, f in $l(M)$ admits a formal series expansion of the form

$$v = [v_o, v_1, v_2, ...]^{tr} = \sum_o^\infty v_n z^n$$

where v_i is in M for all $i \geq 0$. Here we do not distinguish between formal series and its corresponding infinite column vector. Let F be a Schur contraction where F_n is the coefficient of z^n in the power series expansion of F. Then F defines a linear map F_+ from $X_l = l(D_A \oplus D_T)$ into $X_l' = l(D_A \oplus D_{T'})$ by convolution:

$$F_+ v = \sum_{n=0}^\infty \sum_{i=0}^n F_{n-i} v_i z^n \tag{4.13}$$

or in terms of matrices $F_+ v$ corresponds to

$$F_+ v = \begin{bmatrix} F_0 & 0 & 0 & ... \\ F_1 & F_0 & 0 & ... \\ F_2 & F_1 & F_0 & ... \\ F_3 & F_2 & F_1 & ... \\ \vdots & \vdots & \vdots & ... \end{bmatrix} \begin{bmatrix} v_0 \\ v_1 \\ v_2 \\ v_3 \\ \vdots \end{bmatrix} . \tag{4.14}$$

Let N and N' be the linear spaces defined by

$$N = l(D_A) \oplus l(D_{T'}) \oplus H \oplus l(D_T)$$

$$\tag{4.15}$$

$$N' = l(D_A) \oplus l(D_T) \oplus H' \oplus l(D_{T'}) .$$

The notation $M_1 \oplus M_2$ refers to the orthogonal direct sum when M_1 and M_2 are Hilbert spaces. If M_1 and M_2 are linear spaces, then $M_1 \oplus M_2$ is the usual direct sum, that is, $M_1 \oplus M_2$ can be identified with the space formed by the column vectors

$$M_1 \oplus M_2 = \{[m_1, m_2]^{tr} : m_1 \in M_1 \text{ and } m_2 \in M_2\} .$$

As before $H, H', l(D_A), l(D_T)$ and $l(D_{T'})$ are identified with the appropriate subspaces in N and N'. For example $l(D_T)$ is identified with $\{0\} \oplus \{0\} \oplus \{0\} \oplus l(D_T)$ in N and $\{0\} \oplus l(D_T) \oplus \{0\} \oplus \{0\}$ in N'. The space $l(D_A \oplus D_T)$ is identified with $l(D_A) \oplus l(D_T) \oplus \{0\} \oplus \{0\}$ in N' and $l(D_A) \oplus \{0\} \oplus \{0\} \oplus l(D_T)$ in N. Clearly K and K' are subspaces in N and N' under the identification $\{0\} \oplus \{0\} \oplus H \oplus H^2(D_T)$ in N and $\{0\} \oplus \{0\} \oplus H' \oplus H^2(D_{T'})$ in N'. Also for convenience we denote by $l^\infty(M)$ the subspace of $l(M)$ formed by the elements $v = [v_0, v_1, v_2, ...]^{\tau}$ such that $\sup\|v_n\|$ is finite. The linear maps U_T on N and U_A from N to N' and $U_{T'}$ on N' are defined by

$$U_T(d \oplus v' \oplus h \oplus v) = d \oplus v' \oplus Th \oplus (D_T h + zv)$$

$$U_A(d \oplus v' \oplus h \oplus v) = (D_A h + zd) \oplus v \oplus Ah \oplus v' \qquad (4.16)$$

$$U_{T'}(d \oplus v \oplus h' \oplus v') = d \oplus v \oplus T' h' \oplus (D_{T'} h' + zv')$$

where h is in H, v is in $l(D_T)$ and d is in $l(D_A)$, h' is in H' and v' is in $l(D_{T'})$. Obviously $U_T | K$ and $U_{T'} | K'$ are the minimal isometric dilations U and U' of T and T', respectively. Notice that if we set $K_l = H \oplus l(D_T)$ and $K'_l = H' \oplus l(D_{T'})$, then $N = X'_l \oplus K_l$ and $N' = X_l \oplus K'_l$ and U_A maps $X'_l \oplus K_l$ to $X_l \oplus K'_l$. Let P' be the linear map from N' to K'_l and P_A the linear map from X'_l to X_l defined by

$$P'(d \oplus v \oplus h' \oplus v') = 0 \oplus v' \quad \text{and} \quad P_A(d \oplus v') = d \oplus 0. \qquad (4.17)$$

As before Π' is the operator from N' onto $l(D_{T'})$ which picks out the $l(D_{T'})$ component. Finally, in this setting the unilateral shift S on $l(D_A) \oplus l(D_T) \oplus l(D_{T'})$ is formal multiplication by z, that is, $S(d \oplus v \oplus v') = zd \oplus zv \oplus zv'$.

Now consider the Redheffer feedback system in (4.12) extended to the following general linear setting

$$\begin{bmatrix} x \\ k' \end{bmatrix} = \begin{bmatrix} SP_A & D_A\Pi_H \oplus Q \\ P' & A\Pi_H \end{bmatrix} \begin{bmatrix} x' \\ k \end{bmatrix} \qquad (4.18)$$

$$\text{subject to } x' = F_+ x$$

where F is a Schur contraction, $x' \oplus k$ is in $X'_l \oplus K_l = N$ and $x \oplus k'$ is in $X_l \oplus K'_l = N'$. Here Π_H is the linear map from K_l onto H and Q is the linear map from K_l onto $l(D_T)$ defined by

$$\Pi_H(h \oplus v) = h \text{ and } Q(h \oplus v) = v \qquad (4.19)$$

where $h \oplus v$ is in $H \oplus l(D_T)$. It is easy to show that the 2 by 2 matrix in (4.18) is precisely U_A. In other words the Redheffer cascading system in (4.18) is precisely the closed loop system for the scattering map U_A with respect to the feedback F_+ (see Figure 4 in Section 1 where U_A replaces L, F_+ replaces D_1 and the roles of x and x' are reversed). Notice that $I - SP_A F_+$ is an invertible map on X_l. (This follows from the fact that $I - zP_A F_+$ corresponds to an analytic Toeplitz matrix with the identify on the diagonal; see (4.14).) So solving for k' in terms of k in (4.18) we see that $k' = U_F k$ where

U_F is the linear map from K_l to K_l' defined by

$$U_F = A\Pi_H \oplus \Pi'(I - SF_+P_A)^{-1}F_+(D_A\Pi_H \oplus Q) =$$

$$A\Pi_H \oplus \Pi'F_+(I - SP_AF_+)^{-1}(D_A\Pi_H \oplus Q) . \tag{4.20}$$

The second equation in (4.20) follows from the identity

$$F_+(I - SP_AF_+)^{-1} = (I - F_+SP_A)^{-1}F_+ .$$

Now let B_F equal U_F restricted to the Hilbert space K, that is $B_F = U_F | K$. The following shows that B_F is an operator and even more importantly a contractive intertwining lifting of A. (Recall that an operator is a bounded linear map acting between the appropriate Hilbert spaces.)

4.1 THEOREM. *If* F *is a Schur contraction, then* B_F *is a contractive intertwining lifting of* A.

PROOF. Let $h \oplus v$ be in the Hilbert space $H \oplus H^2(D_T) = K$ and let $d \oplus v'$ be the element in $l(D_A) \oplus l(D_{T'}) = X_l'$ defined by

$$d \oplus v' = (I - SF_+P_A)^{-1}F_+(D_Ah \oplus v) . \tag{4.21}$$

Clearly this implies that

$$(I - SF_+P_A)(d \oplus v') = F_+(D_Ah \oplus v)$$

or using the fact that S commutes with F_+ we have

$$F_+((D_Ah + zd) \oplus v) = d \oplus v' . \tag{4.22}$$

Therefore finding $d \oplus v'$ in (4.21) is equivalent to solving (4.22) where h and v are given. Now let d_n and v_n' be the coefficient of z^n in the formal power series expansion of d and v', respectively. Also let v_n and F_n be the coefficient of z^n in the power series expansion of v and F. By using the definition of the map F_+ in (4.13) or (4.14) we see that (4.22) gives

$$
\begin{bmatrix}
F_o & 0 & 0 & \ldots & 0 \\
F_1 & F_o & 0 & \ldots & 0 \\
F_2 & F_1 & F_o & \ldots & 0 \\
\vdots & \vdots & \vdots & \ldots & \vdots \\
F_{n-1} & F_{n-2} & F_{n-3} & \ldots & 0 \\
F_n & F_{n-1} & F_{n-2} & \ldots & F_o
\end{bmatrix}
\begin{bmatrix}
D_A h \oplus v_o \\
d_o \oplus v_1 \\
d_1 \oplus v_2 \\
\vdots \\
d_{n-2} \oplus v_{n-1} \\
d_{n-1} \oplus v_n
\end{bmatrix}
=
\begin{bmatrix}
d_o \oplus v_o' \\
d_1 \oplus v_1' \\
d_2 \oplus v_2' \\
\vdots \\
d_{n-1} \oplus v_{n-1}' \\
d_n \oplus v_n'
\end{bmatrix} . \tag{4.23}
$$

Since F is a Schur contraction the previous matrix is also a contraction. Thus

$$
\|d_n\|^2 + \sum_{i=0}^{n} \|v_i'\|^2 \le \|D_A h\|^2 + \sum_{i=0}^{n} \|v_i\|^2 \le \|D_A h\|^2 + \|v\|^2 . \tag{4.24}
$$

This readily implies that d is in $l^\infty(D_A)$. Moreover, by letting n approach infinity $\|v'\|^2 \le \|D_A h\|^2 + \|v\|$. So using $\|D_A h\|^2 = \|h\|^2 - \|Ah\|^2$ we obtain

$$
\|Ah\|^2 + \|v'\|^2 \le \|h\|^2 + \|v\|^2 . \tag{4.25}
$$

In other words if $h \oplus v$ is in the Hilbert space K and $d \oplus v'$ is given by (4.21), then d is in $l^\infty(D_A)$ and v' is in the Hilbert space $H^2(D_{T'})$.

By (4.20) and (4.21) we see that $Ah \oplus v' = U_F(h \oplus v)$. Equation (4.25) guarantees that $B_F \overset{\Delta}{=} U_F | K$ is a contraction from K to K'. By following the analysis in (4.11) we obtain $U_{T'} U_F = U_F U_T$. This and the fact that $U_T | K = U$ and $U_{T'} | K' = U'$ is the minimal isometric dilation of T and T', respectively implies that $U' B_F = B_F U$. Obviously B_F is a lifting of A. Therefore B_F is a contractive intertwining lifting of A. This completes the proof.

4.2 THEOREM. *Let* F *be a Schur contraction. The map* $F \to B_F$ *is one to one.*

PROOF. Let $B = B_F$ where F is an unspecified Schur contraction. As before let $h \oplus v$ be in the Hilbert space K and $d \oplus v'$ the element in $l^\infty(D_A) \oplus H^2(D_{T'})$ given by (4.21), or equivalently, (4.22). Consulting (4.23) along with the fact that $F|F$ is a constant ($F_o|F = \omega$ and $F_n|F = 0$ for $n \ge 1$) we have

$$
\omega P_F(D_A h \oplus v_o) = P_{F'}(d_o \oplus v_o') \overset{\Delta}{=} \tilde{d}_o \oplus \tilde{v}_o'
$$
$$
\tag{4.26}
$$
$$
\omega P_F(d_{n-1} \oplus v_n) = P_{F'}(d_n \oplus v_n') \overset{\Delta}{=} \tilde{d}_n \oplus \tilde{v}_n'
$$

for all $n \ge 1$ where v_n, d_n and v_n' are the coefficients of z^n in the power series expansion of v, d and v', respectively. This and (4.21) shows that we can calculate $\tilde{d}_o \oplus \tilde{v}_o'$ and $\tilde{d}_n \oplus \tilde{v}_n'$ for $n \ge 1$ given $k = h \oplus v$. Clearly $\tilde{d}_n \oplus \tilde{v}_n' - d_n \oplus v_n'$ is orthogonal to F'. By virtue of the definition of F' in (4.2) we have

$$D_A d_n = D_A \tilde{d}_n + A^* D_{T'}(\tilde{v}_n - v'_n) \qquad (n \geq 0) . \qquad (4.27)$$

Thus, by (4.26), (4.27), $D_A d_o$ is uniquely determined by k and $Bk = Ah \oplus v'$, and therefore since $d_o \in D_A$ so is d_o. By recurrence it easily follows from (4.26), (4.27) that d_n is also uniquely determined by k and Bk for all $n \geq 1$. Now $F_o(D_A h \oplus v_o) = d_o \oplus v'_o$ shows that F_o is uniquely determined by k and B. This with (4.23) and an induction argument shows that F_n is also uniquely determined by k and B. Since $F = \sum F_n z^n$ the proof is complete.

5. MORE ON THE SCHUR REPRESENTATION

In this section we will show that the set *of all* contractive intertwining liftings B of A is given by the representation $B = B_F$ with a suitable Schur contraction F.

The following is main result of this section.

5.1 THEOREM. *The set of all contractive intertwining liftings B of A is given by $B = B_F$ where F is a Schur contraction. Moreover, the mapping $F \rightarrow B_F$ is one to one. In particular, there is a one to one correspondence between the set of all contractive intertwining liftings B of A, and the set of all Schur contractions.*

5.2 REMARK. The previous theorem shows that the set of all contractive intertwining liftings of A is given by $B_F = U_F | K$ where F is a Schur contraction. Since U_F is the closed loop system for the scattering map U_A with respect to the feedback F_+, Figure 9 provides a Redheffer cascading interpretation for the set of all contractive intertwining liftings $B = B_F$ of A.

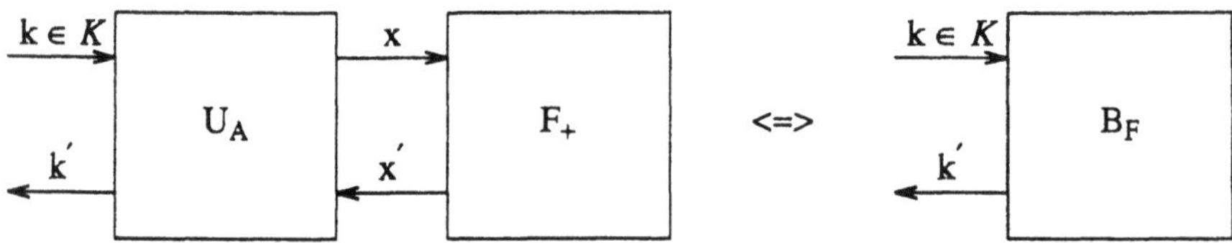

Figure 9

Before presenting a proof of this theorem we will establish some preliminary results. To simplify notation in the proof of this theorem throughout this section we will always assume that $T = U$ is an isometry on H, that is, T is its own minimal isometric dilation. If this is not the case, then one can replace A by AP_H and replace T by U. Then B is a contractive intertwining lifting of A with respect to T and T' if and only if B is a contractive intertwining lifting of AP_H with respect to U and T'. Therefore, obtaining all contractive intertwining liftings of A is equivalent to obtaining all contractive intertwining liftings of AP_H (see Lemma VII.1.1). So for convenience we will assume that $T = U$.

In this setting, the unitary operator ω in (4.3) maps

$$F = \{D_A Th : h \in H\}^- \ \text{onto}\ F' = \{D_A h \oplus D_{T'} Ah : h \in H\}^-$$

and is defined by

$$\omega(D_A Th) = D_A h \oplus D_{T'} Ah \qquad (h \in H) . \tag{5.1}$$

As before ω is an inner and $*$-inner constant function in $H^\infty(F, F')$. Since $T = U$ the spaces N and N' in the previous section become

$$N = l(D_A) \oplus l(D_{T'}) \oplus H \ \text{and}\ N' = l(D_A) \oplus H' \oplus l(D_{T'}) .$$

Let F be an analytic function in D with values in $L(D_A, D_A \oplus D_{T'})$ and $\hat{F}_{n+1}$ the $n + 1$ by $n + 1$ analytic Toeplitz matrix defined by

$$\hat{F}_{n+1} = \begin{bmatrix} F_o & 0 & 0 & \dots & 0 & 0 \\ F_1 & F_o & 0 & \dots & 0 & 0 \\ F_2 & F_1 & F_o & \dots & 0 & 0 \\ . & . & . & \dots & . & . \\ . & . & . & \dots & . & . \\ . & . & . & \dots & F_o & . \\ F_n & F_{n-1} & F_{n-2} & \dots & F_1 & F_o \end{bmatrix} \tag{5.2}$$

where F_n is the coefficient of z^n in the power series expansion of F. Obviously F is a Schur contraction if and only if $\hat{F}_{n+1}$ is a contraction for all $n \geq 0$ and $F_o | F = \omega$ (see Lemma VIII.1.2). According to (4.21), (4.22) and (4.23) adjusted to this setting we see that

$$d \oplus v' = (I - SF_+ P_A)^{-1} F_+ D_A h \qquad (h \in H) \tag{5.3}$$

if and only if

$$\hat{F}_{n+1}[D_A h, d_o, d_1, ..., d_{n-1}]^{tr} = [d_o \oplus v_o', d_1 \oplus v_1', ..., d_n \oplus v_n']^{tr} \quad \text{(for all } n \geq 1) \tag{5.4}$$

where as in the previous section h is in H, d_i and v_i' is the coefficient of z^n in the formal series expansion of d and v', respectively. Obviously $d_i = d_i(D_A h)$ and $v_i' = v_i'(D_A h)$ are linear maps for all i depending on h.

The previous section shows that each B_F is a contractive intertwining lifting of A, and the mapping $F \to B_F$ is one to one. So to complete the proof of Theorem 5.1 we will show by using (5.3) and (5.4) that every contractive intertwining lifting B of A produces a Schur contraction F such that $B = B_F$. To this end we first establish the following result.

5.3 LEMMA. *Let* B *be a contractive intertwining lifting of* A. *Then* B *admits a decomposition of the form*

$$Bh = Ah \oplus YD_A h = Ah \oplus v' \quad (h \in H) \tag{5.5}$$

where Y *is a contraction mapping* D_A *into* $H^2(D_{T'})$ *satisfying*

$$YD_A Th = D_{T'} Ah \oplus YD_A h \quad (h \in H). \tag{5.6}$$

PROOF. Equation (5.5) follows from (4.7) and the fact that in this case $D_T = \{0\}$, or equivalently, $I - P_H = 0$. Equation (5.6) and the matrix form of the minimal isometric dilation U' of T' gives

$$ATh \oplus YD_A Th = BTh = U'(Ah \oplus YD_A h) = T'Ah \oplus (D_{T'} Ah \oplus YD_A h).$$

This yields (5.6) and completes the proof.

Let B be a contractive intertwining lifting of A and Y its corresponding contraction in Lemma 5.3. Throughout the contraction Y_n from D_A to $l^2_{n+1}(D_{T'})$ is the compression of Y to $l^2_{n+1}(D_{T'})$, that is, if $YD_A h = [v_o', v_1', v_2', ...]^{tr}$, then $Y_n D_A h = [v_o', v_1', ..., v_n']^{tr}$. According to the previous lemma and its proof

$$Y_i D_A Th = D_{T'} Ah \oplus Y_{i-1} D_A h \quad (h \in H \text{ and } i \geq 0). \tag{5.7}$$

(We set $Y_{-1} = 0$.)

5.4 LEMMA. *Let* Y_n *be a contraction satisfying (5.7) for* $i \leq n$ *and* $\hat{F}_{n+1}$ *a contractive block Toeplitz matrix of the form (5.2) where* $F_o | F = \omega$. *Assume that* $[h, d_o(D_A h), d_1(D_A h), ..., d_n(D_A h)]^{tr} \oplus Y_n D_A h \quad is \quad the \quad unique \quad element \quad in$

$(H \oplus l^2_{n+1}(D_A)) \oplus l^2_{n+1}(D_{T'})$ *solving*

$$\hat{F}_{n+1}[D_A h, d_0(D_A h), d_1(D_A h), ..., d_{n-1}(D_A h)]^{tr} =$$

$$\tag{5.8}$$

$$[d_0(D_A h), d_1(D_A h), ..., d_n(D_A h)]^{tr} \oplus Y_n D_A h .$$

(The $\oplus$ in (5.8) means that if $Y_n D_A h = [v'_0, ..., v'_n]^{tr}$, then the right hand side of (5.8) equals $[d_0 \oplus v'_0, ..., d_n \oplus v'_n]^{tr}$.) Then

$$d_0(D_A Th) = D_A h \text{ and } d_{i+1}(D_A Th) = d_i(D_A h) \qquad (0 \le i < n \text{ and } h \in H) . \tag{5.9}$$

PROOF. Using $F_0 | F = \omega$ and (5.7) we have

$$\hat{F}_{n+1}[D_A Th, D_A h, d_0(D_A h), ..., d_{n-2}(D_A h)]^{tr} =$$

$$[D_A h \oplus D_{T'} Ah, \hat{F}_n[D_A h, d_0(D_A h), ..., d_{n-2}(D_A h)]^{tr} =$$

$$[D_A h, d_0(D_A h), d_1(D_A h), ..., d_{n-1}(D_A h)]^{tr} \oplus \begin{bmatrix} D_{T'} Ah \\ Y_{n-1} D_A h \end{bmatrix} =$$

$$[D_A h, d_0(D_A h), d_1(D_A h), ..., d_{n-1}(D_A h)]^{tr} \oplus Y_n D_A Th , \tag{5.10}$$

$$\hat{F}_{n+1}[D_A Th, d_0(D_A Th), d_1(D_A Th), ..., d_{n-1}(D_A Th)]^{tr} =$$

$$[d_0(D_A Th), d_1(D_A Th), d_2(D_A Th), ..., d_n(D_A Th)]^{tr} \oplus Y_n D_A Th .$$

However, $Th \oplus [d_0(D_A Th), ..., d_n(D_A Th)]^{tr} \oplus Y_n D_A Th$ is the unique solution of (5.4) where Th replaces h. So comparing the two sets of equations in (5.10) yields (5.9). This completes the proof.

PROOF OF THEOREM 5.1. To prove the remaining part of Theorem 5.1, assume that B is a contractive intertwining lifting of A. Throughout $Bh = Ah \oplus v'$ where h is in H and $Y D_A h = v' = [v'_0, v'_1, v'_2, ...]^{tr}$ is in $H^2(D_{T'})$. By our previous results, we see that $B = B_F$ for some Schur contraction F if and only if there exists a set of operators $\{F_i\}^{\infty}_0$ mapping D_A into $D_A \oplus D_{T'}$ such that

$$\|\hat{F}_{n+1}\| \le 1 , \ F_0 | F = \omega \ \text{ and (5.8) holds for all } n \ge 0 . \tag{5.11}$$

As noted earlier the lower triangular structure of $\hat{F}_{n+1}$ guarantees that d_n in D_A is recursively determined by h and the previous d_i's. The proof of Theorem 4.2 also shows that d_n is uniquely determined by h and $\{v'_i\}^n_0$.

We proceed by induction. The conditions in (5.11) for F_o become

$$\|F_o\| \le 1, \; F_o \,|\, F = \omega \quad \text{and} \quad \Pi' F_o = Y_o \tag{5.12}$$

where Π' is the operator from $D_A \oplus D_{T'}$ onto $D_{T'}$ which picks out the $D_{T'}$ component, that is, $\Pi'(x \oplus y) = y$. This is equivalent to the existence of a contraction of the form

$$F_o = \begin{bmatrix} A_{1,1} & A_{1,2} \\ A_{2,1} & A_{2,2} \end{bmatrix} : F \oplus G \to D_A \oplus D_{T'}$$

satisfying

$$[A_{2,1}, A_{2,2}] = \Pi' F_o = Y_o \text{ and } F_o \,|\, F = \omega, \text{ i.e.} [A_{1,1}^*, A_{2,1}^*]^* = \omega,$$

where $G = D_A \ominus F$. Equations (5.1), (5.7) for $i = 0$ and $F_o \,|\, F = \omega$ give:

$$\Pi' F_o D_A T = D_{T'} A = Y_o D_A T.$$

But $\Pi' \omega = A_{2,1} = \Pi' F_o \,|\, F = Y_o \,|\, F$ and $\Pi' F_o$ and Y_o are compatible on F. Since Y_o and ω are both contractions, Corollary IV.3.6 adjusted to this setting shows that there exists a $A_{1,2}$ such that F_o is a contraction. By the proof of Theorem 4.2, this F_o is unique. This completes the construction of $F_o = \hat{F}_1$.

We proceed by induction and assume that $\{F_i\}_0^n$ (with $n \ge 0$) were constructed such that $\hat{F}_{n+1}$ is a contraction satisfying (5.8). We will now show how to construct F_{n+2}. Namely, we define

$$M = \{[D_A h, d_o(D_A h), ..., d_n(D_A h)]^{\tau} : h \in H\}$$

and set

$$X_{n+1} D_A h = v'_{n+1}(D_A h) - \sum_{i=0}^{n} \Pi' F_{n-i} d_i(D_A h)$$

where $Y_{n+1} D_A h = [v'_o(D_A h), ..., v'_{n+1}(D_A h)]^{\tau}$. Therefore

$$\begin{bmatrix} F_o & 0 & 0 & ... & 0 \\ F_1 & F_o & 0 & ... & 0 \\ F_2 & F_1 & F_o & ... & 0 \\ \vdots & \vdots & \vdots & ... & \vdots \\ F_n & F_{n-1} & F_{n-2} & ... & 0 \\ X_{n+1} & \Pi' F_n & \Pi' F_{n-1} & ... & \Pi' F_o \end{bmatrix} \begin{bmatrix} D_A h \\ d_o(D_A h) \\ d_1(D_A h) \\ \vdots \\ d_{n-1}(D_A h) \\ d_n(D_A h) \end{bmatrix} = \begin{bmatrix} d_o(D_A h) \\ d_1(D_A h) \\ d_2(D_A h) \\ \vdots \\ d_n(D_A h) \\ 0 \end{bmatrix} \oplus \begin{bmatrix} Y_n D_A h \\ v'_{n+1}(D_A h) \end{bmatrix}$$

Obviously this matrix C' is a contraction from M to $X \oplus D_{T'} \triangleq l^2_{n+1}(D_A \oplus D_{T'}) \oplus D_{T'}$.

Also $P_X C' = [\hat{F}_{n+1}\ 0]$ is a contraction. By adjusting Corollary IV.3.6 to this setting there exists a contraction C from $l^2_{n+2}(D_A)$ to $X \oplus D_T$ satisfying $P_X C = [\hat{F}_{n+1}\ 0]$ and $C|M = C'$. Thus

$$C = \begin{bmatrix} F_o & 0 & 0 & \dots & 0 \\ F_1 & F_0 & 0 & \dots & 0 \\ F_2 & F_1 & F_o & \dots & 0 \\ \vdots & \vdots & \vdots & \dots & \vdots \\ F_n & F_{n-1} & F_{n-2} & \dots & 0 \\ C_{n+1} & C_n & C_{n-1} & \dots & C_o \end{bmatrix}$$

where

$$C_{n+1}D_A h + \sum_{i=0}^{n} C_{n-i}d_i(D_A h) = v'_{n+1}(D_A h) \, . \tag{5.13}$$

Since C is a contraction and $F_o D_A Th = \omega D_A Th$ for all h in H, we have $C_i D_A Th = 0$ for all $1 \le i \le n+1$. So using (5.9) and (5.13) we obtain

$$C_o D_A h = C_{n+1} D_A T^{n+1}h + \cdots + C_1 D_A Th + C_o D_A h =$$

$$C_{n+1} D_A T^{n+1}h + C_n d_o(D_A T^{n+1}h) + \cdots + C_o d_n(D_A T^{n+1}h) =$$

$$v'_{n+1}(D_A T^{n+1}h) = v'_o(D_A h) = Y_o D_A h = \Pi' F_o D_A h \, .$$

The second from the last equality follows from (5.6) in Lemma 5.2 and the last equality from (5.12). By a similar argument

$$C_1 D_A h + \Pi' F_o d_o(D_A h) = C_{n+1} D_A T^n h + \cdots + C_1 D_A h + C_o d_o(D_A h) =$$

$$C_{n+1} D_A T^n h + C_n d_o(D_A T^n h) + \cdots + C_o d_n(D_A T^n h) =$$

$$v'_{n+1}(D_A T^n h) = v'_1(D_A h) = \Pi' F_1 D_A h + \Pi' F_o d_o(D_A h) \, .$$

Thus $C_1 = \Pi' F_1$. Proceeding in this fashion $C_i = \Pi' F_i$ for $0 \le i \le n$.

Now consider the matrix $\hat{F}_{n+2}$ written under the form

$$
\hat{F}_{n+2} = \begin{bmatrix}
F_o & 0 & \dots & 0 & 0 \\
F_1 & F_o & \dots & 0 & 0 \\
\vdots & \vdots & \dots & \vdots & \vdots \\
F_n & F_{n-1} & \dots & F_o & 0 \\
X_{n+1} & \Pi' F_n & \dots & \Pi' F_1 & \Pi' F_o \\
A_{2,1} & \Pi'_A F_n & \dots & \Pi'_A F_1 & \Pi'_A F_o
\end{bmatrix} = \begin{bmatrix}
A_{1,1} & A_{1,2} \\
A_{2,1} & A_{2,2}
\end{bmatrix}
$$

where Π'_A is the operator from $D_A \oplus D_{T'}$ onto D_A which picks out the D_A component. Obviously $[A_{1,1}, A_{1,2}] = C$ and $[A_{1,2}, A_{2,2}]^{tr} = [0 \, \hat{F}_{n+1}]^{tr}$ are both contractions. By adjusting Corollary IV.3.6 to this setting there exists a $A_{2,1}$ such that $\hat{F}_{n+2}$ is a contraction. This $\hat{F}_{n+2}$ has the desired properties. The proof is now complete.

6. ANOTHER REDHEFFER CASCADING INTERPRETATION OF THE COMMUTANT LIFTING THEOREM

In this section we will use the Redheffer cascading system in equations (1.23) and (1.24) corresponding to Figure 5, to obtain the Schur type representation for the commutant lifting theorem, which is the natural generalization of the classical Schur fractional representation for the Carathéodory and Nevanlinna-Pick interpolation problem. This representation will be used in the next section to compute the set of all contractive intertwining liftings.

We begin by establishing some notation. As before ω is the unitary operator defined in (4.3), and F, F', G, G' are the spaces in (4.2) and (4.4). The linear maps p from $l(D_A \oplus D_T)$ onto $l(F)$ and q from $l(D_A \oplus D_T)$ onto $l(G)$ are defined by

$$
p(d \oplus v) = \sum_o^\infty P_F(d_n \oplus v_n)z^n \text{ and } q(d \oplus v) = \sum_o^\infty P_G(d_n \oplus v_n)z^n \quad (d \oplus v \in l(D_A \oplus D_T))
$$

where d_n and v_n is the coefficient of z^n in the formal power series expansion of d and v, respectively. The maps p' from $l(D_A \oplus D_{T'})$ onto $l(F')$ and q' from $l(D_A \oplus D_{T'})$ onto $l(G')$ are defined analogously. Now let p_A from $l(D_A)$ to $l(F)$ and q_A from $l(D_A)$ to $l(G)$ be the linear maps defined by $p_A = p \, | \, l(D_A)$ and $q_A = q \, | \, l(D_A)$. As before Π_A is the linear map from $Z = l(D_A) \oplus l(D_T) \oplus l(D_{T'})$ which picks out the $l(D_A)$ component, that is, $\Pi_A(d \oplus v \oplus v') = d$. By identifying $l(D_A \oplus D_T)$ or $l(D_A \oplus D_{T'})$ or $l(F)$ etc. with the appropriate subspace of Z we also use Π_A as the map from any one of these spaces to $l(D_A)$ which picks out the $l(D_A)$ component.

Now consider the Redheffer cascading system in (4.18) decomposed in the following way

$$\begin{bmatrix} f \\ g \\ h' \oplus v' \end{bmatrix} = \begin{bmatrix} Sp_A \Pi_A & Sp_A \Pi_A & p(D_A \oplus Q) \\ Sq_A \Pi_A & Sq_A \Pi_A & q(D_A \oplus Q) \\ P' & P' & A \oplus 0 \end{bmatrix} \begin{bmatrix} f' \\ g' \\ h \oplus v \end{bmatrix} \tag{6.1}$$

$$\text{subject to } f' = \omega_+ f \text{ and } g' = R_+ g$$

where $F = \omega \oplus R$ is a Schur contraction, that is, R is a contractive analytic function in $H^\infty(G, G')$. Here $f \oplus g$ in $l(F) \oplus l(G)$ and $f' \oplus g'$ in $l(F') \oplus l(G')$ is the decomposition of $d \oplus v$ in $l(D_A) \oplus l(D_T)$ and $d \oplus v'$ in $l(D_A) \oplus l(D_{T'})$ given by

$$f \oplus g = p(d \oplus v) \oplus q(d \oplus v) \text{ and } f' \oplus g' = p'(d \oplus v') \oplus q'(d \oplus v'). \tag{6.2}$$

Since (6.1) is a reformulation of the Redheffer cascading system in (4.18) according to the decomposition in (6.2), it follows that the solution to (6.1) is given by $h' \oplus v' = U_F(h \oplus v)$ and $B_F \triangleq U_F \mid K$ is a contractive intertwining lifting of A. It is important to notice that in (6.1) the 3 by 3 matrix and ω_+ are fixed, while R is an arbitrary contractive analytic function in $H^\infty(G, G')$. Therefore by adjusting (1.23) and its solution (1.24) to this setting, it follows that (6.1) is equivalent to the following Redheffer cascading equation

$$\begin{bmatrix} g \\ h' \oplus v' \end{bmatrix} = \tag{6.3}$$

$$\begin{bmatrix} Sq_A(I_A - S\Pi_A \omega_+ p_A)^{-1} \Pi_A & [Sq_A(I_A - S\Pi_A \omega_+ p_A)^{-1} \Pi_A \omega_+ p + q](D_A \oplus Q) \\ P'(I_{X'} - S\omega_+ p_A \Pi_A)^{-1} & P'(I_{X'} - S\omega_+ p_A \Pi_A)^{-1} \omega_+ p(D_A \oplus Q) + A \oplus 0 \end{bmatrix} \begin{bmatrix} g' \\ h \oplus v \end{bmatrix}$$

$$\text{subject to } g' = R_+ g$$

where I_A and $I_{X'}$ is the identity on $l(D_A)$ and $X'_l = l(D_A \oplus D_{T'})$ respectively, and Q is defined in (4.19). To see how to obtain (6.3) from (1.24) notice that the first term in (1.24) adjusted to this setting gives

$$A_{2,1}(I - WA_{1,1})^{-1}WA_{1,2} + A_{2,2} = Sq_A \Pi_A(I_{F'} - S\omega_+ p_A \Pi_A)^{-1} \omega_+ Sp_A \Pi_A + Sq_A \Pi_A =$$

$$Sq_A[(I_A - S\Pi_A \omega_+ p_A)^{-1} S\Pi_A \omega_+ p_A + I]\Pi_A = Sq_A(I_A - S\Pi_A \omega_+ p_A)^{-1} \Pi_A$$

which is precisely the first term in the matrix in (6.3). A similar calculation produces the rest of the entries in (6.3). Finally, it is noted that all the inverses in (6.3) exist in the appropriate $l(\cdot)$ spaces because they all contain the operator S. This corresponds to inverting (between the appropriate $l(\cdot)$ spaces) an analytic Toeplitz matrix with the identity on the diagonal.

Let C_A be the 2 by 2 matrix from $l(G') \oplus (H \oplus l(D_T))$ to $l(G) \oplus (H' \oplus l(D_{T'}))$ in (6.3). Then U_F is precisely the solution $(h' \oplus v' = U_F(h \oplus v))$ to (6.3). In other words U_F is the closed loop system for the scattering map C_A with respect to the feedback R_+. The network interpretation for C_A is given in Figure 10, which shows that C_A is obtained by feeding back ω_+ in U_A.

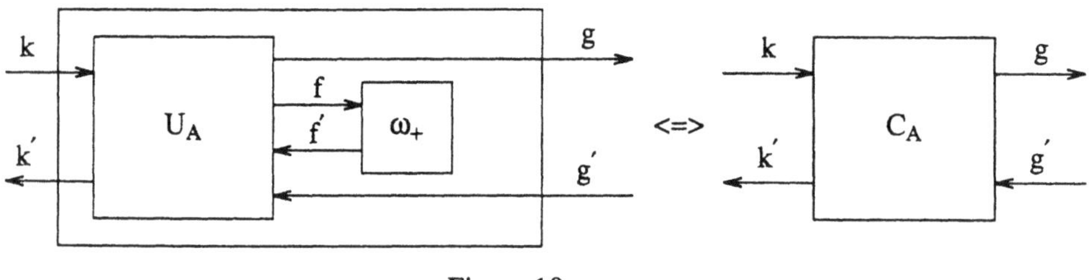

Figure 10

According to Theorem 5.1 the set of all contractive intertwining liftings B of A is given by $B_F = U_F \,|\, K$ where F is a Schur contraction (and $K = H \oplus H^2(D_T)$). However, U_F is also the solution to the Redheffer cascading system in (6.3). In other words U_F is precisely the closed loop system for the scattering map C_A with respect to the feedback R_+. Figure 11 provides a Redheffer scattering interpretation for the set of all contractive intertwining liftings $B = B_F$ of A. It is emphasized that in Figure 11 the operator C_A depends only on the data A, T and T' while R is arbitrary in $H_1^\infty(G, G')$.

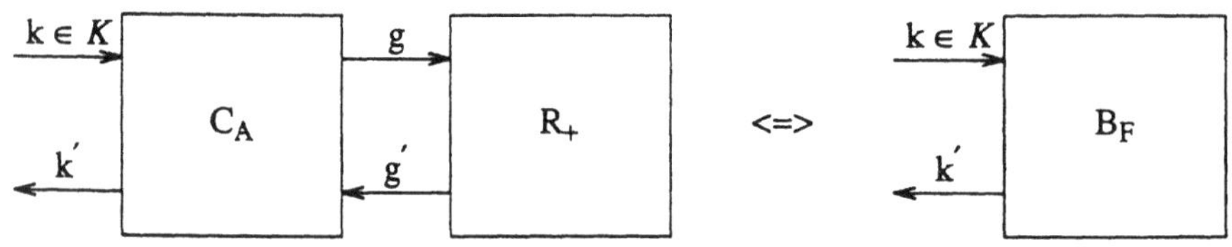

Figure 11

The main advantage of the network interpretation of the set of all contractive intertwining liftings B_F of A in Figure 11 over that in Figure 9, is that in Figure 11 we only have to feedback R in $H_1^\infty(G, G')$ and G and G' may be of low dimension; whereas in Figure 9 we must feedback F in $H_1^\infty(D_A \oplus D_T, D_A \oplus D_{T'})$ and D_A may be of very high dimension.

The following result which will be useful in computing the set of all contractive intertwining liftings shows that $C_A \mid (H^2(G') \oplus K)$ is a contraction.

6.1 THEOREM. *The linear map C_A is a contraction from the Hilbert space $H^2(G') \oplus K$ into the Hilbert space $H^2(G) \oplus K'$.*

PROOF.   Let $x \oplus (h' \oplus v')$ be in $l(D_A \oplus D_T) \oplus (H' \oplus l(D_{T'}))$ and $(d \oplus v') \oplus (h \oplus v)$ in $l(D_A \oplus D_{T'}) \oplus (H \oplus H^2(D_T))$ be a solution to (4.18), that is,

$$x \oplus (h' \oplus v') = ((D_A h + zd) \oplus v) \oplus (Ah \oplus v') \tag{6.4}$$

where $x' = F_+ x$ and $F = \omega \oplus R$ is a Schur contraction. Now let $f \oplus g = px \oplus qx$ and $f' \oplus g' = p'(d \oplus v') \oplus q'(d \oplus v')$. Then $f' = \omega_+ f$ and $g' = R_+ g$. As before let $d_n,\ v_n,\ v'_n,\ f_n$ etc. be the coefficient of z^n in the power series expansion of d, v, v', f etc. respectively. Then

$$\|h\|^2 + \sum_0^{n+1} \|v_i\|^2 + \sum_0^n \|g'_i\|^2 =$$

$$\|Ah\|^2 + \|D_A h\|^2 + \|v_0\|^2 + \sum_1^{n+1} \|v_i\|^2 + \sum_0^n \|g'_i\|^2 =$$

$$\|Ah\|^2 + \|D_A h \oplus v_0\|^2 + \sum_0^n \|d_i \oplus v_{i+1}\|^2 - \sum_0^n \|d_i\|^2 + \sum_0^n \|g'_i\|^2 =$$

$$\|Ah\|^2 + \sum_0^{n+1} \|x_i\|^2 - \sum_0^n \|d_i\|^2 + \sum_0^n \|g'_i\|^2 =$$

$$\|Ah\|^2 + \sum_0^{n+1} \|f'_i\|^2 + \sum_0^{n+1} \|g_i\|^2 - \sum_0^n \|d_i\|^2 + \sum_0^n \|g'_i\|^2 \geq$$

$$\|Ah\|^2 + \sum_0^n \|d_i \oplus v'_i\|^2 + \sum_0^n \|g_i\|^2 - \sum_0^n \|d_i\|^2 =$$

$$\|Ah\|^2 + \sum_0^n \|g_i\|^2 + \sum_0^n \|v'_i\|^2 \; .$$

The third equality follows from the definition of x in (6.4). The fourth equality is a consequence of $f_i \oplus g_i = px_i \oplus qx_i$ and $f'_i = \omega f_i$. The inequality follows from

$f_i' \oplus g_i' = (p' \oplus q')(d_i \oplus v_i')$. Finally, using $h' = Ah$ and letting n approach infinity in the previous computation, we obtain that $\|g \oplus (h' \oplus v')\|^2 \leq \|g' \oplus (h \oplus v)\|^2$. By construction $g \oplus (h' \oplus v') = C_A(g' \oplus (h \oplus v))$. Thus $C_A \,|\, (H^2(G') \oplus K)$ is a contraction. The proof is now complete.

Recall that if T is a $*$-stable contraction, then $L_* \stackrel{\Delta}{=} \overline{(I - UT^*)}H$ is the wandering subspace for its minimal isometric dilation U and $K = M_+(L_*)$; see Section VI.4. In particular, $U|K$ is a unilateral shift with respect to the decomposition $K = M_+(L_*)$. In fact the operator $W = W_T$ from K onto $H^2(D_{T^*})$ defined by

$$W(h \oplus v) = D_{T^*}(I - zT^*)^{-1}h + \Theta_T v \quad (h \oplus v \in H \oplus H^2(D_T)) \tag{6.5}$$

is a unitary operator satisfying $WU = SW$ where S also refers to the unilateral shift on $H^2(D_{T^*})$; see Theorem IX.6.4. In the next section we will use the following remarkable property of the operator W namely

$$W^* d_* = D_{T^*} d_* \oplus (- T^* d_*) \quad (d_* \in D_{T^*}) \tag{6.6}$$

for every constant function d_* in $H^2(D_{T^*})$. To see this notice that for h in H and v in $H^2(D_T)$ we have

$$(h \oplus v, W^* d_*) = (W(h \oplus v), d_*) = (D_{T^*}(I - zT^*)^{-1}h + \Theta_T(z)v(z), d_*) =$$

$$(D_{T^*}h + \Theta_T(0)v(0), d_*) = (h \oplus v, D_{T^*}d_* \oplus (- T^* d_*)) .$$

The last equality follows from $\Theta_T(0) = - T|D_T$ (see (IX.6.1)). Thus (6.6) holds.

Now assume that T' is also $*$-stable and let $W' = W_{T'}$ be the corresponding unitary operator from K' onto $H^2(D_{T'^*})$ for U'. A simple calculation shows that $(S \oplus U_{T'})C_A = C_A(S \oplus U_T)$. In particular, by the previous theorem $(I \oplus W')C_A(I \oplus W^*)$ is a contraction from the Hilbert space $H^2(G') \oplus H^2(D_{T'^*})$ into the Hilbert space $H^2(G) \oplus H^2(D_{T^*})$ which intertwines two unilateral shifts. By Theorem IX.1.1 this implies that C_A uniquely determines a contractive analytic function. This readily proves the following result.

6.2 COROLLARY. *Let* T *and* T' *be two $*$-stable contraction. Then* $C = (I \oplus W')C_A(I \oplus W^*)\,|\,(G \oplus D_{T^*})$ *is a contractive analytic function in*

$H^{\infty}(G^{'} \oplus D_{T^{*}}, G \oplus D_{T^{'*}})$ *satisfying* $C_{+} = (I \oplus W^{'})C_{A}(I \oplus W^{*})$.

7. REDHEFFER CASCADING AND THE
COMPUTATION OF ALL CONTRACTIVE INTERTWINING LIFTINGS

In this section we will use the Redheffer cascading system in (6.3) to compute the set of all contractive intertwining liftings B of A when A is a strict contraction, and both T and $T^{'}$ are $*$-stable. In this case $K = M_{+}(L_{*})$ and $K^{'} = M_{+}(L_{*}^{'})$, and B is a multiplication operator commuting with the unilateral shifts U and $U^{'}$. Using this fact we will present a block diagram which will allow one to design a circuit to actually compute the set of all contractive intertwining liftings. Then in Section 9 we will use Redheffer products to also compute the set of all contractive intertwining liftings of a strictly contractive Hankel operator.

We begin by establishing some notation. Throughout this section A is a strict contraction in $I(T, T^{'})$. As before ω is the unitary operator from F onto $F^{'}$ defined by

$$\omega \begin{bmatrix} D_{A}T \\ D_{T} \end{bmatrix} = \begin{bmatrix} D_{A} \\ D_{T^{'}}A \end{bmatrix} \tag{7.1}$$

where F and $F^{'}$ are the spaces in (4.2). The residue spaces G and $G^{'}$ are defined in (4.4). Let us recall that if C is an operator from X to Y and $C^{*}C$ is invertible, then $C(C^{*}C)^{-\frac{1}{2}}$ is an isometry from X to Y. Moreover, C and $C(C^{*}C)^{-\frac{1}{2}}$ have the same range. Now let C be the operator from H to $D_{A} \oplus D_{T}$ defined by the column matrix $C = [D_{A}T, D_{T}]^{\tau}$. It is easy to verify that $C^{*}C = D_{AT}^{2}$. Since A is a strict contraction, D_{AT} is invertible. So applying the inverse of D_{AT} to both sides of (7.1) we obtain

$$\omega C(C^{*}C)^{-\frac{1}{2}} = \omega \begin{bmatrix} D_{A}T \\ D_{T} \end{bmatrix} D_{AT}^{-1} = \begin{bmatrix} D_{A} \\ D_{T^{'}}A \end{bmatrix} D_{AT}^{-1} \ .$$

Because $C(C^{*}C)^{-\frac{1}{2}}$ is an isometry whose range is F we can apply the adjoint of $C(C^{*}C)^{-\frac{1}{2}}$ to both sides of the previous equation to obtain

$$\omega P_{F} = \begin{bmatrix} D_{A}D_{AT}^{-2}T^{*}D_{A} & D_{A}D_{AT}^{-2}D_{T} \\ D_{T^{'}}AD_{AT}^{-2}T^{*}D_{A} & D_{T^{'}}AD_{AT}^{-2}D_{T} \end{bmatrix} : D_{A} \oplus D_{T} \to D_{A} \oplus D_{T^{'}} \ . \tag{7.2}$$

It is emphasized that in this setting ωP_{F} is a block matrix.

Let $x \oplus y$ in $D_{A} \oplus D_{T}$ be an element in G. Thus $x \oplus y$ is orthogonal to $\{D_{A}Th \oplus D_{T}h : h \in H\}$, or equivalently, $D_{A}x \oplus y$ is orthogonal to $\{Th \oplus D_{T}h : h \in H\}$. Recall that the rotation matrix

$$R_T \triangleq \begin{bmatrix} T & D_{T^\bullet} \\ D_T & -T^* \end{bmatrix} : H \oplus D_{T^\bullet} \to H \oplus D_T$$

is unitary (see Corollary IV.1.4). This implies that $D_A x \oplus y = D_{T^\bullet} d \oplus (-T^* d)$ for some d in $D_{T^\bullet}$. Since D_A is invertible $x \oplus y = D_A^{-1} D_{T^\bullet} d \oplus (-T^* d)$. Clearly $x \oplus y$ is in the range of the operator

$$C_1 = \begin{bmatrix} D_A^{-1} D_{T^\bullet} \\ -T^* \end{bmatrix} : D_{T^\bullet} \to D_A \oplus D_T .$$

It is easy to verify that the range of C_1 is also orthogonal to F. Therefore $C_1 D_{T^\bullet} = G$. Notice that for all d in $D_{T^\bullet}$ we have

$$\|C_1 d\|^2 = \|D_A^{-1} D_{T^\bullet} d\|^2 + \|T^* d\|^2 \geq \delta \|D_{T^\bullet} d\|^2 + \|T^* d\|^2 \geq \delta \|d\|^2$$

where δ is any number $0 < \delta < 1$ satisfying $\|D_A^{-1} h\|^2 \geq \delta \|h\|^2$ for all h in H. Therefore the range of C_1 is closed and $C_1^* C_1$ is invertible. In particular, if we set $G \triangleq C_1 (C_1^* C_1)^{-\frac{1}{2}}$, then G is an isometry whose range is G. Moreover, this G is given by

$$G = \begin{bmatrix} D_A^{-1} D_{T^\bullet} \\ -T^* \end{bmatrix} N^{-\frac{1}{2}} : D_{T^\bullet} \to D_A \oplus D_T \text{ where } N = (TT^* + D_{T^\bullet} D_A^{-2} D_{T^\bullet}) | D_{T^\bullet} . \tag{7.3}$$

(Because T^* maps $D_{T^\bullet}$ into D_T and T maps D_T into $D_{T^\bullet}$ the operator TT^* maps $D_{T^\bullet}$ into $D_{T^\bullet}$.) Finally, it is noted that (7.3) shows that the dimension of G equals $\delta_{T^\bullet}$ the dimension of $D_{T^\bullet}$.

The vector $x \oplus y$ in $D_A \oplus D_{T'}$ is in G' if and only if $x \oplus y$ is orthogonal to $\{D_A h \oplus D_{T'} Ah : h \in H\}$, or equivalently, $D_A x + A^* D_{T'} y = 0$. Since D_A is invertible $x = -D_A^{-1} A^* D_{T'} y$. Thus $x \oplus y$ is in the range of the operator

$$C_2 = \begin{bmatrix} -D_A^{-1} A^* D_{T'} \\ I \end{bmatrix} : D_{T'} \to H \oplus D_{T'}$$

On the other hand since $D_A = H$, it is easy to verify that the range of C_2 is contained in G'. Obviously $C_2^* C_2$ is invertible. So the range of C_2 is closed and equal to G'. Therefore the operator $G' \triangleq C_2 (C_2^* C_2)^{-1}$ is an isometry, and is given by

$$G' = \begin{bmatrix} -D_A^{-1} A^* D_{T'} \\ I \end{bmatrix} N_1^{-\frac{1}{2}} : D_{T'} \to D_A \oplus D_{T'} \text{ where } N_1 = (I + D_{T'} A D_A^{-2} A^* D_{T'}) | D_{T'}$$

$$\tag{7.4}$$

Moreover, the range of G' is G' and the dimension of G' is $\delta_{T'}$ the dimension of $D_{T'}$. Finally, it is noted that a simple calculation shows that

$$N_1 = [T^{'*}T^{'} + D_{T^{'}}D_A^{-2} \cdot D_{T^{'}}]|D_{T^{'}} \ . \tag{7.4a}$$

Therefore N formed by A, T is the dual of N_1 formed by $T^{'}$, A, that is, if we replace A by A^* and T by $T^{'*}$ in the formula for N we obtain N_1.

From now on, throughout this section, we make the supplementary assumption that both T and $T^{'}$ are $*$-stable. As before let $W = W_T$ be the unitary operator from $K = H \oplus H^2(D_T)$ onto $H^2(D_{T^*})$ in (6.5) satisfying $WU = SW$ and $W^{'} = W_{T^{'}}$ the corresponding unitary operator from $K^{'} = H^{'} \oplus H^2(D_{T^{'}})$ onto $H^2(D_{T^{'*}})$ satisfying $SW^{'} = W^{'}U^{'}$. Let Φ_+ be the operator from $H^2(D_{T^{'}}) \oplus H^2(D_{T^*})$ to $H^2(D_{T^*}) \oplus H^2(D_{T^{'*}})$ defined by

$$\Phi_+ = (G_+^* \oplus W^{'})C_A(G_+^{'} \oplus W^*) \ .$$

Corollary 6.2 shows that Φ_+ defines a unique contractive analytic function Φ in $H^\infty(D_{T^{'}} \oplus D_{T^*}, D_{T^*} \oplus D_{T^{'*}})$ by

$$\Phi(z) = \begin{bmatrix} \Phi_{1,1}(z) & \Phi_{1,2}(z) \\ \Phi_{2,1}(z) & \Phi_{2,2}(z) \end{bmatrix} \tag{7.5}$$

where

$$\Phi_{1,1}(z) = zG^*\Pi_A^*(I_A - z\Pi_A\omega P_F\Pi_A^*)^{-1}\Pi_A G^{'}$$

$$\Phi_{1,2}(z) = [zG^*\Pi_A^*(I_A - z\Pi_A\omega P_F\Pi_A^*)^{-1}\Pi_A\omega P_F + G^*](D_A \oplus Q)W^* |D_{T^*}$$

$$\tag{7.5a}$$

$$\Phi_{2,1}(z) = \Theta_{T^{'}}(z)\Pi^{'}(I - z\omega P_F P_{D_A})^{-1}G^{'}$$

$$\Phi_{2,2}(z) = \{\Theta_{T^{'}}(z)\Pi^{'}(I - z\omega P_F P_{D_A})^{-1}\omega P_F(D_A \oplus Q) + [D_{T^{'*}}(I - zT^{'*})^{-1}A, \ 0]\}W^* |D_{T^*}$$

To convert (7.5a) to a more usable form we need the operator T_A on H defined by $T_A \triangleq D_A^2 T D_{AT}^{-2}$. We claim that T_A is similar to a contraction. To see this notice that

$$T_A \triangleq D_A^2 T D_{AT}^{-2} = D_{AT}(D_A D_{AT}^{-1})^*(D_A T D_{AT}^{-1})D_{AT}^{-1} \ .$$

Using $\|D_{AT}h\|^2 = \|D_A h\|^2 + \|D_{T^{'}}Ah\|^2$ and $\|D_{AT}h\|^2 = \|D_T h\|^2 + \|D_A Th\|^2$ for all h in H, a simple computation shows that both $D_A D_{AT}^{-1}$ and $D_A T D_{AT}^{-1}$ are contractions. Thus, T_A is similar to a contraction. Therefore, the operator $(I - zT_A^*)^{-1}$ which plays a basic role in the following results is well defined for all z in D.

By using (7.2) to (7.4) the first entry $\Phi_{1,1}(z)$ in (7.5) becomes

$$\Phi_{1,1}(z) = zG^* \Pi_A^* (I_A - z\Pi_A \omega P_F \Pi_A^*)^{-1} \Pi_A G' =$$

$$- zN^{-\frac{1}{2}} D_{T^\cdot} D_A^{-1} (I - zD_A D_{AT}^{-2} T^* D_A)^{-1} D_A^{-1} A^* D_{T'} N_1^{-\frac{1}{2}} =$$

$$- zN^{-\frac{1}{2}} D_{T^\cdot} (I - zD_{AT}^{-2} T^* D_A^2)^{-1} D_A^{-2} A^* D_{T'} N_1^{-\frac{1}{2}} .$$

So if we use $T_A^* = D_{AT}^{-2} T^* D_A^2$, then

$$\Phi_{1,1}(z) = - zN^{-\frac{1}{2}} D_{T^\cdot} (I - zT_A^*)^{-1} D_A^{-2} A^* D_{T'} N_1^{-\frac{1}{2}} . \tag{7.6}$$

A similar calculation shows that the $\Phi_{1,2}(z)$ term of $\Phi(z)$ is given by

$$\Phi_{1,2}(z) = \{zN^{-\frac{1}{2}} D_{T^\cdot} (I - zT_A^*)^{-1} [T_A^*, \ D_{AT}^{-2} D_T Q] + N^{-\frac{1}{2}} [D_{T^\cdot}, \ - TQ] \} W^* \, | D_{T^\cdot}$$

$$\tag{7.7a}$$

$$= [N^{-\frac{1}{2}} D_{T^\cdot} (I - zT_A^*)^{-1} (\Pi_H + zD_{AT}^{-2} D_T Q) - N^{-\frac{1}{2}} TQ] W^* \, | D_{1^\cdot} .$$

Using (6.6) and $D_T T^* = T^* D_{T^\cdot}$ we have

$$\Phi_{1,2}(z) = N^{-\frac{1}{2}} D_{T^\cdot} (I - zT_A^*)^{-1} (I - zT_A^* D_A^{-2}) D_{T^\cdot} + N^{-\frac{1}{2}} TT^* =$$

$$N^{-\frac{1}{2}} + N^{-\frac{1}{2}} D_{T^\cdot} [(I - zT_A^*)^{-1} (I - zT_A^* D_A^{-2}) - I)] D_{T^\cdot} =$$

$$N^{-\frac{1}{2}} + N^{-\frac{1}{2}} D_{T^\cdot} (I - zT_A^*)^{-1} (I - zT_A^* D_A^{-2} - I + zT_A^*) D_{T^\cdot} =$$

$$N^{-\frac{1}{2}} + zN^{-\frac{1}{2}} D_{T^\cdot} (I - zT_A^*)^{-1} T_A^* (I - D_A^{-2}) D_{T^\cdot} =$$

$$N^{-\frac{1}{2}} - N^{-\frac{1}{2}} D_{T^\cdot} (I - D_A^{-2}) D_{T^\cdot} - N^{-\frac{1}{2}} D_{T^\cdot} (I - zT_A^*)^{-1} D_A^{-2} A^* A D_{T^\cdot} .$$

Finally, using the definition of N in (7.3) we obtain

$$\Phi_{1,2}(z) = N^{\frac{1}{2}} - N^{-\frac{1}{2}} D_{T^\cdot} (I - zT_A^*)^{-1} D_A^{-2} A^* A D_{T^\cdot} . \tag{7.7}$$

Using (7.2) to (7.4) we see that the $\Phi_{2,1}(z)$ term of $\Phi(z)$ is

$$\Phi_{2,1}(z) = \Theta_{T'}(z)\Pi'(I - z\omega P_F P_{D_A})^{-1}G' =$$

$$\Theta_{T'}(z)\Pi'\begin{bmatrix} I - zD_A D_{A'T}^{-2}T^*D_A & 0 \\ -zD_{T'}AD_{A'T}^{-2}T^*D_A & I \end{bmatrix}^{-1} G' = \quad (7.8a)$$

$$\Theta_{T'}(z)N_1^{-\frac{1}{2}} - \Theta_{T'}(z)zD_{T'}AD_{A'T}^{-2}T^*D_A(I - zD_A D_{A'T}^{-2}T^*D_A)^{-1}D_A^{-1}A^*D_{T'}N_1^{-\frac{1}{2}} =$$

$$\Theta_{T'}(z)N_1^{-\frac{1}{2}} - z\Theta_{T'}(z)D_{T'}A(I - zT_A^*)^{-1}D_{A'T}^{-2}T^*A^*D_{T'}N_1^{-\frac{1}{2}} .$$

This implies that

$$\Phi_{2,1}(z) = \Theta_{T'}(z)N_1^{-\frac{1}{2}} - z\Theta_{T'}(z)D_{T'}A(I - zT_A^*)^{-1}T_A^*D_A^{-2}A^*D_{T'}N_1^{-\frac{1}{2}} =$$

$$\Theta_{T'}(z)N_1^{-\frac{1}{2}} + \Theta_{T'}(z)D_{T'}AD_A^{-2}A^*D_{T'}N_1^{-\frac{1}{2}} - \Theta_{T'}(z)D_{T'}A(I - zT_A^*)^{-1}D_A^{-2}A^*D_{T'}N_1^{-\frac{1}{2}} .$$

So using the definition of N_1 in (7.4) we obtain

$$\Phi_{2,1}(z) = \Theta_{T'}(z)N_1^{\frac{1}{2}} - \Theta_{T'}(z)D_{T'}A(I - zT_A^*)^{-1}D_A^{-2}A^*D_{T'}N_1^{-\frac{1}{2}} . \quad (7.8)$$

Finally, a calculation similar to (7.8a) shows that the last term is given by

$$\Phi_{2,2}(z) = \Theta_{T'}(z)D_{T'}A(I - zT_A^*)^{-1}D_{A'T}^{-2}[T^*D_A^2, \ D_T Q]W^* \,|D_{T^\bullet} +$$

$$[D_{T^\bullet}(I - zT'^*)^{-1}A, \ 0]W^* \,|D_{T^\bullet} . \quad (7.9a)$$

Using (6.6) and $D_T T^* = T^* D_{T^\bullet}$ we have

$$\Phi_{2,2}(z) = \Theta_{T'}(z)D_{T'}A(I - zT_A^*)^{-1}T_A^*(I - D_A^{-2})D_{T^\bullet} + D_{T^\bullet}(I - zT'^*)^{-1}AD_{T^\bullet} =$$

$$-\Theta_{T'}(z)D_{T'}A(I - zT_A^*)^{-1}T_A^*D_A^{-2}A^*AD_{T^\bullet} + D_{T^\bullet}(I - zT'^*)^{-1}AD_{T^\bullet} .$$

Therefore the last term in $\Phi(z)$ becomes

$$\Phi_{2,2}(z) = D_{T^\bullet}(I - zT'^*)^{-1}AD_{T^\bullet} - \Theta_{T'}(z)D_{T'}AT_A^*(I - zT_A^*)^{-1}D_A^{-2}A^*AD_{T^\bullet} . \quad (7.9)$$

We note that all the terms $\Phi_{i,j}$ in (7.6) to (7.9) contain the same term $\Lambda(z) \overset{\Delta}{=} (I - zT_A^*)^{-1}D_A^{-2}A^*$. So after computing $\Lambda(z)$ the entries of Φ are given by

$$\Phi_{1,1}(z) = - zN^{-\frac{1}{2}}D_{T^*}\Lambda(z)D_{T'}N_1^{-\frac{1}{2}}$$

$$\Phi_{1,2}(z) = N^{\frac{1}{2}} - N^{-\frac{1}{2}}D_{T^*}\Lambda(z)AD_{T^*}$$

$$(7.10)$$

$$\Phi_{2,1}(z) = \Theta_{T'}(z)N_1^{\frac{1}{2}} - \Theta_{T'}(z)D_{T'}A\Lambda(z)D_{T'}N_1^{-\frac{1}{2}}$$

$$\Phi_{2,2}(z) = D_{T'^*}(I - zT'^*)^{-1}AD_{T^*} - \Theta_{T'}(z)D_{T'}AT_A^*\Lambda(z)AD_{T^*} .$$

Finally, since T_A is similar to a contraction, $\Phi_{i,j}$ is well defined for all z in D and for $1 \leq i, j \leq 2$.

It is emphasized that if both H and H' are finite dimensional, then one can compute the analytic function $\Phi(z)$. To do this one must compute D_T, $D_{T'}$ and the inverses of D_A^2 and D_{AT}^2. In this case one can compute the inverse of $I - zT_A^*$ by using standard techniques from linear algebra involving the Cayley-Hamilton theorem or Jordan matrices. Then $\Phi(z)$ is computed from (7.6) to (7.9).

According to Theorem 5.1 the set of all contractive intertwining liftings B of A is given by $B = B_F$ where $F = \omega P_F \oplus R$ is a Schur contraction. Since G and G' are isometries $R = G'F_1G^*$ where F_1 is a contractive analytic function in $H^\infty(D_{T^*}, D_{T'})$. By converting (6.3) to this setting, that is, by setting

$$g = G_+x, \quad g' = G_+'x', \quad u = W(h \oplus v) \text{ and } y = W'(h' \oplus v') , \qquad (7.11a)$$

then the Redheffer cascading system in (6.3) viewed as an analytic function in D becomes

$$\begin{bmatrix} x(z) \\ y(z) \end{bmatrix} = \begin{bmatrix} \Phi_{1,1}(z) & \Phi_{1,2}(z) \\ \Phi_{2,1}(z) & \Phi_{2,2}(z) \end{bmatrix} \begin{bmatrix} x'(z) \\ u(z) \end{bmatrix} \qquad (7.11)$$

subject to $x'(z) = F_1(z)x(z)$.

So by our previous analysis and Theorem 5.1, the set of all contractive intertwining liftings B of A is given by $B = W'^*CW$ where $Cu(z) = y(z)$ is the solution to (7.11) for u in $H^2(D_{T^*})$ and F_1 is an arbitrary contractive analytic function in $H^\infty(D_{T^*}, D_{T'})$. Because both T and T' are $*$-stable $C = \Psi_+$ where $\Psi(z) = C|D_{T^*}$. These observations with (1.20) and (1.21) adjusted to this setting readily lead to the following result.

7.1 THEOREM. *Assume that both* T *and* T' *are $*$-stable and* A *is a strict contraction. Let* $\Phi_{i,j}$ *for* $1 \leq i, j \leq 2$ *be the functions defined according to (7.6) to (7.9). Then the set of all contractive intertwining liftings* B *of* A *is given by*

$$B = W'^* \Psi_+ W \qquad (7.12)$$

where

$$\Psi(z) = \Phi_{2,1}(z)(I - F_1(z)\Phi_{1,1}(z))^{-1} F_1(z)\Phi_{1,2}(z) + \Phi_{2,2}(z) =$$

$$(7.12a)$$

$$\Phi_{2,1}(z)F_1(z)(I - \Phi_{1,1}(z)F_1(z))^{-1}\Phi_{1,2}(z) + \Phi_{2,2}(z) ,$$

and F_1 *is any contractive analytic function in* $H^\infty(D_{T^*}, D_{T'})$. *Moreover, formulas (7.12), (7.12a) establish a one to one correspondence between the set of all contractive intertwining liftings* B *of* A *and the closed unit ball* $H_1^\infty(D_{T^*}, D_{T'})$.

It is emphasized that the inverse in (7.12a) is well defined for all z in D, because the 2 by 2 matrix $\Phi(z)$ in (7.11) is a contractive analytic function (see Corollary 6.2). The block diagram representation for the set of all contractive intertwining liftings $B(= W'^* \Psi_+ W)$ is given in Figure 12 where u = Wk and y = W'k'. If both H and H' are finite dimensional, then one can compute all the entries in $\Phi(z)$ and design a circuit corresponding to Figure 12. So the set of all contractive intertwining liftings is obtained by the closed loop system formed by cascading the fixed system Φ to an arbitrary feedback F_1 in $H_1^\infty(D_{T^*}, D_{T'})$. Finally, it is noted that if we set $F_1 \equiv 0$, then $B = W'^* \Psi_+ W$ where $\Psi(z) = \Phi_{2,2}(z)$ is a particularly easy contractive intertwining lifting to compute. It is usually referred to as the *maximal entropy* contractive intertwining lifting.

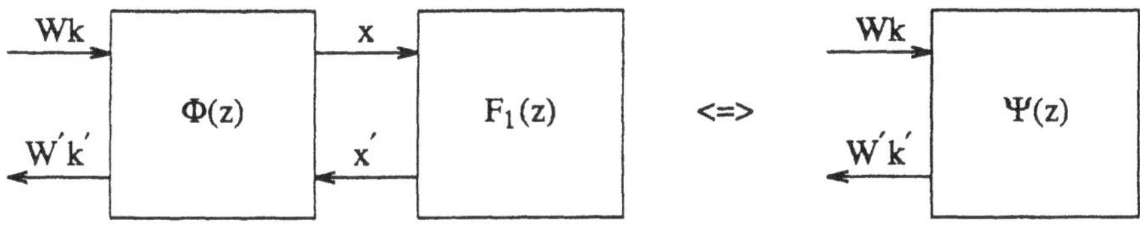

Figure 12

7.2 EXAMPLE. Consider the one step Carathéodory interpolation problem where T = 0 on H, T' = 0 on H' and A is a strict contraction from H to H'. In this case the characteristic function $\Theta = \Theta_T$ for T is zI where I is the identity on H and $\Theta' = \Theta_{T'} = zI'$ where I' is the identity on H'. The normalizing operator N in (7.3) equals D_A^{-2}. Using

(7.4a) it follows that N_1 equals $D_{A^*}^{-2}$. Thus $N^{-\frac{1}{2}} = D_A$ and $N_1^{-\frac{1}{2}} = D_{A^*}$. Substituting these results into (7.6) to (7.9) yields

$$\Phi(z) = \begin{bmatrix} -zA^* & D_A \\ zD_{A^*} & A \end{bmatrix}.$$

So by virtue of (7.12a) the set of all contractive intertwining liftings Ψ_+ of A is given by

$$\Psi(z) = A + zD_{A^*}F_1(I + zA^*F_1)^{-1}D_A \tag{7.13}$$

where F_1 is a contractive analytic function in $H^\infty(H, H')$. Equation (7.13) is precisely the set of all solutions to the one step Carathéodory interpolation problem in Lemma 3.3, as well as Example XIII.3.7.

8. A COMPUTATIONAL PROCEDURE FOR HERMITE-FEJER INTERPOLATION.

In this section, first we particularize Theorem 7.1 to the case when $T = S(mI)$ and $T' = S(mI')$. Then this will be used to compute the set of all solutions to the classical Hermite-Fejér interpolation Problem X.5.1.

Throughout this section m is a nontrivial inner function in H^∞, and consider the particular case when $T = S(mI)$ on $H = H(mI)$ and $T' = S(mI')$ on $H' = H(mI')$ where I is the identity on E and I' the identity on E'. Recall from Section IX.5 that the unilateral shifts S on $H^2(E)$ and S' in $H^2(E')$ are minimal isometric dilations of T and T', respectively. As before A is a strict contraction in $I(T, T')$. Theorem 7.1 provides the set of all contractive intertwining liftings B of A with respect to the minimal isometric dilations U and U' of T and T', respectively. However, the solution to the classical Hermite-Fejér interpolation problem requires the computation of the set of all contractive intertwining liftings with respect to the minimal isometric dilations S and S'. Because all minimal isometric dilations are unitarily equivalent we can convert Theorem 7.1 to the Hermite-Fejér setting. In this section we will provide an explicit identification to accomplish this and solve the classical Hermite-Fejér interpolation problem.

To begin let ϕ and ϕ_* be the functions in H^∞ defined by

$$\phi(z) = \frac{m(z) - m(0)}{z\sqrt{1 - |m(0)|^2}} \quad \text{and} \quad \phi_*(z) = \frac{1 - m(z)\overline{m(0)}}{\sqrt{1 - |m(0)|^2}}. \tag{8.1}$$

It is easy to show that both ϕ and ϕ_* are unit vectors in $H(m)$. Let Π_0 be the operator from H to E defined by $\Pi_0 h = h(0)$ where h is in H. Notice that this h is orthogonal to $mH^2(E)$. Therefore $\overline{m}h$ admits a power series expansion of the form

$$\overline{m}h = e^{-it}h_{-1} + e^{-2it}h_{-2} + e^{-3it}h_{-3} + \cdots .$$

Now let Π_{-1} be the operator from H to E defined by $\Pi_{-1}h = h_{-1}$. If a is in E, then ϕa and $\phi_* a$ are in H. Moreover,

$$\Pi_o \phi_* a = \sqrt{1 - |m(0)|^2}\, a \text{ and } \Pi_{-1}\phi a = \sqrt{1 - |m(0)|^2}\, a .$$

Since m is nontrivial ($|m(0)| \neq 1$) the range of both Π_o and Π_{-1} is closed and equal to E. (This fact also follows from Section IX.5.) It is also useful to notice that

$$\Pi_o^* a = (1 - m\overline{m(0)})a \text{ and } \Pi_{-1}^* a = S^* ma \qquad (a \in E) . \tag{8.2}$$

The first equality follows from the fact that $\Pi_o^* a = P_H a = (1 - m\overline{m(0)})a$. The second equation in (8.2) follows from

$$(\Pi_{-1}h,\, a) = (e^{it}\overline{m}h,\, a) = (Sh,\, ma) = (h,\, S^* ma)$$

where h is in H. Finally, we are ready for the following useful result.

8.1 LEMMA. *Let* m *be a nontrivial inner function in* H^∞ *and set* $T = S(mI)$. *If* h *is in* H, *then*

    (i) $D_T h = \phi \Pi_{-1} h$

    (ii) $D_{T^*} h = \phi_* \Pi_o h$

    (iii) $D_{T^*}(I - zT^*)^{-1}h = \phi_* h(z)$

    (iv) $\Theta_T(z)D_T h = \phi_* m(z)\Pi_{-1} h.$

Moreover, $D_T = \phi E$ *and* $D_{T^*} = \phi_* E$.

    PROOF. To establish part (i) first notice that

$$Th \triangleq P_H zh = zh - P_{mH^2}zh = zh - m\Pi_{-1}h . \tag{8.3}$$

So using the definition of ϕ and $T^* = S^*|H$ we have

$$D_T^2 h = (I - T^* T)h = h - S^*(zh - m\Pi_{-1}h) = \frac{m - m(0)}{z}\Pi_{-1}h . \tag{8.4}$$

Now let C be the operator in H defined by the right hand side of (i), that is, $Ch = \phi\Pi_{-1}h$. Thus

$$C^2 h = C\phi\Pi_{-1}h = \phi\Pi_{-1}\phi\Pi_{-1}h = \phi\sqrt{1 - |m(0)|^2}\,\Pi_{-1}h = D_T^2 h .$$

The last equality follows from (8.4). Finally,

$$(Ch, h) = \frac{(S^* m \Pi_{-1} h, h)}{\sqrt{1 - |m(0)|^2}} = \frac{(\Pi_{-1} h, e^{it} \overline{m} h)}{\sqrt{1 - |m(0)|^2}} = \frac{\|\Pi_{-1} h\|^2}{\sqrt{1 - |m(0)|^2}} .$$

Thus, C is a positive operator satisfying $C^2 = D_T^2$. Since the positive square root is unique $C = D_T$. A similar argument establishes part (ii).

To verify part (iii) let h_n be the coefficient of z^n in the power series expansion of $h(z)$. Using $T^* = S^* | H$ and the power series expansion of $(I - zT^*)^{-1}$ we have

$$\Pi_0 (I - zT^*)^{-1} h = \sum_0^\infty z^n \Pi_0 S^{*n} h = \sum_0^\infty z^n h_n = h(z) .$$

Now part (iii) follows from this and the form of D_{T^*} in part (ii).

Finally, using parts (ii), (iii) with (8.3), (8.4) the definition of Θ_T in (IX.6.1) and $D_{T^*} T = T D_T$ we have

$$\Theta_T(z) D_T h = [z D_{T^*} (I - zT^*)^{-1} D_T - T] D_T h =$$

$$z D_{T^*} (I - zT^*)^{-1} S^* m \Pi_{-1} h - D_{T^*} T h =$$

$$\phi_* (m(z) - m(0)) \Pi_{-1} h - \phi_* \Pi_0 (zh - m \Pi_{-1} h) = \phi_* m(z) \Pi_{-1} h .$$

The last part follows from (i), (ii) and the fact that both Π_{-1} and Π_0 are onto E. This completes the proof.

It is emphasized that in parts (iii) and (iv) the vectors $\phi_* h(z)$ and $\phi_* m(z) \Pi_{-1} h$ are in $D_{T^*} (\subseteq H)$ while $h(z)$ and $m(z) \Pi_{-1} h$ are in E for each z in D.

Let α_* be the operator from E onto D_{T^*} defined by $\alpha_* a = \phi_* a$ where a is in E. Since $\|\phi_* a\| = \|a\|$ Lemma 8.1, shows that α_* is indeed a unitary operator. Using part (ii) of Lemma 8.1, the definition of N in (7.3), (8.3) and $T^* = S^* | H$ we have

$$N \alpha_* a = TT^* \phi_* a + D_{T^*} D_A^{-2} D_{T^*} \phi_* a =$$

$$-T \frac{m - m(0)}{z \sqrt{1 - |m(0)|^2}} \overline{m(0)} a + \phi_* \Pi_0 D_A^{-2} \phi_* \sqrt{1 - |m(0)|^2} \, a = \qquad (8.5)$$

$$\phi_* |m(0)|^2 a + \phi_* \Pi_0 D_A^{-2} \phi_* \sqrt{1 - |m(0)|^2} \, a .$$

So if we define the operator E on E by

$$E \overset{\Delta}{=} |m(0)|^2 I + \Pi_0 D_A^{-2} \Pi_0^* \qquad\qquad (a \in E) , \qquad\qquad (8.6)$$

then (8.2) and (8.5) shows that $N \alpha_* = \alpha_* E$, that is, N is unitarily equivalent to E. In particular, part (ii) of Lemma 8.1 implies that

$$N^{-\frac{1}{2}}D_{T^{\cdot}} = N^{-\frac{1}{2}}\alpha_{*}\Pi_{o} = \alpha_{*}E^{-\frac{1}{2}}\Pi_{o} \ . \tag{8.7}$$

Finally, we let α'_{*} be the unitary operator from E' onto $D_{T^{\cdot}}$ defined by $\alpha'_{*}a' = \phi_{*}a'$ where a' is in E'.

Let Π'_{o} from H' onto E' be the operator corresponding to Π_{o} and Π'_{-1} from H' onto E' be the operator corresponding to Π_{-1}. Let α' be the operator from E' onto $D_{T'}$ defined by $\alpha'a' = \phi a'$ where a' is in E'. As before Lemma 8.1 adjusted to this setting guarantees that α' is unitary. Let E_1 be the operator on E' defined by

$$E_1 \triangleq |m(0)|^2 I' + \Pi'_{-1}D_{A}^{-2\cdot}\Pi'^{*}_{-1} \ . \tag{8.8}$$

A calculation similar to (8.5) involving the form of N_1 in (7.4a) shows that $N_1\alpha' = \alpha'E_1$ that is, N_1 is unitarily equivalent to E_1. So by adjusting part (i) of Lemma 8.1 to this setting, and using (8.2) we have

$$D_{T'}N_1^{-\frac{1}{2}}\alpha' = D_{T'}\alpha'E_1^{-\frac{1}{2}} = \alpha'\sqrt{1 - |m(0)|^2}\,E_1^{-\frac{1}{2}} = \Pi'^{*}_{-1}E_1^{-\frac{1}{2}} \ . \tag{8.9}$$

Finally, using part (iv) of Lemma 8.1 we obtain the following useful result

$$\alpha'^{*}_{*}\Theta_{T'}(z)\alpha'a' = \alpha'^{*}_{*}\Theta_{T'}(z)D_{T'}(1 - |m(0)|^2)^{-1}S^{*}ma' = m(z)a' \qquad (a' \in E') \ . \tag{8.10}$$

Let γ_{*} be the unitary operator from $H^2(E)$ onto $H^2(D_{T^{\cdot}})$ determined by $\gamma_{*}z^{n}a = z^{n}\phi_{*}a$ where a is in E and $n = 0, 1, 2, \cdots$. Let γ'_{*} be the corresponding unitary operator from $H^2(E')$ onto $H^2(D_{T^{\cdot}})$. By construction both γ_{*} and γ'_{*} intertwine unilateral shifts on the appropriate spaces. The definition of $W = W_{T}$ in (6.5) and part (iii) of Lemma 8.1, imply that $\gamma_{*}^{*}Wh = \gamma_{*}^{*}\phi_{*}h(z) = h$ for all h in H. Because $\gamma_{*}^{*}W$ commutes with S and H is cyclic for the minimal isometric dilation S of T we have $\gamma_{*}^{*}W = I$. A similar argument shows that $\gamma'^{*}_{*}W' = I$ where $W' = W_{T'}$. Therefore by Theorem 7.1 the set of all contractive intertwining liftings B of A is given by $\gamma'^{*}_{*}\Psi_{+}\gamma_{*}$ where $\Psi(z)$ is the contractive analytic function computed according to (7.12a) and F_1 is an arbitrary contractive analytic function in $H^{\infty}(D_{T^{\cdot}}, D_{T'})$. Finally, to obtain an explicit formula for $\gamma'^{*}_{*}\Psi(z)\gamma_{*}$ we need the unitary operator γ' from $H^2(E')$ onto $H(D_{T'})$ determined by $\gamma'z^{n}a' = \phi z^{n}a'$ where a' is in E' and $n = 0, 1, 2, \cdots$. Obviously γ' intertwines the unilateral shifts on the appropriate spaces.

Now let Ω be the contractive analytic function in $H^{\infty}(E' \oplus E, E \oplus E')$ defined by the multiplication operator $\Omega_{+} = (\gamma_{*}^{*} \oplus \gamma'^{*}_{*})\Phi_{+}(\gamma' \oplus \gamma_{*})$ where Φ is the contractive analytic function in (7.5). Using $\gamma_{*}|E = \alpha_{*}$, $\gamma'|E' = \alpha'$ and $\gamma'_{*}|E' = \alpha'_{*}$ we see that $\Omega(z)$ admits a matrix decomposition of the form

$$\Omega(z) = \begin{bmatrix} \Omega_{1,1}(z) & \Omega_{1,2}(z) \\ \Omega_{2,1}(z) & \Omega_{2,2}(z) \end{bmatrix} \tag{8.11}$$

where the entries are given by

$$\Omega_{1,1}(z) = \alpha_*^* \Phi_{1,1}(z) \alpha'$$

$$\Omega_{1,2}(z) = \alpha_*^* \Phi_{1,2}(z) \alpha_*$$

$$\Omega_{2,1}(z) = \alpha_*'^* \Phi_{2,1}(z) \alpha' \tag{8.12}$$

$$\Omega_{2,2}(z) = \alpha_*'^* \Phi_{2,2}(z) \alpha_*$$

Using (7.10), (8.7), (8.9), (8.10) and the identity $D_{T^*} \alpha_* = \Pi_o^*$ as well as (iii), (iv) of Lemma 8.1 we have

$$\Omega_{1,1}(z) = - z E^{-1/2} \Pi_o \Lambda(z) \Pi_{-1}'^* E_1^{-1/2}$$

$$\Omega_{1,2}(z) = E^{1/2} - E^{-1/2} \Pi_o \Lambda(z) A \Pi_o^*$$

$$\Omega_{2,1}(z) = m(z) E_1^{1/2} - m(z) \Pi_{-1}' A \Lambda(z) \Pi_{-1}'^* E_1^{-1/2} \tag{8.13}$$

$$\Omega_{2,2}(z) = A \Pi_o^*(z) - m(z) \Pi_{-1}' A T_A^* \Lambda(z) A \Pi_o^* .$$

(The notation $A\Pi_o^*(z)$ means that $A\Pi_o^*(z)a = (A\Pi_o^* a)(z)$ for all a in E and z in D. Recall that $\Lambda(z) = (I - zT_A^*)^{-1} D_A^{-2} A^*$ and $T_A = D_A^2 T D_{AT}^{-2}$.) By combining this with Theorem 7.1 and our previous analysis, we readily obtain the following result.

8.2 THEOREM. *Let* $T = S(mI)$ *and* $T' = S(mI')$ *where* m *is a nontrivial inner function in* H^∞. *Assume that* A *is a strict contraction in* $I(T, T')$. *Let* $\Omega_{i,j}(z)$ *be the functions defined according to (8.13). Then the set of all contractive intertwining liftings* B *of* A *is given by* $B = C_+$ *where*

$$C(z) = \Omega_{2,1}(z)(I - F_1(z)\Omega_{1,1}(z))^{-1} F_1(z)\Omega_{1,2}(z) + \Omega_{2,2}(z) = \tag{8.14}$$

$$\Omega_{2,1}(z)F_1(z)(I - \Omega_{1,1}(z)F_1(z))^{-1}\Omega_{1,2}(z) + \Omega_{2,2}(z)$$

and F_1 *is any contractive analytic function in* $H^\infty(E, E')$. *Moreover, formula (8.14) establishes a one to one correspondence between the set of all contractive intertwining liftings* B *of* A *and the closed unit ball* $H_1^\infty(E, E')$.

Theorem 8.2 can be extended to the case when $T = S(\Theta)$ and $T' = S(\Theta')$ where Θ in $H^\infty(E_1, E)$ and Θ' in $H^\infty(E_1', E')$ are two inner functions such that both $\Theta(0)$ and $\Theta'(0)$ are strict contractions. This is left as an instructive exercise for the reader.

The Redheffer cascading equations corresponding to (8.14), or equivalently, to the set of all contractive intertwining liftings B of A is given by

$$\begin{bmatrix} x(z) \\ y(z) \end{bmatrix} = \begin{bmatrix} \Omega_{1,1}(z) & \Omega_{1,2}(z) \\ \Omega_{2,1}(z) & \Omega_{2,2}(z) \end{bmatrix} \begin{bmatrix} x'(z) \\ u(z) \end{bmatrix}$$

(8.15)

$$\text{subject to } x'(z) = F_1(z)x(z)$$

where F_1 is an arbitrary contractive analytic function in $H^\infty(E, E')$. Its network interpretation is presented in Figure 13.

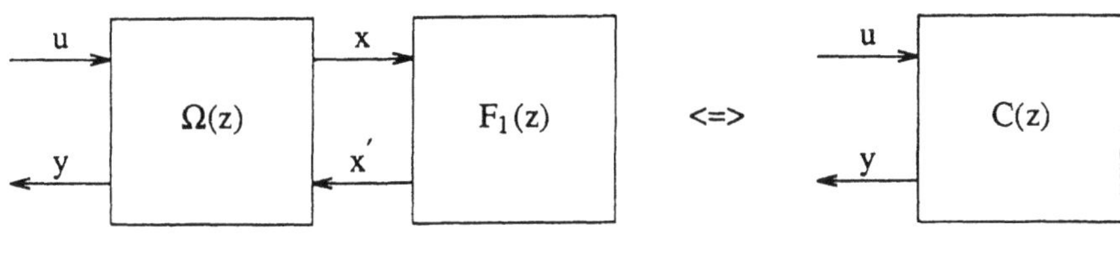

Figure 13

Now we will use the previous theorem along with the matrix identifications for $S(m)$ and A in Chapter X to solve the following classical Hermite-Fejér interpolation Problem X.5.1 restated here for convenience as

8.3 PROBLEM. Given a sequence of operators $\{A_{i,j} : 1 \le i \le n \text{ and } 0 \le j < n_i\}$ in $L(E, E')$ and a set of distinct complex numbers $\alpha_1, \alpha_2, \cdots, \alpha_n$ in D, find the set of all contractive analytic functions F in $H^\infty(E, E')$ satisfying

$$\frac{F^{(j)}(\alpha_i)}{j!} = A_{i,j} \quad (\text{for } 1 \le i \le n \text{ and } 0 \le j < n_i) . \tag{8.16}$$

To obtain a solution to this Hermite-Fejér interpolation problem, let m be the Blaschke product in (X.1.1) whose zeros α_i are repeated n_i times for $1 \le i \le n$. Let M be the matrix in (X.1.19) which is unitarily equivalent to $S(m)$ with respect to the orthogonal basis $\{\phi_j\}_1^d$ in (X.1.18). (Recall that $d = \Sigma n_i$.) Now set $T_m = I \otimes M$ and $T_m' = I' \otimes M$ where the tensor notation is defined in (X.5.8). Obviously T_m on $l_d^2(E)$ and T_m' on $l_d^2(E')$

are the matrix representation of $T = S(mI)$ and $T' = S(mI')$ with respect to the orthonormal splitting $\{\phi_j E\}_1^d$ and $\{\phi_j E'\}_1^d$ (see Section X.5). Now compute the Hermite-Fejér matrix $A_m = P(M)$ from $l_d^2(E)$ to $l_d^2(E')$ which is uniquely determined by the data for Problem 8.3 (see (X.5.9)). Without loss of generality we assume that A_m is a strict contraction. (If A_m is not a contraction there is no solution to the Hermite-Fejér interpolation problem.) Notice that for

$$h = [\phi_1 I, \ \phi_2 I, \ \cdots, \ \phi_n I][a_1, \ a_2, \ \cdots, \ a_d]^{tr} \tag{8.17}$$

where a_j is in E for $1 \le j \le d$ we have

$$Ah = [\phi_1 I', \ \phi_2 I', \ \cdots, \ \phi_d I']A_m[a_1, \ a_2, \ \cdots, \ a_d]^{tr} \ ;$$

similar relations hold for T and T_m as well as for T' and T'_m. Under these identifications the matrix representation M_A for T_A is given by

$$M_A = (I - A_m^* A_m)T_m(I - T_m^* A_m^* A_m T_m)^{-1} \tag{8.18}$$

and the rational matrix $\Lambda_m(z)$ corresponding to $\Lambda(z)$ is

$$\Lambda_m(z) = (I - zM_A^*)^{-1}(I - A_m^* A_m)^{-1}A_m^* \ . \tag{8.19}$$

Since T_A is similar to a contraction, $\Lambda_m(z)$ is well defined for all z in D. For h as in (8.17) we have

$$\Pi_0 h = h(0) = [\phi_1(0)I, \ \phi_2(0)I, \ \cdots, \ \phi_d(0)I][a_1, \ a_2, \ \cdots, \ a_d]^{tr} \ .$$

So that the matrix representation M_0 from $l_d^2(E)$ onto E corresponding to Π_0 with respect to the orthogonal splitting $\{\phi_j E\}_1^d$ is given by

$$M_0 = [\phi_1(0)I, \ \phi_2(0)I, \ \cdots, \ \phi_d(0)I] \ . \tag{8.20}$$

Because ϕ_j is in $H(m)$ and $\overline{m}\phi_j$ is orthogonal to H^2, we see that $\overline{m}\phi_j$ admits a power series expansion of the form

$$(\overline{m}\phi_j)(e^{it}) = \phi_{j,-1}e^{-it} + \phi_{j,-2}e^{-2it} + \ \cdots$$

This and the fact that $|m(e^{it})|^2 = 1$ implies that $z\phi_j(z)/m(z)$ is a rational function analytic in $|z| > 1$. Moreover,

$$\phi_{j,-1} = \lim_{z \to \infty} \frac{z\phi_j(z)}{m(z)} \overset{\Delta}{=} (z\phi_j/m)(\infty) \ .$$

(By replacing z by $1/z$ it also follows that $\phi_{j,-1} = \left[\dfrac{\phi_j(1/z)}{zm(1/z)}\right](0)$). This implies that the matrix representation M_{-1} from $l_d^2(E)$ onto E of Π_{-1} is

$$M_{-1} = [(z\phi_1/m)(\infty)I, \ (z\phi_2/m)(\infty)I, \ \cdots, \ (z\phi_d/m)(\infty)I] \ . \tag{8.21}$$

It is emphasized that one can readily compute the entries $(z\phi_j/m)(\infty)$ of the matrix M_{-1} from the rational functions $z\phi_j(z)/m(z)$. Finally, let M_{-1}' from $l_d^2(E')$ onto E' be the matrix representation for Π_{-1}' corresponding to (8.21). Under these identifications the entries $\Omega_{i,j}$ for $\Omega(z)$ in (8.13) become

$$\Omega_{1,1}(z) = -\, zE^{-\frac{1}{2}}M_o\Lambda_m(z)M_{-1}'^{*}E_1^{-\frac{1}{2}}$$

$$\Omega_{1,2}(z) = E^{\frac{1}{2}} - E^{-\frac{1}{2}}M_o\Lambda_m(z)A_mM_o^{*}$$

$$\tag{8.22}$$

$$\Omega_{2,1}(z) = m(z)E_1^{\frac{1}{2}} - m(z)M_{-1}'A_m\Lambda_m(z)M_{-1}'^{*}E_1^{-\frac{1}{2}}$$

$$\Omega_{2,2}(z) = [\phi_1(z)I', \ \phi_2(z)I', \ \cdots, \ \phi_d(z)I']A_mM_o^{*} - m(z)M_{-1}'A_mM_A^{*}\Lambda_m(z)A_mM_o^{*} \ ,$$

where the matrices E on E in (8.6) and E_1 on E' in (8.8) are given by

$$E = |\,m(0)\,|^2 I + M_o(I - A_m^{*}A_m)^{-1}M_o^{*}$$

$$\tag{8.23}$$

$$E_1 = |\,m(0)\,|^2 I' + M_{-1}'(I - A_mA_m^{*})^{-1}M_{-1}'^{*} \ .$$

Combining this with Theorem 8.2 readily produces the following computational procedure for solving the classical Hermite-Fejér interpolation Problem 8.3.

8.4 PROCEDURE. Let m be a Blaschke product of the form (X.1.1) whose zeros α_i are repeated n_i times for $1 \leq i \leq n$ and specify an orthonormal basis $\{\phi_j\}_1^d$ for $H(m)$ of the form (X.1.18). Compute the matrices T_m, A_m, T_m', M_A and the rational matrix $\Lambda_m(z)$ in (8.19). (Here we have assumed that A_m is a strict contraction. If A_m is not a strict contraction, then one cannot use this procedure.) Next compute M_o, M_{-1}, M_{-1}' the operators E, E_1 and their appropriate square roots. Then the entries $\Omega_{i,j}(z)$ for $1 \leq i, j \leq 2$ of $\Omega(z)$ are computed according to (8.22). By Theorem 8.2 and the results in Section X.5, the set of all solutions $F(z)$ to the classical Hermite-Fejér interpolation Problem 8.3 is given by $F(z) = C(z)$ where $C(z)$ is given by (8.14) and $F_1(z)$ is any contractive analytic function in $H^\infty(E, E')$.

It is noted that the solution $F(z) = \Omega_{2,2}(z)$ is the easiest solution of Problem 8.3 to compute. It is an interesting exercise to apply Procedure 8.4 to the one step Carathéodory interpolation problem in Example 7.2. In this case Procedure 8.4 produces the same solution. Finally, it is noted that one can compute the entries $\Omega_{i,j}(z)$ of $\Omega(z)$. Then one can design a circuit given by the block diagram in Figure 13, which produces the set of all solutions to the classical Hermite-Fejér interpolation problem. This circuit could be incorporated as the compensator for solving the H^∞ control problems discussed in Chapter XII.

By applying Procedure 8.4 to the classical Carthéodory interpolation Problem I.1.2, one can show with some tedious work, that by using Lemma XIII.8.1 to invert $I - zM_A^*$ (where $\phi_i = z^{i-1}$) and following the ideas in Section XIII.9 we obtain (for $\|A_n\| < 1$)

$$\Omega(z) = \frac{1}{y_n^\#} \begin{bmatrix} -zx_n^\# & \sqrt{t_n} \\ z^n\sqrt{t_n} & x_n \end{bmatrix} \tag{8.24}$$

where $t_n = (1 - |r_0|^2)(1 - |r_1|^2) \cdots (1 - |r_{n-1}|^2)$, the $r_i's$ are the Schur numbers for the Carathéodory data, and x_n and y_n are the Schur polymonials defined in Section I.3. So by using (8.14) in Theorem 8.2 along with (I.3.3) and (I.3.4), the set of all solutions f to the classical Carathéodory interpolation Problem I.1.2 is given by

$$f(z) = \frac{x_n(z) + zy_n(z)f_n(z)}{y_n^\#(z) + zx_n^\#(z)f_n(z)} \tag{8.25}$$

where f_n is in H_1^∞. This is precisely the Schur representation (I.3.8) for solving the classical Carathéodory interpolation problem. Finally it is noted that as expected the matrix Ω in (8.24) is inner.

9. THE HANKEL CASE

In this section we will adapt the results in Section 6 and 7 to the case of strictly contractive Hankel operators. Since these proofs are closely related to our previous proofs, some details will be left to the reader as more or less simple exercises.

Throughout this section A is a strictly contractive Hankel operator from $H^2(E)$ to $K_o^2(E')$ of the form $A = P_- M_{Q_o} | H^2(E)$ where Q_o is in $L^\infty(E, E')$ and P_- is the orthogonal projection onto $K_o^2(E')$. In this setting, T' is the compression of V' to $K_o^2(E')$ and $T = S$ is the unilateral shift on $H^2(E)$. (Recall that V' is the bilateral shift on $L^2(E')$.) If h is in $H = H^2(E)$ we define, as before $\Pi_o h = h(0) \in E$; thus by identifying E with the subspace of constant functions in H we now have $D_{T^*} = D_{S^*} = \Pi_o$. However, if h' is in $H' = K_o^2(E')$ we set $\Pi'h' = h'_{-1}$ the Fourier coefficient of e^{-it} of h' in the power series expansion of h'; now we have $D_{T'}h' = e^{-it}\Pi'h'$. In this setting B is a contractive intertwining lifting of A if $B = M_C | H^2(E)$ where C is a function in $L^\infty(E, E')$ satisfying

$A = P_- M_C \,|\, H^2(E)$ and $\|C\|_\infty \leq 1$. In other words, the set of all contractive intertwining liftings B of A is given by $B = M_C$ where C is a function in $L^\infty(E, E')$ of the form $C = Q_o + \Psi$ satisfying $\|Q_o + \Psi\|_\infty \leq 1$ and Ψ is in $H^\infty(E, E')$; see Sections XIII.3 and XIII.8 for further details.

According to Theorem 5.1 the set of all contractive intertwining liftings B of A is given by $B = B_F$ where $F = \omega P_F \oplus R$ is a Schur contraction. Since G and G' are isometries $R = G' \Pi'^* F_1 \Pi_o G^*$ where F_1 is a contractive analytic function in $H^\infty(E, E')$. By converting (6.3) to this setting, that is, by setting $g = (G\Pi_o^*)_+ x$ and $g' = (G'\Pi'^*)_+ x'$ and using the fact that $D_T = D_S = \{0\}$ along with Theorem 6.1, we see that the Redheffer cascading system in (6.3) becomes

$$
\begin{bmatrix} x \\ h' \oplus v' \end{bmatrix} = \left[\begin{matrix} (\Pi_o G^*)_+ S q_A (I_A - S\Pi_A \omega_+ p_A)^{-1} \Pi_A (G'\Pi'^*)_+ \\ (\Pi')_+ P'(I_{X'} - S\omega_+ p_A \Pi_A)^{-1} (G'\Pi'^*)_+ \end{matrix} \right.
\tag{9.1}
$$

$$
\left. \begin{matrix} (\Pi_o G^*)_+ [S q_A (I_A - S\Pi_A \omega_+ p_A)^{-1} \Pi_A \omega_+ p + q] D_A \\ (\Pi')_+ P'(I_{X'} - S\omega_+ p_A \Pi_A)^{-1} \omega_+ p D_A + A \end{matrix} \right] \begin{bmatrix} x' \\ h \end{bmatrix}
$$

$$
\text{subject to } x' = F_{1+} x
$$

where I_A and $I_{X'}$ are the identity operators on $l(D_A)$ and $l(D_A \oplus E')$, respectively. The 2 by 2 matrix is a contraction from $H^2(E') \oplus H = H^2(E') \oplus H^2(E)$ into $H^2(E) \oplus (H' \oplus H^2(E')) = H^2(E) \oplus (K_o^2(E') \oplus H^2(E'))$. The operator $(\Pi')_+$ in (9.1) provides the natural identification from $H^2(D_{T'})$ onto $H^2(E')$. In the sequel we will use the natural identification $L^2(E') = K_o^2(E') \oplus H^2(E')$.

Let N on E, N_1 on E' and T_A on H be the operators defined by

$$
N = \Pi_o D_A^{-2} \Pi_o^*, \quad N_1 = \Pi' D_A^{-2} \cdot \Pi'^* \quad \text{and} \quad T_A = D_A^2 S D_{AS}^{-2} .
\tag{9.2}
$$

By proceeding as in Section 7 (see (7.6), (7.7a), (7.8a), (7.9a)) we obtain that the cascading equations (9.1) take the form

$$
\begin{bmatrix} x \\ k' \end{bmatrix} = \begin{bmatrix} \Psi_{1,1+} & \Psi_{1,2+} \\ \Psi_{2,1+} & M \end{bmatrix} \begin{bmatrix} x' \\ h \end{bmatrix}
\tag{9.3}
$$

$$
\text{subject to } x' = F_{1+} x
$$

where for z in D

$$\Psi_{1,1}(z) = -\, zN^{-1/2}\Pi_0(I - zT_A^*)^{-1}D_A^{-2}A^*\Pi'^*N_1^{-1/2}$$

$$\Psi_{1,2}(z) = N^{-1/2}\Pi_0(I - zT_A^*)^{-1}\Pi_0^* \tag{9.4}$$

$$\Psi_{2,1}(z) = N_1^{-1/2} - z\Pi'A(I - zT_A^*)^{-1}D_{AS}^{-2}S^*A^*\Pi'^*N_1^{-1/2}\;.$$

Moreover, M is the operator in $I(S, V')$ defined by

$$(Mh)(e^{it}) = (Ah)(e^{it}) + \sum_{n=0}^{\infty} e^{int}\Pi'AT_A^{*n+1}h \qquad (h \in H)\;. \tag{9.5}$$

By virtue of Theorem IX.1.1, the operator M is a multiplication operator $M = M_\Phi\,|\,H^2(E)$ from $H^2(E)$ to $L^2(E')$ defined by the operator

$$\Phi(e^{it})a = (A\Pi_0^*a)(e^{it}) + \sum_{n=0}^{\infty} e^{int}\Pi'AT_A^{*n+1}\Pi_0^*a \qquad (a \in E)\;. \tag{9.6}$$

(Note that Π_0^*a for a in E is the consistent notation for the constant function identical with a in $H^2(E)$.) The consistency of the formulas (9.4), (9.5) and (9.6) follows from the fact that by virtue of Theorem 6.1 the matrix

$$\begin{bmatrix} \Psi_{1,1+} & \Psi_{1,2+} \\ \Psi_{2,1+} & M \end{bmatrix} : H^2(E') \oplus H^2(E) \to H^2(E) \oplus L^2(E')$$

is a contraction. In particular, $\Psi_{1,1}$ is a contractive analytic function in $H^\infty(E', E)$. The factor z of $\Psi_{1,1}(z)$ guarantees that $I - F_1(z)\Psi_{1,1}(z)$ is invertible for all z in D when F_1 is a contractive analytic function. This fact along with Theorem 5.1 readily produces the following result.

9.1 THEOREM. *Let* $A = P_-M_{Q_o}\,|\,H^2(E)$ *be a strictly contractive Hankel operator in* $I(S, T')$ *where* Q_o *is in* $L^\infty(E, E')$. *Then the set of all contractive intertwining liftings* B *of* A *is given by* $B = M_C\,|\,H^2(E)$ *where* C *is the contractive function in* $L^\infty(E, E')$ *given a.e. by*

$$C(e^{it}) = [\Psi_{2,1}(I - F_1\Psi_{1,1})^{-1}F_1\Psi_{1,2}](e^{it}) + \Phi(e^{it}) =$$

$$\tag{9.7}$$

$$[\Psi_{2,1}F_1(I - \Psi_{1,1}F_1)^{-1}\Psi_{1,2}](e^{it}) + \Phi(e^{it})$$

and $F_1(z)$ *is any contractive analytic function in* $H^\infty(E, E')$. *Moreover, formula (9.7) establishes a one to one correspondence between the set of all contractive intertwining liftings* B *of* A *and the closed unit ball* $H_1^\infty(E, E')$.

In (9.7) the operator valued function in the square brackets are in $H^\infty(E, E')$ and therefore, their boundary values exists a.e. (see Section IX.1).

9.2 COROLLARY. *Let* $A = P_- M_{Q_0} | H^2(E)$ *be a strictly contractive Hankel operator in* $I(S, T')$ *where* Q_0 *is in* $K_0^\infty(E, E')$. *Then the following properties are equivalent.*

(i) *The operator* $M_C | H^2(E)$ *is a contractive intertwining lifting of* A;

(ii) *The function* $C = Q_0 + \Psi$ *where* Ψ *is a contractive analytic function in* $H^\infty(E, E')$ *given by the formulas*

$$\Psi(z) = \Psi_{2,1}(z)(I - F_1(z)\Psi_{1,1}(z))^{-1}F_1(z)\Psi_{1,2}(z) + \Psi_{2,2}(z) = \qquad (z \in D) \quad (9.8)$$

$$\Psi_{2,1}(z)F_1(z)(I - \Psi_{1,1}(z)F_1(z))^{-1}\Psi_{1,2}(z) + \Psi_{2,2}(z)$$

where $\Psi_{2,2}(z)$ *is defined by*

$$\Psi_{2,2}(z) = \Pi' A(I - zT_A^*)^{-1}T_A^*\Pi_0^* \qquad (z \in D) \qquad (9.9)$$

and where F_1 *is any contractive analytic function in* $H^\infty(E, E')$;

(iii) *The function* $C = Q_0 + \Psi$ *where* Ψ *is a function in* $H^\infty(E, E')$ *satisfying* $\|Q_0 + \Psi\|_\infty \leq 1$.

The proof (which uses the fact that $Ca = M_C a$ for a in E) is left to the reader as a simple exercise.

It is emphasized that the previous Corollary and Theorem XIII.8.3 produce the same contractive intertwining lifting for the same F_1. This follows because both results were derived from the same Schur representation in Theorem XIII.3.4 with the same $F = \omega P_F \oplus R$ and $R = G'\Pi'^* F_1 \Pi_0 G^*$. One can also prove this result directly. However, it is a very tedious exercise.

9.3 REMARK. We note that in many applications such as H^∞ control (see Section XII.1) one is only interested in the analytic part $\Psi(z)$ of the contractive intertwining lifting B of A. This part is obtained by solving the following Redheffer cascading system $(y(z) = \Psi(z)u(z))$

$$\begin{bmatrix} x(z) \\ y(z) \end{bmatrix} = \begin{bmatrix} \Psi_{1,1}(z) & \Psi_{1,2}(z) \\ \Psi_{2,1}(z) & \Psi_{2,2}(z) \end{bmatrix} \begin{bmatrix} x'(z) \\ u(z) \end{bmatrix} \tag{9.10}$$

$$\text{subject to } x'(z) = F_1(z)x(z)$$

where F_1 is a contractive analytic function in $H^\infty(E, E')$. The solution $\Psi_{2,2}(z)$ corresponding to $F_1 = 0$ is particularly easy to compute. It corresponds to the contractive intertwining lifting $B = M_C \,|\, H^2(E)$ where $C = \Psi_{2,2} + Q_o$. If $\Omega(z)$ is the 2 by 2 matrix in (9.10), then Figure 13 in Section 8 with $C(z)$ replaced by $\Psi(z)$ provides the network interpretation of $\Psi(z)$. As before it is emphasized that this circuit could be incorporated as the compensator for an H^∞ control problem.

To obtain a computational procedure for Corollary 9.2, let Q be a rational function in $L^\infty(E, E')$ and $A = P_- M_Q \,|\, H^2(E)$ a strictly contractive Hankel operator in $I(S, T')$. It is easy to see that $Q = Q_o + Q_1$ where Q_o is in $K_o^\infty(E, E')$ and Q_1 is in $H^\infty(E, E')$. Thus, the Hankel operator

$$A = P_- M_Q \,|\, H^2(E) = P_- M_{Q_o} \,|\, H^2(E)$$

satisfies the conditions in Corollary 9.2. As before, let β be the unitary operator from $K_o^2(E')$ onto $H^2(E')$ defined by $\beta = V'^* J \,|\, K_o^2(E')$ where $Jf(e^{it}) = f(e^{-it})$ for all f in $L^2(E')$. Using $V'^* J P_- = P_+ V'^* J$ we have $\beta A = \Gamma(e^{-it} Q_o(e^{-it}))$ where Γ is the Hankel operator in $I(S, S'^*)$ with symbol $Q_2(e^{it}) = e^{-it} Q_o(e^{-it})$ defined in Section IX.3. According to Proposition X.6.3 or (X.7.30) the Hankel operator $\Gamma = \Gamma(Q_2)$ admits a factorization of the form $\Gamma = W_1 A_m W^*$ where W from $\mathbb{C}^n$ to $H^2(E)$ and W_1 from $\mathbb{C}^n$ to $H^2(E')$ are two isometric operators which we can explicitly obtain and A_m is a matrix on $\mathbb{C}^n$. (These operators W and W' below should not be confused with the isometries in (6.5).) Moreover, $W^* S = S_m W^*$ where S_m is also a matrix on $\mathbb{C}^n$. So if W' is the isometry from $\mathbb{C}^n$ to $K_o^2(E')$ defined by $W' = \beta^* W_1 = J V' W_1$, then

$$A = W' A_m W^* \quad \text{and} \quad AS = W' A_m S_m W^* . \tag{9.11}$$

It is emphasized that we can compute W, W', A_m and S_m by using the orthogonal basis method in Section X.6, or the state space techniques in Section X.7. Using this set up we see that

$$D_A^2 = W(I - A_m^* A_m)^{-1} W^* + I - WW^* \quad \text{and}$$

$$D_{AS}^{-2} = W(I - S_m^* A_m^* A_m S_m)^{-1} W^* + I - WW^* . \tag{9.12}$$

Now let M_o be the matrix from $\mathbb{C}^n$ to E defined by $M_o = W(0)$, that is, the entries of the matrix M_o are precisely the entries of $W(z)$ evaluated at $z = 0$. Let M' be the corresponding matrix from C^n to E' defined by $M' = W_1(0)$. From the definitions of Π_o and Π' we have

$$\Pi_o W x = M_o x \quad \text{and} \quad \Pi' W' x = M' x \qquad (x \in C^n) . \qquad (9.13)$$

So the $\Psi_{1,1}$, $\Psi_{2,1}$ terms in (9.4) and N, N_1 in (9.2) become

$$\Psi_{1,1}(z) = - zN^{-\frac{1}{2}}M_o(I - zM_A^*)^{-1}(I - A_m^*A_m)^{-1}A_m^*M'^*N_1^{-\frac{1}{2}}$$

$$\Psi_{2,1}(z) = N_1^{-\frac{1}{2}} - zM'A_m(I - zM_A^*)^{-1}(I - S_m^*A_m^*A_mS_m)^{-1}S_m^*A_m^*M'^*N_1^{-\frac{1}{2}} \qquad (9.14)$$

$$N = M_o(I - A_m^*A_m)^{-1}M_o^* + I_E - M_oM_o^* \quad \text{and}$$

$$N_1 = M'(I - A_mA_m^*)^{-1}M'^* + I_{E'} - M'M'^* .$$

where

$$M_A \triangleq (I - A_m^*A_m)S_m(I - S_m^*A_m^*A_mS_m)^{-1} .$$

The $\Psi_{1,2}$ and $\Psi_{2,2}$ terms are harder to compute. To compute these terms let M be the kernel of W^*, or equivalently, the orthogonal complement of the range of W. Obviously $I - WW^* = P_M$. By (9.12) and $S^*W = WS_m^*$

$$T_A^* = (WD_{A_mS_m}^{-2}W^* + P_M)S^*(WD_{A_m}^2W^* + P_M) = WM_A^*W^* + WD_{A_mS_m}^{-2}W^*S^*P_M + P_MS^*$$

$$(9.15)$$

where we used the fact that WC^n is an invariant suspace for S^*, or equivalently, $P_MS^* = P_MS^*P_M$. Therefore T_A^* admits a matrix representation on $WC^n \oplus M$ of the form

$$T_A^* = \begin{bmatrix} WM_A^*W^* & WD_{A_mS_m}^{-2}W^*S^* \\ 0 & P_MS^* \end{bmatrix} .$$

This matrix form and (9.15) readily shows that

$$(I - zT_A^*)^{-1} = \qquad (9.16)$$

$$W(I - zM_A^*)^{-1}W^* + zW(I - zM_A^*)^{-1}D_{A_mS_m}^{-2}W^*S^*P_M(I - zS^*)^{-1} + P_M(I - zS^*)^{-1} .$$

Notice that for a in E we have $P_M a = (I - WM_o^*)a$. Using this in (9.16) along with $S^*W = WS_m^*$ produces

$$\Psi_{1,2}(z) = N^{-\frac{1}{2}}\Pi_o(I - zT_A^*)^{-1}\Pi_o^* =$$

$$N^{-\frac{1}{2}}M_o(I - zM_A^*)^{-1}M_o^* - zN^{-\frac{1}{2}}M_o(I - zM_A^*)^{-1}D_{A_m}^{-2}S_m S_m^* M_o^* + N^{-\frac{1}{2}}(I - M_o M_o^*) =$$

$$N^{-\frac{1}{2}}[I - zM_o(I - zM_A^*)^{-1}(I - S_m^* A_m^* A_m S_m)^{-1}S_m^* A_m^* A_m M_o^*] . \tag{9.17}$$

A similar calculation involving (9.9), (9.15) and (9.16) yields

$$\Psi_{2,2}(z) = - M' A_m(I - zM_A^*)^{-1}(I - S_m^* A_m^* A_m S_m)^{-1}S_m^* A_m^* A_m M_o^* . \tag{9.18}$$

This readily leads to the following computational procedure for obtaining all contractive intertwining liftings for a rational Hankel operator.

9.4 PROCEDURE. Let $A = P_- M_{Q_o}|H^2(E)$ be a strictly contractive Hankel operator in $I(S, T')$ whose symbol Q_o is a rational function in $K_o^\infty(E, E')$. Using the results in Section X.6 or the state space techniques in Section X.7, compute a factorization for $\Gamma(e^{-it}Q_o(e^{-it})) = \Gamma$ of the form $W_1 A_m W^*$ where W is an isometry from C^n to $H^2(E)$ and W_1 is an isometry from C^n to $H^2(E')$ and A_m is a matrix on C^n. (Note one can even choose A_m to be a diagonal matrix.) Set $W' = V'^* J W_1$. Also compute the contraction S_m on C^n satisfying $S_m W^* = W^* S$. Then compute $\Psi_{i,j}(z)$, N, and N_1 in (9.14), (9.17) and (9.18). (In the scalar setting $N = N_1$; see Corollary XIII.8.4.) By Corollary 9.2, the set of all contractive intertwining lifting B of A is given by $B = M_C|H^2(E)$ where $C = \Psi + Q_o$ and $\Psi(z)$ is computed according to (9.8) with an arbitrary contractive analytic function F_1 in $H^\infty(E, E')$.

As expected the contractive intertwining lifting $B = M_C$ where $F_1 = 0$, that is, $C = \Psi_{2,2} + Q_o$ is particularly easy to compute. Here we did not reduce the computations of $\Psi_{i,j}$ to those involving the operators $\hat{X}_1$ and $\hat{Y}_1$ as in Section XIII.8. This is left to the reader as an exercise. In fact by following the ideas in Section XIII.8 one can show that for Q_o in $K_o^\infty(E, E')$

$$\Psi_{1,1}(z) = - zN^{-\frac{1}{2}}E_o(z)\hat{Y}(z)N_1^{\frac{1}{2}}$$

$$\Psi_{1,2}(z) = N^{-\frac{1}{2}}E_o(z)$$

$$\Psi_{2,1}(z) = N_1^{\frac{1}{2}} + z(P_+ e^{it}Q_o\hat{X})(z)E_o(z)\hat{Y}(z)N_1^{\frac{1}{2}} - (P_+ e^{it}Q_o\hat{Y})(z)N_1^{\frac{1}{2}} \tag{9.19}$$

$$\Phi(e^{it}) = [Q_o(e^{it}) + (AS\hat{X}_1)(e^{it})]E_o(e^{it})$$

where $E_o(z) = (I + z\hat{X}(z))^{-1}$ and $\hat{X}, \hat{Y}, \hat{X}_1$ are the operators defined in Section XIII.8. (The operator $\Psi_{2,2}(z)$ is $- (P_+ e^{it}Q_o\hat{X})(z)E_o(z)$.) By equation (XIII.8.30) one can express

$\hat{X}$, $\hat{Y}$, E_o and Φ in terms of the inverses of D_A^2 and $D_{A^*}^2$. (Notice that $\hat{X} = S^* D_A^{-2} \Pi_o^* N^{-1}$ and $\Pi_o = \Pi_*$). By using the second equation in (9.7) a simple calculation shows that C is indeed given by (XIII.8.26). In other words Theorem 9.1 provides another proof of Theorem XIII.8.3. If Q_o is rational, then one can compute the operators in (9.19) and build a circuit to implement C in (9.7), or $\Psi(z)$ in (9.8). (By following the ideas in Section XIII.9, one can use Procedure 9.4 to also derive the Schur representation in (I.3.8) for solving the classical Carathéodory interpolation problem.) In some problems it may be easier to invert $I - zM_A^*$ rather than invert $I + z\hat{X}(z) + z\hat{Y}(z)N_1^{1/2}F_1(z)N^{-1/2}$ or $I + z\hat{X}(z)$ which is required by Procedure XIII.9.1, or by (9.19). (Sometimes it is easier to invert $I - zM_A^*$ rather than invert a n by n matrix of analytic functions.) In other words for certain problems Procedure 9.4 may be better than Procedure XIII.9.1 or vice versa. For certain problems it may be better to use (9.4), (9.6) rather than (9.19) or vice versa.

XIV.10. NOTES AND COMMENTS

The results in Section 1 concerning Redheffer products are now classical in network theory. We only mention that Redheffer products and their related results are originally due to Redheffer [Re]. The results in Section 2 are an expanded version of [FoF 2]. The relationship between Redheffer products and Carathéodory interpolation in Section 3 is also classical, e.g., [BrK], [DeGK 1], [Dew], [DewD], [DewVK], [DudM], [Ka 2]. As noted in Section I.8, the layer peeling algorithm is given in [Ka 2] and [LeK]. Sections 4 and 5 are an improved version of the papers [FoF 3,4]. The results in Sections 6 and 7 are believed to be new. Let us notice that one can also obtain a layer peeling inverse scattering algorithm for the commutant lifting theorem by peeling off the contraction C_A in (6.3), and then using the layer peeling algorithm in Section 3. For other algorithms to solve the matrical Nevanlinna-Pick problem and Nehari H^∞ optimization problems see [AAK 4], [BGR], [DeGK 2], [Fran 2], [Gl 1], [GoKW 1], [Helt 5], [KuL] and [Y 4].

We will conclude this section summarizing the connection between the two Schur representations for lifting Hankel operators given in Sections XIII.8 and 9. To see this, let us recall the Redheffer scattering interpretation for the set of all contractive intertwining liftings given by Figure 14 below

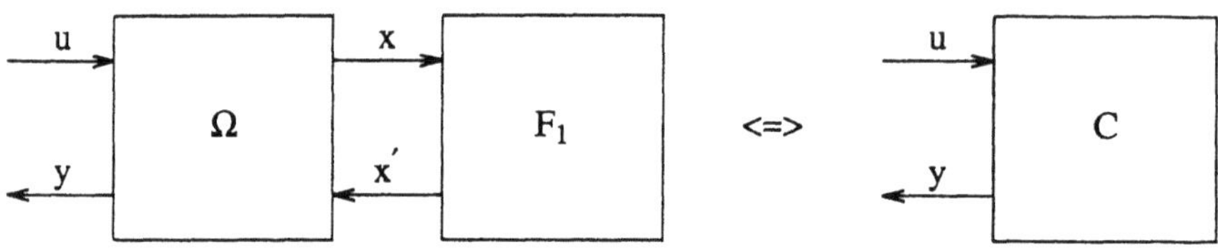

Figure 14

where Ω is the 2 by 2 matrix in (9.3). The Redheffer cascading system corresponding to Figure 14 is

$$\begin{bmatrix} x \\ y \end{bmatrix} = \begin{bmatrix} A_{1,1} & A_{1,2} \\ A_{2,1} & A_{2,2} \end{bmatrix} \begin{bmatrix} x' \\ u \end{bmatrix} \tag{10.1}$$

subject to $x' = F_1 x$,

where F_1 is an arbitrary function in $H_1^\infty(E, E')$. By rearranging the terms in (10.1) and assuming that the appropriate inverses exists, we obtain the following transmission version of (10.1)

$$\begin{bmatrix} u \\ y \end{bmatrix} = \begin{bmatrix} B_{1,1} & B_{1,2} \\ B_{2,1} & B_{2,2} \end{bmatrix} \begin{bmatrix} x \\ x' \end{bmatrix} \tag{10.2}$$

subject to $x' = F_1 x$.

The 2 by 2 matrix Θ in (10.2) is usually referred to as the *transmission matrix*. It transfers the vector $[x, x']^{tr}$ to $[u, y]^{tr}$. In Chapter III, this corresponds to the transmission of the waves in one layer through the interface to the waves in the next layer. By solving for y in (10.2) and assuming that the appropriate inverse exists we have

$$y = [B_{2,1} + B_{2,2}F_1]x = [B_{2,1} + B_{2,2}F_1][B_{1,1} + B_{1,2}F_1]^{-1}u .$$

Comparing the formulas (XIII.8.26), (XIII.8.23) with (9.7) one can see, after some hard work, that the Schur representation (XIII.8.26), (XIII.8.23) is actually the transmission form corresponding to the system whose Schur representation (9.7) is obtained from the Redheffer cascading system (9.3). Many authors prefer to use the transmission matrix, e.g., [AAK 4], [BH], [DewD], [Dy 1,2], [GoKW 1], [Ka 2] and [LeK], because it has algebraic advantages when A is a strict contraction. If the norm of A is one, then the Redheffer scattering approach seems to be preferable.

CHAPTER XV

A GEOMETRIC APPROACH TO POSITIVE DEFINITE SEQUENCES

In this chapter we present a geometric approach for studying positive definite sequences. First, we will obtain the Naimark dilation theorem which is the fundamental result of this chapter. Then we will use this theorem to show that there is a one to one correspondence between the set of all positive definite sequences on H, and the set of all choice sequences initiated on H. We also solve the positive definite Carathéodory interpolation problem in the operator setting, and present a geometric proof for many of the results in Chapter II. Then a recursive inverse scattering algorithm to compute the choice sequence from the positive definite sequence is presented. Further, we use a special matrix representation of the Naimark dilation to demonstrate how this choice sequence occurs in the Levinson algorithm for block Toeplitz matrices. This leads to another inverse scattering algorithm to compute the choice sequence. Also shown is how the Naimark dilation occurs in an operator version of the marine seismology and layered medium models discussed in Chapter III. Finally, we will show that there is a one to one correspondence through the Cayley transform, between the set of all positive definite sequences on H, and the set of all contractive analytic functions in $H^\infty(H, H)$. From this it will follow that the positive definite operator valued Carathéodory interpalation problem is equivalent to the usual operator valued Carathéodory interpalation problem involving n by n contractive analytic Toeplitz matrices.

1. THE NAIMARK DILATION THEOREM

In this section we present the Naimark dilation theorem for positive definite sequences along with some of its simple consequences.

To begin we establish some notation. Throughout this chapter $\{R_i\}_1^\infty$ is a sequence of operators on H and $R_{-n} \overset{\Delta}{=} R_n^*$ for all n with $R_o \overset{\Delta}{=} I$. A sequence of operators $\{R_i\}_1^\infty$ on H is *positive definite* if

$$\sum_{0,0}^{\infty,\infty} (R_{i-j}h_i, h_j) \geq 0 \tag{1.1}$$

for all sequences $\{h_i\}_0^\infty$ in H with finite support. By converting (1.1) to a matrix it is easy to show that $\{R_i\}_1^\infty$ is positive definite if and only if the Toeplitz matrix

$$\hat{R}_n \triangleq \begin{bmatrix} I & R_1 & \cdots & R_{n-2} & R_{n-1} \\ R_{-1} & I & \cdots & R_{n-3} & R_{n-2} \\ R_{-2} & R_{-1} & \cdots & R_{n-4} & R_{n-3} \\ . & . & \cdots & . & . \\ . & . & \cdots & . & . \\ . & . & \cdots & . & . \\ R_{1-n} & R_{2-n} & \cdots & R_{-1} & I \end{bmatrix} \tag{1.2}$$

is positive (≥ 0) for all $n \geq 1$. Finally, recall that a subspace H is *cyclic* for an operator U on K if

$$K = \bigvee_0^\infty U^i H . \tag{1.3}$$

The following is the Naimark dilation theorem which plays a basic role in this chapter.

   1.1 THEOREM. *A sequence of operators $\{R_i\}_1^\infty$ on H is positive definite if and only if there exists an isometry U on $K(\supseteq H)$ such that*

$$R_i = P_H U^i | H \quad \text{(for all } i \geq 0) . \tag{1.4}$$

In this case if H is cyclic for U, then U is unique up to an isomorphism.

   By unique up to an isomorphism we mean that if U' on K' is another isometry satisfying

$$R_i = P_H U'^i | H \quad \text{(for all } i \geq 0) \tag{1.5}$$

where H is cyclic for U', then there exists a unitary operator W mapping K onto K' such that $U'W = WU$ and $W|H = I$. The isometry U on K satisfying (1.3) and (1.4) is called *the minimal isometric dilation of $\{R_i\}_1^\infty$.*

   PROOF. Assume that U is an isometry satisfying (1.4). Let $\{h_i\}_0^\infty$ be a sequence in H with finite support. Using (1.4) and $R_i^* = R_{-i}$ we have

$$0 \le \|\sum_{o}^{\infty} U^i h_i\|^2 = \sum_{0,0}^{\infty,\infty} (U^i h_i, U^j h_j) = \sum_{i \ge j}(U^{i-j}h_i, h_j) + \sum_{i<j}(h_i, U^{j-i}h_j) =$$

$$\sum_{i \ge j}(R_{i-j}h_i, h_j) + \sum_{i<j}(R^*_{j-i}h_i, h_j) = \sum_{0,0}^{\infty,\infty}(R_{i-j}h_i, h_j) . \tag{1.6}$$

Hence $\{R_i\}_1^\infty$ is positive definite.

To prove the other half assume that $\{R_i\}_1^\infty$ is a positive definite sequence on H. Let K_f be the linear space consisting of all sequences $k = \{k_i\}_0^\infty$ in H with finite support. Addition and scalar multiplication is performed component wise. Consider the bilinear functional $<\cdot,\cdot>$ on K_f defined by

$$<k,k'> = \sum_{0,0}^{\infty,\infty}(R_{i-j}k_i, k_j')$$

where k and $k' = \{k_i'\}_0^\infty$ are both in K_f. Using $R_i^* = R_{-i}$ it is easy to verify that $<k,k'> = \overline{<k',k>}$. Since $\{R_i\}_1^\infty$ is positive definite $<k,k> \ge 0$. This implies that the Cauchy-Schwartz inequality holds $|<k,k'>|^2 \le <k,k> <k',k'>$. Thus $K_o = \{k \in K_f : <k,k> = 0\}$ forms a linear subspace of K_f. By identifying elements in K_o with zero, the quotient space K_f/K_o defines an inner product $<\cdot,\cdot>$ on K_f/K_o. By completing K_f/K_o we obtain a Hilbert space K.

To construct an isometry satisfying (1.4), let U be the forward shift operator on K_f defined by

$$U\{k_0, k_1, k_2, ...\} = \{0, k_0, k_1, k_2, ...\}$$

where we identify $\{k_i\}_0^\infty$ with $\{k_0, k_1, k_2, ...\}$. Using $k = \{k_i\}_0^\infty$ we have

$$<Uk,Uk> = \sum_{1,1}^{\infty,\infty}(R_{i-j}k_{i-1}, k_{j-1}) = \sum_{0,0}^{\infty,\infty}(R_{i-j}k_i, k_j) = <k,k> .$$

Therefore U defines an isometry on K_f/K_o. By continuity U extends to an isometry (also denoted by) U on K. Let us identify h and h' in H with $\{h, 0, 0, ...\}$ and $\{h', 0, 0, ...\}$ in K, respectively. Then

$$<P_H U^n h, h'> = <U^n h, h'> = (R_n h, h') \quad \text{(for all } n \ge 0) .$$

This yields (1.4). Obviously H is cyclic for U.

To complete the proof assume that U' on $K'(\supseteq H)$ is another isometry satisfying (1.5) and H is cyclic for U'. Let $\{h_i\}_0^\infty$ be a sequence in H with finite support. By (1.4) (1.5) and (1.6) we have

$$\|\sum_{o}^{\infty} U^i h_i\|^2 = \sum_{0,0}^{\infty,\infty} (U^i h_i, U^j h_j) = \sum_{0,0}^{\infty,\infty} (R_{i-j} h_i, h_j) = \|\sum_{o}^{\infty} U'^i h_i\|^2 .$$

Since H is cyclic for both U and U$'$ this implies that there exists a unitary operator W mapping K onto K' satisfying

$$W \sum_{o}^{\infty} U^i h_i = \sum_{o}^{\infty} U'^i h_i .$$

This readily gives $WU^i h = U'^i h$ for all $i \geq 0$ and h in H, which yields,

$$WU \sum_{o}^{\infty} U^i h_i = U' \sum_{o}^{\infty} U'^i h_i = U' W \sum_{o}^{\infty} U^i h_i .$$

Because H is cyclic $WU = U'W$. Obviously $W|H = I$. This completes the proof.

1.2 COROLLARY. *The sequence of operators $\{R_i\}_1^\infty$ on H is positive definite if and only if there exists a contraction* T *on* $X (\supseteq H)$ *such that*

$$R_i = P_H T^i | H \quad (for\ all\ i \geq 0) . \tag{1.7}$$

PROOF. Assume that (1.7) holds. Let U be the minimal isometric dilation of T. Equation (1.7) and $U^*|X = T$ gives (1.4). The Naimark dilation theorem shows that $\{R_i\}_1^\infty$ is positive definite. The other half follows by choosing $T = U$ in the Naimark dilation theorem.

The previous corollary shows that if T is a contraction on $X = H$, then $\{T^i\}_1^\infty$ is a positive definite sequence. In this case the minimal isometric dilation U of T is precisely the minimal isometric dilation of the positive definite sequence $\{T^i\}_1^\infty$. So the concept of a minimal isometric dilation for positive definite sequences generalizes the notion of the minimal isometric dilation of a contraction.

1.3 COROLLARY. *A sequence of operators $\{R_i\}_1^\infty$ on H is positive definite if and only if there exists a unitary operator* $\hat{U}$ *on* $\hat{K} (\supseteq H)$ *such that*

$$R_i = P_H \hat{U}^i | H \quad (for\ all\ i \geq 0) \tag{1.8}$$

In this case if

$$\hat{K} = \bigvee_{-\infty}^{\infty} \hat{U}^i H , \tag{1.9}$$

then $\hat{U}$ *is unique up to an isomorphism.*

In this case, by *unique up to an isomorphism,* we mean that if $\hat{U}'$ on $\hat{K}'$ is another unitary operator satisfying (1.8) and (1.9) where $\hat{U}'$ on $\hat{K}'$ replaces $\hat{U}$ on $\hat{K}$, then there exists a unitary operator W mapping $\hat{K}$ onto $\hat{K}'$ such that $\hat{U}'W = W\hat{U}$ and $W|H = I$. the unitary operator $\hat{U}$ on $\hat{K}$ satisfying (1.8) and (1.9) is called *the minimal unitary dilation of* $\{R_i\}_1^\infty$.

PROOF. Let U be an isometry on K. By Corollary VI.2.3 the isometry U has a unique minimal unitary extension $\hat{U}$ on K satisfying

$$\hat{U}|K = U \text{ and } \hat{K} = \bigvee_{-\infty}^{\infty} \hat{U}^i K . \tag{1.10}$$

By unique we mean that if $\hat{U}'$ on $\hat{K}'$ is another unitary operator satisfying (1.10) where $\hat{U}'$ on $\hat{K}'$ replaces $\hat{U}$ on $\hat{K}$, then there exists a unitary operator W mapping $\hat{K}$ onto $\hat{K}'$ such that $\hat{U}'W = W\hat{U}$. The corollary follows from this fact (1.10) and the Naimark dilation theorem.

Let $\{r_i\}_1^\infty$ be a positive definite sequence of complex numbers. The previous corollary shows that

$$r_n = (\hat{U}^n h, h) = \int_0^{2\pi} e^{in\lambda} d(E_\lambda h, h)$$

where 1 is identified with h in $\hat{K}$. The last equation follows from the spectral representation for the unitary operator $\hat{U}$ where E_λ is the resolution of the identity for $\hat{U}$ (see RieszSz.-N). Setting $\mu(\lambda) = (E_\lambda h, h)$ we obtain first part of the following result which is Hergoltz's theorem for positive definite sequences.

1.4 COROLLARY. *A sequence of complex numbers* $\{r_i\}_1^\infty$ *is positive definite if and only if there exists positive measure* $\mu(\lambda)$ *on* $[0, 2\pi)$ *such that*

$$r_n = \int_0^{2\pi} e^{in\lambda} d\mu(\lambda) . \tag{1.11}$$

PROOF. To complete the proof it remain to show that if (1.11) holds for some positive measure, then $\{r_i\}_1^\infty$ is positive definite. So if $\{h_i\}_0^\infty$ is a sequence of complex numbers with finite support, then

$$\sum_{0,0}^{\infty,\infty} r_{n-j} h_n \bar{h}_j = \sum_{0,0}^{\infty,\infty} \int_0^{2\pi} e^{i(n-j)\lambda} h_n \bar{h}_j d\mu(\lambda) = \int_0^{2\pi} | \sum_{n=0}^{\infty} e^{int} h_n |^2 d\mu(\lambda) \geq 0 .$$

Therefore $\{r_i\}_1^\infty$ is positive definite. This completes the proof.

2. POSITIVE DEFINITE FUNCTIONS AND CHOICE SEQUENCES

In this section we will show that there is a one to one correspondence between the set of all positive definite functions $\{R_1\}_1^\infty$ on H, and the set of all choice sequences initiated on H. One can obtain the results in this section by using the Naimark dilation theorem along with the theory of non-minimal isometric liftings in Section VI.7. However, because this special case is instructive we will proceed from first principles.

To begin, let $\{R_n\}_1^\infty$ be a positive definite sequence on H and U on K be its minimal isometric dilation. The minimality condition (1.3) implies that U has a matrix representation of the form:

$$U = \begin{bmatrix} \Gamma_1 & * & * & . & . & . & . & . & . \\ B & * & * & . & . & . & . & . & . \\ 0 & * & * & . & . & . & . & . & . \\ 0 & 0 & * & . & . & . & . & . & . \\ 0 & 0 & 0 & . & . & . & . & . & . \\ . & . & . & . & . & . & . & . & . \\ . & . & . & . & . & . & . & . & . \\ . & . & . & . & . & . & . & . & . \end{bmatrix} \tag{2.1}$$

$$\text{on } K = H \oplus D_1 \oplus D_2 \oplus D_3 \oplus \cdots$$

where

$$D_m = H_{m+1} \ominus H_m \text{ and}$$

$$H_m = \bigvee_{n=0}^{m-1} U^n H = H \oplus D_1 \oplus D_2 \oplus \cdots \oplus D_{m-1} \quad (m \geq 1). \tag{2.2}$$

Since U is an isometry the columns must be isometric and orthogonal. The first column in (2.1) is easy to obtain. Equation (1.4) gives $R_1 = P_H U | H = \Gamma_1$. This and the isometric property shows that

$$\|Bh\|^2 = \|h\|^2 - \|\Gamma_1 h\|^2 = \|D_1 h\|^2 \quad (h \in H)$$

where $D_1 = D_{\Gamma_1}$. So there exists a unitary operator W mapping $\overline{BH}$ onto $D_1 = D_{\Gamma_1}$ such that $WB = D_1$. Without loss of generality we can choose $W = I$. This follows because the isometry U in the Naimark dilation theorem is unique up to an isomorphism. Therefore the first column of U is

$$[\Gamma_1^*, D_1, 0, 0, 0, ...]^* .$$

Furthermore $H_2 = H \oplus D_1$ where $D_1 = D_{\Gamma_1}$.

Now let $D_{1*} = D_{\Gamma_1^*}$ and $D_{1*} = D_{\Gamma_1^*}$. Using the fact that the rotation matrix

$$R_{\Gamma_1} = \begin{bmatrix} \Gamma_1 & D_{1*} \\ D_1 & -\Gamma_1^* \end{bmatrix} : H \oplus D_{1*} \to H \oplus D_1$$

of Γ_1 is unitary (see Corollary IV.1.4) we have

$$\begin{bmatrix} H \\ D_1 \end{bmatrix} \ominus \begin{bmatrix} \Gamma_1 \\ D_1 \end{bmatrix} H = \begin{bmatrix} D_{1*} \\ -\Gamma_1^* \end{bmatrix} D_{1*} . \tag{2.3}$$

Notice that the operator $[D_{1*}, -\Gamma_1]^*$ is an isometry. Since the second column in (2.1) is orthogonal to the first column, (2.3) shows that the second column must be of the form:

$$[D_{1*}\Gamma_2, -\Gamma_1^*\Gamma_2, *, 0, 0, ...]^{\mathrm{tr}}$$

where Γ_2 is a *uniquely determined* contraction mapping D_1 into D_{1*}. Because $[D_{1*}, -\Gamma_1]^*$ is an isometry and the second column of U is isometric, the second column of U can be identified with

$$[D_{1*}\Gamma_2, -\Gamma_1^*\Gamma_2, D_2, 0, 0, ...]^{\mathrm{tr}}$$

(As expected $D_n = D_{\Gamma_n}$ and $D_{n*} = D_{\Gamma_n^*}$ and $D_n = D_{\Gamma_n}$ and $D_{n*} = D_{\Gamma_n^*}$ for all n.) Clearly $H_3 = H \oplus D_1 \oplus D_2$. Continuing in this fashion, that is, using (1.3) and the fact that all the columns of U must be isometric and orthogonal shows that U can be identified with a matrix of the form:

$$U = \begin{bmatrix} \Gamma_1 & D_{1*}\Gamma_2 & D_{1*}D_{2*}\Gamma_3 & D_{1*}D_{2*}D_{3*}\Gamma_4 & \cdots \\ D_1 & -\Gamma_1^*\Gamma_2 & -\Gamma_1^*D_{2*}\Gamma_3 & -\Gamma_1^*D_{2*}D_{3*}\Gamma_4 & \cdots \\ 0 & D_2 & -\Gamma_2^*\Gamma_3 & -\Gamma_2^*D_{3*}\Gamma_4 & \cdots \\ 0 & 0 & D_3 & -\Gamma_3^*\Gamma_4 & \cdots \\ 0 & 0 & 0 & D_4 & \cdots \\ 0 & 0 & 0 & 0 & \cdots \\ 0 & 0 & 0 & 0 & \cdots \\ \cdot & \cdot & \cdot & \cdot & \cdots \\ \cdot & \cdot & \cdot & \cdot & \cdots \\ \cdot & \cdot & \cdot & \cdot & \cdots \end{bmatrix} \tag{2.4}$$

$$\text{on } K = H \oplus D_1 \oplus D_2 \oplus D_3 \oplus \cdots$$

Here $\Gamma_1 = R_1$ and Γ_{n+1} is a uniquely determined contraction mapping D_n into D_{n*} for n ≥ 1, or equivalently, $\{\Gamma_i\}_1^\infty$ is a choice sequence initiated in H. (Recall that $\{\Gamma_i\}_1^\infty$ is a choice sequence initiated in H if Γ_1 is a contraction on H and Γ_{n+1} is a contraction mapping D_n into D_{n*} for all n ≥ 1.)

The isometry U on K in (2.4) can be put in a simple form. To see this let U_n be the isometry defined by

$$U_n = I \oplus I \oplus \cdots \oplus I \oplus \begin{bmatrix} \Gamma_n & D_{n*} \\ D_n & -\Gamma_n^* \end{bmatrix} \oplus I \oplus I \oplus \cdots \tag{2.5}$$

where Γ_n is the contraction appearing in the n-th position and I is the identity operator on the appropriate space. Notice that the 2 by 2 matrix appearing in (2.5) is the rotation matrix determined by Γ_n. A simple calculation shows that U in (2.4) is given by

$$U = \prod_{n=1}^\infty U_n = \text{strong } \lim_{i \to \infty} U_1 U_2 U_3 U_4 \cdots U_i \tag{2.6}$$

Clearly the minimal isometric dilation U of $\{R_n\}_1^\infty$ and the choice sequence $\{\Gamma_n\}_1^\infty$ initiated on H uniquely determine each other. Combining the above with the uniqueness of U yields the following result.

2.1 THEOREM. *There is a one to one correspondence between the set of all positive definite sequences $\{R_i\}_1^\infty$ on H and the set of all choice sequences $\{\Gamma_i\}_1^\infty$ initiated on H. In this correspondence the minimal isometric dilation U on K for $\{R_i\}_1^\infty$ is isomorphic to the isometry in (2.4), or equivalently, (2.5), (2.6).*

2.2 REMARK. The above analysis shows that if U is an isometry on K and H is cyclic for U, then U is unitarily equivalent to a matrix of the form (2.4) where $\{\Gamma_i\}_1^\infty$ is a choice sequence initiated on H. In other words there is a one to one correspondence between the set of all isometries U on K where H is cyclic for U, and the set of all choice sequences $\{\Gamma_i\}_1^\infty$ initiated on H. This result is a special case of the theory of non-minimal isometric liftings in Section VI.7. In fact one can easily obtain Theorem 2.1 by using the results in Section VI.7 along with the uniqueness and minimality condition (1.3) for U supplied by the Naimark dilation Theorem.

3. THE CARATHEODORY INTERPOLATION PROBLEM
FOR POSITIVE DEFINITE FUNCTIONS

In this section we will solve an operator version of the positive definite Carathéodory interpolation problem in Chapter II. Then we will use our geometric structure to present another approach to some classical results. Namely we will show that the scalar valued Carathéodory positive extension problem has a unique solution if and only if its Toeplitz matrix is positive and singular. Then we will show that the Levinson polynomial has all its zeros in the unit disc, respectively on the unit circle if and only if its corresponding Toeplitz matrix is strictly positive, respectively positive and singular.

A sequence of operators $\{R_i\}_1^m$ on H has a *positive extension* if there exists a sequence of operators $\{R_i\}_{m+1}^\infty$ on H such that $\{R_i\}_1^\infty$ is positive definite. (Recall that $R_o \stackrel{\Delta}{=} I$.) This leads to the following Carathéodory interpolation problems for positive definite sequences.

3.1 PROBLEM. Given a sequence of operators $\{R_i\}_1^n$ on H, find necessary and sufficient conditions for the existence of a positive extension of $\{R_i\}_1^n$.

3.2 PROBLEM. Give an explicit description of all solutions (if there are any) to Problem 3.1.

Throughout $\hat{R}_{n+1}$ is the Toeplitz matrix in (1.2) where n+1 replaces n. The following is the classical solution to Problem 3.1. The proof is similar to the proof of the Naimark dilation theorem.

3.3 THEOREM. *A sequence of operators* $\{R_i\}_1^n$ *on H admits a positive extension if and only if its Toeplitz matrix* $\hat{R}_{n+1}$ *is positive* (≥ 0).

PROOF. Assume that $\hat{R}_{n+1}$ is positive. To show that $\{R_i\}_1^n$ admits a positive extension we construct a contraction A_n on $X_n(\supseteq H)$ such that

$$R_i = P_H A_n^i | H \quad (0 \leq i \leq n). \tag{3.1}$$

Then by Corollary 1.2 the sequence $\{P_H A_n^i | H\}_1^\infty$ is a positive extension of $\{R_i\}_1^n$. To this end let F_n be the quotient space defined by $l_{n+1}^2(H)/\ker \hat{R}_{n+1}$ with the inner product

$$(f,g)_{F_n} = (\hat{R}_{n+1}f, g) = \sum_{i=0, j=0}^{n,n} (R_{i-j}f_i, g_j) \tag{3.2}$$

where $f = [f_0, f_1, ..., f_n]^{tr}$ and $g = [g_0, g_1, ..., g_n]^{tr}$ are in F_n. Consider the operator A_n on X_n defined by the first n components of F_n.

$$A_n[f_0, f_1, ..., f_{n-1}]^{tr} = P_{X_n}[0, f_0, f_1, ..., f_{n-1}]^{tr} \quad \text{(for } [f_0, ..., f_{n-1}]^{tr} \in X_n) \tag{3.3}$$

where X_n is the subspace of F_n identified with $[l_n^2(H), \{0\}]^{tr}$. By virtue of (3.2), (3.3) we have

$$(A_n^i h, h')_{F_n} = (R_i h, h') \quad \text{(for } 0 \le i \le n)$$

where h and h' are in H. (Note h in H is identified with $[h, 0, 0, ..., 0]^{tr}$.) Hence (3.1) holds. To see that A_n is a contraction notice that the form of $\hat{R}_{n+1}$ in (1.2) along with (3.2) gives

$$\|A_n[f_0, f_1, ..., f_{n-1}]^{tr}\|_{X_n}^2 = \|P_{X_n}[0, f_0, f_1, ..., f_{n-1}]^{tr}\|_{F_n}^2 \le$$

$$\|[0, f_0, f_1, ..., f_{n-1}]^{tr}\|_{F_n}^2 = \|[f_0, f_1, ..., f_{n-1}, 0]^{tr}\|_{X_n}^2 . \tag{3.4}$$

Therefore $\{R_i\}_0^n$ admits a positive extension by Corollary 1.2. The other half of the proof is obvious.

To complete the section we will use the operator A_n on X_n in (3.3) to obtain several classical results for Toeplitz operators $\hat{R}_{n+1}$ on C^{n+1} generated by a sequence of complex numbers $\{R_i\}_1^n$ on $H = C^1$.

3.4 LEMMA. *Let $\hat{R}_{n+1}$ be a positive Toeplitz matrix on C^{n+1}. The following are equivalent:*

(i) $\hat{R}_{n+1}$ is singular.

(ii) There exists a sequence of complex numbers $\{\alpha_i\}_0^{n-1}$ such that

$$\hat{R}_{n+1}[\alpha_0, \alpha_1, ..., \alpha_{n-1}, 1]^{tr} = 0 . \tag{3.5}$$

(iii) The space $X_n = F_n$.

(iv) There exists an $f = [f_0, f_1, ..., f_n]^{tr}$ in F_n where f_n is nonzero and f is in X_n.

PROOF. Assume that $\hat{R}_{n+1}$ is singular. Let i be the first integer such that $\hat{R}_i$ is singular. This implies that there exists an α in C^i of the form $\alpha = [\alpha_0, ..., \alpha_{i-2}, 1]^{tr}$ satisfying $\hat{R}_i \alpha = 0$. (Because $\hat{R}_i$ is a singular expansion of the nonsingular Toeplitz matrix $\hat{R}_{i-1}$ we can choose the last component $\alpha_{i-1} = 1$.) Using $\hat{R}_i \alpha = 0$ we have

$$(\hat{R}_{n+1}(0\oplus\alpha), (0\oplus\alpha)) = (\hat{R}_i\alpha, \alpha) = 0 .$$

Since $\hat{R}_{n+1}$ is positive this implies that $\hat{R}_{n+1}(0\oplus\alpha) = 0$. Thus (ii) holds. Obviously (ii) implies (i). Hence (i) and (ii) are equivalent.

Assume that (ii) holds and let $f = [f_0, f_1, ..., f_n]^{tr}$ be in F_n. Notice that

$$f = [f_0 - \alpha_0 f_n, f_1 - \alpha_1 f_n, ..., f_{n-1} - \alpha_{n-1} f_n, 0]^{tr} + f_n[\alpha_0, \alpha_1, ..., \alpha_{n-1}, 1]^{tr} .$$

The second term is in X_n and by (3.5) the third term is in ker $\hat{R}_{n+1}$. Thus f is in X_n, or equivalently, $F_n \subseteq X_n$. Obviously $X_n \subseteq F_n$. Thus $X_n = F_n$ and (iii) holds. If (iii) holds, then $f = [0, 0, ..., 0, 1]$ is in F_n and $f_n = 1$ is nonzero. Since $X_n = F_n$ this f is in X_n and (iv) holds. If (iv) holds, then f admits a decomposition of the form: $f = [f_0', f_1', ..., f_{n-1}', 0]^{tr} + g$ where g is in the ker $\hat{R}_{n+1}$. Because f_n is nonzero g is nonzero. Thus $\hat{R}_{n+1}$ is singular. This completes the proof.

3.5 COROLLARY. *Let $\hat{R}_{n+1}$ be a positive Toeplitz matrix on C^{n+1} and A_n the contraction on X_n defined in (3.3). Then*

(i) The contraction A_n is unitary if and only if $\hat{R}_{n+1}$ is singular.

(ii) The eigenvalues of A_n are in the open unit disc if and only if $\hat{R}_{n+1}$ is nonsingular. In this case X_n is n dimensional.

(iii) The spectrum of A_n is either on the unit circle or in the open unit disc.

PROOF. If A_n is unitary, then there is equality in (3.4). This implies that there is a nonzero f_{n-1} such that $[0, f_0, f_1, ..., f_{n-1}]$ is in X_n. Part (iv) of the previous lemma shows that $\hat{R}_{n+1}$ is singular. On the other hand if $\hat{R}_{n+1}$ is singular Lemma 3.4 implies that $X_n = F_n$. So there is equality in (3.4) and A_n is unitary. This proves part (i).

If the eigenvalues of A_n are in the open unit disc, then A_n is not unitary. Part (i) implies that $\hat{R}_{n+1}$ is nonsingular. Now assume that $\hat{R}_{n+1}$ is nonsingular. Since A_n is a contraction the spectrum of A_n is in the closed unit disc. We use contradiction and assume that there exists a nonzero eigenvector $f = [f_0, f_1, ..., f_{n-1}]$ in X_n satisfying $A_n f = \lambda f$ where $|\lambda| = 1$. Let f_i be a nonzero component of f. Using $A_n^j f = \lambda^j f$ for all $j \geq 1$ with (3.3) we have

$$\|f\| = \|A_n^{n-i} f\| = \|P_{X_n}[0, 0, ..., 0, f_0, f_1, ..., f_i]^{tr}\|$$

$$\leq \|[0, 0, ..., 0, f_0, f_1, ..., f_i]^{tr}\| = \|f\| .$$

So we have equality and $[0, 0, ..., 0, f_0, f_1, ..., f_i]^{tr}$ is in X_n. Lemma 3.4 part (iv) implies that $\hat{R}_{n+1}$ is singular. This is a contradiction. Therefore $|\lambda| < 1$.

To complete the proof we show that X_n is n dimensional when $\hat{R}_{n+1}$ is nonsingular. Clearly $H(=C^1)$ is cyclic for A_n. If the dimension of X_n is strictly less than n, then $\{A_n^i H\}_{i=0}^{n-2}$ spans X_n. Since A_n is isometric on $\bigvee_0^{n-2} A_n^i H$ this implies that A_n is unitary. By Part (i) this contradicts the nonsingularity of $\hat{R}_{n+1}$. Thus X_n is n dimensional and the proof is complete.

The following is the classical uniqueness result for the positive definite Carathéodory interpolation problem, which is equivalent to Theorem 3.2 in Chapter II.

3.6 THEOREM. *Let $\hat{R}_{n+1}$ be a positive Toeplitz matrix on C^{n+1}. Then $\{R_i\}_1^n$ admits a unique positive extension if and only if the Toeplitz matrix $\hat{R}_{n+1}$ is positive and singular. In this case the unique positive extension of $\{R_i\}_1^n$ is $\{P_H A_n^i | H\}_{i=1}^\infty$ and A_n on X_n is the minimal isometric dilation of this extension.*

PROOF. Assume that $\hat{R}_{n+1}$ is positive and nonsingular. We claim that $\{\hat{R}_i\}_1^n$ does not admit a unique extension. To see this let

$$U_o = \begin{bmatrix} A_n & 0 & 0 & 0 & \cdots \\ D_{A_n} & 0 & 0 & 0 & \cdots \\ 0 & I & 0 & 0 & \cdots \\ 0 & 0 & I & 0 & \cdots \\ 0 & 0 & 0 & I & \cdots \\ \cdot & \cdot & \cdot & \cdot & \cdots \\ \cdot & \cdot & \cdot & \cdot & \cdots \\ \cdot & \cdot & \cdot & \cdot & \cdots \end{bmatrix} \tag{3.6}$$

on $X_n \oplus l^2(D_{A_n})$ be the minimal isometric dilation of A_n. Since H is cyclic for A_n the space H is also cyclic for U_o. (Notice that H is identified with C^1.) Using $U_o^* | X_n = A_n^*$ gives

$$R_i = P_H A_n^i | H = P_H U_o^i | H \qquad (0 \le i \le n) \ . \tag{3.7}$$

By the Naimark dilation theorem $\{P_H U_o^i | H\}_1^\infty$ is a positive extension of $\{R_i\}_1^n$. Corollary 3.5 implies that D_{A_n} is nonzero and U_o is infinite dimensional.

Now consider the unitary operator U_1 on $X_n \oplus X_n$ defined by

$$U_1 = \begin{bmatrix} A_n & D_{A_n^*} \\ D_{A_n} & -A_n^* \end{bmatrix} .$$ (3.8)

Since A_n is isometric on $\bigvee_0^{n-2} A_n^i H$ we have $D_{A_n} A_n^i h = 0$ for $0 \le i \le n-2$. This gives

$$R_i = P_H A_n^i | H = P_H U_1^i | H \qquad (0 \le i \le n) .$$ (3.9)

By the Naimark dilation theorem $\{P_H U_1^i | H\}_1^\infty$ is a positive extension of $\{R_i\}_1^n$. The unitary operator U_1 is not necessarily the minimal isometric dilation of $\{P_H U_1^i | H\}_1^\infty$. The minimal isometric dilation of $\{P_H U_1^i | H\}_1^\infty$ is obtained by restricting U_1 to the invariant subspace generated by $K_1 = \bigvee_0^\infty U_1^i H$.

We claim that $\{P_H U_0^i | H\}_1^\infty$ and $\{P_H U_1^i | H\}_1^\infty$ are two different positive extensions of $\{R_i\}_1^n$. If they were the same extension, then the Naimark dilation theorem implies that their respective isometric dilations U_0 and $U_1 | K_1$ are unitarily equivalent. This is impossible because U_0 is infinite dimensional and U_1 is finite dimensional. Therefore if $\hat{R}_{n+1}$ is positive and nonsingular, then $\{R_i\}_1^n$ admits at least two different positive extensions, or equivalently, if $\{R_i\}_1^n$ admits a unique positive extension, then $\hat{R}_{n+1}$ is positive and singular.

If $\hat{R}_{n+1}$ is positive and singular, then $\{R_i\}_1^n$ admits a unique positive extension. This fact follows from the proof of Theorem 3.2 in Chapter II. The last statement follows from (3.9) and Part (i) of Corollary 3.5. The proof is now complete.

As before let $\hat{R}_{n+1}$ be a positive Toeplitz matrix on $\mathbb{C}^{n+1}$. Consider the Levinson system

$$\hat{R}_{n+1}[\alpha_0, \alpha_1, ..., \alpha_{n-1}, 1]^{\text{tr}} = [0, 0, ..., 0, \epsilon^2]^{\text{tr}}$$ (3.10)

where α_i are complex numbers and ϵ^2 is a positive number. (The positivity of $\hat{R}_{n+1}$ guarantees that ϵ^2 is positive.) Let $\alpha(z)$ be the polynomial defined by

$$\alpha(z) = z^n + \sum_0^{n-1} \alpha_i z^i .$$ (3.11)

This sets the stage for the following classical result.

3.7 THEOREM. *Let $\hat{R}_{n+1}$ be a positive Toeplitz matrix on $\mathbb{C}^{n+1}$.*
(i) If $\hat{R}_{n+1}$ is nonsingular, then $\alpha(z)$ in (3.11) is the minimal polynomial of A_n.
In particular, the zeros of $\alpha(z)$ are contained in the open unit disc.

(ii) If $\hat{R}_n$ is nonsingular and $\hat{R}_{n+1}$ is singular, then $\alpha(z)$ is the minimal polynomial of A_n. In particular, the zeros of $\alpha(z)$ are on the unit circle.

PROOF. Let e_j for $0 \le j \le n$ be the standard $j+1$ unit vector in C^{n+1}, that is, the $j+1$ component of e_j is one and all the other components of e_j are zero. We identify 1 in $H = C^1$ with $h = e_o$ in X_n. Using $A_n^j h = e_j$ for $0 \le j < n$ along with the shift property of A_n we have

$$(A_n^i h, A_n^j h)_{F_n} = (\hat{R}_{n+1} e_i, e_j) = R_{i-j} \quad \text{(for } 0 \le i \le n \text{ and } 0 \le j < n) \tag{3.12}$$

Using this and (3.10) yields

$$(\alpha(A_n)h, A_n^j h) = 0 \quad \text{(for } 0 \le j < n) . \tag{3.13}$$

Since h is cyclic for A_n this implies that $\alpha(A_n)h = 0$. Thus $\alpha(A_n)A_n^k h = 0$ for all k. Applying the cyclic property of h again yields $\alpha(A_n) = 0$. Because h is cyclic for A_n the repeated eigenvalues are all contained in one Jordan block. So the degree of the minimal polynomial for A_n is the dimension of X_n. Obviously the degree of $\alpha(z)$ is n. Therefore $\alpha(z)$ is the minimal polynomial for A_n if and only if the dimension of X_n is n. This and Part (ii) of Corollary 3.5 proves (i).

To prove part (ii) it is sufficient to show that the dimension of X_n is n when $\hat{R}_n$ is nonsingular. We claim that $\{e_i : 0 \le i \le n - 1\}$ is a basis for X_n. To see this assume that

$$\sum_{i=0}^{n-1} \gamma_i e_i = 0 \quad \text{in } X_n .$$

This imples that $f = [\gamma_0, \gamma_1, ..., \gamma_{n-1}, 0]^{tr}$ is in the kernel of $\hat{R}_{n+1}$. Therefore by the structure of the Toeplitz matrix $[\gamma_0, \gamma_1, ..., \gamma_{n-1}]^{tr}$ is in the kernel of $\hat{R}_n$. So $\gamma_i = 0$ for all i and the dimension of X_n is n. This completes the proof.

4. AN INVERSE SCATTERING ALGORITHM

In this section we will give a recursive algorithm to calculate the choice sequence associated with a positive definite sequence $\{R_i\}_1^\infty$ on H.

Let $\{R_i\}_1^\infty$ be a positive definite sequence on H and U on K its minimal isometric dilation. The *n step contraction* T_n on H_n *associated with* $\{U, H\}$ is defined by

$$T_n = P_{H_n} U | H_n \quad \text{where } H_n = \bigvee_0^{n-1} U^i H \quad (n \ge 1) . \tag{4.1}$$

Using (1.4) and $H \subseteq H_n$ with $T_n^j | H = U^j | H$ for $0 \le j < n$ it is easy to verify that

$$R_i = P_H T_n^i | H \qquad (0 \leq i \leq n) . \qquad (4.2)$$

The following shows that two positive definite sequences whose first n operators $\{R_i\}_1^n$ are the same determine the same n step contraction T_n.

4.1 PROPOSITION. *Let $\{R_i\}_1^\infty$ and $\{R_i'\}_1^\infty$ be two positive definite sequences on H with minimal isometric dilations U and U', respectively. Let T_n on H_n and T_n' on H_n' be the n step contractions associated with $\{U, H\}$ and $\{U', H\}$, respectively. Then*

$$R_i = R_i' \qquad (for \ 0 \leq i \leq n) \qquad (4.3)$$

if and only if there exists a unitary operator W mapping H_n onto H_n' satisfying $T_n'W = WT_n$ and $W|H = I$.

PROOF. The proof is similar to the uniqueness part of the proof of the Naimark dilation theorem. Assume that (4.3) holds and let W be the operator mapping H_n onto H_n' defined by

$$W \sum_0^{n-1} T_n^i h_i = \sum_0^{n-1} T_n'^i h_i$$

where h_i is in H for all i. Obviously $W|H = I$. By (4.1) to (4.3) we have

$$\|\sum_0^{n-1} T_n^i h_i\|^2 = \|\sum_0^{n-1} U^i h_i\|^2 = \sum_{0,0}^{n-1,n-1} (R_{i-j} h_i, h_j) = \|\sum_0^{n-1} T_n'^i h_i\|^2 .$$

Therefore W is unitary. Using $WT_n^i h_i = T_n'^i h_i$ for $0 \leq i < n$ with h in H we have:

$$(T_n^n h, T_n^i h_i) = (U^n h, U^i h_i) = (R_{n-i} h, h_i) =$$

$$(T_n'^n h, T_n'^i h_i) = (T_n'^n h, WT_n^i h_i) = (W^* T_n'^n h, T_n^i h_i) .$$

Since H is cyclic for T_n this implies that $WT_n^n h = T_n'^n h$. Hence

$$WT_n T_n^i h = T_n'^{i+1} h = T_n' T_n'^i h = T_n' WT_n^i h$$

for all $0 \leq i < n$. This and the fact that H is cyclic readily yields $WT_n = T_n'W$. The other half of the proof is a simple consequence of (4.2), and the unitary equivalence of T_n and T_n'. This completes the proof.

The following is an immediate consequence of the previous proposition.

4.2 COROLLARY. *Let $\hat{R}_{n+1}$ be the positive Toeplitz matrix generated by the sequence $\{R_i\}_1^n$ on H. Then all positive extensions $\{R_i\}_1^\infty$ of $\{R_i\}_1^n$ determine the same n* *step contraction* T_n *up to a unitary operator. In particular, the n step contraction* T_n *is uniquely determined by* $\{R_i\}_1^n$.

Assume that $\{R_i\}_1^n$ admits a positive extension, or equivalently, $\hat{R}_{n+1}$ is positive. Due to the previous corollary, we say that a contraction T' on H' is *the n step contraction associated* with $\{R_i\}_1^n$ if there exists a unitary operator W mapping H_n onto H' satisfying $WT_n = T'W$ and $W|H = I$ where T_n on H_n is the n step contraction associated with $\{U, H\}$, and U is the minimal isometric dilation of any positive extension $\{R_i\}_1^\infty$ of $\{R_i\}_1^n$.

4.3 COROLLARY. *Let* $\hat{R}_{n+1}$ *be a positive Toeplitz matrix on* $l_{n+1}^2(H)$ *and* A_n *on* X_n *the contraction defined in (3.3). Then* A_n *is precisely the n step contraction associated with* $\{R_i\}_1^n$.

PROOF. To see this let U_o be the minimal isometric dilation of A_n in (3.6). Since H is cyclic for A_n and $A_n | \bigvee_0^{n-2} A_n^i H$ is an isometry we have

$$H_n = \bigvee_0^{n-1} U_o^i H = \bigvee_0^{n-1} A_n^i H = X_n \ .$$

Thus $H_n = X_n$ and A_n is the n step contraction associated with U_o. Because $\{P_H U_o^i | H\}_1^\infty$ is a positive extension of $\{R_i\}_1^n$ the contraction A_n is the n step contraction associated with $\{R_i\}_1^n$. This completes the proof.

Let $\{R_i\}_1^\infty$ be a positive definite sequence on H and U on K its minimal isometric dilation in (2.4). Let T_n be the n step contraction associated with $\{U, H\}$, or equivalently, let T_n on H_n be the n by n matrix contained in the upper n by n left hand corner of U in (2.4), that is,

$$T_n = P_{H_n} U | H_n = P_{H_n} U_1 U_2 \cdots U_n | H_n \tag{4.4}$$

$$H_n = H \oplus D_1 \oplus D_2 \oplus \cdots \oplus D_{n-1} \ .$$

This (1.4) and the special form of U in (2.4) yields

$$R_n = P_H T_{n-1}^n | H + D_{1*} \cdots D_{n-1*} \Gamma_n D_{n-1} \cdots D_1 \qquad (n \geq 2) \ . \tag{4.5}$$

(Recall that $R_1 = \Gamma_1$.) The second term containing T_{n-1} is determined by $\{\Gamma_i\}_1^{n-1}$. The operator Γ_n only appears in the last term. Therefore $\{R_i\}_1^n$ and $\{\Gamma_i\}_1^n$ uniquely determine each other. Moreover, the first n operators in two positive definite sequences are equal if and only if the first n contractions in their corresponding choice sequences are equal.

Given a choice $\{\Gamma_i\}_1^n$ initiated on H equation (4.5) can be used to recursively calculate $\{R_i\}_1^n$. By extending $\{\Gamma_i\}_1^\infty$ to an infinite choice sequence we readily see that $\{R_i\}_1^n$ has a positive definite extension. By Theorem 3.3 the Toeplitz matrix $\hat{R}_{n+1}$ is positive. If $\hat{R}_{n+1}$ is positive, then Theorem 3.3 shows that $\{R_i\}_1^n$ has a positive extension. In this case $\{R_i\}_1^n$ uniquely determines a choice sequence $\{\Gamma_i\}_1^n$. In other words $\{R_i\}_1^n$ admits a positive extension if and only if the operators $\{\Gamma_i\}_1^n$ determined by (4.5) is a choice sequence initiated on H. The following inverse scattering algorithm uses (4.5) to recursively calculate this choice sequence.

4.4 PROCEDURE. Assume that the Toeplitz matrix $\hat{R}_{n+1}$ generated by $\{R_i\}_1^n$ on H is positive. To recursively obtain the choice sequence $\{\Gamma_i\}_1^n$ associated with $\{R_i\}_1^n$ first set $\Gamma_1 = R_1$. Now assume that we have computed $\{\Gamma_i\}_1^{n-1}$. Using $\{\Gamma_i\}_1^{n-1}$ recursively compute $T_{n-1}^n \,|\, H$. This (4.5) and R_n recursively yields Γ_n by inverting the self adjoint operators D_{i*} and D_i in (4.5).

4.5 REMARK. In the previous procedure if Γ_n is not a contraction for any n, then the procedure stops. In this case the sequence $\{R_i\}_1^n$ does not have a positive definite extension, by Theorem 2.1. On the other hand, if $\{\Gamma_i\}_1^n$ computed by the previous procedure is a choice sequence, then the set of all positive definite extensions $\{R_i\}_1^\infty$ of $\{R_i\}_1^n$ is the set of all positive definite sequences generated by the choice sequence $\{\Gamma_i\}_1^\infty$, where $\{\Gamma_i\}_{n+1}^\infty$ is an arbitrary choice sequence initiated from D_n to D_{n*}. In particular, there exists a unique positive definite extension of $\{R_i\}_1^n$ if and only if $\{\Gamma_i\}_1^n$ is a choice sequence and $D_n = \{0\}$ or $D_{n*} = \{0\}$.

4.6 REMARK. Assume that $\{R_i\}_1^n$ on H admits a positive definite extension and $\{\Gamma_i\}_1^n$ is its choice sequence. Obviously all positive definite extensions of $\{R_i\}_1^n$ have the same first n contractions $\{\Gamma_i\}_1^n$ in its choice sequence. So all positive definite extensions determine the same matrix T_n. Since T_n is the n step contraction associated with $\{R_i\}_1^n$

$$R_i = P_H T_n^i \,|\, H \qquad (0 \le i \le n) . \qquad (4.6)$$

Consider the sequence $\{R_i^0\}_1^\infty$ on H generated by

$$R_i^0 = P_H T_n^i \,|\, H \qquad (i \ge 0) . \qquad (4.7)$$

Corollary 1.2 and (4.6) show that $\{R_n^0\}_1^\infty$ is a positive definite extension of $\{R_i\}_1^n$. We claim that $\{R_i^0\}_1^\infty$ is *the maximal entropy extension of* $\{R_i\}_1^n$, that is, the choice sequence for $\{R_i^0\}_1^\infty$ is

$$\{\Gamma_1, \Gamma_2, ..., \Gamma_n, 0, 0, 0, \cdots\} . \tag{4.8}$$

To prove our claim let $U = U_o$ be the isometry in (2.4) generated by the choice sequence in (4.8). Using $U = U_o$ in (2.4) with $\{\Gamma_i\}_{n+1}^{\infty}$ zero shows that H_n is invariant for U_o^*. So U_o is an isometric lifting of T_n. Since H is cyclic for T_n and T_n is contained in the upper left hand corner of U_o we see that H is cyclic for U_o. Thus U_o is the minimal isometric dilation of T_n. This (4.7) and the dilation property gives

$$R_i^o = P_H T_n^i \,|\, H = P_H U_o^i \,|\, H \qquad (i \geq 0) . \tag{4.9}$$

Therefore $\{R_i^o\}_1^{\infty}$ is the maximal entropy extension of $\{R_i\}_1^n$.

By Corollary 4.2 the n step contraction T_n is unique up to a unitary operator. So the above analysis shows that if T_n is any n step contraction associated with $\{R_i\}_1^n$ (or if $T_n = A_n$), then

$$R_i^o = P_H T_n^i \,|\, H \qquad (\text{for } 0 \leq i)$$

is the maximal entropy positive extension of $\{R_i\}_1^n$.

Finally, it is noted that if $\{R_i\}_1^m$ admits a unique positive definite extension ($D_m = \{0\}$ or $D_{m*} = \{0\}$), then this unique extension is given by (4.7) where the n step contraction T_n is computed recursively by Procedure 4.4, and n is the first integer where Γ_n is isometric or co-isometric.

4.7 REMARK. Assume that $\{R_i\}_1^{n-1}$ generates a positive Toeplitz matrix $\hat{R}_n$. Equation (4.5) shows that the set of all positive Toeplitz extension $\hat{R}_{n+1}$ of $\hat{R}_n$ is given by the operator quasi-ball

$$R_n = C_n + L_{n*}\Gamma_n L_n \tag{4.10}$$

where Γ_n is a contraction from D_{n-1} to D_{n-1*}. The center C_n, right radius L_n, and left radius L_{n*} are given by

$$C_n = P_H T_{n-1}^n \,|\, H, \; L_n = D_{n-1} D_{n-2} \cdots D_1 \text{ and } L_{n*} = D_{1*} D_{2*} \cdots D_{n-1*} . \tag{4.11}$$

Finally, it is noted that C_n, L_n and L_{n*} are all uniquely determined by $\hat{R}_n$, or equivalently, by its choice sequence $\{\Gamma_i\}_1^{n-1}$.

4.8 REMARK. Assume that $\{R_i\}_1^n$ on H admits a positive definite extension and $\{\Gamma_i\}_1^n$ is its choice sequence. Consider the matrix Δ_n mapping $l_n^2(H)$ into H_n defined by

$$\Delta_n = [I \,|\, H, \; T_n \,|\, H, \; T_n^2 \,|\, H, ..., T_n^{n-1} \,|\, H] . \tag{4.12}$$

We claim that Δ_n is an upper triangular factorization of the Toeplitz matrix $\hat{R}_n$, defined in (1.2), that is, $\Delta_n^* \Delta_n = \hat{R}_n$. (Recall that an upper triangular factorization Δ_n of $\hat{R}_n$ means that Δ_n is upper triangular and $\Delta_n^* \Delta_n = \hat{R}_n$.)

To prove this result let U on K in (2.4) be the minimal isometric dilation of $\{R_i\}_1^\infty$ where $\{R_i\}_1^\infty$ is any positive definite extension of $\{R_i\}_1^n$. Since T_n is contained in the upper left hand corner of U we have

$$T_n^i \,|\, H = U^i \,|\, H \quad (\text{for } 0 \le i \le n-1) . \tag{4.13}$$

This and the form of U in (2.4) shows that Δ_n is upper triangular. To prove that $\Delta_n^* \Delta_n = \hat{R}_n$ let $[h_i]_0^{n-1}$ be in $l_n^2(H)$. Then (1.4), (4.12) and (4.13) gives

$$(\hat{R}_n [h_i]_0^{n-1}, [h_i]_0^{n-1}) = \sum_{0,0}^{n-1,n-1} (R_{i-j}h_i, h_j) = \|\sum_0^{n-1} U^i h_i\|^2 = \|\Delta_n [h_i]_0^{n-1}\|^2 .$$

Since $[h_i]_0^{n-1}$ is arbitrary this yields $\Delta_n^* \Delta_n = \hat{R}_n$ and verifies our claim. It is emphasized that one can obtain the upper triangular factorization Δ_n of $\hat{R}_n$ in (4.12), recursively by Procedure 4.4. Finally, it is noted that one can always obtain an upper triangular factorization Δ of any positive (not necessarily Toeplitz) matrix R. This is done by simply solving for the upper triangular entries in Δ satisfying $\Delta^* \Delta = R$. However, if the matrix $R = \hat{R}_n$ is Toeplitz, then (4.12) shows how the choice sequence $\{\Gamma_i\}_1^{n-1}$ naturally enters in the elements of Δ_n.

We complete this section by presenting an

ALTERNATE PROOF OF THEOREM 3.6. Since $\{R_i\}_1^n$ are complex numbers the choice sequence $\{\Gamma_i\}_1^n$ corresponding to $\{R_i\}_1^n$ are also complex numbers. Thus $D_n = D_{n*}$. So $\{R_i\}_1^n$ admits a unique extension if and only if $D_n = \{0\}$. Since A_n is unitarily equivalent to T_n the upper left n by n matrix of U in (2.4), the operator A_n is unitary if and only if $D_n = \{0\}$. This follows because T_n is unitary if and only if $T_n = U$, or equivalently, $D_n = \{0\}$. By Corollary 3.5 the sequence $\{R_i\}_1^n$ admits a unique positive extension if and only if $\hat{R}_{n+1}$ is singular. This completes the proof.

5. DUALITY AND CHOICE SEQUENCES

This section is the dual of the previous section. First we will show that $\{\Gamma_i\}_1^n$ is the choice sequence for $\{R_i\}_1^n$ if and only if $\{\Gamma_i^*\}_1^n$ is the choice sequence for $\{R_i^*\}_1^n$. Then we will use Procedure 4.4 to recursively compute a lower triangular factorization Δ_{n*} of $\hat{R}_n$, that is, we recursively compute a lower triangular matrix Δ_{n*} satisfying $\Delta_{n*}^* \Delta_{n*} = \hat{R}_n$.

We begin with the following result, which states that two operator quasi-balls with the same left and right radii, which are equal as sets, have the same center. Since this result is a special case of Theorem XIII.7.1 the proof is omitted.

5.1 LEMMA. *Let* C *and* C' *be two operators on* H. *Let* L^* *from* D *to* H *and* L_* *from* D_* *to* H *be two operators with zero kernel. Then*

$$\{C + L_*\Gamma L:\Gamma \text{ is a contraction from } D \text{ to } D_*\} = \tag{5.1}$$

$$\{C' + L_*\Gamma' L:\Gamma' \text{ is a contraction from } D \text{ to } D_*\}$$

if and only if $C = C'$. *In this case* $C + L_*\Gamma L = C' + L_*\Gamma' L$ *if and only if* $C = C'$ *and* $\Gamma = \Gamma'$.

Let $\hat{I}_n$ be the unitary operator on $l_n^2(H)$ defined by

$$\hat{I}_n = \begin{bmatrix} 0 & 0 & \dots & 0 & I \\ 0 & 0 & \dots & I & 0 \\ \vdots & \vdots & & \vdots & \vdots \\ 0 & I & \dots & 0 & 0 \\ I & 0 & \dots & 0 & 0 \end{bmatrix} \tag{5.2}$$

The identity appears on the southwest to northeast diagonal and zeros appear elsewhere. Let $\hat{R}_n$ be the Toeplitz matrix generated by $\{R_i\}_1^{n-1}$ on H and $\hat{R}_{n*}$ be the Toeplitz matrix generated by $\{R_i^*\}_1^{n-1}$. A simple calculation shows that

$$\hat{R}_{n*} = \begin{bmatrix} I & R_1^* & \dots & R_{n-2}^* & R_{n-1}^* \\ R_1 & I & \dots & R_{n-3}^* & R_{n-2}^* \\ R_2 & R_1 & \dots & R_{n-4}^* & R_{n-3}^* \\ \vdots & \vdots & & \vdots & \vdots \\ R_{n-1} & R_{n-2} & \dots & R_1 & I \end{bmatrix} = \hat{I}_n \hat{R}_n \hat{I}_n. \tag{5.3}$$

This sets the stage for the following result.

5.2 PROPOSITION. *Let* $\hat{R}_{n+1}$ *be the Toeplitz matrix generated by* $\{R_i\}_1^n$ *on* H *and* $\hat{R}_{n+1*}$ *be the Toeplitz matrix generated by* $\{R_i^*\}_1^n$. *Then* $\hat{R}_{n+1}$ *is positive if and only if* $\hat{R}_{n+1*}$ *is positive. In this case* $\{\Gamma_i\}_1^n$ *is the choice sequence for* $\hat{R}_{n+1}$ *if and only if* $\{\Gamma_i^*\}_1^n$ *is the choice sequence for* $\hat{R}_{n+1*}$.

PROOF. The first part concerning the positivity of $\hat{R}_{n+1}$ and $\hat{R}_{n+1*}$ follows from (5.3) with $\hat{I}_n^* = \hat{I}_n$. Since $R_1 = \Gamma_1$ the second part holds for $n = 1$. Now we use induction and assume that $\{\Gamma_i\}_1^{n-1}$ is the choice sequence for $\hat{R}_n$ if and only if $\{\Gamma_i^*\}_1^{n-1}$ is the choice sequence for $\hat{R}_{n*}$. Applying Remark 4.7 to $\hat{R}_{n+1*}$ with the induction hypothesis, shows that $\hat{R}_{n+1*}$ is positive if and only if

$$R_n^* = C_{n*} + L_n^* \Gamma_{n*} L_n^* \tag{5.4}$$

where Γ_{n*} is an n-th choice contraction from D_{n-1*} to D_{n-1} and L_n and L_{n*} are the operators defined in (4.11). Equations (4.10) and (4.11) show that $\hat{R}_{n+1}$ is positive if and only if

$$R_n^* = C_n^* + L_n^* \Gamma_n^* L_{n*}^* \tag{5.5}$$

where Γ_n is an n-th choice contraction from D_{n-1} to D_{n-1*}. Since $\hat{R}_{n+1}$ is positive if and only if $\hat{R}_{n+1*}$ is positive, (5.4) and (5.5) generate the same set as Γ_{n*} and Γ_n^* run through the set of all contractions. By Lemma 5.1 we see that $C_{n*} = C_n^*$ and $\Gamma_{n*} = \Gamma_n^*$. This completes the induction and the proof.

5.3 REMARK. Assume that $\{R_i\}_1^{n-1}$ generates a positive Toeplitz matrix. Then Procedure 4.4 can be used to recursively compute a lower triangular factorization Δ_{n*} of $\hat{R}_n$, that is, a lower triangular matrix Δ_{n*} satisfying $\Delta_{n*}^* \Delta_{n*} = \hat{R}_n$. To obtain Δ_{n*} let Δ_n' be the upper triangular matrix in (4.12) recursively computed by applying Procedure 4.4 to $\hat{R}_{n*}$. A simple calculation shows that $\Delta_{n*} = \hat{I}_n \Delta_n' \hat{I}_n$ is lower triangular. This (5.3) and $\hat{I}_n^2 = \hat{I}_n$ yields:

$$\hat{R}_n = \hat{I}_n \hat{R}_{n*} \hat{I}_n = \hat{I}_n \Delta_n'^* \hat{I}_n \hat{I}_n \Delta_n' \hat{I}_n = \Delta_{n*}^* \Delta_{n*} . \tag{5.6}$$

Therefore Δ_{n*} is a lower triangular factorization of $\hat{R}_n$ and is recursively obtained by applying Procedure 4.4 to $\hat{R}_{n*}$.

6. THE LEVINSON ALGORITHM FOR BLOCK TOEPLITZ MATRICES

In this section we use the special form of U in (2.4) to obtain the Levinson algorithm for block Toeplitz matrices. One can easily generalize the results in Section II.2 to obtain a Levinson algorithm for block Toeplitz matrices. However, the main purpose of this section is to demonstrate how the choice sequence $\{\Gamma_i\}_1^\infty$ in the Naimark dilation U naturally occur in the Levinson algorithm.

As before $\{R_i\}_1^\infty$ on H is a positive definite sequence, $\{\Gamma_i\}_1^\infty$ is its choice sequence initiated on H and U on K is the minimal isometric dilation of $\{R_i\}_1^\infty$ displayed in (2.4). Finally, in this section only we assume that $\{\Gamma_i\}_1^\infty$ is a strictly contractive choice sequence, that is, Γ_i is a strict contraction for all i.

To begin, we define the n step past and future spaces by

$$P_n = \bigvee_{i=0}^{n-1} U^i H \quad \text{and} \quad F_n = \bigvee_{i=1}^{n} U^i H \ . \tag{6.1}$$

(Note H is always identified with $H \oplus \{0\} \oplus \{0\} \oplus \cdots$) The form of U in (2.4) yields

$$P_n = H \oplus D_1 \oplus D_2 \oplus \cdots \oplus D_{n-1} \ , \tag{6.2}$$

$$F_n = \bigvee \{\text{First n columns of U}\} \ .$$

If h is in H, then (2.4) and (6.2) give

$$U^n h - P_{P_n} U^n h = [0, 0, \cdots, 0, D_n D_{n-1} \cdots D_1 h, 0, 0, 0 \cdots]^{tr} = \phi_n \varepsilon_n h \tag{6.3}$$

where $\varepsilon_n \overset{\Delta}{=} D_n D_{n-1} \cdots D_1$ and the nonzero term appears in the n+1 position. Here ϕ_n is the isometry which embeds D_n in the n+1 position. Therefore the "forward prediction error" is

$$\|U^n h - P_{P_n} U^n h\| = \|\varepsilon_n h\| \qquad (h \in H) \ . \tag{6.4}$$

Since all the columns of U are isometric and orthogonal, F_n is orthogonal to the n+1 column of U for all choices of Γ_{n+1}. Thus F_n is orthogonal to the range of the isometry ϕ_{n*} mapping D_{n*} into K defined by

$$\phi_{n*} = \begin{bmatrix} D_{1*}D_{2*} & \cdots & D_{n*} \\ -\Gamma^* D_{2*} & \cdots & D_{n*} \\ -\Gamma_2^* D_{3*} & \cdots & D_{n*} \\ & \vdots & \\ & -\Gamma_n^* & \\ & 0 & \\ & \vdots & \end{bmatrix} \tag{6.5}$$

Using (6.2), (6.5) with the fact that the rotation matrix is unitary it is easy to verify that

$$H \oplus D_1 \oplus D_2 \oplus \cdots \oplus D_n = F_n \oplus \phi_{n*} D_{n*} \ . \tag{6.6}$$

Equation (6.6) also follows from

$$F_n \oplus \phi_{n*} D_{n*} = \operatorname{ran} P_{P_{n+1}} U_1 U_2 U_3 \cdots U_n = H \oplus D_1 \oplus D_2 \oplus \cdots \oplus D_n$$

where U_i for $1 \le i \le n$ are the unitary operators defined in (2.5).

Equation (6.6) implies that $h - P_{F_n} h = \phi_{n*} g$ for some g in D_{n*}. Since ϕ_{n*} is an isometry a simple calculation gives

$$h - P_{F_n} h = \phi_{n*} \varepsilon_{n*} h \qquad (h \in H) . \tag{6.7}$$

where $\varepsilon_{n*} \triangleq D_{n*} D_{n-1*} \cdots D_{1*}$ Therefore the "backward prediction error" is

$$\|h - P_{F_n} h\| = \|\varepsilon_{n*} h\| \qquad (h \in H) . \tag{6.8}$$

Finally, it is emphasized that the form of U in (2.4) easily lead to the forward and backward prediction errors in (6.4) and (6.8).

The operators $\phi_n \varepsilon_n$ and $\phi_{n*} \varepsilon_{n*}$ in (6.3) and (6.7) are the forward and backward error operators, respectively. Using (2.4), (6.3), (6.7) and $D_i \Gamma_i^* = \Gamma_i^* D_{i*}$ we obtain the Levinson recursion:

$$[\phi_{n+1} \varepsilon_{n+1}, \; \phi_{n+1*} \varepsilon_{n+1*}] = [U \phi_n \varepsilon_n, \; \phi_{n*} \varepsilon_{n*}] \begin{bmatrix} I & -\varepsilon_n^{-1} \Gamma_{n+1}^* \varepsilon_{n*} \\ -\varepsilon_{n*}^{-1} \Gamma_{n+1} \varepsilon_n & I \end{bmatrix} \tag{6.9}$$

(The operators ε_n and ε_{n*} are invertible because we have assumed that the choice sequence $\{\Gamma_i\}_1^\infty$ is strictly contractive.) To complete the recursion the following is also needed:

$$P_H U \phi_n \varepsilon_n | H = \varepsilon_{n*}^* \Gamma_{n+1} \varepsilon_n . \tag{6.10}$$

This follows from the definition of ϕ_n in (6.3) and (2.4). To complete this section we will convert (6.9) and (6.10) to the classical form of the Levinson algorithm.

By (6.1) to (6.3) and (6.7) there exists operators $A_{n,i}$ and $B_{n,i}$ on H satisfying

$$U^n h + \sum_{i=0}^{n-1} U^i A_{n,i} h = U^n h - P_{P_n} U^n h = \phi_n \varepsilon_n h$$

$$\tag{6.11}$$

$$h_* + \sum_{i=1}^{n} U^i B_{n,i} h_* = h_* - P_{F_n} h_* = \phi_{n*} \varepsilon_{n*} h_*$$

where h and h_* are in H. We claim that

$$h + P_H \sum_{i=0}^{n-1} U^{*n} U^i A_{n,i} h = \varepsilon_n^* \varepsilon_n h$$

$$(6.12)$$

$$h_* + P_H \sum_{i=1}^{n} U^i B_{n,i} h_* = \varepsilon_{n*}^* \varepsilon_{n*} h_* \ .$$

The first equation follows from the first equation in (6.11) and the following calculation:

$$(\varepsilon_n^* \varepsilon_n h, \ h') = (U^n h + \sum_{i=0}^{n-1} U^i A_{n,i} h, \ U^n h') = (h + P_H \sum_{i=0}^{n-1} U^{*n} U^i A_{n,i} h, \ h')$$

where h and h$'$ are in H. A similar calculation involving the second equation in (6.11) yields the second equation in (6.12). Notice that the first equation in (6.11) is orthogonal to $U^k H$ for $0 \le k < n$, and the second equation in (6.11) is orthogonal to $U^j H$ for $1 \le j \le n$, or equivalently,

$$\sum_{i=0}^{n-1} P_H(U^{*k} U^i A_{n,i} \ h) + P_H U^{*k} U^n h = 0 \quad (0 \le k < n)$$

$$(6.12a)$$

$$P_H U^{*j} h_* + \sum_{i=1}^{n} P_H U^{*j} U^i B_{n,i} \ h_* = 0 \quad (1 \le j \le n) \ .$$

This and (1.4), (6.12) gives the usual relation between the A's and B's:

$$\begin{bmatrix} I & R_1 & R_2 & \cdots & R_n \\ R_{-1} & I & R_1 & \cdots & R_{n-1} \\ R_{-2} & R_{-1} & I & \cdots & R_{n-2} \\ \vdots & \vdots & \vdots & \ddots & \vdots \\ R_{-n} & R_{-n+1} & R_{-n+2} & \cdots & I \end{bmatrix} \begin{bmatrix} A_{n,0} & I \\ A_{n,1} & B_{n,1} \\ A_{n,2} & B_{n,2} \\ \vdots & \vdots \\ I & B_{n,n} \end{bmatrix} = \begin{bmatrix} 0 & \varepsilon_{n*}^* \varepsilon_{n*} \\ 0 & 0 \\ 0 & 0 \\ \vdots & \vdots \\ \varepsilon_n^* \varepsilon_n & 0 \end{bmatrix} \quad (6.13)$$

Equations (6.9) and (6.11) yield the Levinson recursion in its usual form:

$$[A_{n+1,0}, A_{n+1,1}, ..., A_{n+1,n}, I] =$$

$$[0, A_{n,0}, A_{n,1}, ..., A_{n,n-1}, I] - [I, B_{n,1}, ..., B_{n,n}, 0] \varepsilon_{n*}^{-1} \Gamma_{n+1} \varepsilon_n \ ,$$

$$[I, B_{n+1,1}, ..., B_{n+1,n+1}] = \qquad\qquad (6.14)$$

$$-[0, A_{n,0}, A_{n,1}, ..., A_{n,n-1}, I] \varepsilon_n^{-1} \Gamma_{n+1}^* \varepsilon_{n*} + [I, B_{n,1}, ..., B_{n,n}, 0] \ .$$

Equations (1.4), (6.10) and the first equation in (6.11) give

$$\varepsilon_{n*}^{*}\Gamma_{n+1}\varepsilon_{n}h = P_{H}U\phi_{n}\varepsilon_{n}h = R_{n+1}h + \sum_{i=1}^{n} R_{i}A_{n,i-1}h \quad (h \in H) . \qquad (6.15)$$

Equations (6.14) and (6.15) is known as the Levinson algorithm for block Toeplitz matrices. Finally, it is noted that one does not need (6.9) to obtain the Levinson recursion in (6.14). This recursion also follows from (6.13), (6.15) and

$$\varepsilon_{n}^{*}\Gamma_{n+1}^{*}\varepsilon_{n*} = R_{n+1}^{*} + \sum_{i=1}^{n} R_{i}^{*}B_{n,n-i+1}$$

which is the analogue of (6.15). (This identity can be obtained by applying $\hat{I}_{n+1}$ to (6.13) as in (5.3).) Then by Proposition 5.2 the role of $A_{n,i}$ is played by $B_{n,n-i}$ and with Γ_{i} replaced by Γ_{i}^{*}. Now the previous identity follows from (6.15).) Equations (6.14) and (6.15) lead to the following recursive inverse scattering Levinson algorithm for block Toeplitz matrices.

6.1 PROCEDURE (Levinson). Assume that $\hat{R}_{n+2}$ is positive and $\{\Gamma_{i}\}_{1}^{n+1}$ its corresponding choice sequence is strictly contractive. To calculate the choice sequence $\{\Gamma_{i}\}_{1}^{n+1}$ from $\{R_{i}\}_{1}^{n+1}$ first set $\Gamma_{1} = R_{1}$ and $A_{1,0} = -\Gamma_{1}$ and $B_{1,1} = -\Gamma_{1}^{*}$ and

$$\varepsilon_{1} = D_{1} \text{ and } \varepsilon_{1*} = D_{1*} . \qquad (6.16)$$

Using these initial conditions assume that we have calculated the choice sequence $\{\Gamma_{i}\}_{1}^{n}$ and the operators $A_{n,i}$ for $0 \le i < n$ and $B_{n,i}$ for $1 \le i \le n$. The next choice operator Γ_{n+1} is obtained by inverting the operators ε_{n} and ε_{n*} in (6.15). The new operators $A_{n+1,i}$ and $B_{n+1,i}$ are obtained recursively from (6.14) and

$$\varepsilon_{n+1} = D_{n+1}\varepsilon_{n} \text{ and } \varepsilon_{n+1*} = D_{n+1*}\varepsilon_{n*} . \qquad (6.17)$$

This gives the next step and completes the recursion.

6.2 REMARK. If at any step in the previous procedure Γ_{n} is not a contraction, then the procedure stops. In this case $\{R_{i}\}_{1}^{n}$ does not admit a positive extension, or equivalently, $\hat{R}_{n+1}$ is not positive.

6.3 REMARK. If $\{R_{i}\}_{1}^{\infty}$ is a sequence of complex numbers, then its choice sequence $\{\Gamma_{i}\}_{1}^{\infty}$ is also a sequence of complex numbers and the previous algorithm reduces to the Levinson algorithm in Chapter II. In this case $\varepsilon_{n} = \varepsilon_{n*}$ and $A_{n,i} = \overline{B}_{n,n-i}$ for $0 \le i \le n - 1$.

6.4 REMARK. In the next chapter we will show that $\hat{R}_{n+1}$ is a strictly positive Toeplitz matrix if and only if its choice sequence $\{\Gamma_i\}_1^n$ is strictly contractive.

6.5 REMARK. Assume that $\hat{R}_{n+1}$ is strictly positive, or equivalently, its choice sequence $\{\Gamma_i\}_1^n$ is strictly positive. Then one can obtain the maximal entropy extension $\{R_i^o\}_1^\infty$ of $\{R_i\}_1^n$ directly from the $A_{n,i}$ operators calculated in the Levinson algorithm, that is, the maximal entropy extension $\{R_i^o\}_1^\infty$ of $\{R_i\}_1^n$ is given by the following difference equation:

$$R_{n+k+1}^o = -\sum_{i=1}^{n} R_{i+k}^o A_{n,i-1} \qquad (k \geq 0) \qquad (6.18)$$

subject to the initial conditions

$$R_i^o = R_i \qquad (\text{for } 1 \leq i \leq n) . \qquad (6.19)$$

To verify (6.18) notice that by using $\Gamma_i = 0$ for $i > n$ in (6.14) we have

$$[A_{n+k,0}, ..., A_{n+k,n+k-1}] = [0, 0, ..., 0, A_{n,0}, ..., A_{n,n-1}]$$

$$(6.20)$$

$$[B_{n+k,1}, ..., B_{n+k,n+k}] = [B_{n,1}, ..., B_{n,n}, 0, 0, ..., 0] .$$

The operators $A_{n+k,i}$ and $B_{n+k,i}$ correspond to the maximal entropy extension $\{R_i^o\}_1^\infty$ of $\{R_i\}_1^n$. Since $\Gamma_i = 0$ for all $i > n$ Equation (6.15) becomes

$$R_{n+k+1}^o = -\sum_{i=1}^{n+k} R_i^o A_{n+k,i-1} \quad (k \geq 0)$$

The first equation in (6.20) produces (6.18) and (6.19). This completes the proof.

Finally, it is noted that one can easily generalize the techniques in Section II.2 to obtain the block Levinson algorithm in (6.14) and (6.15). We chose to obtain the Levinson algorithm by using the Naimark dilation U of $\{R_i\}_1^\infty$ to show that the choice sequence $\{\Gamma_i\}_1^\infty$ naturally appearing in the Levinson algorithm is precisely the choice sequence for $\{R_i\}_1^\infty$.

7. THE NAIMARK DILATION AND MARINE SEISMOLOGY

In this section we show that the matrix minimal isometric dilation U on K of $\{R_i\}_1^\infty$ in (2.4) naturally occurs in marine seismology, and at the same time, generalize the marine seismology structure in Section III.5 to the operator setting.

To this end consider waves with values (amplitudes) in a Hilbert space H traveling through a layered medium, orthogonal to the interfaces between the layers. The interface between the j-th layer and j+1 layer is called the j-th interface; see Figure 1. Each interface is characterized by the following properties:

0th layer $\Gamma_o = -I$

____ 0th interface Γ_0

    1st layer

____ 1st interface Γ_1

    2nd layer

____ 2nd interface Γ_2

    3rd layer

____ 3rd interface Γ_3

.
.
.

Figure 1

(i) If a downgoing wave of amplitude h in D_{j-1} strikes the j-th interface, then the wave of amplitude $\Gamma_j h$ will be reflected upward into the j-th layer and a wave of amplitude $D_j h$ will be transmitted downward into the j+1 layer; see Figure 2. The operator Γ_j is always a contraction mapping D_{j-1} into D_{j-1*}, and is called the *reflection operator* of the j-th interface; as expected $D_o = D_{o*} = H$ and $\{\Gamma_i\}_1^\infty$ is a choice sequence initiated on H.

(ii) If an upgoing wave of amplitude h' in D_{j-1*} strikes the j-th interface, then the wave of amplitude $-\Gamma_j^* h'$ is reflected downward into the j+1 layer and the wave of amplitude $D_{j*} h'$ is transmitted upward into the j-th layer; see Figure 2.

We always assume that the time it takes for a wave to travel between the j and j+1 interface (or j+1 and j interface) is $\tau/2$ where τ is the same positive constant for all j.

In marine seismology one has the usual layered media situation in Figure 1. However, the zero interface water acts like a perfect reflector $\Gamma_o = -I$. Throughout we assume $\Gamma_o = -I$. Now consider the downgoing waves of amplitudes $w_j(n)$ in the j-th layer immediately below the $j-1$ interface, as shown in Figure 3. The amplitude of w_j at time $n\tau$ is $w_j(n)$. (Since the zero interface acts like a perfect reflector, $w_1 = v_1$ where v_1 is the upgoing wave immediately below the zero interface.) Using the reflection and the

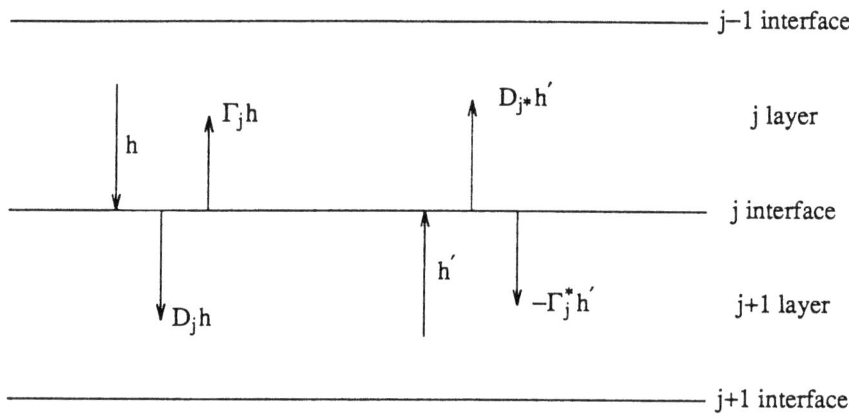

Figure 2

transmission rules in (i), (ii) and taking travel time into account with linearity, we have

$$
\begin{bmatrix}
w_1(n+1) \\
w_2(n+\frac{1}{2}) \\
w_3(n) \\
w_4(n-\frac{1}{2}) \\
\cdot \\
\cdot \\
\cdot
\end{bmatrix}
=
\begin{bmatrix}
\Gamma_1 & D_{1*}\Gamma_2 & D_{1*}D_{2*}\Gamma_3 & \cdots \\
D_1 & -\Gamma_1^*\Gamma_2 & -\Gamma_1^*D_{2*}\Gamma_3 & \cdots \\
0 & D_2 & -\Gamma_2^*\Gamma_3 & \cdots \\
0 & 0 & D_3 & \cdots \\
0 & 0 & 0 & \cdots \\
\cdot & \cdot & \cdot & \cdots \\
\cdot & \cdot & \cdot & \cdots \\
\cdot & \cdot & \cdot & \cdots
\end{bmatrix}
\begin{bmatrix}
w_1(n) \\
w_2(n-\frac{1}{2}) \\
w_3(n-1) \\
w_4(n-\frac{3}{2}) \\
\cdot \\
\cdot \\
\cdot
\end{bmatrix}
\tag{7.1}
$$

This is precisely the isometry U in (2.4).

One of the problems in marine seismology is to compute the reflection operators $\{\Gamma_j\}_1^\infty$, without drilling to the j-th interface. To accomplish this one sets off an explosion $w_1(0) = h \in H$ immediately below the zero interface. By (7.1) this produces the wave

$$
w_1(i) = P_H U^i h = R_i h \qquad (h \in H \text{ and } i > 0)
\tag{7.2}
$$

immediately below the zero interface where H is identified with $H \oplus \{0\} \oplus \{0\} \oplus \ldots$ Since $\{\Gamma_j\}_1^\infty$ is a choice sequence the operator U in (7.1) is an isometry. By the Naimark

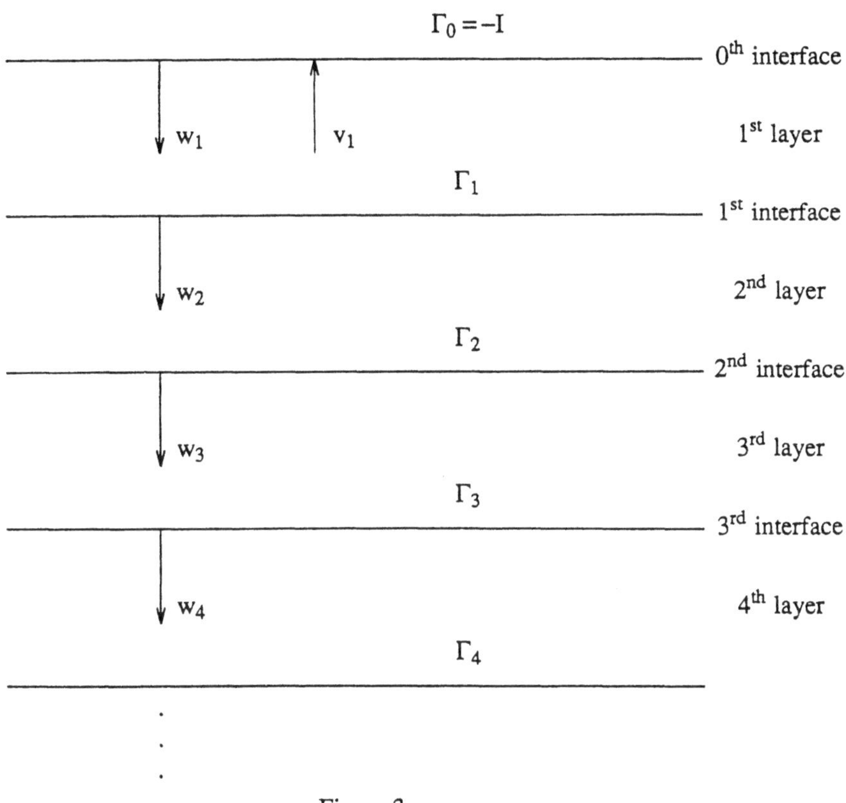

Figure 3

dilation theorem the operators $\{R_i\}_1^{\infty}$ are positive definite. Therefore the choice sequence $\{\Gamma_j\}_1^{\infty}$ defined in Theorem 2.1 for $\{R_i\}_1^{\infty}$ is precisely the set of reflection operators in the marine seismology problem. Summing up we obtain the following result.

7.1 REMARK. By Theorem 2.1 there is a one to one correspondence between the set of all positive definite functions $\{R_i\}_1^{\infty}$ and the marine seismology structure in Figure 3. In this correspondence the choice sequence for $\{R_i\}_1^{\infty}$ is precisely the reflection operators in Figure 3. Moreover, $\{R_i\}_1^{\infty}$ is obtained by (7.2) where $w_1(0) = h \in H$ is the explosion set off immediately below the zero interface.

7.2 REMARK. To compute the reflection operators let $w_1(0) = h$ run through an orthonormal basis in H. Then record the resulting values of $w_1(i)$ and obtain $\{R_i\}_1^\infty$ by (7.2). (Notice that $w_1(i)$ is a linear operator on H, and $w_1(i, h) = w_1(i) = R_i h$ for all $i \geq 0$ and h in H. So one can compute R_i from $w_1(i)$.) At this point one can use Procedure 4.4 or the Levinson algorithm Procedure 6.1, to recursively compute the reflection operators $\{\Gamma_i\}_1^\infty$.

Finally, recall that $\{R_i\}_1^n$ and its reflection operators $\{\Gamma_i\}_1^n$ uniquely determine each other. This fact also follows from the marine seismology structure in Figure 3 by observing that it takes $\tau/2$ units of time to travel between adjacent interfaces.

8. THE INVERSE CAYLEY TRANSFORM OF $\hat{R}_n$.

In this section we compute the inverse Cayley transform $\hat{A}_n$ of a positive Toeplitz matrix $\hat{R}_n$. It is shown that there is a one to one correspondence between the set of all positive definite Toeplitz matrices $\hat{R}_n$ and the set of all n by n contractive analytic Toeplitz matrices $\hat{A}_n$ whose diagonal term A_o is zero. In particular, the operator version of the Carathéodory interpolation problems in Section VIII.1 is shown to be equivalent to the positive definite Carathéodory interpolation problems in Section 3. Then we will generalize the results in Section II.1 to the operator setting, that is, we will show that an analytic function with values in $L(E, E)$ is positive-real if and only if its inverse Cayley transform is a contractive analytic function in $H^\infty(E, E)$.

Throughout the rest of this chapter we assume that H is finite dimensional. Although this assumption is not necessary, it does simplify the presentation in certain instances. Let $\hat{R}_n$ be a (not necessarily positive) Toeplitz matrix on $l_n^2(H)$, and let L_n^* be the strictly upper triangular matrix defined by

$$L_n^* = \begin{bmatrix} 0 & R_1 & R_2 & R_3 & \cdots & R_{n-1} \\ 0 & 0 & R_1 & R_2 & \cdots & R_{n-2} \\ 0 & 0 & 0 & R_1 & \cdots & R_{n-3} \\ \vdots & & & & & \\ 0 & 0 & 0 & 0 & \cdots & R_1 \\ 0 & 0 & 0 & 0 & \cdots & 0 \end{bmatrix}. \tag{8.1}$$

The elements on the diagonal and below the diagonal of L_n^* are zero. Obviously L_n is strictly lower triangular. The elements on the diagonal and above the diagonal of L_n are zero. It is easy to verify that

$$\hat{R}_n = L_n + I + L_n^* = (I+L_n)^*(I+L_n) - L_n^*L_n \ . \tag{8.2}$$

This implies that $\hat{R}_n$ is positive if and only if

$$0 \le (\hat{R}_n x,\ x) = \|(I+L_n)x\|^2 - \|L_n x\|^2 \quad (\text{for all x in } l_n^2(H)) \ . \tag{8.3}$$

Therefore $\hat{R}_n$ is positive if and only if there exists a contraction $\hat{A}_n$ on $l_n^2(H)$ satisfying

$$\hat{A}_n(I+L_n) = L_n \ , \text{or equivalently,} \ \hat{A}_n = L_n(I+L_n)^{-1} \ . \tag{8.4}$$

The strictly lower triangular structure of L_n guarantees that $I + L_n$ is invertible and $\hat{A}_n$ is defined on all of $l_n^2(H)$. The operator $\hat{A}_n$ in (8.4) is the *inverse analytic Cayley transform of* L_n. Since $\hat{R}_n$ and L_n uniquely determine each other, by a slight abuse of terminology, we call $\hat{A}_n$ the *inverse Cayley transform of* $\hat{R}_n$. Because L_n is strictly lower triangular $\hat{A}_n$ is a strictly lower triangular matrix of the form

$$\hat{A}_n = \begin{bmatrix} 0 & 0 & 0 & \cdots & 0 & 0 \\ A_1 & 0 & 0 & \cdots & 0 & 0 \\ A_2 & A_1 & 0 & \cdots & 0 & 0 \\ A_3 & A_2 & A_1 & \cdots & 0 & 0 \\ \cdot & \cdot & \cdot & \cdots & \cdot & \cdot \\ \cdot & \cdot & \cdot & \cdots & \cdot & \cdot \\ \cdot & \cdot & \cdot & \cdots & \cdot & \cdot \\ A_{n-1} & A_{n-2} & A_{n-3} & \cdots & A_1 & 0 \end{bmatrix} \tag{8.5}$$

where A_i are operators on H for $1 \le i < n$. In other words $\hat{A}_n$ is a n by n analytic Toeplitz matrix whose diagonal entry A_o is zero.

On the other hand, if $\hat{A}_n$ is an operator of the form (8.5). Then $I - \hat{A}_n$ is invertible and

$$L_n = (I-\hat{A}_n)^{-1}\hat{A}_n \tag{8.6}$$

defines a strictly lower triangular matrix whose adjoint is of the form (8.1). (The operator L_n was obtained by solving for L_n in the first equation in (8.4).) We define the matrix $\hat{R}_n$ on $l_n^2(H)$ by

$$\hat{R}_n = L_n + I + L_n^* = (I+L_n)^*(I-\hat{A}_n^*\hat{A}_n)(I+L_n) \ . \tag{8.7}$$

The first equality shows that $\hat{R}_n$ is a Toeplitz matrix. The second equality follows by using $\hat{A}_n(I+L_n) = L_n$ in

$$L_n + I + L_n^* = (I+L_n)^*(I+L_n) - L_n^*L_n =$$

$$(I+L_n)^*(I+L_n) - (I+L_n)^*\hat{A}_n^*\hat{A}_n(I+L_n) = (I+L_n)^*(I-\hat{A}_n^*\hat{A}_n)(I+L_n) \, .$$

Since $(I+L_n)$ is invertible, (8.7) implies that $\hat{R}_n$ is positive if and only if $\hat{A}_n$ is a contraction. The strictly lower triangular matrix L_n in (8.6) is called the *analytic Cayley transform of* $\hat{A}_n$. Since L_n and $\hat{R}_n$ uniquely determine each other by (8.7), we call $\hat{R}_n$ the *Cayley transform* of $\hat{A}_n$. Obviously $\hat{R}_n$ is uniquely determined by $\hat{A}_n$. Summing up the above analysis yields the following basic result.

8.1 PROPOSITION. *The operator $\hat{A}_n$ is a contraction of the form (8.5) if and only if its Cayley transform $\hat{R}_n$ is a positive Toeplitz matrix. In this case $\hat{A}_n$ and $\hat{R}_n$ uniquely determine each other by (8.1), (8.2), (8.4) and (8.7).*

In this setting the Carathéodory interpolation problems in Section VIII.1 becomes

8.2 PROBLEM. Given a matrix $\hat{A}_n$ of the form (8.5) find necessary and sufficient conditions for the existence of an infinite contractive analytic Toeplitz matrix $\hat{A}_\infty$ lifting $\hat{A}_n$.

8.3 PROBLEM. Give an explicit description of all solutions (if there are any) to Problem 8.2.

The results in Chapters VIII to XIV completely solve these interpolation problems, which are a special case of the interpolation problems involving the commutant lifting theorem and generalized Hankel matrices. Proposition 8.1 also shows that the previous Carathéodory interpolation problems involving $\hat{A}_n$ are equivalent to the positive definite Carathéodory interpolation Problems 3.1 and 3.2 involving the Toeplitz matrix $\hat{R}_n$. Therefore one can use Proposition 8.1 along with our previous results in this chapter to also solve the previous Carathéodory interpolation problems involving $\hat{A}_n$. In particular, since $\hat{R}_n$ admits a positive Toeplitz expansion $\hat{R}_\infty$ if and only if $\hat{R}_n$ is positive, the operator $\hat{A}_n$ admits an infinite contractive analytic Toeplitz lifting $\hat{A}_\infty$ (or equivalently, Problem 8.1 is solvable) if and only if $\hat{A}_n$ (the inverse Cayley transform of $\hat{R}_n$) is a contraction. In this case there is a one to one correspondence through the Cayley transform between the set of all positive Toeplitz expansions $\hat{R}_j$ of $\hat{R}_n$ and the set of all j by j contractive analytic Toeplitz liftings $\hat{A}_j$ of $\hat{A}_n$ for all $j > n$. Moreover, since $\hat{R}_n$ is uniquely determined by a choice sequence $\{\Gamma_i\}_1^{n-1}$ initiated on H, its inverse Cayley transform $\hat{A}_n$ is also uniquely determined by the same choice sequence. In fact the set of all infinite contractive analytic Toeplitz liftings $\hat{A}_\infty$ of $\hat{A}_n$ is uniquely determined by a choice sequence $\{\Gamma_i\}_n^\infty$ initiated from D_{n-1} to D_{n-1*}.

Equation (8.7) shows that $\hat{R}_n$ is strictly positive if and only if $\hat{A}_n$ is a strict contraction. Now let $H = C^1$ and $\hat{R}_n$ be a Toeplitz matrix on C^n. In this case, recall that $\hat{R}_n$ admits a unique positive extension if and only if $\hat{R}_n$ is positive and singular. By Proposition 8.1 the operator $\hat{A}_n$ in (8.5) admits a unique infinite contractive analytic Toeplitz lifting A_∞ (or equivalently, Problem 8.2 has a unique solution) if and only if the norm of $\hat{A}_n$ is one.

In this setting we say that $R_p(z)$ is *positive-real* if $z^{-1}R_p(z)$ is analytic in D with values in H and

$$R(z) \overset{\Delta}{=} R_p(z) + I + R_p(z)^* \geq 0 \quad \text{(for all z in D)} . \tag{8.8}$$

(Since we have chosen $R_o = I$ we have modified the definition of positive-real in Section II.1 to suit our needs in this chapter.) A sequence $\{R_i\}_1^\infty$ on H *generates a positive-real function* $R_p(z)$ if

$$R_p(z) = \sum_1^\infty R_i z^i \tag{8.9}$$

is positive-real. The following is an operator version of Theorem 1.2 in Chapter II. The proof of Theorem II.1.2 can be easily extended to the operator setting. Here we will give an alternate proof of this result.

8.4 THEOREM. *Let $\{R_i\}_1^\infty$ be a sequence of operators on H. Then $\{R_i\}_1^\infty$ generates a positive-real function $R_p(z)$ by (8.9) if and only if $\hat{R}_n$ is a positive Toeplitz matrix for all* $n \geq 0$.

PROOF. Assume that the Toeplitz matrix $\hat{R}_n$ is positive for all n. Let us identify $l^2(H)$ with $H^2(H)$ in the usual way. In this case the infinite analytic Toeplitz operator $\hat{A}_\infty$ which is the strong limit of $\hat{A}_n$ (see Lemma VII.2.1) is unitarily equivalent to $(zF_1)_+$ on $H^2(H)$ where

$$F_1(z) = \sum_1^\infty A_i z^{i-1}, \text{ and } R_p(z) = \sum_1^\infty R_i z^i \tag{8.10}$$

is analytic in D. The last statement follows because R_i is a contraction for all i. Proposition 8.1 implies that $\hat{A}_n$ (the inverse Cayley transform of $\hat{R}_n$) is a contraction for all $n \geq 0$. Thus $(zF_1)_+$ is a contraction, or equivalently, zF_1 is in $H_1^\infty(H,H)$; see Theorem L.1.1. In particular, F_1 is in $H_1^\infty(H,H)$. Letting $n \to \infty$ in (8.6) along with (8.10) gives

$$R_p(z) = z(I-zF_1(z))^{-1}F_1(z) \quad (z \in D) \ . \tag{8.11}$$

The function $R_p(z)$ is well defined and analytic in D because F_1 is in $H_1^\infty(H,H)$. Using (8.11) we have

$$R(z) = R_p(z) + I + R_p(z)^* = (I-zF_1)^{-1}(I-z\bar{z}F_1F_1^*)^{-1}(I-\bar{z}F_1^*)^{-1} \geq 0$$

for all z in D. Therefore $R_p(z)$ is positive-real, because F_1 is in $H_1^\infty(H,H)$.

On the other hand, let $R_p(z)$ be positive-real. Solving for zF_1 in (8.11) yields

$$zF_1(z) = R_p(z)(I+R_p(z))^{-1} \quad (z \in D) \ . \tag{8.12}$$

Because $R_p(z)$ is positive-real, condition (8.8) implies that $F_1(z)$ is well defined for z in D. If $R_p(z)h = -h$ for some z in D and h in H, then (8.8) implies that $-(h, h) \geq 0$, or equivalently, $h = 0$. So $I + R_p(z)$ is invertible and its inverse is analytic in D, since H is finite dimensional. Hence

$$(I-\bar{z}zF_1^*(z)F_1(z)) = I - (I+R_p^*)^{-1}R_p^*R_p(I+R_p)^{-1} =$$

$$(I+R_p^*)^{-1}(R_p+I+R_p^*)(I+R_p)^{-1} \geq 0$$

for all z in D. Therefore zF_1 is analytic and even in $H_1^\infty(H,H)$. Thus $(zF_1)_+$ is a contraction. So $\hat{A}_n$ is a contraction for all $n \geq 0$. Equations (8.4), (8.10) and (8.12) show that $\hat{R}_n$ is the Cayley transform of $\hat{A}_n$. Hence $\hat{R}_n$ is positive for all n. This completes the proof.

The previous analysis shows that F_1 is in $H_1^\infty(H,H)$ if and only if its *Cayley transform* $R_p(z)$ defined in (8.11) is positive-real. Equivalently $R_p(z)$ is positive-real if and only if its *inverse Cayley transform* F_1 defined in (8.12) is in $H_1^\infty(H,H)$. (We have used the definition of the Cayley transform to describe both the relationship between $\hat{R}_n$ and $\hat{A}_n$, and their corresponding analytic functions $R_p(z)$ and $zF_1(z)$.) Obviously F_1 and R_p uniquely determine each other. So there is a one to one correspondence through the Cayley transform (8.11), (8.12) between the set of all functions zF_1 in $H_1^\infty(H,H)$, and the set of all positive-real functions. The positive definite Carathéodory interpolation Problems 3.1 and 3.2 stated in terms of positive-real functions become:

8.5 PROBLEM. Given a sequence of operators $\{R_i\}_1^n$ on H find necessary and sufficient conditions for the existence of a positive-real function $R_p(z)$ of the form:

$$R_p(z) = \sum_1^\infty R_i z^i \ .$$

8.6 PROBLEM. Given an explicit description of all solutions (if there are any) to Problem 8.5.

Theorem 8.4 shows that Problems 8.2 and 8.3 are equivalent to Problems 8.5 and 8.6. Therefore our previous results in this chapter provide a solution to these problems. In particular, $\{R_i\}_1^n$ admits a positive-real extension, or equivalently, Problem 8.5 is solvable if and only if the Toeplitz matrix $\hat{R}_{n+1}$ is positive.

9. THE NAIMARK DILATION FOR $H_1^\infty(H, H)$

In this section we use the Cayley transform in (8.11), (8.12) to obtain a Naimark dilation theorem for functions F_1 in $H_1^\infty(H,H)$, which is precisely Lemma XIII.3.1. Then we obtain an operator Moebious transform. This transform is used to obtain the Schur contractions for F_1.

The following is a Naimark dilation theorem for $H_1^\infty(H,H)$, which is precisely Lemma XIII.3.1 (see also Theorem XIII.2.7).

9.1 THEOREM. *Assume that* F_1 *admits a formal power series of the form (8.10). Then* F_1 *is in* $H_1^\infty(H,H)$ *if and only if there exists an isometry* U *on* $K(\supseteq H)$ *such that*

$$A_i = PU(QU)^{i-1}|H \quad (\textit{for all } i \geq 1) \tag{9.1}$$

where $P = P_H$ *and* $Q = I - P$. *In this case, if* H *is cyclic for* U, *then* U *is unique up to an isomorphism, and* U *is the minimal isometric dilation of* $\{R_i\}_1^\infty$ *corresponding to* $\hat{R}_\infty$ *which is the Cayley transform of* $\hat{A}_\infty$.

PROOF. Assume that F_1 is in $H_1^\infty(H,H)$. Let R_p be the Cayley transform of F_1 defined in (8.11). Obviously R_p is positive-real and $\{R_i\}_1^\infty$ obtained from (8.10) forms a positive definite sequence. Let U be the minimal isometric dilation of $\{R_i\}_1^\infty$. Using (1.4) in (8.10) yields

$$R_p(z) = \sum_1^\infty R_i z^i = z\sum_0^\infty PUU^i z^i |H = zPU(I-zU)^{-1}|H \tag{9.2}$$

for all z in D. Notice that

$$(I{-}zQU)(I{-}zU)^{-1}\,|\,H = (I{-}zQU{-}zPU{+}zPU)(I{-}zU)^{-1}\,|\,H =$$

$$(9.3)$$

$$(I{+}zPU(I{-}zU)^{-1})\,|\,H = (I{+}R_p(z))\ .$$

Because $I + R_p(z)$ is invertible, $(I{-}zQU)(I{-}zU)^{-1}$ maps H onto H. Therefore its inverse also maps H onto H. Substituting (9.2) and (9.3) into (8.12) gives

$$F_1(z) = z^{-1}R_p(I{+}R_p)^{-1} = PU(I{-}zQU)^{-1}\,|\,H\ . \qquad (9.4)$$

This and (8.10) produces (9.1).

Now assume that (9.1) holds. Since U is an isometry

$$F_1(z) = \sum_1^\infty A_i z^{i-1} = \sum_0^\infty PU(QU)^i z^i\,|\,H = PU(I{-}zQU)^{-1}\,|\,H \qquad (9.5)$$

is analytic in D. Notice that

$$(I{-}zU)(I{-}zQU)^{-1}\,|\,H = (I{-}zQU{-}zPU)(I{-}zQU)^{-1}\,|\,H =$$

$$(I{-}zPU(I{-}zQU)^{-1})\,|\,H = (I{-}zF_1)\ . \qquad (9.6)$$

In particular, $(I{-}zU)(I{-}zQU)^{-1}$ maps H into H. Since H is finite dimensional, its inverse maps H onto H, and $(I - zF_1(z))^{-1}$ exists and is analytic in D. Substituting (9.5) and (9.6) into (8.11) yields

$$R_p(z) = z(I{-}zF_1)^{-1}F_1 = zF_1(I{-}zF_1)^{-1} = zPU(I{-}zU)^{-1}\,|\,H\ . \qquad (9.7)$$

This and (8.10) produces (1.4). Theorem 8.4 and the Naimark dilation theorem shows that $R_p(z)$ is positive-real. So its inverse Cayley transform $F_1(z)$ is in $H_1^\infty(H,H)$. The uniqueness aspect of the Naimark dilation theorem shows that U is unique up to an isomorphism. This completes the proof.

Let $\hat{A}_\infty$ be an infinite contractive analytic Toeplitz matrix whose diagonal term is zero, or equivalently, let F_1 in (8.10) be in $H_1^\infty(H, H)$. Recall that R_p the Cayley transform of F_1 determines a positive definite sequence $\{R_i\}_1^\infty$. Since R_p and F_1 are uniquely determined by the same isometry, Theorem 2.1 shows that one can always choose U to be given by (2.4) where $\{\Gamma_i\}_1^\infty$ is a choice sequence initiated on H. As noted earlier $\{\Gamma_i\}_1^n$, $\{R_i\}_1^n$ and $\{A_i\}_1^n$ all uniquely determine each other. As before let T_n be the n step contraction determined by U, that is, let T_n be the contraction given by the n by n matrix in the upper left hand corner of U. Then (9.1) and the form of U in (2.4) gives

$$A_{n+1} = PT_n(QT_n)^n \,|\, H + D_{1*}D_{2*} \, \cdots \, D_{n*}\Gamma_{n+1}D_n \, \cdots \, D_2 D_1 \qquad (9.8)$$

where Γ_{n+1} is a contraction mapping D_n into D_{n*} and $A_1 = \Gamma_1$. Notice that the second term in (9.8) only depends on $\hat{A}_{n+1}$, or equivalently, $\{\Gamma_i\}_1^n$. This observation leads to the following recursive inverse scattering algorithm to determine the choice sequence $\{\Gamma_i\}_1^\infty$ directly from $\{A_i\}_1^\infty$

9.2 PROCEDURE. Assume that $\hat{A}_{n+2}$ is a contraction of the form (8.5) where n+2 replaces n. To recursively obtain the choice sequence $\{\Gamma_i\}_1^{n+1}$ associated with $\hat{A}_{n+2}$ first set $\Gamma_1 = A_1$. Now assume we have computed $\{\Gamma_i\}_1^n$ and T_n. This (9.8) and A_{n+1} recursively yields Γ_{n+1} by inverting the self adjoint operators D_{i*} and D_i in (9.8). Once Γ_{n+1} is known we compute T_{n+1} and continue as before.

The previous procedure is equivalent to Procedure 2.9 in Chapter XIII with $T = T' = 0$.

9.3 REMARK. In the previous procedure if Γ_i is not a contraction for some i, then the procedure stops. In this case $\hat{A}_{i+1}$ is not a contraction and for $n \geq i$ and $\hat{A}_{n+1}$ does not have a contractive lifting. On the other hand, if $\{\Gamma_i\}_1^n$ computed by the previous procedure is a choice sequence, then the set of all infinite contractive analytic Toeplitz liftings $\hat{A}_\infty$ of $\hat{A}_{n+1}$ is generated by (9.1) where U is the isometry in (2.4) determined by the choice sequence $\{\Gamma_i\}_1^\infty$ and $\{\Gamma_i\}_{n+1}^\infty$ is an arbitrary choice sequence initiated from D_n to D_{n*}. In particular, there exists a unique infinite contractive analytic Toeplitz lifting $\hat{A}_\infty$ of $\hat{A}_{n+1}$ if and only if $D_n = \{0\}$ or $D_{n*} = \{0\}$.

9.4 REMARK. Assume that $\hat{A}_{n+1}$ is a contraction on $l_{n+1}^2(H)$ and let $\{\Gamma_i\}_1^n$ be its corresponding choice sequence, computed according to Procedure 9.2. Let T_n be the n by n matrix formed by the upper n by n left hand corner of U in (2.4). Probability the easiest contractive infinite analytic Toeplitz lifting $\hat{A}_\infty$ of $\hat{A}_{n+1}$ to compute is the one given by setting $\Gamma_i = 0$ for all $i > n$. By the above analysis its corresponding analytic function F_1 in (8.10) is given by

$$F_1(z) = PT_n(I - zQT_n)^{-1}\,|\, H \quad \text{and} \quad R_p(z) = zPT_n(I - zT_n)^{-1}\,|\, H$$

is its Cayley transform. It is emphasized that both $F_1(z)$ and $R_p(z)$ can be computed by employing Procedure 9.2 and then computing the appropriate inverses. Finally, it is noted that one can also obtain a rational form for this $R_p(z)$ by using the Levinson algorithm and (6.18). Infact it is a simple exercise to show that this $R_p(z)$ is given by

$R_p(z) = N(z)A(z)^{-1}$ where

$$A(z) = I + zA_{n,n-1} + z^2 A_{n,n-2} + \cdots + z^n A_{n,o}$$

is the polymonial formed by the solution $[A_{n,o}, ..., A_{n,n-1}, I]^{tr}$ to the Levinson system in (6.13), and $N(z)$ is the polymonial given by

$$N(z) = zR_1 + z^2(R_2 + R_1 A_{n,n-1}) + z^n(R_n + R_{n-1}A_{n,n-1} + \cdots + R_1 A_{n,1}) \ .$$

To complete this section we will use an operator Moebius transform on $F_1(z)$ in $H^\infty(H,H)$ to obtain the choice sequence $\{\Gamma\}_1^\infty$ determined by F_1. Since these results are presented in Section XIV.3, they will only be sketched here. Let Γ_j be a contraction mapping D_{j-1} into D_{j-1*} where $D_o = D_{o*} = H$. Let F_{j+1} be in $H_1^\infty(D_j, D_{j*})$ and F_j the function defined by

$$F_j(z) = \Gamma_j + zD_{j*}F_{j+1}(z)(I+z\Gamma_j^* F_{j+1}(z))^{-1} D_j \ . \tag{9.9}$$

Since Γ_j is a contraction F_j is well defined and analytic in D. We claim that F_j is $H_1^\infty(D_{j-1}, D_{j-1*})$.

To see this consider the unitary system defined by

$$\begin{bmatrix} x' \\ y \end{bmatrix} = \begin{bmatrix} -\Gamma_j^* & D_j \\ D_{j*} & +\Gamma_j \end{bmatrix} \begin{bmatrix} x \\ u \end{bmatrix}$$

$$\text{subject to } x = zF_{j+1}x' \ . \tag{9.10}$$

Solving this equation produces $y = F_j u$ where F_j is defined in (9.9). Moreover, since this 2 by 2 matrix is unitary and zF_{j+1} is a contraction we have (see Corollary XIV.1.3)

$$\|y\|^2 + \|x'\|^2 = \|x\|^2 + \|u\|^2 = \|zF_{j+1}x'\|^2 + \|u\|^2 \le \|x'\|^2 + \|u\|^2 \ .$$

This implies that $\|y\|^2 \le \|u\|^2$ and $F_j(z)$ is a contraction for all z in D. Therefore F_j is in $H_1^\infty(D_{j-1}, D_{j-1*})$ which proves our claim.

Notice that $F_j(0) = \Gamma_j$ in (9.9). Now assume that F_j is a function in $H_1^\infty(D_{j-1}, D_{j-1*})$ satisfying $F_j(0) = \Gamma_j$ and F_{j+1} is the function defined by

$$F_{j+1}(z) = \frac{1}{z} (D_{j*}F_j(z)(I-\Gamma_j^* F_j(z))^{(-1)}D_j - \Gamma_j) \ , \tag{9.11}$$

where $A^{(-1)}$ is the psuedo inverse of an operator A, that is, if $Ax = y$ and x is orthogonal to the kernel of A, then $x = A^{(-1)}y$. According to Theorem IX.5.3, the operator $\Gamma_j^* F_j(z)$ is a strict contraction mapping D_{j-1} into D_{j-1}, because H is finite dimensional. So the inverse in (9.11) makes since. Using $F_j(0) = \Gamma_j$ we have

$$D_{j*}F_j(0)(I-\Gamma_j^*F_j(0))^{(-1)}D_j = \Gamma_j D_j(I-\Gamma_j^*\Gamma_j)^{(-1)}D_j = \Gamma_j \ .$$

This and Schwartz's lemma implies that F_{j+1} is analytic in D. We claim that F_{j+1} is in $H_1^\infty(D_j, D_{j*})$. Moreover, F_j is given by (9.9), that is, F_j and F_{j+1} uniquely determine each other by (9.9) and (9.11).

To see this consider the unitary system defined by

$$\begin{bmatrix} x \\ u \end{bmatrix} = \begin{bmatrix} -\Gamma_j & D_{j*} \\ D_j & +\Gamma_j^* \end{bmatrix} \begin{bmatrix} x' \\ y \end{bmatrix}$$

$$\text{subject to } y = F_j u \ . \tag{9.12}$$

Solving this equation produces $x = zF_{j+1}x'$ where F_{j+1} is defined in (9.11). Moreover, since the 2 by 2 matrix in (9.12) is unitary, and F_j is a contraction we have (see Corollary XIV.1.3)

$$\|x\|^2 + \|u\|^2 = \|x'\|^2 + \|y\|^2 = \|x'\|^2 + \|F_j u\|^2 \le \|x'\|^2 + \|u\|^2 \ .$$

This implies that $\|x\| \le \|x'\|$ and $F_{j+1}(z)$ is a contraction for all z in D. Therefore F_{j+1} is in $H_1^\infty(D_j, D_{j*})$. Notice that the unitary matrix in (9.10) which produces F_j in (9.9), is the adjoint of the unitary matrix in (9.12) which produces F_{j+1} in (9.11). Thus F_j and F_{j+1} in (9.9) and (9.11) uniquely determine each other. This proves our claim.

Let F_1 be in $H_1^\infty(H,H)$ and let F_{j+1} be the functions in $H_1^\infty(D_j, D_{j*})$ recursively defined in (9.11) for $j \ge 1$ where

$$\Gamma_j = F_j(0) \quad \text{for } j \ge 1 \ . \tag{9.13}$$

Our previous analysis guarantees that F_{j+1} is in $H_1^\infty(D_j, D_{j*})$. By the maximum principle Γ_j is a contraction mapping D_{j-1} into D_{j-1*} for all $j \ge 1$, that is, $\{\Gamma_i\}_1^\infty$ defined by F_j in (9.13) is a choice initiated on H. The choice sequence defined by (9.13) is called *the Schur choice sequence for* F_1. Obviously this choice sequence is uniquely determined by F_1. By using the Cayley transform, Theorem 9.1 or Procedure 9.2 we also obtain a choice sequence for a F_1 in $H_1^\infty(H, H)$. In the following sections we use the marine seismology structure developed in Section 7 to show that the choice sequence determined by Theorem 9.1 is precisely the Schur choice sequence.

The transformations in (9.9) and (9.11) are operator Moebius transformations. If $H = C^1$, then these transformations reduce to usual Moebius transforms used in studying the Carathéodory interpolation problem in Chapter I, that is, if $H = C^1$, then (9.9) and (9.11) reduce to

$$F_j = \frac{\Gamma_j + zF_{j+1}}{1 + z\Gamma_j^* F_{j+1}} \quad \text{and} \quad F_{j+1} = \frac{1}{z}\,\frac{F_j - \Gamma_j}{1 - \Gamma_j^* F_j} \tag{9.14}$$

respectively. In this case Γ_j in (9.13) is precisely the j-th Schur number for F_1 defined in Chapter I.

10. THE NAIMARK DILATION AND LAYERED MEDIUM

In this section we present a geophysical or scattering interpretation for the Naimark representation of a contractive analytic function F_1 in Theorem 9.1, or equivalently, for the representation of A_i in (9.1).

To this end consider the following layered medium model: The reflection operator for the i-th interface is Γ_i. Each interface satisfies the usual scattering properties (i) and (ii) described in Section 7. As before the time it takes for a wave to travel between adjoining interfaces is $\tau/2$ where τ is a fixed positive constant. The input u(n) and output y(n) are waves with values in H at time $n\tau$. This leads to the following inverse scattering problem: Given the input u(n) and output y(n) in Figure 4 below, find the reflection operators $\{\Gamma_i\}_1^\infty$. This problem naturally occurs in oil exploration on land. In this case, the first interface is the surface of the earth and there is no zero interface. Once the reflection operators are known, then one can determine if there is oil under this particular spot of the earth.

To solve the previous inverse scattering problem first notice that the reflection and transmission rules (i) and (ii) in Section 7 yield:

$$
\begin{bmatrix}
y(n) \\
x_1(n) \\
x_2(n-\tfrac{1}{2}) \\
x_3(n-1) \\
x_4(n-\tfrac{3}{2}) \\
\cdot \\
\cdot \\
\cdot
\end{bmatrix}
=
\begin{bmatrix}
\Gamma_1 & D_{1*}\Gamma_2 & D_{1*}D_{2*}\Gamma_3 & \cdots \\
D_1 & -\Gamma_1^*\Gamma_2 & -\Gamma_1^*D_{2*}\Gamma_3 & \cdots \\
0 & D_2 & -\Gamma_2^*\Gamma_3 & \cdots \\
0 & 0 & D_3 & \cdots \\
0 & 0 & 0 & \cdots \\
\cdot & \cdot & \cdot & \cdots \\
\cdot & \cdot & \cdot & \cdots \\
\cdot & \cdot & \cdot & \cdots
\end{bmatrix}
\begin{bmatrix}
u(n) \\
x_1(n-1) \\
x_2(n-\tfrac{3}{2}) \\
x_3(n-2) \\
x_4(n-\tfrac{5}{2}) \\
\cdot \\
\cdot \\
\cdot
\end{bmatrix}
\tag{10.1}
$$

where $x_i(n)$ is the downgoing wave immediately below the i-th interface at times $n\tau$ (see Figure 4). The infinite matrix in (10.1) is precisely the isometry U in (2.4).

Let x(n) be the column vector formed by the transpose of

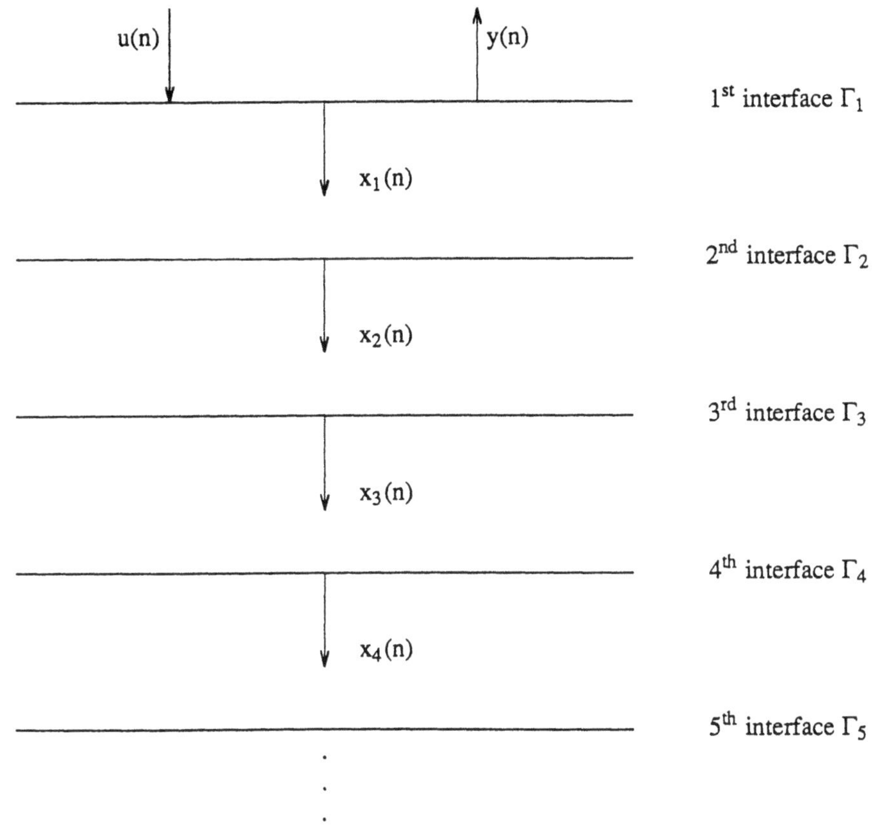

Figure 4

$$[x_1(n-1),\ x_2(n-\tfrac{3}{2}),\ x_3(n-2),\ x_4(n-\tfrac{5}{2}),\ ...]\ .$$

In this case (10.1) becomes

$$\begin{bmatrix} y(n) \\ x(n+1) \end{bmatrix} = [U] \begin{bmatrix} u(n) \\ x(n) \end{bmatrix} \tag{10.2}$$

where U is the isometry in (10.1), or equivalently, U is the isometry on $K = H \oplus D_1 \oplus D_2 \oplus D_3 \oplus \cdots$ in (2.4). As in the previous section $P = P_H$ and $Q = I - P$. Using

$$A = QU|H^\perp,\ \ B = QU|H,\ \ C = PU|H^\perp\ \text{and}\ D = PU|H = \Gamma_1 \tag{10.3}$$

we see that (10.2) admits a state variable representation of the form:

$$x(n+1) = Ax(n) + Bu(n)$$
$$y(n) = Cx(n) + Du(n) .$$

(10.4)

The state variable is $x(n)$. Throughout, it is always assumed that the initial condition $x(0)$ is zero.

By recursively solving for $x(n)$ in the first equation in (10.4) with $x(0) = 0$ we have

$$x(n) = \sum_{i=0}^{n-1} A^{n-i-1} Bu(i) .$$

(10.5)

Substituting this into the second equation in (10.4) yields

$$y(n) = \sum_{i=0}^{n} F'_{n-i} u(i)$$

(10.6)

where $\{F'_i\}_0^\infty$ are operators on H defined by

$$F'_0 = D \text{ and } F'_i = CA^{i-1}B \quad (\text{if } i \geq 1) .$$

(10.7)

Equation (10.6) shows that the output $y(n)$ is obtained by convolution.

The *z-transform* $u(z)$ of $u(n)$ and $y(z)$ of $y(n)$ are the formal series obtained by identifying z^n with $n\tau$, that is,

$$u(z) = \sum_0^\infty u(n)z^n \text{ and } y(z) = \sum_0^\infty y(n)z^n .$$

(10.8)

Since $y(n)$ is obtained by convoluting F'_n with $u(n)$, a simple calculation shows that $y(z) = F(z)u(z)$ where $F(z)$ is the formal series defined by

$$F(z) = \sum_0^\infty F'_i z^i = D + \sum_{i=1}^\infty CA^{i-1}Bz^i = D + zC(I-zA)^{-1}B .$$

(10.9)

The second equality follows from (10.7). The function $F(z)$ is the *transfer function* for the layered media model in Figure 4, or equivalently, for its state variable representation (10.4).

Since A is a contraction the last equality in (10.9) shows that $F(z)$ is analytic in D. The following provides an even stronger result.

10.1 THEOREM. *Let* U *be an isometry and* $F(z)$ *be the transfer function defined in (10.9) where* A, B, C *and* D *are defined in (10.3). Then* F *is a contractive analytic function in* $H^\infty(H,H)$.

PROOF. Equations (10.3) and (10.7) yield

$$F_i' = PU(QU)^i | H \quad (\text{for } i \geq 0) \tag{10.10}$$

Theorem 9.1 with $A_i = F_{i-1}'$ implies that $F(z)$ is in $H_1^\infty(H,H)$. This completes the proof.

If the input $u(z)$ is in $H^2(H)$, then the previous theorem shows that the output $y(z)$ $= F(z)u(z)$ is also in $H^2(H)$. In this case the formal series for $u(z)$ and $y(z)$ generate analytic functions $u(z)$ and $y(z)$ in $H^2(H)$. Finally, it is noted that Theorem 9.1 also implies that any F in $H_1^\infty(H,H)$ produces a layered media model of the form given in Figure 4. In this case the choice sequence for F is the sequence of reflection operators for its layered media model.

10.2 REMARK. Let F_i be the transfer function for the i-th interface, that is, F_i is the transfer function for the layered medium model in Figure 4 where the choice sequence $\{\Gamma_1, \Gamma_2, \Gamma_3, ...\}$ initiated on H is replaced by the choice sequence $\{\Gamma_i, \Gamma_{i+1}, \Gamma_{i+2}, ...\}$ initiated from D_{i-1} to D_{i-1*}. We claim that F_i is in $H_1^\infty(D_{i-1}, D_{i-1*})$ for all $i \geq 1$ where $D_0 = D_{0*} = H$. To see this embed both D_{i-1} and D_{i-1*} in a larger space H. Let $\{\hat{\Gamma}_i, \hat{\Gamma}_{i+1}, \hat{\Gamma}_{i+2}, \cdots \}$ be the choice sequence on H where $\hat{\Gamma}_j | D_{j-1} = \Gamma_j$ and $\hat{\Gamma}_j$ is zero on the orthogonal complement of D_{j-1}. Then the transfer function $\hat{F}_i$ generated by $\{\hat{\Gamma}_j\}_i^\infty$ is zero on $H \ominus D_{i-1}$ and $\hat{F}_i | D_{i-1} = F_i$. Theorem 10.1 implies that $\hat{F}_i$ is in $H_1^\infty(H,H)$. Therefore F_i is in $H_1^\infty(D_{i-1}, D_{i-1*})$.

11. THE LAYER PEELING ALGORITHM REVISITED

In this section we present a layer peeling algorithm to recursively obtain the reflection operators $\{\Gamma_i\}$ in Figure 4 from the input $u(n)$ and output $y(n)$. This algorithm is also discussed in Section I.1, Procedure III.4.2 and Procedure XIV.3.7.

To this end, let $u_i(n)$ be the ingoing wave immediately above the i-th interface, and $u_i'(n)$ be the ingoing wave immediately below the i-th interface at time $n\tau$. The outgoing wave immediately above the i-th interface is $y_i(n)$ and the outgoing wave immediately below the i-th interface is $y_i'(n)$ at time $n\tau$ (see Figure 5). Notice that $u_1(n) = u(n)$ and $y_1(n) = y(n)$ where $u(n)$ and $y(n)$ are the input and output waves in Figure 4. Finally, $y_i'(n) = x_i(n)$ for all $i \geq 1$.

The reflection and transmission rules (i) and (ii) in Section 7 applied to Figure 5 yields

$$\begin{bmatrix} y_i(n) \\ y_i'(n) \end{bmatrix} = \begin{bmatrix} \Gamma_i & D_{i*} \\ D_i & -\Gamma_i^* \end{bmatrix} \begin{bmatrix} u_i(n) \\ u_i'(n) \end{bmatrix}. \tag{11.1}$$

The 2 by 2 matrix in (11.1) is the *Scattering matrix* for the i-th interface. This scattering

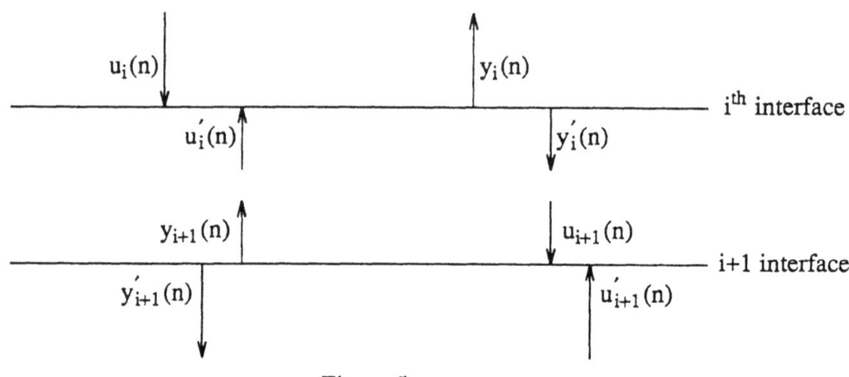

Figure 5

matrix is the unitary matrix which expresses the outgoing waves $y_i(n)$ and $y'_i(n)$ in terms of the incoming waves $u_i(n)$ and $u'_i(n)$. Moreover, this scattering matrix is precisely the rotation matrix determined by the reflection operator Γ_i. Because the scattering matrix in (11.1) is unitary we have

$$\begin{bmatrix} u_i(n) \\ u'_i(n) \end{bmatrix} = \begin{bmatrix} \Gamma_i^* & D_i \\ D_{i*} & -\Gamma_i \end{bmatrix} \begin{bmatrix} y_i(n) \\ y'_i(n) \end{bmatrix} . \tag{11.2}$$

We will now obtain the matrix version of the layer peeling algorithm in Section I.1. Throughout this section we assume that the choice sequence $\{\Gamma_i\}_1^\infty$ associated with Figure 4, or an F in $H_1^\infty(H,H)$ is strictly contractive. This implies that both D_i and D_{i*} are invertible for all i. By inverting D_i and solving for y'_i in the first equation of (11.2), we obtain the first of the following two equations:

$$y'_i(n) = D_i^{-1} u_i(n) - D_i^{-1} \Gamma_i^* y_i(n) \tag{11.3}$$

$$u'_i(n) = -D_{i*}^{-1} \Gamma_i u_i(n) + D_{i*}^{-1} y_i(n) .$$

The second equation follows by inverting D_{i*} and solving for u'_i in the first equation of (11.1). Consulting Figure 5 along with the fact that it takes $\tau/2$ units of time for a wave to travel between adjoining interfaces gives

$$u_{i+1}(n+\tfrac{1}{2}) = y'_i(n) \text{ and } y_{i+1}(n-\tfrac{1}{2}) = u'_i(n) \quad (\text{for } i \geq 1) . \tag{11.4}$$

Substituting this into (11.3) produces

$$\begin{bmatrix} u_{i+1}(n+\frac{1}{2}) \\ y_{i+1}(n-\frac{1}{2}) \end{bmatrix} = \begin{bmatrix} D_i^{-1} & -D_i^{-1}\Gamma_i^* \\ -D_{i*}^{-1}\Gamma_i & D_{i*}^{-1} \end{bmatrix} \begin{bmatrix} u_i(n) \\ y_i(n) \end{bmatrix}. \tag{11.5}$$

The 2 by 2 matrix in (11.5) is the *transmission matrix* of the i-th interface. It expresses the waves u_i' and y_i' immediately below the i-th interface (or equivalently, the waves u_{i+1} and y_{i+1} immediately above the i+1-interface) in terms of the waves u_i and y_i immediately above the i-th interface.

Throughout the rest of this section only it is assumed that $n = 0, 1, 2, 3, \ldots$ and that the input $u(j) = u_1(j)$ is zero for $j < 0$. Obviously this implies that $y(j) = y_1(j)$ is zero for $j < 0$. Because it takes $\tau/2$ units of time to travel between adjacent interfaces, this also implies that both $u_i(n)$ and $y_i(n)$ are zero for $n < (i-1)/2$. Let $u_{i,n}$ and $y_{i,n}$ be the input and output wave of the i-th interface with the time index shifted back to the origin, that is,

$$u_{i,n} = u_i(n+(i-1)/2) \text{ and } y_{i,n} = y_i(n+(i-1)/2) \qquad \text{(for } i \geq 1) \tag{11.6}$$

where $n = 0, 1, 2, 3, \ldots$ Substituting (11.6) into (11.5) with $y_{i+1,-1} = 0$ yields

$$\begin{bmatrix} u_{i+1,0}, & u_{i+1,1}, u_{i+1,2}, u_{i+1,3}, & \cdots \\ 0 & y_{i+1,0}, y_{i+1,1}, y_{i+1,2}, & \cdots \end{bmatrix} = \Theta_i \begin{bmatrix} u_{i,0}, u_{i,1}, u_{i,2}, u_{i,3}, & \cdots \\ y_{i,0}, y_{i,1}, y_{i,2}, y_{i,3}, & \cdots \end{bmatrix} \tag{11.7}$$

where Θ_i is the 2 by 2 transmission matrix in (11.5). Equation (11.7) relates the input and output waves immediately above the i+1-interface to the input and output waves immediately above the i-th interface. The time index for all of these waves starts at the origin $n = 0$.

We claim that if $u_{i,0}^\alpha$ for α in A spans a dense set in D_{i-1}, then $u_{i+1,0}^\alpha$ for α in A spans a dense set in D_i where $D_o = H$. To see this notice that $y_{i,0}^\alpha = \Gamma_i u_{i,0}^\alpha$. Substituting this into (11.7) yields $u_{i+1,0}^\alpha = D_i u_{i,0}^\alpha$. Hence $u_{i+1,0}^\alpha$ runs through a dense set in D_i when $u_{i,0}^\alpha$ runs through a dense set in D_{i-1}. In particular, if $u^\alpha(0) = u_{1,0}^\alpha$ for α in A spans a dense set in H, then $u_{j,0}^\alpha$ for α in A spans a dense set in D_{j-1} for all $j \geq 1$. This observation along with (11.7) gives the following layer peeling recursive inverse scattering algorithm.

11.1 PROCEDURE. Assume that F is in $H_1^\infty(H,H)$, or equivalently, F is the transfer function of the layered medi model in Figure 4. Choose an input $u^\alpha(n) = u_{1,n}^\alpha$ where $u_{1,0}^\alpha$ for α in A spans a dense set in H. Then compute the output $y^\alpha(n) = y_{1,n}^\alpha$ by convolving F with $u^\alpha(n)$ for all α in A. (In the layered media model $y^\alpha(n)$ is obtained by reading the output for $u^\alpha(n)$.) The first choice sequence Γ_1 is obtained by solving $y^\alpha_{1,0} = \Gamma_1 u^\alpha_{1,0}$. Since $u^\alpha_{1,0}$ for α in A spans a dense set in H, this equation has a unique solution. Now assume that we have computed the i-th choice contraction Γ_i, the i-th input $u^\alpha_{i,n}$ and the i-th output $y^\alpha_{i,n}$ for all α in D. Now compute Θ_i and the i+1 input $u^\alpha_{i+1,n}$ and the i+1 output $y^\alpha_{i+1,n}$ for all α in A by using (11.7). The next choice contraction Γ_{i+1} is obtained by solving $y^\alpha_{i+1,0} = \Gamma_{i+1} u^\alpha_{i+1,0}$. Since $u^\alpha_{i,n}$ for α in A spans a dense set in D_{i-1} our previous analysis shows that $u^\alpha_{i+1,0}$ for α in A spans a dense set in D_i. Thus $y_{i+1,0} = \Gamma_{i+1} u^\alpha_{i+1,0}$ admits a unique solution and completes the recursion.

11.2 REMARK. If at any step in the previous algorithm Γ_i is not a contraction, then the procedure stops. In this case F is not in $H^\infty_1(H,H)$.

11.3 REMARK. Let $\{R_i\}^\infty_1$ be a positive definite sequence on H. The previous procedure can be used to recursively obtain the choice sequence $\{\Gamma_i\}^\infty_1$ associated with $\{R_i\}^\infty_1$. To this end notice that the marine seismology structure in Figure 3 contains the layered medium model in Figure 4 with a zero interface and reflection operator $\Gamma_0 = -I$. Recall that Section 7 shows that $\{R_i\}^\infty_1$ is generated by the marine seismology structure in Figure 3, where $u^h(0) = u^h_{1,0} = h$ is in H and $y^h(n) = y^h_{1,n} = R_n h$ for all $n > 0$. Since $\Gamma_0 = -I$ this implies that the marine structure in Figure 3 is equivalent to the layered media structure in Figure 4 where

$$u^h_{1,0} = h \text{ and } u^h_{1,n} = y^h_{1,n-1} = R_n h \quad \text{(for h in } H \text{ and } n > 0) . \tag{11.8}$$

Using the initial conditions in (11.8), Procedure 11.1 recursively yields the choice sequence $\{\Gamma_i\}^\infty_1$ associated with $\{R_i\}^\infty_1$. As before if at any step in this procedure Γ_i is not a contraction, then $\hat{R}_{i+1}$ is not a positive Toeplitz matrix and $\{R_i\}^\infty_1$ is not positive definite.

12. CHOICE SEQUENCES AND SCHUR NUMBERS

In this section we use the layered media model in Figure 4 to show that the choice sequence for F_1 in $H^\infty_1(H,H)$ is precisely the Schur choice sequence for F_1 defined in (9.13).

To begin, let F_i be the transfer function for the i-th interface, that is, F_i is the transfer function for the layered media model in Figure 4 where $\{\Gamma_1, \Gamma_2, \Gamma_3, \cdots\}$ is replaced by $\{\Gamma_i, \Gamma_{i+1}, \Gamma_{i+2}, \cdots\}$. Remark 10.2 shows that F_i is in $H^\infty_1(D_{i-1}, D_{i-1*})$ for all $i \geq 1$. Let us identify the times $n\tau$ with z^n. For instance a wave of amplitude hz^n or $hz^{n/2}$ means that at time $n\tau$ or $n\tau/2$ the amplitude of the wave is h, respectively. In terms of z the input waves $u_i(n)$, $u'_i(n)$ and output waves $y_i(n)$, $y'_i(n)$ of the i-th interface become

$$u_i(z) = \sum_n u_i(n)z^n \text{ and } u_i'(z) = \sum_n u_i'(n)z^n$$

$$(12.1)$$

$$y_i(z) = \sum_n y_i(n)z^n \text{ and } y_i'(z) = \sum_n y_i'(n)z^n$$

where n will vary over the integers for odd i and n will vary over odd integer multiples of 1/2 for even i. If $u_i(z)$ is the downgoing wave immediately above the i-th interface, then $y_i(z) = F_i(z)u_i(z)$ is the upgoing wave produced by $u_i(z)$ immediately above the i-th interface (see Figure 5). The output $y_i(z)$ is obtained by formally multiplying $F_i(z)$ by $u_i(z)$.

Because it takes $\tau/2$ units of time for a wave to travel between adjacent interfaces, Figure 5 shows that

$$u_i'(z) = zF_{i+1}(z)y_i'(z) \,. \tag{12.2}$$

Substituting this into the second equation in (11.1) along with (12.1) gives

$$y_i' = D_i u_i - z\Gamma_i^* F_{i+1} y_i' \text{ and } y_i' = (I+z\Gamma_i^* F_{i+1})^{-1} D_i u_i \,.$$

The inverse is well defined for z in D because F_{i+1} is in $H_1^\infty(D_i, D_{i*})$. Using this in the first equation in (11.1) along with (12.1) and (12.2) yields

$$y_i(z) = \Gamma_i u_i + D_{i*} u_i' = \Gamma_i u_i + zD_{i*} F_{i+1} y_i' =$$

$$[\Gamma_i + zD_{i*} F_{i+1}(I+z\Gamma_i^* F_{i+1})^{-1} D_i u_i(z) = F_i(z)u_i(z) \,.$$

Therefore

$$F_i(z) = \Gamma_i + zD_{i*} F_{i+1}(z)(I+z\Gamma_i^* F_{i+1}(z))^{-1} D_i \tag{12.3}$$

which is precisely equation (9.9).

Using $y_i(z) = F_i(z)u_i(z)$ in the first equation in (11.2) along with (12.1) gives

$$u_i(z) = (I - \Gamma_i^* F_i(z))^{(-1)} D_i y_i'(z) \,.$$

The psuedo inverse is well defined for z in D because F_i is in $H_1^\infty(D_{i-1}, D_{i-1*})$ and $\Gamma_i^* F_i$ is a strict contraction on D_{i-1} for z in D (see Theorem IX.5.3). Substituting this into the second equation in (11.2) with $y_i = F_i u_i$ gives

$$u_i' = [D_{i*} F_i(I-\Gamma_i^* F_i)^{(-1)} D_i - \Gamma_i]y_i' \,.$$

This and (12.2) produces the result we have been looking for, that is,

$$F_{i+1}(z) = \frac{1}{z}\,[D_{i*}F_i(z)(I-\Gamma_i^*F_i(z))^{(-1)}D_i - \Gamma_i] \tag{12.4}$$

which is precisely Equation (9.11). We sum up the above analysis in the following remark.

12.1 REMARK. Equations (12.3) and (12.4) show that the transfer functions F_i for the i-th interface and F_{i+1} for the i+1-interface are related by an operator Moebius transformation. (As noted earlier if H is scalar valued, then these transformations reduce to the Moebious transformations (9.14) used in Chapter I to solve the Carathéodory interpolation problem.) Moreover, the Schur contractions $\Gamma_i = F_i(0)$ are precisely the reflection operators of the i-th interace, or equivalently, the choice sequence for F_1.

To complete this section, we will show that the positive definite sequence $\{R_i\}_1^\infty$ on H in (7.2) determined by the Marine seismology structure in Figure 3 in Section 7, is the Cayley transform of the transfer function F_1 determined by the first interface in Figure 3. Since it takes $\tau/2$ units of time to travel between adjoining interfaces and $\Gamma_0 = -I$, Figure 3 implies that $v_1 = zF_1w_1$ where $v_1 = v_1(z)$ and $w_1 = w_1(z)$ are the waves in Figure 3. Recall that in marine seismology one sets off an explosion $w_1(0) = h$ in H immediately below the zero interface. (Notice also that w_{i+1} in Figure 3 becomes y_i' in Figure 5.) This and $\Gamma_0 = -I$ gives $w_1 = h + v_1$. Therefore $w_1 = h + zF_1w_1$, or equivalently,

$$w_1 = (I-zF_1)^{-1}h \text{ and } v_1 = zF_1(I-zF_1)^{-1}h . \tag{12.5}$$

The inverse is well defined for z in D because F_1 is in $H_1^\infty(H,H)$. Notice that $\Gamma_0 = -I$ implies that $v_1(n) = w_1(n)$ for all $n > 0$. This with (7.2) and the second equation in (12.5) yields

$$R_p(z) = \sum_1^\infty R_n z^n = zF_1(I-zF_1)^{-1} = z(I-zF_1)^{-1}F_1 . \tag{12.6}$$

This and (8.11) shows that $\{R_i\}_1^\infty$ is the Cayley transform of F_1. Combining the above analysis produces the following remark.

12.2 REMARK. Let $\{R_i\}_1^\infty$ on H be the positive definite sequence generated by the marine seismology structure in Figure 3. The inverse Cayley transform of $\{R_i\}_1^\infty$ is F_1 where F_1 is the transfer function of the first interface in Figure 3. Furthermore the choice sequence for $\{R_i\}_1^\infty$, or equivalently, the sequence of reflection operators for the marine seismology problem in Figure 3, is precisely the sequence of Schur contractions for F_1.

XV.13. NOTES AND COMMENTS

The Naimark dilation theorem is due to Naimark [Na]. Our presentation was taken from [Sz.-NF 10]. The proof of Hergoltz's theorem was taken from [Ha 5]. The Hessenberg matrix representation for the Naimark dilation in (2.4) was developed in [Con 2] and independently in [Frz 5]. The matrix U in (2.4) is really a special case of the matrix W in (XIII.2.19) presented in [ArsCF 2], which naturally occurs in characterizing all contractive intertwining liftings. This is demonstrated in Section XIII.3. In the scalar case Theorem 2.1 is due to Schur [Schur 1]. The matrix case is in [DeGK 1]. It also follows from the matrix Levinson algorithm in [MoVK] (see [DudM] for related results and historical comments in this direction). Our approach to Theorem 2.1 via the Naimark dilation is in [Con 2] and [Frz 5,7]. Sz.-Nagy and Koranyi were the first to study operator versions of classical interpolation problems [Sz.-NK 1,2] (see also [Sz.–N 7,8] for related results). All the results in Section 3 are classical [GrS], and date back to Carathéodory [Ca 1,2] and Toeplitz [To 4]. For example, Corollary 3.5 is a well known result in orthogonal polynomials [Szegö], [Ge]. Here we have tried to present a geometric approach to some of these results. The inverse scattering algorithms in Section 4 are taken from [Con 2] and [Frz 7] (see also [ArsCo] and [Con 1,3]). Actually these algorithms are just a positive version of Procedure XIII.2.9 which was obtained in [ArsCF 2]. It is also noted that many of the results in this chapter concerning the Naimark dilation are special cases of the stochastic bilinear systems theory in [Frz 7]. There are many fast efficient algorithm to obtain an upper triangular factorization for a Toeplitz matrix, for example see [Ba], [GoS], [LeRG], [Mar], [Mo] and [Riss]. The main purpose of Remark 4.8 was to show how choice sequences naturally occurs in upper triangular factorizations. The Levinson algorithm for block Toeplitz matrices was first obtained in [Wh], [WigR] (see [DudM] further results and historical comments). One can easily obtain a matrix version of the Levinson algorithm by extending the original proof of the Levinson algorithm in Corollary II.2.3 to the matrix setting [MoVK]. Our approach is rather unorthodox. Here we have presented a proof of the Levinson algorithm based on the matrix form of the Naimark dilation in (2.4). This was done mainly to show how the choice sequence $\{\Gamma_i\}$ uniquely determined by the Naimark dilation naturally occurs in the block Levinson algorithm. The norms $\|\varepsilon_n h\|$ and $\|\varepsilon_{n*} h\|$ are usually referred to as the forward and backward prediction errors [DudM], [MoVK]. The results in Section 7 providing a layered medium interpretation for the Naimark dilation is presented in [KaB]; see also [BrK]. The results in Section 8 are classical applications of the Cayley transform and dates back to [Schur 1]. The results in the rest of this chapter uses the Cayley transform to show that the choice sequence for the Naimark dilation is precisely the Schur operators for its inverse Cayley transform. The layer peeling algorithm in Section 11 was taken from [Ka 2] and adjusted to this setting (see also [DewVK] and [LeK 1,2]). Finally it is noted that the main emphasis of this chapter was the connection between the Naimark dilation, choice sequences and positive Toeplitz matrices. For a nice derivation and characterization of all solutions to the positive definite Carathéodory interpolation problem see [GoKW 1]. In connection with this chapter see [Ak], [AlpD], [Ar], [ArC], [ArCL], [ArsCo], [Bur], [Cl], [Con 1,2,3], [CoS], [DeGK], [Des], [Dew], [DewD], [FoF 1], [Ge], [GoKW], [GrS], [KaBM], [KaVM], [Kan], [KoT], [Kor], [KovP], [KrK], [KrN], [Lev], [Lo], [Ma], [Mas], [RoT], [Sz.-NK] and [WiM].

CHAPTER XVI

POSITIVE DEFINITE BLOCK MATRICES

This chapter is a positive matrix version of Chapter IV. First we give a complete characterization of all 2 by 2 and 3 by 3 positive block matrices. Then this is used to obtain some standard results for positive Toeplitz matrices and the Levinson algorithm. We also show that there is a one to one correspondence between the set of all n by n positive block matrices, and a set of n positive operators combined with $n(n-1)/2$ contractions. As a special case of this result, we also establish the fact that there is a one to one correspondence between the set of all positive block Toeplitz matrices on $l_n^2(H)$, and the set of all choice sequences $\{\Gamma_i\}_1^{n-1}$ initiated on H.

1. POSITIVE 2 by 2 BLOCK MATRICES

In this section we present a complete classification of all 2 by 2 positive block matrices, and their upper and lower triangular factorizations.

To this end consider the block matrix T defined by

$$T = \begin{bmatrix} A & B \\ B^* & C \end{bmatrix} \quad \text{on} \quad H \oplus G \tag{1.1}$$

where A, B and C are operators acting between the appropriate spaces. Finally, notice that for any positive operator Q on H we have $\overline{QH} = \overline{Q^{1/2}H}$, that is, the closed range of Q equals the closed range of $Q^{1/2}$. (Recall that positive means that $(Qh, h) \geq 0$ for all h.) The following result is a positive definite version of Lemma IV.2.1, which gave a complete characterization of all lower triangular contractive block matrices.

1.1 THEOREM. *The operator* T *defined in (1.1) is positive if and only if* A *and* C *are both positive and there exists a contraction* Γ *mapping* $\overline{CG}$ *into* $\overline{AH}$ *satisfying*

$$B = A^{1/2}\Gamma C^{1/2} \qquad (\|\Gamma\| \leq 1). \tag{1.2}$$

PROOF. Assume that T in (1.1) is a positive matrix. Obviously A and C are both positive. For h in H and g in G we have

$$0 \leq \left(\begin{bmatrix} A & B \\ B^* & C \end{bmatrix} \begin{bmatrix} h \\ g \end{bmatrix}, \begin{bmatrix} h \\ g \end{bmatrix} \right) = \|A^{1/2}h\|^2 + 2\mathrm{Re}(Bg,h) + \|C^{1/2}g\|.^2 \qquad (1.3)$$

We claim that there exists a constant M such that

$$\|Bg\| \leq M\|C^{1/2}g\| \quad \text{(for all g in } G). \qquad (1.4)$$

If this is not the case, then there exists a sequence g_n in G for all $n \geq 1$ such that $n\|C^{1/2}g_n\| \leq \|Bg_n\| \neq 0$. By a suitable normalization we can assume that $\|Bg_n\| = 1$. Using $g = -ng_n$ and $h = Bg_n$ in (1.3) yields

$$0 \leq \|A^{1/2}h\|^2 - 2n\,\mathrm{Re}(Bg_n,h) + n^2\|C^{1/2}g_n\|^2 \leq \|A^{1/2}\| - 2n\|Bg_n\|^2 + 1 .$$

For large n the last equation becomes negative which is a contradiction to the positivity of T. Thus (1.4) holds. This implies that there exists an operator Q mapping $\overline{CG}$ into H satisfying $B = QC^{1/2}$.

Using $B = QC^{1/2}$ in (1.3) yields

$$0 \leq \|A^{1/2}h\|^2 + 2\mathrm{Re}(C^{1/2}g, Q^*h) + \|C^{1/2}g\|^2 =$$

$$ \qquad (1.5)$$

$$\|C^{1/2}g + Q^*h\|^2 + \|A^{1/2}h\|^2 - \|Q^*h\|^2 .$$

Since $\overline{CG} \supseteq Q^*H$ we can always choose a sequence g_n in G such that $C^{1/2}g_n \to Q^*h$. This implies that $\|Q^*h\| \leq \|A^{1/2}h\|$ for all h in H. So there exists a contraction Γ_1 mapping $\overline{AH}$ into $\overline{CG}$ satisfying $Q^* = \Gamma_1 A^{1/2}$. This and $B = QC^{1/2}$ proves (1.2) where Γ is now the adjoint of Γ_1.

To prove the other half assume that (1.2) holds. Then

$$(T(h \oplus g), h \oplus g) = \|A^{1/2}h\|^2 + 2\mathrm{Re}(A^{1/2}\Gamma C^{1/2}g, h) + \|C^{1/2}g\|^2 =$$

$$\|A^{1/2}h + \Gamma C^{1/2}g\|^2 - \|\Gamma C^{1/2}g\|^2 + \|C^{1/2}g\|^2 = \qquad (1.6)$$

$$\|A^{1/2}h + \Gamma C^{1/2}g\|^2 + \|D_\Gamma C^{1/2}g\|^2 \geq 0 .$$

This completes the proof.

1.2 COROLLARY. *Let* T *be a positive operator for the form (1.1) and* Γ *the contraction defined in (1.2). Then the operators* W *from* $\overline{\mathrm{ran}(T)}$ *onto* $\overline{AH} \oplus D_\Gamma$ *and* W* *from* $D_{\Gamma^*} \oplus \overline{CG}$ *onto* $\overline{\mathrm{ran}(T)}$ *defined by*

$$\Delta = \begin{bmatrix} A^{1/2} & \Gamma C^{1/2} \\ 0 & D_\Gamma C^{1/2} \end{bmatrix} = W T^{1/2} \tag{1.7}$$

and

$$\Delta_* = \begin{bmatrix} A^{1/2} D_{\Gamma^*} & A^{1/2}\Gamma \\ 0 & C^{1/2} \end{bmatrix} = T^{1/2} W_* \tag{1.8}$$

are unitary.

PROOF. Equations (1.1), (1.2) and (1.7) give $\Delta^* \Delta = T = T^{1/2} T^{1/2}$. This implies that $\|\Delta x\|^2 = \|T^{1/2} x\|^2$ for all x in $H \oplus G$. Thus (1.7) defines a unitary operator W. Equations (1.1), (1.2) and (1.8) give $\Delta_* \Delta_*^* = T = T^{1/2} T^{1/2}$. This implies that the equation $\Delta_*^* = W_1 T^{1/2}$ defines a unitary operator W_1. Taking the adjoint and setting W_* equal to the adjoint of W_1 yields (1.8). This completes the proof.

1.3 REMARK. Equation (1.7) provides an upper triangular factorization of T, that is, $T = \Delta^* \Delta$ where Δ is upper triangular. On the other hand (1.8) provides a lower triangular factorization of T, that is, $T = \Delta_1^* \Delta_1$ where $\Delta_1 = \Delta_*^*$ is lower triangular.

1.4 COROLLARY. *If* T *is a positive operator of the form (1.1), then* T *admits a factorization of the form:*

$$T = \begin{bmatrix} A^{1/2} & 0 \\ 0 & C^{1/2} \end{bmatrix} \begin{bmatrix} I & \Gamma \\ \Gamma^* & I \end{bmatrix} \begin{bmatrix} A^{1/2} & 0 \\ 0 & C^{1/2} \end{bmatrix}. \tag{1.9}$$

PROOF. A simple matrix multiplication shows that the left hand side of (1.9) equals T in (1.1) where B is given by (1.2). This and Theorem 1.1 completes the proof.

Recall that an operator Q is *strictly positive* if Q is a positive operator satisfying $Q \geq \delta I > 0$ where δ is a positive constant. Notice that Q is strictly positive if and only if Q is positive and invertible. This terminology sets the stage for the following result.

1.5 COROLLARY. *The operator* T *in (1.1) is strictly positive if and only if* A *and* C *are both strictly positive and (1.2) holds where* Γ *is a strict contraction.*

PROOF. If T is strictly positive, then both A and C are strictly positive. So without loss of generality we assume that both A and C are strictly positive. Equation (1.7) shows that $T^{1/2}$ is invertible if and only if D_Γ is invertible, or equivalently, Γ is a strict contraction. Thus $T = T^{1/2} T^{1/2}$ is invertible if and only if Γ is a strict contraction.

This completes the proof.

2. LEVINSON SYSTEMS

In this section we will obtain a Levinson type equation for all 2 by 2 positive block matrices. Then we will use this system of equations to obtain necessary and sufficient conditions for a block Toeplitz matrix to be strictly positive. We will also use the Levinson system to provide an upper and lower triangle factorization for a strictly positive block Toeplitz matrix.

To begin, consider the following linear system of equations

$$\begin{bmatrix} A & B \\ B^* & C \end{bmatrix} \begin{bmatrix} -F \\ G \end{bmatrix} = \begin{bmatrix} Q \\ E \end{bmatrix} \tag{2.1}$$

where F, G, E and Q are operators acting between the appropriate spaces. As before T is the operator defined by A, B, and C in (1.1).

2.1 LEMMA. *Assume that* A *is invertible and* G, Q *are known operators. Then there is a unique solution* F, E *to* (2.1) *namely:*

$$F = A^{-1}BG - A^{-1}Q$$
$$E = CG - B^*A^{-1}BG + B^*A^{-1}Q . \tag{2.2}$$

PROOF. By substituting (2.2) into (2.1) it is easy to verify that (2.2) is a solution to (2.1). (This solution was obtained by first solving for F in the first row of (2.1), then substituting this F into the last row to obtain E). We will now prove uniqueness. Because of linearity, we can assume without loss of generality that the known operators G and Q are both zero. Since A is invertible, this implies that $F = 0$. Thus $E = 0$ and the proof is complete.

The Levinson system for the 2 by 2 matrix in (1.1) is:

$$\begin{bmatrix} A & B \\ B^* & C \end{bmatrix} \begin{bmatrix} -F \\ I \end{bmatrix} = \begin{bmatrix} 0 \\ E \end{bmatrix} \tag{2.3}$$

where E is an operator on G and F maps G into H. If A is invertible, then the previous lemma shows that

$$F = A^{-1}B \quad \text{and} \quad E = C - B^*A^{-1}B \tag{2.4}$$

is the unique solution to (2.3).

Finally, we say that an operator R is *singular* if the kernel of R is nonzero. This sets the stage for the following basic result.

2.2 COROLLARY. *Let* A *and* C *both be strictly positive. The matrix* T *in (1.1) is positive (strictly positive) (positive and singular) if and only if the solution* E *to the Levinson system in (2.3) is positive (strictly positively) (positive and singular), respectively.*

PROOF. For all g and G Equation (2.3) gives

$$(T(-Fg \oplus g), \, (-Fg \oplus g)) = (Eg, g) \, .$$

Therefore, if T is positive (strictly positive) this implies that E is positive (strictly positive), respectively.

On the other hand assume that E is positive. By (2.4)

$$\|C^{1/2}g\|^2 - \|A^{-1/2}Bg\|^2 = (Eg, g) \, .$$

This implies that $\|C^{1/2}g\| \geq \|A^{-1/2}Bg\|$. Hence there exists a contraction Γ mapping $\overline{CG}$ into $\overline{AH}$ satisfying $\Gamma C^{1/2} = A^{-1/2}B$, or equivalently, $B = A^{1/2}\Gamma C^{1/2}$. By Theorem 1.1 the operator T is positive.

Substituting $B = A^{1/2}\Gamma C^{1/2}$ into (2.4) yields $E = C^{1/2}D_\Gamma^2 C^{1/2}$. Since C is invertible, this shows that E is strictly positive if and only if D_Γ^2 is strictly positive, or equivalently, Γ is a strict contraction. By Corollary 1.5 the operator T is strictly positive if and only if E is strictly positive.

If E is positive and singular, then the previous part of this corollary shows that T is positive. Equation (2.3) implies that T is singular. On the other hand, if T is positive and singular, then E is positive and there exists a nonzero $h \oplus g$ in $H \oplus G$ such that $T(-h \oplus g) = 0$. Since A and C are both invertible, h and g are both nonzero. Solving for h in terms of g gives $h = A^{-1}Bg$, or equivalently, by (2.4) we have $h = Fg$. Thus $0 = T(-Fg \oplus g) = 0 \oplus Eg$ where g is nonzero. Therefore E is singular and the proof is complete.

2.3 REMARK. Let T be an operator of the form (1.1) where A and C are both strictly positively. Theorem 1.1 shows that T is positive if and only if $\Gamma = A^{-1/2}BC^{-1/2}$ is a contraction. However, it is easier to check the positively of T by the previous corollary, that is, T is positive if and only if E is positive. Equation (2.4) shows that E only involves the inverse of A. We do not have to calculate the inverse and square roots of both A and C required by Theorem 1.1. For the same reason the previous corollary provides a more efficient method than Corollary 1.5 for determining the strict positivity of T.

The following is a simple consequence of Theorem 1.1, Corollary 2.2 and (2.4).

2.4 COROLLARY. *Let* T *be a positive operator of the form (1.1) where* A *and* C *are both strictly positive and* Γ *be the unique contraction defined by* $B = A^{1/2}\Gamma C^{1/2}$. *Then the unique solution to the Levinson system in (2.3) is:*

$$F = A^{-1/2}\Gamma C^{1/2} \text{ and } E = C^{1/2}D_\Gamma^2 C^{1/2} .$$ (2.5)

Moreover, T *is strictly positive (positive and singular) if and only if* Γ *is a strict contraction* $(D_\Gamma$ *is singular), respectively.*

To complete this section we will present the dual of the previous Levinson system. To this end consider the following linear system of equations

$$\begin{bmatrix} A & B \\ B^* & C \end{bmatrix} \begin{bmatrix} G_* \\ -F_* \end{bmatrix} = \begin{bmatrix} E_* \\ Q_* \end{bmatrix}$$ (2.6)

where F_*, G_*, E_* and Q_* are operators acting between the appropriate spaces. The following is the dual of Lemma 2.1. The proof is omitted because it is similar to the proof of Lemma 2.1.

2.5 LEMMA. *Assume that* C *is invertible and* G_*, Q_* *are known operators. Then there is a unique solution* F_*, E_* *to (2.6) namely:*

$$\begin{aligned} F_* &= C^{-1}B^*G_* - C^{-1}Q_* \\ E_* &= AG_* - BC^{-1}B^*G_* + BC^{-1}Q_* . \end{aligned}$$ (2.7)

The dual of the Levinson system in (2.3) is:

$$\begin{bmatrix} A & B \\ B^* & C \end{bmatrix} \begin{bmatrix} I \\ -F_* \end{bmatrix} = \begin{bmatrix} E_* \\ 0 \end{bmatrix}$$ (2.8)

where E_* is an operator on H and F_* maps H into G. If C is invertible, then the previous lemma shows that

$$F_* = C^{-1}B^* \text{ and } E_* = A - BC^{-1}B^*$$ (2.9)

is the unique solution to (2.8). The dual to Corollary 2.2 is given by the following result.

2.6 COROLLARY. *Let* A *and* C *both be strictly positive. The matrix* T *in (1.1) is positive (strictly positive) (positive and singular) if and only if the solution* E_* *to the Levinson system in (2.8) is positive (strictly positive) (positive and singular),* *respectively.*

The proof is omitted because it is similar to the proof of Corollary 2.2. (By applying the operator $\hat{I}$ from $H \oplus G$ to $G \oplus H$ defined by

$$\hat{I} = \begin{bmatrix} 0 & I \\ I & 0 \end{bmatrix}$$

to (2.8), one can convert the Levinson system in (2.8) to the Levinson system in (2.3) where A, B and C are replaced by C, B^* and A, respectively. So the previous corollary can also be obtained from Corollary 2.2, and the fact that T is positive (strictly positive) (positive and singular) if and only if $\hat{I}T\hat{I}^*$ is positive (strictly positive) (positive and singular), respectively.)

2.7 REMARK. Let T be an operator of the form (1.1) where A and C are both strictly positive. As noted in Remark 2.3, Theorem 1.1 shows that T is positive if and only if $\Gamma = A^{-1/2}BC^{-1/2}$ is a contraction. As expected it is easier to check the positivity of T by the previous corollary, that is, T is positive if and only if E_* is positive. Equation (2.9) shows that E_* only involves the inverse of C. We do not have to calculate the inverse and square roots of both A and C required by Theorem 1.1. For the same reasons the previous corollary provides a more efficient method than Corollary 1.5 for determining the strict positivity of T. Finally, it is noted that Corollary 2.2 might be easier to use than Corollary 2.6 or vice versa depending upon which operator A or C is easier to invert.

The following is a dual of Corollary 2.4. The proof is a simple consequence of Theorem 1.1 and (2.9).

2.8 COROLLARY. *Let T be a positive operator of the form (1.1) where A and C are both strictly positive and Γ be the unique contraction defined by $B = A^{1/2}\Gamma C^{1/2}$. Then the unique solution to the Levinson system in (2.8) is:*

$$F_* = C^{-1/2}\Gamma^* A^{1/2} \text{ and } E_* = A^{1/2}D_{\Gamma^*}^2 A^{1/2} . \tag{2.10}$$

Moreover, T is strictly positive (positive and singular) if and only if Γ is a strict contraction (D_{Γ^*} is singular), respectively.

To complete this section we will use Corollary 2.2 to prove a standard result on the strict positivity of a Toeplitz matrix. To this end let $\{R_i\}_1^n$ be a sequence of operators on

H and $\hat{R}_{n+1}$ its Toeplitz matrix on $l^2_{n+1}(H)$ defined in (XV.1.2) where n+1 replaces n. Consider the following Levinson system

$$\hat{R}_{i+1} \begin{bmatrix} A_{i,0} \\ A_{i,1} \\ \cdot \\ \cdot \\ \cdot \\ A_{i,i-1} \\ I \end{bmatrix} = \begin{bmatrix} 0 \\ 0 \\ \cdot \\ \cdot \\ \cdot \\ 0 \\ E_i \end{bmatrix} \qquad (2.11)$$

where $A_{i,j}$ and E_i are all operators on H. This sets the stage for the following result.

2.9 THEOREM. *Let $\{R_i\}_1^n$ be a sequence of operators on H. The Toeplitz matrix $\hat{R}_{n+1}$ is strictly positive if and only if the error operators E_i are strictly positive for all $1 \le i \le n$.*

PROOF. For $i = 1$ the Levinson system in (2.11) reduces to

$$\hat{R}_2 \begin{bmatrix} A_{1,0} \\ I \end{bmatrix} = \begin{bmatrix} I & R_1 \\ R_1^* & I \end{bmatrix} \begin{bmatrix} A_{1,0} \\ I \end{bmatrix} = \begin{bmatrix} 0 \\ E_1 \end{bmatrix} .$$

Corollary 2.2 shows that $\hat{R}_2$ is strictly positive if and only if E_1 is strictly positive. Now we use induction and assume that the Toeplitz matrix $\hat{R}_n$ is strictly positive if and only if E_i is strictly positive for $1 \le i < n$. In this case $\hat{R}_{n+1}$ admits matrix decomposition of the form

$$\hat{R}_{n+1} = \begin{bmatrix} \hat{R}_n & B \\ B^* & I \end{bmatrix}$$

where $\hat{R}_n$ is strictly positive. Corollary 2.2 implies that $\hat{R}_{n+1}$ is strictly positive if and only if E_n is strictly positive. This completes the induction and the proof.

Notice that if $\{R_i\}_1^n$ is a sequence of complex numbers, that is, $H = C^1$, then $\hat{R}_{n+1}$ is a Toeplitz matrix on C^{n+1} and the error operators E_i are all complex numbers. This observation leads to the following result which is precisely Theorem 2.1 in Chapter II.

2.10 COROLLARY. *Let $\hat{R}_{n+1}$ on C^{n+1} be the Toeplitz matrix generated by a sequence of complex numbers $\{R_i\}_1^n$. Then $\hat{R}_{n+1}$ is strictly positive if and only if the* *scalar valued error operators* $E_i > 0$ *for all* $1 \le i \le n$.

The dual of the Levinson system in (2.11) is

$$
\hat{R}_{i+1}
\begin{bmatrix}
I \\
B_{i,1} \\
\cdot \\
\cdot \\
\cdot \\
B_{i,i-1} \\
B_{i,i}
\end{bmatrix}
=
\begin{bmatrix}
E_{i*} \\
0 \\
\cdot \\
\cdot \\
\cdot \\
0 \\
0
\end{bmatrix}
\tag{2.12}
$$

where $B_{n,i}$ and E_{i*} are operators on H. The dual of Theorem 2.9 is given by:

2.11 THEOREM. *Let* $\{R_i\}_1^n$ *be a sequence of operators on* H. *The Toeplitz matrix* $\hat{R}_{n+1}$ *is strictly positive if and only if the error operators* E_{i*} *are strictly positive for all* $1 \le i \le n$.

The proof is omitted because it is identical to the proof of the previous theorem. (The only difference is that one uses Corollary 2.6 instead of Corollary 2.2.) Finally, the dual to Corollary 2.10 is given by:

2.12 COROLLARY. *Let* $\hat{R}_{n+1}$ *on* $\mathbb{C}^{n+1}$ *be the Toeplitz matrix generated by a sequence of complex numbers* $\{R_i\}_1^n$. *Then* $\hat{R}_{n+1}$ *is strictly positive if and only if the scalar valued error operators* $E_{i*} > 0$ *for all* $1 \le i \le n$.

The operators E_i and E_{i*} can be recursively computed by the Levinson algorithm in Section XV.6. Therefore, the Levinson algorithm along with the previous two theorems provides us with a recursive method of determining the strict positivity of $\hat{R}_{n+1}$.

2.13 REMARK. By a slight abuse of notation, let $\hat{R}_{n+1}$ be a block matrix on $H_o \oplus H_1 \oplus \cdots \oplus H_n$ with block entries $[R_{i,j} : 0 \le i \le n$ and $0 \le j \le n]$ and assume that its diagonal entries $R_{i,i}$ are all strictly positive ($0 \le i \le n$). Let $A_{i,j}$ from H_i to H_j and E_i on H_i be operators satisfying (2.11) for $1 \le i \le n$ where $\hat{R}_{i+1}$ is the i+1 by i+1 block matrix contained in the north west corner of $\hat{R}_{n+1}$. Then by replacing I by the appropriate $R_{i,i}$ in the proof of Theorem 2.9, we see that $\hat{R}_{n+1}$ is strictly positive if and only if the error operators E_i are all strictly positive for $1 \le i \le n$. In particular, if $\hat{R}_{n+1}$ is a matrix on

C^{n+1}, that is, if $H_i = C^1$ for all i, then $\hat{R}_{n+1}$ is strictly positive if and only if the complex numbers E_i are strictly positive for $1 \le i \le n$. However, in this case because $\hat{R}_{n+1}$ is not necessarily a Toeplitz matrix, one can not use the Levinson algorithm to recursively compute the error operators E_i. One can also convert Theorem 2.11 and Corollary 2.12 to the case when $\hat{R}_{n+1}$ is an arbitrary block matrix. In this case $\hat{R}_i$ in (2.12) must be the i by i matrix contained in the southeast corner of $\hat{R}_{n+1}$. The details are left to the reader as an instructive exercise.

To complete this section we will use the Levinson system in (2.3) and (2.8) to obtain a lower and upper triangular factorization for the inverse of the block Toeplitz matrix $\hat{R}_{n+1}$. As before we assume that T in (1.1) is strictly positive, or equivalently, that A, C, E and E_* are all strictly positive. To begin let

$$\Lambda = \begin{bmatrix} A^{-1/2} & -FE^{-1/2} \\ 0 & E^{-1/2} \end{bmatrix} \tag{2.13}$$

where F and E is the solution to the Levinson system in (2.3). Using $F = A^{-1}B$ and $E = C - B^*A^{-1}B$ a simple calculation shows that $\Lambda^*T\Lambda = I$, or equivalently, $I = \Lambda^{-1}T^{-1}\Lambda^{*-1}$. Therefore $\Lambda\Lambda^* = T^{-1}$ and Λ^* is a lower triangular factorization of T^{-1}. This implies that there exists a unitary operator W on $H \oplus G$ satisfying $WT^{-1/2} = \Lambda^*$.

To obtain an upper triangular factorization Λ_*^* of T^{-1} let Λ_* be the matrix defined by

$$\Lambda_* = \begin{bmatrix} E_*^{-1/2} & 0 \\ -F_*E_*^{-1/2} & C^{-1/2} \end{bmatrix} \tag{2.14}$$

where F_* and E_* is the solution to the Levinson system in (2.8); see (2.9). Using $F_* = C^{-1}B^*$ a simple calculation shows that $\Lambda_*^*T\Lambda_* = I$, or equivalently, $I = \Lambda_*^{-1}T^{-1}\Lambda_*^{*-1}$. Therefore $\Lambda_*\Lambda_*^* = T^{-1}$ is an upper triangular factorization of T^{-1}. This implies that there exists a unitary operator W_* on $H \oplus G$ satisfying $W_*T^{-1/2} = \Lambda_*^*$. Summing up the above analysis yields the following result.

2.14 PROPOSITION. *Let* T *in (1.1) be strictly positive and* F, E *and* F_*, E_* *be the unique solution to the Levinson systems in (2.3) and (2.8). Then*

$$\Lambda\Lambda^* = T^{-1} = \Lambda_*\Lambda_*^* . \tag{2.15}$$

Moreover, there exist two unitary operators W *and* W_* *on* $H \oplus G$ *satisfying*

$$\Lambda W = T^{-1/2} \; and \; \Lambda_* W_* = T^{-1/2} \; . \qquad (2.16)$$

Assume that $\{R_i\}_1^n$ on H generates a strictly positive Toeplitz matrix $\hat{R}_{n+1}$ (see (XV.1.2)). Let $F_{i,j}$ be the operators defined by

$$[F_{i,0}, F_{i,1}, ..., F_{i,i-1}, F_{i,i}] = [A_{i,0}, A_{i,1}, ..., A_{i,i-1}, I]E_i^{-1/2} \qquad (2.17)$$

where $A_{i,j}$ and E_i form the solution to the Levinson system in (2.11). Let Λ_{n+1} be the matrix defined by

$$\Lambda_{n+1} = \begin{bmatrix} I & F_{1,0} & F_{2,0} & \cdots & \cdots & F_{n-1,0} & F_{n,0} \\ 0 & F_{1,1} & F_{2,1} & \cdots & \cdots & F_{n-1,1} & F_{n,1} \\ 0 & 0 & F_{2,2} & \cdots & \cdots & F_{n-1,2} & F_{n,2} \\ 0 & 0 & 0 & \cdots & \cdots & F_{n-1,3} & F_{n,3} \\ . & . & . & \cdots & \cdots & . & . \\ . & . & . & \cdots & \cdots & . & . \\ . & . & . & \cdots & \cdots & . & . \\ 0 & 0 & 0 & \cdots & \cdots & F_{n-1,n-1} & F_{n,n-1} \\ 0 & 0 & 0 & \cdots & \cdots & 0 & F_{n,n} \end{bmatrix} \qquad (2.18)$$

By recursively applying the previous proposition (and using $A^{-1/2} = (\hat{R}_i)^{-1/2} = \Lambda_i W_i$ where W_i is unitary) to

$$T = \begin{bmatrix} \hat{R}_i & B \\ B^* & I \end{bmatrix} = \hat{R}_{i+1}$$

it is easy to verify that Λ_{n+1}^* is a lower triangular factor of $\hat{R}_{n+1}^{-1}$, that is,

$$\Lambda_{n+1}\Lambda_{n+1}^* = \hat{R}_{n+1}^{-1} \; . \qquad (2.19)$$

The Levinson algorithm provides a fast recursive algorithm for computing the lower triangular square root Λ_{n+1}^* of R_{n+1}^{-1}.

To obtain the upper triangular factorization Λ_{n+1*}^* of $\hat{R}_{n+1}^{-1}$ let $G_{i,j}$ be the operators defined by

$$[G_{i,0}, G_{i,1}, ..., G_{i,i-1}, G_{i,i}] = [I, B_{i,1}, ..., B_{i,i-1}, B_{i,i}]E_{i*}^{-1/2} \qquad (2.20)$$

where $B_{i,j}$ and E_{i*} form the solution to the Levinson system in (2.12). Let Λ_{n+1*} be the matrix defined by

$$\Lambda_{n+1*} = \begin{bmatrix} G_{n,0} & 0 & 0 & \cdots & 0 & 0 \\ G_{n,1} & G_{n-1,0} & 0 & \cdots & 0 & 0 \\ G_{n,2} & G_{n-1,1} & G_{n-2,0} & \cdots & 0 & 0 \\ \cdot & \cdot & \cdot & \cdots & \cdot & \cdot \\ \cdot & \cdot & \cdot & \cdots & \cdot & \cdot \\ \cdot & \cdot & \cdot & \cdots & \cdot & \cdot \\ G_{n,n-1} & G_{n-1,n-2} & G_{n-2,n-3} & \cdots & G_{1,0} & 0 \\ G_{n,n} & G_{n-1,n-1} & G_{n-2,n-2} & \cdots & G_{1,1} & I \end{bmatrix} \tag{2.21}$$

By recursively applying Proposition 2.14 (and using $C^{-\frac{1}{2}} = (\hat{R}_i)^{-\frac{1}{2}} = \Lambda_{i*} W_{i*}$ where W_{i*} is unitary) to

$$T = \begin{bmatrix} I & B \\ B^* & \hat{R}_i \end{bmatrix} = \hat{R}_{i+1}$$

it is easy to verify that Λ_{n+1*}^* is an upper triangular factorization of $\hat{R}_{n+1}^{-1}$, that is,

$$\Lambda_{n+1*} \Lambda_{n+1*}^* = \hat{R}_{n+1}^{-1} . \tag{2.22}$$

The Levinson algorithm also provides a fast recursive algorithm for computing the upper triangular square root Λ_{n+1*}^* of $\hat{R}_{n+1}^{-1}$.

2.15 REMARK. Let $\hat{R}_{n+1}$ be a strictly positive block matrix on $F = H_o \oplus H_1 \oplus \cdots \oplus H_n$ with block entries $[R_{i,j}]$. As in Remark 2.13, let $A_{i,j}$ and E_i be the operators defined according to (2.11). Consider the inner product defined by

$$<[f_j]_o^n, [g_i]_o^n> = \sum_{o,o}^{n,n} (R_{i,j} f_j, g_i)_{H_i} = (\hat{R}_{n+1}[f_j]_o^n, [g_i]_o^n)_F \quad (f_i \text{ and } g_i \in H_i \text{ for all i}) .$$

Since $\hat{R}_{n+1}$ is strictly positive $<\cdot, \cdot>$ defines a Hilbert space denoted by G for n+1 tuples of the form $[f_o, f_1, ..., f_n]^{tr}$ where f_i is in H_i for all i. Let Φ_i for $0 \le i \le n$ be the operators mapping H_i into F defined by $\Phi_o = [R_{o,o}^{-\frac{1}{2}}, 0, 0, ..., 0]^{tr}$ and

$$\Phi_i = [A_{i,o}, A_{i,1}, ... A_{i,i-1}, I, 0, 0, ..., 0]^{tr} E_i^{-\frac{1}{2}} \quad (\text{for } 1 \le i \le n) .$$

Using (2.11) it is easy to verify that

$$< \Phi_i f_i, \Phi_j g_j > = \delta_{i,j}(f_i, g_j) \quad (\text{for all } f_i \in H_i \text{ and } g_j \in H_j) \tag{2.23}$$

where $\delta_{i,j} = 1$ for $i = j$ and zero otherwise. Moreover, since E_i is invertible $\{\Phi_i H_i\}$ forms a spanning set for G, that is,

$$G = \Phi_0 H_0 \oplus \Phi_1 H_1 \oplus \cdots \oplus \Phi_n H_n .$$

Now let Λ_{n+1} be the operator on F defined by $\Lambda_{n+1} = [\Phi_0, \Phi_1, ..., \Phi_n]$. Notice that Λ_{n+1} is precisely the operator in (2.18) when $\hat{R}_{n+1}$ is Toeplitz. If f and g are in F, then (2.23) implies that

$$(f, g) = \; < \Lambda_{n+1} f, \; \Lambda_{n+1} g > \; = (\hat{R}_{n+1} \Lambda_{n+1} f, \; \Lambda_{n+1} g) = (\Lambda_{n+1}^* \hat{R}_{n+1} \Lambda_{n+1} f, \; g) .$$

The previous equation readily implies that $\Lambda_{n+1}^* \hat{R}_{n+1} \Lambda_{n+1} = I$. From this it easily follows that $\hat{R}_{n+1}^{-1} = \Lambda_{n+1} \Lambda_{n+1}^*$ and Λ_{n+1}^* provides a lower triangular factorization of $\hat{R}_{n+1}$. One can also recursively use Proposition 2.14 to obtain this result. However, in this case $\hat{R}_{n+1}$ is not necessarily a Toeplitz operator, and thus one can not use the Levinson algorithm to compute Λ_{n+1}. One can provide a similar argument using $< \cdot , \; \cdot >$ to show that Λ_{n+1*}^* also provides an upper triangular factorization of $\hat{R}_{n+1}$. The details as left to the reader.

3. POSITIVE 3 by 3 BLOCK MATRICES

In this section we give a complete classification of all 3 by 3 positive block matrices. Then we use this classification to present another proof of Theorem XV.3.3 which solves the positive Carathéodory interpolation Problem XV.3.1.

Throughout this section T is the positive operator in (1.1) and Γ mapping $\overline{CG}$ into $\overline{AH}$ is the unique contraction defined by $B = A^{1/2} \Gamma C^{1/2}$. The operator T' is the positive operator defined by

$$T' = \begin{bmatrix} C & B' \\ B'^* & D \end{bmatrix} \text{ on } G \oplus D \tag{3.1}$$

and Γ' mapping $\overline{DD}$ into $\overline{CG}$ is the unique contraction defined by $B' = C^{1/2} \Gamma' D^{1/2}$. Finally, T_3 is the operator defined by

$$T_3 = \begin{bmatrix} A & B & R \\ B^* & C & B' \\ R^* & B'^* & D \end{bmatrix} \text{ on } H \oplus G \oplus D . \tag{3.2}$$

The following which is the main result of this section is the positive definite version Theorem IV.3.1, which gave a complete classification of all 2 by 2 block contractive matrices.

3.1 THEOREM. *The operator T_3 in (3.2) is positive if and only if both T and T' are positive and*

$$R = A^{1/2}D_{\Gamma^*}\Gamma_R D_{\Gamma'}D^{1/2} + A^{1/2}\Gamma\Gamma'D^{1/2} \tag{3.3}$$

where Γ_R is a contraction mapping $D_{\Gamma'}$ into D_{Γ^*}.

PROOF. If T_3 is positive, then both T and T' are positive. So without loss of generality, we assumed that both T and T' are positive. By Theorem 1.1 and (1.8) we see that T_3 is positive if and only if there exists a contraction $[C_1, C_2]^{tr}$ mapping $\overline{DD}$ into $D_{\Gamma^*} \oplus \overline{CG}$ satisfying:

$$\begin{bmatrix} R \\ B' \end{bmatrix} = T^{1/2}W_* \begin{bmatrix} C_1 \\ C_2 \end{bmatrix}D^{1/2} = \begin{bmatrix} A^{1/2}D_{\Gamma^*} & A^{1/2}\Gamma \\ 0 & C^{1/2} \end{bmatrix}\begin{bmatrix} C_1 \\ C_2 \end{bmatrix}D^{1/2} \tag{3.4}$$

This and $B' = C^{1/2}\Gamma'D^{1/2}$ implies that $C_2 = \Gamma'$. According to Lemma 1.1 in Chapter IV, the set of all contractions of the form $[C_1, \Gamma']^{tr}$ is given by $C_1 = \Gamma_R D_{\Gamma'}$ where Γ_R is a contraction mapping $D_{\Gamma'}$ into D_{Γ^*}. Substituting this into (3.4) yields (3.3) and completes the proof.

3.2 REMARK. Let T and T' both be fixed positive operators. The previous theorem shows that there exists an R such that T_3 is positive. In fact the set of all such $R^{,s}$ is given by (3.3). Formula (3.3) shows that the set of all R such that T_3 is positive is given by a quasi-ball with center $A^{1/2}\Gamma\Gamma'D^{1/2}$, left radius $A^{1/2}D_{\Gamma^*}$ and right radius $D_{\Gamma'}D^{1/2}$ (see section XIII.7 for some related results on quasi-balls). Moreover, there exists a unique R such that T_3 is positive if and only if $D_{\Gamma'} = \{0\}$ or $D_{\Gamma^*} = \{0\}$. Notice that if $A = 0$, then Γ^* mapping $\{0\}$ into $\overline{CG}$ is an isometry and $D_{\Gamma^*} = \{0\}$. If $D = 0$, then Γ' mapping $\{0\}$ into $\overline{CG}$ is an isometry and $D_{\Gamma'} = \{0\}$. In these last two cases there exists a unique positive extension $R = 0$.

Recall that if $\{R_i\}_1^n$ is a sequence of operators on H, then $\hat{R}_{n+1}$ its corresponding block Toeplitz matrix on $l_{n+1}^2(H)$ is given by (XV.1.2) where $n+1$ replaces n. The previous theorem allows us to give an alternate proof of Theorem XV.3.3, which solves the positive Carathéodory interpolation problem, restated here for conveniences as

3.3 COROLLARY. *A sequence of operators $\{R_i\}_1^n$ on H admits a positive extension if and only if its Toeplitz matrix $\hat{R}_{n+1}$ is positive.*

PROOF. Assume that $\hat{R}_{n+1}$ is positive. Setting $C = \hat{R}_n$ and $T = \hat{R}_{n+1} = T'$ the previous remark shows that there exists a $R = R_{n+1}$ such that the Toeplitz matrix $T_3 = \hat{R}_{n+2}$ is positive. Continuing in this fashion there exists a positive Toeplitz matrix

$\hat{R}_{n+k}$ expanding $\hat{R}_{n+1}$ for all $k > 0$. This implies that $\{R_i\}_1^n$ admits a positive extension. The other half of the proof is obvious. This completes the proof.

3.4 COROLLARY. *Let* T *and* T' *both be strictly positive. Then* T_3 *is strictly positive (positive and singular) if and only if* R *is given by (3.3) where* Γ_R *is a strict contraction (is a contraction and* D_{Γ_R} *is singular), respectively.*

PROOF. Since T' is strictly positive, Corollary 1.5 implies that Γ' is a strict contraction. By replacing A by T and C by D in Corollary 1.5 equations (1.8) and (3.4) imply that T_3 is strictly positive if and only if

$$\begin{bmatrix} C_1 \\ C_2 \end{bmatrix} = \begin{bmatrix} \Gamma_R D_{\Gamma'} \\ \Gamma' \end{bmatrix}$$

is a strict contraction. Because Γ' is a strict contraction, Corollary 1.2 in Chapter IV shows that $[C_1, C_2]^{\pi}$ is a strict contraction if and only if Γ_R is a strict contraction.

Corollary 2.4 implies that T_3 is positive and singular if and only if

$$(I - [C_1^*, C_2^*][C_1, C_2]^{\pi}) = D_{\Gamma'} D_{\Gamma_R}^2 D_{\Gamma'}$$

is singular and Γ_R is a contraction. Since Γ' is a strict contraction T_3 is positive and singular if and only if Γ_R is a contraction and D_{Γ_R} is singular. This completes the proof.

4. LEVINSON SYSTEMS FOR 3 by 3 BLOCK MATRICES

In this section we will obtain the Levinson algorithm for the positive matrix T_3 in (3.2). Then we will use this algorithm to obtain the classical Levinson algorithm for block Toeplitz matrices.

Throughout this section, it is assumed that both T in (1.1) and T' in (3.1) are strictly positive. Moreover, F_*, E_*, F' and E' is the unique solution to the following Levinson systems:

$$\begin{bmatrix} A & B \\ B^* & C \end{bmatrix} \begin{bmatrix} I \\ -F_* \end{bmatrix} = \begin{bmatrix} E_* \\ 0 \end{bmatrix}$$

$$\begin{bmatrix} C & B' \\ B'^* & D \end{bmatrix} \begin{bmatrix} -F' \\ I \end{bmatrix} = \begin{bmatrix} 0 \\ E' \end{bmatrix}. \tag{4.1}$$

Because T and T' are both strictly positive, Corollaries 2.2 and 2.6 guarantee that both E_* and E' are invertible. By applying (2.5) to T' we have

$$F' = C^{-1/2}\Gamma'D^{1/2} \text{ and } E' = D^{1/2}D_{\Gamma'}^2 D^{1/2} \ . \tag{4.2}$$

The Levinson equations for T_3 are

$$T_3 \begin{bmatrix} -F_1 & I \\ -F_2 & -F_{1*} \\ I & -F_{2*} \end{bmatrix} = \begin{bmatrix} 0 & E_{3*} \\ 0 & 0 \\ E_3 & 0 \end{bmatrix} \ . \tag{4.3}$$

To solve this system in terms of the solution to (4.1), notice that (4.1) gives

$$T_3 \begin{bmatrix} 0 \\ -F' \\ I \end{bmatrix} = \begin{bmatrix} R - BF' \\ 0 \\ E' \end{bmatrix} \quad \text{and} \quad T_3 \begin{bmatrix} I \\ -F_* \\ 0 \end{bmatrix} = \begin{bmatrix} E_* \\ 0 \\ R^* - B^{'*}F_* \end{bmatrix}$$

Multiplying the second equation by $E_*^{-1}(R - BF')$ on the right and subtracting it from the first yields the following equation

$$T_3 \left[\begin{bmatrix} 0 \\ -F' \\ I \end{bmatrix} - \begin{bmatrix} I \\ -F_* \\ 0 \end{bmatrix} E_*^{-1}(R - BF') \right] = \begin{bmatrix} 0 \\ 0 \\ E' - (R^* - B^{'*}F_*)E_*^{-1}(R - BF') \end{bmatrix}$$

This and (4.3) gives

$$\begin{bmatrix} -F_1 \\ -F_2 \\ I \end{bmatrix} = \begin{bmatrix} 0 \\ -F' \\ I \end{bmatrix} - \begin{bmatrix} I \\ -F_* \\ 0 \end{bmatrix} E_*^{-1}(R - BF')$$

$$\tag{4.4}$$

$$E_3 = E' - (R^* - B^{'*}F_*)E_*^{-1}(R - BF') \ .$$

A similar calculation yields

$$\begin{bmatrix} I \\ -F_{1*} \\ -F_{2*} \end{bmatrix} = \begin{bmatrix} I \\ -F_* \\ 0 \end{bmatrix} - \begin{bmatrix} 0 \\ -F' \\ I \end{bmatrix} E'^{-1}(R^* - B^{'*}F_*)$$

$$\tag{4.5}$$

$$E_{3*} = E_* - (R - BF')E'^{-1}(R^* - B^{'*}F_*) \ .$$

Using (1.2), (3.3), (4.2) and (3.3), (2.10) respectively:

$$R - BF' = A^{1/2}D_{\Gamma^*}\Gamma_R D_{\Gamma'}D^{1/2} = \varepsilon_*^*\Gamma_R\varepsilon'$$

$$\tag{4.6}$$

$$R^* - B'^*F_* = D^{1/2}D_{\Gamma'}\Gamma_R^*D_{\Gamma^*}A^{1/2} = \varepsilon'^*\Gamma_R^*\varepsilon_* \ .$$

where ε_* and ε' are the operators defined by

$$\varepsilon_* = D_{\Gamma^*}A^{1/2} \text{ and } \varepsilon' = D_{\Gamma'}D^{1/2} \ . \tag{4.7}$$

Notice that (2.10) and (4.2) show that ε_* and ε' provide a factorization of E_* and E', that is,

$$E_* = \varepsilon_*^*\varepsilon_* \text{ and } E' = \varepsilon'^*\varepsilon' \ . \tag{4.8}$$

Using (4.6) to (4.8) in (4.4) and (4.5) produces the following Levinson recursion:

$$\begin{bmatrix} -F_1 \\ -F_2 \\ I \end{bmatrix} = \begin{bmatrix} 0 \\ -F' \\ I \end{bmatrix} - \begin{bmatrix} I \\ -F_* \\ 0 \end{bmatrix}\varepsilon_*^{-1}\Gamma_R\varepsilon'$$

$$\tag{4.9}$$

$$\begin{bmatrix} I \\ -F_{1*} \\ -F_{2*} \end{bmatrix} = \begin{bmatrix} I \\ -F_* \\ 0 \end{bmatrix} - \begin{bmatrix} 0 \\ -F' \\ I \end{bmatrix}\varepsilon'^{-1}\Gamma_R^*\varepsilon_*$$

This generalizes the Levinson recursion in (XV.6.14). Using (4.6) to (4.8) in (4.4) and (4.5) produces the following update of E_3, E_{3*} and ε'_3, ε_{3*}

$$E_3 = \varepsilon'^*D_{\Gamma_R}^2\varepsilon' \quad , \quad E_{3*} = \varepsilon_*^*D_{\Gamma_R^*}^2\varepsilon_* \quad , \quad \varepsilon'_3 = D_{\Gamma_R}\varepsilon' \text{ and } \varepsilon_{3*} = D_{\Gamma_R^*}\varepsilon_* \ . \tag{4.10}$$

The previous analysis is summarized in the following proposition, which is the Levinson algorithm moving from a 2 by 2 positive block matrix to a 3 by 3 positive block matrix. By redefining the block matrices this contains the general case.

4.1 PROPOSITION. *Assume that both* T *and* T' *are strictly positive and* F_*, F', ε_* *and* ε' *are the operators obtained from the Levinson system in (4.1), (4.7). Then*
(i) The set of all R *such that* T_3 *in (3.2) is positive is given by*

$$R = \varepsilon_*^*\Gamma_R\varepsilon' + BF' \tag{4.11}$$

where Γ_R *is a contraction from* $\overline{\varepsilon'D}$ *to* $\overline{\varepsilon_*H}$
(ii) The operator T_3 *in (3.2) is positive (strictly positive) if and only if*

$$\Gamma_R = \varepsilon_*^{*-1}(R - BF')\varepsilon'^{-1} \tag{4.12}$$

is a contraction (strict contraction), respectively.

(iii) The solution to the Levinson system for T_3 in (4.3) is given by (4.9), (4.10).

PROOF. Part (i) follows from (4.6) and Theorem 3.1. Part (ii) follows from (4.6) and Theorem 3.1 (Corollary 3.4), respectively. (Corollary 1.5 guarantees that both of the operators ε_* and ε' in (4.12) are invertible.) Finally, Part (iii) follows from (4.9) and (4.10). This completes the proof.

The previous Levinson algorithm reduces to the usual Levinson algorithm in Procedure XV.6.1 when T_3 is Toeplitz. To see this assume that $\{R_i\}_1^n$ on H generates a strictly positive Toeplitz matrix $\hat{R}_{n+1}$. We set

$$T = T' = \hat{R}_{n+1} \text{ and } C = \hat{R}_n \text{ and } A = D = I \text{ on } H$$

$$R = R_{n+1} \text{ and } \Gamma_R = \Gamma_{n+1} \text{ and } T_3 = \hat{R}_{n+2} . \tag{4.13}$$

In this case the Levinson system in (4.1) reduces to

$$\hat{R}_{n+1} \begin{bmatrix} F_n & I \\ I & F_{n*} \end{bmatrix} = \begin{bmatrix} 0 & E_{n*} \\ E_n & 0 \end{bmatrix} . \tag{4.14}$$

Here $-F' = F_n$ and $-F_* = F_{n*}$ are both n column matrices of the form:

$$-F' = F_n = [A_{n,0}, A_{n,1}, ..., A_{n,n-1}]^{tr}$$

$$-F_* = F_{n*} = [B_{n,1}, B_{n,2}, ..., B_{n,n}]^{tr} \tag{4.15}$$

where $A_{n,i}$ for $0 \le i < n$ and $B_{n,i}$ for $1 \le i \le n$ are operators on H. Since $\hat{R}_{n+1}$ is strictly positive Corollaries 2.2 and 2.6 guarantee that both E_n and E_{n*} are strictly positive. Finally, let $\varepsilon' = \varepsilon_n$ and $\varepsilon_* = \varepsilon_{n*}$ be factors of $E' = E_n$ and $E_* = E_{n*}$, that is,

$$E_n = \varepsilon_n^* \varepsilon_n \text{ and } E_{n*} = \varepsilon_{n*}^* \varepsilon_{n*} . \tag{4.16}$$

Corollaries 2.2 and 2.6 guarantee that both ε_n and ε_{n*} are invertible.

Equations (4.8), (4.9), (4.14), (4.15) and (4.16) give the following recursion:

$$F_{n+1} = [A_{n+1,0}, A_{n+1,1}, ..., A_{n+1,n}]^{tr} =$$

$$[0, A_{n,0}, A_{n,1}, ..., A_{n,n-1}]^{tr} - [I, B_{n,1}, B_{n,2}, ..., B_{n,n}]^{tr} \varepsilon_{n*}^{-1} \Gamma_{n+1} \varepsilon_n ,$$

$$(4.17)$$

$$F_{n+1*} = [B_{n+1,1}, B_{n+1,2}, ..., B_{n+1,n+1}]^{tr} =$$

$$[B_{n,1}, B_{n,2}, ..., B_{n,n}, 0]^{tr} - [A_{n,0}, A_{n,1}, ..., A_{n,n-1}, I]^{tr} \varepsilon_n^{-1} \Gamma_{n+1}^* \varepsilon_{n*} .$$

This recursion is precisely (XV.6.14). Equations (4.11), (4.13), (4.15) and

$$B = [R_1, R_2, ..., R_n]$$

give

$$\varepsilon_{n*}^* \Gamma_{n+1} \varepsilon_n = R_{n+1} + \sum_{i=1}^{n} R_i A_{n,i-1} , \qquad (4.18)$$

which is precisely (XV.6.15) in the Levinson algorithm. Finally, since $A = D = I$ Equation (4.10) and (4.16) produce the following update of ε_n and ε_{n*}

$$\varepsilon_{n+1} = D_{\Gamma_{n+1}} \varepsilon_n \text{ and } \varepsilon_{n+1*} = D_{\Gamma_{n+1}^*} \varepsilon_{n*} . \qquad (4.19)$$

Equation (4.17) to (4.19) form the Levinson algorithm in Procedure XV.6.1.

The initial conditions for (4.17) and (4.19) are obtained by solving the following Levinson system

$$\begin{bmatrix} I & R_1 \\ R_1^* & I \end{bmatrix} \begin{bmatrix} A_{1,0} & I \\ I & B_{1,1} \end{bmatrix} = \begin{bmatrix} 0 & E_{1*} \\ E_1 & 0 \end{bmatrix} .$$

Corollaries 2.4 and 2.8 or a direct calculation provide these initial conditions, that is,

$$\Gamma_1 = R_1 \text{ and } A_{1,0} = -R_1 \text{ and } B_{1,1} = -R_1^*$$

$$(4.20)$$

$$\text{and } \varepsilon_1 = D_{\Gamma_1} \text{ and } \varepsilon_{1*} = D_{\Gamma_1^*} .$$

Therefore the Levinson algorithm generated by (4.17) to (4.19) with the initial conditions in (4.20) is precisely the Levinson algorithm presented in Procedure XV.6.1. Both methods recursively produce the same choice sequence $\{\Gamma_i\}_1^n$ for $\{R_i\}_1^n$. Moreover, by Theorems 2.9 and 2.11 and (4.19), (4.20) we obtain the following result.

4.2 PROPOSITION. *Let* $\hat{R}_{n+1}$ *be the Toeplitz matrix generated by* $\{R_i\}_1^n$ *on H. Then* $\hat{R}_{n+1}$ *is strictly positive if and only if its choice sequence* $\{\Gamma_i\}_1^n$ *is strictly contractive.*

5. MORE RESULTS ON POSITIVE 3 by 3 BLOCK MATRICES

In this section we will study the extension problem for 3 by 3 matrices when R is on C^1, that is,

$$T_3 = \begin{bmatrix} A & B & R \\ B^* & C & B^{'} \\ R^* & B^{'*} & D \end{bmatrix} \text{ on } C^1 \oplus G \oplus C^1$$

(5.1)

$$B^* \alpha = \alpha b \text{ and } B^{'} \alpha = \alpha b^{'} \text{ for all } \alpha \text{ in } C^1$$

where b and $b^{'}$ are "column vectors" in G. In this case $A = [a]$ and $D = [d]$ and $R = [r]$ where a, d and r are all complex numbers in C^1. As before T on $C^1 \oplus G$ and $T^{'}$ on $G \oplus C^1$ are the operators defined in (1.1) and (3.1), respectively.

We being with the following result.

5.1 PROPOSITION. *The matrix T_3 in (5.1) is positive for an adequate choice of* r *if and only if both* T *and* $T^{'}$ *are positive. In this case, there exists an* α *in* C^1 *and a* $\tau \geq 0$ *such that the set of all positive* T_3 *(with the same* T *and* $T^{'}$*) is given by*

$$r = \alpha + \tau\rho \quad (with \ \rho \in \overline{D}).$$

(5.2)

Moreover, α *and* τ *are both uniquely determined by* T *and* $T^{'}$*.*

PROOF. The first assertion is a restatement of Theorem 3.1. The form of T_3 in (5.1) shows that the contraction Γ (uniquely determined by T) maps $\overline{CG}$ into C^1 and the contraction $\Gamma^{'}$ (uniquely determined by $T^{'}$) maps C^1 into $\overline{CG}$. Thus $A^{1/2}\Gamma\Gamma^{'}D^{1/2} = [\alpha]$ where α is a complex number. Obviously α is uniquely determined by T and $T^{'}$. Moreover, $D_{\Gamma^*} = [d_*]$ and $D_{\Gamma^{'}} = [d^{'}]$ where d_* and $d^{'}$ are complex numbers. Equation (3.3) in Theorem 3.1 produces (5.2) where $\Gamma_R = [\rho]$ and $\tau = \sqrt{a}\, d_* d^{'} \sqrt{d}$. Obviously τ is uniquely determined by T and $T^{'}$. This completes the proof.

5.2 REMARK. Assume that T and $T^{'}$ are both fixed positive matrices. The representation (5.2) expresses the fact that T_3 is positive if and only if r belongs to the closed disc $\alpha + \tau\overline{D}$ with center α and radius τ. In particular, this implies that α and τ are uniquely determined. Moreover, T_3 is is the unique positive operator of the form (5.1) expanding T and $T^{'}$ if and only if a = 0 or $d_* = 0$ or $d^{'} = 0$ or d = 0.

5.3 PROPOSITION. *Let* T *and* T$'$ *both be positive and assume that* C *is strictly positive. Then* $D_{\Gamma^*} = \{0\}(D_{\Gamma'} = \{0\})$ *if and only if* T(T$'$) *is singular, respectively.*

PROOF. If a = 0, then Γ^* is an isometry mapping $\{0\}$ into $\overline{CG}$. Thus $D_{\Gamma^*} = \{0\}$ and B = 0 and T is singular. A similar argument shows that if d = 0, then $D_{\Gamma'} = \{0\}$ and T$'$ is singular. So without loss of generality we assume that both a and d are nonzero. Now the proposition follows from Corollary 2.8 (Corollary 2.4) and the fact that $D_{\Gamma^*}(D_{\Gamma'})$ is scalar valued, that is, on C^1, respectively.

5.4 REMARK. If C is not strictly positive, then the conclusion of the previous proposition might not be true. Indeed if

$$a = 1,\ B = [.5,\ 0] \text{ and } C = \begin{bmatrix} 1 & 0 \\ 0 & 0 \end{bmatrix},$$

then T is singular and $\Gamma = .5$.

The following is a simple consequence of the previous proposition and Theorem 3.1.

5.5 COROLLARY. *Let* T *and* T$'$ *both be positive and assume that* C *is strictly positive. Then there exists a unique positive* T_3 *in* (5.1) *extending* T *and* T$'$ *if and only if* T *is singular or* T$'$ *is singular.*

5.6 COROLLARY. *Let* T *and* T$'$ *both be strictly positive. Then the operator* T_3 *in* *(5.1) extending* T *and* T$'$ *is positive and singular if and only if* $|\rho| = 1$ *where* $\Gamma_R = [\rho]$.

PROOF. This follows from Corollary 3.4 and the fact that $\Gamma_R = [\rho]$ is an operator on C^1.

6. SOME CLASSICAL RESULTS FOR TOEPLITZ MATRICES ON C^n REVISITED

In this section we will use the previous results to obtain some classical results for positive Toeplitz matrices on C^n.

Throughout this section $\{R_i\}_1^n$ is a sequence on C^1 and $\hat{R}_{n+1}$ is its corresponding Toeplitz on C^{n+1}. We begin with the following result.

6.1 PROPOSITION. *Let $\hat{R}_j$ be a fixed positive Toeplitz matrix on C^j for some $j \geq 1$. Then*

(i) There exists a constant R_j in C^1 such that $\hat{R}_{j+1}$ is a positive Toeplitz expansion of $\hat{R}_j$.

(ii) The set of all R_j such that $\hat{R}_{j+1}$ is a positive Toeplitz expansion of $\hat{R}_j$ is given by

$$R_j = \alpha_j + \tau_j \rho_j \quad (\text{with } \rho_j \in \overline{D}) \tag{6.1}$$

where α_j is in C^1 and $\tau_j \geq 0$ are constants which only depend on $\hat{R}_j$.

(iii) If $\hat{R}_j$ is strictly positive, then $\hat{R}_{j+1}$ is positive and singular if and only if $|\rho_j| = 1$.

(iv) There exists a unique positive Toeplitz expansion $\hat{R}_{j+1}$ of $\hat{R}_j$ if and only if $\hat{R}_j$ is singular.

(v) There exists a unique positive extension of $\{R_i\}_1^n$ if and only if the Toeplitz matrix $\hat{R}_{n+1}$ is positive and singular. In other words the scalar valued version of the Carathéodory interpolation Problem XV.3.1 has a unique solution if and only if $\hat{R}_{n+1}$ is positive and singular.

PROOF. Parts (i) and (ii) follows from Proposition 5.1 with $T = T' = \hat{R}_j$. Part (iii) follows from Corollary 5.6 with $T = T' = \hat{R}_j$. For part (iv) assume that the positive expansion $\hat{R}_{j+1}$ of $\hat{R}_j$ is unique. If $\hat{R}_j$ is strictly positive, then $C = \hat{R}_{j-1}$ is strictly positive. This produces a contradiction to Corollary 5.5. Therefore $\hat{R}_j$ is singular. On the other hand if $\hat{R}_j$ is singular, the proof of Theorem 3.2 in Chapter II shows that $\hat{R}_j$ admits a unique positive Toeplitz expansion $\hat{R}_{j+1}$.

For part (v) notice that if $\{R_i\}_1^n$ admits a unique positive extension, then $\hat{R}_{n+1}$ has a unique positive Toeplitz expansion $\hat{R}_{n+2}$. Part (iv) implies that $\hat{R}_{n+1}$ is positive and singular. On the other hand, if $\hat{R}_{n+1}$ is positive and singular, then Part (iv) implies that $\hat{R}_{n+1}$ admits a unique positive expansion $\hat{R}_{n+2}$. Since $\hat{R}_{n+1}$ is singular, Theorem 1.1 with $C = \hat{R}_{n+1}$ and $A = 1$ implies that $\hat{R}_{n+2}$ is singular. (If $Cg = 0$, then $B = \Gamma C^{1/2} g = 0$. Thus $T(0 \oplus g) = 0$ and T is singular.) By Part (iv) $\hat{R}_{n+2}$ admits a unique positive Toeplitz expansion $\hat{R}_{n+3}$. Moreover, $\hat{R}_{n+3}$ is singular. Continuing in this fashion $\hat{R}_{n+1}$ admits a unique positive Toeplitz expansion $\hat{R}_{n+k}$ for all $k \geq 1$. This completes the proof.

Assume that the Toeplitz matrix $\hat{R}_n$ formed by $\{R_i\}_1^{n-1}$ on C^n is positive. Proposition 6.1 part (ii) implies that each R_j is in the closed disc $\alpha_j + \tau_j \overline{D}$ with center α_j and radius τ_j. The first "data" point R_1 is in $\overline{D}$, that is, $\alpha_1 = 0$ and $\tau_1 = 1$. By convention we set $\rho_j = 0$ if $\tau_j = 0$ in (6.1). Recall that α_j and τ_j are uniquely determined by $\hat{R}_j$ (see

Proposition 6.1). This and $R_1 = \rho_1$ implies that the complex numbers $\{\rho_j\}_1^{n-1}$ in (6.1) are uniquely determined by $\hat{R}_n$. In other words the set $\{R_j\}_1^{n-1}$ uniquely determines a set of $n-1$ complex numbers $\{\rho_j\}_1^{n-1}$ in $\overline{D}$ by (6.1). These numbers $\{\rho_j\}_1^{n-1}$ are called the *choice sequence* for $\hat{R}_n$. We are now ready for the following result.

6.2 PROPOSITION. *Let $\hat{R}_n$ be a positive Toeplitz matrix on $\mathbf{C}^n$ with choice sequence $\{\rho_i\}_1^{n-1}$. Then*

(i) The matrix $\hat{R}_n$ is a strict contraction if and only if $|\rho_j| < 1$ for all $1 \le j < n$.

(ii) The matrix $\hat{R}_n$ is singular if and only if there exists an integer j such that

$$|\rho_i| < |\rho_j| = 1 \text{ for } 1 \le i < j, \text{ and } \rho_i = 0 \text{ for } j < i < n . \tag{6.2}$$

PROOF. Obviously $\hat{R}_j$ is a strictly positive for all $j \le n$ if and only if $\hat{R}_n$ is a strictly positive. Thus (i) follows from Proposition 6.1 part (iii). For part (ii) assume that $\hat{R}_n$ is singular. Let $j + 1$ be the first integer such that $\hat{R}_{j+1}$ is singular. Then (i) and Proposition 6.1 part (iii) implies that $|\rho_i| < |\rho_j| = 1$ for $1 \le i < j$. By Proposition 6.1 part (iv) the Toeplitz matrix $\hat{R}_{j+1}$ has a unique positive Toeplitz expansion. (This expansion must be $\hat{R}_n$). Hence $\tau_i = 0$ for $i > j$. By our convention $\rho_i = 0$ for $i > j$ and (6.2) holds. On the other hand if (6.2) holds, then (i) and Proposition 6.1 part (iii) implies that $\hat{R}_{j+1}$ is singular. So $\hat{R}_n$ is singular and the proof is complete.

Using $\alpha_1 = 0$ and $\tau_1 = 1$ one can show that there is a one to one correspondence between the set of all positive Toeplitz matrices $\hat{R}_n$ and the set of all choice sequences $\{\rho_j\}_1^{n-1}$ satisfying either $|\rho_j| < 1$ for all $1 \le j < n$, or (6.2). We have already given several algorithm to recursively calculate this choice sequence. The following shows that the set of Schur numbers for a Toeplitz matrix defined in Section II.4 is precisely its choice sequence.

6.3 COROLLARY. *Let $\hat{R}_n$ be a positive Toeplitz matrix on $\mathbf{C}^n$. Let $\{\rho_i\}_1^{n-1}$ be the choice sequence for $\hat{R}_n$ and $\{r_i\}_1^{n-1}$ be the Schur numbers associated with $\hat{R}_n$. Then $\rho_i = r_i$ for $1 \le i < n$.*

PROOF. We proceed by induction. Since $\rho_1 = R_1 = r_1$ we assume that $\rho_i = r_i$ for $1 \le i < j$ and $j > 1$. According to Chapter II the set of all R_j such that $\hat{R}_{j+1}$ is positive is given by a closed disc $c_j + \gamma_j \overline{D}$ with center c_j and radius γ_j. (This follows by substituting (I.3.16) into the first equation of (II.1.6); see also Remark II.2.5.) By Part (iii) of Proposition 6.1 the set of all R_j such that $\hat{R}_{j+1}$ is positive is given by $\alpha_j + \tau_j \overline{D}$ with center α_j and radius τ_j. In both cases R_j describes the same set $\alpha_j + \tau_j \overline{D} = c_j + \gamma_j \overline{D}$. Since the center and radius of a closed disc is uniquely determined, $\alpha_j = c_j$ and $\tau = \gamma_j$ for $1 \leq j < n$. Recalling that $R_j = \alpha_j + \tau_j \rho_j = c_j + \gamma_j r_j$ yields $r_j = \rho_j$ for all j. This completes the proof.

7. POSITIVE n by n BLOCK MATRICES

In this section we will present a complete characterization of all n by n block positive matrices. It is shown that there is a one to one correspondence between the set of all n by n positive block matrices, and n positive operators with $n(n-1)/2$ contractions.

Throughout T_n is the positive matrix defined by

$$T_n = \begin{bmatrix} A_{1,1} & A_{1,2} & \cdots & A_{1,n} \\ A_{1,2}^* & A_{2,2} & \cdots & A_{2,n} \\ \cdot & & & \\ \cdot & & & \\ \cdot & & & \\ A_{1,n}^* & A_{2,n}^* & \cdots & A_{n,n} \end{bmatrix} \quad \text{on } \overset{n}{\underset{1}{\oplus}} H_i \tag{7.1}$$

and $\Gamma_{i,j}$ for $1 \leq i < j$ and $2 \leq j \leq n$ are a set of contractions satisfying

$$\Gamma_{j-1,j} : \overline{A_{j,j}H_j} \to \overline{A_{j-1,j-1}H_{j-1}} \quad (\text{for } 2 \leq j \leq n)$$

$$\Gamma_{i,j} : D_{i+1,j} \to D_{i,j-1*} \quad (\text{for } 1 \leq i < j-1 \text{ and } 3 \leq j \leq n) \tag{7.2}$$

$$D_{i,j} = D_{\Gamma_{i,j}} \text{ and } D_{i,j*} = D_{\Gamma_{i,j}^*} \text{ and } D_{i,j} = D_{\Gamma_{i,j}} \text{ and } D_{i,j*} = D_{\Gamma_{i,j}^*} .$$

The index on i,j may seem a bit funny. However, the contraction $\Gamma_{i,j}$ will correspond to the $A_{i,j}$ operator in T_n. Let Q_{j*} be the column matrix contraction defined by

$$Q_{j*} = [\Gamma_{1,j}D_{2,j} \cdots D_{j-1,j}, \Gamma_{2,j}D_{3,j} \cdots D_{j-1,j}, ..., \Gamma_{j-2,j}D_{j-1,j}, \Gamma_{j-1,j}]^{tr}$$

$$\tag{7.3}$$

$$Q_{j*}^* = [D_{j-1,j}D_{j-2,j} \cdots D_{2,j}\Gamma_{1,j}^*, D_{j-1,j}D_{j-2,j} \cdots D_{3,j}\Gamma_{2,j}^*, ..., D_{j-1,j}\Gamma_{j-2,j}^*, \Gamma_{j-1,j}^*] .$$

Corollary 1.3 in Chapter XIII guarantees that Q_{j*} is a contraction. The operator Q_{j*} will correspond to the j-th column of T_n. We begin with the following result which will be used in obtaining the lower triangular factorization Δ_{n*} of T_n.

7.1 LEMMA. *Let $\Gamma_{i,j}$ be a set of contractions satisfying (7.2). The equation*

$$
D_{Q_{j}^{\cdot}} W_{j^{\cdot}} = \begin{bmatrix}
D_{1,j^{*}} & -\Gamma_{1,j}\Gamma_{2,j}^{*} & -\Gamma_{1,j}D_{2,j}\Gamma_{3,j}^{*} & \cdots & -\Gamma_{1,j}D_{2,j}D_{3,j}\cdots D_{j-2,j}\Gamma_{j-1,j}^{*} \\
0 & D_{2,j^{*}} & -\Gamma_{2,j}\Gamma_{3,j}^{*} & \cdots & -\Gamma_{2,j}D_{3,j}D_{4,j}\cdots D_{j-2,j}\Gamma_{j-1,j}^{*} \\
0 & 0 & D_{3,j} & \cdots & -\Gamma_{3,j}D_{4,j}D_{5,j}\cdots D_{j-2,j}\Gamma_{j-1,j}^{*} \\
0 & 0 & 0 & \cdots & -\Gamma_{4,j}D_{5,j}D_{6,j}\cdots D_{j-2,j}\Gamma_{j-1,j}^{*} \\
\cdot & \cdot & \cdot & \cdots & \cdot \\
\cdot & \cdot & \cdot & \cdots & \cdot \\
\cdot & \cdot & \cdot & \cdots & \cdot \\
0 & 0 & 0 & \cdots & D_{j-1,j^{*}}
\end{bmatrix} \quad (7.4)
$$

defines a unitary operator $W_{j^{*}}$ *mapping* $D_{1,j^{*}} \oplus D_{2,j^{*}} \oplus \cdots \oplus D_{j-1,j^{*}}$ *onto* $D_{Q_{j}^{\cdot}}$.

PROOF. We prove this result for the case when $j = 4$. The general case follows by exactly the same reasoning. By rearranging the rows and columns in Corollary 1.4 in Chapter XIII, the equation

$$
W_{j^{*}}^{*}D_{Q_{j}^{\cdot}} = \begin{bmatrix}
D_{1,4^{*}} & 0 & 0 \\
-\Gamma_{2,4}\Gamma_{1,4}^{*} & D_{2,4^{*}} & 0 \\
-\Gamma_{3,4}D_{2,4}\Gamma_{1,4}^{*} & -\Gamma_{3,4}\Gamma_{2,4}^{*} & D_{3,4^{*}}
\end{bmatrix}
$$

defines a unitary operator $W_{j^{*}}^{*}$ mapping $D_{Q_{j}^{\cdot}}$ onto $D_{1,4^{*}} \oplus D_{2,4^{*}} \oplus D_{3,4^{*}}$. Taking the adjoint yields (7.4) for $j = 4$. This completes the proof.

The upper triangular operator $\Delta_{j^{*}}$ mapping $D_{1,j^{*}} \oplus \cdots \oplus D_{j-1,j^{*}} \oplus \overline{A_{j,j}H_{j}}$ into $\overset{j}{\underset{1}{\oplus}} H_{i}$ is defined by the difference equation:

$$
\Delta_{j^{*}} = \begin{bmatrix} \Delta_{j-1} & 0 \\ 0 & I \end{bmatrix} \begin{bmatrix} D_{Q_{j}^{\cdot}} W_{j^{*}} & Q_{j^{*}} \\ 0 & A_{j,j}^{1/2} \end{bmatrix} \qquad (2 \leq j \leq n) \qquad (7.5)
$$

with the initial condition

$$
\Delta_{1^{*}} = A_{1,1}^{1/2} \text{ and } \Delta_{2^{*}} = \begin{bmatrix} A_{1,1}^{1/2} & 0 \\ 0 & I \end{bmatrix} \begin{bmatrix} D_{1,2^{*}} & \Gamma_{1,2} \\ 0 & A_{2,2}^{1/2} \end{bmatrix}. \qquad (7.6)
$$

Notice that $D_{Q_{j}^{\cdot}} W_{j^{*}}$ in (7.5) is given by the right hand side of (7.4). This sets the stage for the following result

7.2 THEOREM. *The matrix* T_{n} *in (7.1) is positive if and only if* $A_{i,i}$ *for* $1 \leq i \leq n$ *are all positive and there exists a set of contractions* $\Gamma_{i,j}$ *of the form (7.2) satisfying*

$$\Delta_{j-1*} Q_{j*} A_{j,j}^{1/2} = [A_{1,j}, A_{2,j}, ..., A_{j-1,j}]^{tr} \quad (2 \le j \le n) \tag{7.7}$$

Moreover, the equation

$$\Delta_{n*} = T_n^{1/2} \phi_{n*} \tag{7.8}$$

defines a unitary operator ϕ_{n*} mapping $\overline{\mathrm{ran}(\Delta_{n*}^*)}$ onto $\overline{\mathrm{ran} T_n}$.

PROOF. Corollary 1.2 shows that

$$\Delta_{2*} = \begin{bmatrix} A_{1,1}^{1/2} & 0 \\ 0 & A_{2,2}^{1/2} \end{bmatrix} \begin{bmatrix} D_{1,2*} & \Gamma_{1,2} \\ 0 & I \end{bmatrix} = T_2^{1/2} \phi_{2*}$$

defines a unitary operator ϕ_2, or equivalently, Δ_{2*}^* is a lower triangular factorization of $T_2 (= \Delta_{2*} \Delta_{2*}^*)$. This and Theorem 1.1 with $A = T_2$ and $C = A_{3,3}$ shows that T_3 is positive if and only if there exists a contraction Q_{3*} satisfying

$$\Delta_{2*} Q_{3*} A_{3,3}^{1/2} = [A_{1,3}, A_{2,3}]^{tr} = B .$$

Lemma 1.1 in Chapter XIII yields the form of Q_{3*} in (7.3) for $j = 3$. Applying Corollary 1.2 along with (7.6) and Lemma 7.1 we see that

$$T_3^{1/2} \phi_{3*} = \begin{bmatrix} \Delta_{2*} & 0 \\ 0 & I \end{bmatrix} \begin{bmatrix} D_{Q_{3*}^*} W_{3*} & Q_{3*} \\ 0 & A_{3,3}^{1/2} \end{bmatrix} = \Delta_{3*}$$

defines a unitary operator ϕ_{3*}, that is, Δ_{3*}^* is a lower triangular factorization of T_3. Thus (7.8) holds for $n = 3$.

Now we use induction and assume that the theorem is true for $n-1$. Choosing $A = T_{n-1}$ and $C = A_{n,n}$, Theorem 1.1 shows that T_n is positive if and only if there exists a contraction $Q_n (= \phi_{n-1*}^* \Gamma)$ such that (7.7) holds for $j = n$. As before Lemma XIII.1.1 yields the form of Q_{n*} in (7.3). Applying Corollary 1.2 along with (7.8) for $j = n-1$ and Lemma 7.1 we have

$$T_n^{1/2} \phi_{n*} = \begin{bmatrix} \Delta_{n-1*} & 0 \\ 0 & I \end{bmatrix} \begin{bmatrix} D_{Q_n^*} W_{n*} & Q_{n*} \\ 0 & A_{n,n}^{1/2} \end{bmatrix} = \Delta_{n*}$$

where ϕ_{n*} is unitary. This completes the proof.

The proof of the previous theorem along with Corollary 1.5 and Corollary XIII.1.2 give the following result.

7.3 COROLLARY. *Let* T_n *be the positive operator in (7.1) and* $\Gamma_{i,j}$ *in (7.2) its corresponding contractions defined according to (7.7). Then* T_n *is strictly positive if and only if* $A_{i,i}$ *are strictly positive for all* $1 \leq i \leq n$ *and all* $\Gamma_{i,j}$ *are strict contractions.*

7.4 REMARK. The previous theorem shows that any n by n positive block matrix is uniquely determined by n positive operators $\{A_{i,i}\}_1^n$ and $n(n-1)/2$ contractions $\Gamma_{i,j}$.

7.5 REMARK. Let T_n in (7.1) be positive. Then Δ_{n*} in (7.8) obtained recursively from (7.5) and (7.7) provides a lower triangular factorization of T_n, that is, Δ_{n*}^* is lower triangular and $\Delta_{n*}\Delta_{n*}^* = T_n$. Finally, it is noted that one can always obtain a lower triangular factorization L^*L of T_n by solving for the entries of L in $L^*L = T_n$ starting with the southeast corner. However, the lower triangular factorization Δ_{n*}^* of T_n displays the role of the contractions $\Gamma_{i,j}$ in a lower triangular factorization.

8. DUALITY AND POSITIVE n by n BLOCK MATRICES

This section is the dual of the previous section. Here we present a complete characterization of all n by n block positive matrices, by obtaining an upper triangular factorization. As before, it is shown that there is a one to one correspondence between the set of all n by n block positive matrices, and n positive operators with $n(n-1)/2$ contractions. Throughout this section T_n is the positive matrix defined by

$$T_n = \begin{bmatrix} A_{n,n} & A_{n,n-1} & \cdots & A_{n,2} & A_{n,1} \\ \cdot & \cdot & \cdots & \cdot & \cdot \\ \cdot & \cdot & \cdots & \cdot & \cdot \\ \cdot & \cdot & \cdots & \cdot & \cdot \\ A_{2,n} & A_{2,n-1} & \cdots & A_{2,2} & A_{2,1} \\ A_{1,n} & A_{1,n-1} & \cdots & A_{1,2} & A_{1,1} \end{bmatrix} \text{ on } \overset{1}{\underset{n}{\oplus}} H_i \qquad (8.1)$$

where $A_{i,j} = A_{j,i}^*$. We have rearranged the indices on the $A_{i,j}$ terms to simplify some of the following notation.

Throughout this section $\Gamma_{i,j}$ for $2 \leq i \leq n$ and $1 \leq j < i$ are a set of contractions satisfying

$$\Gamma_{j+1,j} : \overline{A_{j,j}H_j} \to \overline{A_{j+1,j+1}H_{j+1}} \qquad \text{(for } 1 \leq j < n\text{)}$$

$$\Gamma_{i,j} : D_{i-1,j} \to D_{i,j+1*} \quad \text{(for } 3 \leq i \leq n \text{ and } 1 \leq j < i-1\text{)} \qquad (8.2)$$

$$D_{i,j} = D_{\Gamma_{i,j}} \text{ and } D_{i,j*} = D_{\Gamma_{i,j}^*} .$$

As before the contraction $\Gamma_{i,j}$ will correspond to $A_{i,j}$ in T_n. Let Q_i be the row matrix defined by

$$Q_i = [\Gamma_{i,i-1}, D_{i,i-1*}\Gamma_{i,i-2}, D_{i,i-1*}D_{i,i-2*}\Gamma_{i,i-3}, ..., D_{i,i-1*}D_{i,i-2*} \cdots D_{i,2*}\Gamma_{i,1}] \quad (8.3)$$

Corollary 1.4 in Chapter XIII shows that the equation

$$W_i D_{Q_i} = \begin{bmatrix} D_{i,i-1} & -\Gamma_{i,i-1}^*\Gamma_{i,i-2} & \cdots & -\Gamma_{i,i-1}^*D_{i,i-2*} \cdots D_{i,2*}\Gamma_{i,1} \\ 0 & D_{i,i-2} & \cdots & -\Gamma_{i,i-2}^*D_{i,i-3*} \cdots D_{i,2*}\Gamma_{i,1} \\ 0 & 0 & \cdots & -\Gamma_{i,i-3}^*D_{i,i-4*} \cdots D_{i,2*}\Gamma_{i,1} \\ 0 & 0 & \cdots & -\Gamma_{i,i-4}D_{i,i-5*} \cdots D_{i,2*}\Gamma_{i,1} \\ \cdot & \cdot & \cdots & \cdot \\ \cdot & \cdot & \cdots & \cdot \\ \cdot & \cdot & \cdots & \cdot \\ 0 & 0 & \cdots & D_{i,1} \end{bmatrix} \quad (8.4)$$

defines a unitary operator W_i mapping D_{Q_i} into $D_{i,i-1*} \oplus \cdots \oplus D_{i,1*}$. Equation (8.4) will be used in obtaining the upper triangular factorization Δ_n of T_n.

The upper triangular Δ_i is defined by the following difference equation:

$$\Delta_i = \begin{bmatrix} A_{i,i}^{1/2} & Q_i \\ 0 & W_i D_{Q_i} \end{bmatrix} \begin{bmatrix} I & 0 \\ 0 & \Delta_{i-1} \end{bmatrix} \quad (8.5)$$

with the initial condition

$$\Delta_1 = A_{1,1}^{1/2} \text{ and } \Delta_2 = \begin{bmatrix} A_{2,2}^{1/2} & \Gamma_{2,1}A_{1,1}^{1/2} \\ 0 & D_{2,1}A_{1,1}^{1/2} \end{bmatrix} \quad (8.6)$$

Notice that $W_i D_{Q_i}$ in (8.5) is given by the right hand side of (8.4). This sets the stage for the following result which is the dual of Theorem 7.2.

8.1 THEOREM. *The matrix T_n in (8.1) is positive if and only if $A_{i,i}$ for $1 \le i \le n$ are all positive and there exists a set of contractions $\Gamma_{i,j}$ of the form (8.2) satisfying*

$$A_{i,i}^{1/2}Q_i\Delta_{i-1} = [A_{i,i-1}, A_{i,i-2}, ..., A_{i,1}] \quad (2 \le i \le n) \quad (8.7)$$

Moreover, the equation

$$\Delta_n = \phi_n T_n^{1/2} \tag{8.8}$$

defines a unitary operator ϕ_n from $\overline{\mathrm{ran}T_n}$ onto $\overline{\mathrm{ran}\Delta_n}$.

PROOF. Corollary 1.2 shows that

$$\Delta_2 = \begin{bmatrix} I & \Gamma_{2,1} \\ 0 & D_{2,1} \end{bmatrix} \begin{bmatrix} A_{2,2}^{1/2} & 0 \\ 0 & A_{1,1}^{1/2} \end{bmatrix} = \phi_2 T_2^{1/2}$$

defines a unitary operator ϕ_2. This and Theorem 1.1 with $A = A_{3,3}$ and $C = T_2$ shows that T_3 is positive if and only if there exists a contraction $Q_3 (= \Gamma \phi_2^*)$ satisfying

$$A_{3,3}^{1/2} Q_3 \Delta_2 = [A_{3,2}, A_{3,1}] = B .$$

Lemma 1.1 in Chapter XIII yields the form of Q_3 in (8.3) for $i = 3$. Applying Corollary 1.2 along with (8.4) and (8.5) we have

$$\phi_3 T_3^{1/2} = \begin{bmatrix} A_{3,3}^{1/2} & Q_3 \\ 0 & W D_{Q_3} \end{bmatrix} \begin{bmatrix} I & 0 \\ 0 & \Delta_2 \end{bmatrix} = \Delta_3$$

defines a unitary operator ϕ_3. This produces (8.8) for $n = 3$. As in the proof of Theorem 7.2 an induction argument finishes the proof.

The proof of the previous theorem along with Corollary 1.5 and Corollary XIII.1.2 give the following result.

8.2 COROLLARY. *Let T_n be the positive operator in (8.1) and $\Gamma_{i,j}$ in (8.2) its corresponding contractions defined according to (8.7). Then T_n is strictly positive if and only if $A_{i,i}$ are strictly positive for all $1 \leq i \leq n$ and all $\Gamma_{i,j}$ are strict contractions.*

8.3 REMARK. The previous theorem shows that any n by n positive block matrix is uniquely determined by n positive operators $\{A_{i,i}\}_1^n$ and $n(n-1)/2$ contractions $\Gamma_{i,j}$.

8.4 REMARK. Let T_n in (8.1) be positive. Then Δ_n in (8.8) obtained recursively from (8.5) and (8.7) provides an upper triangular factorization of T_n, that is, Δ_n is upper triangular and $\Delta_n^* \Delta_n = T_n$. As in Remark 7.5 one can always obtain an upper triangular factorization $U^* U$ of T_n by solving for the entries of the upper triangular matrix U in $U^* U = T_n$. However, the upper triangular factorization Δ_n of T_n displays the role of the contractions $\Gamma_{i,j}$ in an upper triangular factorization of T_n.

9. CHOICE SEQUENCES AND POSITIVE BLOCK TOEPLITZ MATRICES REVISITED

The previous section shows that there is a one to one correspondence between the set of all n by n positive block matrices and n positive operators with $n(n-1)/2$ contractions. Recall that an n by n positive block Toeplitz matrix (with $R_o = I$) is uniquely determined by a choice sequence of length $n-1$. In this section we use the ideas in the previous section to present another proof of this result.

To this end let $\hat{R}_n$ be the positive Toeplitz matrix in (XV.1.2) generated by $\{R_i\}_1^{n-1}$. Let $\{\Gamma_i\}_1^{n-1}$ be a choice sequence initiated on H. As before, let $D_i = D_{\Gamma_i}$ and $D_{i*} = D_{\Gamma_i^*}$ and $D_i = D_{\Gamma_i}$ and $D_{i*} = D_{\Gamma_i^*}$ for all i. The row matrix Q_{i+1} is defined by

$$Q_{i+1} = [\Gamma_1, D_{1*}\Gamma_2, D_{1*}D_{2*}\Gamma_3, ..., D_{1*} \cdots D_{i-1*}\Gamma_i] . \tag{9.1}$$

Corollary 1.4 in Chapter XIII implies that there exists a unitary operator W_{i+1} mapping $D_{Q_{i+1}}$ onto $D_1 \oplus D_2 \oplus \cdots \oplus D_i$ satisfying

$$W_{i+1}D_{Q_{i+1}} = \begin{bmatrix} D_1 & -\Gamma_1^*\Gamma_2 & \cdots & -\Gamma_1^*D_{2*}\cdots D_{i-1*}\Gamma_i \\ 0 & D_2 & \cdots & -\Gamma_2^*D_{3*}\cdots D_{i-1*}\Gamma_i \\ 0 & 0 & \cdots & -\Gamma_3^*D_{4*}\cdots D_{i-1*}\Gamma_i \\ . & . & \cdots & . \\ . & . & \cdots & . \\ . & . & \cdots & . \\ 0 & 0 & \cdots & D_i \end{bmatrix} \tag{9.2}$$

In this setting the upper triangular matrix Δ_i is defined by the following difference equation:

$$\Delta_i = \begin{bmatrix} I & Q_i \\ 0 & W_iD_{Q_i} \end{bmatrix} \begin{bmatrix} I & 0 \\ 0 & \Delta_{i-1} \end{bmatrix} \tag{9.3}$$

with the initial condition

$$\Delta_1 = I \quad \text{and} \quad \Delta_2 = \begin{bmatrix} I & \Gamma_1 \\ 0 & D_1 \end{bmatrix} . \tag{9.4}$$

The matrix $W_iD_{Q_i}$ is given by the right hand side of (9.2) where $i-1$ replaces i. We begin with the following useful result.

9.1 LEMMA. *Let Δ_i be the upper triangular matrix in (9.3), (9.4) generated by a choice sequence $\{\Gamma_i\}_1^\infty$ initiated on H. Then Δ_i admits an upper triangular* *decomposition of the form:*

$$\Delta_i = \begin{bmatrix} \Delta_{i-1} & x_i \\ 0 & L_i \end{bmatrix} \quad (i \geq 2).$$

(9.5)

Moreover, x_i from H to $H \oplus D_1 \oplus \cdots \oplus D_{i-2}$ and L_i from H to D_{i-1} are the operators recursively computed by the following difference equation

$$\begin{bmatrix} x_{i+1} \\ L_{i+1} \end{bmatrix} = x'_{i+1} = \begin{bmatrix} Q_{i+1} \\ W_{i+1}D_{Q_{i+1}} \end{bmatrix} \begin{bmatrix} x_i \\ L_i \end{bmatrix} \quad (i \geq 2)$$

(9.6)

where x_{i+1} is the operator from H to $H \oplus D_1 \oplus \ldots D_{i-1}$ contained in the first i components of x'_{i+1} and L_{i+1} is the operator from H to D_i contained in the last component of x'_{i+1} with the initial conditions $(x_1 = I, L_1 = 0)$ $x_2 = \Gamma_1$ and $L_2 = D_1$.

PROOF. Equation (9.4) shows that (9.5) holds for $i = 2$. Now we use induction and assume that (9.5) holds for i. By (9.2), (9.3) and (9.5) we have

$$\Delta_{i+1} = \begin{bmatrix} I & Q_{i+1} \\ 0 & W_{i+1}D_{Q_{i+1}} \end{bmatrix} \begin{bmatrix} I & 0 & 0 \\ 0 & \Delta_{i-1} & x_i \\ 0 & 0 & L_i \end{bmatrix} =$$

$$\begin{bmatrix} I & Q_i & * \\ 0 & W_iD_{Q_i} & * \\ 0 & 0 & * \end{bmatrix} \begin{bmatrix} I & 0 & 0 \\ 0 & \Delta_{i-1} & 0 \\ 0 & 0 & 0 \end{bmatrix} + \begin{bmatrix} I & Q_{i+1} \\ 0 & W_{i+1}D_{Q_{i+1}} \end{bmatrix} \begin{bmatrix} 0 & 0 & 0 \\ 0 & 0 & x_i \\ 0 & 0 & L_i \end{bmatrix} = \begin{bmatrix} \Delta_i & x_{i+1} \\ 0 & L_{i+1} \end{bmatrix}.$$

The last equality follows from (9.3) and the definition of x_{i+1} and L_{i+1} in (9.6). (The top row of zeros in the sixth matrix multiplies the first column in the fifth matrix.) This completes the induction and the proof.

The main result of this section is the following result.

9.2 THEOREM. *Let $\hat{R}_{n+1}$ be the Toeplitz matrix generated by the sequence $\{R_i\}_1^n$ on H. Then $\hat{R}_{n+1}$ is positive if and only if there exists a choice sequence $\{\Gamma_i\}_1^n$ initiated on H satisfying $\Gamma_1 = R_1$ and*

$$[R_1, R_2, ..., R_j] = Q_{j+1}\Delta_j \quad (1 \leq j \leq n).$$

(9.7)

Moreover, the equation

$$\alpha_j \hat{R}_j^{1/2} = \Delta_j \quad (1 \leq j \leq n+1)$$

(9.8)

defines a unitary operator α_j mapping $\overline{\mathrm{ran}\hat{R}_j}$ onto $\overline{\mathrm{ran}\Delta_j}$.

PROOF. Applying Theorem 1.1 to $\hat{R}_2$ yields the first choice contraction $\Gamma = R_1 = \Gamma_1$. Corollary 1.2 with $T = \hat{R}_2$ and Δ_2 in (9.4) proves (9.8) for $j = 2$. Applying Theorem 1.1 again with $A = I$ and $C = \hat{R}_2$ we see that $\hat{R}_3$ is positive if and only if there exists a contraction $Q_3 = [C_1, C_2]$ such that

$$B = [R_1, R_2] = Q_3\Delta_2 = [C_1, C_2] \begin{bmatrix} I & \Gamma_1 \\ 0 & D_1 \end{bmatrix}$$

This and $R_1 = \Gamma_1$ readily yields $C_1 = \Gamma_1$. By Corollary 1.3 in Chapter XIII, $\hat{R}_3$ is positive if and only if $C_2 = D_{1*}\Gamma_2$ where Γ_2 is a contraction mapping D_1 into D_{1*}, or equivalently, (9.7) holds for $j = 2$.

Now we use induction and assume that $\hat{R}_j$ is positive and (9.7), (9.8) holds for j. By Theorem 1.1 along with (9.8), $A = I$ and $C = \hat{R}_j$ and B in (9.7), we see that $\hat{R}_{j+1}$ is positive if and only if Q_{j+1} is a contraction, or equivalently, by Corollary 1.3 in Chapter XIII the operators $\{\Gamma_i\}_1^j$ forms a choice sequence. In this case Corollary 1.2 along with (9.2), (9.3) and (9.8) give

$$\alpha_{j+1}\hat{R}_{j+1}^{1/2} = \begin{bmatrix} I & Q_{j+1}\Delta_j \\ 0 & W_{j+1}D_{Q_{j+1}}\Delta_j \end{bmatrix} = \Delta_{j+1}$$

where α_{j+1} is unitary. This proves (9.8) for $j+1$.

By Theorem 1.1 and (9.8) with $A = I$ and $C = \hat{R}_{j+1}$ we see that $\hat{R}_{j+2}$ is positive if and only if there exists a contraction $Q_{j+2} = [Q_{j+1}, C_{j+1}]$ satisfying

$$B = [R_1, R_2, ..., R_{j+1}] = Q_{j+2}\Delta_{j+1} =$$

$$[Q_{j+1}, C_{j+1}] \begin{bmatrix} \Delta_j & x_{j+1} \\ 0 & L_{j+1} \end{bmatrix} = [[R_1, R_2, ..., R_j], R_{j+1}]$$

The third equality follows from (9.5). The last equality follows from the induction hypothesis in (9.7). This shows that Q_{j+2} must be in the form (9.1) with i replaced by $j + 1$. Corollary 1.3 in Chapter XIII implies that $Q_{j+2} = [Q_{j+1}, C_{j+1}]$ is a contraction if and only if $C_{j+1} = D_{1*} \cdots D_{j*}\Gamma_{j+1}$ where Γ_{j+1} is a contraction mapping D_j into D_{j*}. Therefore $\hat{R}_{j+2}$ is positive if and only if (9.7) holds for $j + 1$. This completes the induction on (9.7) and finishes the proof.

As before the sequence of contractions $\{\Gamma_i\}_1^n$ obtained in Theorem 9.2 from a positive Toeplitz matrix $\hat{R}_{n+1}$ is called the *choice sequence associated with* $\hat{R}_{n+1}$. The previous theorem also shows that there is a one to one correspondence between the set of all positive Toeplitz matrices $\hat{R}_{n+1}$ on $l^2_{n+1}(H)$ and the set of all choice sequences $\{\Gamma_i\}_1^n$ initiated on H.

Notice that

$$Q_{n+1} = [Q_n, \; D_{1*}D_{2*} \; \cdots \; D_{n-1*}\Gamma_n]$$

Using (9.5), (9.6) and (9.7):

$$R_n = Q_{n+1}\Delta_n[0, \, 0, \, ..., \, 0, \, I]^* = Q_n x_n + D_{1*}D_{2*} \; \cdots \; D_{n-1*}\Gamma_n L_n \; .$$

Equation (9.6) with the form of $W_{i+1}D_{Q_{i+1}}$ in (9.2) shows that

$$L_n = D_{n-1}D_{n-2} \; \cdots \; D_1 \; .$$

This along with the previous theorem proves the following result.

9.3 COROLLARY. *Let $\hat{R}_n$ be a positive Toeplitz matrix and $\{\Gamma_i\}_1^{n-1}$ be its choice sequence. Then $\hat{R}_{n+1}$ is a positive Toeplitz expansion of $\hat{R}_n$ if and only if*

$$R_n = C_n + L_{n*}\Gamma_n L_n \tag{9.9}$$

where Γ_n is a contraction mapping D_{n-1} into D_{n-1*} and

$$C_n = Q_n x_n \; , \; L_n = D_{n-1}D_{n-2} \; \cdots \; D_1 \; and \; L_{n*} = D_{1*}D_{2*} \; \cdots \; D_{n-1*} \; . \tag{9.10}$$

9.4 REMARK. Let $\hat{R}_n$ be a positive Toeplitz matrix. The previous corollary shows that the set of all positive Toeplitz expansions $\hat{R}_{n+1}$ of $\hat{R}_n$ is determined by the set of all R_n in the quasi-ball (9.9) with center C_n, right radius L_n and left radius L_{n*} all uniquely determined by $\{\Gamma_i\}_1^{n-1}$, or equivalently, by $\hat{R}_n$, and Γ_n is an arbitrary contraction from D_{n-1} to D_{n-1*}.

The previous corollary leads to the following inverse scattering algorithm for determining the choice sequence $\{\Gamma_i\}_1^n$ from $\{R_i\}_1^n$.

9.5 PROCEDURE. Let $\hat{R}_{n+1}$ be a positive Toeplitz matrix with choice sequence $\{\Gamma_i\}_1^n$. Set $\Gamma_1 = R_1$ and assume that we have calculated $\{\Gamma_i\}_1^{n-1}$ and x_n recursively from $\{R_i\}_1^{n-1}$. The next choice contraction Γ_n is given by

$$\Gamma_n = L_{n*}^{(-1)}(R_n - C_n)L_n^{(-1)} \tag{9.11}$$

where the inverse is the psuedo inverse, and L_{n*}, L_n and C_n are given in (9.10). Once Γ_n is known, calculate x_{n+1}, from (9.6) and L_{n+1*}, L_{n+1} from (9.10), and proceed as before.

9.6 REMARK. If at any step in the previous procedure Γ_n is not a contraction, then this procedure stops. In this case $\hat{R}_{n+1}$ is not positive.

9.7 REMARK. Let $\hat{R}_n$ be a positive Toeplitz matrix. Then Δ_n is an upper triangular factorization of $\hat{R}_n$, that is, Δ_n is upper triangular and (9.8) gives $\Delta_n^* \Delta_n = \hat{R}_n$. Moreover, Procedure 9.5 provides a recursive algorithm to calculate Δ_n. This is done by solving for x_i and L_i in (9.6). Then (9.5) gives

$$\Delta_n = \begin{bmatrix} I & x_1 & x_2 & \cdots & x_{n-1} & x_n \\ 0 & L_1 & L_2 & \cdots & L_{n-1} & L_n \\ 0 & 0 & 0 & \cdots & 0 & \end{bmatrix} \tag{9.12}$$

where a zero column vector of the appropriate length has been added to the column vector $[x_i, L_i]^{tr}$ to make Δ_n a square matrix. There are no zeros added to the last column in (9.12). Since $[Q_i, W_i D_{Q_i}]^{tr}$ in (9.1), (9.2) is precisely the upper $i-1$ by i matrix for U in (XV.2.4). Equation (9.6) implies that $[x_{i+1}, L_{i+1}]^{tr} = U^i | H$. Therefore the upper triangular factorizations Δ_n in (9.12) and (XV.4.12) of $\hat{R}_n$ are identical. From this it follows that the previous procedure is equivalent to Procedure XV.4.4.

The previous proof along with Corollary 1.2 in Chapter XIII and Corollary 1.5 yield the following result.

9.8 COROLLARY. *Let $\hat{R}_{n+1}$ be a positive Toeplitz matrix. Then $\hat{R}_{n+1}$ is strictly positive if and only if its choice sequence $\{\Gamma_i\}_1^n$ is strictly contractive.*

9.9 REMARK. Let $\{\Gamma_i\}_1^\infty$ be a strictly contractive choice sequence. Then $\hat{R}_n (= \Delta_n^* \Delta_n)$ and Δ_n are both invertible for all n.

10. DUALITY AND CHOICE SEQUENCES FOR POSITIVE BLOCK TOEPLITZ MATRICES

This section is the dual of the previous section. Here we will recursively use Theorem 1.1 and (1.8) to obtain the choice sequence $\{\Gamma_i\}_1^{n-1}$ associated with a positive Toeplitz matrix $\hat{R}_n$.

Let $\{\Gamma_i\}_1^{n-1}$ be a choice sequence initiated on H and Q_{i+1*} the column matrix defined by

$$Q_{i+1*} = [\Gamma_i D_{i-1} \cdots D_1, \Gamma_{i-1} D_{i-2} \cdots D_1, ..., \Gamma_2 D_1, \Gamma_1]^{tr}. \tag{10.1}$$

Lemma 7.1 shows that the equation

$$D_{Q_{i+1*}^*} W_{i+1*} = \begin{bmatrix} D_{i*} & -\Gamma_i \Gamma_{i-1}^* & -\Gamma_i D_{i-1} \Gamma_{i-2}^* & \cdots & -\Gamma_i D_{i-1} \cdots D_2 \Gamma_1^* \\ 0 & D_{i-1*} & -\Gamma_{i-1} \Gamma_{i-2}^* & \cdots & -\Gamma_{i-1} D_{i-2} \cdots D_2 \Gamma_1^* \\ 0 & 0 & D_{i-2} & \cdots & -\Gamma_{i-2} D_{i-3} \cdots D_2 \Gamma_1^* \\ 0 & 0 & 0 & \cdots & -\Gamma_{i-3} D_{i-4} \cdots D_2 \Gamma_1^* \\ \cdot & \cdot & \cdot & \cdots & \cdot \\ \cdot & \cdot & \cdot & \cdots & \cdot \\ \cdot & \cdot & \cdot & \cdots & \cdot \\ 0 & 0 & 0 & \cdots & D_{1*} \end{bmatrix} \tag{10.2}$$

defines a unitary operator mapping $D_{i*} \oplus \cdots \oplus D_{1*}$ onto $D_{Q_{i+1*}^*}$. This result can also be obtained by rearranging the rows and columns of (XIII.1.8), (XIII.1.9) in Corollary XIII.1.4.

In this setting the upper triangular matrix Δ_{i*} is defined by the following difference equation

$$\Delta_{i*} = \begin{bmatrix} \Delta_{i-1*} & 0 \\ 0 & I \end{bmatrix} \begin{bmatrix} D_{Q_{i*}^*} W_{i*} & Q_{i*} \\ 0 & I \end{bmatrix} \tag{10.3}$$

with the initial condition

$$\Delta_{1*} = I \quad \text{and} \quad \Delta_{2*} = \begin{bmatrix} D_{1*} & \Gamma_1 \\ 0 & I \end{bmatrix} \tag{10.4}$$

The matrix $D_{Q_{i*}^*} W_{i*}$ is given by the right hand side of (10.2) where i is replaced by i−1. The following useful result is the dual of Lemma 9.1.

10.1 LEMMA. *Let Δ_{i*} be the upper triangular matrix in (10.3), (10.4) generated by a choice sequence $\{\Gamma_i\}_1^\infty$ initiated on H. Then Δ_{i*} admits an upper triangular decomposition of the form:*

$$\Delta_{i*} = \begin{bmatrix} L_{i*} & y_i \\ 0 & \Delta_{i-1*} \end{bmatrix} \quad (i \geq 2) \tag{10.5}$$

where y_i from $D_{i-2*} \oplus \cdots \oplus D_{1*} \oplus H$ to H and L_{i*} from D_{i-1*} to H are the operators recursively computed by the following difference equation

$$[L_{i+1*}, y_{i+1}] = y'_{i+1} = [L_{*i}, y_i][D_{Q^*_{i+1*}} W_{i+1*}, Q_{i+1*}] \quad (i \geq 2) \tag{10.6}$$

where L_{i+1*} is the first and y_{i+1} is the remaining components of y'_{i+1} with the initial conditions $L_{2*} = D_{1*}$ and $y_2 = \Gamma_1$.

The proof is omitted because it is similar to the proof of Lemma 9.1. This sets the stage for the following result, which is the dual of Theorem 9.2.

10.2 THEOREM. *Let $\hat{R}_{n+1}$ be the Toeplitz matrix generated by the sequence $\{R_i\}_1^n$ on H. Then $\hat{R}_{n+1}$ is positive if and only if there exists a choice sequence $\{\Gamma_i\}_1^n$ initiated on H satisfying $\Gamma_1 = R_1$ and*

$$[R_j, R_{j-1}, ..., R_1]^{tr} = \Delta_{j*} Q_{j+1*} \quad (1 \leq j \leq n) . \tag{10.7}$$

Moreover, the equation

$$\Delta_{j*} = \hat{R}_{j*}^{1/2} \alpha_{j*} \quad (1 \leq j \leq n+1) \tag{10.8}$$

defines a unitary operator α_{j*} mapping $\overline{\text{ran } \Delta_{j*}^*}$ onto $\overline{\text{ran } \hat{R}_j}$.

PROOF. The proof is similar to the proof of Theorem 9.2 and is only sketched. Applying Theorem 1.1 to $\hat{R}_2$ yields the first choice contraction $\Gamma = R_1 = \Gamma_1$. Corollary 1.2 with $T = \hat{R}_2$ and Δ_{2*} in (10.4) proves (10.8) for $j = 2$. Applying Theorem 1.1 again with $A = \hat{R}_2$ and $C = I$, we see that $\hat{R}_3$ is positive if and only if there exists a contraction $Q_{3*} = [C_2, C_1]^{tr}$ such that

$$B = \begin{bmatrix} R_2 \\ R_1 \end{bmatrix} = \Delta_{2*} Q_{3*} = \begin{bmatrix} D_{1*} & \Gamma_1 \\ 0 & I \end{bmatrix} \begin{bmatrix} C_2 \\ C_1 \end{bmatrix} .$$

Using $R_1 = \Gamma_1$ readily yields $C_1 = \Gamma_1$. This and Lemma 1.1 in Chapter IV shows that $\hat{R}_3$ is positive if and only if $C_2 = \Gamma_2 D_1$ where Γ_2 is a contraction mapping D_1 into D_{1*}, or equivalently, (10.7) holds for $j = 2$.

Equation (1.8) in Corollary 1.2 along with (10.2) and (10.8) give

$$\hat{R}_3^{1/2} \alpha_* = \begin{bmatrix} \Delta_{2*} & 0 \\ 0 & I \end{bmatrix} \begin{bmatrix} D_{Q_{3*}^*} W_{3*} & Q_{3*} \\ 0 & I \end{bmatrix}$$

where α_* is unitary. This proves (10.8) for $j = 3$. Now the proof follows by induction.

The previous theorem also shows that there is a one to one correspondence between the set of all positive Toeplitz matrices $\hat{R}_{n+1}$ on $l_{n+1}^2 (H)$ and the set of all choice sequences $\{\Gamma_i\}_1^n$ initiated on H. By following an argument similar to the proof of Corollary 9.3 we have

10.3 COROLLARY. *Let $\hat{R}_n$ be a positive Toeplitz matrix and $\{\Gamma_i\}_1^{n-1}$ be its choice sequence. Then $\hat{R}_{n+1}$ is a positive Toeplitz expansion of $\hat{R}_n$ if and only if*

$$R_n = C_n + L_{n*}\Gamma_n L_n \tag{10.9}$$

where Γ_n is a contraction mapping D_{n-1} into D_{n-1*} and

$$C_n = y_n Q_{n*}, \ L_n = D_{n-1}D_{n-2} \cdots D_1 \ and \ L_{n*} = D_{1*}D_{2*} \cdots D_{n-1*}. \tag{10.10}$$

10.4 REMARK. Let $\hat{R}_n$ be a positive Toeplitz matrix. The previous corollary shows that the set of all positive Toeplitz expansions $\hat{R}_{n+1}$ of $\hat{R}_n$ is determined by the quasi-ball in (10.9) with center C_n, right radius L_n, left radius L_{n*} all uniquely determined by $\{\Gamma_i\}_1^{n-1}$, or equivalently, by $\hat{R}_n$, and Γ_n is an arbitrary contraction from D_{n-1} to D_{n-1*}.

The previous corollary leads to the following inverse scattering algorithm for determining the choice sequence $\{\Gamma_i\}_1^n$ from $\{R_i\}_1^n$.

10.5 PROCEDURE. Let $\hat{R}_{n+1}$ be a positive Toeplitz matrix with choice sequence $\{\Gamma_i\}_1^n$. Set $\Gamma_1 = R_1$ and assume that we have calculated $\{\Gamma_i\}_1^{n-1}$ and y_n and Δ_{n*} recursively from $\{R_i\}_1^{n-1}$. The next choice contraction Γ_n is given by

$$\Gamma_n = L_{n*}^{(-1)}(R_n - C_n)L_n^{(-1)} \tag{10.11}$$

where the inverse is the psuedo inverse and L_n, L_{n*} and C_n are given in (10.10). Once Γ_n is known calculate y_{n+1} from (10.6) and L_{n+1}, L_{n+1*} from (10.10), and proceed as before.

10.6 REMARK. If at any step in the previous procedure Γ_n is not a contraction, then the previous procedure stops. In this case $\hat{R}_{n+1}$ is not positive.

10.7 REMARK. Let $\hat{R}_n$ be a positive Toeplitz matrix. Then Δ_{n*}^* is a lower triangular factorization of $\hat{R}_n$, that is, Δ_{n*}^* lower triangular and (10.8) for $j = n$ gives $\Delta_{n*}\Delta_{n*}^* = \hat{R}_n$. Moreover, Procedure 10.5 provides a recursive algorithm to calculate Δ_{n*}.

The proof of Theorem 10.2 along with Corollary 1.2 in Chapter XIII and Corollary 1.5 yield the following result.

10.8 COROLLARY. *Let* $\hat{R}_n$ *be a positive Toeplitz matrix. Then* $\hat{R}_n$ *is strictly positive if and only if its choice sequence* $\{\Gamma_i\}_1^{n-1}$ *is strictly contractive.*

10.9 REMARK. Let $\{\Gamma_i\}_1^\infty$ be a strictly contractive choice sequence, then $\hat{R}_n(=\Delta_{n*}\Delta_{n*}^*)$ and Δ_{n*} are both invertible for all n.

10.10 REMARK. Assume that $\{R_i\}_1^n$ generates a positive Toeplitz matrix $\hat{R}_{n+1}$. Then all the inverse scattering algorithms in this chapter and the previous chapter produce the same choice sequence $\{\Gamma_i\}_1^n$. The proof of this is similar to the proof of Proposition XV.5.2, that is, all the algorithms start with $\Gamma_1 = R_1$. By induction assume that all these algorithms generate the same choice sequence $\{\Gamma_i\}_1^{n-1}$. This implies that R_n is in a ball with center C_n, right radius L_n and left radius L_{n*} where L_n and L_{n*} are defined in (10.10). All these algorithms generate the set of all positive extensions, or equivalently, the same quasi-ball. By Lemma XV.5.1 all these algorithms produce the same center C_n and the same n-th choice contraction Γ_n. This completes the induction and the proof of our claim. This result also follows from our results on quasi-balls in Section XIII.7.

XVI.11. NOTES AND COMMENTS

Theorem 1.1 is a classical result for positive block matrices, which dates back to Schur. It is also a positive version of Lemma IV.2.1 in [Sz.-NF 6]. Theorem 1.1 plays a fundamental role in probability. To see this let x and y be two mean zero finite dimensional real random vectors. Let T in (1.1) be the covariance matrix generated by the random vector $[x, y]^{tr}$, that is, $A = E[xx^{tr}] \triangleq \sigma_x^2$, is the *covariance matrix for* x, $C = E[yy^{tr}] \triangleq \sigma_y^2$ is the *covariance matrix for* y and $B = E[xy^{tr}] \triangleq \sigma_{xy}$ is the *covariance matrix between* x *and* y, where E denotes the expectation. Obviously this T is a positive matrix and in probabilistic terminology $\Gamma \triangleq \rho_{xy}$ is known as the *correlation matrix between* x *and* y. In other words, Theorem 1.1 produces the classical result $\sigma_{xy} = \sigma_x \rho_{xy} \sigma_y$. If x and y are scalar valued, then ρ_{xy} is called the *correlation coefficient* and is given by $\rho_{xy} = \sigma_{xy}/(\sigma_x \sigma_y)$. In geometric terms $\rho_{xy} = \cos(\theta)$ where θ is the angle between x and y with respect to the inner product generated by the expectation. To generalize this concept to the matrix setting, let X and Y be two subspaces of a Hilbert space K such that $\dim X \geq \dim Y = m$. The *principle angles* $\theta_1, \theta_2, \ldots, \theta_m$ between X and Y are recursively defined by [GolVaL]

$$\cos(\theta_j) = \sup(u, v) = (u_j, v_j) \, ,$$

where the supremum is taken over unit vectors u in X and v in Y, subject to the constraint $(u, u_i) = 0$ and $(v, v_i) = 0$ for $i = 1, 2, \ldots, j - 1$. It is emphasized that $0 \leq \theta_1 \leq \theta_2 \leq \ldots \leq \theta_m \leq \pi/2$. The cosine of the principle angles are precisely the singular values of $P_Y | X$. Now let K be the Hilbert space generated by vectors of the form $[h, g]^{tr}$, where h is in H and g is in G, whose inner product is given by $< [h, g]^{tr}, [h', g']^{tr} > = (T[h, g]^{tr}, [h', g']^{tr})$. Let X and Y be the subspaces generated by vectors of the form $[h, 0]^{tr}$ and $[0, g]^{tr}$, where h is in H and g in G, respectively. Then it follows that cosine of the angles between X and Y are precisely the singular values of the contraction Γ in Theorem 1.1. In probabilistic language if X and Y are the linear subspaces generated by the random vectors x and y, respectively, then the cosine of the principle angles between X and Y are precisely the singular values of the correlation matrix ρ_{xy}. These singular values determine the amount of information between x and y.

The results in Section 2 are standard results for block Levinson systems [Des], [DudM], [MoVK], [Wh], [WigR]. Here we have tried to emphasize the role of Theorem 1.1 in these Levinson systems. Theorem 3.1 is taken from [FrzK] and is implicitly contained in [Con 3]. It is a positive version of Theorem IV.3.1 due to [DaKW] and [ArsG]. The existence aspect of Theorem 3.1 for scalar valued Hermitian matrices appears in [JoR 1]. For some nice results concerning the completion of partial Hermitian matrices and contractive matrices see [GroJSW] and [JoR 1,2]. The contraction Γ_R in Theorem 3.1 has an interesting interpretation in terms of principle angles. To see this let K be the Hilbert space generated by vectors of the form $[h, g, d]^{tr}$, where h is in H, g in G and d in D whose inner product is given by

$$<[h, g, d]^{tr}, [h', g', d']^{tr}> = (T_3[h, g, d]^{tr}, [h', g', d']^{tr})$$

and T_3 is the positive operator in (3.2). Now let $X_{1,2}$ and G_1 and $X_{2,3}$ be the subspace generated, respectively by vectors of the form $[h, g, 0]^{tr}$ and $[0, g, 0]^{tr}$ and $[0, g, d]^{tr}$, where h is in H, g in G and d is in D. Then one can show that the cosine of the angles between $X_{1,2} \ominus G_1$ and $X_{2,3} \ominus G_1$ are precisely the singular values of Γ_R. In probabilistic terminology the singular values of Γ_R represented the amount of information between $X_{1,2}$ and $X_{2,3}$ not already contained in their common part G_1.

As noted in Section XV.13, the matrix version of the Levinson algorithm was originally presented in [Wh], [WigR] (see [DudM] for further results and historical comments). Here we have used Theorem 3.1 to obtain the Levinson algorithm. All the results in Section 6 are classical and date back to the papers [Ca 1,2], [To 4] and [Schur 1]. Here we have tried to use the special forms of 2 by 2 and 3 by 3 positive matrices to prove these results. The characterization of all n by n positive block matrices in terms of positive operators and a set of contractions is in [Con 3]. This result is a positive version Theorem XIII.1.5 due to Davis [Da 2]. For a solution to the Carathéodory interpolation problem based on expanding positive matrices see [KovP]. For a solution to this problem based on the band method see [GoKW 1]. The upper triangular factorizations and the inverse scattering algorithms in the rest of this chapter are essentially contained in [ArsCo], [Con 1,2]. As noted in Section XV.13 there are many fast algorithm to obtain an upper triangular factorization for a Toeplitz matrix [Ba], [GoS], [LeRG], [Mar], [Mo], [Riss], [Tr]. Our purpose of presenting upper triangular factorizations in this chapter was to demonstrate how the choice sequence naturally occurs in these factorizations rather than presenting fast algorithms. For further results on fast algorithms to factor other kinds of matrices see [LeK 2].

CHAPTER XVII

A PHYSICAL BASIS FOR THE LAYERED MEDIUM MODEL

In this chapter we sketch a physical framework for the layered medium model in Chapter III. First we introduce some elementary notions from the mechanics of elastic bodies. In particular, Poisson's primary and secondary waves. We show that the analysis in Chapter III is relevant to the study of vertically moving horizontal primary waves through a layered medium, with isotropic perfectly elastic layers welded along horizontal interfaces. We illustrate the connection between primary waves in such a medium and the theory of Carathéodory interpolation as developed in Chapters I, II, III, VIII, XIII and XIV.

1. WAVES IN AN ELASTIC MEDIUM

In this section we present the rigorous physical framework for the concepts of reflection and transmission coefficients introduced in Chapter III.

The mechanics of a continuous medium is governed by Newton's laws as adapted by Cauchy. Namely, the movement of a "material point" of the medium placed at $x = (x_1, x_2, x_3) \in R^3$ is given by

$$\rho a_j = \sum_{k=1}^{3} \frac{\partial}{\partial x_k} T_{jk} + F_j \qquad (j = 1, 2, 3) \qquad (1.1)$$

where (a_1, a_2, a_3) is the acceleration of the material point at x, ρ is the density of the medium at x, and $F = (F_1, F_2, F_3)$ is the body force acting at x while $T = (T_{jk})_{j,k=1}^3$ is the stress tensor at x. We recall that (T_{j1}, T_{j2}, T_{j3}) is the traction at x on any surface with normal $e_j = (\varepsilon_{j1}, \varepsilon_{j2}, \varepsilon_{j3})$ of the "portion" of the medium on the "positive side" of the surface onto that on the "negative side." If $u = (u_1(x, t), u_2(x, t), u_3(x, t))$ represents the displacement of the "material point" placed at x at time t, then the constitutive law of the medium is a functional relation expressing the stress T in terms of the displacement u. Assuming that u_j, $\dfrac{\partial u_j}{\partial x_k}$, etc. are very small and that the medium is isotropic (invariant to rotation) and perfectly elastic (linear in deformation) one has

$$T_{jk} = \lambda \delta_{jk} \, \text{div } u + 2\mu e_{jk} \qquad (1 \le j, \, k \le 3) \qquad\qquad (1.1a)$$

where δ_{jk} is the usual Kronecker delta, $\text{div } u = \sum\limits_{k=1}^{3} \dfrac{\partial}{\partial x_k} u_k$ and $(e_{jk})_{j,k=1}^{3}$ is the strain tensor, that is,

$$e_{jk} = \frac{1}{2} \left[\frac{\partial u_j}{\partial x_k} + \frac{\partial u_k}{\partial x_j} \right] \qquad (1 \le j, \, k \le 3) \, ; \qquad\qquad (1.1b)$$

finally λ, μ are the so called Lamé elastic parameters at x. Also $a_j = \dfrac{d^2}{dt^2} u_j$ can be replaced with $\dfrac{\partial^2}{\partial t^2} u_j$ $(1 \le j \le 3)$ again because in $\dfrac{d}{dt} = \dfrac{\partial}{\partial t} + \sum\limits_{j=1}^{3} u_j \dfrac{\partial}{\partial x_j}$ the contribution of the last sum is negligible. We will further assume that the medium is homogeneous, that is, ρ, λ and μ are constant. Introducing the above relations in (1.1) we obtain

$$\rho \frac{\partial^2}{\partial t^2} u_j = \lambda \frac{\partial}{\partial x_j} \, \text{div } u + \mu \sum_{k=1}^{3} \left[\frac{\partial^2}{\partial x_k^2} u_j + \frac{\partial^2}{\partial x_k \partial x_j} u_k \right] + F_j \qquad\qquad (1.2)$$

so that, by taking $F = 0$ (that is, neglecting the body forces)

$$\rho \frac{\partial^2 u}{\partial t^2} = (\lambda + \mu) \, \text{grad} \, (\text{div } u) + \mu \Delta u \qquad\qquad (1.3)$$

where grad is the gradient and $\Delta = \sum\limits_{k=1}^{3} \dfrac{\partial^2}{\partial x_k^2}$ is the Laplacian. There are two remarkable consequences of these equations due to Poisson. Namely, applying the "divergence operator" to (1.3) and using the fact that div grad $= \Delta$, we obtain for $\psi = \text{div } u$ the relation

$$\rho \frac{\partial^2 \psi}{\partial t^2} = (\lambda + 2\mu) \Delta \psi \, , \qquad\qquad (1.4)$$

which is a wave equation with wave speed

$$\alpha = \sqrt{\frac{\lambda + 2\mu}{\rho}} \, , \qquad\qquad (1.4a)$$

while applying to (1.3) the "curl operator" and using curl grad $= 0$ we obtain for $\phi = \text{curl } u$

$$\rho \frac{\partial^2 \phi}{\partial t^2} = \mu \Delta \phi \, , \qquad\qquad (1.5)$$

a wave equation with wave speed

$$\beta = \sqrt{\frac{\mu}{\rho}} \ . \tag{1.5a}$$

For the sake of the reader recall that

$$\phi = \text{curl } u = \begin{bmatrix} \dfrac{\partial u_3}{\partial x_2} - \dfrac{\partial u_2}{\partial x_3} \\[2ex] \dfrac{\partial u_1}{\partial x_3} - \dfrac{\partial u_3}{\partial x_1} \\[2ex] \dfrac{\partial u_2}{\partial x_1} - \dfrac{\partial u_1}{\partial x_2} \end{bmatrix}$$

In seismology the solutions to the equation (1.4) are called *primary waves*, while those of equation (1.5) are called *secondary waves*. It is found that for many rocks $\mu \sim \lambda$ so that $\alpha \sim \beta\sqrt{3}$, that is, the primary seismic waves travel almost 3/2 times faster than the secondary seismic waves.

In order to avoid the technical complications present in the mathematical study of the equations (1.3), (1.4) and (1.5) we shall consider only plane waves. We start by considering vertically moving horizontal plane waves of the form

$$u = (0, \, 0, \, w(z, \, t)) \tag{1.6}$$

where we denoted by z the coordinate x_3 in order to emphasize its role; also we shall always assume that the positive direction of the $0z$ axis is pointing downwards. The equation (1.3) yields

$$\rho \frac{\partial^2 w}{\partial t^2} = (\lambda + 2\mu)\frac{\partial^2 w}{\partial z^2} \ . \tag{1.7}$$

It is well known (and easy to prove) that the solutions of (1.7) are of the form

$$w(z, \, t) = f(t + \frac{z}{\alpha}) + g(t - \frac{z}{\alpha}) \tag{1.8}$$

where α is given by (1.4a) and $f, \, g$ are smooth arbitrary functions. It is clear (see Figure 1) that $f(t + \frac{z}{\alpha})$ and $g(t - \frac{z}{\alpha})$ represent upgoing and respectively downgoing waves with speed α. Obviously the corresponding u in (1.5) is a primary wave; also it is easy to check that if u is a vertically moving horizontal plane primary wave, then necessarily it is of the form (1.6).

Now assume that we have a layered medium with a horizontal interface at $z = 0$ and with homogeneous isotropic perfectly elastic layers of infinite width above and

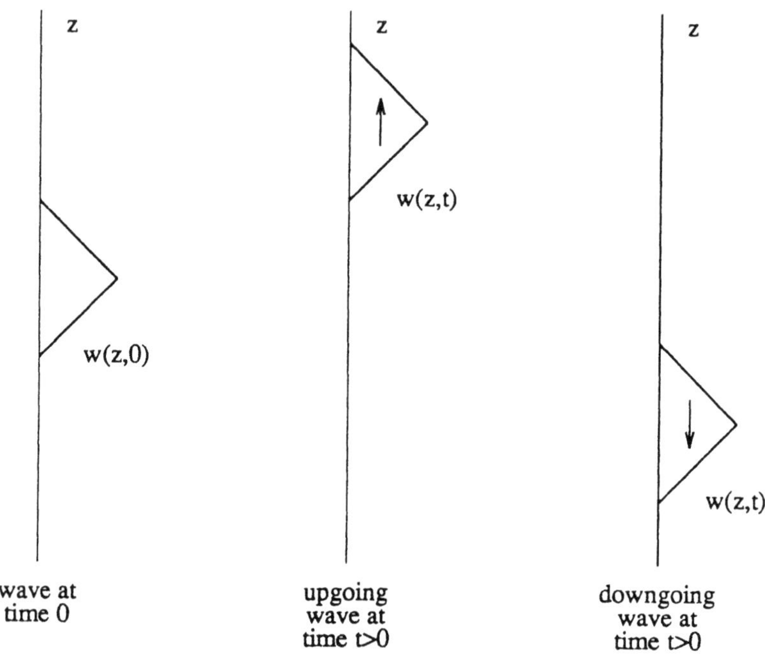

Figure 1

below the interface (see Figure 2). The upper (lower) layer will be labeled as the 0-layer
(respectively the 1-layer). We denote by ρ_i, λ_i, μ_i and $\alpha_i = \sqrt{(\lambda_i + 2\mu_i)/\rho_i}$ $(i = 0, 1)$
the corresponding physical constants of the two layers. For a plane wave of the form
(1.6) we will thus have the equation

$$\rho_i \frac{\partial^2}{\partial t^2} w = (\lambda_i + 2\mu_i) \frac{\partial^2}{\partial z^2} w$$

in the i-th layer (i=0, 1). Hence (see (1.8))

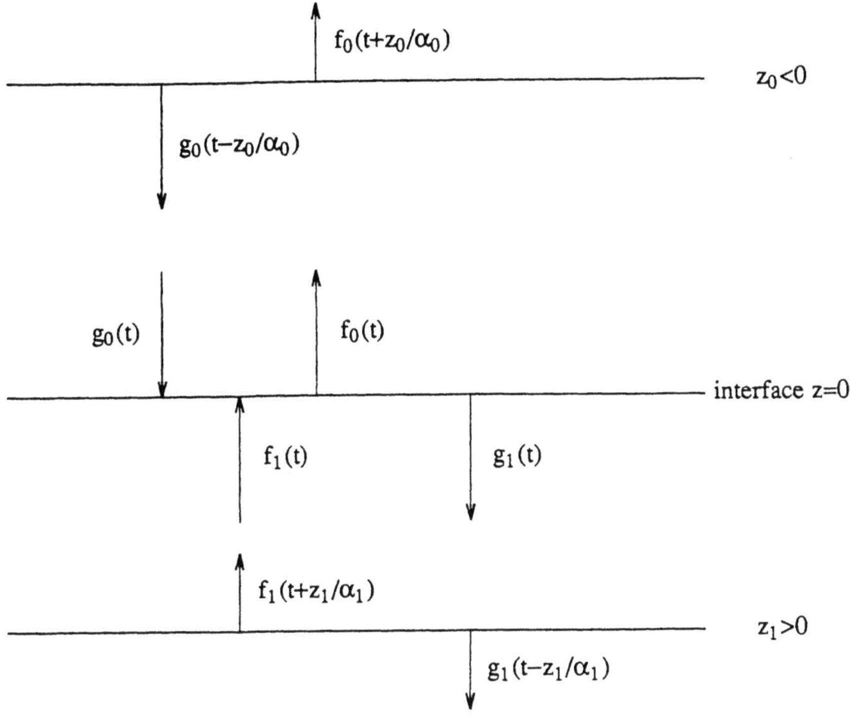

Figure 2

$$w(z,\,t) = \begin{cases} f_0(t + \dfrac{z}{\alpha_0}) + g_0(t - \dfrac{z}{\alpha_0}) \text{ for } z < 0 \\[2ex] f_1(t + \dfrac{z}{\alpha_1}) + g_1(t - \dfrac{z}{\alpha_1}) \text{ for } z > 0 \end{cases} \qquad (1.9)$$

where f_0, g_0 f_1, g_1 are arbitrary (smooth) functions. Clearly (1.9) must be supplemented with the boundary conditions across the interface. These conditions follow from the natural assumption that the layers have a welded contact along the interface. Indeed in this case, the displacements and the traction on the interface must be continuous across the interface. This means that

$$u(+\,0,\,t) = u(-\,0,\,t)$$
$$T_{3j}(+\,0,\,t) = T_{3j}(-\,0,\,t) \qquad (1 \le j \le 3)$$

for all times t, or equivalently,

$$f_1(t) + g_1(t) = w(+0, t) = w(-0, t) = f_0(t) + g_0(t) \tag{1.10a}$$

and by (1.1a)

$$\frac{\lambda_1 + 2\mu_1}{\alpha_1}(f_1'(t) - g_1'(t)) = T_{33}(+0, t) =$$

$$\tag{1.10b}$$

$$= T_{33}(-0, t) = \frac{\lambda_0 + 2\mu_0}{\alpha_0}(f_0'(t) - g_0'(t))$$

for all t where prime denotes the derivative. It is convenient to introduce now the physical constant

$$Z_j = \frac{\lambda_j + 2\mu_j}{\alpha_j} = \sqrt{(\lambda_j + 2\mu_j)\rho_j} = \alpha_j\rho_j \qquad (j = 1, 2) \tag{1.11}$$

to which we will refer as the (longitudinal) *impedance* of the j-th layer. Then taking the derivatives in (1.10a), solving for f_0' and g_1' in (1.10) and integrating we easily obtain

$$f_0 = \frac{Z_0 - Z_1}{Z_0 + Z_1}g_0 + \frac{2Z_1}{Z_0 + Z_1}f_1 + a$$

$$g_1 = \frac{2Z_0}{Z_0 + Z_1}g_0 + \frac{Z_1 - Z_0}{Z_0 + Z_1}f_1 + a$$

where a is a constant which can be incorporated in the definition of f_0 and g_1. Thus we obtained the following basic fact.

1.1 LEMMA. *The vertically moving horizontal primary waves in the two-layered medium are of the form* u = (0, 0, w) *where*

$$w(z, t) = \begin{cases} g_0(t - \dfrac{z}{\alpha_0}) + r_0 g_0(t + \dfrac{z}{\alpha_0}) + (1 - r_0)f_1(t + \dfrac{z}{\alpha_0}) & \text{for } z < 0 \\[4mm] (1 + r_0)g_0(t - \dfrac{z}{\alpha_1}) - r_0 f_1(t - \dfrac{z}{\alpha_1}) + f_1(t + \dfrac{z}{\alpha_1}) & \text{for } z > 0 \end{cases} \tag{1.12}$$

and

$$r_0 = \frac{Z_0 - Z_1}{Z_0 + Z_1}. \tag{1.13}$$

It is important to observe (see Figure 2) that in the 0-layer $g_0(t - \frac{z}{\alpha_0}) + r_0 g_0(t + \frac{z}{\alpha_0})$ represents a downgoing wave plus its part reflected by the interface as an upgoing wave, while $(1 + r_0)g_0(t - \frac{z}{\alpha_1})$ is its part, still a downgoing wave, which passed through the interface. Conversely $f_1(t + \frac{z}{\alpha_1}) - r_0 f_1(t - \frac{z}{\alpha_1})$ is an upgoing wave plus its part reflected by the interface as a downgoing wave and $(1 - r_0)f_1(t + \frac{z}{\alpha_0})$ is the part of the upgoing wave which passed through the interface. These observations and the formulas (1.12), (1.13) now explain the definitions of the reflection coefficients in Chapter III. Indeed we see that the reflection coefficient r_0 (respectively $- r_0$) represents the fraction of the amplitude of the wave reflected in the 0-layer (respectively 1-layer) while $1 + r_0$ (respectively $1 - r_0$) represents the fraction of the amplitude of the wave transmitted through the interface, downward (respectively upwards).

Before concluding this section, it is worth mentioning that the density of the mechanical energy at x in an isotropic perfectly elastic medium associated with the solution $u = (u_1, u_2, u_3)$ of (1.3) is given by the formula

$$E = \frac{1}{2}[\rho \sum_{j=1}^{3} \left[\frac{\partial u_j}{\partial t}\right]^2 + \lambda(\text{div } u)^2 + 2\mu \sum_{j,k=1}^{3} e_{jk}^2] . \tag{1.14}$$

Computing this for the wave (1.6) we obtain

$$E = \frac{1}{2}[\rho \left[\frac{\partial w}{\partial t}\right]^2 + (\lambda + 2\mu)\left[\frac{\partial w}{\partial z}\right]^2] . \tag{1.15}$$

So by (1.4a) the density of energy associated with the wave (1.12) in the two layered medium at the upperside, respectively lowerside, of the interface at $z = 0$ is

$$E = \rho_0[g_0'(t)^2 + (r_0 g_0'(t) + (1 - r_0)f_1'(t))^2] \tag{1.16a}$$

respectively

$$E = \rho_1[f_1'(t)^2 + ((1 + r_0)g_0'(t) - r_0 f_1'(t))^2] . \tag{1.16b}$$

Thus the energy of the incoming wave (that is, waves moving into the interface) per unit area of the interface at time t is

$$[\rho_0 g_0'(t)^2 + \rho_1 f_1'(t)^2]dt \tag{1.17a}$$

while that of the outgoing waves (that is, waves emerging from the interface) is

$$[\rho_0(r_0 g_0'(t) + (1 - r_0)f_1'(t))^2 + \rho_1((1 + r_0)g_0'(t) - r_0 f_1'(t))^2]dt \ . \tag{1.17b}$$

The fact that (1.17b) equals (1.17a) is the mathematical facet of the conservation of energy through the interface, namely the energy emerging from the interface equals the energy entering the interface. This also shows that the amplitudes considered in Chapter III (see especially Section III.6) are actually the amplitudes of the velocity fields $\frac{\partial w}{\partial t}$. We remark for further applications that the upgoing (respectively downgoing) wave component of this field is given in the j-th layer by

$$\frac{1}{2}\left[\frac{\partial}{\partial t} + \alpha_j \frac{\partial}{\partial z}\right]w \ \ \text{(respectively} \ \frac{1}{2}\left[\frac{\partial}{\partial t} - \alpha_j \frac{\partial}{\partial z}\right]w)$$

as shown immediately by a trivial calculation.

2. WAVES IN A LAYERED MEDIUM

In this section we consider the propagation of vertically moving horizontal primary waves in a layered medium, and establish the connection with the Carathéodory interpolation problem.

We start by considering a layered medium with (n+1)-horizontal interfaces and n isotropic perfectly elastic layers of width $d_1, d_2, ..., d_n$. The upper (0-layer) and the lower layer ((n+1)-th layer) are of infinite width and all contacts along the interfaces are welded. For the j-th layer the physical constants considered in Section 1 will be denoted by ρ_j, λ_j, μ_j, α_j, Z_j, etc. The origin is chosen on the 0-interface separating the 0-layer and the 1-layer. Thus for $j = 1, 2, ... n$ the j-th interface (that is, the lower border of the j-th layer) will pass through $(0, 0, z_j)$ where $z_j = d_1 + \cdots + d_j$ and $z_0 = 0$ for the 0-interface. As in Section 1 it is easy to prove that any vertically moving horizontal primary wave has the form

$$u = (0, 0, w(z, t)) \tag{2.1}$$

where

$$\rho_j \frac{\partial^2}{\partial t^2} w = (\lambda_j + 2\mu_j)\frac{\partial^2}{\partial z^2} w \quad \text{in the j-th layer}, \tag{2.2}$$

and w satisfies at the j-th interface the boundary conditions

$$\begin{cases} w(z_j + 0, t) = w(z_j - 0, t) \\ (\lambda_{j+1} + 2\mu_{j+1})\dfrac{\partial}{\partial z}w(z, t)|_{z=z_j+0} = (\lambda_j + 2\mu_j)\dfrac{\partial}{\partial z}w(z, t)|_{z=z_j-0} \end{cases} \tag{2.3}$$

for all $j = 0, 1, 2, \ldots, n$ and all times t. In order to describe all such waves we will study in more detail their behavior in the layers above and below the j-th interface. By virtue of the remark at the end of the preceding section it is natural to introduce the following functions

$$\begin{cases} U_j(z, t) = \dfrac{1}{2}\left[\dfrac{\partial}{\partial t} + \alpha_j\dfrac{\partial}{\partial z}\right]w(z,t) \\ D_j(z, t) = \dfrac{1}{2}\left[\dfrac{\partial}{\partial t} - \alpha_j\dfrac{\partial}{\partial z}\right]w(z, t) \end{cases} \quad \text{(for } z_{j-1} < z < z_j) \tag{2.4}$$

where $z_{-1} = -\infty$, $z_{n+1} = \infty$ and the time is arbitrary. (Notice that if $w(z, t) = f_j(t + z/\alpha_j) + g_j(t - z/\alpha_j)$ in the j-th layer, then $U_j(z, t) = f_j'(t + z/\alpha_j)$ and $D_j(z, t) = g_j'(t - z/\alpha_j)$.) Then

$$\begin{cases} \left[\dfrac{\partial}{\partial t} - \alpha_j\dfrac{\partial}{\partial z}\right]U_j(z, t) = 0 \\ \left[\dfrac{\partial}{\partial t} + \alpha_j\dfrac{\partial}{\partial z}\right]D_j(z, t) = 0 \end{cases} \quad \text{(for } z_{j-1} < z < z_j \text{ and all } t)$$

and hence for $j = 1, 2, \ldots, n$, $z_{j-1} < z < z_j$ and all t we have

$$U_j\left[z_j - 0, t - \frac{z_j - z}{\alpha_j}\right] = U_j(z, t) = U_j\left[z_{j-1} + 0, t + \frac{z - z_{j-1}}{\alpha_j}\right]$$

$$\tag{2.5}$$

$$D_j\left[z_j - 0, t + \frac{z_j - z}{\alpha_j}\right] = D_j(z, t) = D_j\left[z_{j-1} + 0, t - \frac{z - z_{j-1}}{\alpha_j}\right].$$

Moreover, U_j is an upgoing wave while D_j is a downgoing wave in the j-th layer. Both have the same speed α_j. For $j = 0$ (respectively $j = n + 1$) only the first (respectively second) equalities in (2.5) hold. In order to express the boundary conditions only in terms of these upgoing and downgoing waves, we introduce now the definitions of the reflection coefficients considered in Sections III.1, III.6, and the previous Section

$$r_j = \frac{Z_j - Z_{j+1}}{Z_j + Z_{j+1}} \qquad (0 \le j \le n) . \qquad (2.6)$$

The boundary conditions (2.3) across the j-th interface now yield

$$U_j(z_j - 0,\ t) = \frac{1}{2}\left[\frac{\partial}{\partial t} w(z_j - 0,\ t) + \alpha_j \frac{\partial}{\partial z} w(z_j - 0,\ t)\right] =$$

$$= \frac{1}{2}\left[\frac{\partial}{\partial t} w(z_j + 0,\ t) + \frac{1}{Z_j}(\lambda_{j+1} + 2\mu_{j+1})\frac{\partial}{\partial z} w(z_j + 0,\ t)\right] =$$

$$= \frac{1}{2}(U_{j+1}(z_j + 0,\ t) + D_{j+1}(z_j + 0,\ t)) +$$

$$+ \frac{1}{2}\frac{Z_{j+1}}{Z_j}(U_{j+1}(z_j + 0,\ t) - D_{j+1}(z_j + 0,\ t)) =$$

$$= \frac{1}{1 + r_j}U_{j+1}(z_j + 0,\ t) + \frac{r_j}{1 + r_j}D_{j+1}(z_j + 0,\ t)$$

and similarly

$$D_j(z_j - 0,\ t) = \frac{r_j}{1 + r_j}U_{j+1}(z_j + 0,\ t) + \frac{1}{1 + r_j}D_{j+1}(z_j + 0,\ t)$$

for all t, that is,

$$[D_j(z_j - 0,\ t), U_j(z_j - 0,\ t)] =$$

$$[D_{j+1}(z_j + 0,\ t),\ U_{j+1}(z_j + 0,\ t)]\begin{bmatrix} 1 & r_j \\ r_j & 1 \end{bmatrix}\frac{1}{1 + r_j} . \qquad (2.7)$$

This formula is similar to the formula (4.3) in Chapter III. In order to make (2.7) almost identical to (III.4.3) let us assume that the layered medium has a peculiar geometry, namely that a wave passes through a layer in 1/2 units of time, that is,

$$2d_j/\alpha_j = 1 \ \text{time unit} \ \ (1 \le j \le n) . \qquad (2.8)$$

In this case, for all t, we have

$$[D_{j+1}(z_j + 0,\ t),\ U_{j+1}(z_j + 0,\ t)] =$$
$$[D_{j+1}(z_{j+1} - 0,\ t + 1/2),\ U_{j+1}(z_{j+1} - 0,\ t - 1/2)] \qquad (0 \le j \le n-1) \qquad (2.9)$$

and therefore

$$[D_j(z_j - 0, t), U_j(z_j - 0, t)] = \qquad\qquad (0 \leq j \leq n-1)$$

(2.10)

$$[D_{j+1}(z_{j+1} - 0, t + 1/2), U_{j+1}(z_{j+1} - 0, t - 1/2)] \begin{bmatrix} 1 & r_j \\ r_j & 1 \end{bmatrix} \frac{1}{1 + r_j} .$$

Introducing the auxiliary functions

$$\phi_j(t) = D_j(z_j - 0, t + j/2) \text{ and } \psi_j(t) = U_j(z_j - 0, t + j/2) ,$$

and denoting by τ the translation operator $\psi(t) \to \psi(t - 1)$, we can rewrite (2.10) as

$$[\phi_j, \psi_j] = [\phi_{j+1}, \tau\psi_{j+1}] \begin{bmatrix} 1 & r_j \\ r_j & 1 \end{bmatrix} \frac{1}{1 + r_j}$$

(2.10a)

(for $0 \leq j \leq n-1$). On the other hand in the case when $U_{n+1} \equiv 0$, the formula (2.7) gives

$$\psi_n = r_n \phi_n .$$

(2.10b)

Comparing (2.10a) to (III.4.4) and using (2.10b) we readily obtain

$$\psi_j = R_j(\tau)\phi_j$$

where $R_j(z)$ is the scattering function at the j-th interface (see Section III.4). Thus we can easily conclude with the following result.

2.1 PROPOSITION. *Under the above assumptions, if*

$$R_j(z) = \sum_{k=0}^{\infty} a_{jk} z^k$$

(2.11)

denotes the scattering function at the j-th interface, then

$$U_j(z_j - 0, t) = \sum_{k=0}^{\infty} a_{jk} D_j(z_j - 0, t - k) .$$

(2.12)

We notice that this $R_j(z)$ is precisely the function in H_1^∞ with Schur numbers $r_j, r_{j+1}, ..., r_n, 0, 0, 0, ...$

If the conditions in (2.8) do not hold, then one can assume with great precision that the numbers $2d_j/\alpha_j$ for $1 \leq j \leq n$ are all rational. Then denoting by m their common denominator, we can divide the j-th layer into $2md_j/\alpha_j$ ficticious layers of width $\alpha_j/2m$ for all $1 \leq j \leq n$. Notice that the waves cross each of the new ficticious layers in exactly $1/2m$ units of time. Obviously the reflection coefficients of each of these new ficticious interfaces are zero. Changing the time scale we now see that the new layered medium has the geometrical structure corresponding to the Carathéodory interpolation problem in Proposition 2.1.

XVII.3 NOTES AND COMMENTS

The aim of this chapter was only to sketch the well known physical justification to the layered medium model introduced in Chapter III. The mathematical theory of wave propagation through a layered medium is well developed and deep; see for instance [AkiR], [Ke] and [New]. Part of this theory can be connected to some of the results and techniques in this book. In particular, the reflection matrices of plane interfaces for seismic waves (see Chapter 5 in [Ke]) can be viewed as the contractive free parameters in the contractive matrix expansions treated in Chapter IV and elsewhere in this book. For further developments in geophysics see [Cl], [Kan], [Ro 1] and [RoT 1].

REFERENCES

[AA] Adamjan, V.M. and D.Z. Arov,

[1] A class of scattering operators and of characteristic operator-functions of contractions, *Doklady,* **160** (1965) pp. 1-5.

[2] Scattering operators and contraction semigroups in Hilbert space, *Doklady,* **165** (1965) pp. 1377-1380.

[3] On the unitary couplings of isometric operators, Mat. Issled. Kisinev, **1** (1966) pp. 3-66 (Russian).

[AAK] Adamjan, V.M., Arov, D.Z. and M.G. Krein,

[1] Infinite Hankel matrices and generalized problems of Carathéodory - Fejér and I. Schur, *Functional Anal. i Prilozen,* **2** (1968) pp. 1-19 (Russian).

[2] On bounded operators commuting with contractions of class C_{oo} of unit rank of non-unitarity class, *Funkcional. Anal. i Prilozen,* **3** (1969) pp. 86-87 (Russian).

[3] Analytic properties of Schmidt pairs for a Hankel operator and the generalized Schur-Takagi problem, *Math USSR Sbornick,* **15** (1971) pp. 31-73.

[4] Infinite Hankel block matrices and related extension problems, *Izv. Akad. Nauk. Armjan SSR, Matematika,* **6** (1971) pp. 87-112 (English Translation *Amer. Math. Soc. Transl.,* **III** (1978) pp. 133-156.)

[Ag] Agler, J.,

[1] Rational dilations on an annulus, *Ann. of Math.,* **121** (1985) pp. 537-564.

[Ak] Akhiezer, N.I.,

[1] *The Classical Moment Problem,* Olivier and Boyd, Edinburgh, Scotland, 1965.

[AkG] Akhiezer, N.I. and I. M. Glazman,

[1] *Theory of Linear Operators in Hilbert Space,* Frederick Ungar Publishing, New York, 1966.

[AkiR] Aki, K. and P.G. Richards,

[1] *Quantitative Seismology Theory and Methods*, W.H. Freeman and Company, New York, 1980.

[AllY] Allison, A.C. and N.J. Young,

[1] Numerical algorithms for the Nevalinna-Pick problem, *Numerische Mathematik,* **42** (1983) pp. 125-145.

[AlpD] Alpay, D. and H. Dym,

[1] On applications of reproducing kernel spaces to the Schur algorithm and rational J unitary factorizations, *I. Schur Methods in Operator Theory and Signal Processing; Operator Theory: Advances and Applications,* **18**, Ed. I. Gohberg (1986) pp. 89-159.

[An] Ando, T.,

[1]   On a pair of commutative contractions, *Acta Sci. Math.*, **24** (1963) pp. 88-90.

[AnCF] Ando, T., Ceausescu, Z. and C. Foias.,

[1]   On intertwining dilations II, *Acta Sci. Math.*, **39** (1977) pp. 3-14.

[Ar] Arocena, R.,

[1]   Generalized Toeplitz kernels and dilations of intertwining operators, *Intergal Equations and Operator Theory*, **6** (1983) pp. 759-778.

[2]   On the parameterization of Adamjan, Arov and Krein, *Publ. Math. Orsay*, **83** (1983) pp. 7-23.

[3]   On generalized Toeplitz kernels and their relation with a paper of Adamjan, Arov and Krein, *Functional Analysis Homomorphy and Approximation Theory Math Studies*, **86** North-Holland Amsterdam (1984) pp. 1-22.

[4]   A theorem of Naimark, linear systems and scattering operators, *J. Functional Analysis*, **69** (1986) pp. 281-288.

[5]   Unitary extensions of isometries and contractive intertwining dilations, *The Gohberg Anniversary Collection; Operator Theory: Advances and Applications*, **41**, Ed H. Dym, S. Goldberg, M. A. Kaashoek and P. Lancaster (1989) pp. 13-23.

[ArC] Arocena, R. and M. Cotlar,

[1]   Dilations of generalized Toeplitz kernels and some vectorial moment and weighted problems, *Lecture Notes in Math.*, **908** (1982) pp. 169-188.

[ArCL] Arocena, R., Cotlar, M. and J. Leon,

[1]   Toeplitz kernels, scattering structures and covariance systems, *North-Holland Math. Lib.*, **34** (1985) pp. 1-19.

[Arov] Arov, D.Z.,

[1]   Darlington's method for dissipative systems, *Sov. Phys. Dokl.*, **16** (1972) pp. 954-956.

[2]   On unitary couplings with losses (scattering theory with losses), *Functional Anal. Appl.*, **8** (1974) pp. 280-294.

[3]   Scattering theory with dissipation of energy (with appendix), *Soviet. Math. Doklady*, **15** (1974) pp. 848-854.

[ArovG] Arov, D.Z. and Z. Grossman,

[1]   Scattering matrices in the theory of dilation of isometric operators, *Soviet Math. Doklady*, **27** (1983) pp. 518-522.

[ArsC] Arsene, Gr. and Z. Ceausescu,

[1]   On intertwining dilations IV, *Tokoku Math. J.*, **30** (1978) pp. 423-438.

[ArsCF] Arsene, Gr., Ceausescu, Z. and C. Foias,

[1]   On intertwining dilations VII, *Proc. Coll. Complex Analysis, Joensuu, Lecture Notes in Math.*, **747** (1979) pp. 24-45.

[2]   On intertwining dilations VIII, *J. Operator Theory*, **4** (1980) pp. 55-91.

[ArsCo] Arsene, Gr. and T. Constantinescu,

[1]   The structure of the Naimark dilation and Gaussian stationary processes, *Integral Equations and Operator Theory,* **8** (1985) pp. 181-204.

[ArsG] Arsene, Gr. and A. Gheondea,

[1]   Completing matrix contractions, *J. Operator Theory,* **7** (1982) pp. 179-189.

[Arv] Arveson, W.B.,

[1]   Interpolation problems in nest algebras, *J. Functional Analysis,* **20** (1975) pp. 208-233.

[BC] Ball, J.A. and N. Cohen,

[1]   De Branges - Rovnyak operator models and system theory: a survey, preprint.

[BFHT] Ball, J.A., Foias, C., Helton, J.W. and A. Tannenbaum,

[1]   On a local nonlinear commutant lifting theorem, *Indiana University Math. Journal,* **36** (1987) pp. 693-707.

[BG] Ball, J.A. and I. Gohberg,

[1]   A commutant lifting theorem for triangular matrices with diverse applications, *Intergal Equations and Operator Theory,* **8** (1985) pp. 205-267.

[BGR] Ball, J.A., Gohberg, I. and L. Rodman.

[1]   *Interpolation for Rational Matrix Functions,* Birkhäuser, to appear.

[BH] Ball, J.A. and J.W. Helton,

[1]   A Beurling-Lax Theorem for the Lie group U(m,n) which contains most classical interpolation theory, *J. Operator Theory,* **9** (1983) pp. 107-142.

[BK] Ball, J.A. and T. L. Kriete III,

[1]   Operator-valued Nevanlinna-Pick kernels and their functional models for contraction operators, *Integral Equations and Operator Theory,* **10** (1987) pp. 17-61.

[BL] Ball, J.A. and A. Lubin,

[1]   On a class of contractive perturbations of restricted shifts, *Pacific Journal of Math.,* **63** (1976) pp. 309-323.

[BR] Ball, J.A. and A.C.M. Ran,

[1]   Hankel norm approximation of a rational matrix function in terms of its realization, *Modeling Identification and Robust Control,* Eds. E. I. Byrnes and A. Lindquist, Elsevier Science Publishers B.V. North Holland (1986) pp. 285-296.

[2]   Optimal Hankel norm model reductions and Wiener-Hopf Factorizations I: The canonical case, *SIAM J. Contr. and Optimization,* **25** (1987) pp. 362-382.

[3]   Optimal Hankel norm model reductions and Wiener-Hopf factorizations II: The non-canonical case, *Integral Equations and Operator Theory,* **10** (1987) pp. 416-436.

[Ba] Bareiss, E.H.,

[1] Numerical solution of linear equations with Toeplitz and vector Toeplitz matrices, *Numer. Math.*, **13** (1969) pp. 404-424.

[Ber] Bercovici, H.,

[1] *Operator Theory and Arithmetic in* H$^\infty$, American Mathematical Society, Providence, Rhode Island, 1988.

[BerFT] Bercovici, H., Foias, C. and A. Tannenbaum,

[1] On skew Toeplitz operators I, *Topics in Operator Theory and Interpolation, Operator Theory: Advances and Applications*, **28** Ed. I. Gohberg (1988) pp. 21-43.

[Beu] Beurling, A.,

[1] On two problems concerning linear transformations in Hilbert space, *Acta Math.*, **81** (1949) pp. 239-255.

[Br] de Branges, L.,

[1] Some Hilbert spaces of analytic functions II, *J. Math. Anal. Appl.*, **11** (1966) pp. 44-72.

[2] *Hilbert Spaces of Entire Functions*, Prentice-Hall Englewood Cliffs, New Jersey 1968.

[3] The Carathéodory-Fejér extension theorem, *Integral Equations and Operator Theory*, **5** (1982) pp.160-183.

[4] Unitary linear systems whose transfer functions are Reimann mapping functions, *Operator Theory Advances and Applications*, **19** (1986) pp. 105-124.

[5] *Square Summable Power Series*, Springer-Verlag, to appear.

[BrR] de Branges, L. and J. Rovnyak,

[1] Canonical models in quantum scattering theory, *Perturbation theory and its applications in quantum mechanics*, ed. by C.H. Wilcox (New York-London-Sidney, 1966) pp. 295-392.

[2] *Square Summable Power Series*, Holt Richard and Winson, New York, 1966.

[BroH] Brown, A. and P.R. Halmos,

[1] Algebraic properties of Toeplitz operators, *J. Reine Angew. Math.*, **231** (196?) pp. 89-102.

[BrK] Bruckstein, A.M. and T., Kailath,

[1] Inverse Scattering for Discrete Transmission-line Models, *SIAM REVIEW*, 9 (1987) pp. 359-389.

[Bu] Bunce, J.W.,

[1] Models for N-touples of noncommuting operators, *J. Funct. Anal,* **57** (1984) pp. 21-30.

[Bur] Burg, J.P.,

[1]   Maximal entropy spectral analysis, paper presented at *The 37th Annual International Meeting of the Society of Exploration Geophysicists*, Oklahoma City, Oklahoma (1967).

[2]   The relationship between maximum entropy spectra and maximum likelihood spectra, *Geophysics*, **37** (1972) pp. 375-376.

[3]   *Maximum Entropy Spectral Analysis*, Ph.D. thesis, Department of Geophysics, Stanford University (1975).

[Ca] Carathéodory, C.,

[1]   Über den Variabilitätsbereich der Koeffizienten von Potenzreihen, die gegebene Werte nicht annehmen, *Math. Ann.*, **64** (1907) pp. 95-115.

[2]   Über den Variabilitätsbereich der Fourierschen Konstanten von positiven harmonischen Funktionen, *Rend. Circ. Mat. Palermo*, **32** (1911) pp. 193-217.

[CaFe] Carathéodory, C. and L., Fejér,

[1]   Über den Zusammenhang der Extremen von harmonischen Funktionen mit ihren Koeffizienten ünd uber den Picard-Landauschen Satz, *Rend. Circ. Mat. Palermo*, **32** (1911) pp. 218-239.

[CarS] Carswell, J.G.W. and C. F. Schubert,

[1]   Lifting of operators that commute with shifts, *Michigan Math. J.*, **22** (1975) pp. 65-69.

[Ce] Ceausescu, Z.,

[1]   On intertwining dilations, *Acta. Sci. Math.*, **38** (1976) pp. 281-290.

[2]   Lifting of a contraction intertwining two isometries, *Michigan Math. J.*, **26** (1979) pp. 231-241.

[CeF] Ceausescu, Z. and C. Foias,

[1]   On intertwining liftings III, *Revue Roum. Math. Pures et Appl.*, **22** (1977) pp. 1387-1396.

[2]   On intertwining dilations *V.*, *Acta Sci. Math.*, **40** (1978) pp. 9-32; see also Letter to the Editor, *Acta Sci. Math.* **41** (1979) pp. 457-459.

[3]   On intertwining dilations VI, *Rev. Roumaine Math. Pures Appl.* **23** (1978) pp. 1471-1482.

[CeS] Ceausescu, Z. and I. Suciu,

[1]   Isometric dilations of commuting contractions I, *J. Operator Theory*, **12** (1984) pp. 65-88.

[2]   Isometric dilations of commuting contractions II, *Dilation Theory Toeplitz Operators and Other Topics; Operator Theory Advances and Applications*, **11** (1982) pp. 53-80.

[3]   Extreme points in the set of contractive intertwining dilations, *Topics in Operator Theory Constantin Apostol Memorial Issue; Operator Theory Advances and Application*, **32**, Ed. I. Gohberg (1988) pp. 57-65.

[ChDL] Chu, C.C., Doyle, J.C., and E.B. Lee,

[1]   The general distance problem in H_∞ optimal control theory, *Int. J. Control*, **44** (1986) pp.565-596.

[Cl] Claerbout, J.F.,

[1]   *Fundamentals of Geophysical Data Processing*, McGraw-Hill, New York, 1976.

[ClG] Clancey, K. and I. Gohberg,

[1]   *Factorization of Matrix Functions and Singular Integral Operators*, Birkhauser, Basel, Switzerland, 1981.

[Co] Cohn, A.,

[1]   Üher die Anzahl der Wurzeln einer algebraischen Gleichung in Kreise, *Math. Z.*, **14** (1922) pp. 110-148.

[Con] Constantinescu, T.,

[1]   On the structure of positive Toeplitz forms, *Dilation Theory Toeplitz Operators and Other Topics; Operator Theory: Advances and Applications*, **11**, Birkhauser Verlag (1983) pp. 127-149.

[2]   On the structure of the Naimark Dilation, *J. Operator Theory*, **12** (1984) pp. 159-175.

[3]   Schur analysis of positive block matrices, *I. Schur Methods in Operator Theory and Signal Processing; Operator Theory: Advances and Applications*, **18**, Ed. I. Gohberg (1986) pp. 191-206.

[4]   A maximum entropy principle for contractive intertwining dilations, *Operator theory: Advances and Applications*, **24** (1987) pp. 69-85.

[Conway] Conway, J.B.,

[1]   *Functions of One Complex Variable*, Springer - Verlag, New York, 1978.

[CoS] Cotlar, M. and C. Sadosky,

[1]   On the Helson-Szegö theorem and a related class of modified Toeplitz kernels, *Proc. Symp. Pure Math. AMS*, **35:1** (1979) pp. 383-407.

[2]   Generalized Toeplitz kernels, stationarity and harmonizability, *J. Analyse Math.*, **44** (1985) pp. 117-123.

[3]   Prolongements des formes de Hankel généralisées en formes de Toeplitz, *C.R. Acad. Sci. Paris*, **305** *Serie I* (1987) pp. 167-170.

[Da] Davis, C.,

[1]   An external problem for extensions of a sesquilinear form, *Linear Algebra and its Applications*, **13** (1976) pp. 91-102.

[2]   A factorization of an arbitrary $m \times n$ contractive operator, *Proceedings of the Toeplitz Centennial Conference, Tel-Aviv*, Birkhäuser-Verlag (1982) pp. 217-232.

[3]   Completing a matrix so as to minimize the rank, *Topics in Operator Theory and Interpolation; Operator Theory: Advances and Aplications*, **29** , Ed. I. Gohberg (1988) pp. 87-95.

[DaKW] Davis, C., Kahan, W.M. and H.F. Weinberger,

[1]   Norm-preserving dilations and their applications to optimal error bounds, *SIAM J. Numer. Anal.*, **19** (1982) pp. 445-469.

[DeGK] Delsarte, P., Genin, Y. and Y. Kamp,

[1]   Schur parameterization of positive definite block, Toeplitz systems, *SIAM J. Appl. Math.*, **36** (1979) pp. 34-46.

[2]   The Nevanlinna-Pick problem for matrix-valued functions, *SIAM J. Appl. Math.*, **36** (1979) pp. 47-61.

[Des] Desai, U.B.,

[1]   *Lecture Notes on Estimation Realization and Identification*, Lecture notes (personal copy).

[Dev] Devinatz, A.,

[1]   The factorization of operator valued functions, *Ann. of Math.*, **73** (1961) pp. 458-495.

[DevS] Devinatz, A. and M. Shinbrot,

[1]   General Wiener-Hopf operators, *Trans. American Math. Soc.*, **145** (1969) pp. 467-494.

[Dew] Dewilde, P.,

[1]   The lossless inverse scattering problem in the network - theory context, *Topics in Operator Theory Systems and Networks; Operator Theory Advances and Applications*, **12**, Ed. H. Dym and I. Gohberg (1984) pp. 109-127.

[DewD] Dewilde, P., and H. Dym.,

[1]   Lossless Inverse scattering, digital filters, and estimation theory, *IEEE Trans. on Information Theory*, **30** (1984) pp. 644-662.

[DewVK] Dewilde, P., Vieira, A.C., and T. Kailath,

[1]   On a generalized Szegö-Levinson realization algorithm for optimal linear predictors based on a network synthesis approach, *IEEE Trans. on Circuit and Systems*, **25** (1978) pp. 663-675.

[Do] Douglas, R.G.,

[1]   On majorization, factorization, and range inclusion of operators in Hilbert space, *Proc. Amer. Math. Soc.*, **17** (1966) pp. 413-415.

[2]   On factoring positive operator functions, *J. Math. and Mech.*, **16** (1966) pp. 119-126.

[3]   Structure theory of operators I, *J. reine agnew. Math.*, **232** (1968) pp. 180-193.

[4]   On the operator equation $S^*XT = X$ and related topics, *Acta Sci. Math.*, **30** (1969) pp. 19-32.

[5]   On the hyperinvariant subspaces for isometries, *Math. Z.*, **107** (1969) pp. 297-300.

[6]   Canonical models, *Topics in Operator Theory*, Ed. C. Pearcy, Mathematical Surveys, **13**, The American Mathematical Society, Providence, 1974 pp. 161-218.

[DoF] Douglas, R.G. and C. Foias,

[1] Subisometric dilations and the commutant lifting theorem, *Topics in Operator Theory Systems and Networks; Operator Theory Advances and Applications*, **12**, Ed. H. Dym and I. Gohberg (1984) pp. 129-139.

[DoH] Douglas, R.G. and J.W., Helton,

[1] Inner dilations of analytical matrix functions and Darlington synthesis, *Acta Sci. Math.*, **34** (1973) pp. 301-310.

[DoMP] Douglas, R.G., Muhly, P.S. and C. Pearcy,

[1] Lifting commuting operators, *Michigan Math. J.*, **15** (1968) pp. 385-395.

[DoP] Douglas, R.G. and V.I. Pausen,

[1] *Hilbert Modules Over Function Algebra*, Pitman Research Notes in Mathematics, **217**, Longman Group U.K. Liminated, 1989, New York.

[DoSS] Douglas, R.G., Shapiro, H.S. and A.L. Shields,

[1] Cyclic vectors and invariant subspaces for the backward shift, *Ann. Inst. Fourier, Grenoble* **20** (1971) pp. 37-76.

[Doy] Doyle, J.C.,

[1] *Lecture Notes on a Honeywell Workshop in Advances in Multivariable Control*, Minneapolis, (1984).

[DoyF] Doyle, J.C. and B.A. Francis,

[1] Linear control theory with an H_∞ optimality criterion, *SIAM J. Control and Optimization*, **25** (1987) pp. 815-844.

[DoyFT] Doyle, J., Francis, B.A. and A. Tannenbaum,

[1] *Classical Feedback Control Theory*, Lecture notes (personal copy).

[DoyGKF] Doyle, J.C., Glover, K., Khargonekar, P. and B. Frances,

[1] State space solutions to standard H_2 and H_∞ control problems, *Proceedings of the 1988 Conference on Decision and Control* pp. 1691-1696.

[Dr] Dritschel, M.A.,

[1] A lifting theorem for bicontractions on Krein spaces, *J. Functional Analysis*, to appear.

[DudM] Dudgeon, D.E. and R.M., Mersereau,

[1] *Multidimensional Digital Signal Processing*, Prentice Hall, Englewood Cliffs, New Jersey (1984).

[DuS] Dunford, N. and J. Schwartz,

[1] *Linear operators, Part I: General Theory*, Interscience, New York, 1958.

[Duren] Duren, P.L.,

[1] *Theory of H^p Spaces*, Academic Press, New York, 1970.

[DursSz.-N] Durszt, E. and B. Sz.-Nagy,

[1] Remarks on a paper of A. E. Frazho, Models for noncommuting operators, *J. Functional Analysis,* **52** (1983) pp. 146-147.

[Dy] Dym, H.J.,

[1] *J Contractive Matrix Functions, Reproducing Kernel Hilbert Spaces and Interpolation,* CBMS Regional Conference series, **71**, American Mathematical Society, Providence, Rhode Island, 1989.

[2] On reproducing kernel spaces, J unitary matrix functions, interpolation and displacement rank, *The Gohberg Anniversity Collection; Operator Theory: Advances and Applications*, **41**, Ed. H. Dym, S. Goldberg, M. A. Kaashoek and P. Lancaster (1989) pp. 173-239.

[DyG] Dym, H. and I. Gohberg,

[1] Extension of kernels of Fredholm operators *J. d Analyse Math.,* **42** (1982) pp. 83-125.

[2] Unitary interpolants, factorization indices and infinte block Hankel matrices, *J. Functional Analysis,* **54** (1983) pp. 229-289.

[3] A maximum entropy principle for contractive interpolants, *J. Functional Analysis,* **65** (1986) pp. 83-125.

[4] A new class of contractive interpolants and maximum entropy principles, *Topics in Operator Theory and Interpolation, Operator Theory: Advances and Applications,* **29** , Ed. I. Gohberg (1988) pp. 117-150.

[Eg] Egervary, E.,

[1] On the contractive linear transformations of n-dimensional vector space, *Acta Sci. Math.,* **15** (1954) pp. 178-182.

[ElL] Ellis, R.L. and D.C. Lay,

[1] Rank - preserving extensions of band matrices, preprint.

[Er] Erwe, F.,

[1] Koeffizientenproblem für beschränkte Systeme holomorpher Funktionen und Verallgemeinerung eines Satzes von Carathéodory-Fejér, *Math. Z.,* **93** (1966) pp. 210-215.

[Fed] Fedcina, I.P.,

[1] A criterion for the solvability of the Nevalinna Pick tangent problem, *Mat. Issled.,* **7** (1972) pp. 213-227.

[2] The tangential Nevanlinna-Pick problem with multiple points, *Akad. Nauk Armjan. SSR Dokl.,* **61** (1975) pp. 214-218.

[FeFr] Feintuch, A. and B.A. Francis,

[1] Uniformally optimal control of linear feedback systems, *Automatica,* **21** (1985) pp. 563-574.

[2] Distance formulas for operator algebras arising in optimal control problems, *Topics in Operator Theory and Interpolation; Operator Theory: Advances and Applications,* **29**, Ed. I Gohberg (1988) pp. 151-170.

[Fejér] Fejér, L.

[1]　Über Weierstrass'sche approximation, besonders durch Hermitesche interpolation, *Mathemtische Annalen*, **102** (1930), pp. 707-720.

[2]　Die Abschätzung cinco Polynoms in einesm Intervalle, wenn Schranken für seine werte and ersten Ableitungswerte in einzelnen Punkten des Intervalles gegeben sind, und ihre Anwendung auf die Konvergenzfrage Kermitischer Interpolationseihen, *Mathematische Zeitschrift*, **32** (1930), pp. 426-457.

[3]　On the characterization of some remarkable systems of points of interpolation by means of conjugate points, *American Mathematical Monthly*, **41** (1934), pp. 1-14.

[Fi] Fillmore, P.A.,

[1]　*Notes on Operator Theory,* Van Nostrand Reinhold, New York, 1968.

[Fo] Foias, C.,

[1]　La mesure harmonique-spectrale et la théorie spectrale des opérateurs généraux d'un espace de Hilbert, *Bull. Soc. Math. France,* **85** (1957) pp. 263-282.

[2]　Sur certains theorèmes de J. von Neumann concernant les ensembles spectraux, *Acta Sci. Math.,* **18** (1957) pp. 15-20.

[3]　Certaines applications des ensembles spectraux I. Mesure harmonique-spectrale, *studii cercetari mat.,* **10** (1959) pp. 365-401 (Roumanian, with Russian and French summaries).

[4]　On Hille's spectral theory and operational calculus for semi-groups of operators in Hilbert space, *Composito Math.,* **14** (1959) pp. 71-73.

[5]　A remark on the universal model of G. C. Rota for contractions, *Com. Acad. R. P. Romane,* **13** (1963) pp. 343-352 (Roumanian, with Russian and English summaries).

[6]　La maximalite′ de l'espace H^∞ dans le calcul fonctionnel, *Analele Univ. Timisoara* (ser. mat.-fiz.), **2** (1964) pp. 77-81 (Roumanian, with Russian and French summaries).

[7]　A classification of doubly cyclic operators, *Colloqua Mathematical Societatis Janos Bolyai*, (1970) pp. 155-161.

[8]　Some applications of structural models for operators on Hilbert spaces, *Proc. Intern. Congr. of Math.,* Nice, Sept. 1970, Tom. 2, Gauthier-Villars, Paris, 1971 pp. 433-440.

[9]　On the Lax-Phillips nonconservative scattering theory, *J. Functional Anal.,* **19** (1975) pp. 273-301.

[10]　Contractive intertwining dilations and waves in layered media, *Proceedings of the International Congress of Mathematicians,* Helsinki (1978) pp. 605-613.

[11]　On an interpolation problem of Dym and Gohberg, *Integral Equations and Operator Theory*, **11** (1988) pp. 769-775.

[FoF] Foias, C. and A.E. Frazho,

[1]　Markov subspaces and minimal unitary dilations, *Acta Sci. Math.,* **45** (1983) pp. 165-175.

[2]　Redheffer products and the lifting of contractions on Hilbert space, *J. Operator Theory,* **11** (1984) pp. 193-196.

[3]   On the Schur representation in the Commutant Lifting Theorem I, *Schur Methods in Operator Theory and signal processing, Operator Theory Advances and Applications*, **18**, Ed. I. Gohberg (1986) pp. 207-217.

[4]   On the Schur representation in the commutant lifting theorem II, *Topics in Operator Theory and Interpolation; Operator Theory: Advances and Applications*, **29**, Ed. I Gohberg (1988) pp. 171-179.

[FoT] Foias, C. and A. Tannenbaum,

[1]   On the four block problem I, *Operator Theory: Advances and Applications* **32** (1988) pp. 93-112, On the four block problem II, The singular system, *Integral Equations and Operator Theory*, **11** (1988), pp. 727-767.

[2]   Some remarks on optimal interpolation, *Systems and Control Letters*, **11** (1988) pp. 259-264.

[3]   The strong Parrott Theorem, *Proceedings of the American Math. Soc.*, **106** (1989) pp. 777-784.

[FoTZ] Foias, C., Tannenbaum, A. and G. Zames,

[1]   On the H^∞-optimal sensitivity problem for systems with delays, *SIAM J. Control and Optimization*, **25** (1987) pp. 686-706.

[2]   Some explicit formulae for the singular values of certain Hankel operators with factorizable symbol, *SIAM J. Math. Anal.*, **19** pp. 1081-1089.

[Fran] Francis, B.A.,

[1]   Optimal disturbance attenuation with control weighting, *Lecture Notes in Control and Inf. Sci.*, **66** (1985) Springer-Verlag. Proc. 1984 Twente Workshop on Systems and Optimization.

[2]   *A Course in H^∞ Control Theory*, Lecture Notes in Control and Information Science, Springer, New York, 1987.

[FranHZ] Francis, B.A., Helton, J.W., and G. Zames,

[1] H_∞-optimal feedback controllers for linear multivariable systems, *IEEE Trans. Auto. Cont.*, **29** (1984) pp. 888-900.

[FranZ] Francis, B.A. and G. Zames,

[1]   On H_∞-optimal sensitivity theory for SISO feedback systems, *IEEE Trans. Auto. Cont.*, **29** (1984) pp. 9-16.

[Frz] Frazho, A.E.,

[1]   A shift operator approach to bilinear system theory, *SIAM J. Control*, **18** (1980) pp. 640-658.

[2]   A shift operator approach to state affine system theory, *IEEE Trans. Automat. Control*, **17** (1982) pp. 117-122.

[3]   Models for noncommuting operators, *J. Functional Analysis*, **48** (1982) pp. 1-11.

[4]   Complements to models for noncommuting operators, *J. Functional Analysis*, **59** (1984) pp. 445-461.

[5]   Schur contractions and bilinear stochastic systems, *Proceedings of the 1984 Conference on Information Sciences and Systems*, Princeton University pp. 190-196.

[6] On Uniqueness and the Lifting Theorem, *Acta Sci. Math.*, **49** (1985) pp. 353-355.

[7] On stochastic bilinear systems, in *Modeling and Application of Stochastic Processes*, Ed U. B. Desai (1986) pp. 215-241.

[8] Three inverse scattering algorithms for the lifting theorem, *I. Schur Methods in Operator Theory and Signal Processing; Operator Theory: Advances and Applications*, **18**, Ed. I. Gohberg (1986) pp. 219-248.

[FrzK] Frazho, A.E. and A.M. King,

[1] An approach to recursive q-Markov covariance equivalent realizatons, *Control Theory and Advanced Technology*, **4** (1988) pp. 447-463.

[FrzL] Frazho, A.E. and D.K. Lee

[1] A pole placement procedure via H^∞ optimization, *The 26 Annual Allerton Conference*, Sept. (1988) pp. 872-880.

[Fu] Fuhrmann, P.A.,

[1] A functional calculus in Hilbert space based on operator valued analytic functions, *Israel J. Math.*, **6** (1968) pp. 267-278.

[2] On the Corona problem and its application to spectral problems in Hilbert space, *Trans. Amer. Math. Soc.*, **132** (1968) pp. 55-66.

[3] *Linear Systems and Operators in Hilbert Space*, McGraw-Hill, New York, 1981.

[Ga] Garnett, J.B.,

[1] *Bounded Analytic Functions*, Academic Press, New York, 1981.

[Geo] Georgiou, T.T.,

[1] *Partial Realization of Covariance Sequences*, Ph.D. Thesis, The University of Florida, 1983.

[Ge] Geronimus, Ya.L.,

[1] *Orthogonal Polynomials*, New York Consultants Bureau, 1961.

[Gl] Glover, K.,

[1] All optimal Hankel-norm approximations of linear multivariable systems and their L_∞-error bounds, *Int.J. Cont.*, **39** (1984) pp. 1115-1193.

[2] Robust stabilization of linear multivariable systems: relations to approximation, *Int.J. Control*, **43** (1986) pp. 741-766.

[Go] Gohberg, I.,

[1] *I. Schur Methods in Operator Theory and Signal Processing; Operator Theory: Advances and Applications*, **18**, Birkhauşer, Boston, 1986.

[2] *Topics in Operator Theory and Interpolation; Operator Theory: Advances and Applications*, **29,** Birkhauser, Boston, 1988.

[GoG] Gohberg, I., and S. Goldberg,

[1] *Basic Operator Theory*, Birkhauser, Boston, 1981.

[GoKW] Gohberg, I., Kaashoek, M.A., and H.J. Woerdeman,

[1] The band method for positive and strictly contractive extension problems: an alternative version and new applications, *Integral Equations and Operator Theory*, **12** (1989) pp. 343-382.

[2] A maximum entropy principle in the general framework of the band method, preprint.

[GoS] Gohberg, I. and A.A. Semencul,

[1] On the inversion of finite Toeplitz matrices and their continuous analogs, *Mat. Issled.*, **2** (1972) pp. 201-233.

[GolVaL] Golub, G.H. and C.F. Van Loan,

[1] *Matrix Computations*, Johns Hopkins University Press, Baltimore, 1983.

[GrS] Grenander, U. and G. Szegö,

[1] *Toeplitz Forms and Their Applications*, University of California Press, Berkley, California, 1958.

[GroJSW] Grone, R., Johnson, C.R., Sá, E.M. and H. Wolkowicz,

[1] Positive definite completions of partial Hermitian matrices, *Linear Algebra and its Applications*, **58** (1984) pp. 109-124.

[Ha] Halmos, P.R.,

[1] Normal dilations and extensions of operators, *Summa Brasil. Math.*, **2** (1950) pp. 125-134.

[2] Shifts on Hilbert spaces, *J. reine angew. Math.*, **208** (1961) pp. 102-112.

[3] On Foguel's answer to Nagy's question, *Proc. Amer. Math. Soc.*, **15** (1964) pp. 791-793.

[4] *Positive definite sequences and the miracle of w,* (A talk before the functional analysis seminar at the University of Michigan, 8 July 1965) 17 pages.

[5] *A Hilbert space problem book,* Springer-Verlag, New York, 1982.

[Heins] Heins, M.,

[1] A bibliography on Pick-Nevanlinna interpolation and cognate questions, 1984.

[He] Helson, H.,

[1] *Lectures on Invariant Subspaces,* New York: Academic Press, 1964.

[HeL] Helson, H. and D. Lowdenslager,

[1] Prediction theory and Fourier series in several variables, *Acta Math.*, **99** (1958) pp. 165-202.

[2] Prediction theory and Fourier series in several variables II, *Acta Math.*, **106** (1961) pp. 175-213.

[Helt] Helton, J.W.,

[1] Discrete time systems, operator models and scattering theory, *J. Functional Analysis*, **16** (1974) pp. 15-38.

[2] Orbit structure of the Möbius transformation semi-group acting in H$^\infty$ (broadband matching), *Topics in Functional Analysis (essays dedicated to M. G. Krein on the occasion of his 70th birthday) Advances in Math. suppl. stud.,* **3** Academic press, New York (1978) pp. 129-157.

[3] The distance of a Function to H$^\infty$ in the Poincare´ Metric; Electrical Power Transfer, *J. Functional Analysis,* **38** (1980) pp. 273-314.

[4] Worst case analysis in the frequency-domain; an H$_\infty$ approach to control," *IEEE Trans. Auto. Control,* **30** (1985) pp. 1154-1170.

[5] *Operator Theory, Analytic Functions, Matrices, and Electrical Engineering,* CBMS Regional Conference Series in Math., **68**, Amer. Math. Soc., Providence, Rhode Island, 1987.

[Hi] Hille, E.,

[1] *Functional analysis and semi-groups,* American Mathematical Society Providence, Rhode Island, 1948.

[Ho] Hoffman, K.,

[1] *Banach Spaces of Analytic Functions,* Englewood Cliffs, Prentice Hall, New Jersey, 1962.

[JoR] Johnson, C.R. and L. Rodman,

[1] Inertia possibilities for completions of partial Hermitian matrices, *Linear and Multilinear Algebra,* **16** (1984) pp. 179-195.

[2] Completion of partial matrices to contractions, *J. Functional Analysis,* **69** (1986) pp. 260-267.

[Ju] Julia, G.,

[1] Sur la représentation analytique des opérateurs bornés ou fermés de l'espace hilbertien, *C.R. Acad. Sci. Paris,* **219** (1944) pp. 225-227.

[2] Les projections des systèmes orthonormaux de l'espace hilbertien et les opérateurs bornés, *C.R. Acad. Sci. Paris,* **219** (1944) pp. 8-11.

[3] Sur les projections des systèmes orthonormaux de l'espace hilbertien, *C. R. Acad. Sci. Paris,* **218** (1944) pp. 892-895.

[Jury] Jury, E.I.,

[1] A stability test for linear discrete systems using simple division, *Proceedings of the IRE* (1961) pp. 1948-1949.

[JuryB] Jury, E.I. and J. Blanchard,

[1] A stability test for linear discrete systems in table form, *Proceedings of the IRE* (1961) pp. 1947-1948.

[Ka] Kailath, T.,

[1] *Linear Systems,* Englewood Cliffs: Prentice-Hall, Inc., New Jersey, 1980.

[2] A theorem of I. Schur and its impact on modern signal processing, *I. Schur Methods in Operator Theory and Signal Processing; Operator Theory: Advances and Applications,* **18,** Ed. I. Gohberg (1986) pp. 9-30.

[KaB] Kailath, T. and A.M. Bruckstein,

[1] Naimark dilations, state-space generators and transmission lines, *Advances in Invariant Subspaces and other results of Operator theory; Operator theory: Advances and Applications* (1984) pp. 173-186.

[KaBM] Kailath, T., Bruckstein, A.M. and D. Morgan,

[1] Fast matrix factorizations via discrete transmission lines, *Linear Algebra and Its Applications*, **75** (1986) pp. 1-25.

[KaVM] Kailath, T., Vieira A. and M. Morf,

[1] Inverses of Toeplitz operators innovations and orthogonal polynomials, *SIAM Review*, **20** (1978) pp. 106-119.

[Kan] Kanasewich, E.R.,

[1] *Time Sequence Analysis in Geophysics*, The University of Alberta Press, Edmonton, 1973.

[Ke] Kennett, B.L.N.,

[1] *Seismic Wave Propagation in Stratified Media*, Cambridge University Press, Cambridge, 1983.

[Ki] Kimura, H.,

[1] Robust stabilization for a class of transfer functions, *IEEE Trans. Auto. Cont.*, **29** (1984) pp. 788-793.

[2] On interpolation-minimization problems in H_∞, *Control-Theory and Advanced Technology*, **2** (1986) pp. 1-25.

[KoT] Koehler, F. and Taner, M.T.

[1] Direct and inverse problems relating refraction coefficients and reflection response for horizontally layered media, *Geophysics*, **42** (1977) pp. 1199-1206.

[Kol] Kolmogorov, A.,

[1] Sur l'interpolation et l'exprapolation des suites stationnaires, *C. R. Acad. Sci (Paris)* **208** (1939) pp. 2043-2045.

[2] Stationary sequences in Hilbert space. (Russian) *Bull. Math. Univ., Moscow* **2** (1941) 40 pp. (English translation by Natasha Artin.)

[Kor] Koranyi, A.,

[1] On a theorem of Löwner and its connections with resolvents of self adjoint transformations, *Acta Sci. Math.*, **17** (1956) pp. 63-70.

[KovP] Kovalishina, I.V. and V.P. Potapov,

[1] *Integral Representation of Hermitian Positive Functions*, Private Translation by T. Ando, Division of Applied Math., Research Institute of Applied Electricity, Hokkaido University, Sappora, Japan.

[Kr] Krein, M.G.,

[1] Analytical problems and results in the theory of linear opeators on Hilbert spaces, *Proc. Internat. Congress. Math., Moscow* - August 1966, Izd. Mir. Moscow (1968) pp. 189-216.

[KrK] Krein, M. and M. A. Krasnoselski,

[1] Fundamental theorems on the extensions of Hermitian operators and some of their applications to the theory of orthogonal polynomials and the moment problem, *Uspekhi mat. Nauk,* **2** (1947).

[KrN] Krein, M.G. and A.A. Nudel'man,

[1] *The Markov Moment Problem and External Problems,* Transl. Math. Monographs **50**, American Math. Society, Providence, Rhode Island, 1977.

[KrR] Krein, M.G. and P.G., Rehtman,

[1] On the problem of Nevanlinna-Pick, *Trudi Odes'kogo Derz. Univ. Mat.,* **2** (1938) pp. 63-68.

[Kri] Kriete III, T. L.

[1] Similarity of canonical models, *Bull Amer. Math. Soc.,* **76** (1970) pp. 326-330.

[KuL] Kung, S. and D.W. Lin,

[1] Optimal Hankel norm model reductions: multivariable systems, *IEEE Trans. Automatic Control,* **26** (1981) pp. 832-852.

[La] Lax, P.,

[1] Translation invariant spaces, *Acta Math.,* **101** (1959) pp. 163-178.

[2] Translation invariant spaces, *Proc. Internat. Symp. Linear Spaces, Jerusalem 1960,* pp. 299-306 (New York, 1961).

[LaP] Lax, P. and R. S. Phillips,

[1] Scattering theory, *Bull. Amer. Math. Soc.,* **70** (1964) pp. 130-142.

[2] *Scattering theory,* New York, 1967.

[Lee] Leech, R.B.

[1] Factorization of analytic functions and operator inequalities.

[LeRG] LeRoux, J. and Gueguen, C.,

[1] A fixed point computation of partial correlation coefficients, *IEEE Trans. Acoust. Speech an Signal Processing,* **25** (1977) pp. 257-259.

[LeK] Lev-Ari, H. and T. Kailath,

[1] Lattice filter parameterization and modeling of nonstationary processes, *IEEE Trans. on Information Theory,* **30** (1984) pp. 2-16.

[2] Triangular Factorization of Structured Hermitian Matrices, *I. Schur Methods in Operator Theory and Signal Processing; Operator Theory: Advances and Applications,* **18**, Ed. I. Gohberg (1986) pp. 301-324.

[Lev] Levinson, N.,

[1]   The Wiener RMS (root mean square) error criterion in filter design and prediction, *J. Math. Phys.,* **25** (1947) pp. 261-278.

[Li] Li, K.Y.,

[1]   *Applications of deBranges Theory of Contractively Contained Spaces,* Ph.D. Thesis, University of California, Berkley, 1989.

[LiY] Livsic, M.S. and A.A. Yantsevich,

[1]   *Operator Colligations in Hilbert Space,* John Wiley and Sons, New York, 1979.

[Lo] Lowdenslager, D.B.,

[1]   On factoring matrix valued functions, *Ann. of Math.,* **78** (1963) pp. 450-454.

[Lu] Lubin, A.,

[1]   Isometries of *-invariant subspaces, *Trans. American Math. Soc.,* **190** (1974) pp. 405-415.

[Ma] Marden, M.,

[1]   *The Geometry of the zeros of a Polymonial in a Complex Variable,* American Math. Sci., 1949.

[Mar] Marple, S.L.,

[1]   *Digital Spectral Analysis with Applications,* Prentice Hall, Englewood Cliffs, New Jersey 1987.

[Mas] Masani, P.,

[1]   The prediction theory of multivariate stochastic processes III, *Acta Math.,* **104** (1960) pp. 141-162.

[2]   On isometric flows on Hilbert space, *Bull. Amer. Math.,* **68** (1962) pp. 624-632.

[3]   Shift invariant spaces and prediction theory, *Acta Math.,* **107** (1962) pp. 275-290.

[4]   On the representation theorem of scattering, *Bull. Amer. Math.,* **74** (1968) pp. 618-624.

[Morán] Morán, M.D.,

[1]   On intertwining Dilations, *J. Math. Anal. and Appl.,* **141** (1989) pp. 219-234.

[Mo] Morf, M.,

[1]   *Fast algorithms for multivariable systems,* Ph.D. Dissertation, Dept. of Electrical Engineering, Stanford University , Stanford, California, 1974.

[MoVK] Morf, M., Vieira, A. and T. Kailath,

[1]   Covariance characterization by partial autocorrelation matrices, *The Annals of Statistics,* **6** (1978) pp. 643-648.

[Mu] Muhly, P.S.,

[1]   Commutants containing a compact operator, *Bull. Amer. Math. Soc.,* **75** (1969) pp. 313-356.

[Na] Naimark, M.A.,

[1]   Self-adjoint extensions of the second kind of a symmetric operator, *Bulletin (Szvestiya) Acad. Sci. URSS (ser. Math.)* **4** (1940) pp. 53-104 (Russian, with English summary.)

[2]   Positive definite operator functions on a commutative group, *Bulletin (Izvestiya) Acad. Sci. URSS* (ser. math)., **7** (1943) pp. 237-244. (Russian, with English summary.)

[Ne] Nehari, Z.,

[1]   On bounded bilinear forms, *Ann. of Math.,* **65** (1957) pp. 153-162.

[Neu] Neumann, J. von,

[1]   Allgemeine Eigenwerttheorie Hermitescher Funkitionaloperatoren, *Math. Ann.,* **102** (1929) pp. 49-131.

[2]   Die Eindeutigkeit der Schrodingerschen Operatoren, *Math. Ann.,* **104** (1931) pp. 570-578.

[3]   Uber einen Satz von Herrn M. H. Stone, *Ann. of Math.,* **33** (1932) pp. 567-573.

[4]   Eine Spektraltheorie fur allgemeine Operatoren eines unitaren Raumes, *Math. Nachr.,* **4** (1951) pp. 258-281.

[Nev] Nevanlinna, R.,

[1]   Über beschränkte Funktionen, die in gegebenen Puntken vorgeschriebene Werte annehmen, *Ann. Acad. Sci. Fenn,* **13:1** (1919), 71 pp.

[2]   Kriterien für die Randwerte beschränkter Funktionen, *Math. Z.,* **13** (1922) pp. 1-9.

[3]   Asymptotische Entwicklungen beschränkter Funktionen und das Stieltjessche Momentenproblem, *Ann. Acad. Scie. Fenn. Ser. A* **18,** (1922) 53 pp.

[4]   Über beschränkte analytische Funktionen, *Ann. Acad. Sci. Fenn. Ser. A,* **32** (1929) 75 pp.

[5]   Beschränkte Potentiale, *Math Nachr.,* **4** (1951) pp. 489-501.

[New] Newton, R.G.,

[1]   Inversion of reflection data for layered media: a review of exact methods, *Geophys. J.R. astr. Soc.,* **65** (1981) pp. 191-215.

[Ni] Nikolskii, N.K.,

[1]   *Treatise on the Shift Operator,* Springer-Verlag, New York, 1986.

[NiV] Nikolskii, N.K. and V. I. Vasyunin,

[1]   A unified approach to functional models, and the transcription problem, *The Gohberg Anniversary Collection; Operator Theory Advances and Applications,* **41,** Ed. H. Dym, S. Goldberg, M. A. Kaashoek, P. Lancaster (1989) pp. 405-434.

[Nu] Nudelman, A.A.,

[1]   On a new problem of moment type, *Dokl. Akad. Nauk SSSR,* **233** (1977) pp. 792-795, *Soviet Math. Dokl.,* **18** (1977) pp. 507-510.

[2]   "On a generalization of classical interpolation problems," *Dokl. Akad. Nauk SSR,* **256** (1981) pp. 790-793, *Soviet Math. Dokl.,* **23** (1981) pp. 125-128.

[Pag] Page, L.B.,

[1]   Bounded and compact vectorial Hankel operators, *Trans. Amer. Math. Soc.,* **150** (1970) pp. 529-534.

[2]   Applications of the Sz.-Nagy and Foias Lifting Theorem, *Indiana University Math. Journal,* **20** (1970) pp. 135-145.

[3]   Operator that commute with a unilateral shift on an invariant subspace, Pacific J. Math., **36** (1971) pp. 787-794.

[Par] Parks, P.C.,

[1]   Liapunov and the Schur-Cohn stability criterion, *IEEE Trans. Automatic Contr.,* **19** (1964) p. 121.

[Parr] Parrott, S.,

[1]   Unitary dilations for commuting contractions, *Pacific J. Math.,* **34** (1970) pp. 481-490.

[2]   On a quotient norm and the Sz.-Nagy-Foias lifting theorem, *J. Functional Analysis,* **30** (1978) pp. 311-328.

[Pau] Paulsen, V. I.,

[1]   *Completely bounded maps and dilations,* Pitman Research Notes in Mathematics Series, **146**, John Wiley and Sons, New York, 1986.

[Pi] Pick, G.,

[1]   Über die Beschränkungen analytischer Functionen, welche durch vorgegebene Funktionswerte bewirkt sind., *Math. Ann.,* **77** (1916) pp. 7-23.

[2]   Über die Beschränkungen analytischer Funktionen, welche durch vorgegebene, funktionswerte bewirkt sind, *Math. Ann.,* **78** (1918) pp. 270-275.

[Po] Power, S. C.,

[1]   *Hankel Operators on Hilbert Space,* Pitman, London, 1982.

[PtV] Pták, V. and P. Vrbová,

[1]   Lifting intertwining relations, *Integral Equations and Operator Theory,* **11** (1988) pp. 128-147.

[2]   Extension of intertwining relations, *Integral Equations and Operator Theory,* **12** (1989) pp. 227-240.

[PtY] Pták, V. and N. J. Young,

[1]   A generalization of the zero locations theorem of Schur and Cohn, *IEEE Trans. Automatic Contr.,* **25** (1980) pp. 978-980.

[2]   Functions of operators and the spectral radius, *Linear Algebra and its applications,* **29** (1980) pp. 357-392.

[Re] Redheffer, R.M.,

[1]   On a certain linear fractional transformation, *J. Math. Phys.*, **39** (1960) pp. 269-286.

[2]   On the relation of transmission - line theory to scattering and transfer, *J. Math. Phys.*, **41** (1962) pp. 1-41.

[RieszSz.-N] Riesz, F. and B. Sz.-Nagy,

[1]   Über Krontraktionen des Hilbertschen Raumes, *Acta Sci. Math.*, **10** (1943) pp. 202-205.

[2]   *Functional Analysis* (New York, 1955); translation of *Lecons d'analyse fonctionnelle,* 2nd ed. (Budapest, 1953).

[Riss] Rissanen, J.,

[1]   Algorithms for triangular decomposition of block Hankel and Toeplitz matrices with application to factoring positive matrix polynomials, *Math. Comput.*, **27** (1973) pp. 147-154.

[Ro] Robinson, E.A.,

[1]   *Statistical Communications and Detections with Special References to Digital Data Processing of Radar and Seismic Signals,* Chules Griffin and Co., London, 1967.

[2]   Dynamic predictive deconvolution, *Geophysical Prospecting,* **23** (1975) pp. 779-797.

[RoT] Robinson, E.A. and S. Treitel,

[1]   *Geophysical Signal Analysis,* Prentice-Hall, Englewood Cliffs, New Jersey, 1980.

[2]   Maximum entropy and the relationship of the partial autocorrelation to the reflection coefficients of a layered system, *IEEE Trans. Acoust. Speech, Signal Process,* **28** (1980) pp. 224-235.

[Ros] Rosenblum, M.,

[1]   A Corona theorem for countably many functions, *Integral Equations and Operator Theory,* **3** (1980) pp. 125-137.

[RosR] Rosenblum, M. and J. Rovnyak,

[1]   An operator-theoretic approach to theorems of the Pick-Nevanlinna and Loewner types I, *Integral Equations and Operator Theory,* **3** (1980) pp. 408-436.

[2]   An operator-theoretic approach to theorems of the Pick-Nevanlinna and Loewner types II, *Integral Equations and Operator Theory,* **5** (1982) pp. 870-887.

[3]   *Hardy Classes and Operator Theory,* Oxford Univ. Press, New York, 1985.

[Rota] Rota, G.C.,

[1]   On models for linear operators, *Comm. Pure Appl. Math.,* **13** (1960) pp. 468-472.

[Ru] Rudin, W.,

[1]   *Real and Complex Analysis,* McGraw-Hill, New York, 1966.

[Rugh] Rugh, W.J.,

[1]   *Nonlinear System Theory, The Volterra Wiener Approach,* The Johns Hopkins University Press, Baltimore, Maryland, 1981.

[Sa] Sarason, D.,

[1]   On spectral sets having connected complement, *Acta Sci. Math.,* **26** (1965) pp. 289-299.

[2]   Generalized interpolation in H^∞, *Trans. American Math. Soc.,* **127** (1967) pp. 179-203.

[3]   Invariant subspaces, *Topics in Operator Theory,* Ed. C. Pearcy, Mathematical Surveys, **13**, The American Mathematical Society, Providence 1974.

[4]   Shift invariant spaces from the Bragesian point of view, The Bieberbach conjecture, Proceedings on the occasion of the proof, *Math. Surveys,* **21**, American Math. Society, Providence, Rhode Island, (1986), pp. 153-166.

[Sc] Schäffer, J.J.,

[1]   On unitary dilations of contractions, *Proc. Amer. Math. Soc.,* **6** (1955) p. 322.

[Schur] Schur, I.,

[1]   On power series which are bounded in the interior of the unit circle I, *J. für die Reine und Angewandte Mathematik,* **147** (1917) pp. 205-232, English translation in *I. Schur Methods in Operator Theory and Signal Processing; Operator Theory: Advances and Applications,* **18**, Ed. I. Gohberg (1986) pp. 31-59.

[2]   On power series which are bounded in the interior of unit circle II., *J. für die Reine and Angewandte Mathematik,* **148** (1918) pp. 122-145 (German). English translation in *I. Schur Methods in Operator Theory and Signal Processing; Operator Theory: Advances and Applications,* **18**, Ed. I. Goherg (1986) pp. 61-88.

[Sk] Skelton, R.E.,

[1]   *Dynamic Systems Control,* John Wiley and Sons, New York, 1988.

[Sm] Smuljan, Ju. L.,

[1]   Operator Balls, *Theor. Funkcii, Funkcional Anal. i. Prilozen,* **6** (1968) pp. 68-81, (Russian).

[Szegö] Szegö, G.,

[1]   *Orthogonal Polynomials,* Colloquium Publications, **23**, American Mathematical Society, Providence, Rhode Island, 1939.

[Sz.-N] Sz.-Nagy, B.,

[1]   Sur les contraction de l'espace de Hilbert, *Acta Sci. Math.,* **15** (1953) pp. 87-92.

[2]   Transformations de l'espace de Hilbert, fonctions de type positif sur un groupe, *Acta Sci. Math.,* **15** (1954) pp. 104-114.

[3] Sur les contractions de l'espace de Hilbert II, *Acta Sci. Math.*, **18** (1957) pp. 1-15.

[4] Extensions of linear transformations in Hilbert space which extend beyond this space, *Appendix to F. Riesz and B. Sz.-Nagy: Functional analysis* (New York, 1960). Translation of "Prolongement des transformations de l'espace de Hilbert qui sortent de cet espace" (Budapest, 1955).

[5] On Schäffer's construction of unitary dilations, *Ann. Univ. Budapest* (sect. math.), **3-4** (1960/61) pp. 343-346; announced in: *Deuxiaeme congres math. hongrois,* Budapest, 24-31 août 1960.

[6] Un calcul fonctionnel pour les contractions. - Sur la structure des dilatations unitaires des operateurs de l'espace de Hilbert, *Seminari dell'Istituto Nazionale de Alta Matematica 1962-63,* pp. 525-534.

[7] Positive definite kernels generated by operator-valued analytic functions, *Acta Sci. Math.*, **26** (1965) pp. 191-192.

[8] Positive-definite, durch Operatoren erzeugte Funktionen, *Wiss. Zeitschur. d. Techn. Univ. Dresden,* **15** (1966) pp. 219-222.

[9] Hilbertraum-Operatoren der Klasse C_0, *Abstract Spaces and Approximation* (Basel, 1969) pp. 72-81.

[10] *Unitary dilations of Hilbert space operators and related topics.* CBMS Regional Conference Series in Math., **19**, Amer. Math. Soc., Providence, Rhode Island 1974.

[Sz.-NF] Sz.-Nagy, B. and C. Foias,

[1] Modèles fonctionnels des contractions de l'espace de Hilbert. La fonction caracteristique, *C. R. Acad. Sci. Paris,* **256** (1963) pp. 3236-3239.

[2] Proprietés des fonctions caractéristiques, modeles triangulaires et une classification des contractions, *Acad. Sci Paris,* **258** (1963) pp. 3413-3415.

[3] Sur les contractions de l'espace de Hilbert, VII. Triangulations canoniques. Fonctions minimum, *Acta Sci. Math.*, **25** (1964) pp. 12-37.

[4] Sur les contractions de l'espace de Hilbert, VIII. Fonctions caractéristiqaues. Modeles fonctionnels, *Acta Sci. Math.*, **25** (1964) pp. 38-71.

[5] Une caractérisation des sous-espaces invariants pour une contraction de l'espace de Hilbert, *C.R. Acad. Sci. Paris,* **258** (1964) pp. 3426-3429.

[6] Forme triangulaire d'une contraction et factorisation de la fonction caracteristique, *Acta Sci. Math.*, **28** (1967) pp. 201-212.

[7] Dilatation des commutants d'opérateurs, *C.R. Acad. Sci. Paris, Serie A,* **266** (1968) pp. 493-495.

[8] Commutants de certains operatéurs, *Acta Sci. Math.*, **29** (1968) pp. 1-17.

[9] Modèle de Jordan pour une classe d'opérateurs de l'espace de Hilbert, *Acta Sci. Math.*, **31** (1970) pp. 93-117.

[10] *Harmonic analysis of operators on Hilbert space,* North Holland Publishing Co., Amsterdam-Budapest, 1970.

[11] The "lifting theorem" for intertwining operators and some new applications, *Indiana University Math. Journal,* **20** (1971) pp. 901-904.

[12] On the structure of intertwining operators, *Acta Sci. Math.*, **35** (1973) pp. 225-254.

[13] Regular factorizations of contractions, *Proceedings of the American Mathematical Society*, **43** (1974) pp. 91-93.

[14] On contractions similar to isometries and Toeplitz operators, *Ann. Acad. Sci. Finn. Math.*, **2** (1976) pp. 553-564.

[Sz.-NK] Sz.-Nagy, B. and A. Koranyi,

[1] Relations d'un problème de Nevanlinna et Pick avec la theorie des opérateurs de l'espace Hilbertien, *Acta Sci. Math.*, **7** (1956) pp. 295-302.

[2] Operator theoretische Behandlung and Verallgemeinerung eines Problemkreises in der komplexen funktionentheorie, *Acta Math.*, **100** (1958) pp. 171-202.

[Ta] Tannenbaum, A.,

[1] Feedback stablization of linear dynamical plants with uncertainty in the gain factor, *Int. J. Control*, **32** (1980) pp. 1-16.

[2] Modified Nevanlinna-Pick interpolation of linear plants with uncertainty in the gain factor, *Int. J. Control*, **36** (1982) pp. 331-336.

[To] Toeplitz, O.,

[1] Zur Transformationen der Scharen bilinearen Formen von unendlichvielen Veränderlichen, *Nachr. Akad, Wiss. Göttingen: Math. Phys. Kl.*, (1907) pp. 110-115.

[2] Zur Theorie der quadratischen Formen von unednlichvielen Variablen, *Nachr. Akad. Wiss. Göttingen: Math. Phys. Kl.*, (1910) pp. 489-506.

[3] Zur Theorie der quadratischen und bilinearen Formen von unednlichvielen Veränderlichen, *Math. Ann.*, **70** (1911) pp. 191-192.

[4] Über die Fourier'sche entwicklung positiven funktionen, *Rend. Cuc. Mat. Palermo*, **32** (1911) pp. 191-192.

[Tol] Tolokonnikov, V.A.,

[1] Estimates in the Carleson Corona theorem ideals of the Algebra H^∞, a Sz.-Nagy problem, *Investigations in Linear Operators and the Theory of Functions*, **9** (1981) pp. 178-198, (in Russian, published by the Leningrad division of the Steklov institute of Mathematics).

[Tr] Trench, W.F.,

[1] An algorithm for the inversion of finite Toeplitz matrices, *SIAM J. Appl. Math.*, **13** (1964) pp. 515-522.

[V] Vidyasagar, M.,

[1] *Control System Synthesis*, The MIT Press, Cambridge, Massachusetts, 1985.

[VK] Vieira, A. and T. Kailath,

[1] On another approach to the Schur-Cohn criterion, *IEEE Trans. on Circuits and Systems*, (1977) pp. 218-230.

[Wa] Walsh, J.L.,

[1] *Interpolation and Approximation*, Amer. Math. Soc. Coll. Publ., **20**, New York, 1935.

[Wh] Whittle, P.,

[1] On the fitting of multivariable autoregressions and the approximate canonical factorization of a spectral density matrix, *Biometrika*, **50** (1963) pp. 129-134.

[Wi] Wiener, N.,

[1] *Smoothing of Stationary Time Series*, Wiley, New York, 1949. (Originally a report in February, 1942).

[WiM] Wiener, N. and P. Masani,

[1] The prediction theory of multivariate stochastic processes I, *Acta Math.*, **98** (1957) pp. 111-150.

[2] The prediction theory of multivariate stochastic processes II, The linear predictor, *Acta Math.*, **99** (1958) pp. 93-139.

[WigR] Wiggins, R.A. and E.A., Robinson,

[1] Recursive solution to the multichannel filtering problem, *J. Geophys. Res.*, **70** (1965) pp. 1885-1891.

[Woe] Woerdeman, H.J.,

[1] *Matrix and Operator Extension*, Ph.D. Thesis, Vrije Universiteit te Amsterdam, 1989.

[Wo] Wold, H.,

[1] *A study in the Analysis of Stationary Time Series*, Stockholm, 1938; 2nd ed. 1954.

[Won] Wong, W.S.,

[1] *Operator theoretic methods in nonlinear systems*, Ph.D. thesis, Harvard University, Cambridge, Massachusetts, 1980.

[Y] Young, N.J.,

[1] Norms of certain rational functions of a matrix and Schur's criteria for polynomials, *Comment math. University Carolina*, **19** (1978) pp. 673-688.

[2] A maximum principle for interpolation in H^∞, *Acta Sci. Math.*, **43** (1981) pp. 147-152.

[3] The singular-value decomposition of an infinite Hankel matrix, *Linear Algebra and its Applications*, 50 (1983) pp. 639-656.

[4] The Nevanlinna-Pick problem for matrix valued functions, *J. Operator Theory*, **15** (1986) pp. 239-265.

[Z] Zames, G.,

[1] "Optimal sensitivity and feedback: weighted seminorms, approximate inverses, and plant invariant schemes," *Proc. Allerton Conf.*, 1979.

[2] "Feedback and optimal sensitivity: model reference transformations, multiplicative seminorms, and approximate inverses," *IEEE Trans. Auto. Cont.*, **26** (1981) pp. 301-320.

[ZF] Zames, G. and B. A. Francis,

[1] "Feedback, minimax sensitivity, and optimal robustness," *IEEE Trans. Auto. Cont.,* **28** (1983) pp. 585-601.

NOTATION

A^*	the adjoint of the operator A
$A^{1/2}$	the positive square root of the positive operator A
$\det A$	the determinant of A
$[a_{i,j}]_{1,1}^{n,m}$	a matrix with entries $a_{i,j}$
$A\,|\,M$	the restriction of the operator A to the subspace M
C^n	the n-dimensional linear space of complex n-tuples
D	the open unit disc
D_T	the positive square root of $I - T^*T$
D_{T^*}	the positive square root of $I - TT^*$
$\mathcal{D}_T$	the closed range of D_T
$\mathcal{D}_{T^*}$	the closed range of D_{T^*}
F_+	the operator from $H^2(\cdot)$ to $H^2(\cdot)$ defined by $F_+x = Fx$ when F is in the appropriate $H^\infty(\cdot, \cdot)$ space and x is in $H^2(\cdot)$
$F(A{\cdot}T)$	equals $\{D_A Th \oplus D_T h : h \in H\}^-$
$G(A{\cdot}T)$	equals $(\mathcal{D}_A \oplus \mathcal{D}_T) \ominus F(A{\cdot}T)$
H^2	the Hardy space of analytic functions on the unit disc
$H^2(E)$	the Hardy space of analytic functions on the unit disc with values in the Hilbert space E
H^∞	the space of all bounded analytic functions on D
H_1^∞	the closed unit ball in H^∞
$H^\infty(E, E')$	the space of all bounded analytic functions on D with values in $L(E, E')$.
$H_1^\infty(E, E')$	the closed unit ball of $H^\infty(E, E')$
$H(\Theta)$	$[H^2(E') \oplus \overline{\Delta_\Theta L^2(E)}] \ominus \{\Theta u \oplus \Delta_\Theta u : u \in H^2(E)\}$

$H(m)$	$H^2 \ominus mH^2$ if m is an inner function in H^∞
I	the identity operator
I_H	identity on H
$I(T, T')$	the set of all operators A satisfying $T'A = AT$
$I_1(T, T')$	the set of all contractions A satisfying $T'A = AT$
J	the unitary operator on $L^2(\cdot)$ defined by $(Jf)(e^{it}) = f(e^{-it})$
J_{n+1}	the operator which reverses and conjugates the coefficients of an n-th degree polymonial $J_{n+1}p \overset{\Delta}{=} z^n\,\overline{p(1/\bar{z})}$
ker A	the kernel of A
K_o^2	$L^2 \ominus H^2$
$K_o^2(E)$	$L^2(E) \ominus H^2(E)$
K_o^∞	the set of all f in L^∞ of the form $f = \sum_1^\infty f_n e^{-int}$
$K_o^\infty(E, E')$	the set of all F in $L^\infty(E, E')$ of the form $F = \sum_1^\infty F_n e^{-int}$
$L(E, E')$	the set of all operators from E to E'
L^2	the set of all square integrable Lebesgue measurable functions in $[0, 2\pi)$
$L^2(E)$	the set of all square integrable Lebesgue measurable functions in $[0, 2\pi)$ with values in E.
L^∞	the set of all Lebesgue measurable function uniformally bounded a.e. in $[0, 2\pi)$
$L^\infty(E, E')$	the set of all strongly Lebesgue measurable functions uniformally bounded a.e. in $[0, 2\pi)$ with values in $L(E, E')$
l^2	the set of all square summable unilateral sequences $[x_o, x_1, x_2, ...]^{tr}$
$l^2(E)$	the set of all square summable unilateral sequences $[x_o, x_1, x_2, ...]^{tr}$ with

	values in E, that is , $x_i \in E$ for all i
$l_n^2(E)$	$\overset{n}{\underset{1}{\oplus}} E$ which is identified with the set of all vectors of the form $[x_1, x_2, ..., x_n]^{tr}$ where $x_i \in E$ for all i
$\hat{l}^2$	the set of all square summable bilateral sequences
$\hat{l}^2(E)$	the set of all square summable bilateral sequences with values in E
$l(E)$	the space of all unilateral formal series $(\overset{\infty}{\underset{0}{\sum}} f_n z^n)$ with values in E
m_T	the minimal function for T
$\overline{M}$ or M^-	the closure of the set M
M_F	the operator from $L^2(\cdot)$ to $L^2(\cdot)$ defined by $M_F x = Fx$ when F is in the appropriate $L^\infty(\cdot, \cdot)$ space and x is in $L^2(\cdot)$
$p^\#$	the reverse polymonial of p
P_H	the orthogonal projection onto the subspace H
P_+	the orthogonal projection onto $H^2(\cdot)$
P_-	the orthogonal projection onto $K_0^2(\cdot)$
ran A	the range of the operator A
R_A	the rotation matrix defined by A
S	the unilateral shift on $H^2(\cdot)$
$S(\Theta)$	the compression of multiplication by e^{it} to $H(\Theta)$
$S(m)$	the compression of S to $H(m)$ if m is an inner function in H^∞
tr	transpose
$[x_i]_1^n$	a vector in C^n identified with $[x_1, x_2, ..., x_n]^{tr}$
$[A_1, A_2, ..., A_n]^{tr}$	equals $[A_1^*, A_2^*, ..., A_n^*]^*$ when the $A_i's$ are operators

δ_T the rank of D_T

δ_{T^*} the rank of D_{T^*}

Θ_T the characteristic function of T

Δ_Θ the positive square root of $I - \Theta(e^{it})^*\Theta(e^{it})$ a.e.

$\tilde{\Theta}$ equals $\Theta(e^{-it})^*$

(x, y) the inner product between x and y

$\|x\|$ the norm of x

$\|h\|_\infty$ the $L^\infty(\cdot)$ norm of h

$\oplus$ the orthogonal direct sum

$\bigoplus_m^n x_i \in \bigoplus_m^n X_i$ the vector x_i is in the Hilbert space X_i for $m \le i \le n$ and the space X_i is orthogonal to X_j for all $i \ne j$; the vector $\bigoplus_m^n x_i$ is identified with $[x_m, x_{m+1}, ..., x_n]^{tr}$.

$K \ominus M$ the orthogonal complement of M in K.

$M^\perp$ the orthogonal complement of M

$x \perp y$ x is orthogonal to y

$\triangleq$ definition

INDEX

If you have any concerns about our products,
you can contact us on
ProductSafety@springernature.com

In case Publisher is established outside the EU,
the EU authorized representative is:
Springer Nature Customer Service Center GmbH
Europaplatz 3, 69115 Heidelberg, Germany

Printed by Libri Plureos GmbH
in Hamburg, Germany